NORMED ALGEBRAS

NORMED ALGEBRAS

BY

M. A. NAIMARK

Steklov Institute of Mathematics,
Academy of Sciences, U.S.S.R.

TRANSLATED FROM THE SECOND RUSSIAN EDITION

BY

LEO F. BORON

University of Idaho, Moscow, U.S.A.

WOLTERS-NOORDHOFF PUBLISHING GRONINGEN
THE NETHERLANDS

ISBN: 90 01 61475 2

Library of Congress Catalog Card Number: 74–151038

Third completely revised American edition of "Normed Rings", translated from the second Russian edition, published 1968, Moscow.

Printing history: 1st American edition of "Normed Rings" published in 1959; Reprinted in 1970.

Printed in the Netherlands
by Drukkerij Dijkstra Niemeyer bv, Groningen

DEDICATION

THIS TRANSLATION IS DEDICATED

TO

PROF. DR. MARK ARONOVICH NAIMARK

L.F.B.

CONTENTS

CHAPTER I

BASIC IDEAS FROM TOPOLOGY AND FUNCTIONAL ANALYSIS

CHAPTER V
SOME SPECIAL ALGEBRAS

CHAPTER VI
GROUP ALGEBRAS

CHAPTER VII
ALGEBRAS OF OPERATORS IN HILBERT SPACE

CHAPTER VIII
DECOMPOSITION OF AN ALGEBRA OF OPERATORS INTO IRREDUCIBLE ALGEBRAS

FOREWORD TO THE THIRD AMERICAN EDITION

This third American edition is close to the second Soviet edition which appeared in November 1968.

In comparison with the second (= first revised) American edition, most of the chapters (especially Chapters III and VIII) have been significantly revised and extended by the discussion of new results; the bibliography has also been significantly extended. In the third American edition the term *ring* has been replaced by the now universally used term *algebra*; this accounts for the corresponding change in the title of the book.

In the technical part of the preparation of the third American edition, I was helped by my wife, L. P. Naimark; I use the present occasion to express my profound gratitude.

Inasmuch as the third American edition is coming off the press several years afterthe second Soviet edition, there have been included in the galley proofs the statement of several of the most essential theoretical results that have been obtained recently, and the bibliography has consequently been extended by several papers; I wish to express my profound gratitude to Dr. A. I. Shtern for the great help rendered in connection with the preparation of this work and in the reading of galley proofs. I also thank Professor L. F. Boron for translating the revised and supplementary material and for editing this American edition.

Moscow, USSR M. A. NAIMARK

March 1972

FOREWORD TO THE SECOND (=FIRST REVISED) AMERICAN EDITION

In this edition we have corrected all misprints and inaccuracies, we have extended the literature list and we have made a number of further references concerning the most essential recent results in the theory of normed rings. Moreover, the last chapter, i.e. Chapter VIII, has been rewritten.

Since it turned out that the theory of Tomita is valid only under the additional assumptions of separability type, the author considered it expedient to give a discussion in Chapter VIII which is closer to the initial simpler theory of von Neumann for the separable case; in this discussion, the articles NAIMARK [3] and NAIMARK and FOMIN [1] were used.

For the convenience of the reader, Appendices II-IV were added to the book – these are necessary for understanding the last chapter.

The author considers it his pleasant duty to express his gratitude to Prof. Leo F. Boron who undertook the task of preparing the revised American edition of this book and to the publisher NOORDHOFF who made possible the appearance of the second edition and enabled the author to introduce the above-mentioned modifications and additions.

Moscow M. A. NAIMARK

August 1963

FOREWORD TO THE SECOND SOVIET EDITION

In this second edition the initial text has been worked over again and improved, certain portions have been completely rewritten; in particular, Chapter VIII has been rewritten in a more accessible form. The changes and extensions made by the author in the Japanese, German, first and second (= first revised) American, and also in the Romanian (lithographed) editions, were hereby taken into account.

Appendices II and III, which are necessary for understanding Chapter VIII, have been included for the convenience of the reader.

The book discusses many new theoretical results which have been developing intensively during the decade after the publication of the first edition. Of course, limitations on the volume of the book obliged the author to make a tough selection and in many cases to limit himself to simply a formulation of the new results or to pointing out the literature. The author was also compelled to make a choice of the exceptionally extensive collection of new works in extending the literature list. Monographs and survey articles on special topics of the theory which have been published during the past decade have been included in this list and in the literature pointed out in the individual chapters.

The author is very much indebted to all his colleagues who pointed out misprints and errors contained in the first edition.

D. A. Raikov and M. G. Sonis carefully edited the manuscript and enabled me by their remarks to improve individual parts of the book; furthermore, M. G. Sonis wrote subsections 3, 5 of §9, subsection 7 of §11, IX in subsection 2, §18, corollaries 1 – 4 to I, subsection 3 – §23 and IV, V, subsection 3, §23.

The author considers it to be his pleasant duty to express his profound gratitude to D. A. Raikov and M. G. Sonis. The author is grateful to A. Z. Rybkin for his careful consideration of the manuscript and expresses his indebtedness to D. P. Zhelobenko for his aid in reading the galley proofs.

Moscow, M. A. NAIMARK

February 1967

FROM THE FOREWORD TO THE FIRST SOVIET EDITION

The theory of normed algebras, although of recent origin, has developed into an extensive branch of functional analysis; it now has numerous applications in various other branches of mathematics.

The first cycle of papers devoted to concrete normed algebras, i.e. algebras of bounded linear operators in Hilbert space, was begun in 1930 by VON NEUMANN [1] and then continued in the works of MURAY and VON NEUMANN [1]. The advantage of considering algebras of operators was already apparent in these works. However, it was the abstract point of view which turned out to be most fruitful; from this viewpoint, the nature of the elements of the algebra does not play any role, so that a normed algebra is simply an arbitrary set of elements which forms an algebra in the algebraic sense and which is, furthermore, provided with a norm which satisfies simple requirements.

This point of view was systematically developed by GELFAND [1 – 7] in his theory of commutative normed algebras. The discovery by Gelfand of the role of maximal ideals, the construction of a compact space of maximal ideals, and the representation of the elements of a semisimple algebra in the form of an algebra of continuous functions on this space were of decisive significance in this connection. Even the first applications showed the power of the theory of normed algebras. Thus, with the aid of normed algebras, a simple proof of the Wiener theorem on trigonometric series was unexpectedly obtained (see WIENER [1]), simple proofs and generalizations of many theorems of Tauberian type were also obtained, and so on.

An essential role in the development of these applications was played by a large cycle of works by SHILOV [1] which are devoted to the investigation of various classes of commutative normed algebras and the structure of ideals in them.

The application of the theory of commutative normed algebras to the theory of locally compact commutative groups which led to the construction of a harmonic analysis on such groups (by GELFAND and M. KREIN [1], M. KREIN [6], and RAIKOV [2 – 5]) and, in particular, to a simple analytic proof of the Pontryagin duality theorem by RAIKOV [4] is especially important.

Another important class of algebras which are no longer commutative, namely algebras with involution (see §10), was considered in the work by GELFAND and NAIMARK [1]. In this work, it was shown that under certain natural conditions, every such algebra can be isomorphically mapped into an algebra of bounded linear operators in Hilbert space such that the involution operation goes over into the operation $A \to A^*$ (where A^* is the adjoint operator of A), and the norm goes over into the operator norm.

Here, an important role was played by the concept of a positive functional, i.e. a linear functional f in the algebra satisfying the condition $f(x^*x) \geqq 0$. The methods worked out in this paper, in particular the concept of a positive functional, were used later in the works of Gelfand and in numerous works of other authors in the investigation of algebras with involution and in the construction of the theory of representations of such algebras; for the particular case of group algebras, these methods were used in the investigation of unitary representations of topological groups.

A second construction of the representations of locally compact groups with the aid of positive definite functions was first given by GELFAND and RAIKOV [2]; in particular, they proved the completeness of the system of all irreducible unitary representations of a locally compact group.

Later, these results of Gelfand and Raikov were repeated in part independently and then developed in the works of GODEMENT [3].

Despite the presence of a large number of results, the theory of normed algebras, especially of non-commutative algebras, cannot be considered complete and many interesting problems in this theory remain open at the present time.

Of special interest is the further development of the theory of characters and harmonic analysis on locally compact groups, constructed in the works of GELFAND and NAIMARK [1 – 8] for the complex classical groups and then carried over, in numerous works of several authors, to other classes of locally compact groups. Furthermore, there remain unsolved a number of problems connected with the decomposition of a given representation of a group or algebra into irreducible representations.

Despite the importance of the theory of normed algebras in many applications and the large number of results which have already been obtained in this theory, there are, at the present time, very few books devoted to this theory. Thus, in the foreign literature there is the book by LOOMIS [1] in which, however, primary attention is given to the theory of commutative and Hilbert algebras and its application to harmonic analysis on locally compact commutative groups and on compact non-commutative groups. Further, some problems in the theory of normed algebras are discussed in the book *Functional Analysis and Semigroups* by Hille.

In the present book, the theory of normed algebras and also of some (commutative as well as non-commutative) topological algebras and its various applications, mainly to the theory of representations of locally compact groups, are discussed. For the convenience of the reader, the first chapter contains the information from functional analysis which will be needed in the remainder of the book.

The author expresses profound gratitude to D. A. RAIKOV who read the book in manuscript form and made a number of valuable observations. The author also expresses his sincere thanks to I. M. GELFAND and G. E. SHILOV for valuable advice.

Moscow M. A. NAIMARK

August 1955

CHAPTER I

BASIC IDEAS FROM TOPOLOGY AND FUNCTIONAL ANALYSIS

§ 1. Linear spaces

1. Definition of a linear space. A set R is called a *linear*, or *vector*, *space* if:

a) the *sum* $x+y$ of any two elements x, y in R is defined, and $x+y$ is an element in the set R;

b) the *product* αx of an arbitrary real or complex number α and an arbitrary element x in R is defined, and αx is an element in the set R;

c) the above operations of addition of elements and multiplication of an element by a number satisfy the following conditions:

a_1) $x+y = y+x$;

a_2) $(x+y)+z = x+(y+z)$;

a_3) there exists an element 0 in R such that $x+0 = x$ for arbitrary element x in R;

a_4) for every element x in R there exists an element $-x$ such that $x+(-x) = 0$;

b_1) $1 \cdot x = x$;

b_2) $\alpha(\beta x) = (\alpha\beta)x$;

b_3) $(\alpha+\beta)x = \alpha x+\beta x$;

b_4) $\alpha(x+y) = \alpha x+\alpha y$.

The set R is called a *real* linear space if, in this connection, multiplication in R is defined only for real numbers; but if multiplication in R is defined for arbitrary complex numbers, then R is called a *complex* linear space.

Obviously, *every complex linear space R can also be considered as a real linear space.*

The elements of a linear space R are usually called *vectors.*

We note that the nature of the elements in the space R and also the manner of defining the operations of addition and multiplication by a number can be taken perfectly arbitrarily; it is important only that conditions a_1)$-a_4$) and b_1)$-b_4$) be satisfied.

EXAMPLES. 1. We denote the set of all ordered n-tuples $x = \langle \xi_1, \xi_2, \cdots, \xi_n \rangle$, where $\xi_1, \xi_2, \cdots, \xi_n$ are complex numbers, by C^n. We define the operations of addition and multiplication by a complex number by the formulas

$$\langle \xi_1, \xi_2, \cdots, \xi_n \rangle + \langle \eta_1, \eta_2, \cdots, \eta_n \rangle = \langle \xi_1+\eta_1, \xi_2+\eta_2, \cdots, \xi_n+\eta_n \rangle,$$
$$\alpha\langle \xi_1, \xi_2, \cdots, \xi_n \rangle = \langle \alpha\xi_1, \alpha\xi_2, \cdots, \alpha\xi_n \rangle.$$

Clearly, conditions a_1) $-a_4$) and b_1) $-$ b_4) are satisfied and so C^n is a complex linear space.

But if we regard ξ_j and η_j as restricted to the reals, then we obtain the real linear space R^n.

In particular, C^1 is simply the set of all complex numbers and R^1 is the set of all real numbers, with the usual operations of addition and multiplication.

2. Let $\mathfrak{P}_n$ be the set of all polynomials $p(x) = c_0+c_1x+\cdots+c_nx^n$ of degree $\leqq n$ with complex coefficients; we define the operations of addition and multiplication by a complex number in $\mathfrak{P}_n$ in the usual way. It is easily seen that then $\mathfrak{P}_n$ becomes a complex linear space.

3. Let $C[a, b]$ be the set of all continuous complex-valued functions $x = x(t)$ on the fixed interval $[a, b]$; we define the operations of addition and multiplication by a complex number in the usual way. One can easily verify that $C[a, b]$ then becomes a complex linear space.

2. Linear dependence and independence of vectors. A *linear combination* of the vectors $x_1, x_2, \cdots, x_k$ is any sum of the form $\alpha_1 x_1+\alpha_2 x_2+\cdots+\alpha_k x_k$. The vectors $x_1, x_2, \cdots, x_k$ are said to be *linearly independent* if and only if any linear combination $\alpha_1 x_1+\alpha_2 x_2+\cdots+\alpha_k x_k$ is zero inplies $\alpha_1 = \alpha_2 = \cdots = \alpha_k = 0$, and *linearly dependent* otherwise.

A space R is said to be *finite-dimensional* (namely, *n-dimensional*) if there exists only a finite number (precisely n) linearly independent vectors in R; otherwise, the space R is called *infinite-dimensional*. In the case of an n-dimensional space R, any set of n linearly independent vectors is called a *basis* in R.

The space C^n in Example 1, subsection 1, is n-dimensional; the set of vectors $x_1 = \langle 1, 0, 0 \ldots, 0\rangle$, $x_2 = \langle 0, 1, 0, \cdots, 0\rangle, \cdots, x_n = \langle 0, 0, \cdots, 0, 1\rangle$ is, for instance, a basis in C^n.

The space $C[a, b]$ in Example 3, subsection 1, is infinite-dimensional since this space contains an arbitrary number of linearly independent functions, for instance,

$$1, t, \cdots, t^N \qquad (N = 1, 2, 3, \cdots).$$

If R is n-dimensional and $\{x_1, x_2, \cdots, x_n\}$ is a basis in R, then every vector x in R can be uniquely represented in the form

$$x = \xi_1 x_1+\xi_2 x_2+\cdots+\xi_n x_n. \tag{1}$$

In fact, since R is n-dimensional, the vectors $x, x_1, x_2, \cdots, x_n$ are linearly dependent; this means that there exist numbers $c, c_1, \cdots, c_n$, not all equal to zero, for which

$$cx+c_1 x_1+\cdots+c_n x_n = 0. \tag{2}$$

Here, $c \neq 0$, for $c = 0$ would imply that $c_1 x_1+\cdots+c_n x_n = 0$, which in virtue of the linear independence of the vectors $x_1, \cdots, x_n$ is possible only when $c_1 = c_2 = \cdots = c_n = 0$. But then it follows from (2) that

$$x = -\frac{c_1}{c}x_1-\cdots-\frac{c_n}{c}x_n,$$

i.e. formula (1) with coefficients $\xi_i = -c_i/c$. If also $x = \xi'_1 x_1+\xi'_2 x_2+\cdots+\xi'_n x_n$, then, subtracting this equality from (1), we obtain:

$$0 = (\xi_1-\xi'_1)x_1+(\xi_2-\xi'_2)x_2+\cdots+(\xi_n-\xi'_n)x_n.$$

In virtue of the linear independence of the vectors $x_1, x_2, \cdots, x_n$ this is possible only when $\xi_1-\xi_1' = \xi_2-\xi_2' = \cdots = \xi_n-\xi_n' = 0$, i.e. when $\xi_1 = \xi_1', \xi_2 = \xi_2', \cdots, \xi_n = \xi_n'$. This proves the uniqueness of representation (1).

The numbers $\xi_1, \xi_2, \cdots, \xi_n$ are called the *coordinates* of the vector x with respect to the basis $\{x_1, x_2, \cdots, x_n\}$.

Two linear spaces R and R' are said to be *isomorphic* if there exists a one-to-one correspondence $x \leftrightarrow x'$ between their sets of vectors having the following properties:

1) if $x \leftrightarrow x'$, then $\alpha x \leftrightarrow \alpha x'$;

2) if $x \leftrightarrow x'$ and $y \leftrightarrow y'$, then $x+y \leftrightarrow x'+y'$.

The correspondence $x \leftrightarrow x'$ itself is called an *isomorphism* of the spaces R and R'.

EXERCISES. 1. Prove that the correspondence $x \leftrightarrow \langle \xi_1, \xi_2, \cdots, \xi_n \rangle$ between the vectors of the n-dimensional space R and their coordinates $\xi_1, \xi_2, \cdots, \xi_n$ with respect to a fixed basis is an isomorphism between the spaces R and C^n (see Example 1, subsection 1).

2. Prove that the vectors

$$\begin{aligned} y_1 &= \xi_{11}x_1+\xi_{12}x_2+\cdots+\xi_{1n}x_n, \\ y_2 &= \xi_{21}x_1+\xi_{22}x_2+\cdots+\xi_{2n}x_n, \\ &\cdots\cdots\cdots\cdots\cdots\cdots \\ y_k &= \xi_{k1}x_1+\xi_{k2}x_2+\cdots+\xi_{kn}x_n \end{aligned}$$

of the n-dimensional space R with basis $\{x_1, x_2, \cdots, x_n\}$ are linearly independent if and only if the rank of the matrix

$$\begin{Vmatrix} \xi_{11} & \xi_{12} & \cdots & \xi_{1n} \\ \xi_{21} & \xi_{22} & \cdots & \xi_{2n} \\ \cdot & \cdot & \cdot & \cdot \\ \xi_{k1} & \xi_{k2} & \cdots & \xi_{kn} \end{Vmatrix}$$

is k. Recall that the *rank* of a matrix is the maximum of the orders of all possible non-zero determinants obtained from the matrix by striking out rows and columns.

3. Subspaces. A subset $\mathfrak{M}$ of the linear space R is called a *subspace* of R if: a) the sum of any two elements in the set $\mathfrak{M}$ also belongs to $\mathfrak{M}$; b) the product of any element in $\mathfrak{M}$ by an arbitrary number also belongs to $\mathfrak{M}$. Clearly, $\mathfrak{M}$ also forms a linear space if we retain the same definitions of the operations of addition and multiplication by a number in $\mathfrak{M}$ as in the entire space.

EXAMPLES. 1. The set of all ordered n-tuples $x = \langle 0, \xi_2, \cdots, \xi_n \rangle$ in C^n is a subspace of C^n.

2. The set of all functions $x(t)$ in $C[a, b]$ which are equal to zero at a fixed point $t_0 \in [a, b]$ is a subspace of $C[a, b]$.

We note that R itself and also the set (0), which consists of the single element 0, are subspaces of R. We shall call them *trivial subspaces*.

Obviously, *the intersection of an arbitrary number of subspaces in R is a subspace in R*. In particular, the intersection of all subspaces which contain a given set

$\mathfrak{S} \subset R$ is the *minimal subspace containing* $\mathfrak{S}$; this minimal subspace is called the *linear hull of the set* $\mathfrak{S}$ or the *subspace enveloping* $\mathfrak{S}$.

I. *The linear hull of the set* $\mathfrak{S}$ *is the set of all finite linear combinations* $\alpha_1 x_1 + \cdots + \alpha_k x_k$ *of elements* x_i *in* $\mathfrak{S}$.

In fact, the set of all such linear combinations is a subspace containing $\mathfrak{S}$; on the other hand, every subspace containing $\mathfrak{S}$ must contain all these linear combinations.

A particular case of a linear hull is the sum of a finite number of subspaces $\mathfrak{M}_1, \mathfrak{M}_2, \cdots, \mathfrak{M}_k$. The *sum* of the subspaces $\mathfrak{M}_1, \mathfrak{M}_2, \cdots, \mathfrak{M}_k$ is the set of all sums $x_1+x_2+\cdots+x_k$, $x_i \in \mathfrak{M}_i$. Clearly, this set is the minimal subspace containing all the vectors of the subspaces $\mathfrak{M}_i$, $i = 1, 2, \cdots, k$, i.e. it is the linear hull of the set of all the vectors of these subspaces.

The sum of the subspaces $\mathfrak{M}_1, \mathfrak{M}_2, \cdots, \mathfrak{M}_k$ will be denoted by $\mathfrak{M}_1+\mathfrak{M}_2+\cdots+\mathfrak{M}_k$.

The subspaces $\mathfrak{M}_1, \mathfrak{M}_2, \cdots, \mathfrak{M}_k$ are said to be *linearly independent* provided the equality $x_1+x_2+\cdots+x_k = 0$, where $x_i \in \mathfrak{M}_i$, is possible if and only if $x_1 = x_2 = \cdots = x_k = 0$.

II. *If the subspaces* $\mathfrak{M}_1, \mathfrak{M}_2, \cdots, \mathfrak{M}_k$ *are linearly independent, then every vector* x *in the sum* $\mathfrak{M}_1+\mathfrak{M}_2+\cdots+\mathfrak{M}_k$ *can be uniquely represented in the form*

$$x = x_1+x_2+\cdots+x_k, \; x_i \in \mathfrak{M}_i.$$

In fact, if we also have $x = x_1'+x_2'+\cdots+x_k'$, $x_i' \in \mathfrak{M}_i$, then $0 = (x_1-x_1')+(x_2-x_2')+\cdots+(x_k-x_k')$, $x_i-x_i' \in \mathfrak{M}_i$, from which, in virtue of the linear independence of the subspaces $\mathfrak{M}_i$,

$$x_1-x_1' = x_2-x_2' = \cdots = x_k-x_k' = 0, \; x_1 = x_1', \; x_2 = x_2', \cdots, x_k = x_k'.$$

4. Quotient space. Let $\mathfrak{M}$ be fixed subspace in R. Two vectors x_1, x_2 will be called *equivalent modulo* $\mathfrak{M}$ if $x_1-x_2 \in \mathfrak{M}$; this is written symbolically as

$$x_1 \equiv x_2 (\text{mod } \mathfrak{M}). \tag{1}$$

The concept of equivalence with respect to a modulus has all the general properties of equivalence, namely:

1) $x \equiv x$ (mod $\mathfrak{M}$);

2) if $x_1 \equiv x_2$ (mod $\mathfrak{M}$), then $x_2 \equiv x_1$ (mod $\mathfrak{M}$);

3) if $x_1 \equiv x_2$ (mod $\mathfrak{M}$) and $x_2 \equiv x_3$ (mod $\mathfrak{M}$), then $x_1 \equiv x_3$ (mod $\mathfrak{M}$).

In fact, $x-x = 0 \in \mathfrak{M}$, and therefore 1) holds. Further, if $x_1 \equiv x_2$ (mod $\mathfrak{M}$), then $x_1-x_2 \in \mathfrak{M}$; then also $x_2-x_1 = -(x_1-x_2) \in \mathfrak{M}$, i.e. $x_2 \equiv x_1$ (mod $\mathfrak{M}$); consequently 2) holds. Finally, if $x_1 \equiv x_2$ (mod $\mathfrak{M}$), $x_2 \equiv x_3$ (mod $\mathfrak{M}$), then $x_1-x_2 \in \mathfrak{M}$, $x_2-x_3 \in \mathfrak{M}$; but then we also have $x_1-x_3 = (x_1-x_2)+(x_2-x_3) \in \mathfrak{M}$, i.e. $x_1 \equiv x_3$ (mod $\mathfrak{M}$) and 3) also holds.

We shall denote the set of all vectors equivalent to a fixed vector x with respect to the modulus $\mathfrak{M}$ by ξ_x; in virtue of properties 2) and 3) all vectors in ξ_x are mutually equivalent; ξ_x is called a *class of equivalent vectors*, and each vector in ξ_x is called a *representative of the class*. Clearly, a class is completely determined by any one of its representatives; in other words, if $y \in \xi_x$, then $\xi_y = \xi_x$. From this it follows that two classes ξ_x, ξ_y either are disjoint (when $y \notin \xi_x$) or they coincide (when $y \in \xi_x$). Therefore the entire space R decomposes into classes ξ_x of mutually equivalent vectors. [The foregoing discussion is of a general nature and is in fact applicable to any set $\mathfrak{A}$ in which there is defined some equivalence relation satisfying conditions 1)–3). Then the set $\mathfrak{A}$ decomposes into classes of mutually equivalent elements.]

We shall consider these classes as elements of a new linear space where the operations of addition of classes and the multiplication of a class by a number will be defined by the formulas

$$\xi_x + \xi_y = \xi_{x+y}, \qquad \alpha\xi_x = \xi_{\alpha x}. \tag{2}$$

This definition does not depend on the choice of representatives x, y of the classes ξ_x, ξ_y, respectively. In fact, if $\xi_{x_1} = \xi_x$ and $\xi_{y_1} = \xi_y$, then $x_1 - x \in \mathfrak{M}$, $y_1 - y \in \mathfrak{M}$, and therefore also $(x_1 + y_1) - (x + y) = (x_1 - x) + (y_1 - y) \in \mathfrak{M}$, i.e. $\xi_{x_1+y_1} = \xi_{x+y}$; further, $\alpha x_1 - \alpha x = \alpha(x_1 - x) \in \mathfrak{M}$, and therefore $\xi_{\alpha x_1} = \xi_{\alpha x}$. It is easy to verify that conditions a_1)–a_4) and b_1)–b_4) in the definition of a vector space (see subsection 1) are satisfied; in this connection, the zero vector is the class ξ_0, which contains the zero vector, i.e. the subspace $\mathfrak{M}$ itself; consequently, with these definitions of the operations of addition and multiplication by a number the set of all classes forms a linear space. This space is called the *quotient space R* modulo $\mathfrak{M}$ and is denoted by $R/\mathfrak{M}$.

EXAMPLE. Let R be the set of all vectors in real three-dimensional space, emanating from the origin of coordinates, and let $\mathfrak{M}$ be the set of those vectors in this set which lie on a fixed line l. Then every class ξ_x consists of vectors whose endpoints lie on the straight line l' which passes through the endpoint of the vector x parallel to l. Thus, to the classes of equivalent vectors there correspond in this case straight lines parallel to the line l.

5. Linear operators. Let R and R' be linear spaces and let $\mathfrak{D}$ and $\mathfrak{R}$ be certain sets in R and R', respectively. If there is assigned to each vector x in $\mathfrak{D}$ a vector y in $\mathfrak{R}$, where every vector y in $\mathfrak{R}$ is the image of at least one vector x in $\mathfrak{D}$, then we shall say that *an operator $y = A(x)$ from R into R' with domain $\mathfrak{D}$ and range $\mathfrak{R}$ is defined.*

If $R' = R$, then A is called an *operator in R*.

Thus, the concept of operator is a natural generalization of the concept of function to the case when the argument and value of the function are elements of vector spaces. In the sequel, we shall omit the parentheses and write simply $y = Ax$; the letter A denotes the operator itself, i.e. that rule according to which vectors y in $\mathfrak{R}$ are assigned to vectors x in $\mathfrak{D}$. If it is necessary to emphasize that $\mathfrak{D}$ and $\mathfrak{R}$ are the domain and range of the operator A, then we write $\mathfrak{D}_A$ and $\mathfrak{R}_A$. Two operators A and B from R into R' are considered to be equal if $\mathfrak{D}_A = \mathfrak{D}_B$ and $Ax = Bx$ for all vectors x in $\mathfrak{D}_A = \mathfrak{D}_B$.

If, in particular, we take C^1 as R', so that the values of the operator are complex numbers, then the operator is called a *functional*; the usual notation for a function is retained for a functional and we write $f(x)$ instead of Ax.

An operator B is called an *extension* of the operator A, and the operator A is called a *restriction* of the operator B, if $\mathfrak{D}_A \subset \mathfrak{D}_B$ and $Ax = Bx$ for all $x \in \mathfrak{D}_A$. If B is an extension of the operator A (which means that A is a restriction of the operator B), then we write $B \supset A$ or $A \subset B$. If A is an operator from R into R' and $M \subset \mathfrak{D}_A$, then AM denotes the set of all Ax, where x runs over M.

EXAMPLES. 1. $R = \mathfrak{D}_A = C^n$, $R' = \mathfrak{R}_A = C^m$ (see Example 1, subsection 1),

$$A\langle \xi_1, \xi_2, \cdots, \xi_n \rangle = \langle \xi_1', \xi_2', \cdots, \xi_m' \rangle, \tag{1}$$

where

$$\xi_j' = \sum_{k=1}^{n} a_{jk} \xi_k \qquad (j = 1, 2, \cdots, m) \tag{2}$$

and the a_{jk} are certain constants. Thus, A is an operator from C^n into C^m; when $m = n$, A is an operator in C^n. The matrix

$$a = \begin{Vmatrix} a_{11} & a_{12} \cdots a_{1n} \\ a_{21} & a_{22} \cdots a_{2n} \\ \cdot & \cdot \quad \cdot \quad \cdot \\ a_{m1} & a_{m2} \cdots a_{mn} \end{Vmatrix}$$

is called the *matrix of the operator* A.

2. $R = R' = C[a, b]$, $\mathfrak{D}_B = C'[a, b]$, $Bx = dx/dt$ where $C[a, b]$ is the set of all continuous functions on the interval $[a, b]$ (see Example 3, subsection 1), and $C'[a, b]$ is the set of all continuously differentiable functions on this interval. It is clear that here $\mathfrak{R}_B = C[a, b]$ and that B is an operator in $C[a, b]$.

3. $R = R' = C[a, b]$, $\mathfrak{D}_A$ is the set of all functions $x(t)$ in $C'[a, b]$ which satisfy the conditions $x(a) = x(b) = 0$; $Ax = dx/dt$. A is an operator in R. Clearly, A is a restriction of the operator B from Example 2, and B is an extension of the operator A.

4. $R = R' = C[a, b]$, $\mathfrak{D}_B = C'[a, b]$, $Bx = x\, dx/dt$; B is an operator in $C[a, b]$.

5. $R = C[a, b]$, $\mathfrak{D}_f = C'[a, b]$, $f(x) = \int_a^b (dx/dt)^2 dt$; $f(x)$ is a functional in $C[a, b]$.

6. $R = C[a, b]$, $\mathfrak{D}_f = C[a, b]$, $f(x) = \int_a^b x(t) dt$; $f(x)$ is a functional in $C[a, b]$.

7. $R = R' = \mathfrak{D}_A = C[a, b]$; $Ax(t) = \int_a^b K(t, s)x(s)ds$, where $K(t, s)$ is a continuous function in both variables t, s in the square $a \leqq t, s \leqq b$; A is an operator in $C[a, b]$. It is called an *integral operator* and the function $K(t, s)$ is called the *kernel* of this operator.

An operator A is said to be *linear* if $\mathfrak{D}_A$ is subspace of R (which may coincide with the entire space) and

$$A(\alpha x) = \alpha A x, \quad A(x+y) = Ax + Ay \tag{3}$$

for all $x, y \in \mathfrak{D}_A$ and all numbers α. In particular, a functional f is said to be *linear* if $\mathfrak{D}_f$ is a subspace and

$$f(\alpha x) = \alpha f(x), \; f(x+y) = f(x)+f(y) \tag{4}$$

for all $x, y \in \mathfrak{D}_f$ and all numbers α.

If it is necessary to emphasize that f is a linear functional in the real space R, so that $f(\alpha x) = \alpha f(x)$ for all real α, then f is said to be *real linear*; analogously, f is said to be *complex linear* if $f(\alpha x) = \alpha f(x)$ for all complex α. In the sequel, we shall consider primarily linear operators and functionals.

The operator A in Examples 1, 2, 3, 7 is linear; nonlinear in Example 4; the functional f is linear in Example 6; nonlinear in Example 5.

We shall consider as an example the case of a linear functional f in the finite-dimensional space C^n, having domain all of C^n. Let $\{e_1, e_2, \cdots, e_n\}$ be a basis in C^n; setting $x = \xi_1 e_1 + \xi_2 e_2 + \cdots + \xi_n e_n$ and $c_k = f(e_k)$, $k = 1, 2, \cdots, n$, we have:

$$f(x) = \xi_1 f(e_1) + \cdots + \xi_n f(e_n) = c_1 \xi_1 + \cdots + c_n \xi_n.$$

Thus, every linear functional f in C^n is given by the formula

$$f(x) = c_1 \xi_1 + \cdots + c_n \xi_n, \tag{5}$$

where $\xi_1, \cdots, \xi_n$ are the coordinates of the vector x with respect to the fixed basis, and $c_1, \cdots, c_n$ are fixed numbers.

It is clear that, also conversely, every functional defined by formula (5) is linear.

In particular, every linear functional in real three-dimensional space R^3 having domain all of R^3 is given by the formula $f(x) = c_1 \xi_1 + c_2 \xi_2 + c_3 \xi_3$ and therefore the equation $f(x) = c$ with fixed c defines a plane in R^3.

Suppose now that f is an arbitrary linear functional in an arbitrary linear space X and let $\mathfrak{D}_f = X$. In analogy with the case of the space R^3 we shall call the set of all vectors x in X which satisfy the equation $f(x) = c$ for fixed c a *hyperplane* in X.

The hyperplane $f(x) = c$ in the real space R divides the entire space X into two parts defined by the inequalities $f(x) \leqq c$ and $f(x) \geqq c$, respectively, and intersecting in the hyperplane itself.

If a set M lies entirely in some one of these parts then we say that it lies *on one side* of the hyperplane $f(x) = c$.

The operator T_a defined by the formula $T_a x = x+a$, where a is a fixed vector, is called a *translation operator*. Obviously, *under a translation the hyperplane* $f(x) = c$ *goes over into the hyperplane* $f(x) = f(a)+c$.

EXERCISE. Prove that every linear operator from C^n into C^m with domain $\mathfrak{D}_A = C^n$ has the form given in Example 1.

6. Operator calculus. a) *Addition of operators.* The *sum* $A+B$ of two operators A, B from R into R' is an operator from R into R' defined by the conditions

$$\mathfrak{D}_{A+B} = \mathfrak{D}_A \cap \mathfrak{D}_B, \tag{1}$$

$$(A+B)x = Ax+Bx \text{ for } x \in \mathfrak{D}_A \cap \mathfrak{D}_B. \tag{2}$$

It is easy to verify that

$$A+B = B+A, \ A+(B+C) = (A+B)+C. \tag{3}$$

We shall denote by O the operator from R into R' defined by the conditions $\mathfrak{D}_O = R$ and $Ox = 0$, where the 0 in the right member is the zero vector in R'. The operator O is called the *zero operator*; it is easy to see that

$$A+O = A. \tag{4}$$

b) *Multiplication of an operator by a number.* The *product* αA *of the operator* A from R into R' *by a number* α is the operator from R into R' defined by the conditions

$$\mathfrak{D}_{\alpha A} = \mathfrak{D}_A, \tag{5}$$

$$(\alpha A)x = \alpha(Ax) \text{ for all } x \in \mathfrak{D}_A. \tag{6}$$

In this connection,

$$\alpha(\beta A) = (\alpha\beta)A, \tag{7}$$

$$(\alpha+\beta)A = \alpha A+\beta A, \tag{8}$$

$$\alpha(A+B) = \alpha A+\alpha B, \tag{9}$$

$$1 \cdot A = A, \tag{10}$$

$$0 \cdot A \subset O \tag{11}$$

[in relation (11) 0 is a number (on the left) and O is the zero operator (on the right)].

We shall verify, say, relation (9); the verification of the remaining relations (7), (8), (10), and (11) is left to the reader. We have:

$$\mathfrak{D}_{\alpha(A+B)} = \mathfrak{D}_{A+B} = \mathfrak{D}_A \cap \mathfrak{D}_B,$$

$$\mathfrak{D}_{\alpha A+\alpha B} = \mathfrak{D}_{\alpha A} \cap \mathfrak{D}_{\alpha B} = \mathfrak{D}_A \cap \mathfrak{D}_B,$$

so that

$$\mathfrak{D}_{\alpha(A+B)} = \mathfrak{D}_{\alpha A+\alpha B}; \tag{12}$$

moreover, for $x \in \mathfrak{D}_{\alpha(A+B)}$,

$$\alpha(A+B)x = \alpha(Ax+Bx) = \alpha Ax+\alpha Bx = (\alpha A+\alpha B)x. \tag{13}$$

By the definition of the equality of operators, (12) and (13) mean that

$$\alpha(A+B) = \alpha A + \alpha B.$$

c) *Product of operators.* Let B be an operator from R into R' and let A be an operator from R' into R''. The *product* AB of the operators A and B is the operator from R into R'' whose domain $\mathfrak{D}_{AB}$ consists of those and only those vectors $x \in \mathfrak{D}_B$ for which $Bx \in \mathfrak{D}_A$, and which is defined by the equality

$$(AB)x = A(Bx) \text{ for } x \in \mathfrak{D}_{AB}.$$

It is easy to show that

$$A(BC) = (AB)C, \tag{14}$$

$$(A+B)C = AC+BC, \tag{15}$$

and, if the operator A is linear, that

$$A(B+C) \supset AB+AC. \tag{16}$$

The proof of these relations will be left to the reader.

We denote by 1_R the operator in R with domain $\mathfrak{D}_1 = R$, defined by the condition

$$1_R x = x \text{ for all } x \in R; \tag{17}$$

1_R is called the *identity operator* in R. The index R is usually omitted whenever this does not lead to any misunderstanding. If A is an operator from R into R' then, obviously,

$$1_{R'} \cdot A = A 1_R = A. \tag{18}$$

c) *Powers of an operator*; *operator polynomials.* Let A be an operator in R; the product $A \cdot A$ is called the *square* of the operator A and is denoted by A^2. Further, the product $A^2 \cdot A$ is called the *cube* of the operator A and is denoted by A^3. By induction, one can define in an analogous manner the *power* A^n of A for arbitrary natural number n by setting

$$A^{n+1} = A^n \cdot A. \tag{19}$$

Furthermore, we shall define the 0-th power of A by

$$A^0 = 1_R. \tag{20}$$

We have:

$$A^n A^m = A^{n+m}. \tag{21}$$

In fact, when $m = 1$ this relation coincides with (19); applying induction and making use of the associativity property (14), it is easy to check that relation (21) is valid for any natural number m.

Now suppose $p(\lambda)$ is any polynomial

$$p(\lambda) = c_0 + c_1\lambda + \cdots + c_n\lambda^n; \tag{22}$$

the *polynomial* $p(A)$ in the operator A in R is the operator

$$p(A) = c_0 1 + c_1 A + \cdots + c_n A^n. \tag{23}$$

Applying the basic properties of the operations of addition, multiplication by a number and the multiplication of operators (in partuclar, relation (21)), we conclude that:

For a fixed linear operator A the correspondence $p(\lambda) \to p(A)$ possesses the following properties:

1) *if* $p(\lambda) = \alpha_1 p_1(\lambda) + \alpha_2 p_2(\lambda)$, *then*

$$p(A) = \alpha_1 p_1(A) + \alpha_2 p_2(A);$$

2) *if* $p(\lambda) = p_1(\lambda)p_2(\lambda)$, *then* $p(A) = p_1(A)p_2(A)$.

d) *Inverse operator*. The operator B from R' into R is called the *inverse* of the operator A from R into R' if $\mathfrak{D}_B = \mathfrak{R}_A$ and

$$BAx = x \text{ for all } x \in \mathfrak{D}_A. \tag{24}$$

We note that then we also have $\mathfrak{D}_A = \mathfrak{R}_B$ and

$$ABy = y \text{ for all } y \in \mathfrak{D}_B, \tag{25}$$

so that A is also the inverse of B.

In fact, when x runs through $\mathfrak{D}_A$ then Ax runs through all of $\mathfrak{R}_A = \mathfrak{D}_B$ and therefore $x - BAx$ runs through exactly $\mathfrak{R}_B$; in other words, $\mathfrak{D}_A = \mathfrak{R}_B$. Further, if we apply the operator A to both sides of (24) and set $y = Ax$, we obtain $ABy = y$, where $y = Ax$ runs through all of $\mathfrak{D}_B = \mathfrak{R}_A$.

The operator inverse to A is denoted by A^{-1}. Thus, by the very definition of the inverse operator, we have

$$\mathfrak{D}_{A^{-1}} = \mathfrak{R}_A, \ \mathfrak{R}_{A^{-1}} = \mathfrak{D}_A \tag{26}$$

and

$$A^{-1}Ax = x, \ AA^{-1}y = y \text{ for all } x \in \mathfrak{D}_A, \ y \in \mathfrak{D}_{A^{-1}}. \tag{27}$$

The operators A and A^{-1} appear symmetrically in these relations and therefore A (as was already pointed out above) is the inverse of A^{-1}:

$$(A^{-1})^{-1} = A. \tag{28}$$

Obviously, relations (27) mean that

$$A^{-1}A \subset 1_R, \ AA^{-1} \subset 1_{R'}. \tag{29}$$

If A is an operator in R then it is possible to define negative integral powers of the operator A by setting

$$A^{-n} = (A^{-1})^n \tag{30}$$

for $n = 1, 2, 3, \cdots$. One must, however, note that the inverse operator does not exist for every operator A and, consequently, it is not true that negative integral powers A^{-n} exist for every operator A in R.

I. *If A is a linear operator and A^{-1} exists, then A^{-1} is also linear.*

In fact, if $y_1, y_2 \in \mathfrak{D}_{A^{-1}} = \mathfrak{R}_A$, then

$$A(\alpha_1 A^{-1}y_1 + \alpha_2 A^{-1}y_2) = \alpha_1 AA^{-1}y_1 + \alpha_2 AA^{-1}y_2 = \alpha_1 y_1 + \alpha_2 y_2;$$

therefore, $\alpha_1 A^{-1}y_1 + \alpha_2 A^{-1}y_2 = A^{-1}(\alpha_1 y_1 + \alpha_2 y_2)$.

II. *A linear operator A has an inverse if and only if the equation*

$$Ax = 0, \text{ where } x \in \mathfrak{D}_A, \tag{31}$$

holds only for $x = 0$.

Proof. If the operator A has an inverse, then, applying A^{-1} to both sides of (31), we obtain, in virtue of (27), that $x = A^{-1}Ax = 0$; consequently, the given condition is necessary. Conversely, suppose this condition is satisfied. We define the operator B by setting $\mathfrak{D}_B = \mathfrak{R}_A$ and

$$By = x \text{ for } y = Ax, \ x \in \mathfrak{D}_A. \tag{32}$$

These conditions define the operator B uniquely. In fact, if $y = Ax_1 = Ax_2$, then $A(x_1 - x_2) = 0$, and consequently, in virtue of the condition, we have $x_1 - x_2 = 0$, $x_1 = x_2$, i.e. x is determined uniquely by y in the formula $y = Ax$. It follows from (32) that $x = By = BAx$ for all $x \in \mathfrak{D}_A$, i.e. B is the operator inverse to A.

Let A be a linear operator in the linear space X. The number λ is called a *characteristic value* of the operator A if there exists a vector $x \neq 0$, $x \in \mathfrak{D}_A$, for which $Ax = \lambda x$; the vector x is then called a *characteristic vector of the operator A corresponding to the characteristic value λ*, and the subspace

$$\mathfrak{M}_\lambda = \{x : x \in \mathfrak{D}_A, \ Ax = \lambda x\}$$

is called the *characteristic subspace of the operator A corresponding to the characteristic value λ.*

It follows from Proposition II that *the operator $A - \lambda 1$ has an inverse if, and only if, λ is not a characteristic value of the operator A*; in particular, A has an inverse if, and only if, 0 is not a characteristic value of the operator A.

EXERCISES. Prove that a) the addition of two linear operators from C^n into C^m with domain C^n corresponds to the addition of their matrices (see Exercise 1 in subsection 5); b) the multiplication of such an operator by a number corresponds to the multiplication of its matrix by that number; c) the multiplication of two such operators from C^n into C^m and from C^m into C^p corresponds to the multiplication of their matrices (more precisely, if B is an operator from C^n into C^m and A is an operator from C^m into C^p, then the matrix of the operator AB is the product in the same order of the matrices of the operators A and B); d) a linear operator from C^n into C^m with domain C^n has an inverse if, and only if, the rank of the matrix of the operator A equals n; e) the product A_1A_2 of two integral operators in $C[a, b]$ with kernels K_1 and K_2 respectively (see Example 7, subsection 5) is the integral operator with kernel

$$K(t_1, t_2) = \int_a^b K_1(t_1, t)K_2(t, t_2)dt. \tag{33}$$

The kernel K is called the *composition* of the kernels K_1 and K_2; f) if A is an everywhere defined linear operator in the space C^n, then the number λ is a characteristic value of the operator A if, and only if, the determinant of the matrix of the operator $A-\lambda_1$ equals zero.

7. Invariant subspaces. Let A be a linear operator in R with domain $\mathfrak{D}_A = R$. A subspace $\mathfrak{M} \subset R$ is said to be *invariant* with respect to A if the operator A maps every vector in $\mathfrak{M}$ into a vector in $\mathfrak{M}$, i.e.

$$\xi \in \mathfrak{M} \text{ implies } A\xi \in \mathfrak{M}. \tag{1}$$

If $\mathfrak{M}$ is invariant with respect to the operator A, then A can also be considered as an operator in $\mathfrak{M}$. Specifically, one can define the operator $A_{\mathfrak{M}}$ in $\mathfrak{M}$ by setting $A_{\mathfrak{M}}\xi = A\xi$ for $\xi \in \mathfrak{M}$.

Let us now consider the quotient space $R/\mathfrak{M}$ (see subsection 4); if $\mathfrak{M}$ is invariant with respect to the operator A then one can define an operator $A^{\frown}$ in $R/\mathfrak{M}$ by setting $A^{\frown}\xi_x = \xi_y$ for $Ax = y$. This definition does not depend on the choice of the representative $x \in \xi_x$. In fact, if $\xi_{x_1} = \xi_x$ then $x_1 - x \in \mathfrak{M}$, from which, in virtue of the invariance of $\mathfrak{M}$, we have $Ax_1 - Ax = A(x_1 - x) \in \mathfrak{M}$, so that the vectors $y_1 = Ax_1$ and $y = Ax$ define the same class ξ_y.

8. Convex sets and Minkowski functionals. The *segment* $[x_1, x_2]$ *connecting the vectors* x_1 *and* x_2 is the set of all vectors $x = (1-t)x_1 + tx_2$, where $0 \leqq t \leqq 1$; the vectors x_1, x_2 themselves are called the *endpoints* of the segment $[x_1, x_2]$ and all the remaining points of this segment are called its *interior points*. If $x_1 = x_2$, then the segment $[x_1, x_2]$ consists of only one point, $x_1 = x_2$.

A set K in R is said to be *convex* if together with any two of its vectors x_1, x_2 it contains the entire segment which connects them.

It follows directly from this definition that the intersection of convex sets is also a convex set.

Moreover, if A is a linear operator from R into R' and $K \subset \mathfrak{D}_A$ is a convex set, then AK is a convex set in R'.

I. *If* $x_1, x_2, \cdots, x_n$ *belong to a convex set* K, *then also* $t_1x_1 + t_2x_2 + \cdots + t_nx_n \in K$ *for* $t_1 \geqq 0, \cdots, t_n \geqq 0$ *and* $t_1 + t_2 + \cdots + t_n = 1$.

For $n = 2$, this assertion coincides with the definition of a convex set (since $t_1x_1+t_2x_2$ belongs to the segment $[x_1, x_2]$), and for larger n this is proved by induction. Thus, for example, if $t_1+t_2 > 0$ and $t_1+t_2+t_3 = 1$, then

$$\frac{t_1}{t_1+t_2}x_1+\frac{t_2}{t_1+t_2}x_2 \in K,$$

and therefore also

$$t_1x_1+t_2x_2+t_3x_3 = (t_1+t_2)\left(\frac{t_1}{t_1+t_2}x_1+\frac{t_2}{t_1+t_2}x_2\right)+t_3x_3 \in K.$$

Let K be a convex set containing not less than two points. A point x_0 is called a *boundary point* of the set K if there exists a segment $[x_1, x_2]$, containing x_0 in its interior, such that all interior points of the segment $[x_1, x_0]$, belong, but no interior points of the segment $[x_0, x_2]$ belong to the set K. If K consists of one point x_0 then we consider x_0 to be a boundary point of the set K. But in general a boundary point x_0 may or may not belong to the set K. The set of all boundary points of a convex set is called its *boundary*.

This concept of a boundary does not coincide with the concept of topological boundary, defined below in subsection 3, § 2. An example, for instance, is a convex set in the plane (with the usual definitions of the topology and operations of a linear space; see Example 1, subsection 1, § 1 and Example 1, in subsection 2, § 2), consisting of all the points of some segment.

II. *Under the translation $T_ax = x+a$, a convex set K goes over into the convex set $K+a$ and the boundary of the set K goes over into the boundary of the set $K+a$.* [M_1+M_2, where M_1, M_2 are arbitrary sets in the linear space X, is the set of all vectors $x+y$, $x \in M_1$, $y \in M_2$; in particular, M_1+a denotes the set of all vectors $x+a$, $x \in M_1$. Analogously, αM denotes the set of all vectors αx, $x \in M$, and if $\mathfrak{A}$ is a set of numbers, then $\mathfrak{A}M$ is the set of all vectors αx, $\alpha \in \mathfrak{A}$, $x \in M$.]

This assertion follows directly from the fact that under the indicated translation every segment $[x_1, x_2]$ goes over into the segment $[x_1+a, x_2+a]$.

A convex set K in R is called *absorbing* if for every $x \in R$ there exists a positive number α such that $\alpha x \in K$, i.e. $x \in (1/\alpha)K$. It is clear that $0 \in K$ and if $x \in \beta K$ and $\gamma > \beta$, then $x \in \gamma K$.

III. *Let K be an absorbing convex set in R. Then the formula*

$$p(x) = \inf\{\beta : \beta > 0, x \in \beta K\} \tag{1}$$

defines a functional p on R which possesses the following properties:

1) $p(x) \geqq 0$ *for all* $x \in R$;
2) $p(x+y) \leqq p(x)+p(y)$ *for all* $x, y \in R$;
3) $p(\alpha x) = \alpha p(x)$ *for all* $\alpha \geqq 0$ *and all* $x \in R$;
4) *if* $x \in K$, *then* $p(x) \leqq 1$;
5) *if* $p(x) < 1$, *then* $x \in K$.

Proof. We note first of all that by the definition of an absorbing convex set the numbers β occurring in formula (1) exist for every $x \in R$ so that p is defined on all of R. By its very definition, $p(x) \geqq 0$, and for every $\varepsilon > 0$

$$x \in [p(x)+\varepsilon]K. \tag{2}$$

We shall prove that condition 2) is satisfied. To this end, we set $t = (p(y)+\varepsilon)/\{p(x)+p(y)+2\varepsilon\}$; consequently,

$$1-t = \frac{p(x)+\varepsilon}{p(x)+p(y)+2\varepsilon}. \tag{3}$$

We deduce from (2) that

$$\frac{1}{p(x)+\varepsilon}\, x \in K, \quad \frac{1}{p(y)+\varepsilon}\, y \in K;$$

consequently, in virtue of the convexity of the set K,

$$(1-t)\,\frac{1}{p(x)+\varepsilon}\, x + t\,\frac{1}{p(y)+\varepsilon}\, y \in K.$$

Taking (3) into account, we obtain

$$\frac{1}{p(x)+p(y)+2\varepsilon}(x+y) \in K,$$

i.e. $x+y \in [p(x)+p(y)+2\varepsilon]K$.

By the definition of $p(x+y)$ this means that $p(x+y) \leqq p(x)+p(y)+2\varepsilon$, from which 2) follows since the number $\varepsilon > 0$ is arbitrary. Further, for $\alpha > 0$, $p(\alpha x) = \inf\{\beta : \beta > 0, \alpha x \in \beta K\} = \inf\{\beta : \beta > 0, x \in (\beta/\alpha)K\} = \inf\{\alpha\beta : \beta > 0, x \in \beta K\} = \alpha p(x)$; for $\alpha = 0$, $p(0x) = \inf\{\beta : \beta > 0, 0x \in \beta K\} = \inf\{\beta : \beta > 0\} = 0$; thus we have 3). Finally, if $x \in K$, then $1 \in \{\beta : x \in \beta K\}$, from which 4) follows. And if $p(x) < 1$, then, setting $\varepsilon = 1-p(x)$ in (2), we obtain that $x \in K$.

The functional p, constructed in Proposition III, is called the *Minkowski functional* or *gauge* of the set K.

IV. *Every functional p on R, satisfying conditions* 1)–3) *of Proposition III, is the Minkowski functional of an absorbing convex set.*

Proof. Let $K = \{x : x \in R, p(x) \leqq 1\}$. K is convex. In fact, if $x_1, x_2 \in K$ and $0 \leqq t \leqq 1$, then $p((1-t)x_1+tx_2) \leqq (1-t)p(x_1)+tp(x_2) \leqq (1-t)+t = 1$; consequently, $[x_1, x_2] \in K$. Furthermore, for every $x \in R$ and $\varepsilon > 0$ we have that $[1/(p(x)+\varepsilon)]x \in K$, so that K is an absorbing set. Finally,

$$p(x) = \inf\{\beta : \beta > 0, p(x) \leqq \beta\} = \inf\{\beta : \beta > 0, x \in \beta K\},$$

i.e. p is the Minkowski functional of the set K.

Motivated by this, we shall call every functional p on R possessing properties 1)–3) of Proposition III a *Minkowski functional.*

We note, however, that p will be a Minkowski functional not only of the set $K = \{x: p(x) \leqq 1\}$ but also of the set $K_0 = \{x: p(x) < 1\}$, and also of every convex set included between K_0 and K.

We note that if p is a Minkowski functional, then αp is also a Minkowski functional for arbitrary $\alpha \geqq 0$.

V. *If $p(x)$ is a Minkowski functional, then for arbitrary $c > 0$ and arbitrary $a \in R$ the set K of all vectors x satisfying the condition*

$$p(x-a) \leqq c$$

is convex and its boundary consists of precisely those vectors x for which $p(x-a) = c$. If moreover $a = 0$, then K is an absorbing set.

Proof. Without loss of generality, we may assume that $a = 0$; in virtue of II, the general case reduces to this case by means of the translation $T_a x = x+a$. Let x_1, $x_2 \in K$, i.e. $p(x_1) \leqq c$, $p(x_2) \leqq c$. Then, in virtue of conditions 2) and 3), we have, for $0 \leqq t \leqq 1$, that

$$\begin{aligned} p((1-t)x_1+tx_2) &\leqq p((1-t)x_1)+p(tx_2) = (1-t)p(x_1)+tp(x_2) \\ &\leqq (1-t)c+tc = c; \end{aligned}$$

consequently, the entire segment $[x_1, x_2]$ also belongs to K, i.e. K is convex.

We shall now prove the second assertion. Suppose $p(x_0) = c$. We set $x_1 = \alpha x_0$, $x_2 = \beta x_0$, where $0 < \alpha < 1$ and $\beta > 1$. Then, for $0 < t < 1$,

$$\begin{aligned} p((1-t)x_0+tx_1) &= p((1-t+\alpha t)x_0) = [1-(1-\alpha)t]c < c, \\ p((1-t)x_0+tx_2) &= p((1-t+\beta t)x_0) = [1+(\beta-1)t]c > c, \end{aligned}$$

so that all interior points of the segment $[x_1, x_0]$, but none of the interior points of the segment $[x_0, x_2]$, belong to K; consequently, x_0 is a boundary point of the set K.

Conversely, let x_0 be a boundary point of the set K and suppose all interior points of the segment $[x_1, x_0]$, but that none of the interior points of the segment $[x_0, x_2]$, belong to K. This means that for $0 < t < 1$, we have

$$p((1-t)x_0+tx_1) \leqq c, \quad p((1-t)x_0+tx_2) > c.$$

But then

$$\begin{aligned} (1-t)p(x_0) = p((1-t)x_0) &= p((1-t)x_0+tx_1-tx_1) \\ &\leqq p((1-t)x_0+tx_1)-tp(-x_1) \leqq c+tp(-x_1) \end{aligned}$$

and

$$c < p((1-t)x_0)+p(tx_2) = (1-t)p(x_0)+tp(x_2),$$

i.e.

$$(1-t)p(x_0) \leqq c+tp(-x_1) \text{ and } c < (1-t)p(x_0)+tp(x_2).$$

Passing to the limit as $t \to 0$ in these inequalities, we obtain

$$p(x_0) \leqq c \text{ and } c \leqq p(x_0),$$

from which it follows that $p(x_0) = c$.

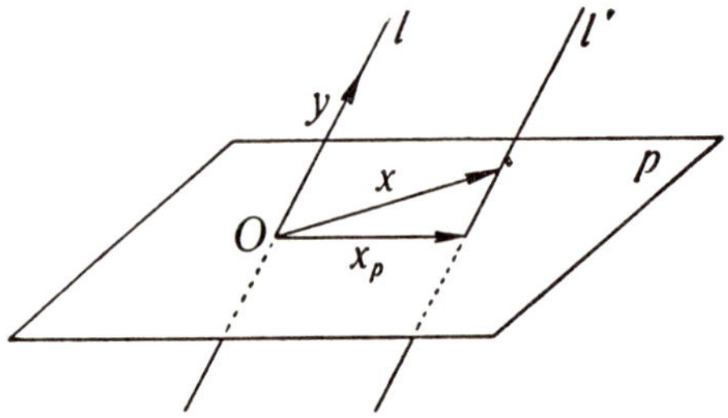

Fig. 1.

The last assertion of Proposition V is obvious.

REMARK. Proposition V remains valid for the set of all vectors x, satisfying an inequality of the form $p(x-a) < c$.

EXAMPLES. 1. In real three-dimensional space R^3 the endpoints of the vectors $x = (1-t)x_1 + tx_2$, $0 \leqq t \leqq 1$, form a segment which connects the endpoints of the vectors x_1 and x_2 (Fig. 1).

2. In real two-dimensional space R^2, we set

$$p(x) = \sqrt{a_1 \xi_1^2 + a_2 \xi_2^2},$$

where $x = (\xi_1, \xi_2)$, $a_1 > 0$, $a_2 > 0$. It is easy to verify that p is a Minkowski functional. The endpoints of all vectors x which satisfy the condition $p(x) < c$ fill the interior of the ellipse

$$a_1 \xi_1^2 + a_2 \xi_2^2 = c^2.$$

9. Theorems on the extension of a linear functional.

LEMMA. *Let $\mathfrak{M}$ be a subspace of a real linear space R and let $\mathfrak{M}'$ be the subspace enveloping $\mathfrak{M}$ and a vector $x_0 \notin \mathfrak{M}$. Further, let p be a Minkowski functional in $\mathfrak{M}'$ and f a linear functional in $\mathfrak{M}$ which satisfies the condition*

$$f(x) \leqq p(x) \textit{ for all } x \in \mathfrak{M}. \tag{1}$$

Then f can be extended to a linear functional f' defined in all of $\mathfrak{M}'$ and satisfying the condition

$$f'(x) \leqq p(x) \textit{ for all } x \in \mathfrak{M}'. \tag{2}$$

Proof. In virtue of (1), for $y', y'' \in \mathfrak{M}$,

$$f(y')-f(y'') = f(y'-y'') \leqq p(y'-y'') = p((y'+x_0)-(y''+x_0))$$
$$\leqq p(y'+x_0)+p(-y''-x_0),$$

from which it follows that

$$-p(-y''-x_0)-f(y'') \leqq p(y'+x_0)-f(y').$$

Therefore,

$$m = \sup_{y \in \mathfrak{M}} \{-p(-y-x_0)-f(y)\} \text{ and } M = \inf_{y \in \mathfrak{M}} \{p(y+x_0)-f(y)\}$$

are finite and $m \leqq M$. Let c_0 be any number between m and M:

$$m \leqq c_0 \leqq M;$$

then, for all $y \in \mathfrak{M}$,

$$-p(-y-x_0)-f(y) \leqq c_0 \leqq p(y+x_0)-f(y). \tag{3}$$

By definition, $\mathfrak{M}'$ is the set of all vectors x of the form

$$x = y+\alpha x_0, \tag{4}$$

where $y \in \mathfrak{M}$ and α is a real number. We define a functional f' in $\mathfrak{M}'$ by setting

$$f'(x) = f(x)+\alpha c_0$$

for a vector x of the form (4). We note that every vector x in $\mathfrak{M}'$ is uniquely representable by formula (4) and therefore f' is uniquely defined by the vector $x \in \mathfrak{M}'$. In fact, if $x = y_1+\alpha_1 x_0$ and $x = y_2+\alpha_2 x_0$, then

$$y_1-y_2 = (\alpha_2-\alpha_1)x_0; \tag{5}$$

but $y_1-y_2 \in \mathfrak{M}$, and $(\alpha_2-\alpha_1)x_0 \notin \mathfrak{M}$ with $\alpha_2-\alpha_1 \neq 0$; therefore, (5) is possible only when $\alpha_2-\alpha_1 = 0$, $y_1-y_2 = 0$. It is easy to verify that the functional f' is linear; it remains to prove that condition (2) is satisfied, i.e. that

$$f(y)+\alpha c_0 \leqq p(y+\alpha x_0) \tag{2'}$$

for all $y \in \mathfrak{M}$ and all real α. When $\alpha = 0$ this inequality coincides with condition (1); therefore, it is sufficient to consider the case $\alpha \neq 0$. Suppose $\alpha > 0$; replacing y by $(1/\alpha)y$ in the right inequality of (3), we obtain

$$c_0 \leqq p\left(\frac{y}{\alpha}+x_0\right)-f\left(\frac{y}{\alpha}\right).$$

It follows that

$$\alpha f\left(\frac{y}{\alpha}\right)+\alpha c_0 \leqq \alpha p\left(\frac{y}{\alpha}+x_0\right),$$

i.e.

$$f(y)+\alpha c_0 \leqq p(y+\alpha x_0).$$

Now let $\alpha < 0$; replacing y by $(1/\alpha)y$ in the left inequality of (3), we obtain

$$-p\left(-\frac{y}{\alpha}-x_0\right)-f\left(\frac{y}{\alpha}\right) \leqq c_0.$$

It follows that

$$(-\alpha)p\left(-\frac{y}{\alpha}-x_0\right) \geqq \alpha c_0+\alpha f\left(\frac{y}{\alpha}\right).$$

i.e.

$$p(y+\alpha x_0) \geqq f(y)+\alpha c_0.$$

Thus, condition (2) is satisfied in both cases, and the lemma is proved.

THEOREM 1 (HAHN [1]-BANACH [1]). *Suppose p is a Minkowski functional defined in the real linear space R and that f is a linear functional defined in some subspace $\mathfrak{M} \subset R$ and satisfying the condition*

$$f(x) \leqq p(x) \textit{ for all } x \in \mathfrak{M}. \tag{6}$$

Then f can be extended to a linear functional F defined in the entire space R and satisfying the condition

$$F(x) \leqq p(x) \textit{ for all } x \in R. \tag{7}$$

Proof. Let F_p be the set of all linear functionals f' which are extensions of the functional f and satisfy the inequality $f'(x) \leqq p(x)$ for all $x \in \mathfrak{D}_{f'}$. If f'_1, f'_2 are two functionals in F_p we shall agree to write $f'_1 \prec f'_2$ if f'_2 is an extension of the functional f'_1. Then F_p is a partially ordered set which satisfies the condition of Zorn's lemma (see Appendix I); namely, an upper bound of the linearly ordered set $F'_p \subset F_p$ is a functional $f^\frown$ defined on $\bigcup_{f' \in F'_p} \mathfrak{D}_{f'}$ by means of the equality $f^\frown(x) = f'(x)$ where $x \in \mathfrak{D}_{f'}, f' \in F'_p$. Consequently, F_p has a maximal element F which, in virtue of the preceding lemma, is defined on the entire space R.

COROLLARY 1. *If p is a Minkowski functional in the real linear space R, then for arbitrary vector $x_0 \in R$ there exists a linear functional F, defined on all of R and satisfying the conditions*

$$F(x_0) = p(x_0), \tag{8}$$

$$F(x) \leqq p(x) \textit{ for all } x \in R. \tag{7}$$

Proof. We shall denote by $\mathfrak{M}$ the set of all vectors αx_0, where α runs through all real numbers; $\mathfrak{M}$ is a subspace of R. We define a linear functional f in $\mathfrak{M}$ by means of the formula

$$f(\alpha x_0) = \alpha p(x_0), \tag{9}$$

so that

$$f(x_0) = p(x_0). \tag{10}$$

We shall prove that f satisfies condition (6); on the basis of Theorem 1 it can then be extended to a linear functional F defined on all of R and satisfying condition (7); in virtue of (10) condition (8) will then be satisfied also, and Corollary 1 will be proved.

But condition (6) for f reduces to the satisfaction of the inequality

$$f(\alpha x_0) \leqq p(\alpha x_0) \tag{11}$$

for all real α. It is satisfied (and reduces to an equality) for $\alpha \geqq 0$ because in virtue of (9) we have

$$f(\alpha x_0) = \alpha p(x_0) = p(\alpha x_0).$$

In order to prove (11) for $\alpha < 0$, we first note that, since $p(x) \geqq 0$ for every $x \in R$,

$$-p(x_0) \leqq p(-x_0).$$

It follows that when $\alpha < 0$,

$$f(\alpha x_0) = \alpha p(x_0) \leqq -\alpha p(-x_0) = p(\alpha x_0).$$

REMARK. The assertions of the lemma, Theorem 1 and its corollary remain valid for functionals p which satisfy only conditions 2 and 3 of Proposition III. In fact, in the proofs of the lemma and Theorem 1, the condition $p(x) \geqq 0$ was not used; and in the proof of the corollary it suffices to note that $0 = p(x-x) \leqq p(x)+p(-x)$ and therefore $-p(-x) \leqq p(x)$ for all $x \in R$.

A Minkowski functional p (in a real or complex linear space R) is called a *seminorm* if $p(\alpha x) = |\alpha| p(x)$ for all (real respectively complex) numbers α and all $x \in R$. A seminorm p is called a *norm* if $p(x) > 0$ for $x \neq 0$; to denote a norm we usually write $||x||$ or $|x|$ instead of $p(x)$. A set M in the (real or complex) space R is said to be *symmetric* if $x \in M$ and $|\alpha| = 1$ imply $\alpha x \in M$.

I. *If p is a seminorm, then*:

1) *the set $\mathfrak{M}$ of all vectors x in R which satisfy the condition $p(x) = 0$ is a subspace in R*;

2) *for fixed $c > 0$, the set K of all vectors x in R, which satisfy the inequality $p(x) \leqq c$, is a symmetric absorbing convex set in R. Conversely, if K is a symmetric absorbing convex set, then its Minkowski functional p is a seminorm.*

In fact, if $x, y \in \mathfrak{M}$, then $p(x) = 0$, $p(y) = 0$ and therefore $p(\alpha x) = |\alpha| p(x) = 0$, $0 \leqq p(x+y) \leqq p(x)+p(y) = 0$; consequently, αx, $x+y \in \mathfrak{M}$ also, i.e. $\mathfrak{M}$ is a subspace.

Further, if $x \in K$, $|\alpha| = 1$, then $p(\alpha x) = |\alpha| p(x) = p(x) \leqq c$ and therefore $\alpha x \in K$; consequently, K is symmetric. The remaining properties of K were proved in subsection 8.

Conversely, let K be a symmetric absorbing convex set and p its Minkowski functional. Let $|\alpha| = 1$. Then $\alpha^{-1}K = K$ and hence

$$\begin{aligned} p(\alpha x) &= \inf \{\beta \colon \beta > 0,\ \alpha x \in \beta K\} \\ &= \inf \{\beta \colon \beta > 0,\ x \in \beta(\alpha^{-1}K)\} \\ &= \inf \{\beta \colon \beta > 0,\ x \in \beta K\} = p(x). \end{aligned}$$

It follows that, for arbitrary $\alpha = re^{i\theta}$, where $r \geqq 0$ and θ is real, $p(\alpha x) = p(re^{i\theta}x) = p(rx) = rp(x) = |\alpha| p(x)$.

THEOREM 2 (SUKHOMLINOV [1]). (This result was also obtained, independently, by BOHNENBLUST and SOBCZYK [1].) *Suppose p is a seminorm in the complex linear space R and that f is a linear functional defined in some subspace $\mathfrak{M} \subset R$ and satisfying the condition*

$$|f(x)| \leqq p(x) \text{ for all } x \in \mathfrak{M}. \tag{12}$$

Then f can be extended to a linear functional F defined on the entire space R, and satisfying the condition

$$|F(x)| \leqq p(x) \text{ for all } x \in R. \tag{13}$$

Proof. We shall consider R as a real space and set

$$f(x) = f_1(x) + if_2(x), \tag{14}$$

where $f_1(x)$ and $f_2(x)$ are real. Clearly, f_1 and f_2 are real linear functionals in $\mathfrak{M}$ considered as a real space; since

$$i[f_1(x) + if_2(x)] = if(x) = f(ix) = f_1(ix) + if_2(ix),$$

we have

$$f_1(ix) = -f_2(x) \text{ for all } x \in \mathfrak{M}. \tag{15}$$

It follows from (12) that $f_1(x) \leqq p(x)$ for all $x \in \mathfrak{M}$, so that on the basis of Theorem 1, f_1 can be extended to a real linear functional F_1, defined on all of R and satisfying the condition

$$F_1(x) \leqq p(x) \text{ for all } x \in R. \tag{16}$$

We set

$$F(x) = F_1(x) - iF_1(ix); \tag{17}$$

in virtue of (14) and (15), $F(x) = f(x)$ on $\mathfrak{M}$.

It follows from (17) that $F(ix) = iF(x)$; consequently, F is a complex linear functional in the complex space R. To complete the proof, it remains to establish that F satisfies condition (13). This is clear when $F(x) = 0$. Suppose $F(x) \neq 0$ and set $\theta = \arg F(x)$. Then, in virtue of (16),

$$|F(x)| = F(e^{-i\theta}x) = F_1(e^{-i\theta}x) \leqq p(e^{-i\theta}x) = p(x),$$

and the theorem is proved.

COROLLARY 2. *If p is a seminorm in the complex linear space R, then for arbitrary vector $x_0 \in R$, there exists a linear functional F, defined on all of R and satisfying the conditions*

$$F(x_0) = p(x_0), \tag{18}$$

$$|F(x)| \leqq p(x) \text{ for all } x \in R. \tag{13}$$

Proof. Denote by $\mathfrak{M}$ the set of all vectors αx_0, where α runs through all complex numbers, and define the linear functional f in $\mathfrak{M}$ by means of the formula

$$f(\alpha x_0) = \alpha p(x_0). \tag{19}$$

Condition (12) will be satisfied for f because $|f(\alpha x_0)| = |\alpha| p(x_0) = p(\alpha x_0)$. On the basis of Theorem 2, f can be extended to a linear function F which satisfies condition (13). It follows from (19) that condition (18) will also be satisfied.

The propositions proved above have a simple geometric interpretation. We shall explain, for instance, the geometric meaning of Corollary 1. Suppose K is a convex set defined by the inequality $p(x) < c$ with fixed $c > 0$, x_0 a boundary point of the set K, and P the hyperplane defined by the equation $F(x) = c$. Since $F(x_0) = p(x_0) = c$, this hyperplane passes through the boundary point x_0. On the other hand, $F(x) \leqq p(x) < c$ for $x \in K$; consequently, K lies on one side of the hyperplane $F(x) = c$. A hyperplane passing through a boundary point of the convex set K is called a *hyperplane of support* to K if K lies on one side of this hyperplane. The preceding discussion shows that *if p is a Minkowski functional in a real linear space R, then through any boundary point of the convex set K, defined by an inequality of the form $p(x-a) < c$ (or $p(x-a) \leqq c$), a hyperplane of support to K can be drawn.*

We shall now assume that some vector x_1 does not belong to either the set K or to its boundary. Without loss of generality, we may assume that $a = 0$; then $p(x_1) > c$. We set $t = c/p(x_1)$ and $x_2 = tx_1$; then $p(x_2) = c$, and consequently x_2 belongs to the boundary of K. Suppose $f(x) = c$ is a hyperplane of support to K passing through x_2, so that $f(x_2) = c$. Then $f(x_1) = f\{(1/t)x_2\} = (1/t)f(x_2) = c/t = p(x_1) > c$. But this means that K and x_1 lie on different sides of the hyperplane $f(x) = c$. We shall say in this case that the hyperplane $f(x) = c$ *separates* x_1 from K. Thus, *if p is a Minkowski functional and x_1 is a vector which belongs neither to the convex set K, defined by the inequality $p(x-a) < c$, nor to its boundary, then there exists a hyperplane of support to K separating x_1 from K.*

§ 2. Topological spaces

1. Definition of a topological space. A set X is called a *topological space* if X contains a family $\mathfrak{U}$ of subsets U which has the following properties:

1) $\phi \in \mathfrak{U}$, $X \in \mathfrak{U}$;

2) *the union of an arbitrary family of sets, belonging to* $\mathfrak{U}$, *also belongs to* $\mathfrak{U}$;

3) *the intersection of an arbitrary finite number of sets, belonging to* $\mathfrak{U}$, *also belongs to* $\mathfrak{U}$.

(Here, and in the sequel, ϕ denotes the void set.)

Then the sets $U \in \mathfrak{U}$ are called *open sets* of the topological space X, and the elements $x \in X$ are called *points* of this space. We also say that the family of sets $\mathfrak{U}$ *defines a topology T in the set X.*

Different families $\mathfrak{U}$ may be given in one and the same set X; they then define different topologies in X. We shall say that the topology T_1, defined by the family $\mathfrak{U}_1$, is *weaker* than the topology T_2, defined by the family $\mathfrak{U}_2$, and we shall write $T_1 < T_2$ or $T_2 > T_1$, if $\mathfrak{U}_1 \subset \mathfrak{U}_2$ and $\mathfrak{U}_1 \neq \mathfrak{U}_2$; in this case we shall also say that the topology T_2 is *stronger* than the topology T_1. But if it is only known that $\mathfrak{U}_1 \subset \mathfrak{U}_2$, then we shall write $T_1 \leqq T_2$ or $T_2 \geqq T_1$ and say that T_2 *majorizes* T_1. A family $\mathfrak{B}$ of open sets will be called a *basis* of the topological space X if every nonvoid open set in X is the union of sets $U \in \mathfrak{B}$.

EXAMPLES. 1. Suppose $X = R^1$ is the set of all real numbers. We shall agree to consider all possible unions of open intervals and the empty set ϕ to be open sets in X. Obviously, conditions 1–3 will be satisfied and R^1 is then a topological space. The open intervals themselves form a basis in the space.

2. Suppose $X = C^1$ is the set of all complex numbers. We shall consider a basis in C^1 to be the collection of all sets $|z-z_0| < \varepsilon$ with all possible $z_0 \in C^1$ and $\varepsilon > 0$. Then C^1 will be a topological space.

2. Interior of a set; neighborhoods. Let M be any subset in X. The *interior of the set M* is the union of all open sets contained in M. The interior of the set M is denoted by *Int M*. Clearly, Int M is the maximal open set which is contained in M.

A *neighborhood $U(x)$ of the point x* is any open set U containing x. Clearly, the intersection of two neighborhoods of the point x is also a neighborhood of this point. A family of neighborhoods $W(x)$ of the point x is called a *neighborhood basis* of this point if every neighborhood $U(x)$ also contains some neighborhood $W(x)$ of the family under consideration.

In N. Bourbaki's *General Topology* a neighborhood of the point x_0 is any set M for which $x \in \text{Int } M$. Using this definition does not modify the topological concepts we state.

It is directly clear from this definition that a neighborhood basis must have the following properties:

1) $x \in W(x)$; 2) every intersection $W_1(x) \cap W_2(x)$ of neighborhoods from the basis contains a neighborhood belonging to the basis; 3) if $y \in W(x)$, then there exists a neighborhood $W(y) \subset W(x)$ in the neighborhood basis of the point y.

Conversely, if every point $x \in X$ is assigned a family of sets $W(x)$ which satisfies conditions 1–3, then a topology can be defined in X by taking ϕ and all possible unions of sets $W(x)$ to be the open sets in X. The sets $W(x)$ themselves form a basis of this topological space. Therefore X can also be topologized by selecting a neighborhood basis for every point $x \in X$.

Thus, in Example 1 of subsection 1, all possible open intervals $(x_0-\varepsilon, x_0+\varepsilon)$, $\varepsilon > 0$, can be taken as a neighborhood basis of the number x_0.

Clearly, two families of neighborhoods $\{W'\}$, $\{W''\}$ define the same topological space if and only if every neighborhood W' contains some neighborhood W'', and conversely, every neighborhood W'' contains some neighborhood W'.

Further, a topology defined by the family $\{W'\}$ is weaker than the topology defined by the family $\{W''\}$ if and only if these topologies are distinct and every neighborhood W' contains some neighborhood W''.

EXAMPLES. 1. We define a topology in real n-dimensional space R^n by taking as a neighborhood basis of every point $x_0 = \langle \xi_1^0, \xi_2^0, \cdots, \xi_n^0 \rangle$ the family of sets $W(x)$ defined by the inequalities

$$|\xi_k - \xi_k^0| < \varepsilon, \varepsilon > 0, \qquad k = 1, 2, \cdots, n.$$

It is easily seen that conditions 1–3 will then be satisfied so that R^n becomes a topological space.

2. A topology is defined in the complex space C^n analogously.

3. Closed sets; closure of a set. The complements of open sets are called *closed sets*. It follows from the properties of open sets (see subsection 1) that:

1) *the void set and the entire space are closed*;

2) *the intersection of an arbitrary family of closed sets is closed*;

3) *the union of a finite number of closed sets is closed.*

The intersection of all closed sets containing a given set $M \subset X$ is the minimal closed set containing M. It is called the *closure* of the set M and is denoted by $\overline{M}$. Obviously $\overline{M} = M$ if and only if M is closed. Furthermore, it is clear that:

1′) $M \subset \overline{M}$;

2′) $\overline{\overline{M}} = \overline{M}$ (because $\overline{M}$ is closed);

3′) $\overline{M_1 \cup M_2 \cup \cdots \cup M_n} = \overline{M}_1 \cup \overline{M}_2 \cup \cdots \cup \overline{M}_n$ for an arbitrary finite number of sets $M_1, M_2, \cdots, M_n$;

4′) $\overline{\phi} = \phi$.

One can also define a topological space as a set on which a closure operation is given satisfying conditions 1′–4′. This definition is equivalent to the original definition. In fact, it suffices to take the open sets to be the complements of closed sets, i.e. the sets M satisfying the condition $\overline{M} = M$. Then conditions 1–3 of subsection 1 are satisfied and the topology given by these open sets defines a closure operation equivalent to the initial definition. We leave the verification of these assertions to the reader.

Further, $x \in \overline{M}$ if and only if $x \notin \operatorname{Int}(X-M)$, i.e. if and only if every neighborhood $U(x)$ contains at least one point of the set M. Every point $x \in \overline{M}$ is called a *contact point* of the set M. The concept of limit point is closely related to the concept of contact point. A point x is called a *limit point* of the set M if every neighborhood $U(x)$ contains a point in the set M which is distinct from x. Clearly, every contact point of the set M which does not belong to M is a limit point of M.

A set M is said to be *perfect* if it coincides with the set of all its limit points. The set $\overline{M} - \operatorname{Int} M$ is called the *boundary* of the set M.

The concept of limit of a sequence is also closely related to the notion of a contact point. A point x is called the *limit of the sequence* $x_1, x_2, x_3, \cdots$ and it is denoted by $\lim_{n\to\infty} x_n$, if every neighborhood $U(x)$ contains all the terms of this sequence starting with some one of them; a sequence having a limit is said to be *convergent*.

Clearly, the only possible cases are the following:

1) All the terms of the sequence $x_1, x_2, x_3, \cdots$, beginning with some one of them, coincide; then they coincide with the limit x of this sequence.

2) There is an infinite set of distinct terms in the sequence; then the limit of the sequence is a limit point of the set of its points.

A set M in the topological space X is said to be *dense* in X if $\overline{M} = X$. A space X is said to be *separable* if it contains a countable set which is dense in X.

If two distinct topologies T_1 and T_2 are given on the set X, then T_1 is weaker than T_2 if and only if the closure of every set in the topology T_2 is contained in its closure in the topology T_1; this follows directly from the definition given in subsection 1.

4. Subspaces. Every subset Y of a topological space X can be transformed into a topological space by assuming the open sets in Y to be the intersections of the open sets in X with Y. A space Y with a topology defined in it in this manner is called *a subspace* of the topological space X and we say that the topology in Y is *induced* by the topology of the space X. It follows directly from this definition that

1) *if $\{U\}$ is a basis of open sets in X, then $\{U \cap Y\}$ is a basis of open sets in Y*;

2) *if $\{W\}$ is a neighborhood basis in X, then $\{W \cap Y\}$ is a neighborhood basis in Y*;

3) *every closed set in Y is the intersection with Y of some closed set in X*;

4) *the closure in Y of every set $M \subset Y$ is the intersection with Y of the closure of the set M in X.*

5. Mappings of topological spaces. Let X and Y be two arbitrary sets. We shall say that a *mapping f of the set X into the set Y* is defined if every point $x \in X$ is assigned a point $y \in Y$.

The point y is called the *image* of the point x under the mapping f; we then write

$$y = f(x).$$

In general, the set of all images of the points $x \in M$, where M is an arbitrary subset of X, is called the *image of the set M under the mapping f* and is denoted by $f(M)$. The set of all points $x \in X$ for which $f(x) \in N$, where $N \subset Y$, is called the *inverse image* of the set N and denoted by $f^{-1}(N) = f^{-1}(N \cap f(X)$. If the image $f(X)$ of the set X the entire is space Y, then the mapping f is said to be a mapping of *X onto Y.*

A mapping of X into Y is said to be *one-to-one* if the inverse image of every point $y \in f(X)$ consists of one point. A mapping of X onto X under which every image coincides with its inverse image is called the *unit*, or *identity, mapping.*

Suppose f is a mapping of X into Y and that φ is a mapping of Y into Z; the *product* or *composition* φf of the mapping φ with the mapping f is the mapping consisting of the successive application first of the mapping f and then of the mapping φ.

Let f be a mapping from X into Y. The mapping φ of the set $f(X)$ into X is called the *inverse* of f and is denoted by f^{-1} if φf is the identity mapping in X. In this case, it follows from the equations $(\varphi f)(x) = x$, $(f\varphi f)(x) = f(x)$ that φ is a mapping onto X and that $f\varphi$ is the identity mapping in $f(X)$; consequently, f, considered as a mapping in $f(X)$, is inverse to $\varphi = f^{-1}$.

I. *The inverse mapping f^{-1} exists if and only if the mapping f is one-to-one.*

In fact, the necessity of the condition is obvious. Conversely, if the mapping f is one-to-one, then, assigning to each point $y \in Y$ its inverse image under the mapping f, we obtain the inverse mapping f^{-1}.

Till now we have been considering mappings of arbitrary sets. Now we shall assume that f is a mapping of the topological space X into the topological space Y. By analogy with the usual definition of a continuous function, the mapping f is said to be *continuous at the point* $x_0 \in X$ if the inverse image of every neighborhood $V(y_0)$ of the point $y_0 = f(x_0)$ contains some neighborhood $U(x_0)$ of the point x_0. A mapping f of the space X into the space Y is said to be *continuous* if it is continuous at every point x of the space X. It follows directly from this definition that:

II. *A mapping f of the space X into the space Y is continuous if and only if the inverse image of every open set in Y is an open set* (*or if and only if the inverse image of every closed set in Y is a closed set*).

Moreover:

III. *If a continuous mapping of the space X into the space Y maps the set $M \subset X$ into the set $N \subset Y$, then it also maps $\overline{M}$ into $\overline{N}$.*

In fact, the inverse image of the set $\overline{N}$ is closed and contains M and therefore it also contains $\overline{M}$.

A mapping f of the space X onto the space Y is said to be a *homeomorphism* if:

1) f is one-to-one;

2) the mappings f and f^{-1} are continuous.

A mapping f of the space X into the space Y is called a *homeomorphism into* if f is a homeomorphism of X onto $f(X)$, considered as a topological subspace of the space Y.

Two topological spaces X and Y are said to be *homeomorphic* if there exists a homeomorphism of X onto Y.

It follows directly from Proposition II that

IV. *Under a homeomorphism open sets map into open sets, closed sets map into closed sets, and the closure of every set maps into the closure of the image of this set.*

Thus, the properties of a set of being closed, open, or the closure of some set are preserved under a homeomorphism.

The properties which are left invariant under homeomorphisms are said to be *topological*, and that branch of mathematics which studies topological properties is called *topology*.

6. Compact sets. A family $\{G\}$ of sets is called a *covering* of the set M if the union of all the sets G contains M.

A topological space X is said to be *compact* if each of its coverings $\{G\}$ by open sets G contains a finite number of sets $\{G_1, G_2, \cdots, G_n\}$ which also form a covering of X.

This is the usual English terminology; see, for instance, N. Bourbaki's *General Topology* [3], Ch. I, § 10, subsection 2. Recently, Soviet literature has also been using *compact* instead of *bicompact* and *sequentially compact* instead of *compact* — see § 2, subsection 14, below.

A set $M \subset X$ is said to be *compact* if, considered as a subspace of X, it is compact.

Going over to the complementary sets, we conclude that:

I. *A space X is compact if, and only if, every family $\{F\}$ of closed sets with void intersection in X contains a finite number of sets $F_1, F_2, \cdots, F_n$ with void intersection.*

It follows from I that

II. *A closed subset of a compact space is compact.*

One can give another, equivalent, definition of a compact space. A family $\{M\}$ of sets is said to be *centralized* (or to *have the finite intersection property*) if any finite number of sets M of this family have nonvoid intersection.

In virtue of I, *a space X is compact if and only if every centralized family of closed sets in X has nonvoid intersection.*

This proposition can also be given the following form.

III. *A space X is compact if and only if every centralized family of sets M in X has at least one contact point, common to all M, in X.*

Proof. Suppose X is compact and that $\{M\}$ is a centralized family of sets in X. Then $\{\overline{M}\}$ is a centralized family of closed sets in X and, consequently, $\{\overline{M}\}$ has nonvoid intersection; any point of this intersection is a common contact point of the sets M.

Conversely, suppose every centralized family of sets in X has a common contact point. In particular, every centralized family of closed sets must have a common contact point; but in virtue of the fact that these sets are closed, it belongs to their intersection; consequently, X is compact.

IV. *The continuous image of a compact space is a compact space.*

Proof. Suppose f is a continuous mapping of the compact space X onto the space Y and let $\{G'\}$ be a covering of the space Y by open sets G'. Then the inverse images G of the sets G' are open and form a covering $\{G\}$ of the space X. In virtue of the compactness of X, $\{G\}$ contains a finite covering $\{G_1, \cdots, G_n\}$. But then $\{G'_1, G'_2, \cdots, G'_n\}$ is a finite covering of the space Y which is contained in the covering $\{G'\}$. Thus, every covering $\{G'\}$ of the space Y by open sets contains a finite subcovering, i.e. Y is compact.

7. Hausdorff spaces. A topological space X is said to be *Hausdorff* or *separated* if it satisfies the following separation axiom: every pair of distinct points in X have disjoint neighborhoods.

Thus, the spaces R^n and C^n in the examples of subsection 2 are Hausdorff.

I. *If F is a compact set in the Hausdorff space X and $x \notin F$, then there exist disjoint open sets U and V which contain x and F, respectively.*

Proof. If y is an arbitrary point in F, then there exist disjoint open sets U_y, V_y, which contain x and y, respectively. Since F is compact, the set of V_y's contains a finite subset $V_{y_1}, \cdots, V_{y_n}$ which forms a covering of F. Then the sets $U = \bigcap_{k=1}^n U_{y_k}$, $V = \bigcup_{k=1}^n V_{y_k}$ satisfy the requirements of the proposition.

II. *A compact set in a Hausdorff space is closed.*

Proof. Suppose F is compact and that $x \notin F$. In virtue of I, we also have $x \notin \bar{F}$; consequently, $\bar{F} \subset F$, and therefore $\bar{F} = F$.

As an example, we consider compact sets in the space R^n (Example 1, subsection 2). On the basis of the well-known Borel-Lebesgue lemma, every closed bounded set in R^n is a compact space. It is easily verified that the converse is also true.

In fact, in virtue of II, a compact set M in R^n is closed. We shall prove that it is bounded. To this end, we cover every point of M by a ball of radius 1; in virtue of the compactness of M this covering has a finite subcovering. A ball containing this subcovering covers all of M.

Thus:

III. *A subspace in R^n is compact if and only if it is a closed bounded set in R^n.*

IV. *Under a continuous mapping of a compact space into a Hausdorff space, the images of closed sets are closed.*

Proof. Since a closed subset M of a compact space X is compact (see II, subsection 6), the continuous image of the space M is also compact (see IV, subsection 6) and therefore it is closed in the Hausdorff space which contains it (see II).

V. *A one-to-one continuous mapping f of a compact space X into a Hausdorff space is a homeomorphism.*

Proof. By I, subsection 5, the inverse mapping f^{-1} exists; so it suffices to prove it is continuous, i.e. to establish that the inverse image of every closed set in X under the mapping f^{-1} is a closed set (see II, subsection 5). But this assertion coincides with IV, because it means that the image of a closed set under the mapping f into a Hausdorff space is a closed set.

VI. *If a set X is a Hausdorff space in the topology T_1 and compact in the topology T_2, and if $T_1 \leqq T_2$, then $T_1 = T_2$.*

Proof. Suppose X_1, X_2 are the set X with the topologies T_1, T_2, respectively. Because of the condition $T_1 \leqq T_2$, the identity mapping $f(x) = x$ of the compact space X_2 onto the Hausdorff space X_1 is continuous and therefore (see V) it is a homeomorphism. But this means that $T_1 = T_2$.

We now apply Proposition IV, subsection 6, to the case when f is a continuous mapping of the space X into the space R^1 of all real numbers (Example 1, subsection 1). In this case, f is called a *continuous real-valued function* on X. If X is compact, then in virtue of IV, subsection 6, its image in R^1 is a compact space and therefore (see III) it is a closed bounded set. But a closed bounded set in R^1 contains a maximal and a minimal element. This means that the function f attains its maximal and minimal values on X. Thus:

VII. *A continuous real-valued function on a compact space attains, on this space, its maximal and minimal values.*

VIII. *If f and φ are two continuous mappings of the space X into the Hausdorff space Y, then the set M of all points x for which $f(x) = \varphi(x)$ is closed.*

Proof. We shall prove that the complement CM of the set M, $\{x: f(x) \neq \varphi(x)\}$, is open. Let $x_0 \in CM$; then $f(x_0) \neq \varphi(x_0)$. Consequently, there exist disjoint neighborhoods $U = U(f(x_0))$, $V = V(\varphi(x_0))$. To them there corresponds a neighborhood $W(x_0)$ whose images under the mappings f and φ are contained in U and V respectively. Since U and V are disjoint, $W(x_0) \subset CM$, i.e. CM is open and M is closed.

IX. *If two continuous mappings f and φ of the space X into the Hausdorff space Y coincide on a set $N \subset X$, then they also coincide on its closure $\overline{N}$.*

Proof. Suppose M is the set in VIII. Then $N \subset M$, and therefore $\overline{N} \subset \overline{M} = M$; consequently, $f(x) = \varphi(x)$ on $\overline{N}$.

8. Normal spaces. A topological space X is said to be *normal* if for any two disjoint closed sets $F_1, F_2 \subset X$ there exist disjoint open sets U_1, U_2, containing F_1 and F_2, respectively. This condition is equivalent to the following; for any closed set F and open set $U \supset F$ there exists an open set V such that $F \subset V$ and $\overline{V} \subset U$. In fact, it suffices to consider the closed set $X - U$, which does not intersect F.

I. *A compact Hausdorff space X is normal.*

Proof. Suppose F_1, F_2 are closed (and hence compact) disjoint sets in X. By I, subsection 7, for any point $y \in F_2$ there exist disjoint open sets U_y, V_y, which contain F_1 and y, respectively. Among the sets V_y there is a finite number $V_{y_1}, V_{y_2}, \cdots, V_{y_n}$ which form a covering of F_2. Then

$$U = \bigcap_{k=1}^{n} U_{y_k}, \quad V = \bigcup_{k=1}^{n} V_{y_k}$$

are disjoint open sets, containing F_1 and F_2, respectively.

II (Urysohn's Lemma). *For any two disjoint closed subsets F_0, F_1 of a normal space X there exists a continuous real-valued function f on X satisfying the conditions*:

1) $0 \leqq f(x) \leqq 1$;

2) $f(x) = 0$ *on* F_0;

3) $f(x) = 1$ *on* F_1.

Proof. The assertion is trivial if one of the sets F_0, F_1 is void. In fact, if, e.g., F_0 is a void set, it suffices to set $f(x) = 1$ for all $x \in X$. Therefore, we shall assume that F_0 and F_1 are nonvoid sets. We set $V_1 = X - F_1$. Since the space X is normal, there exists an open set – which we shall denote by V_0 – such that $F_0 \subset V_0$, $\overline{V}_0 \subset V_1$.

Analogously, there exists an open set – we denote it by $V_{\frac{1}{2}}$ – such that $\overline{V}_0 \subset V_{\frac{1}{2}}$ and $\overline{V}_{\frac{1}{2}} \subset V_1$. Continuing this process, we define, for every number r of the form $m/2^n$, $0 \leqq m \leqq 2^n$, an open set V_r such that $F_0 \subset V_r$, $\overline{V}_{r_1} \subset V_{r_2}$ with $r_1 < r_2$.

We now set

$$V_t = \bigcup_{r \leqq t} V_r$$

for arbitrary real number t in the interval $0 < t < 1$, and $V_t = \phi$ for $t < 0$, $V_t = X$ for $t > 1$. This defines the open sets V_t for all real values of t. Moreover,

$$\overline{V}_{t_1} \subset V_{t_2} \text{ if } t_1 < t_2. \tag{1}$$

In fact, for $0 \leqq t_1 < t_2 \leqq 1$ there exist rational numbers r_1, r_2 of the form $m/2^n$, satisfying the conditions $t_1 < r_1 < r_2 < t_2$, and therefore $\overline{V}_{t_1} \subset \overline{V}_{r_1} \subset V_{r_2} \subset V_{t_2}$.

And if $t_1 < 0$ or $t_2 > 1$, relation (1) is obvious, because then $V_{t_1} = \phi$, $V_{t_2} = X$, respectively.

We set

$$f(x) = \inf\{t : x \in V_t\} \tag{2}$$

and prove that f satisfies all the requirements. For $t > 1$, the set $V_t = X$ contains an arbitrary point x; therefore $\{t : x \in V_t\} \supset (1, \infty)$ from which, in virtue of (2), it follows that $f(x) \leqq 1$. For $t < 0$, the set $V_t = \phi$ contains no point x. Therefore the set $\{t : x \in V_t\}$ contains no points $t < 0$ so that, in virtue of (2), $f(x) \geqq 0$. If $x \in V_0$, then $0 \in \{t : x \in V_t\}$; consequently, $f(x) \leqq 0$. In combination with the inequality $f(x) \geqq 0$ this yields $f(x) = 0$ on V_0; in particular, $f(x) = 0$ on $F_0 \subset V$. Analogously, from $x \notin V_1$, i.e. $x \in F_1$ it follows that $1 \notin \{t : x \in V_t\}$ from which, in virtue of (1), $f(x) \geqq 1$. On the other hand, by what was proved above, $f(x) \leqq 1$. Therefore, $f(x) = 1$ on F_1. It remains to prove the continuity of the function f. Let $x_0 \in X$ and $\varepsilon > 0$; we set $y_0 = f(x_0)$. In virtue of (1), $x_0 \in V_t$ for $t > y_0$ and $x_0 \notin \overline{V}_t$ for $t < y_0$. In particular, $x_0 \in U(x_0) = V_{y_0+\varepsilon} - \overline{V}_{y_0-\varepsilon}$, so that, being an open set, $U(x_0)$ is a neighborhood of the point x_0. If $x \in V_{y_0+\varepsilon} - \overline{V}_{y_0-\varepsilon}$, then $y_0 + \varepsilon \in \{t : x \in V_t\}$ and $y_0 - \varepsilon \in \{t : x \notin V_t\}$. Therefore, in virtue of (1) and (2), $y_0 - \varepsilon \leqq f(x_0) \leqq y_0 + \varepsilon$, i.e.

$$f(x_0) - \varepsilon \leqq f(x) \leqq f(x_0) + \varepsilon. \tag{3}$$

This means that f is continuous.

9. Locally compact spaces. A topological space X is called *locally compact* if every point $x \in X$ has a neighborhood whose closure is compact.

I. *A locally compact space X which is not compact can be made into a compact space by adjoining one more element $x_\infty \notin X$.*

In fact, we shall set $X_\infty = X \bigcup x_\infty$ and consider the open sets in X_∞ to be all open sets in X and also all sets of the form $U \bigcup x_\infty$ where U is an open set in X, whose complement is compact.

It is easy to verify that X_∞ is a compact space. Moreover, if X is Hausdorff then X_∞ is also Hausdorff. In fact, two distinct points x_1, x_2 have disjoint neighborhoods by assumption provided both of them are elements in X. But if $x_1 = x_\infty$, $x_2 \in X$, then there exists a neighborhood U of the point x_2 whose closure is compact. Then $(X - \bar{U}) \bigcup x_\infty$ and U are disjoint neighborhoods of the points x_∞ and x_2 respectively.

Combining this result with the Urysohn lemma, we obtain:

II. *If U is a given open set and $F \subset U$ is a given compact set in a locally compact Hausdorff space X, then there exists a continuous real-valued function f on X satisfying the conditions*:

$$0 \leqq f(x) \leqq 1, f(x) = 1 \textit{ on } F, f(x) = 0 \textit{ outside of } U.$$

In fact, it suffices to apply Urysohn's lemma to the closed sets F and $X_\infty - U$ in the compact space X_∞.

If X itself is compact, then we take $X_\infty = X$ by definition. In this case Proposition II coincides with Urysohn's Lemma.

If f is any numerical function on a locally compact space X which is not compact, then the equality $\lim_{x\to\infty} f(x) = A$ will mean that for arbitrary $\varepsilon > 0$ the set $\{x: |f(x) - A| \geqq \varepsilon\}$ is compact; it is easy to see that this condition defines the number A uniquely, provided it exists. In particular, the equality $\lim_{x\to\infty} f(x) = 0$ will mean that for any $\varepsilon > 0$, the set $\{x: |f(x)| \geqq \varepsilon\}$ is compact; in this case we shall say that f *vanishes at infinity*.

10. Stone's theorem. Let Q be an arbitrary set. We introduce the notation

$$(f_1 \cup f_2 \cup \cdots \cup f_n)(q) = \max\{f_1(q), f_2(q), \cdots, f_n(q)\},$$
$$(f_1 \cap f_2 \cap \cdots \cap f_n)(q) = \min\{f_1(q), f_2(q), \cdots, f_n(q)\}$$

for real-valued functions f on Q. We shall say that a set $\mathfrak{A}$ of real-valued functions on Q forms a *lattice* if together with every pair of functions f_1, f_2 it also contains $f_1 \cup f_2$ and $f_1 \cap f_2$; clearly, in this case, it also contains $f_1 \cup f_2 \cup \cdots \cup f_n$ and $f_1 \cap f_2 \cap \cdots \cap f_n$ whenever it contains $f_1, \cdots, f_n$. A set $\mathfrak{A}$ of real-valued functions on Q is called a *real linear function algebra* if together with every function it contains its product by an arbitrary real number and together with every pair of functions it contains their sum and product. In the sequel we shall consider only linear algebras and the term "algebra" will always denote a linear algebra. [Concerning these terms see the beginning of Ch. II.]

A real function algebra $\mathfrak{A}$ is said to be *uniformly closed* if the limit of every sequence of functions $f_n \in \mathfrak{A}$, uniformly convergent on Q, belongs to $\mathfrak{A}$. If $\mathfrak{A}$ is not uniformly closed, then, adjoining all such limits to it, we obtain a new set of functions, containing $\mathfrak{A}$ and, as can easily be seen, is a uniformly closed algebra. This algebra is called the *uniformly closed hull of the algebra* $\mathfrak{A}$ and is denoted by $\overline{\mathfrak{A}}$.

An example of a uniformly closed algebra of functions is the set $C^r(T)$ of all continuous real-valued functions on a given topological space T, and also the set $C^r_{t_0}(T)$ of all continuous real-valued functions on T, which are equal to zero at the point t_0.

I. *Every real uniformly closed algebra R of bounded functions, containing all constants, is a lattice.*

Proof. It suffices to prove that if $x \in R$ then also $|x| \in R$, because then $x, y \in R$ implies that the functions

$$x \cup y = \tfrac{1}{2}(x+y+|x-y|),\ x \cap y = \tfrac{1}{2}(x+y-|x-y|)$$

also belong to the algebra R.

Suppose $x \in R$ and $|x(t)| \leqq c$ for all $t \in T$. Then

$$|x(t)| = \sqrt{c^2-[c^2-x^2(t)]} = c\sqrt{1-\left(1-\frac{x^2(t)}{c^2}\right)}$$

$$= c\left\{1-\sum_{n=1}^{\infty}\frac{1\cdot 1\cdot 3\cdots(2n-3)}{2\cdot 4\cdot 6\cdots 2n}\left(1-\frac{x^2(t)}{c^2}\right)^n\right\},$$

where the series converges uniformly on T because $0 \leqq 1-x^2(t)/c^2 \leqq 1$. Consequently, $|x| \in R$.

REMARK. Proposition I is also true for an algebra of functions which does not contain nonzero constants. In order to verify this, it suffices to consider the set R_1 of all functions of the form $y(t) = c+x(t)$, $x \in R$. R_1 is a uniformly closed algebra of functions which contains all constants.

In fact, we must show that if $c_n+x_n(t)$, where $x_n \in R$, converges uniformly to $y(t)$, then $y(t) = c+x(t)$, where $x \in R$. We note first of all that the sequence c_n must be bounded; otherwise, we would select a subsequence (denoting it again by $c_n+x_n(t)$), for which $c_n \to \infty$ and hence $1+(1/c_n)\,x_n(t)$ converges uniformly to zero so that, contrary to assumption, $1 \in R$, as the uniform limit of the function $-(1/c_n)x_n(t) \in R$. But if c_n is bounded, then we can select a subsequence (denoting it again by $c_n+x_n(t)x_n(t)$) for which c_n converges. Let $c_n \to c$; then $x_n(t)$ converges uniformly to $y(t)-c$. Consequently, since R is complete, $x(t) = y(t)-c \in R$ and $y(t) = c+x(t)$.

Now if $x \in R$, then, according to what we proved in I, $|x| \in R_1$. Suppose $|x(t)| = c + x_1(t)$,

$x_1 \in R$. Then $x^2(t) = c^2 + 2cx_1(t) + x_1^2(t)$ and $c^2 = x^2(t) - 2cx_1(t) - x_1^2(t) \in R$; consequently, $c^2 = 0$, $c = 0$ and $|x| = x_1 \in R$.

II. *Suppose* $\mathfrak{A}$ *is the set of continuous real-valued functions* x *on a compact space* T, *satisfying the conditions*:

1) $\mathfrak{A}$ *is a lattice*;

2) *for any two distinct points* $\tau, \sigma \in T$ *and arbitrary real numbers* a, b *there exists a function* $x_{\tau\sigma} \in \mathfrak{A}$ *such that* $x_{\tau\sigma}(\tau) = a$, $x_{\tau\sigma}(\sigma) = b$.

Then every continuous real-valued function on T *is the limit of a uniformly convergent sequence of functions* $x_n \in \mathfrak{A}$.

Proof. Suppose x is an arbitrary continuous real-valued function on T, ε is any positive number, and $x_{\tau\sigma}$ is a function in $\mathfrak{A}$ which satisfies condition 2 with $a = x(\tau)$, $b = x(\sigma)$. We shall denote by $U_{\tau\sigma}$, $V_{\tau\sigma}$ the sets of points $t \in T$ on which $x_{\tau\sigma}(t) < x(t) + \varepsilon$ and $x_{\tau\sigma}(t) > x(t) - \varepsilon$ respectively; $U_{\tau\sigma}$ and $V_{\tau\sigma}$ are open sets, containing τ and σ respectively. For fixed σ the sets $U_{\tau\sigma}$ form a covering of the compact space T; selecting a finite covering $\{U_{\tau_1\sigma}, U_{\tau_2\sigma}, \cdots, U_{\tau_n\sigma}\}$ from this covering and setting $y_\sigma = x_{\tau_1\sigma} \cap x_{\tau_2\sigma} \cap \cdots \cap x_{\tau_n\sigma}$, we obtain a function $y_\sigma \in \mathfrak{A}$ satisfying the conditions

$$y_\sigma(t) < x(t) + \varepsilon \text{ on all of } T,$$
$$y_\sigma(t) > x(t) - \varepsilon \text{ for } t \in V_\sigma = \bigcap_{j=1}^{n} V_{\tau_j\sigma}.$$

Selecting now a finite covering $\{V_{\sigma_1}, \cdots, V_{\sigma_m}\}$ from the covering $\{V_\sigma\}$ of the space T and setting $z = y_{\sigma_1} \cup y_{\sigma_2} \cup \cdots \cup y_{\sigma_m}$, we obtain a function $z \in \mathfrak{A}$, satisfying the inequalities $x(t) - \varepsilon < z(t) < x(t) + \varepsilon$ on all of T; this completes the proof.

A set $\mathfrak{A}$ of functions on T is said to *separate points in* T if for any two distinct points $t_1, t_2 \in T$ there exists a function $x \in \mathfrak{A}$ such that $x(t_1) \neq x(t_2)$.

THEOREM 1 (STONE [3]). *Let* R *be a real algebra of continuous functions on the compact space* T *which separates points. Then the uniformly closed hull* $\bar{R}$ *of the algebra* R *either coincides with* $C^r(T)$ *or it coincides with* $C^r_{t_0}(T)$ *for some* $t_0 \in T$.

Proof. a) We shall first assume that for every $t_0 \in T$ there exists a function x in the algebra R such that $x(t_0) \neq 0$. Then for $t_1, t_2 \in T$, $t_1 \neq t_2$ there exists a function x in the algebra R such that

$$x(t_1) \neq 0, \; x(t_1) \neq x(t_2). \tag{1}$$

In fact, there exist functions $y \in R$, $z \in R$, satisfying the conditions $y(t_1) \neq y(t_2)$, $z(t_1) \neq 0$; we set

$$x(t) = \begin{cases} y(t) & \text{if } y(t_1) \neq 0, \\ z(t) & \text{if } y(t_1) = 0, \; z(t_1) \neq z(t_2), \\ y(t) + z(t) & \text{if } y(t_1) = 0, \; z(t_1) = z(t_2). \end{cases} \tag{2}$$

Moreover, we may assume that $x(t_2) = 0$; otherwise, we could replace $x(t)$ by the function

$$u(t) = \frac{1}{x(t_2)}x(t) - \left[\frac{1}{x(t_2)}x(t)\right]^2.$$

But then, setting $x_1(t) = \{1/x(t_1)\}x(t)$, we obtain a function $x_1 \in R$, satisfying the conditions $x_1(t_1) = 1$, $x_1(t_2) = 0$. Analogously, there exists a function $x_2 \in R$ such that $x_2(t_1) = 0$, $x_2(t_2) = 1$. Consequently, if we further set $y = ax_1 + bx_2$, we conclude that for arbitrary $t_1, t_2 \in T$, $t_1 \neq t_2$ and any real a, b there exists a function y in R such that $y(t_1) = a$, $y(t_2) = b$. According to the remark to I, the algebra $\bar{R}$ satisfies all the conditions of Proposition II and therefore it coincides with $C^r(T)$.

b) If case a) does not occur, all functions on R vanish at some point $t_0 \in T$. Then for the algebra R' of all functions of the form $y(t) = c + x(t)$, $x \in R$ case a) holds. Therefore for an arbitrary continuous real-valued function z – in particular, for such a function which vanishes at the point t_0 – and arbitrary $\varepsilon > 0$, there exists a function $y(t) = c + x(t)$, $x \in R$ such that $|z(t) - [c + x(t)]| < \varepsilon$ for all $t \in T$. If we set $t = t_0$ in this inequality, we obtain that $|c| < \varepsilon$, and therefore $|z(t) - x(t)| < 2\varepsilon$. But this means that $\bar{R} = C^r_{t_0}(T)$.

THEOREM 2 (WEIERSTRASS' THEOREM). *Let T be a closed bounded set in real n-dimensional space R^n; then every continuous real-valued function $f(x_1, x_2, \cdots, x_n)$ on T is the limit of a sequence of polynomials in the variables $x_1, \cdots, x_n$ with real coefficients, which is uniformly convergent on T.*

To prove this theorem, it suffices to apply Stone's theorem to the algebra $\mathfrak{A}$ of all polynomials in $x_1, \cdots, x_n$ with real coefficients, considered as functions on T. The case $\bar{\mathfrak{A}} = C^r_{t_0}(T)$ is impossible here because $\mathfrak{A}$ contains all constants.

11. Weak topology, defined by a family of functions. Let Q be an arbitrary set. One way of defining a topology in Q consists in the following.

Suppose $\{f_\alpha\}$ is a family of functions on Q with values, each in its topological space X_α. We define a basis in Q to be the set of all possible intersections of a finite number of sets $f_\alpha^{-1}(U_\alpha)$, where the U_α are open sets in X_α. The topology in Q defined by this basis is called the *weak topology in Q defined by the family* $\{f_\alpha\}$. Clearly:

I. *The weak topology defined by the family $\{f_\alpha\}$ is the weakest of the topologies in which all the functions f_α are continuous.*

II. *Suppose F is a family of continuous complex-valued functions on the locally compact space X, satisfying the following conditions:*

1) *all the functions in F vanish at infinity;*

2) *F separates points of the space X;*

3) *there are no points in X at which all functions in F vanish.*

Then the weak topology on X defined by the family F coincides with the initial topology.

Proof. Clearly, the functions in F can be considered as continuous functions on X_∞, which are equal to zero at the point x_∞. Suppose T_1 is the weak topology on X_∞, defined by the family F, and that T_2 is the topology on X_∞ defined by the initial topology on X. By I, we have $T_1 \leqq T_2$; in virtue of conditions 2) and 3), the family F separates points in X_∞, and therefore the topology T_1 is Hausdorff. On the basis of VI, subsection 7, we conclude that $T_1 = T_2$ on X_∞, and hence also on X.

12. Topological product of spaces. Suppose $X_1, X_2, \cdots, X_n$ are topological spaces. We shall denote the set of all n-tuples $x = \langle x_1, x_2, \cdots, x_n \rangle$ where $x_i \in X_i$, $i = 1, 2, \cdots, n$ by $X_1 \times X_2 \times \cdots \times X_n$; in general, if $M_1, M_2, \cdots, M_n$ are arbitrary sets in $X_1, X_2, \cdots, X_n$ respectively, then $M_1 \times M_2 \times \cdots \times M_n$ will denote the set of all n-tuples $x = \langle x_1, x_2, \cdots, x_n \rangle$ which are obtained when $x_1, x_2, \cdots, x_n$ independently run through $M_1, M_2, \cdots, M_n$, respectively.

We introduce a topology in $X_1 \times X_2 \times \cdots \times X_n$ by considering a neighborhood basis of the point $x^0 = \langle x_1^0, x_2^0, \cdots, x_n^0 \rangle$ to be the set of all $U_1(x_1^0) \times U_2(x_2^0) \times \cdots \times U_n(x_n^0)$, where the $U_i(x_i^0)$ are arbitrary neighborhoods of the points x_i^0, $i = 1, 2, \cdots, n$. Clearly, the neighborhood basis axioms (see subsection 2) are satisfied, so that $X_1 \times X_2 \times \cdots \times X_n$ becomes a topological space; this space is called the *topological product of the spaces* $X_1, X_2, \cdots, X_n$.

The definition of the topological product can be extended to the case of an infinite number of spaces. Namely, let $\{X_\alpha\}$ be a given family of topological spaces X_α, where the index α runs through an arbitrary fixed set $\mathfrak{A}$ (of any power). We shall denote by $\prod_{\alpha \in \mathfrak{A}} X_\alpha$ the set of all functions $x = \langle x_\alpha \rangle$ defined for all $\alpha \in \mathfrak{A}$ and with values $x_\alpha \in X_\alpha$. A neighborhood basis of the point $x^0 = \langle x_\alpha^0 \rangle$ is defined to be the family of sets $U(x^0)$ in $\prod_{\alpha \in \mathfrak{A}} X_\alpha$ which is obtained in the following way (A. N. Tikhonov): we choose an (arbitrary) finite number n of elements $\alpha_1, \alpha_2, \cdots, \alpha_n$, take a fixed neighborhood $U(x_{\alpha_i}^0)$ for each point $x_{\alpha_i}^0$, $i = 1, 2, \cdots, n$, and denote by $U(x^0)$ the set of all points $x = \langle x_\alpha \rangle$, which are obtained when the x_{α_i} independently run through the $U(x_{\alpha_i}^0)$, $i = 1, 2, \cdots, n$, and all the remaining x_α run through X_α. If we let n run through all natural values, $\alpha_1, \alpha_2, \cdots, \alpha_n$ through all possible collections of n elements of the set $\mathfrak{A}$ and $U(x_{\alpha_1}^0), \cdots, U(x_{\alpha_n}^0)$ through all possible neighborhoods of the points $x_{\alpha_1}^0, \cdots, x_{\alpha_n}^0$, we obtain a family of sets $\{U(x^0)\}$ which we shall consider to be a neighborhood basis of the point x^0.

It is again easy to verify that with this definition of a neighborhood basis, $\prod_{\alpha \in \mathfrak{A}} X_\alpha$ becomes a topological space. It is called the *topological product of the spaces* X_α.

If $x = \langle x_\alpha \rangle$, then x_α is called the α-th coordinate of the point x and the *projection of the point* x *on* X_α. The correspondence $x \to x_\alpha$ is a mapping of the space $X =$

$\prod_{\alpha \in \mathfrak{A}} X_\alpha$ onto X_α; the image of a set $M \subset X$ under this mapping is called the *projection of M on* X_α.

It is clear from the definition given above of a neighborhood basis in the topological product $X = \prod_{\alpha \in \mathfrak{A}} X_\alpha$ that the topology in X is the weak topology defined by the family of functions $f_\alpha(x) = x_\alpha$. Therefore, the functions $f_\alpha(x) = x_\alpha$ are continuous and under the mapping $f_\alpha(x) = x_\alpha$ the inverse images of open sets are open and the inverse images of closed sets are closed.

I. *If Y is compact, then the projection on X of any closed set* $F \subset X \times Y$ *is closed.*

Proof. Suppose M is the projection of the set F on X and that $x_0 \in \overline{M}$. Then every neighborhood $U(x_0)$ has nonvoid intersection with M; consequently, the sets

$$N_U = \{y : x \times y \in F, \ x \in U(x_0)\}$$

form a centralized family in the compact space Y and therefore they have a common contact point y_0(see III, subsection 6). But then $x_0 \times y_0 \in \bar{F} = F$; consequently, $x_0 \in M$. Thus, if $x_0 \in \overline{M}$, then $x_0 \in M$; this means that M is closed.

II (Tikhonov [1]). *The topological product of an arbitrary family of compact topological spaces is compact.*

Proof. (The following proof is due to Chevalley and Frink [1].) Suppose $X = \prod X_\alpha$ is the topological product of the compact spaces X_α. According to III, subsection 6, it suffices to show that every centralized family of sets in X has at least one contact point in common. Suppose $\{M^\lambda\}$ is a centralized family of sets in X (where λ denotes an index which runs through some set Λ). Without loss of generality, we may assume that $\{M^\lambda\}$ is the maximal centralized family of sets, i.e. that the family $\{M^\lambda\}$ cannot be augmented by adding new sets and still remain centralized. In fact, if the family $\{M^\lambda\}$ is not maximal, then applying Zorn's lemma (see Appendix I), we can extend $\{M^\lambda\}$ to a maximal centralized family. [The set of all centralized families in X is a set, partially ordered with respect to set inclusion, satisfying the condition of Zorn's lemma; an upper bound of a linearly ordered set of centralized families is a centralized family, obtained by forming the union of all centralized families of this set.] Obviously, it suffices to prove the existence of a common contact point for this maximal family, because this point will also be a common contact point of the initial family $\{M^\lambda\}$. Thus, suppose $\{M^\lambda\}$ is a maximal centralized family in X. We shall denote the projection of the set M^λ on X_α by M_α^λ. Then for fixed α the sets $\{M_\alpha^\lambda\}$ form a centralized family in X_α; in view of the fact that X_α is compact, the sets M_α^λ have a common contact point (see III, subsection 6); we denote it by x_α^0. We shall prove that $x^0 = \langle x_\alpha^0 \rangle$ is a common contact point of the family $\{M^\lambda\}$, i.e. that every neighborhood $U(x^0)$ in the neighborhood basis of the point x^0 intersects every set M^λ.

By the definition of the topology in X, every such neighborhood $U(x^0)$ is given by a finite collection $\alpha_1, \alpha_2, \cdots, \alpha_n$ of indices α and neighborhoods $U(x_{\alpha_i}^0)$, $i = 1, 2, \cdots, n$, satisfying the conditions

$$x_{\alpha_i} \in U(x^0_{\alpha_i}), \qquad i = 1, 2, \cdots, n. \tag{1}$$

Suppose $U_i(x^0)$ denotes the neighborhood of the point x^0 defined by only one of the conditions (1) for fixed i, i.e. the set of all points $x = \langle x_\alpha \rangle$ such that $x_{\alpha_i} \in U(x^0_{\alpha_i})$, and all the remaining x_α are arbitrary. Then $U(x^0)$ is the intersection of the neighborhoods $U_i(x^0)$, $i = 1, 2, \cdots, n$. Since $x^0_{\alpha_i}$ is a contact point of every $M^\lambda_{\alpha_i}$, $U(x^0_{\alpha_i})$ intersects each of the $M^\lambda_{\alpha_i}$; this means that $U_i(x^0)$ intersects each of the M^λ. But the intersection of a finite number of sets M^λ is also a set in the family $\{M^\lambda\}$: for otherwise the adjunction of such an intersection would lead to a centralized family, which is an extension of the family $\{M^\lambda\}$, and this contradicts the maximality of $\{M^\lambda\}$. Therefore, we may also say that $U_i(x^0)$ intersects the intersection of an arbitrary finite number of sets M^λ. However, it follows from this that the neighborhoods $U_i(x^0)$, $i = 1, 2, \cdots, n$, must belong to the family $\{M^\lambda\}$ because in the contrary case the adjunction of the neighborhoods $U_i(x^0)$ would lead to a centralized family which is an extension of the family $\{M^\lambda\}$, and this contradicts the fact that $\{M^\lambda\}$ is maximal. But then the intersection $U(x^0)$ of the neighborhoods $U_i(x^0)$, $i = 1, 2, \cdots, n$, also belongs to the family $\{M^\lambda\}$ and therefore it intersects each of the M^λ. This completes the proof of Proposition II.

III. *The topological product of Hausdorff spaces is Hausdorff.*

Proof. If $x, y \in \prod_{\alpha\in\mathfrak{A}} X_\alpha$ and $x \neq y$, then $x_\alpha \neq y_\alpha$ for some α. Since X_α is Hausdorff, there exist disjoint neighborhoods U_α, V_α of the points x_α, y_α. Their inverse images under the mapping $x \to x_\alpha$ will be disjoint neighborhoods of the points x and y.

IV. *If F and G are a closed and open set, respectively, in the topological product $X \times Y$ of the topological spaces X, Y, and Q is a compact set in X then the set $\bigcup_{x\in Q}\{y\colon x \times y \in F\}$ is closed, and the set $\bigcap_{x\in Q}\{y\colon x \times y \in G\}$ is open.*

Proof. The first assertion follows from the fact that $\bigcup_{x\in Q}\{y\colon x \times y \in F\}$ is the projection on Y of the closed set $(Q \times Y) \cap F$ (see I); the second assertion is obtained from the first by going over to the complementary sets.

V. *Let $f(x, y)$ be a continuous function on the topological product $X \times Y$ of the topological spaces X, Y with values in the topological space Z, and let Q be a compact set in X, and G an open set in Z. Then the set $W = \{y\colon f(x, y) \in G \text{ for all } x \in Q\}$ is open in Y.*

Proof. Suppose $\mathfrak{G}$ is the inverse image of G under the mapping $z = f(x, y)$; then $\mathfrak{G}$ is open in $X \times Y$ and

$$W = \bigcap_{x\in Q} \{y : x \times y \in \mathfrak{G}\},$$

and we may now apply IV.

13. Metric spaces. A set X is called a *metric space* if each pair x, y in X is assigned a nonnegative number $\rho(x, y)$(called the distance from x to y), satisfying the following conditions:

1) (identity axiom) $\rho(x, y) = 0$ if and only if $x = y$;
2) (symmetry axiom) $\rho(x, y) = \rho(y, x)$;
3) (triangle axiom) $\rho(x, y)+\rho(y, z) \geqq \rho(x, z)$.

$\rho(x, y)$ is called a *metric* in X. [Note that actually the nonnegativeness of $\rho(x, y)$ follows from axioms 1)–3).]

EXAMPLES. 1. The set R^1 of all real numbers forms a metric space if the distance between x, y is defined by the formula $\rho(x, y) = |x-y|$.

2. The set of all points in three-dimensional space is, obviously, a metric space if $\rho(x, y)$ is understood to be the usual distance between points x and y in R^3. The triangle axiom means that one side of a triangle is not greater than the sum of its other two sides, where in a given case, the equality sign is possible if, and only if, the triangle degenerates into a segment of a straight line, i.e. when the points x, y, z lie on the same straight line.

3. The set $C[a, b]$ of all continuous functions x in the interval $[a, b]$ is a metric space if the distance between x and y is defined by the formula

$$\rho(x, y) = \max_{a \leqq t \leqq b} |x(t)-y(t)|.$$

4. Suppose $\mathfrak{B}$ is the set of all sequences $x = \langle n_1, n_2, n_3, \cdots \rangle$ of natural numbers. We define distance in $\mathfrak{B}$ by setting $\rho(x, x) = 0$ and $\rho(x, y) = 1/k$, where k is the first of the indices for which x_k is different from y_k. It is easily verified that axioms 1) –3) are satisfied where axiom 3) holds in the following strengthened form:

$$\rho(x, z) \leqq \max \{\rho(x, y), \rho(y, z)\}.$$

The metric space $\mathfrak{B}$ is called the *Baire null-dimensional space*.

The set of all points x satisfying one of the conditions

$$\rho(x, x_0) = r, \rho(x, x_0) < r, \rho(x, x_0) \leqq r \qquad (r > 0),$$

is called a *sphere*, an *open ball*, and a *closed ball*, respectively, with center x_0 and radius r. The open (closed) ball with center x_0 and radius r will be denoted by $B(x_0, r)(\overline{B(x_0, r)})$.

The least upper bound (which may be infinite) of the distances between all possible pairs of points of a given set is called the *diameter* of this set.

A topology may be introduced in the metric space X by taking the neighborhood basis of the point $x_0 \in X$ to be the set of all open balls with center x_0. It is easily verified that axioms 1)–3) for a neighborhood basis and the separation axiom (see subsections 2 and 7) are satisfied, so that X is a Hausdorff topological space. In this case, we shall say that the topology in X is *defined by the metric* $\rho(x, y)$.

A topological space X is said to be *metrizable* if a metric can be introduced in X which defines a topology in X, coinciding with the initial topology.

[For further information concerning conditions that a space be metrizable, see HAUSDORFF [1], § 26 and YU. SMIRNOV [1].]

A mapping f of a metric space X into the metric space X' is said to be *isometric* if it preserves distances, i.e. if the distance $\rho(x, y)$ between any two points $x, y \in X$

equals the distance $\rho(x', y')$ between their images in X'. Clearly, *an isometric mapping of X into X' is a homeomorphism.* Two metric spaces X, X' are said to be *isometric* if there exists an isometric mapping of X onto X'. Obviously, *isometric spaces are homeomorphic.*

A sequence $\langle x_n \rangle$ of elements in a metric space is said to be *fundamental* (or *Cauchy*) if for every positive number ε there exists an index $N(\varepsilon)$ such that

$$\rho(x_m, x_n) < \varepsilon, \text{ for } m, n > N(\varepsilon). \tag{1}$$

Every convergent sequence is fundamental. In fact, if $\lim_{n\to\infty} x_n = x_0$, then beginning with some one of them, say with x_N, all the terms of the sequence $\langle x_n \rangle$ are in the neighborhood $\rho(x, x_0) < \varepsilon/2$. Thus, we have $\rho(x_n, x_0) < \varepsilon/2$, $\rho(x_m, x_0) < \varepsilon/2$ for $n, m > N$ and, consequently,

$$\rho(x_n, x_m) \leqq \rho(x_n, x_0) + \rho(x_0, x_m) < \frac{\varepsilon}{2} + \frac{\varepsilon}{2} = \varepsilon.$$

The converse is not true in general. If, for example, X is the set of all rational numbers, considered as a subspace in R^1, then every sequence of numbers in X, which converges to an irrational number, is fundamental, but it does not converge in X.

A metric space is said to be *complete* if every fundamental sequence converges to a point in the space. The well-known Cauchy convergence criterion asserts that R^1 is a complete space.

A metric space X which is not complete can be embedded in a complete metric space in the following way, which generalizes Cantor's construction of the set of all real numbers.

We shall denote the set of all fundamental sequences $\tilde{x} = \langle x_1, x_2, x_3, \cdots \rangle$, $x_n \in X$ by $\tilde{X}$; here, we agree not to distinguish two fundamental sequences

$$\tilde{x} = \langle x_1, x_2, x_3, \cdots \rangle \text{ and } \tilde{y} = \langle y_1, y_2, y_3, \cdots \rangle$$

if, and only if,

$$\lim_{n\to\infty} \rho(x_n, y_n) = 0.$$

[Specifically, two fundamental sequences $\tilde{x} = \langle x_1, x_2, \cdots \rangle$ and $\tilde{y} = \langle y_1, y_2, \cdots \rangle$ will be called equivalent if $\lim_{n\to\infty} \rho(x_n, y_n) = 0$; the elements of the space $\tilde{X}$ are classes of mutually equivalent fundamental sequences.]

We define a metric in $\tilde{X}$ by setting

$$\rho(\tilde{x}, \tilde{y}) = \lim_{n\to\infty} \rho(x_n, y_n),$$

where $\tilde{x} = \langle x_1, x_2, x_3, \cdots \rangle$, $\tilde{y} = \langle y_1, y_2, y_3, \cdots \rangle$. This limit exists; in fact,

$$\rho(x_m, y_m) \leqq \rho(x_m, x_n) + \rho(x_n, y_n) + \rho(y_n, y_m);$$

consequently,

$$\rho(x_m, y_m) - \rho(x_n, y_n) \leqq \rho(x_m, x_n) + \rho(y_n, y_m).$$

Interchanging the roles of n and m, we conclude that

$$|\rho(x_m, y_m) - \rho(x_n, y_n)| \leqq \rho(x_m, x_n) + \rho(y_n, y_m),$$

so that the numbers $\rho(x_n, y_n)$ form a fundamental sequence in R^1 which converges inasmuch as R^1 is complete.

It is easy to verify the distance axioms 1)–3) for $\rho(\tilde{x}, \tilde{y})$, so that $\tilde{X}$ is a metric space. It contains, in particular, all sequences $\tilde{x} = \langle x, x, x, \cdots \rangle$, where the correspondence $x \to \langle x, x, x, \cdots \rangle$ is an isometric mapping of X into $\tilde{X}$. We shall therefore agree not to distinguish between x and the sequence $\langle x, x, x, \cdots \rangle$. Then X can be considered as a subspace of $\tilde{X}$. In particular, we can talk about the distance $\rho(\tilde{x}, y)$ between the points $\tilde{x} \in \tilde{X}$ and $y \in X$: by the definition of distance in $\tilde{X}$,

$$\rho(\tilde{x}, y) = \lim_{n \to \infty} \rho(x_n, y) \text{ with } \tilde{x} = \langle x_1, x_2, x_3, \cdots \rangle.$$

Therefore, if $\tilde{x} = \langle x_1, x_2, x_3, \cdots \rangle$ is an arbitrary fundamental sequence, then, passing to the limit as $m \to \infty$ in (1), we obtain

$$\rho(\tilde{x}, x_n) \leqq \varepsilon \text{ for } n > N(\varepsilon), \tag{2}$$

so that $\lim_{n\to\infty} \rho(\tilde{x}, x_n) = 0$. Thus, *for arbitrary* $\tilde{x} \in \tilde{X}$ *and* $\varepsilon > 0$, *there exists a point* $x \in X$ *such that* $\rho(\tilde{x}, x) < \varepsilon$.

The space $\tilde{X}$ *is complete*. In fact, suppose $\{\tilde{x}_n\}$ is a fundamental sequence in $\tilde{X}$. By what we proved above, for every $\tilde{x}_n$ there exists a point $x_n \in X$ such that

$$\rho(\tilde{x}_n, x_n) < \frac{1}{n} \tag{3}$$

and therefore $\tilde{x} = \langle x_1, x_2, x_3, \cdots \rangle$ is a fundamental sequence. But then, in virtue of (2), $\lim_{n\to\infty} \rho(\tilde{x}, x_n) = 0$. Combining this result with (3), we conclude that $\lim_{n\to\infty} \rho(\tilde{x}_n, \tilde{x}) = 0$, i.e. that $\lim_{n\to\infty} \tilde{x}_n = \tilde{x}$.

The space $\tilde{X}$ is called the *completion of the space X*.

I. $\tilde{X}$ *is the minimal of all complete metric spaces which contain X, i. e. if Y is a complete space containing X, then there exists an isometric mapping of* $\tilde{X}$ *into Y leaving all points* $x \in X$ *invariant.*

In fact, we obtain this mapping by assigning to each fundamental sequence $\tilde{x} = \langle x_1, x_2, x_3, \cdots \rangle$ its limit in Y.

II. *In a complete metric space, a nonincreasing sequence* $F_1 \supset F_2 \supset F_3 \supset \cdots$ *of bounded nonvoid closed sets* $F_1, F_2, F_3, \cdots$, *whose diameters* $d(F_n)$ *tend to zero, has a nonvoid intersection consisting of a single point.*

Proof. We select an element x_n in each F_n; then for $m > n$ we have $x_m, x_n \in F_n$, and consequently, $\rho(x_m, x_n) \leq d(F_n)$. Therefore $\langle x_n \rangle$ is a fundamental sequence;

its limit (which exists in virtue of the fact that the space is complete) is also the limit of every sequence $\langle x_n, x_{n+1}, x_{n+2}, \cdots \rangle$ and therefore it belongs to every F_n. There cannot be more than one point belonging to all the F_n because $d(F_n) \to 0$ when $n \to \infty$.

A set M in the metric space X is said to be *nowhere dense* if every open ball in X contains an open ball which does not intersect M. A set M is called a set of the *first category* if it is the union of a countable number of nowhere dense sets. Every set M which is not a set of the first category is called a set of the *second category*.

III. *A complete metric space is a set of the second category.*

Proof. Let us assume the contrary; suppose $X = \bigcup_{n=1}^{\infty} M_n$, where M_n is nowhere dense in X, is a complete metric space. We consider an arbitrary open ball S_0 of radius 1. Since M_1 is nowhere dense there exists a ball S_1 of radius $r_1 < \frac{1}{2}$ in S_0, whose closure $\bar{S}_1$ does not intersect M_1. Since M_2 is nowhere dense, there exists a ball S_2 of radius $r_2 < (\frac{1}{2})^2$ in S_1, whose closure $\bar{S}_2$ does not intersect M_2. Repeating this argument, we obtain a decreasing sequence of closed balls $\bar{S}_1 \supset \bar{S}_2 \supset \bar{S}_3 \supset \cdots$, to which Proposition II is applicable. Suppose x_0 is the point belonging to all the balls $\bar{S}_n$, $n = 1, 2, 3, \cdots$. By construction, x_0 does not belong to any of the M_n, which, however, contradicts the equality $X = \bigcup_{n=1}^{\infty} M_n$.

IV. *A separable metric space X has a countable neighborhood basis.*

Proof. If $\{x_1, x_2, x_3, \cdots\}$ is a countable dense set in X, then the open balls

$$\rho(x, x_k) < \frac{1}{n}, \quad k, n = 1, 2, 3, \cdots$$

form a countable basis in X. In fact, suppose $U(x_0, \varepsilon)$ is an arbitrary neighborhood defined by the inequality $\rho(x_0, x) < \varepsilon$ and $x' \in U(x_0, \varepsilon)$; then $\rho(x_0, x') < \varepsilon - \delta$ for some $\delta > 0$. We choose n_0 so that $1/n_0 < \delta/2$ and then an x_{k_0} for which $\rho(x', x_{k_0}) < 1/n_0$. Then the open ball defined by the inequality $\rho(x, x_{k_0}) < 1/n_0$ lies entirely in $U(x_0, \varepsilon)$ and contains x'. Thus, every neighborhood $U(x_0, \varepsilon)$, and hence also every open set in X, is the union of the balls $\rho(x, x_k) < 1/n$, i.e. these balls form a neighborhood basis.

Clearly, the converse assertion is also true.

V. *A metric space X with a countable neighborhood basis is separable.*

In fact, if U_k is a countable neighborhood basis in X, then, choosing an x_k in each U_k, we obtain a countable set $\{x_k\}$. It is dense in X because for each $x_0 \in X$ and $\varepsilon > 0$ there exists a $U_k \subset U(x_0, \varepsilon)$. Then $x_k \in U(x_0, \varepsilon)$, i.e. $\rho(x_0, x_k) < \varepsilon$.

Propositions IV and V imply

VI. *Every subset X_1 of a separable metric space X is separable.*

In fact, if U_k is countable neighborhood basis in X, then $U_k \cap X$ is a countable neighborhood basis in X_1 and it remains to apply Proposition V.

14. Compact sets in metric spaces. A set Q in a metric space X is called *precompact* if every sequence of elements in Q contains a fundamental subsequence. [Another definition of a precompact set – furthermore in an arbitrary topological space – which coincides with this definition in the case of a metric space, can be found in N. Bourbaki [3], Ch. II, § 4, subsection 4.] A set Q in a metric space is called *sequentially compact* if every sequence of elements in Q contains a subsequence which converges to some element in Q. [It can be shown that a set Q in a metric space is sequentially compact if and only if it is compact; we leave the proof of this fact as a useful exercise to the reader.]

Obviously, every sequentially compact set is precompact, and a closed precompact set in a complete metric space is sequentially compact.

A set E of points in a metric space X is said to be an *ε-net* for the set $M \subset X$ if for arbitrary element $x \in M$ there exists an element y in the set E whose distance from x is less than ε.

I (HAUSDORFF [1]). *A subset M of a metric space X is precompact if and only if X contains a finite ε-net for the set M for every $\varepsilon > 0$.*

Proof. a) *Necessity.* Suppose M is precompact and let $x_1 \in M$. If $\rho(x, x_1) < \varepsilon$ for all $x \in M$, then the set consisting of the single point x_1 is already a finite ε-net. But if this is not the case, there exists a point $x_2 \in M$ such that $\rho(x_1, x_2) \geqq \varepsilon$. Then if for arbitrary point $x \in M$ either $\rho(x, x_1) < \varepsilon$ or $\rho(x, x_2) < \varepsilon$, the set $\{x_1, x_2\}$ will be a finite ε-net. In the contrary case, there exists a point x_3 such that $\rho(x_1, x_3) \geqq \varepsilon$, $\rho(x_2, x_3) \geqq \varepsilon$. Repeating this argument, we construct points x_i such that $\rho(x_i, x_j) \geqq \varepsilon$ with $i \neq j$. *A priori*, two cases are possible: either this process terminates at some finite, say the n-th, step, and then $\{x_1, x_2, \cdots, x_n\}$ is a finite ε-net, or this process is infinite, and then we obtain an infinite sequence

$$x_1, x_2, x_3, \cdots \tag{1}$$

of elements in the M such that

$$\rho(x_i, x_j) \geqq \varepsilon \text{ for } i \neq j,\ i, j = 1, 2, 3, \cdots. \tag{2}$$

But this is impossible because in virtue of (2) no subsequence of the sequence (1) is fundamental, contradicting the fact that the set M is precompact.

b) *Sufficiency.* Suppose, for arbitrary $\varepsilon > 0$, there exists in X a finite ε-net for the set M. We choose a sequence ε_n, $\varepsilon_n \to 0$ as $n \to \infty$, and for every ε_n we construct a finite ε_n-net

$$\{x_1^{(n)}, x_2^{(n)}, \cdots, x_{k_n}^{(n)}\}$$

for the set M. Now suppose $T = \{x_n\}$ is a sequence of elements in M; we shall prove that it contains a fundamental subsequence. This will prove the precompactness of the set M.

We describe a ball of radius ε_1 about each of the points in the ε_1-net $\{x_1^{(1)}, x_2^{(1)}, \cdots, x_{k_1}^{(1)}\}$.

Each of the points in the set T falls into one of these balls; consequently, at least one of these balls contains (an infinite) *subsequence* of T. We denote this subsequence by T_1. We describe a ball of radius ε_2 about each of the points in the ε_2-net $\{x_1^{(2)}, x_2^{(2)}, \cdots, x_{k_2}^{(2)}\}$; then at least one of these balls will contain a subsequence T_2 of T_1. Repeating this argument, we obtain a sequence of nested sequences $T \supset T_1 \supset T_2 \supset T_3 \supset \cdots$, where each T_n is contained in a ball of radius ε_n. We now take a point $x_{n_1} \in T_1$, a point $x_{n_2} \in T_2$, with $n_2 > n_1$, a point $x_{n_3} \in T_3$, with $n_3 > n_2$, and so on. We thus obtain a subsequence $x_{n_1}, x_{n_2}, \cdots$ of T. This subsequence is fundamental. In fact, $x_{n_p} \in T_p$, $x_{n_{p+q}} \in T_{p+q} \subset T_p$, and therefore both points $x_{n_p}, x_{n_{p+q}}$ lie in a ball of radius ε_p, from which it follows that

$$\rho(x_{n_{p+q}}, x_{n_p}) \leqq 2\varepsilon_p.$$

II. *A subset M of a metric space X is precompact if and only if for arbitrary $\varepsilon > 0$ there exists in X a precompact ε-net for the set M.*

Proof. The necessity is obvious because, according to I, there even exists a finite ε-net for the set M. We shall prove the sufficiency. Suppose E is a precompact $\varepsilon/2$-net for M; by I, there exists a finite $\varepsilon/2$-net F for the set E. Then F is a finite ε-net for M. In fact, if $x \in M$, then there exists an element $y \in E$ such that $\rho(x, y) < \varepsilon/2$, and an element $z \in F$ such that $\rho(y, z) < \varepsilon/2$; consequently, $\rho(x, z) < \varepsilon$. On the basis of I, we conclude from this that M is precompact.

15. Topological product of metric spaces. Let X_α be metric spaces, $X = \prod_\alpha X_\alpha$ their topological product (see subsection 12), and ρ_α a metric in X_α.

I. *The topological product of a finite or countable number of metric spaces X_j is metrizable.*

Proof. In the case of a finite number of spaces $X_1, \cdots, X_n$ it is clear that a topology in $X = \prod_{j=1}^n X_j$ is defined by the metric $\rho(x, y) = 1.\ \rho_1(x_1, y_1) + \cdots + \rho_n(x_n, y_n)$ where $x = \langle x_1 \cdots, x_n\rangle$ and $y = \langle y_1, \cdots, y_n\rangle$. In the case of a countable number of spaces $X_1, X_2, \cdots$, we set

$$\rho(x, y) = \sum_{n=1}^{\infty} \frac{1}{2^n} \frac{\rho_n(x_n, y_n)}{1+\rho_n(x_n, y_n)}, \tag{1}$$

where $x = \langle x_n\rangle$, $y = \langle y_n\rangle$, $x_n, y_n \in X$. It is not difficult to verify directly that $\rho(x, y)$ satisfies all the distance axioms. We shall prove that the topology in $X = \prod_{n=1}^\infty X_n$ is defined by the metric $\rho(x, y)$. By the definition of the topology in X, a neighborhood basis in X is formed by all possible sets $V(x_1^0, \cdots, x_n^0; \varepsilon_1, \cdots, \varepsilon_n)$ defined by the inequalities

$$\rho_j(x_j, x_j^0) < \varepsilon_j, \qquad j = 1, \cdots, n, \tag{2}$$

for arbitrary fixed n, $\varepsilon_j > 0$ and $x_j^0 \in X_j$, $j = 1, \cdots, n$. Each such neighborhood contains the ball $S(x_0, \varepsilon)$ defined by the inequality

$$\rho(x, x_0) < \varepsilon, \tag{3}$$

$\varepsilon = \min_{1 \leq k \leq n}\{(\frac{1}{2})^k[\varepsilon_k/(1+\varepsilon_k)]$ and $x_0 = \{x_j^0\}$, where x_j^0 is arbitrary for $j > n$. In fact, if $\rho(x, x_0) < \varepsilon$, then, in virtue of (1),

$$\frac{1}{2^k}\frac{\rho_k(x_k, x_k^0)}{1+\rho_k(x_k, x_k^0)} < \varepsilon \leqq \frac{1}{2^k}\frac{\varepsilon_k}{1+\varepsilon_k} \text{ for } k = 1, \cdots, n,$$

from which it follows that $\rho_k(x_k, x_k^0) \leqq \varepsilon_k$ for $k = 1, \cdots, n$. Conversely each ball $S(x, \varepsilon)$ contains some neighborhood $V(x_1^0, \cdots, x_n^0; \varepsilon_1, \cdots, \varepsilon_n)$. In fact, we choose n such that $\sum_{k=n+1}^{\infty} (\frac{1}{2})^k < \varepsilon/2$ and set $\varepsilon_1 = \cdots = \varepsilon_n = \varepsilon/2$; then it will follow from the inequalities

$$\rho_k(x_k, x_k^0) < \frac{\varepsilon}{2}, \quad k = 1, \cdots, n$$

that

$$\rho(x, y) = \sum_{k=1}^{n} \frac{1}{2^k}\frac{\rho_k(x_k, x_k^0)}{1+\rho_k(x_k, x_k^0)} + \sum_{k=n+1}^{\infty} \frac{1}{2^k}\frac{\rho_k(x_k, x_k^0)}{1+\rho_k(x_k, x_k^0)}$$
$$< \sum_{k=1}^{n} \frac{1}{2^k}\cdot\frac{\varepsilon}{2} + \sum_{k=n+1}^{\infty} \frac{1}{2^k} < \frac{\varepsilon}{2} + \frac{\varepsilon}{2} = \varepsilon.$$

But this means that $V(x_1^0, \cdots, x_n^0; \varepsilon/2, \cdots, \varepsilon/2) \subset S(x_0, \varepsilon)$.

II. *If the metric spaces $X_1, X_2, \cdots$ are complete, then $\prod_{j=1}^{\infty} X_j$ is complete.*

[Proposition II is also valid for the topological product of a finite number of metric spaces; in this case it is trivial.]

Proof. Let $x^{(n)} = \{x_k^{(n)}\}$ be a fundamental sequence in $X = \prod_{j=1}^{\infty} X_j$. It follows easily from (1) that for every fixed k $x_k^{(n)}$ is a fundamental sequence in X_k. In virtue of the completeness of the space X_k, $\lim_{n\to\infty} x_k^{(n)} = x_k^0 \in X_k$ exists. We set $x^0 = \{x_k^0\}$. Then $\rho(x^{(n)}, x^0) \to 0$ as $n \to \infty$. In fact, for given $\varepsilon > 0$ we first choose N_1 so that $\sum_{k=N_1+1}^{\infty} (\frac{1}{2})^k < \varepsilon/2$ and then N so that $\rho(x_k^{(n)}, x_k^0) < \varepsilon/2$ for $k = 1, \cdots, N_1$ and $n > N$.

Then

$$\rho(x^{(n)}, x^0) \leqq \sum_{k=1}^{N_1} \frac{1}{2^k}\frac{\rho_k(x_k^{(n)}, x_k^0)}{1+\rho_k(x_k^{(n)}, x_k^0)} + \sum_{k=N_1+1}^{\infty} \frac{1}{2^k} < \frac{\varepsilon}{2}\sum_{k=1}^{N_1}\frac{1}{2^k} + \frac{\varepsilon}{2} < \varepsilon$$

for $n > N$.

III. *The topological product of a finite or countable number of separable metric spaces is separable.*

Proof. Let S_k be a countable dense set in X_k. In the case of a finite number of spaces $X_1, \cdots, X_n$ the set of all $x = \langle x_1, \cdots, x_n\rangle$, $x_1 \in S_1, \cdots, x_n \in S_n$ forms a countable dense set in $X_1 \times \cdots \times X_n$. In the case of a countable number of spaces $X_1, X_2, \cdots$ the set of all $x = \langle x_1, \cdots, x_n, 0, 0, \cdots\rangle$, $x_k \in S_k$, $k = 1, \cdots, n$, $n = 1, 2, 3, \cdots$, forms a countable set which is dense in $\prod_{k=1}^{\infty} X_k$.

We agree to denote by X^N, where X is a metric space with metric $\rho^1(x, y)$, the product $\prod_{n=1}^{\infty} X_n$ in which all $X_n = X$ and the metric is defined by the formula

$$\rho(x, y) = \sum_{n=1}^{\infty} \frac{1}{2^n} \frac{\rho_n(x_n, y_n)}{1+\rho_n(x_n, y_n)},$$

$x = \langle x_n \rangle$, $y = \langle y_n \rangle$, $x_n, y_n \in X_n$, where ρ_n is the metric in X_n coinciding with ρ^1.

IV. *The topological product of a finite or countable number of sequentially compact metric spaces is sequentially compact.*

Proof. [The assertion follows from Tikhonov's theorem (II, subsection 12); here we give an independent proof.] Let X_j be sequentially compact and let $x^{(n)} = \langle x_1^{(n)}, x_2^{(n)}, \cdots \rangle$ be a sequence in $X = \prod_j X_j$. Then $\langle x_1^{(n)} \rangle$ is a sequence in the sequentially compact space X_1 and hence there exists a subsequence $x^{(n_1)}$ such that $x_1^{(n_1)}$ converges in X_1. Further, $x_2^{(n_1)}$ is a sequence in the sequentially compact space X_2; therefore, there exists a subsequence $x^{(n_2)}$ of the sequence $x^{(n_1)}$ such that $x_2^{(n_2)}$ converges in X_2; moreover, $x_1^{(n_2)}$ also converges in X_1.

If the number of spaces X_j is finite (and equal to p), then after p steps we obtain the sequence $x^{(n_p)}$ which converges in $X = \prod_{j=1}^{p} X_j$. And if the number of spaces X_j is countable, then for each natural number p we obtain a sequence $x^{(n_p)}$ possessing the following properties:

1) $\langle x^{(n_p)} \rangle \supset \langle x^{(n_{p+1})} \rangle$ and 2) $x^{(n_p)}$ converges in X_j for $j = 1, 2, \cdots, p$. Then the "diagonal" sequence $\langle x_1^{(n_1)}, x_2^{(n_2)}, x_3^{(n_3)}, \cdots \rangle$ will converge in $X = \prod_{j=1}^{\infty} X_j$.

§ 3. Topological linear spaces

1. Definition of a topological linear space. A set X is called a *topological linear space* if:

1) X is a linear space;

2) X is a Hausdorff topological space;

3) the sum $x+y$ of the vectors x, y is continuous in both variables x, y;

4) the product αx of the vector x by the number α is continuous in both variables α, x.

X can be either a real or a complex linear space. (We shall consider primarily complex spaces; unless explicitly stated to the contrary, we shall always consider the topological linear space under consideration to be complex.) To avoid considering these cases separately, we shall agree to denote the set of all real numbers or the set of all complex numbers (depending on whether we are considering a real or a complex space) with their natural topology by $\mathfrak{C}$ (see Examples 1 and 2, subsection 1, § 2). Condition 4) means that the mapping $\langle \alpha, x \rangle \to \alpha x$ of the topological product $\mathfrak{C} \times X$ into X is continuous; analogously, condition 3) means that the mapping $\langle x, y \rangle \to x+y$ of the topological product $X \times Y$ into X is continuous.

If X is a topological linear space, then in virtue of condition 3 the mapping $x \to x+x_0$ is continuous for arbitrary $x_0 \in X$.

This mapping is called a *translation* in X. Since the inverse mapping $x \to x-x_0$ exists and is also continuous, every translation $x \to x+x_0$ is a homeomorphism. Therefore, if $\{U(0)\}$ is a neighborhood basis of the point 0, then $\{U(0)+x_0\}$ is a neighborhood basis of the point x_0. In particular, if $\{U(0)\}$ is the set of all neighborhoods of the point 0, then $\{U(0)+x_0\}$ is the set of all neighborhoods of the point x_0.

I. *The closure $\bar{K}$ of a convex subset K of a topological linear space X is convex.*

In fact, for each λ, $0 \leqq \lambda \leqq 1$, the function $f(x, y) = (1-\lambda)x+\lambda y$ is continuous on $X \times X$ and maps $K \times K$ into K and consequently also $\bar{K} \times \bar{K}$ into $\bar{K}$ (see III, subsection 5, § 2 and subsection 12, § 2). But this means that $(1-\lambda)x+\lambda y \in \bar{K}$ for all $x, y \in \bar{K}$.

The mapping f of the topological linear space X into the topological linear space Y is called an *isomorphism* of X into Y if:

1) f is an isomorphic mapping of the linear space X into the linear space Y;

2) f is a homeomorphic mapping of the topological space X into the topological space Y.

If, moreover, $f(X) = Y$, then f is called an *isomorphism of X onto Y.*

Two topological linear spaces X and Y are said to be *isomorphic* if there exists an isomorphism of X onto Y.

A topological linear space X is said to be *locally convex* if there exists in X a basis of neighborhoods of zero which are convex sets (see subsections 8 and 9, § 1). In the sequel, we shall consider only locally convex topological linear spaces which we shall call, briefly, *locally convex spaces.* We note, however, that some results, in particular all the results of subsection 2, remain valid for general topological linear spaces.

EXAMPLES. 1. The topological space C^n of Example 2, subsection 2, § 2 can also be considered as a linear space (see Example 1, subsection 1, § 1); it is easy to see that then C^n is a complex topological linear space. It is locally convex because a basis neighborhood $U(a, \varepsilon)$ in it is the set of all x which satisfy the condition $p(x-a) < \varepsilon$ where $p(x) = (|x_1|^2+|x_2|^2+\cdots+|x_n|^2)^{\frac{1}{2}}$ is a norm in C^n; therefore $U(a, \varepsilon)$ is a convex set in C^n (see subsection 8, § 1). Analogously, the space R^1 in Example 1, subsection 1, § 2, can be considered as a real linear space (by defining addition as the addition of numbers, and multiplication by a real number as ordinary multiplication); then R^1 is a locally convex real topological linear space.

2. Suppose X is an arbitrary locally compact (in particular, a compact) space. We denote the set of all continuous complex-valued functions on X by $C(X)$. We define addition in $C(X)$ as the addition of functions and multiplication by a complex number as the multiplication of functions by a number. Then $C(X)$ becomes a linear space. Further, we define a topology in $C(X)$ in the following way. Suppose K is an arbitrary compact subset in X and that ε is an arbitrary positive number; a neighborhood $U(f_0; K; \varepsilon)$ of the function $f_0 \in C(X)$ is the set of all functions f in $C(X)$ satisfying the inequality

$$|f(x) - f_0(x)| < \varepsilon \quad \text{for all } x \in K.$$

If we let K run through all compact subsets in X and ε all positive numbers, we obtain a family of neighborhoods which we shall consider to be a neighborhood basis of the function f_0. It is easily verified that conditions 3)–4) in the definition of a topological linear space are satisfied and that $U(f_0; K; \varepsilon)$ is a convex set in $C(X)$; consequently, $C(X)$ is a locally convex complex topological linear space.

2. Closed subspaces in topological linear spaces. A set $\mathfrak{M}$ of elements in a topological linear space X is said to be a *closed subspace* of the space X if:

1) $\mathfrak{M}$ is a linear subspace of the linear space X;

2) $\mathfrak{M}$ is a closed subset of the topological space X.

Clearly:

I. *The intersection of an arbitrary set of closed subspaces in X is also a closed subspace in X* (see subsection 3, § 1 and subsection 3, § 2).

In particular, the intersection of all closed subspaces in X which contain a given set $M \subset X$ is the minimal closed subspace in X which contains M; it is called the *closed linear hull* of the set M; we shall denote it by $[M]$. In general, if a finite, countable or noncountable family of sets $M_\alpha, M_\beta, \cdots$ is given, then $[M_\alpha, M_\beta, M_\gamma, \cdots]$ will denote the closed linear hull of their union. We note that

II. *If $\mathfrak{M}$ (which is in general not closed) is a linear subspace of X, then its closure $\overline{\mathfrak{M}}$ is also a linear subspace and, consequently, it is a closed linear subspace in X.*

Proof. For fixed $y \in \mathfrak{M}$ the continuous mapping $f(x) = x+y$ maps $\mathfrak{M}$ into $\mathfrak{M}$ and therefore $\overline{\mathfrak{M}}$ into $\overline{\mathfrak{M}}$ (see III, subsection 5, § 2); this means that $x+y \in \overline{\mathfrak{M}}$ for $x \in \overline{\mathfrak{M}}$, $y \in \mathfrak{M}$. But then for fixed $x \in \overline{\mathfrak{M}}$, the continuous mapping $f(y) = x+y$ maps $\mathfrak{M}$ into $\overline{\mathfrak{M}}$, and therefore also $\overline{\mathfrak{M}}$ into $\overline{\mathfrak{M}}$; in other words, if $x, y \in \overline{\mathfrak{M}}$ then $x+y \in \overline{\mathfrak{M}}$. It is proved analogously that if $x_0 \in \overline{\mathfrak{M}}$, then also $\alpha x_0 \in \overline{\mathfrak{M}}$.

It follows from II that *the closed linear hull of a set M is the closure of the set of all finite linear combinations of elements in M.*

We shall call a topological linear space X_1 a *subspace of the topological linear space X* if 1) X_1 is a linear subspace of the space X and 2) X_1 is a topological subspace of the topological space X.

3. Convex sets in locally convex spaces. Now let X be a locally convex space. Consider in X a convex set K which contains 0 as an interior point. Then K contains a convex symmetric neighborhood of zero $U(0)$. For arbitrary fixed $x \in X$ the product αx is a continuous function of α which equals zero for $\alpha = 0$; consequently, there exists a positive number δ such that $\alpha x \in U(0)$ for $|\alpha| < \delta$ and hence also $\alpha x \in K$ for $|\alpha| < \delta$. This means that

$$x \in \frac{1}{\alpha} K \text{ for } |\alpha| < \delta. \tag{1}$$

Thus (see subsection 8, § 1):

I. *Every convex set K in the space X, containing* 0 *as an interior point, is absorbing.*

According to Proposition III, subsection 8, § 1, to such a set K there corresponds a Minkowski functional p such that if $x \in K$ then $p(x) \leqq 1$ and if $p(x) < 1$ then $x \in K$. From this and V, subsection 8, § 1, we conclude that the boundary of the set K consists of those and only those vectors x for which $p(x) = 1$.

II. *Let K be a convex set in X which contains* 0 *as an interior point. Then the Minkowski functional p of the set K is continuous.*

Proof. Let $U(0)$ be a neighborhood of the point 0, contained in K: $U(0) \subset K$. Then $p(x) \leqq 1$ for $x \in U(0)$ and hence $p(x) \leqq \varepsilon$ for $x \in \varepsilon U(0)$. Consequently, for $x - x_0 \in \varepsilon U(0)$ we have

$$|p(x) - p(x_0)| \leqq p(x - x_0) \leqq \varepsilon.$$

Now suppose that K is a convex set containing the interior point x_0. Then $K - x_0$ is a convex set containing the interior point 0 and therefore also a convex symmetric neighborhood of zero. Applying the preceding results to the convex set $K - x_0$ we conclude that

III. *For every convex set K containing the interior point x_0, there exists a continuous Minkowski functional p such that:*

1) $p(x - x_0) \leqq 1$ *for all* $x \in K$;

2) $p(x - x_0) = 1$ *for all boundary points x of the set K and only for these;*

3) $p(x - x_0) > 1$ *for arbitrary vector x which belongs neither to K nor to its boundary.*

From this and the results of subsection 9, § 1, it follows that

IV. *If K is a convex set in a real locally convex space containing an interior point, then*

1) *a hyperplane of support to K can be drawn through every boundary point of the set K;*

2) *if x_0 does not belong to K nor to its boundary then there exists a hyperplane of support to K which separates x_0 from K.*

We note further the following proposition.

V. *If K is a convex closed set then all the boundary points of K belong to K.*

Proof. Suppose x_0 is a boundary point of the set K, and $[x_1, x_0]$ is a segment all of whose interior points belong to K. Then for $0 < t_n < 1$ and $t_n \to 1$, the points $x_n = (1 - t_n)x_1 + t_n x_0$ belong to K and form a sequence which converges to x_0. Since K is closed, it follows that $x_0 \in K$.

4. Defining a locally convex topology in terms of seminorms. We now apply the preceding construction to the convex set $K = U(0)$, where $U(0)$ is a symmetric convex neighborhood of zero. Let p be the corresponding Minkowski functional. In virtue of I, subsection 9, § 1, p is a seminorm. If $x \in U(0)$ then $p(x) < 1$. In fact, since $U(0)$ is an open set and the product αx is continuous for $\alpha = 1$, there exists an $\varepsilon > 0$ such that $\alpha x \in U(0)$ for $|\alpha - 1| < \varepsilon$. Therefore the set $\{\beta: \beta > 0, x \in \beta U(0)\}$ contains numbers which are less than 1, and $p(x) < 1$.

Conversely, if $p(x) < 1$, then $x \in U(0)$ in virtue of III, subsection 8, § 1.

I. *Every symmetric convex neighborhood of zero* $U(0)$ *is assigned a seminorm* p *such that* $U(0)$ *is the set of all those vectors* x *for which* $p(x) < 1$.

Thus, the neighborhood basis of zero $\{U(0)\}$ is assigned a family of seminorms $\{p\}$. This family of seminorms has the following property: *for every* $x_0 \neq 0$ *there exists a seminorm* p *in the family* $\{p\}$ *such that* $p(x_0) \neq 0$.

In fact, if $x_0 \neq 0$, then there exists a neighborhood $U(0)$ in $\{U(0)\}$ which does not contain x_0; consequently, the seminorm p, corresponding to this neighborhood $U(0)$, satisfies the condition $p(x_0) \geqq 1$, so that we have *a fortiori*, that $p(x_0) > 0$.

We shall say that a family $\{p\}$ of seminorms in the linear space X is *sufficient* if for every vector $x_0 \neq 0$ in X there exists a seminorm p in this family which is such that $p(x_0) > 0$.

We have proved

II. *To the basis* $\{U(0)\}$ *of symmetric convex neighborhoods of zero in* X *there corresponds a sufficient family* $\{p\}$ *of seminorms.*

Conversely, suppose an arbitrary sufficient family $\{p\}$ of seminorms is given in the linear space X. We introduce a topology in X by considering as a neighborhood of the vector x_0 every set of all vectors x which satisfy the conditions $p_k(x-x_0) < \varepsilon$, $k = 1, 2, \cdots, n$, for fixed $p_k \in \{p\}$, $k = 1, 2, \cdots, n$ for fixed n and $\varepsilon > 0$, and taking the set of neighborhoods obtained for all possible $n = 1, 2, 3, \cdots$, all possible p_k in $\{p\}$ and all possible $\varepsilon > 0$, as the neighborhood basis of the vector x_0. Clearly, these neighborhoods are obtained from the corresponding neighborhoods of zero by means of the translations $x \to x+x_0$, where the neighborhoods of zero are symmetric convex sets in X (see subsection 9, § 1).

We shall prove that *with this definition of a neighborhood basis, X is a topological linear space*, locally convex in virtue of the above discussion.

We must first of all verify that conditions 1)–3) in the definition of a neighborhood basis (see subsection 2, § 2) are satisfied. Condition 1) is obviously satisfied. We shall verify that condition 2) is satisfied.

Let $W_1(x_0)$ be defined by the inequalities

$$p_1(x-x_0) < \varepsilon_1, \cdots, p_m(x-x_0) < \varepsilon_1,$$

and $W_2(x_0)$ by the inequalities

$$p_{m+1}(x-x_0) < \varepsilon_2, \cdots, p_n(x-x_0) < \varepsilon_2,$$

where $p_j \in \{p\}$ $(j = 1, \cdots, n)$ and $\varepsilon_1 > 0$, $\varepsilon_2 > 0$.

We set $\varepsilon = \min(\varepsilon_1, \varepsilon_2)$; then the neighborhood $W(x_0)$ defined by the inequalities $p_1(x-x_0) < \varepsilon, \cdots, p_n(x-x_0) < \varepsilon$ will be contained in $W_1(x_0) \cap W_2(x_0)$. Finally, we shall verify condition 3). Let x_1 be contained in the neighborhood $U(x_0)$ defined by

$$p_k(x-x_0) < \varepsilon, k = 1, 2, \cdots, n$$

so that $p_k(x_1-x_0) < \varepsilon$, $k = 1, 2, \cdots, n$. We denote by $U_1(x_1)$ the neighborhood of the point x_1, defined by the conditions

$$p_k(x-x_1) < \varepsilon_1, k = 1, 2, \cdots, n,$$

where ε_1 is the smallest of the numbers $\varepsilon-p_k(x_1-x_0)$, $k = 1, 2, \cdots, n$. If $x \in U_1(x_1)$, then

$$\begin{aligned} p_k(x-x_0) = p_k(x-x_1+x_1-x_0) &\leqq p_k(x-x_1)+p_k(x_1-x_0) \\ &< \varepsilon_1+p_k(x_1-x_0) \leqq \varepsilon-p_k(x_1-x_0)+p_k(x_1-x_0) = \varepsilon; \end{aligned}$$

consequently, $U_1(x_1) \subset U(x_0)$.

We shall now verify that the separation axiom is satisfied. Suppose $x_1 \neq x_2$; then $x_1-x_2 \neq 0$ and therefore there exists a seminorm $p_0 \in \{p\}$ such that $p_0(x_1-x_2) > 0$. Set $\varepsilon = \frac{1}{2}p_0(x_1-x_2)$; then the neighborhoods $U_1(x_1)$ and $U_2(x_2)$ defined respectively by the inequalities

$$p_0(x-x_1) < \varepsilon \text{ and } p_0(x-x_2) < \varepsilon$$

are disjoint. In fact, any common point x would have to satisfy both of the above inequalities, from which it would follow that

$$p_0(x_1-x_2) = p_0(x_1-x+x-x_2) \leqq p_0(x_1-x)+p_0(x-x_2) < 2\varepsilon;$$

but the last inequality contradicts the definition of the number ε.

We have thus proved that X is a Hausdorff topological space. It remains to prove that conditions 3) and 4) in the definition of a topological linear space are satisfied, i.e. that the sum $x+y$ and the product αx are continuous.

Let $U(x_0+y_0)$ be the neighborhood of the point x_0+y_0, defined by the conditions

$$p_k(x-x_0-y_0) < \varepsilon, \qquad k = 1, 2, \cdots, n.$$

If x and y lie in the neighborhoods $U(x_0)$ and $U(y_0)$, defined respectively by the inequalities

$$p_k(x-x_0) < \frac{\varepsilon}{2}, \quad k = 1, 2, \cdots, n$$

and

$$p_k(y-y_0) < \frac{\varepsilon}{2}, \quad k = 1, 2, \cdots, n$$

then, as can easily be verified, $x+y \in U(x_0+y_0)$; this also proves the continuity of the sum.

Further, let $U(\alpha_0, x_0)$ be the neighborhood of the point $\alpha_0 x_0$, defined by the conditions

$$p_k(x-\alpha_0 x_0) < \varepsilon, \qquad k = 1, 2, \cdots, n.$$

If α and x lie in the neighborhoods $U(\alpha_0)$ and $U(x_0)$, defined respectively by the inequalities

$$|\alpha-\alpha_0| < \varepsilon_1,$$
$$p_k(x-x_0) < \varepsilon_1, \qquad k = 1, 2, \cdots, n,$$

then

$$p_k(\alpha x-\alpha_0 x_0) = p_k(\alpha x-\alpha x_0+\alpha x_0-\alpha_0 x_0) \leqq |\alpha| p_k(x-x_0) + |\alpha-\alpha_0| p_k(x_0) < (|\alpha_0|+\varepsilon_1)\varepsilon_1+\varepsilon_1 p_k(x_0).$$

But the last expression is $< \varepsilon$ for sufficiently small ε_1; this proves that the product αx is continuous.

Thus:

III. *Every sufficient family $\{p\}$ of seminorms in the linear space X defines a topology in X in which X is a locally convex topological linear space and in which the basis neighborhoods in X are defined by inequalities of the form*

$$p_k(x-x_0) < \varepsilon, \qquad k = 1, 2, \cdots, n,$$

where $p_k \in \{p\}$ and $\varepsilon > 0$.

We now extend the sufficient family $\{p\}$ by adjoining to it all possible seminorms $g = \sup(p_1, \cdots, p_n)$, $(p_k \in \{p\})$, $\alpha g (\alpha > 0)$ and all seminorms g^1 majorized by them (i.e. $g^1 \leqq \alpha g$); we obtain a new sufficient family of seminorms g which defines the same topology as $\{p\}$ where every symmetric convex neighborhood $U(0)$ of the point 0 is defined by a single inequality $g(x) < 1$. In the sequel we shall always assume that such an extension of the family $\{p\}$ has already been made, i.e. that each symmetric convex neighborhood $U(0)$ of the point 0 can be given by one inequality, $p(x) < 1$, $p \in \{p\}$.

5. The case of a finite-dimensional space. We shall now apply the results of subsection 4 to a finite-dimensional space. Suppose C^n is n-dimensional space and that $\{e_1, e_2, \cdots, e_n\}$ is any basis in C^n. We set

$$p_0(x) = \sqrt{|\xi_1|^2+|\xi_2|^2+ \cdots +|\xi_n|^2}$$

for an arbitrary element $x = \xi_1 e_1+\xi_2 e_2+ \cdots +\xi_n e_n$; it is easy to see that p_0 is a norm in C^n, so that p_0 alone already constitutes a sufficient family and therefore it defines a locally convex topology in C^n. In this topology, a neighborhood $U_0(x_0)$ of the vector $x_0 = \xi_1^0 e_1+\xi_2^0 e_2+ \cdots +\xi_n^0 e_n$ is the set of all vectors x that satisfy the inequality

$$p_0(x-x_0) = \sqrt{|\xi_1-\xi_1^0|^2+|\xi_2-\xi_2^0|^2+ \cdots +|\xi_n-\xi_n^0|^2} < \varepsilon, \ \varepsilon > 0.$$

I. *Any other locally convex topology in* C^n *coincides with the topology defined by the norm* p_0.

Proof. Suppose $\{p\}$ is a sufficient family of seminorms which defines a locally convex topology in C^n, and that $U(x_0)$ is a neighborhood in this topology, defined by the inequalities

$$p_k(x-x_0) < \varepsilon, \qquad k = 1, 2, \cdots, n,$$

where $p_k \in \{p\}$. We set

$$c = \max_{k,\, j=1,\cdots,n} \{p_k(e_j)\}. \tag{1}$$

Then the neighborhood $U_0(x_0)$, defined by the inequality

$$p_0(x-x_0) < \frac{\varepsilon}{cn}, \tag{2}$$

is contained in $U(x_0)$ because it follows from (1) and (2) that

$$p_k(x-x_0) = p_k((\xi_1-\xi_1^0)e_1 + \cdots + (\xi_n-\xi_n^0)e_n)$$
$$\leqq |\xi_1-\xi_1^0| p_k(e_1) + \cdots + |\xi_n-\xi_n^0| p_k(e_n) \leqq cn\frac{\varepsilon}{cn} = \varepsilon.$$

In order to complete the proof of Proposition I it remains to show that, conversely, every neighborhood $U_0(x)$ contains some neighborhood $U(x_0)$. Suppose x_1 is an arbitrary nonzero vector in C^n; then there exists a seminorm $p_1 \in \{p\}$ such that $p_1(x_1) > 0$. We denote by $\mathfrak{M}_1$ the subspace of all vectors x for which $p_1(x) = 0$; then $x_1 \notin \mathfrak{M}_1$, and therefore the dimension of $\mathfrak{M}_1$ is less than n. If $\mathfrak{M}_1 \neq (0)$, then let x_2 be an arbitrary nonzero vector in $\mathfrak{M}_1$; then there exists a seminorm $p_2 \in \{p\}$ such that $p_2(x_2) > 0$. We denote by $\mathfrak{M}_2$ the subspace in $\mathfrak{M}_1$ which consists of all vectors $x \in \mathfrak{M}_2$ for which $p_2(x) = 0$; then $x_2 \notin \mathfrak{M}_2$, and therefore the dimension of $\mathfrak{M}_2$ is less than the dimension of $\mathfrak{M}_1$ and consequently it is less than $n-1$. Repeating this argument, we arrive at the subspaces

$$\mathfrak{M}_1 \supset \mathfrak{M}_2 \supset \cdots \supset \mathfrak{M}_k = (0), \qquad 1 \leqq k \leqq n,$$

and the corresponding seminorms $p_1, p_2, \cdots, p_k \in \{p\}$, where $\mathfrak{M}_k$ is the set of all vectors $x \in C^n$, satisfying the conditions

$$p_1(x) = p_2(x) = \cdots = p_k(x) = 0. \tag{3}$$

Consequently every vector x, satisfying conditions (3), is equal to zero. We set

$$p(x) = \max \{p_1(x), p_2(x), \cdots, p_k(x)\}; \tag{4}$$

clearly, p is a seminorm in C^n, for which $p(x) = 0$ implies that $x = 0$. Thus p is a norm. We shall prove that

$$p_0(x) \leqq cp(x), \tag{5}$$

where c is some constant. Let us assume the contrary. Then there exists a sequence of vectors x_k in C^n for which

$$p_0(x_k) > kp(x_k), \qquad k = 1, 2, 3, \cdots. \tag{6}$$

We set

$$y_k = \frac{1}{p_0(x_k)} x_k;$$

then

$$p_0(y_k) = 1 \tag{7}$$

and it follows from (6) that

$$p(y_k) < \frac{1}{k}.$$

Suppose

$$y_k = \eta_{1k}e_1 + \cdots + \eta_{nk}e_n;$$

then (7) means that

$$|\eta_{1k}|^2 + \cdots + |\eta_{nk}|^2 = 1, \tag{8}$$

and therefore each of the sequences $\langle\eta_{1k}\rangle, \cdots, \langle\eta_{nk}\rangle$ is bounded. But then by selecting a subsequence which we shall again denote by $\langle y_k\rangle$, we can attain the convergence of the sequences $\langle\eta_{1k}\rangle, \cdots, \langle\eta_{nk}\rangle$ to certain limits $\eta_1, \cdots, \eta_n$. It follows from (8) that also $|\eta_1|^2 + \cdots + |\eta_n|^2 = 1$ and therefore the vector $y = \eta_1 e_1 + \cdots + \eta_n e_n$ is different from zero. Passing to the limit as $k \to \infty$ in the inequality $p(y) \leqq p(y-y_k)+p(y_k) \leqq p(e_1)|\eta_1 \quad \eta_{1k}| + \cdots + p(e_n)|\eta_n - \eta_{nk}| + 1/k$ we conclude that $p(y) = 0$ for $y \neq 0$, and this contradicts the fact that p is a norm.

Thus, the inequality (5) holds. But then, by (4), the set of all vectors x, satisfying the inequality $p(x-x_0) < \varepsilon/c$, is the neighborhood $U(x_0)$ defined by the inequalities

$$p_1(x-x_0) < \frac{\varepsilon}{c}, \cdots, p_k(x-x_0) < \frac{\varepsilon}{c}$$

and contained in the given neighborhood $p_0(x-x_0) < \varepsilon$.

From Proposition I, proved above, we conclude that:

II. *All finite-dimensional locally convex spaces having the same dimension are isomorphic.*

6. Continuous linear functionals. Suppose X is a locally convex space and that $\{p\}$ is a sufficient family of seminorms which corresponds to the basis $\{U(0)\}$ of symmetric convex neighborhoods of zero in X. We shall consider linear functionals with domain X.

I. *A linear functional f in X is continuous at $x = 0$ if and only if the inequality*

$$|f(x)| \leqq p_0(x) \text{ for all } x \in X \tag{1}$$

is satisfied for some p_0 in the family $\{p\}$. In this case, f is continuous for all $x \in X$.

Proof. Suppose f is continuous at $x = 0$. Then there exists a neighborhood $U(0)$ such that

$$|f(x_1)| < 1 \text{ for } x_1 \in U(0). \tag{2}$$

Suppose p_0 is a seminorm in $\{p\}$ corresponding to the neighborhood $U(0)$. Then $U(0)$ is the set of all x_1 such that $p_0(x_1) < 1$; therefore (2) can be written in the form

$$|f(x_1)| < 1 \text{ for } p_0(x_1) < 1. \tag{3}$$

Now suppose x is an arbitrary vector in X. We set

$$x_1 = \frac{1}{p_0(x)+\varepsilon} x,$$

where $\varepsilon > 0$. Then

$$p_0(x_1) = \frac{1}{p_0(x)+\varepsilon} p_0(x) < 1$$

and therefore, by (3), we also have $|f(x_1)| < 1$, i.e.

$$\left| f\left(\frac{1}{p_0(x)+\varepsilon} x\right) \right| < 1.$$

It follows that

$$|f(x)| < p_0(x)+\varepsilon;$$

and therefore

$$|f(x)| \leqq p_0(x)$$

since ε is arbitrary.

Conversely, let (1) be satisfied for some $p_0 \in \{p\}$. We shall prove that f is continuous at an arbitrary point $x_0 \in X$.

We must prove that for arbitrary $\varepsilon > 0$ there exists a neighborhood $U(x_0)$ in which $|f(x)-f(x_0)| < \varepsilon$. But such a neighborhood is the neighborhood defined by the inequality $p_0(x-x_0) < \varepsilon$, because it follows from (1) that

$$|f(x)-f(x_0)| = |f(x-x_0)| \leqq p_0(x-x_0).$$

II. *For every element $x_0 \neq 0$ in a locally convex space X there exists a continuous linear functional f such that $f(x_0) \neq 0$.*

Proof. If $x_0 \neq 0$, then there exists a $p_0 \in \{p\}$ such that $p_0(x_0) > 0$. On the other hand, there exists a linear functional f such that

$$|f(x)| \leqq p_0(x), \qquad f(x_0) = p_0(x_0)$$

(see Corollary 2 to Theorem 2, subsection 9, § 1); f is continuous in virtue of I and $f(x_0) = p_0(x_0) > 0$.

III. *If $\mathfrak{M}$ is a closed subspace of a locally convex X and x_0 is a vector in X which does not belong to $\mathfrak{M}$, then there exists a continuous linear functional f in X satisfying the conditions*:

$$f(x) = 0 \text{ on } \mathfrak{M}, f(x_0) = 1. \tag{4}$$

Proof. Since x_0 does not belong to the closed set $\mathfrak{M}$ there exists a neighborhood $U(x_0)$, which does not intersect $\mathfrak{M}$. By the definition of a neighborhood basis in X, there exists a seminorm $p_0 \in \{p\}$ such that $U(x_0)$ is the set of all vectors $x \in X$, satisfying the condition $p_0(x-x_0) < 1$. Therefore,

$$p_0(y-x_0) \geqq 1 \text{ for all } y \in \mathfrak{M}. \tag{5}$$

Replacing y by $-y$, we can rewrite (5) in the form

$$p_0(x_0+y) \geqq 1 \text{ for all } y \in \mathfrak{M}. \tag{6}$$

We denote the set of all vectors

$$x = y+\alpha x_0, y \in \mathfrak{M} \tag{7}$$

by L; clearly L is a linear subspace in X. We define a linear functional f_1 in L by means of the formula

$$f_1(x) = f_1(y+\alpha x_0) = \alpha; \tag{8}$$

this formula defines f_1 uniquely because $x_0 \notin \mathfrak{M}$ and therefore the representation of vectors x in L in the form (7) is unique. It follows from (8) that

$$f_1(x_0) = 1, f_1(y) = 0 \text{ for } y \in \mathfrak{M}. \tag{9}$$

Moreover,

$$|f_1(x)| \leqq p_0(x) \text{ for all } x \in L,$$

because, by (6) and (8),

$$p_0(x) = p_0(y+\alpha x_0) = |\alpha| p_0\left(x_0+\frac{1}{\alpha}y\right) \geqq |\alpha| = |f_1(x)|.$$

On the basis of Theorem 2, subsection 9, § 1, f_1 can be extended to a linear functional f defined on all of X and also satisfying the inequality $|f(x)| \leqq p_0(x)$.

Then f is continuous, in virtue of I, and by (9) it satisfies conditions (4).

IV. *A linear subspace L of a locally convex space X is dense in X if and only if every continuous linear functional in X, equal to zero on L, is equal to zero on X.*

Proof. A continuous linear functional f which is equal to zero on L is also equal to zero on its closure $\bar{L}$. Therefore, if L is dense in X so that $\bar{L} = X$ then f equals zero on X. But if L is not dense in X then $\bar{L}$ is a closed subspace in X which does not coincide with X. Consequently, there exists a vector x_0 in X which does not belong to $\bar{L}$; by III, there then exists a continuous linear functional f in X which equals zero on $\bar{L}$(and hence also on L) and equal to unity at x_0 and therefore it is not equal to zero on X.

7. Conjugate space. We shall denote the set of all continuous linear functionals f on the locally convex space X by X'. Clearly, the sum f_1+f_2 of two continuous linear functionals f_1, f_2 and the product αf of a continuous linear functional f with the number α are also continuous linear functions; consequently, X' is a linear space. This space is called the *conjugate* to X.

Suppose x is a fixed vector in X. By setting

$$F_x(f) = f(x), \tag{1}$$

we assign to each element $f \in X'$ a number $f(x)$ such that F_x is a functional in X'. It follows from the very definition of the operations in X' that F_x is a linear functional, and consequently, every vector $x \in X$ defines a linear functional in X' according to formula (1).

Further, if we set

$$p_x(f) = |F_x(f)| = |f(x)|, \tag{2}$$

we obtain a seminorm p_x in X' because

$$\begin{aligned} p_x(f_1+f_2) = |F_x(f_1+f_2)| &= |F_x(f_1)+F_x(f_2)| \\ &\leqq |F_x(f_1)|+|F_x(f_2)| = p_x(f_1)+p_x(f_2) \end{aligned}$$

and

$$p_x(\alpha f) = |F_x(\alpha f)| = |\alpha|\,|F_x(f)| = |\alpha| p_x(f).$$

Thus, every vector $x \in X$ defines a seminorm p_x in X' according to formula (2).

We now define a topology in X' so that X' becomes a locally convex space. By III, subsection 4, such a topology is given by a sufficient family of seminorms in X'. It turns out that there are various sufficient families of seminorms which determine essentially distinct topologies in X'. We shall consider only one of these topologies, the so-called *weak topology* in X' which will be just what we shall need and which is defined by the family $\{p_x\}$ where $p_x(f) = |f(x)|$ and x runs through all vectors of the space X. This family is sufficient because by (2) the equality $p_x(f) = 0$ for all $x \in X$ means that $f(x) = 0$ for all $x \in X$, i.e. that f is the zero functional.

Thus (see subsection 4) the set of all functionals $f \in X'$, satisfying the inequalities $|f(x_k)-f_0(x_k)| < \varepsilon (k = 1, 2, \cdots, n)$ for arbitrary fixed n, $x_1, \cdots, x_n \in X$ and $\varepsilon > 0$, is a weak neighborhood of the functional f_0 and the set of all such neighborhoods for all possible n, $x_1, \cdots, x_n \in X$, $\varepsilon > 0$ and $f_0 \in X'$ is a neighborhood basis for the weak topology in X'.

If it is necessary to indicate the vectors $x_1, \cdots, x_n$ and the number $\varepsilon > 0$ which define the prescribed neighborhood of the functional f_0 we shall write

$$U(f_0; x_1, \cdots, x_n; \varepsilon)$$

instead of $U(f_0)$.

In the sequel, unless stated explicitly otherwise, we shall consider X' as a locally convex space with a (weak) topology defined in it in this manner.

I. *The functionals* $F_x(f) = f(x)$ *are continuous in* X'.

This follows directly from I, subsection 6, because $|F_x(f)| = p_x(f)$.

We note that the topology we defined in X' is the weakest topology in which the functionals $F_x(f) = f(x)$ are continuous. We leave the proof of this assertion to the reader.

As an example, we shall find the space conjugate to the finite-dimensional space C^n.

Suppose $\{e_1, e_2, \cdots, e_n\}$ is a basis in C^n and that $\xi_1, \xi_2, \cdots, \xi_n$ are the coordinates of the vector $x \in C^n$ with respect to this basis. Then every linear functional f in C^n has the form $f(x) = c_1\xi_1 + c_2\xi_2 + \cdots + c_n\xi_n$(see subsection 5, § 1) and therefore it is continuous in C^n (see subsections 5 and 6). Thus, the correspondence $f \rightarrow \langle c_1, c_2, \cdots, c_n \rangle$ establishes an isomorphism of $C^{n'}$ with C^n and, consequently, $C^{n'}$ is an n-dimensional space. As a basis in $C^{n'}$ one can take the functionals f_k defined by the formulas

$$f_k(x) = \xi_k, \qquad k = 1, 2, \cdots, n.$$

Then every other linear functional in $C^{n'}$ can be written in the form $f = c_1 f_1 + \cdots + c_n f_n$, where $c_k = f(e_k)$.

We shall now find all linear functionals in $C^{n'}$. Suppose F is such a functional. We set $\alpha_k = F(f_k)$ and $x_0 = \alpha_1 e_1 + \cdots + \alpha_n e_n$; then

$$\begin{aligned} F(f) &= F(c_1 f_1 + \cdots + c_n f_n) = c_1 F(f_1) + \cdots + c_n F(f_n) \\ &= c_1\alpha_1 + \cdots + c_n\alpha_n = \alpha_1 f(e_1) + \cdots + \alpha_n f(e_n) \\ &= f(\alpha_1 e_1 + \cdots + \alpha_n e_n) = f(x_0). \end{aligned}$$

Consequently:

II. *Every linear functional* F *in* $C^{n'}$ *is given by the formula*

$$F(f) = f(x_0),$$

where x_0 *is a fixed vector in* C^n.

III. *The set* Q_p^C *of all functionals* f *in* X' *which satisfy the inequality*

$$|f(x)| \leqq Cp(x),$$

where p *is a fixed seminorm in a sufficient family* $\{p\}$, *is a compact set in* X'. (Let us recall that X' is considered as a topological space with the weak topology defined above.)

Proof. Without loss of generality, we may take $C = 1$ because the correspondence $f \to Cf$ is a homeomorphism into X' which maps Q_p^1 onto Q_p^C. Therefore it suffices to prove the assertion for Q_p^1. For definiteness, let X be a real space.

In this case, when X is a complex space, the segment I_x must be replaced by a closed disc of the complex plane with center at zero and radius $p(x)$. The rest of the proof remains the same.

We assign to each element $x \in X$ a closed interval $I_x = [-p(x), p(x)]$ and we denote by I the topological product of all intervals I_x, $x \in R$. According to the Tikhonov theorem (see II, subsection 12, § 2), I, being the topological product of compact spaces I_x, is also a compact space.

If $f \in Q_p^1$ then $|f(x)| \leqq p(x)$ and, consequently, the number $f(x)$ belongs to the interval I_x. Therefore the correspondence $f \to \{f(x)\}$ is, clearly, a one-to-one mapping of the set Q_p^1 onto some subset I' of the space I. [We recall that every element of the topological product I is a function $f(x)$ with values in the intervals $[-p(x), p(x)]$ (see subsection 12, § 2).] Upon comparison of the weak neighborhoods in X' and the neighborhoods in I, we see immediately that this mappping is a homeomorphism. Therefore, it suffices to prove that I' is closed in I (see II, subsection 6, § 2).

Suppose f_0 is a limit point of the set I'. We shall prove that f_0 is a continuous linear functional in X which belongs to the set Q_p^1. This will also prove that the set I' is closed. We choose x_1, x_2 and $\lambda_1 x_1 + \lambda_2 x_2$ as elements which define the neighborhood $U(f_0; x_1, x_2, \lambda_1 x_1 + \lambda_2 x_2; \varepsilon)$ of the point f_0 in I. Since f_0 is a limit point of the set I', there exists in this neighborhood an element $f \in I'$, i.e.

$$|f(x_1)-f_0(x_1)| < \varepsilon, |f(x_2)-f_0(x_2)| < \varepsilon, |f(\lambda_1 x_1+\lambda_2 x_2)-f_0(\lambda_1 x_1+\lambda_2 x_2)| < \varepsilon.$$

The last inequality can be rewritten in the form

$$|\lambda_1 f(x_1)+\lambda_2 f(x_2)-f_0(\lambda_1 x_1+\lambda_2 x_2)| < \varepsilon.$$

From this and the first two inequalities, it follows that

$$\begin{aligned} |\lambda_1 f_0(x_1)+\lambda_2 f_0(x_2)-f_0(\lambda_1 x_1+\lambda_2 x_2)| &\leqq |\lambda_1|\,|f_0(x_1)-f(x_1)| \\ &+|\lambda_2|\,|f_0(x_2)-f(x_2)|+|\lambda_1 f(x_1)+\lambda_2 f(x_2)-f_0(\lambda_1 x_1+\lambda_2 x_2)| \\ &< \varepsilon(|\lambda_1|+|\lambda_2|+1). \end{aligned}$$

Since ε is arbitrarily small, we have

$$f_0(\lambda_1 x_1+\lambda_2 x_2) = \lambda_1 f_0(x_1)+\lambda_2 f_0(x_2),$$

so that f_0 is a linear functional.

We shall prove that $f_0 \in I'$. Every neighborhood $U(f_0; x; \varepsilon)$ contains an element $f \in I'$, i.e.

$$|f_0(x)-f(x)| < \varepsilon.$$

It follows that

$$|f_0(x)| < |f(x)|+\varepsilon \leqq p(x)+\varepsilon,$$

because $f \in Q_p^1$ by assumption and therefore $|f(x)| \leqq p(x)$. Since ε is arbitrarily small, we have $|f_0(x)| \leqq p(x)$, i.e. $f_0 \in Q_p^1, f_0 \in I'$.

8. Convex sets in a finite-dimensional space.

If K is a convex set in the real finite-dimensional space R^n and x_0 is a vector in this space which belongs neither to K nor to its boundary, then there exists a hyperplane of support to K in R^n which separates x_0 from K.

Proof. Without loss of generality, we may assume that $0 \in K$ because this situation can always be attained by a translation. Suppose $\mathfrak{M}$ is a subspace (which might coincide with R^n) enveloping K; then there exist vectors $x_1, x_2, \cdots, x_m$ in K which form a basis in $\mathfrak{M}$. Since, furthermore, $0 \in K$, every vector x of the form

$$\begin{aligned} x &= t_1 x_1 + t_2 x_2 + \cdots + t_m x_m + (1 - t_1 - t_2 - \cdots - t_m)\, 0 \\ &= t_1 x_1 + t_2 x_2 + \cdots + t_m x_m, \end{aligned}$$

where $t_k \geqq 0$, $t_1 + t_2 + \cdots + t_m \leqq 1$, also belongs to K; obviously, every such vector with $t_k > 0$, $k = 1, 2, \cdots, m$ and $t_1 + t_2 + \cdots + t_m < 1$ is an interior point of the set K, considered as a convex set in $\mathfrak{M}$.

We consider the following (only possible) two cases:

1) $x_0 \in \mathfrak{M}$. On the basis of III, subsection 3, there exists a hyperplane of support $f(x) = c$ to the set K in $\mathfrak{M}$, which separates x_0 from K. If $\mathfrak{M} \neq R^n$, then, extending f in any way to a linear functional F in R^n, we obtain a hyperplane of support $F(x) = c$ which also separates x_0 from K.

2) $x_0 \notin \mathfrak{M}$. We denote the subspace enveloping $\mathfrak{M}$ and x_0 by $\mathfrak{M}_1$, i.e. $\mathfrak{M}_1$ is the set of all vectors of the form $y = x + t x_0$, $x \in \mathfrak{M}$; suppose $f(x) = c$ is any hyperplane in $\mathfrak{M}$, which is a hyperplane of support to K. We shall assume, for definiteness, that $f(x) \leqq c$ for all $x \in K$. We shall extend the functional f to a linear functional f_1 in $\mathfrak{M}_1$ by setting

$$f_1(x + t x_0) = f(x) + t c_1,$$

where c_1 is a fixed number, greater than c. Then $f_1(x_0) = c_1 > c$, and therefore the hyperplane $f_1(x) = c$ in the space $\mathfrak{M}_1$ will be a hyperplane of support to K, and it will separate K from x_0. If we extend f_1 in any way to a linear functional F in R^n, we obtain a hyperplane in R^n, which has the same property.

9. Convex sets in the conjugate space. Let X' be the space conjugate to the real locally convex space X, K a convex compact set in X', and x_0 a fixed element in X. The function $F_{x_0}(f) = f(x_0)$ is continuous and, consequently, it attains its maximal value on the compact set K (see VII, § 2, subsection 7); we denote it by M_{x_0}. We now consider the equation

$$F_{x_0}(f) = M_{x_0}.$$

The set P_{x_0} of all elements f which satisfy this equation form a hyperplane in X'. The hyperplane P_{x_0} decomposes the entire space X' into two parts (half-spaces), defined by the inequalities

$$F_{x_0}(f) \leqq M_{x_0}, \quad F_{x_0}(f) > M_{x_0}$$

respectively. By the definition of the number M_{x_0}, $F_{x_0}(f) \leqq M_{x_0}$ for all $f \in K$, where equality is attained for some of the elements $f \in K$. This means that the set K lies entirely in the first half-space, i.e. on one side of the hyperplane P_{x_0} and has points in common with this hyperplane. Therefore, *P_{x_0} is a hyperplane of support to the set K.*

I. *The intersection $K \cap P_{x_0}$ is also a convex set.*

In fact, the hyperplane P_{x_0} is a convex set and the intersection $K \cap P_{x_0}$ of the convex sets K and P_{x_0} is convex.

II. *If an interior point of the segment $[f_1, f_2]$, lying in K, belongs to the intersection $K \cap P_{x_0}$, then the entire segment $[f_1, f_2]$ also belongs to this intersection.*

Since $K \cap P_{x_0}$ is a convex set, it suffices to prove that the endpoints f_1, f_2 of this segment belong to the hyperplane P_{x_0}. Let us assume the contrary; let us suppose, for instance, that $f_1 \notin P_{x_0}$, i.e.

$$F_{x_0}(f_1) < M_{x_0}.$$

Moreover, since $F_{x_0}(f_2) \leqq M_{x_0}$, we have

$$F_{x_0}[(1-t)f_1 + tf_2] = (1-t)F_{x_0}(f_1) + tF_{x_0}(f_2) < M_{x_0}$$

for $0 < t < 1$. This means that no interior point of the segment $[f_1, f_2]$ lies in the hyperplane P_{x_0}. But this contradicts our assumption.

III. *Every convex compact set K in X′ is the intersection of all half-spaces, bounded by the hyperplanes of support to K, containing K, of the form P_{x_0}, such that for every $f_0 \in X'$ not belonging to K there exists a hyperplane of support P_{x_0}, separating f_0 from K, i.e. such that*

$$\max_{f \in K} f(x_0) < f_0(x_0).$$

Proof. We denote by K' the intersection of all half-spaces bounded by the hyperplanes of support to K and which contain K. Clearly, $K' \supset K$. We must prove that $K' = K$.

Suppose $f_0 \in K'$. We shall prove that an arbitrary neighborhood

$$U(f_0; x_1, \cdots, x_n; \varepsilon)$$

of the point f_0 contains elements of the set K. Since K is closed, it will follow from this that $f_0 \in K$ and hence Proposition III will be proved. Let $\mathfrak{K}$ be the image of the set K under the mapping

$$F: f \to \langle f(x_1), \cdots, f(x_n) \rangle$$

of the space X' into R^n. Clearly, F is continuous and $\mathfrak{K}$ is a convex set in the space R^n. Being the continuous image of a compact space K, the set $\mathfrak{K}$ is compact and hence it is a bounded closed set in R^n (see III, subsection 7, § 2). We shall prove that the point $\langle \eta_1, \cdots, \eta_n \rangle = F(f_0)$ belongs to the set $\mathfrak{K}$. For, in the contrary case, on the basis of the result in subsection 8, there exists in R^n a hyperplane of support to $\mathfrak{K}$ which separates the point $\langle \eta_1, \cdots, \eta_n \rangle$ from $\mathfrak{K}$. In other words (see subsection 7), there exist numbers $\lambda_1, \cdots, \lambda_n$ such that

$$\sum_{k=1}^{n} \lambda_k \xi_k < \sum_{k=1}^{n} \lambda_k \eta_k$$

for all points $\langle \xi_1, \cdots, \xi_n \rangle$ of the set $\mathfrak{K}$. This inequality then means that

$$f(\sum_{k=1}^{n} \lambda_k x_k) < f_0(\sum_{k=1}^{n} \lambda_k x_k)$$

for all $f \in K$. But the last inequality contradicts the fact that $f_0 \in K'$, because it shows that f_0 and K lie on different sides of the hyperplane of support

$$P_x, \text{ where } x = \sum_{k=1}^{n} \lambda_k x_k .$$

Thus, the point $\langle \eta_1, \cdots, \eta_k \rangle$ belongs to the set $\mathfrak{K}$. In other words, there exists a functional $f^{\frown} \in K$ such that

$$f^{\frown}(x_1) = f_0(x_1), \cdots, f^{\frown}(x_n) = f_0(x_n).$$

But then, of course, the inequalities

$$|f^{\frown}(x_k) - f_0(x_k)| < \varepsilon, \qquad k = 1, 2, \cdots, n$$

are satisfied, i.e. the point $f^{\frown}$ in K belongs to the neighborhood $U(f_0; x_1, \cdots, x_n; \varepsilon)$. This completes the proof of Proposition III.

The concept of a hyperplane of support can be generalized in the following way. (Everywhere in the sequel, we shall denote a convex set in X' which is compact in the weak topology by K.)

A *manifold of support* to K is any set V having the following properties:

1° the intersection $V \cap K$ is not void and contains no interior point of K;

2° if an interior point of the segment $[f_1, f_2]$ in K belongs to the intersection $V \cap K$ then the entire segment $[f_1, f_2]$ also belongs to this intersection;

3° V is the set of all points f satisfying the equations

$$f(x) = \alpha_x,$$

where the element x run through some (not necessarily finite) set S, and the α_x are prescribed numbers.

A hyperplane of support is, obviously, a particular case of a manifold of support, namely that case in which the set S in condition 3° consists of one element x and $\alpha_x = M_x$.

IV. *Every manifold of support V is closed.*

In fact, condition 3° means that V is the intersection of the hyperplanes $f(x) = \alpha_x$, $x \in S$. But each of the latter is a closed set because the function $F_x(f) = f(x)$ is continuous.

If a manifold of support consists of one point f_0, then it is called a manifold of support of dimension zero. This point f_0 cannot be an interior point of any segment in K. In fact, by condition 2°, an entire such segment would then belong to the intersection $V \cap K$ and consequently V would consist of more than one point.

A point $f_0 \in K$ is called an *extreme point* of the set K if it is not an interior point of any segment in K.

It follows from the immediately preceding assertions that *a manifold of support of dimension zero is an extreme point.* Conversely, *an extreme point is a manifold of support of dimension zero.* In fact, conditions 1° and 2° are satisfied trivially. In order to satisfy condition 3°, it suffices to take the whole space X for the set S, and set $\alpha_x = f_0(x)$.

V. *Every set $\{V_\alpha\}$ of manifolds of support which is linearly ordered by inclusion has nonvoid intersection which is also a manifold of support.*

Proof. We set $K_\alpha = K \cap V_\alpha$; K_α is the intersection of a compact set K and a closed set V_α; this intersection is nonvoid in virtue of condition 1°. Consequently, K_α is compact. Moreover, the family $\{K_\alpha\}$ is linearly ordered by inclusion. Therefore, the intersection of all K_α is nonvoid (see III, subsection 6, § 2). Then the intersection V of all the V_α is also nonvoid. The intersection $V \cap K$ coincides with the intersection of all the K_α and, consequently, it also is nonvoid. Thus, V satisfies condition 1°. Further, let $[f_1, f_2]$ be a segment in K and let an interior point f of this segment belong to the set V so that $f \in V_\alpha$ for all α. But then, by condition 2°, also $[f_1, f_2] \subset V_\alpha$ for all α and, consequently, $[f_1, f_2] \subset V$. This means that V also satisfies condition 2°. Finally, condition 3° is satisfied trivially. Namely, it suffices to take the union of all sets S_α corresponding to the manifolds V_α as the set S for V.

Thus, V satisfies conditions 1°, 2°, 3° and, consequently, it is a manifold of support. A manifold of support is said to be *minimal* if it contains no other manifold of support as a proper subset.

VI. *Every minimal manifold of support is a manifold of dimension zero, i.e. it is an extreme point.*

Proof. Suppose V is the minimal manifold of support to K. We shall assume that V contains more than one point. Suppose, for example, that $f_1, f_2 \in V$ and $f_1 \neq f_2$. Then there exists an element x_0 such that $f_1(x_0) \neq f_2(x_0)$. We set $K' = V \cap K$; K' is a convex compact set. The element x_0 defines a hyperplane of support P_{x_0} to K' which is distinct from V. In fact, all the elements $f \in P_{x_0}$ satisfy a condition of the form $f(x_0) = M$ and since $f_1(x_0) \neq f_2(x_0)$, at least one of the elements f_1, $f_2 \in V$ does not satisfy this condition. Thus, $P_{x_0} \neq V$; consequently, the intersection $V' = P_{x_0} \cap V$ is a proper subset of the manifold V.

We shall prove that V' is a manifold of support to K. This will contradict the fact that V is minimal, and our assertion will be proved.

Since

$$V' \cap K = P_{x_0} \cap V \cap K = P_{x_0} \cap K' \neq \phi,$$

condition 1° is satisfied. Condition 3° is clearly also satisfied. We shall verify that condition 2° is satisfied; suppose $[f_1, f_2]$ is a segment in K and that f is an interior point of this segment which belongs to the set V'. Then $f \in V$ and consequently $[f_1, f_2] \subset V$. But then also $[f_1, f_2] \subset V \cap K = K'$. Furthermore, $f \in P_{x_0}$ and consequently $[f_1, f_2] \subset P_{x_0}$. Hence $f \in [f_1, f_2] \subset V \cap P_{x_0} = V'$. This proves that condition 2° is also satisfied for V'. Consequently, V' is a manifold of support.

VII. *Every manifold of support to a convex set K contains a minimal manifold of support, i.e. an extreme point of K.*

Proof. Suppose V is a manifold of support to K. The set of all manifolds of support contained in V forms a partially ordered set, ordered by inclusion. By Proposition V, this partially ordered set satisfies the condition of Zorn's lemma and therefore has a minimal element which is also the minimal manifold of support contained in V.

It follows from Proposition VII, in particular, that every convex compact set in X' has at least one extreme point.

THEOREM 1 (M. KREIN and MILMAN [1]). *Every convex compact set $K \subset X'$ contains extreme points and is the smallest convex closed set which contains all the extreme points of the set K.*

Proof. Suppose K' is the smallest convex closed set which contains all the extreme points of the set K. Clearly, $K' \subset K$. We shall assume that $K' \neq K$. Suppose $f_0 \in K - K'$. Then, by III, there exists an element x_0 such that

$$f(x_0) < f_0(x_0) \text{ for all } f \in K'.$$

We set $M = \max_{f \in K} f(x_0)$; then $M \geqq f_0(x_0)$ and, consequently,

$$f(x_0) < M \text{ for all } f \in K'. \tag{1}$$

The equality $f(x_0) = M$ defines a hyperplane of support P_{x_0} to K. According to Proposition VII, this hyperplane contains at least one extreme point f_1 of the set K.

Consequently, the intersection $P_{x_0} \cap K'$ is nonvoid, namely, it contains the point f_1. But this last assertion contradicts inequality (1). This completes the proof of Theorem 1.

We note that, generally speaking, the set of extreme points is not closed. A simple example is the following convex set.

Suppose A, B are two points on the perpendicular to some plane $\mathfrak{P}$, which intersects this plane in an interior point C of the segment AB, and that $\mathfrak{C}$ is a circle in the plane $\mathfrak{P}$ which passes through the point C (Fig. 2).

Consider the smallest convex set enveloping $\mathfrak{C}$ and the points A, B. Its extreme points are all the points of the circle $\mathfrak{C}$, except the point C, and the points A, B. Thus, the set of these extreme points is not closed.

REMARK. Indeed, the basic results of subsection 9, including the Krein–Milman theorem, are valid for convex compact sets in an arbitrary locally convex space (see, e.g., N. BOURBAKI [4], Ch. II, § 4, subsection 2).

However, we have no need for these results in such a general form.

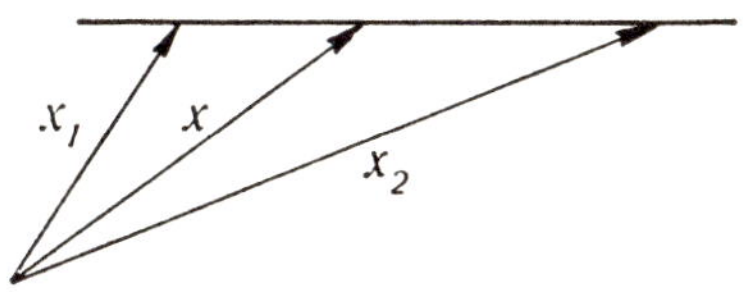

Fig. 2.

10. Cones. A set K in the real linear space X is called a *cone* if:

1) $x \in K$ and $\alpha \geqq 0$ imply that $\alpha x \in K$;

2) $x, y \in K$ implies that $x+y \in K$;

3) $x \in K$, $x \neq 0$ imply that $-x \notin K$.

In virtue of conditions 1)–2), *a cone is a convex set.*

A linear functional f is said to be *positive* (with respect to the cone K) if $f(x) \geqq 0$ for all $x \in K$. If $\mathfrak{M}$ is a subspace in X, then, clearly, $K \cap \mathfrak{M}$ is a cone in $\mathfrak{M}$. A linear functional f defined only in $\mathfrak{M}$ is said to be positive if it is positive with respect to the cone $K \cap \mathfrak{M}$.

In the sequel, an important role is played by the following theorem on the extension of a positive functional.

THEOREM 2 (M. KREIN [1]). *Suppose K is a cone in the real locally convex space X, assume K contains interior points, and let $\mathfrak{M}$ be a subspace in X which contains at least one interior point of K. Then every positive linear functional f in $\mathfrak{M}$ can be extended to a positive linear functional F in X.*

Proof. The cone K is a convex set; by assumption, K contains an interior point $x_0 \in \mathfrak{M}$, and therefore (see III, subsection 3) there exists a seminorm p which satisfies the condition that $p(x-x_0) \leqq 1$ for all $x \in K$. But if $x \in K$, then also $\alpha x \in K$ for all $\alpha > 0$, and therefore also $p(\alpha x - x_0) \leqq 1$, from which it follows that $p\{x-(1/\alpha)x_0\} \leqq 1/\alpha$. If we pass to the limit as $\alpha \to \infty$ in the last inequality and make use of the fact that the seminorm p is continuous (see II, subsection 3), we conclude that $p(x) \leqq 0$ for all $x \in K$. On the other hand, $p(x) \geqq 0$. Therefore,

$$p(x) = 0 \text{ for all } x \in K. \tag{1}$$

Suppose now that f is a linear functional in $\mathfrak{M}$ which is nonnegative on $\mathfrak{M} \cap K$. By the definition of the seminorm p, for arbitrary $x \in K$ and arbitrary $\varepsilon > 0$, we have

$$\frac{1}{p(x)+\varepsilon}\, x \in K - x_0, \text{ and consequently, } \frac{1}{p(x)+\varepsilon}\, x + x_0 \in K.$$

It follows that for any $x \in \mathfrak{M}$,

$$f\left(\frac{1}{p(x)+\varepsilon}\, x + x_0\right) = \frac{1}{p(x)+\varepsilon} f(x) + f(x_0) \geqq 0,$$

so that

$$-f(x) \leqq f(x_0)[p(x)+\varepsilon]$$

and in view of the fact that the number $\varepsilon > 0$ is arbitrary,

$$-f(x) \leqq f(x_0)p(x) \text{ for all } x \in \mathfrak{M}.$$

Since $f(x_0)p(x)$ is a seminorm in X, $-f$ can be extended, on the basis of the Hahn-Banach theorem (see subsection 9, § 1), to a linear functional $-F$ in X so that the inequality

$$-F(x) \leqq f(x_0)p(x)$$

is satisfied for all $x \in X$. Then F is an extension of the functional f and

$$F(x) \geqq -f(x_0)p(x) \text{ for all } x \in X.$$

But it then follows from this, in virtue of (1), that $F(x) \geqq 0$ for all $x \in K$, and hence the theorem is proved.

COROLLARY. *If K is a cone in X with interior point x_0, then there exists a linear functional, positive with respect to K, which equals unity at x_0.*

Proof. Suppose $\mathfrak{M}$ is the one-dimensional subspace in X, consisting of all vectors tx_0, $-\infty < t < \infty$. If we set $f(tx_0) = t$, we obtain a positive linear functional f on $\mathfrak{M}$ which satisfies the condition $f(x_0) = 1$. According to Theorem 2, f can be extended to a positive linear functional F on X and $F(x_0) = f(x_0) = 1$.

REMARK. A set K in a real linear space is called a *wedge* if it satisfies conditions 1) and 2) in the definition of a cone. Theorem 2 and the corollary to it are also valid for a wedge with an interior point; in fact, condition 3) was not used in the proof of Theorem 2.

11. Annihilators in the conjugate space. The *annihilator* of the set $S \subset X$ in the space X' is by definition the set of all functionals $f \in X'$ which satisfy the condition

$$f(x) = 0 \text{ for all } x \in S.$$

I. *The annihilator of any set $S \subset X$ is a closed subspace in X'.*

Proof. Suppose $\mathfrak{N}$ is the annihilator of the set $S \subset X$; $\mathfrak{N}$ is linear because

$$f_1(x) = 0, \ f_2(x) = 0 \text{ for all } x \in S$$

implies that

$$\lambda_1 f_1(x) + \lambda_2 f_2(x) = 0 \text{ for all } x \in S.$$

$\mathfrak{N}$ is closed. In fact, for fixed $x \in R$, the set $\mathfrak{N}_x$ of all $f \in X'$ which satisfy the condition $f(x) = 0$ is closed because $f(x)$ is a continuous function of f. Therefore, $\mathfrak{N}$, being the intersection of the sets $\mathfrak{N}_x$, $x \in S$, is also closed.

II. *Every closed subspace $\mathfrak{N}$ in X' is the annihilator of some closed subspace $S \subset X$.*

Proof. Suppose S is the set of all $x \in X$ which are such that

$$f(x) = 0 \text{ for all } f \in \mathfrak{N}.$$

It follows from the continuity of the linear functionals f that S is a closed subspace in X. We denote the annihilator of the set S by $\mathfrak{N}'$. Clearly, $\mathfrak{N} \subset \mathfrak{N}'$. We shall prove that $\mathfrak{N}' = \mathfrak{N}$, i.e. that if $f_0 \in \mathfrak{N}'$, then $f_0 \in \mathfrak{N}$. Since $\mathfrak{N}$ is closed, it will suffice to prove that f_0 is a contact point of the set $\mathfrak{N}$, i.e. that an arbitrary neighborhood of the functional f_0 contains elements $f \in \mathfrak{N}$.

Let $U(f_0; x_1, \cdots, x_n; \varepsilon)$ be such a neighborhood. We shall consider $\langle f(x_1), \cdots, f(x_n)\rangle$ as a vector in the n-dimensional space R^n. When f runs through $\mathfrak{N}$, the set of all such vectors forms a subspace in R^n. We denote this subspace by $\mathfrak{M}$.

We shall prove that the vector $\xi_0 = \langle f_0(x_1), \cdots, f_0(x_n)\rangle$ belongs to the subspace $\mathfrak{M}$. Let us assume the contrary. Then there exists a vector in $R^{n'}$ which is orthogonal to $\mathfrak{M}$ (e.g. $= 0$ on $\mathfrak{M}$) but is not orthogonal to ξ_0. In other words, there exist numbers $\lambda_1, \cdots, \lambda_n$ such that

$$\lambda_1 f(x_1) + \cdots + \lambda_n f(x_n) = 0 \text{ for all } f \in \mathfrak{N}, \tag{1}$$

$$\lambda_1 f_0(x_1) + \cdots + \lambda_n f_0(x_n) \neq 0. \tag{2}$$

Condition (1) signifies that $\lambda_1 x_1 + \cdots + \lambda_n x_n \in S$. But then, by (2), the functional $f_0 \in \mathfrak{N}'$ is different from zero on the element $\lambda_1 x_1 + \cdots + \lambda_n x_n$ of the set S, which contradicts the definition of $\mathfrak{N}'$.

Hence, $\xi_0 \in \mathfrak{M}$. This means that there exists a functional $\bar{f} \in \mathfrak{N}$ which satisfies the conditions

$$\bar{f}(x_k) = f_0(x_k), \ k = 1, \cdots, n.$$

Consequently, $\bar{f} \in U(f_0; x_1, \cdots, x_n; \varepsilon)$, which was to be proved.

III. *If $\mathfrak{N}$ is the annihilator of a closed subspace $S \subset X$, then S is the set of all $x \in X$ such that $f(x) = 0$ for all $f \in \mathfrak{N}$.*

Proof. If $x_0 \in S$, then $f(x) = 0$ for all $f \in \mathfrak{N}$. But if $x_0 \notin S$, then there exists a functional $f_0 \in X'$ such that $f_0(x_0) \neq 0$ and $f_0(x) = 0$ for all $x \in S$. Thus, $f_0 \in \mathfrak{N}$ and $f_0(x_0) \neq 0$.

Now suppose S_1 and S_2 are closed linear subspaces in X, and that $\mathfrak{N}_1$ and $\mathfrak{N}_2$ are their annihilators in X'.

IV. *If* $S_1 \subset S_2$, *then* $\mathfrak{N}_1 \supset \mathfrak{N}_2$, *and conversely. If, moreover,* $S_1 \neq S_2$, *then* $\mathfrak{N}_1 \neq \mathfrak{N}_2$, *and conversely.*

These assertions follow directly from the definition of the annihilators $\mathfrak{N}_1$ and $\mathfrak{N}_2$, and from Proposition III.

12. Analytic vector-valued functions. Suppose G is a region in the complex plane and that $x(\lambda)$ is a function defined for all $\lambda \in G$, whose values are vectors in a fixed locally convex space X. The function $x(\lambda)$ is said to be *analytic* in G if

$$x'(\lambda_0) = \lim_{\lambda \to \lambda_0} \frac{x(\lambda)-x(\lambda_0)}{\lambda-\lambda_0} \tag{1}$$

exists for arbitrary $\lambda_0 \in G$.

I. *If* $x(\lambda)$ *is analytic in the region* G, *then* $f(x(\lambda))$ *is an analytic numerical function in the region* G *for arbitrary functional* $f \in X'$.

In fact, in virtue of the continuity of the functional f it follows from (1) that

$$\lim_{\lambda \to \lambda_0} \frac{f(x(\lambda))-f(x(\lambda_0))}{\lambda-\lambda_0} = f(x'(\lambda_0))$$

also exists for $\lambda_0 \in G$.

The function $x(\lambda)$ is said to be *analytic at infinity* if

1) $x(\lambda)$ is defined in some neighborhood $|\lambda| > N$ of the point at infinity;

2) $\lim_{\lambda \to 0} x(1/\lambda)$ exists;

3) the function $y(\lambda)$, defined for $|\lambda| < 1/N$ by the conditions

$$y(\lambda) = x\left(\frac{1}{\lambda}\right) \text{ for } \lambda \neq 0,$$

$$y(0) = \lim_{\lambda \to 0} x\left(\frac{1}{\lambda}\right),$$

is analytic in some neighborhood of the point $\lambda = 0$.

II (LIOUVILLE'S THEOREM). *If* $x(\lambda)$ *is analytic in the entire complex plane, including the point at infinity, then* $x(\lambda)$ *is a constant.*

Proof. Suppose $x(\lambda)$ is analytic in the entire complex plane, including the point at infinity. Then for arbitrary $f \in X'$, the numerical function $f(x(\lambda))$ is analytic in the entire complex plane, including the point at infinity. On the basis of the usual Liouville theorem, we conclude from this that $f(x(\lambda))$ is a constant.

Thus, $f(x(\lambda_1)) = f(x(\lambda_2))$ and

$$f(x(\lambda_1)-x(\lambda_2)) = 0 \text{ for arbitrary } \lambda_1, \lambda_2.$$

In view of the fact that the functional $f \in X'$ is arbitrary, this is possible only when

$$x(\lambda_1) - x(\lambda_2) = 0 \text{ for arbitrary } \lambda_1, \lambda_2$$

(see II, subsection 6), i.e. when $x(\lambda)$ is a constant.

13. Complete locally convex spaces. We have already defined the concept of a complete metric space (see subsection 13, § 2). This concept admits the following generalization.

A net $\{x_\alpha\}$ in the topological linear space X is said to be *convergent to the element* x in X if for every neighborhood $U(x)$ there exists an index α_0 such that $x_\alpha \in U(x)$ for all indices $\alpha \geqq \alpha_0$. (See Appendix I.)

A net $\{x_\alpha\} \subset X$ is said to be *fundamental* if every neighborhood $U(0)$ of the zero element $0 \in X$ is assigned an index α_0 such that $x_\alpha - x_\beta \in U(0)$ for all indices α, $\beta \geqq \alpha_0$. A topological linear space X is said to be *complete* if every fundamental net in X converges to some element $x \in X$.

It turns out that *every locally convex space X can be embedded as a subspace in a complete locally convex space $\tilde{X}$, in which X forms a dense set.* (For the proof of this proposition and also those formulated below, the reader is referred to the literature list in DIEUDONNÉ [2]; for another definition of completeness, the proof of the existence of the completion and a detailed discussion of the theory of topological linear spaces, see the book by N. BOURBAKI [2].)

The space $\tilde{X}$ is called the *completion* of the space X.

One can weaken the condition put on X and require only that every ordinary fundamental sequence in X converge to an element $x \in X$; a space X, satisfying this condition, is said to be *sequentially complete.* For metric spaces, these two definitions of completeness are equivalent; but in the general case not every sequentially complete space X is complete.

§ 4. Normed spaces

1. Definition of a normed space. Recall (see § 1, subsection 9) that a norm $|x|$ on a linear space X is a seminorm which satisfies the condition $|x| > 0$ for $x \neq 0$. Thus, a norm $|x|$ on X is a functional which satisfies the following conditions:

1° $|x| \geqq 0$;

2° $|x| = 0$ if and only if $x = 0$;

3° $|\alpha x| = |\alpha|\,|x|$;

4° $|x+y| \leqq |x| + |y|$ (triangle inequality).

A linear space with a norm defined on it is called a *normed space.*

Condition 2° signifies that the set $\{|x|\}$, consisting of the single norm $|x|$, is a sufficient family of seminorms in X and therefore it defines a topology in X relative to which X is a locally convex space (see subsection 4, § 3).

The set of all vectors x, satisfying the inequality

$$|x-x_0| < \varepsilon \tag{1}$$

for fixed $\varepsilon > 0$, is a neighborhood of the element x_0 and the set of all such neighborhoods for all possible fixed $\varepsilon > 0$ and $x_0 > 0$ is a neighborhood basis in X.

The topology so defined in a normed space X is called the *strong topology.*

I. *If p is an arbitrary seminorm in the linear space X and $\mathfrak{M}$ is the subspace of all vectors $x \in X$ on which $p(x) = 0$* (see subsection 9, § 1), *then the formula*

$$|\xi| = p(x), \textit{ where } x \in \xi,$$

defines a norm in the quotient space $X/\mathfrak{M}$.

In fact, if $x, y \in \xi$, then $y-x \in \mathfrak{M}$, and therefore $p(y) = p(x+y-x) \leqq p(x) +p(y-x) = p(x)$, and analogously $p(x) \leqq p(y)$. Thus, $p(x) = p(y)$ and $|\xi|$ does not depend on the choice of the representative $x \in \xi$. Clearly, $|\xi|$ is a seminorm and if $|\xi| = 0$, then $p(x) = 0$ for all $x \in \xi$; but then $\xi = \mathfrak{M}$, i.e. ξ is the zero element in $X/\mathfrak{M}$. Consequently, $|\xi|$ is a norm in $X/\mathfrak{M}$.

One can also introduce a metric in the normed space X by setting

$$\rho(x, y) = |x-y|. \tag{2}$$

It is easy to verify that the metric axioms (see subsection 13, § 2) will then be satisfied. Furthermore, the neighborhoods (1) coincide with the neighborhoods defined by the metric (2).

In other words, *the strong topology in a normed space coincides with the topology defined by the metric $\rho(x, y) = |x-y|$.*

Every linear subspace of a normed space X is also a normed space for the same definition of the norm; it is called a *normed subspace* of the space X. The mapping $x \to x^1$ of a normed space X into a normed space X^1 is called an *isometry* if 1) $x \to x^1$ is an isomorphism of the linear space X into the linear space X^1 and 2) $|x| = |x^1|$ for $x \to x^1$. Two normed spaces X, X^1 are called *isometric* if there exists an isometry of X onto X^1.

A normed space is said to be *complete* if it is a complete metric space in the metric $\rho(x, y) = |x-y|$ (see subsection 13, § 2). A complete normed space is also called a *Banach space.* (This in honor of S. Banach (1892–1945), the well-known Polish mathematician who was one of the founders of modern functional analysis.)

II. *Every non-complete normed space can be completed, i.e. it can be embedded as a normed subspace in a complete normed space.*

In fact, let $\tilde{X}$ denote the completion of the metric space X in the metric $\rho(x, y) = |x-y|$ (see subsection 13, § 2). We shall prove that the norm in X and the operations of addition and scalar multiplication can be defined in $\tilde{X}$ in such a way that $\tilde{X}$ is a normed space.

Suppose $\langle x_n \rangle$, $\langle y_n \rangle$ are fundamental sequences in X, defining the elements x, $y \in \tilde{X}$, respectively. From the relations

$$|\alpha x_n - \alpha x_m| = |\alpha||x_n - x_m|,$$
$$|(x_n + y_n) - (x_m + y_m)| = |(x_n - x_m) + (y_n - y_m)| \leqq |x_n - x_m| + |y_n - y_m|$$

we conclude that $\langle \alpha x_n \rangle$ and $\langle x_n + y_n \rangle$ are also fundamental sequences; we denote the elements which they define in the space $\tilde{X}$ by αx and $x + y$.

This also defines the operations of scalar multiplication and addition in $\tilde{X}$; it is easily verified that the linear space axioms (see subsection 1, § 1) will be satisfied. Furthermore, it follows from (2) that

$$\rho(\alpha x, \alpha y) = |\alpha| \rho(x, y), \; \rho(x + z, y + z) = \rho(x, y) \tag{3}$$

for all $x, y, z \in X$. Passing to the limit, we convince ourselves that relations (3) are also valid for all $x, y, z \in \tilde{X}$. But then the function

$$|x| = \rho(x, 0), \; x \in \tilde{X}$$

satisfies all the axioms $1° - 4°$ for a norm. Thus, for example, in virtue of the second of relations (3), we have

$$|x + y| = \rho(x + y, 0) \leqq \rho(x + y, y) + \rho(y, 0) = \rho(x, 0) + \rho(y, 0) = |x| + |y|.$$

Thus, $\tilde{X}$ is a normed space. Moreover, we have

$$\rho(x, y) = \rho(x - y, 0) = |x - y| \text{ for all } x, y \in \tilde{X}$$

and by construction $\tilde{X}$ is complete in the metric $\rho(x, y)$; consequently, $\tilde{X}$ is a complete normed space.

It follows from I, subsection 13, § 2 that $\tilde{X}$ *is a minimal complete normed space, containing X as a normed subspace, and X is dense in $\tilde{X}$*; $\tilde{X}$ is called the *completion* of the normed space X. It also follows from I, subsection 13, § 2 that the completion is defined uniquely up to an isometry which leaves the points of X fixed.

III. *A finite-dimensional subspace of a normed space is complete and hence closed.*

Proof. According to I, subsection 5, § 3, every norm in the n-dimensional subspace $\mathfrak{M}_n$ is topologically equivalent to the norm

$$|\xi_1 e_1 + \cdots + \xi_n e_n| = \sqrt{\xi_1^2 + \cdots + \xi_n^2},$$

where $\{e_1, \cdots, e_n\}$ is any basis in $\mathfrak{M}_n$, and in this norm $\mathfrak{M}_n$ is clearly complete.

IV. *If $\mathfrak{M}$ is a closed subspace in a normed space X which does not coincide with X, then for arbitrary $\varepsilon > 0$ there exists a vector $y \in X$ such that*

$$|y| = 1, \; |x - y| > 1 - \varepsilon \textit{ for all } x \in \mathfrak{M}.$$

Proof. Suppose y_0 is an element in X which does not belong to $\mathfrak{M}$; we set

$$d = \inf_{x \in \mathfrak{M}} |x - y_0|;$$

then $d > 0$ because if $d = 0$ we should have $y_0 \in \overline{\mathfrak{M}} = \mathfrak{M}$. For arbitrary $\delta > 0$ there exists an element $x_0 \in \mathfrak{M}$ such that $d \leqq |x_0 - y_0| < d + \delta$. If we set

$$y = \frac{1}{|y_0 - x_0|}(y_0 - x_0),$$

we have $|y| = 1$, and for $x \in \mathfrak{M}$,

$$|y-x| = \frac{1}{|y_0 - x_0|}|y_0 - (x_0 + |y_0 - x_0|x)| \geqq \frac{d}{|y_0 - x_0|} > \frac{d}{d+\delta} = 1 - \frac{\delta}{d+\delta} > 1 - \varepsilon,$$

if we choose $\delta > 0$ so that $\delta/(d+\delta) < \varepsilon$.

V. *The unit ball in an infinite-dimensional normed space X is not precompact.*

Proof. We construct by induction an infinite sequence $x_n \in X$ which satisfies the conditions

$$|x_n| = 1, |x_n - x_m| > \tfrac{1}{2} \text{ for } n \neq m.$$

For x_1 we take an arbitrary element in X such that $|x_1| = 1$. If $x_1, \cdots, x_n$ have already been constructed, let $\mathfrak{M}_n$ be the subspace enveloping $x_1, \cdots, x_n$; this subspace is closed in virtue of III. Since X is infinite-dimensional, we have $\mathfrak{M} \neq X$, so that in virtue of IV there exists a vector x_{n+1} satisfying the conditions $|x_{n+1}| = 1$, $|x - x_{n+1}| > \frac{1}{2}$ for all $x \in \mathfrak{M}_n$; in particular, $|x_k - x_{n+1}| > \frac{1}{2}$ for $k = 1, 2, \cdots, n$. The sequence $\langle x_n \rangle$ thus constructed by induction belongs to the unit ball and does not contain a fundamental subsequence; consequently, the unit sphere and hence also the unit ball are not precompact.

A topological linear space X is said to be *normable* if a norm can be introduced in X such that the topology in X defined by this norm will coincide with the initial topology in X. KOLMOGOROV [1] proved that *a topological linear space X is normable if and only if it contains a bounded convex neighborhood of zero.* In this connection, Kolmogorov calls a set M in the topological linear space X *bounded* if $\varepsilon_n x_n \to 0$ for arbitrary elements $x_n \in M$ and numbers $\varepsilon_n \to 0$. It can be shown that in the case of a locally convex space X, this is equivalent to the requirement that every seminorm p in the sufficient family $\{p\}$ which defines the topology in X be bounded on M.

EXAMPLES. 1. *The space $M(T)$.* Suppose T is an arbitrary set. We denote by $M(T)$ the set of all bounded complex-valued functions $x = x(t)$ on T. We define the operations of scalar multiplication and addition in $M(T)$ in the usual way, and the norm by means of the formula

$$|x| = \sup_{t \in T} |x(t)|.$$

It is easily verified that $M(T)$ is then a complete normed space.

2. *The space $C(T)$.* Suppose T is a topological space. We denote by $C(T)$ the set of all bounded continuous complex-valued functions on T. We define in $C(T)$ the operations of scalar multiplication and addition and the norm in the same way as in $M(T)$. It is easy to verify that $C(T)$ is then a complete normed space. Clearly, $C(T)$ is a closed subspace in $M(T)$. If T is compact, then in virtue of VII, subsection 7, § 2, $C(T)$ is the set of all continuous complex-valued functions on X.

3. *The space l^2.* We denote by l^2 the set of all sequences $x = \langle x_n \rangle$, where the x_n are complex numbers which satisfy the condition

$$\sum_{n=1}^{\infty} |x_n|^2 < \infty.$$

We define a norm and the operations of scalar multiplication and addition in l^2 by

$$|x| = (\sum_{n=1}^{\infty} |x|^2)^{\frac{1}{2}}, \quad \alpha x = \langle \alpha x_n \rangle, \quad x+y = \langle x_n + y_n \rangle,$$

where $x = \langle x_n \rangle$, $y = \langle y_n \rangle$. This definition of the operations is legitimate because $x, y \in l^2$ implies that also $\alpha x \in l^2$ and $x+y \in l^2$. In fact, the first assertion is obvious, and the second follows from the inequality

$$|x_n + y_n|^2 \leqq 2(|x_n|^2 + |y_n|^2).$$

We leave to the reader the verification of the fact that all the norm axioms are satisfied (see also subsection 1, § 5) so that l^2 is a normed space, and that $|\sum_{n=1}^{\infty} x_n y_n|^2 \leqq \sum_{n=1}^{\infty} |x_n|^2 \sum_{n=1}^{\infty} |y_n|^2$ for all $\langle x_n \rangle, \langle y_n \rangle \in l^2$.

The space l^2 is complete. In fact, let

$$x^{(k)} = \langle x_n^{(k)} \rangle, \qquad k = 1, 2, 3, \cdots$$

be a fundamental sequence in l^2, so that

$$|x^{(k)} - x^{(p)}|^2 = \sum_{n=1}^{\infty} |x_n^{(k)} - x_n^{(p)}|^2 < \varepsilon^2 \text{ for } k, p > N(\varepsilon). \tag{4}$$

Then, also

$$|x_n^{(k)} - x_n^{(p)}| < \varepsilon \text{ for } k, p > N(\varepsilon);$$

consequently,

$$x_n = \lim_{k \to \infty} x_n^{(k)}$$

exists. Moreover, it follows from (4) that for arbitrary M, we have

$$\sum_{n=1}^{M} |x_n^{(k)} - x_n^{(p)}|^2 < \varepsilon^2 \text{ for } k, p > N(\varepsilon). \tag{5}$$

Passing to the limit as $k \to \infty$ in (5), we obtain that

$$\sum_{n=1}^{M} |x_n - x_n^{(p)}|^2 \leqq \varepsilon^2 \text{ for } p > N(\varepsilon). \tag{6}$$

Since $N(\varepsilon)$ does not depend on M, one can pass to the limit as $M \to \infty$ in (6) and obtain that

$$\sum_{n=1}^{\infty} |x_n - x^{(p)}|^2 \leqq \varepsilon^2 \text{ for } p > N(\varepsilon).$$

This means that $\langle x_n - x_n^{(p)} \rangle \in l^2$ and hence also $\langle x_n \rangle \in l^2$, and that $\langle x_n \rangle$ is the limit of the sequence $\langle x_n^{(p)} \rangle$, $p = 1, 2, 3, \cdots$ in the space l^2.

We note further that l^2 is separable; namely, the set of all sequences $\langle x_n \rangle$ with rational x_n, in which only a finite number of terms x_n are different from zero, form a denumerable dense set in l^2.

Other examples of normed spaces will be given below in § 6.

2. Series in a normed space. A series

$$x_1+x_2+x_3+\cdots \tag{1}$$

of elements x_n in the normed space X is said to be *convergent* if its partial sums

$$s_n = x_1+x_2+\cdots+x_n,\ n = 1, 2, 3, \cdots$$

form a convergent sequence in X; the limit of this sequence is called the *sum* of the series (1). Series (1) is said to be *absolutely convergent* if the numerical series

$$|x_1|+|x_2|+|x_3|+\cdots \tag{2}$$

converges.

I. *A normed space X is complete if and only if every absolutely convergent series in X converges in X.*

Proof. Assume X is complete and that series (2) converges. Then for $m > n$, $|s_m-s_n| = |x_{n+1}+\cdots+x_m| \leqq |x_{n+1}|+\cdots+|x_m|$; consequently, in virtue of the convergence of the series (2) $\langle s_n \rangle$ is a fundamental sequence in the complete space X and hence has a limit.

Conversely, suppose that in the normed space X every absolutely convergent series converges and that $\langle x_n \rangle$ is a fundamental sequence in X. Then $\langle x_n \rangle$ contains a subsequence $\langle x_{n_k} \rangle$ such that

$$|x_{n_{k+1}}-x_{n_k}| < \frac{1}{2^{k+1}}. \tag{3}$$

It follows from (3) that the series $x_{n_1}+(x_{n_2}-x_{n_1}) + (x_{n_3}-x_{n_1})+\cdots$ converges absolutely and thus by assumption it converges to some element $x \in X$. This means that

$$x = \lim_{k\to\infty} [x_{n_1}+(x_{n_2}-x_{n_1})+\cdots+(x_{n_k}-x_{n_{k-1}})] = \lim_{k\to\infty} x_{n_k}.$$

But then from the inequality $|x-x_n| \leqq |x-x_{n_k}|+|x_{n_k}-x_n|$ and the fact that the sequence $\langle x_n \rangle$ is fundamental, we conclude that also $x = \lim_{n\to\infty} x_n$. Consequently, X is complete.

The "sufficiency" part of Proposition I can easily be carried over to locally convex spaces.

Let $\{p\}$ be a sufficient family of seminorms in the locally convex space X. A series $x_1+x_2+x_3+\cdots$ in X is said to be *absolutely convergent* if the series $p(x_1)+p(x_2)+p(x_3)+\cdots$ converges for every $p \in \{p\}$. *In a sequentially complete space X every absolutely convergent series converges* (see subsection 13, § 3).

In fact, it is easy to see that the partial sums $s_n = x_1+\cdots+x_n$ of an absolutely convergent series form a fundamental sequence in X.

3. Quotient spaces of a Banach space. Suppose X is a normed space and that $\mathfrak{M}$ is a closed subspace in X; we consider the quotient space $X/\mathfrak{M}$, i.e. the space whose elements are classes ξ modulo $\mathfrak{M}$ (see subsection 4, § 1). In virtue of the fact that the subspace $\mathfrak{M}$ is closed, all these classes are closed. We define a norm in $X/\mathfrak{M}$ by means of the formula

$$|\xi| = \inf_{x \in \xi} |x|. \tag{1}$$

We shall prove that all the norm axioms (see subsection 1) are then satisfied. Clearly, axioms 1° and 3° are satisfied so that in the proof we need only consider axioms 2° and 4°.

If $\xi = 0$, then ξ coincides with $\mathfrak{M}$ and therefore it contains the zero element of the space X; consequently, it follows from (1) that $|\xi| = 0$. Conversely, suppose $|\xi| = 0$; by (1), it follows from this that the class ξ contains a sequence $\langle x_n \rangle$ for which $|x_n| \to 0$ when $n \to \infty$. But this means that 0 is a contact point of the class ξ and therefore belongs to ξ; consequently, $\xi = \mathfrak{M}$ and therefore it is the zero element in $X/\mathfrak{M}$. This shows that axiom 2° is satisfied.

We now go over to axiom 4°. Suppose $\xi, \eta \in X/\mathfrak{M}$; by definition (1) of the norm, for arbitrary $\varepsilon > 0$, there exist vectors $x \in \xi$, $y \in \eta$ such that

$$|x| < |\xi| + \varepsilon, \; |y| < |\eta| + \varepsilon;$$

consequently,

$$|x+y| \leqq |x| + |y| < |\xi| + |\eta| + 2\varepsilon. \tag{2}$$

On the other hand, $x+y \in \{\xi+\eta\}$, and therefore $|\xi+\eta| \leqq |x+y|$. Combining this inequality with (2), we obtain

$$|\xi+\eta| \leqq |\xi| + |\eta| + 2\varepsilon.$$

It follows that

$$|\xi+\eta| \leqq |\xi| + |\eta|$$

since the number $\varepsilon > 0$ is arbitrary.

The space $X/\mathfrak{M}$, normed by formula (1), will be called a *normed quotient space.*

If X is a Banach space and $\mathfrak{M}$ is a closed subspace in X then the normed quotient space $X/\mathfrak{M}$ is also a Banach space.

Proof. Let $\xi_1 + \xi_2 + \cdots$ be an absolutely convergent series, so that the series

$$|\xi_1| + |\xi_2| + \cdots \tag{3}$$

converges. From the definition (1) of the norm in $X/\mathfrak{M}$ we deduce that there exists an element $x_n \in \xi_n$ such that

$$|x_n| < |\xi_n| + \frac{1}{2^n}. \tag{4}$$

It follows from (4) and the convergence of the series (3) that the series $x_1 + x_2 + \cdots$

converges absolutely and therefore converges to some element $x \in X$, since X is complete. Let ξ be the class containing x. Then, in virtue of (1),

$$|\xi-(\xi_1+\cdots+\xi_n)| \leqq |x-(x_1+\cdots+x_n)| \to 0$$

as $n \to \infty$; consequently, the series $\xi_1+\xi_2+\cdots$ converges to the element $\xi \in X/\mathfrak{M}$ and $X/\mathfrak{M}$ is complete on the basis of I, subsection 2.

4. Bounded linear operators. An operator A from the normed space X into the normed space Y is said to be *bounded* if there exists a constant C such that

$$|Ax| \leqq C|x| \text{ for all } x \in \mathfrak{D}_A. \tag{1}$$

I. *Every linear operator, continuous at the point* $x = 0$, *is bounded*; *conversely, every bounded linear operator is continuous.*

Proof. Suppose the operator A is continuous at the point $x = 0$. This means that to the neighborhood $|y| < \varepsilon$ in Y there corresponds a neighborhood $|x| < \delta$ in X such that $|Ax| < \varepsilon$ for $|x| < \delta, x \in \mathfrak{D}_A$. If we set $z = \{\delta/2|x|\}x$ for arbitrary $x \in \mathfrak{D}_A$, $(x \neq 0)$, we have $|z| < \delta$; therefore, $|Az| < \varepsilon$, i.e. $\{\delta/2|x|\}|Ax| < \varepsilon$, $|Ax| < (2\varepsilon/\delta)|x|$ for all $x \in \mathfrak{D}_A$. And this means that A is bounded. Conversely, if A is bounded, then its continuity follows from the inequality

$$|Ax-Ax_0| = |A(x-x_0)| \leqq C|x-x_0|,$$

which is true for all x, $x_0 \in \mathfrak{D}_A$.

II. *If the space* Y *is complete, then the bounded linear operator* A *is extendible by continuity with preservation of inequality* (1) *to a bounded linear operator with* $\overline{\mathfrak{D}}_A$ *as its domain.*

Proof. Suppose $x_0 \in \overline{\mathfrak{D}}_A$, so that there exists a sequence of vectors $x_n \in \mathfrak{D}_A$ such that

$$|x_n-x_0| \to 0 \text{ as } n \to \infty.$$

Then from the inequality

$$|Ax_n-Ax_m| = |A(x_n-x_m)| \leqq C|x_n-x_m|$$

it follows that $\langle Ax_n \rangle$ is a fundamental sequence in the complete space Y and therefore it has a limit. We set

$$\tilde{A}x_0 = \lim_{n\to\infty} Ax_n; \tag{2}$$

this definition does not depend on the choice of the sequence $x_n \to x_0$. In fact, if also $x'_n \to x_0$, $x'_n \in \mathfrak{D}_A$, then $x'_n - x_n \to 0$ as $n \to \infty$; from this and (1) it follows that

$$Ax'_n - Ax_n \to 0 \text{ as } n \to \infty.$$

Thus, formula (2) yields the operator $\tilde{A}$ from X into Y, with domain $\overline{\mathfrak{D}}_A$, which is an extension of the operator A. It is easily verified (by passing to the limit) that $\tilde{A}$ is linear and that $|\tilde{A}x| \leqq C|x|$ for all $x \in \overline{\mathfrak{D}}_A$.

If, in particular, $\mathfrak{D}_A$ is dense in X then $\tilde{A}$ is a bounded operator, defined on all of X.

We shall now consider the bounded linear operators from X into Y which are defined on all of X; the set of all such operators will be denoted by $L(X, Y)$.

If $A \in L(X, Y)$, then

$$|Ax| \leqq C|x| \text{ for all } x \in X \tag{3}$$

for some C. In particular,

$$|Ax| \leqq C$$

for all vectors $x \in X$ with norm $\leqq 1$. Therefore,

$$\sup_{|x| \leqq 1} |Ax|$$

exists. This least upper bound is called the *norm* of the bounded operator A and is denoted by $|A|$. Thus, by definition,

$$|A| = \sup_{|x| \leqq 1} |Ax|.$$

It is easily seen that also

$$|A| = \sup_{x \neq 0} \frac{|Ax|}{|x|}.$$

In fact, setting $x/|x| = y$, we have $|y| = 1$, and consequently,

$$\sup_{x \neq 0} \frac{|Ax|}{|x|} = \sup_{|y|=1} |Ay| = \sup_{|y| \leqq 1} |Ay| = |A|.$$

It also follows from this that $|A|$ *is the smallest of the numbers* C *for which inequality* (3) *is satisfied.* Clearly, $|A| = 0$ only if $A = O$.

III. *If* A, $B \in L(X, Y)$, *then* $\alpha A \in L(X, Y)$, $A+B \in L(X, Y)$, *and*

$$|\alpha A| = |\alpha|\,|A|,\ |A+B| \leqq |A|+|B|.$$

The assertions follow directly from the relations

$$\frac{|\alpha Ax|}{|x|} = |\alpha| \frac{|Ax|}{|x|},\ \frac{|(A+B)x|}{|x|} \leqq \frac{|Ax|}{|x|} + \frac{|Bx|}{|x|} \leqq |A|+|B|.$$

Proposition III means that $L(X, Y)$ *is a normed space.*

IV. *If* Y *is complete, then the space* $L(X, Y)$ *is also complete.*

Proof. Suppose $\langle A_n \rangle$ is a fundamental sequence in $L(X, Y)$; for arbitrary $\varepsilon > 0$, $|A_n - A_m| < \varepsilon$ for all $n, m > N(\varepsilon)$. It follows that

$$|A_n x - A_m x| = |(A_n - A_m)x| < \varepsilon |x| \tag{4}$$

for $n, m > N(\varepsilon)$ and arbitrary $x \in X$, so that $\langle A_n x\rangle$ is a fundamental sequence in the complete space Y and therefore it has a limit. We set

$$Ax = \lim_{n\to\infty} A_n x.$$

This formula defines an operator from X into Y. It is easily verified that A is linear. We shall prove that $A \in L(X, Y)$ and that A is the limit of the sequence $\langle A_n\rangle$; this will also prove that the space $L(X, Y)$ is complete.

Passing to the limit as $n \to \infty$ in (4), we obtain

$$|Ax - A_m x| \leqq \varepsilon|x| \text{ for } m > N(\varepsilon);$$

consequently, $A - A_m \in L(X, Y)$ and

$$|A - A_m| \leqq \varepsilon \text{ for } m > N(\varepsilon). \tag{5}$$

But then $A = A_m + (A - A_m) \in L(X, Y)$; moreover, (5) means that $|A - A_m| \to 0$ as $m \to \infty$. This completes the proof of Proposition IV.

We shall now consider the case $Y = X$, i.e. the case of bounded operators in X, defined on all of X. The set of all such operators will be denoted by $\mathfrak{B}(X)$, so that $\mathfrak{B}(X) = L(X, X)$. Then, in addition to Propositions III and IV we have

V. *If $A, B \in \mathfrak{B}(X)$, then also $AB \in \mathfrak{B}(X)$ and*

$$|AB| \leqq |A|\,|B|.$$

This assertion follows directly from the inequalities

$$\frac{|ABx|}{|x|} \leqq \frac{|A||Bx|}{|x|} \leqq \frac{|A||B||x|}{|x|} = |A||B|.$$

VI (Banach [1]). *If the bounded linear operator A maps the Banach space X in a one-to-one fashion onto the Banach space Y, then the inverse operator A^{-1} is bounded on Y.*

Proof. Let K be a closed ball in X with center at zero and radius 1. We note first of all that AK cannot be nowhere dense in Y. In fact, if we assume the contrary we deduce that also each of the sets $A(nK) = nAK$ ($n = 1, 2, 3, \cdots$) is nowhere dense in Y. But since $Y = \bigcup_n nAK$, it turns out that Y is of the first category, which contradicts the completeness of Y (see III, subsection 13, § 2). Thus AK is not nowhere dense; consequently, in Y there exists a ball $\bar{B} = \overline{B(y_0, r)}$ in which $AK \cap \bar{B}$ is dense, from which it follows that $\overline{B(y_0, r)} \subset \overline{AK}$. Since $\overline{AK}$ is symmetric, we also have $\overline{B(-y_0, r)} \subset \overline{AK}$. Let $\bar{B}_r = \overline{B(0, r)}$ be a ball in Y with radius r and center 0. If $y \in \bar{B}_r$, then $y_0 + y \in \overline{B(y_0, r)} \subset \overline{AK}$ and $-y_0 + y \in \overline{B(-y_0, r)} \subset \overline{AK}$. But then also $y = \frac{1}{2}(-y_0 + y) + \frac{1}{2}(y_0 + y) \in \overline{AK}$, because $\overline{AK}$ is convex (see I, subsection I, § 3). Thus,

$$\bar{B}_r \subset \overline{AK} \tag{6}$$

and, consequently,

$$\bar{B}_{\varepsilon r} \subset \overline{A(\varepsilon K)} \text{ for every } \varepsilon > 0. \tag{7}$$

We shall now prove that

$$\bar{B}_r \subset 2AK. \tag{8}$$

Let ε_n be a sequence of positive numbers such that $\sum_{n=1}^{\infty} \varepsilon_n \leqq 1$ and let $y \in \bar{B}_r$. By (6), there exists an element $y_1 \in AK$ such that $|y-y_1| \leqq \varepsilon_1 r$. Let x_1 be an element in K for which $y_1 = Ax_1$. Applying the same reasoning to the element $y-y_1 \in \overline{B_{\varepsilon_1 r}}$ and using (7), we conclude that there exists an element $y_2 = Ax_2$, where $x_2 \in \varepsilon_1 K$, such that $|y-(y_1+y_2)| \leqq \varepsilon_2 r$. Continuing this reasoning, we obtain sequences $x_n \in \varepsilon_n K$, $y_n = Ax_n$ such that

$$x_n \in \varepsilon_{n-1} K, \text{ i.e. } |x_n| \leqq \varepsilon_{n-1}, \tag{9}$$

and

$$|y-(y_1+ \cdots +y_n)| \leqq \varepsilon_n r. \tag{10}$$

By (9) and the completeness of the space X, the series $\sum_{k=1}^{\infty} x_k$ converges; let x be its sum. Then

$$|x| \leqq |x_1|+\sum_{k=2}^{\infty} |x_k| \leqq 1+\sum_{k=1}^{\infty} \varepsilon_{k-1} \leqq 2, \text{ i.e., } x \in 2K$$

and

$$Ax = \sum_{k=1}^{\infty} Ax_k = \sum_{k=1}^{\infty} y_k,$$

so that, in virtue of (10), $Ax = y$. Thus, for every $y \in \bar{B}_r$, there exists an element $x \in 2K$ such that $Ax = y$, which proves (8). Now let y be an arbitrary non-zero element in Y. Set $y^1 = (r/|y|)y$. Then $y^1 \in \bar{B}_r$; by (8), $y^1 = Ax^1$, where $|x^1| \leqq 2$. Setting $x = (|y|/r)x^1$, we have $Ax = y$, $x = A^{-1}y$ and

$$|A^{-1}y| = |x| = \frac{|y|}{r}|x^1| \leqq \frac{2}{r}|y|,$$

which proves that the operator A^{-1} is bounded.

VII. *If a linear space X is complete with respect to the norm $|x|$ as well as with respect to the norm $|x|_1$ and if*

$$|x|_1 \geqq |x| \textit{ for all } x \in X, \tag{11}$$

then there exists a constant C such that

$$|x|_1 \leqq C|x|.$$

Proof. We denote by X, X_1 the linear space X considered as a Banach space with norm $|x|$, $|x|_1$, respectively. The operator A from X_1 into X, defined by the equality $Ax = x$, is a one-to-one linear mapping of X_1 onto X; in virtue of (11) A is bounded. Therefore, it remains to apply Proposition VI.

Two norms $|x|$ and $|x|_1$ in the linear space X are said to be *equivalent* if there exist constants C_1 and C_2 such that

$$C_1|x| \leqq |x|_1 \leqq C_2|x|.$$

Clearly, equivalent norms define the same topology in X. It is easily shown that also, conversely, two norms which define the same topology in X are equivalent. Proposition VII means that the norms $|x|$ and $|x|_1$ which satisfy the conditions of this proposition are equivalent.

5. Bounded linear functionals; conjugate space. A linear functional f in the normed space X is said to be *bounded* if there exists a constant C such that

$$|f(x)| \leqq C|x| \text{ for all } x \in \mathfrak{D}_f.$$

Clearly, a bounded linear functional is a particular case of a bounded linear operator, namely an operator from X into Y, where $Y = \mathfrak{C}$, the space of scalars (see subsection 1, § 3). Therefore, the results of subsection 4 concerning bounded operators are applicable to bounded functionals. In particular, the set $L(X, \mathfrak{C})$ of all bounded linear functionals in X, defined on all of X, is a normed linear space, where the norm $|f|$ of the bounded functional f is defined by the relations

$$|f| = \sup_{|x| \leqq 1} |f(x)| = \sup_{x \neq 0} \frac{|f(x)|}{|x|}.$$

This space is called the space *conjugate to* X and is denoted by X'.

The space conjugate to X' is called the *second conjugate space to* X and is denoted by X''. The third, fourth, $\cdots$ conjugate spaces can be defined analogously.

Since the space $\mathfrak{C}$ of all scalars is complete, then in virtue of IV, subsection 4, we have

I. *The conjugate space is complete.*

Further, applying Theorem 2, subsection 9, § 1, and its corollary to $p(x) = |x|$, we conclude that

II. *Every bounded linear functional defined on a subspace of a normed space X can be extended with preservation of norm to the entire space X.*

III. *For every vector $x \in X$ there exists a functional $f \in X'$ such that*

$$f(x) = |x|, |f| = 1.$$

Therefore,

IV. *If $f(x) = 0$ for arbitrary functional $f \in X'$, then $x = 0$.*

For fixed $x \in X$, we set

$$F_x(f) = f(x), f \in X';$$

clearly, F_x is linear functional in X'. This functional is bounded because

$$|F_x(f)| = |f(x)| \leqq |x|\,|f|, \tag{1}$$

i.e. $F_x \in X''$.

It follows from (1) that $|F_x| \leqq |x|$. On the other hand, according to III the equality sign holds in relation (1) for certain f. Consequently, $|F_x| = |x|$. Thus,

V. *The correspondence $x \to F_x$ is an isometric mapping of X into X''.*

Therefore, one can identify the vector x with the functional F_x generated by it and consider X to be a subset of X''. If, in this connection, $X = X''$, then the space X is called *reflexive*. Thus, reflexivity of the space X means that every bounded linear functional F in X' can be written in the form

$$F(f) = f(x),$$

where x is some vector in X.

6. Compact (or completely continuous) operators. A linear operator A in a Banach space X is said to be *compact* (or *completely continuous*) if it maps every bounded set into a precompact set or, equivalently, if every bounded sequence $\langle x_n \rangle$ in X possesses a subsequence $\langle x_{n_k} \rangle$ such that $\langle Ax_{n_k} \rangle$ converges.

An operator A is called *finite-dimensional* if its domain is finite-dimensional. *A bounded finite-dimensional operator A is compact.* In fact, such an operator maps every bounded set in X into a bounded subset of a finite-dimensional space; and the latter set is precompact by virtue of the Bolzano-Weierstrass theorem.

I. *Every compact operator is bounded.*

In fact, otherwise there would exist a sequence $\langle x_n \rangle$ such that

$$|x_n| = 1, \; |Ax_n| > n,$$

and the operator A would map a bounded set $\langle x_n \rangle$ into the set $\langle Ax_n \rangle$ which obviously is not precompact.

We note that the converse assertion is not true. Thus, the identity operator in an infinite-dimensional space is bounded but it is not compact because it maps the unit ball onto itself, i.e. onto a set which is not precompact (see V, subsection 1).

II *If the operator A is compact and the operator B is defined everywhere in X and bounded, then the operators AB and BA are compact.*

Proof. Suppose $\langle x_n \rangle$ is a bounded sequence. Since A is compact, $\langle x_n \rangle$ contains a subsequence $\langle x'_n \rangle$ such that $\langle Ax'_n \rangle$ converges; then in view of the boundedness of the operator B the sequence $\langle BAx'_n \rangle$ also converges. Consequently, the operator BA is compact. Further, the sequence $\langle Bx_n \rangle$ is bounded; therefore, the sequence $\langle ABx_n \rangle$ is precompact and this means that the operator AB is compact.

In a manner analogous to II, one can prove that

III. *If A_1, A_2 are compact operators then $\alpha_1 A_1 + \alpha_2 A_2$ is also a compact operator.*

IV. *If $\langle A_n \rangle$ is a sequence of compact operators such that*

$$|A-A_n| \to 0 \ as \ n \to \infty,$$

then the operator A is also compact.

Proof. Suppose $K = K_c$ is the closed ball in X defined by the inequality $|x| \leqq c$; we set $\varepsilon_n = |A-A_n|c$. Then, for $x \in K$, we have

$$|(A-A_n)x| \leqq |A-A_n|c = \varepsilon_n;$$

consequently, the precompact set A_nK forms an ε_n-net for the set AK. Since $\varepsilon_n \to 0$ as $n \to \infty$, then, on the basis of II, subsection 14, § 2, the set AK is precompact; this means that the operator A is compact.

In the following proposition we give an example of a compact operator.

V. *If*

$$\sum_{i,k=1}^{\infty} |a_{ik}|^2 < +\infty,$$

then the operator A in the space l^2, defined by $A\langle x_i \rangle = \langle y_i \rangle$, where

$$y_i = \sum_{k=1}^{\infty} a_{ik} x_k, \quad i = 1, 2, 3, \cdots, \tag{1}$$

is compact. (See Example 3, subsection 1.)

Proof. We note first of all that A is bounded. In fact, by (1),

$$|Ax|^2 = \sum_{i=1}^{\infty} |y_i|^2 = \sum_{i=1}^{\infty} \sum_{k=1}^{\infty} |a_{ik} x_k|^2 \leqq \sum_{i=1}^{\infty} \sum_{k=1}^{\infty} |a_{ik}|^2 \sum_{k=1}^{\infty} |x_k|^2 = \Big(\sum_{i,k=1}^{\infty} |a_{ik}|^2 \Big) |x|^2. \tag{2}$$

We denote by A_n the operator which maps every vector $x = \langle x_1, x_2, x_3, \cdots \rangle$ into the vector $y^n = \langle y_1, \cdots, y_n, 0, 0, \cdots \rangle$, where the numbers y_i are defined by formula (1). The operator A_n is finite-dimensional and as is clear from (2) it is bounded; consequently, the operator A_n is compact.

On the other hand,

$$|(A-A_n)x|^2 = \sum_{i=n+1}^{\infty} |y_i|^2 = \sum_{i=n+1}^{\infty} |\sum_{k=1}^{\infty} a_{ik} x_k|^2 \leqq \sum_{i=n+1}^{\infty} \sum_{k=1}^{\infty} |a_{ik}|^2 \sum_{k=1}^{\infty} |x_k|^2;$$

it follows that

$$|A-A_n| \leqq \Big(\sum_{i=n+1}^{\infty} \sum_{k=1}^{\infty} |a_{ik}|^2 \Big)^{\frac{1}{2}} \to 0$$

as $n \to \infty$, and it remains to apply Proposition IV.

7. Analytic vector-valued functions in a Banach space. The concept of an analytic vector-valued function in a locally convex space was defined above (see subsection 12, § 3); in particular, it is also by the same token defined in a normed space. Thus, a vector-valued function $x(\lambda)$ with values in the normed space X is said to be *analyt-*

ic in the region G of the complex λ-plane if

$$x'(\lambda) = \lim_{\Delta\lambda \to 0} \frac{x(\lambda+\Delta\lambda)-x(\lambda)}{\Delta\lambda}$$

exists for every point $\lambda \in G$, in the sense of the topology defined by the norm in X.

I (Cauchy's Integral Theorem). *If the vector-valued function $x(\lambda)$ is analytic in the interior of a region bounded by a rectifiable Jordan curve Γ and is continuous on the closure of this region, then*

$$\int_{\Gamma} x(\lambda)d\lambda = 0.$$

Proof. Applying the usual reasoning and using the fact that the space X is complete, it is easily shown that the integral of a continuous vector-valued function $x(\lambda)$ over the rectifiable curve Γ exists in the sense of the topology in X. We set

$$y = \int_{\Gamma} x(\lambda)d\lambda.$$

Then for arbitrary bounded linear functional f, we have

$$f(y) = \int_{\Gamma} f(x(\lambda))d\lambda = 0, \tag{1}$$

because $f(x(\lambda))$ is a numerical function, analytic in the interior of the region bounded by Γ and continuous on the closure of this region, and therefore the usual Cauchy theorem is applicable to $f(x(\lambda))$. In virtue of the fact that the functional f is arbitrary, we conclude from (1) that $y = 0$ (see IV, subsection 5).

In an analogous manner, we prove

II (Cauchy's Integral Formula). *If the vector-valued function $x(\lambda)$ in the complete normed space X is analytic in the interior of a region bounded by a rectifiable Jordan curve Γ and continuous on the closure of this region, then for arbitrary point λ in the interior of the region bounded by Γ, we have*

$$x(\lambda) = \frac{1}{2\pi i} \oint_{\Gamma} \frac{x(\xi)}{\xi-\lambda} d\xi. \tag{2}$$

Applying the usual reasoning to formula (2), we conclude:

III. *Every vector-valued function $x(\lambda)$ in the Banach space X, analytic in the interior of the disc $|\lambda-\lambda_0| < r$, has derivatives of all orders in the interior of this disc and is representable there in the form of the sum of an absolutely convergent Taylor series*:

$$x(\lambda) = x_0+(\lambda-\lambda_0)x_1+(\lambda-\lambda_0)^2x_2+ \cdots, \tag{3}$$

where

$$x_n = \frac{1}{n!}\, x^{(n)}(\lambda_0). \tag{4}$$

The radius R of absolute convergence of this series, on the one hand, equals the distance from λ_0 to the nearest point of non-analyticity of the function $x(\lambda)$, and on the other hand, on the basis of the Cauchy-Hadamard theorem,

$$R = \frac{1}{\overline{\lim\limits_{n\to\infty}} \sqrt[n]{|x_n|}}\,. \tag{5}$$

In fact, if $x(\lambda)$ is analytic in the disc $|\lambda-\lambda_0| < r$, then it is continuous on an arbitrary circle $|\lambda-\lambda_0| = r_1 < r$; therefore, the numerical function $|x(\lambda)|$ is continuous, and this means it is bounded on the circle $|\lambda-\lambda_0| = r_1 < r$. From this we conclude that for $|\lambda-\lambda_0| < r_1 < r$ the series

$$\frac{x(\xi)}{\xi-\lambda} = \frac{x(\xi)}{\xi-\lambda_0}\cdot\frac{1}{1-\dfrac{\lambda-\lambda_0}{\xi-\lambda_0}} = \sum_{n=0}^{\infty}\frac{x(\xi)}{(\xi-\lambda_0)^{n+1}}(\lambda-\lambda_0)^n$$

converges uniformly on the circle $|\xi-\lambda_0| = r_1$. Integrating this series term by term and making use of formula (2), and also of the formulas obtained from (2) by differentiation, we obtain (3)–(4), where the series in formula (3) converges absolutely for $|\lambda-\lambda_0| < r$.

Conversely, if (3)–(4) hold in the disc $|\lambda-\lambda_0| < r$, then by a direct term-by-term integration we obtain that formula (2) is valid for arbitrary Jordan curve Γ, which lies entirely in the interior of this disc. The analyticity of the function $x(\lambda)$ in the interior of the disc $|\lambda-\lambda_0| < r$ follows easily from this.

It is easily seen that Propositions I–III remain valid in an arbitrary sequentially complete locally convex space X. Formula (5) must then be replaced by the formula

$$R = \inf_p \frac{1}{\overline{\lim\limits_{n\to\infty}} \sqrt[n]{p(x_n)}}, \tag{6}$$

where the greatest lower bound is taken over all seminorms p of a sufficient family, which defines the topology in X.

§ 5. Hilbert space

1. Definition of Hilbert space. A vector space R will be called a *pre-Hilbert* space if a function of two variables x and y, usually denoted by (x, y), is defined in R, satisfying the following conditions:

1) $(x, x) \geqq 0$; $(x, x) = 0$ if and only if $x = 0$;
2) $(y, x) = \overline{(x, y)}$;
3) $(\lambda x, y) = \lambda(x, y)$;
4) $(x_1+x_2, y) = (x_1, y)+(x_2, y)$.

This function (x, y) is called the *inner product* of the elements x and y.

It follows from properties 2)–4) that

$$(x, \lambda y) = \bar{\lambda}(x, y),\ (x, y_1+y_2) = (x, y_1)+(x, y_2).$$

The inequality

$$|(x, y)|^2 \leqq (x, x)(y, y) \tag{1}$$

holds in any pre-Hilbert space; it is called the *Cauchy-Bunyakovsky inequality*.

When $y = 0$, this inequality is obvious; when $y \neq 0$, it follows directly from the inequality

$$(x, x)-\bar{\lambda}(x, y)-\lambda\overline{(x, y)}+\lambda\bar{\lambda}(y, y) = (x-\lambda y, x-\lambda y) \geqq 0, \tag{1'}$$

if one sets $\lambda = (x, y)/(y, y)$. It is also clear from this that the equality sign in (1) holds if, and only if, $x = \lambda y$, or $y = 0$.

One can introduce a norm in a pre-Hilbert space by setting

$$|x| = \sqrt{(x, x)}. \tag{2}$$

Then all the norm axioms will be satisfied, i.e.

$|x| \geqq 0$;
$|x| = 0$ if and only if $x = 0$;
$|\lambda x| = |\lambda|\,|x|$;
$|x+y| \leqq |x|+|y|$

(see subsection 1, § 4). The first three properties are obvious and the last follows from the Cauchy-Bunyakovsky inequality; namely,

$$\begin{aligned}|x+y|^2 = (x+y, x+y) = (x, x)+(x, y)+(y, x)+(y, y)\\ \leqq (x, x)+2\sqrt{(x, x)(y, y)}+(y, y) = (|x|+|y|)^2.\end{aligned}$$

Consequently, if we define the norm by means of formula (2), R is a normed space.

In certain cases one must consider a function (x, y) in the linear space R which satisfies the conditions 2)–4) but which perhaps does not satisfy condition 1). Such a function is called a *Hermitian form* in R. A Hermitian form (x, y) is said to be *positive definite* if $(x, x) \geqq 0$. Inequality (1) remains valid for an arbitrary positive definite Hermitian form (x, y) and hence $p(x) = \sqrt{(x, x)}$ is a seminorm. In fact, if $(y, y) > 0$, then we repeat the reasoning in the deduction of inequality (1). If $(y, y) = 0$, but $(x, x) > 0$, then we interchange the roles of x and y in this reasoning. Finally, if $(x, x) = (y, y) = 0$, then, setting $\lambda = (x, y)$ in the following inequality, obtained from (1'),

$$-\bar{\lambda}(x, y)-\lambda\overline{(x, y)} \geqq 0,$$

we conclude that $-2|(x, y)|^2 \geqq 0$; consequently, $(x, y) = 0$ and inequality (1), obviously, is satisfied.

Therefore, applying Proposition I, subsection 1, § 4, to R we arrive at the following result.

I. *If (x, y) is a positive definite Hermitian form in the linear space R and $\mathfrak{M}$ is the subspace of all vectors $x \in R$ for which $(x, x) = 0$, then the formula*

$$(\xi, \eta) = (x, y), \text{ where } x \in \xi,\ y \in \eta,$$

defines an inner product in $R/\mathfrak{M}$.

In actuality we proved that the inequality

$$|b|^2 \leqq ac \tag{2'}$$

is a necessary condition that the form

$$F(\xi, \eta) = a|\xi|^2 + b\xi\bar{\eta} + \bar{b}\bar{\xi}\eta + c|\eta|^2$$

be nonnegative for arbitrary complex ξ and η. This condition together with the inequalities $a \geqq 0$, $c \geqq 0$ are also sufficient that the form $F(\xi, \eta)$ be nonnegative. In fact, if e.g., $a = 0$, then by (2′) also $b = 0$ and $F(\xi, \eta) = c|\eta|^2 \geqq 0$. And if $a > 0$, then by (2′),

$$F(\xi, \eta) = \left| \sqrt{a}\xi + \frac{\bar{b}}{\sqrt{a}}\eta \right|^2 + \frac{1}{a}(ac - |b|^2)|\eta|^2 \geqq 0.$$

Again, let (x, y) be an inner product.

Clearly, inequality (1) can be rewritten in the form

$$|(x, y)| \leqq |x|\,|y|.$$

From this it follows that *the inner product (x, y) is a continuous function with respect to both variables* x, y (in the topology defined by the norm in R). In fact,

$$\begin{aligned} |(x+\Delta x, y+\Delta y) - (x, y)| &= |(x, \Delta y) + (\Delta x, y) + (\Delta x, \Delta y)| \\ &\leqq |x|\,|\Delta y| + |y|\,|\Delta x| + |\Delta x|\,|\Delta y|. \end{aligned} \tag{3}$$

A *Hilbert space* is a pre-Hilbert space which is complete relative to the norm $|x| = \sqrt{(x, x)}$.

We shall usually denote a Hilbert space by the letter $\mathfrak{H}$.

II. *Every pre-Hilbert space R can be completed to a Hilbert space.*

In fact, R is a normed space with norm $|x| = \sqrt{(x, x)}$ and therefore it can be completed to a complete normed space $\tilde{R}$ (see II, subsection 1, § 4). We shall prove that the inner product (x, y) can be defined in $\tilde{R}$ in such a way that the formula $|x| = \sqrt{(x, x)}$ is preserved; this will also prove our proposition.

Suppose $\langle x_n \rangle$, $\langle y_n \rangle$ are fundamental sequences in R which converge to $x_0, y_0 \in \tilde{R}$ respectively. If we set $x = x_n$, $y = y_n$, and $x + \Delta x = x_m$, $y + \Delta y = y_m$ in (3) we conclude that $\langle(x_n, y_n)\rangle$ is a fundamental numerical sequence and therefore it has a limit. We set

$$(x_0, y_0) = \lim_{n\to\infty}(x_n, y_n);$$

it follows easily from (3) that (x_0, y_0) does not depend on the choice of the fundamental sequences $\langle x_n\rangle$, $\langle y_n\rangle$ which converge to x_0, y_0. Passing to the limit in the relations

$$(y_n, x_n) = \overline{(x_n, y_n)}, \; (\lambda x_n, y_n) = \lambda(x_n, y_n),$$
$$(x_n+y_n, z_n) = (x_n, z_n)+(y_n, z_n), \text{ where } z_n \in R \text{ and } z_n \to z_0 \in \tilde{R},$$

and

$$|x_n| = \sqrt{(x_n, x_n)},$$

we verify that (x_0, y_0) satisfies conditions 2)–4) and that $|x_0| = \sqrt{(x_0, x_0)}$ for all $x_0, y_0 \in \tilde{R}$. Since $(x_0, x_0) = |x_0|^2$, condition 1) is also satisfied.

If $\mathfrak{M}$ is a closed subspace in the Hilbert space $\mathfrak{H}$, then the inner product, defined on all of $\mathfrak{H}$, is defined in particular in $\mathfrak{M}$ and $\mathfrak{M}$ is complete with respect to the norm defined by this inner product. Consequently,

III. *Every closed subspace of a Hilbert space is also a Hilbert space.*

EXAMPLE. The space l^2 (see Example 3, subsection 1, § 4) is a Hilbert space if the inner product of any two of its elements $x = \langle x_n\rangle$, $y = \langle y_n\rangle$ is defined by means of the formula

$$(x, y) = \sum_{n=1}^{\infty} x_n \bar{y}_n; \tag{4}$$

the convergence of the series in the right member of formula (4) follows from the inequality

$$|x_n\bar{y}_n| \leqq \tfrac{1}{2}(|x_n|^2 + |y_n|^2).$$

Other examples of Hilbert spaces will be given later on in subsection 4 and in § 6.

2. Projection of a vector on a subspace. Two vectors in the pre-Hilbert space R are said to be *orthogonal* if their inner product vanishes.

The fact that the vectors x and y are orthogonal is denoted by $x \perp y$. Two sets in the pre-Hilbert space R are said to be *mutually orthogonal* if every vector of one set is orthogonal to each of the vectors in the other set. The orthogonality of two sets $S_1, S_2 \subset R$ is denoted by $S_1 \perp S_2$. It is easily verified that *the set of all vectors which are orthogonal to some set $S \subset R$ is a closed subspace in R.* This subspace is called the *orthogonal complement* of S in R and is denoted by $R \ominus S$.

I (THE BEPPO LEVI INEQUALITY). *Suppose $\mathfrak{M}$ is a subspace of the pre-Hilbert space R and that x is a vector in R which is at a distance d from $\mathfrak{M}$; then for any two vectors $y_1, y_2 \in \mathfrak{M}$ we have*

$$|y_1-y_2| \leqq \sqrt{|x-y_1|^2-d^2}+\sqrt{|x-y_2|^2-d^2}. \tag{1}$$

Proof. We set $z_1 = x-y_1$, $z_2 = x-y_2$; for arbitrary complex $\lambda \neq 1$, the vector

$$\frac{y_1-\lambda y_2}{1-\lambda}$$

belongs to $\mathfrak{M}$ and therefore

$$\left| x-\frac{y_1-\lambda y_2}{1-\lambda} \right|^2 \geqq d^2,$$

i.e.

$$|z_1-\lambda z_2|^2 \geqq d^2|1-\lambda|^2, \tag{2}$$

$$[(z_1, z_1)-d^2]-\bar{\lambda}[(z_1, z_2)-d^2]-\lambda[(z_2, z_1)-d^2]+\lambda\bar{\lambda}[(z_2, z_2)-d^2] \geqq 0. \tag{3}$$

Clearly, (2) and hence (3) also hold for $\lambda = 1$. Hence, repeating the line of reasoning followed in subsection 1, above, we obtain

$$|(z_1, z_2)-d^2|^2 \leqq [(z_1, z_1)-d^2][(z_2, z_2)-d^2]. \tag{4}$$

But

$$\begin{aligned} |y_1-y_2|^2 &= |z_1-z_2|^2 = (z_1-z_2, z_1-z_2) = |z_1|^2+|z_2|^2-(z_1, z_2)-(z_2, z_1) \\ &= [|z_1|^2-d^2]+[|z_2|^2-d^2]-[(z_1, z_2)-d^2]-[(z_2, z_1)-d^2] \\ &\leqq [|z_1|^2-d^2]+[|z_2|^2-d^2]+2|(z_1, z_2)-d^2|; \end{aligned}$$

consequently, by (4),

$$\begin{aligned} |y_1-y_2|^2 \leqq [|z_1|^2-d^2]+[|z_2|^2-d^2] \\ +2\sqrt{[|z_1|^2-d^2][|z_2|^2-d^2]} = (\sqrt{|z_1|^2-d^2}+\sqrt{|z_2|^2-d^2})^2, \end{aligned}$$

and inequality (1) follows from this.

II. *If $\mathfrak{M}$ is a closed subspace in the Hilbert space $\mathfrak{H}$ and x is an arbitrary vector in $\mathfrak{H}$, then $\mathfrak{M}$ contains a unique vector x' such that $x-x' \perp \mathfrak{M}$.*

Proof. We set

$$d = \inf_{y \in \mathfrak{M}} |x-y|;$$

there exists a sequence $y_n \in \mathfrak{M}$ such that

$$d = \lim_{n\to\infty} |x-y_n|. \tag{5}$$

The Beppo Levi inequality

$$|y_n-y_m| \leqq \sqrt{|x-y_n|^2-d^2}+\sqrt{|x-y_m|^2-d^2}$$

shows that $\langle y_n \rangle$ is a fundamental sequence and therefore it has a limit; we denote it by x'. Clearly, $x' \in \mathfrak{M}$ because $\mathfrak{M}$ is closed. Passing to the limit in (5), we obtain

$$d = |x-x'|.$$

We shall prove that $x-x' \perp \mathfrak{M}$, i.e. that

$$(x-x', y) = 0 \text{ for all } y \in \mathfrak{M}.$$

We set $x-x' = x''$ and

$$t = -\frac{(x'', y)}{(y, y)}. \tag{6}$$

Since $x'-ty \in \mathfrak{M}$, we have

$$\begin{aligned} d^2 \leqq |x-(x'-ty)|^2 &= |x''+ty|^2 = (x''+ty, x''+ty) \\ &= (x'', x'')+\bar{t}(x'', y)+t(y, x'')+t\bar{t}(y, y). \end{aligned}$$

If we now substitute the value of t from (6) and recall that $(x'', x'') = d^2$, we obtain that

$$-\frac{|(x'', y)|^2}{(y, y)} \geqq 0;$$

consequently, $(x'', y) = 0$.

It remains to prove the uniqueness of the vector x'.

We assume that also $x = y'+y''$, where $y' \in \mathfrak{M}$, $y'' \perp \mathfrak{M}$. Then $x'+x'' = y'+y''$ and consequently

$$x'-y' = y''-x'',$$

where $x'-y' \in \mathfrak{M}$, $y''-x'' \perp \mathfrak{M}$. But in this case $x'-y'$ is orthogonal to itself, i.e. $(x'-y', x'-y') = 0$, and therefore $x'-y' = 0$, $x' = y'$.

The vector $x' \in \mathfrak{M}$, for which $x-x' \perp \mathfrak{M}$ is called the *projection of the vector* x *on the subspace* $\mathfrak{M}$.

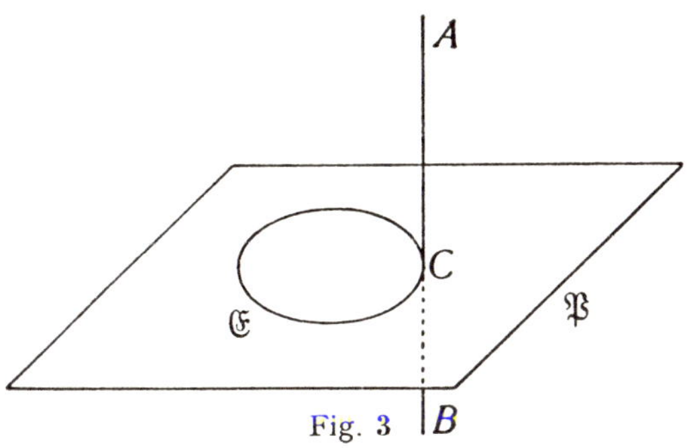

Fig. 3

The concept of projection admits a particularly simple interpretation in three-dimensional space.

Suppose, for instance, that $\mathfrak{M}$ is the set of all vectors emanating from the origin of coordinates and lying in the same plane π (Fig. 3). Then the formula $x = x'+x''$ represents a decomposition of an arbitrary vector x into two components x' and x'' of which the first lies in the plane π and the second is $\perp \pi$.

Consequently, in this case the definition we gave of a projection coincides with the usual definition of the projection of a vector on a plane.

Clearly, Proposition II can be formulated in the following way.

II′. *If* $\mathfrak{M}$ *is a closed subspace in the Hilbert space* $\mathfrak{H}$, *then every vector* $x \in \mathfrak{H}$ *can be uniquely represented in the form* $x = x'+x''$, *where* $x' \in \mathfrak{M}$, $x'' \in \mathfrak{H} \ominus \mathfrak{M}$.

If, in particular, $x \perp \mathfrak{H} \ominus \mathfrak{M}$, then also $x'' = x-x' \perp \mathfrak{H} \ominus \mathfrak{M}$; since, on the other hand, $x'' \in \mathfrak{H} \ominus \mathfrak{M}$, we have $x'' = 0$ and $x = x' \in \mathfrak{M}$. In other words,

III. *A closed subspace* $\mathfrak{M}$ *is the orthogonal complement of its orthogonal complement* $\mathfrak{H} \ominus \mathfrak{M}$:

$$\mathfrak{H} \ominus (\mathfrak{H} \ominus \mathfrak{M}) = \mathfrak{M}. \tag{7}$$

We note further the following important corollary to Proposition II.

IV. *A subspace* $\mathfrak{M}$ *is dense in* $\mathfrak{H}$ *if and only if* $\mathfrak{H}$ *does not contain a nonzero vector which is orthogonal to* $\mathfrak{M}$.

Proof. The closure of $\mathfrak{M}$ is a closed subspace in $\mathfrak{H}$ which coincides with $\mathfrak{H}$ if and only if $\mathfrak{M}$ is dense in $\mathfrak{H}$. Consequently, if $\mathfrak{M}$ is dense in $\mathfrak{H}$ then every vector x which is orthogonal to $\mathfrak{M}$ is also orthogonal to $\overline{\mathfrak{M}} = \mathfrak{H}$ and therefore it equals zero. Conversely, if every vector x which is orthogonal to $\mathfrak{M}$ (and hence also to $\overline{\mathfrak{M}}$) equals zero, then $\mathfrak{H} \ominus \overline{\mathfrak{M}} = \{0\}$ and formula (7) yields $\overline{\mathfrak{M}} = \mathfrak{H} \ominus (\mathfrak{H} \ominus \overline{\mathfrak{M}}) = \mathfrak{H} \ominus \{0\} = \mathfrak{H}$.

3. Bounded linear functionals in Hilbert space.

F. Riesz's Theorem. *Every bounded linear functional* f *in the Hilbert space* $\mathfrak{H}$ *can be uniquely represented in the form*

$$f(x) = (x, y), \tag{1}$$

where y *is a fixed vector in* $\mathfrak{H}$ *and*

$$|f| = |y|. \tag{2}$$

Conversely, (1) *defines a bounded linear functional in* $\mathfrak{H}$ *for arbitrary* $y \in \mathfrak{H}$.

Proof. We denote by $\mathfrak{M}$ the set of all vectors x for which $f(x) = 0$; clearly, $\mathfrak{M}$ is a closed subspace in $\mathfrak{H}$. If $\mathfrak{M} = \mathfrak{H}$, then $f = 0$ and (1) is satisfied for $y = 0$. We therefore consider the case when $\mathfrak{M} \neq \mathfrak{H}$. By III, subsection 2, in this case we have $\mathfrak{H} \ominus \mathfrak{M} \neq \{0\}$ and therefore there exists a vector $z \neq 0$ in $\mathfrak{H} \ominus \mathfrak{M}$. Clearly, $f(z) \neq 0$ because otherwise the vector z would belong to $\mathfrak{M}$. The vector $u = x - \{f(x)/f(z)\}z$ belongs to $\mathfrak{M}$ because

$$f(u) = f(x) - \frac{f(x)}{f(z)} f(z) = 0;$$

consequently, $u \perp z$. This means that $(u, z) = 0$, i.e.

$$(x, z) - \frac{f(x)}{f(z)} (z, z) = 0.$$

It follows that

$$f(x) = \left(x, \frac{\overline{f(z)}z}{(z, z)}\right),$$

so that the vector $y = \{\overline{f(z)}/(z, z)\}z$ satisfies relation (1).

The uniqueness of the vector y is obvious, because if we also have $f(x) = (x, y')$, then $(x, y-y') = 0$ for all $x \in \mathfrak{H}$; if we take $x = y-y'$, we obtain that $y-y' = 0$, i.e. $y' = y$.

It remains to prove formula (2). But it follows from (1) that

$$|f(x)| = |(x, y)| \leqq |x|\,|y|,$$

where the equality sign is attained for $x = y$, and consequently $|f| = |y|$. The last assertion of the theorem is obvious (see the preceding inequality).

The theorem just proved means that there exists a correspondence $f \leftrightarrow y$ between the bounded linear functionals f in $\mathfrak{H}$ and the vectors $y \in \mathfrak{H}$ which preserves norms. It is not difficult to verify that

$$\text{if } f_1 \leftrightarrow y_1 \text{ and } f_2 \leftrightarrow y_2, \text{ then } \alpha_1 f_1 + \alpha_2 f_2 \leftrightarrow \bar{\alpha}_1 y_1 + \bar{\alpha}_2 y_2;$$

such a correspondence is called *anti-linear.*

Let $\mathfrak{H}_1$ be a subspace in $\mathfrak{H}$. A function $(x, y)_1$, defined for all $x, y \in \mathfrak{H}_1$, is called a *bilinear form* if

1) $(x_1+x_2, y)_1 = (x_1, y)_1+(x_2, y)_1$,
2) $(x, y_1+y_2)_1 = (x, y_1)_1+(x, y_2)_1$,
3) $(\alpha x, y)_1 = \alpha(x, y)_1$,
4) $(x, \alpha y)_1 = \bar{\alpha}(x, y)_1$

for all $x, y, x_1, y_1, x_2, y_2 \in \mathfrak{H}_1$ and all numbers α.

The expression "bilinear form" is not exactly appropriate because of condition 4); nevertheless we shall use it inasmuch as it is in general use. In the book by N. Bourbaki [4] the expression "sesquilinear form" is used in place of it.

Clearly, a bilinear form is Hermitian if and only if $(y, x)_1 = \overline{(x, y)_1}$.

A *bilinear* form $(x, y)_1$, defined on a subspace $\mathfrak{H}_1 \subset \mathfrak{H}$, is said to be *bounded* if

$$|(x, y)_1| \leqq C|x|\,|y|, \tag{3}$$

where C is some constant. It follows from F. Riesz's theorem that

Corollary. *Every bounded bilinear form $(x, y)_1$, defined on a subspace $\mathfrak{H}_1$, which is dense in $\mathfrak{H}$, is uniquely representable in the form*

$$(x, y)_1 = (x, Ay),\ x, y \in \mathfrak{H}_1, \tag{4}$$

where A is a bounded operator in $\mathfrak{H}$ with domain $\mathfrak{D}_A = \mathfrak{H}$. Consequently, $(x, y)_1$ is *extendible with preservation of inequality* (3) *to a bounded bilinear form in the entire space* $\mathfrak{H}$.

Proof. It follows from inequality (3) that, for fixed y, the form $(x, y)_1$ is a bounded linear functional in $\mathfrak{H}_1$ and therefore it is uniquely extendible to a bounded linear functional in $\mathfrak{H}$ (see II, subsection 4, § 4). On the basis of F. Riesz's theorem, $(x, y)_1 = (x, y')$, where $y' \in \mathfrak{H}$ and

$$|y'| \leqq C|y|. \tag{5}$$

We set $y' = Ay$; clearly, the operator A is linear and by (5) it is bounded, with $\mathfrak{D}_A = \mathfrak{H}_1$. Consequently, it is extendible with preservation of inequality (5) to a bounded linear operator with domain $\overline{\mathfrak{D}}_A = \mathfrak{H}$. Moreover,

$$(x, Ay) = (x, y') = (x, y)_1 \text{ for all } x, y \in \mathfrak{H}_1.$$

The last assertion of the corollary follows from the fact that (x, Ay) is defined for all $x, y \in \mathfrak{H}$ and $|A| \leqq C$.

4. Orthogonal systems of vectors in Hilbert space. A set of vectors in a Hilbert space is called an *orthogonal system* if any two distinct vectors of this set are orthogonal.

I. *If* $x_1, x_2, \cdots, x_n$ *is an orthogonal system, then*

$$|x_1+x_2+\cdots+x_n|^2 = |x_1|^2+|x_2|^2+\cdots+|x_n|^2.$$

In fact, $(x_j, x_k) = 0$ for $j \neq k$, and therefore

$$|x_1+x_2+\cdots+x_n|^2 = (x_1+x_2+\cdots+x_n, x_1+x_2+\cdots+x_n)$$
$$= (x_1, x_1)+(x_2, x_2)+\cdots+(x_n, x_n) = |x_1|^2+|x_2|^2+\cdots+|x_n|^2.$$

II. *If* $\{x_n\}$ *is a denumerable orthogonal system, then the series*

$$x_1+x_2+x_3+\cdots \tag{1}$$

converges if and only if the series

$$|x_1|^2+|x_2|^2+|x_3|^2+\cdots \tag{2}$$

converges.

Proof. Suppose s_n is the sum of the first n terms of series (1), and that σ_n is the sum of the first n terms of series (2). Then in virtue of I we have

$$|s_{n+p}-s_n|^2 = |x_{n+1}+\cdots+x_{n+p}|^2 = |x_{n+1}|^2+\cdots+|x_{n+p}|^2 = \sigma_{n+p}-\sigma_n,$$

so that $\langle s_n \rangle$ is a fundamental sequence in $\mathfrak{H}$ if and only if $\langle \sigma_n \rangle$ is a fundamental numerical sequence.

The vector x is said to be *normalized* if $|x| = 1$. If $x \neq 0$, then the vector $y = (1/|x|)x$ is normalized. The passage from x to y is called *normalizing the vector x*. An orthogonal system, all vectors of which are normalized, is called an *orthonormal system.*

If e is an element of an orthonormal system, then the inner product

$$\alpha = (x, e)$$

is called *the Fourier coefficient of the vector x relative to the element e.*

III. *If $e_1, e_2, \cdots, e_n$ are vectors of an orthonormal system and*

$$\alpha_k = (x, e_k), \qquad k = 1, 2, \cdots, n \tag{3}$$

are the corresponding Fourier coefficients, then

$$\sum_{k=1}^{n} |\alpha_k|^2 \leqq |x|^2. \tag{4}$$

Proof. We set

$$y = x - \sum_{k=1}^{n} \alpha_k e_k .$$

It easily follows from (3) that $y \perp e_k$, $k = 1, 2, \cdots, n$, and therefore the vectors $\alpha_1 e_1, \cdots, \alpha_k e_k, y$ form an orthogonal system; consequently, in virtue of I, we have

$$|x|^2 = \sum_{k=1}^{n} |\alpha_k e_k|^2 + |y|^2 = \sum_{k=1}^{n} |\alpha_k|^2 + |y|^2,$$

from which it follows that

$$\sum_{k=1}^{n} |\alpha_k|^2 \leqq |x|^2.$$

Inequality (4) is called *Bessel's inequality*. It follows directly from this inequality that

IV. *Every vector x has at most a denumerable number of nonzero Fourier coefficients with respect to a fixed orthonormal system.*

Further,

V. *If $e_1, e_2, e_3, \cdots$ is a denumerable orthonormal system in $\mathfrak{H}$ and $\alpha_k = (x, e_k)$, $k = 1, 2, 3, \cdots$ are the corresponding Fourier coefficients of the vector $x \in \mathfrak{H}$, then the series*

$$\sum_{k=1}^{\infty} \alpha_k e_k \tag{5}$$

converges in $\mathfrak{H}$ and the difference $x - \sum_{k=1}^{\infty} \alpha_k e_k$ is a vector which is orthogonal to all the vectors $e_1, e_2, \cdots$.

Proof. It follows from II and (4) that series (5) converges. Further,

$$(x - \sum_{k=1}^{n} \alpha_k e_k , e_p) = 0 \text{ for } n > p;$$

passing to the limit in this equality and making use of the continuity of the inner product (see subsection 1), we conclude that

$$(x - \sum_{k=1}^{\infty} \alpha_k e_k, e_p) = 0 \text{ for } p = 1, 2, 3, \cdots.$$

Now suppose $\{e_\nu\}$ is an arbitrary orthonormal system in $\mathfrak{H}$ (ν is a subscript which runs through an arbitrary set). By IV, for every $x \in \mathfrak{H}$, $\{e_\nu\}$ contains at most a denumerable number of elements such that $(x, e_\nu) \neq 0$; suppose these are the elements e_{ν_k}, $k = 1, 2, 3, \cdots$. We shall agree to write $\sum_\nu \alpha_\nu e_\nu$ in place of $\sum_k \alpha_{\nu_k} e_{\nu_k}$; by V, this series converges and the vector $x - \sum_\nu \alpha_\nu e_\nu$ is orthogonal to all the e_ν.

An orthogonal system is said to be *complete* if there does not exist a nonzero vector which is orthogonal to all the vectors of this system. In other words, completeness of the system means that this system cannot be augmented to a more extensive orthonormal system by adjoining new elements.

By IV, subsection 2, the following proposition is true.

VI. *An orthonormal system is complete if and only if its enveloping subspace is dense in* $\mathfrak{H}$.

VII. *Every nonzero Hilbert space contains a complete orthonormal system.*

Proof. We consider all possible orthonormal systems in $\mathfrak{H}$; such systems exist for $\mathfrak{H} \neq \{0\}$, for example the systems consisting of one normalized element. For two such systems $\{e_\nu\}$, $\{e'_\mu\}$ we agree to write $\{e_\nu\} \prec \{e'_\mu\}$ if every e_ν is some e'_μ. Then the set of all orthonormal systems in $\mathfrak{H}$ becomes a partially ordered set which satisfies the condition of the Zorn lemma (see Appendix I); namely, an upper bound of a linearly ordered set of orthonormal systems is the orthonormal system obtained by forming the union of all the orthonormal systems of this set. On the basis of the Zorn lemma, $\mathfrak{H}$ contains a maximal orthonormal system. It is a complete orthonormal system in $\mathfrak{H}$.

We note that in a separable Hilbert space $\mathfrak{H}$ one can do without the application of Zorn's lemma. In fact, suppose $\langle x_n \rangle$ is a sequence which is dense in $\mathfrak{H}$. We discard from this sequence every vector x_n, which is a linear combination of the preceding vectors $x_1, \cdots, x_{n-1}$; we obtain a new finite or infinite sequence, which we denote by $\langle y_n \rangle$, in which no vector y_n is a linear combination of the preceding, and whose linear envelope contains $\langle x_n \rangle$ and which is therefore dense in $\mathfrak{H}$.

Suppose $\mathfrak{M}_n$ is the subspace enveloping $y_1, y_2, \cdots, y_n$; we denote by y'_{n+1} the projection of the vector y_{n+1} on $\mathfrak{M}_n$ and set

$$e_1 = \frac{1}{|y_1|} y_1, \quad e_{n+1} = \frac{1}{|y_{n+1} - y'_{n+1}|}(y_{n+1} - y'_{n+1}), \; n = 1, 2, \cdots. \tag{6}$$

The construction of the vectors e_n by formulas (6) is called the *orthogonalization process*. We leave it to the reader to prove the following formulas:

$$e_n = \frac{1}{\sqrt{D_{n-1}D_n}} \begin{vmatrix} y_1 & y_2 & \cdots & y_n \\ (y_1, y_1) & (y_2, y_1) & \cdots & (y_n, y_1) \\ \cdot & \cdot & \cdot & \cdot \\ (y_1, y_{n-1}) & (y_2, y_{n-1}) & \cdots & (y_n, y_{n-1}) \end{vmatrix},$$

where

$$D_n = \begin{vmatrix} (y_1, y_1) & (y_2, y_1) & \cdots & (y_n, y_1) \\ (y_1, y_2) & (y_2, y_2) & \cdots & (y_n, y_2) \\ \cdot & \cdot & \cdot & \cdot \\ (y_1, y_n) & (y_2, y_n) & \cdots & (y_n, y_n) \end{vmatrix}.$$

Then the vectors $\{e_n\}$ form a finite or infinite orthonormal system whose linear envelope coincides with the linear envelope of the vectors y_n and therefore is dense in $\mathfrak{H}$; consequently, by VI, the system $\{e_n\}$ is complete in $\mathfrak{H}$. We have at the same time proved that

VIII. *In a separable Hilbert space there exists a finite or denumerable complete orthonormal system.*

Clearly, the converse assertion is also true. Namely, *if there exists a finite or denumerable complete orthonormal system $\{e_n\}$ in the Hilbert space $\mathfrak{H}$, then $\mathfrak{H}$ is separable.* In fact, the finite linear combinations of vectors e_n with rational coefficients form a denumerable set which is dense in $\mathfrak{H}$.

IX. *The following assertions are equivalent for an orthonormal system $\{e_\nu\}$ in $\mathfrak{H}$:*

1) *$\{e_\nu\}$ is dense in $\mathfrak{H}$;*

2) *every vector $x \in \mathfrak{H}$ is representable in the form*

$$x = \sum_\nu \alpha_\nu e_\nu, \textit{ where } \alpha_\nu = (x, e_\nu); \tag{7}$$

3) *for every vector $x \in \mathfrak{H}$, we have*

$$|x|^2 = \sum_\nu |\alpha_\nu|^2; \tag{8}$$

4) *for every pair of vectors $x, y \in \mathfrak{H}$,*

$$(x, y) = \sum_\nu \alpha_\nu \bar{\beta}_\nu, \textit{ where } \alpha_\nu = (x, e_\nu), \beta_\nu = (y, e_\nu). \tag{9}$$

Proof. Suppose $\{e_\nu\}$ is complete; by what was proved in V, the vector $x - \sum_\nu \alpha_\nu e_\nu$ is orthogonal to all the e_ν and therefore it is equal to zero; consequently, 1) implies 2). Now suppose 2) holds; forming the inner product of both members of equation (7) with y and making use of the continuity of the inner product, we obtain (9); hence, 2) implies 4).

Suppose 4) holds; setting $y = x$ in (9), we obtain (8); consequently, 4) implies 3). Suppose 3) holds and let the vector x be orthogonal to all the e_ν; then all the $\alpha_\nu = 0$ so that $|x|^2 = 0$, and therefore $x = 0$. Thus, every vector x which is orthogonal to all the e_ν is equal to zero, i.e. $\{e_\nu\}$ is complete; this means that 3) implies 1) and this completes the proof of Proposition IX.

X. *All complete orthonormal systems in a given Hilbert space have the same power.*

Proof. In finite-dimensional space, a complete orthonormal system is a basis for this space and, consequently, the number of its elements equals the dimension of the space. Therefore, we can assume the given Hilbert space $\mathfrak{H}$ to be infinite-dimensional.

Let $\{e_\nu\}$ and $\{e'_\mu\}$ be two complete orthonormal systems in $\mathfrak{H}$ and suppose $\mathfrak{a}$ and $\mathfrak{b}$ are their powers. For fixed ν, the inner product (e_ν, e'_μ) is different from zero for at least one μ (because $\{e'_\mu\}$ is complete) and, in virtue of IV, it is different from zero for at most a denumerable set of values μ. To each e_ν we assign an e'_μ such that $(e_\nu, e'_\mu) \neq 0$; this correspondence is an at least single-valued and at most denumerable-valued mapping of $\{e_\nu\}$ onto $\{e'_\mu\}$. Consequently, $\mathfrak{a} \geqq \mathfrak{b}$. Interchanging the roles of $\mathfrak{a}$ and $\mathfrak{b}$, we obtain that $\mathfrak{b} \geqq \mathfrak{a}$, and hence $\mathfrak{a} = \mathfrak{b}$.

The power of a complete orthonormal system in $\mathfrak{H}$ is called the *dimension* of $\mathfrak{H}$ and is denoted by *dim* $\mathfrak{H}$.

A linear operator A from the Hilbert space $\mathfrak{H}_1$ into the Hilbert space $\mathfrak{H}_2$ is said to be *isometric* if it leaves the inner product invariant, i.e. if $(Ax, Ay) = (x, y)$ for all $x, y \in \mathfrak{D}_A$. An isometric operator A in $\mathfrak{H}$ is said to be *unitary* if $\mathfrak{D}_A = \mathfrak{R}_A = \mathfrak{H}$. Two Hilbert spaces $\mathfrak{H}_1$, $\mathfrak{H}_2$ are said to be *isometric* if there exists an isometric operator A with $\mathfrak{D}_A = \mathfrak{H}_1$ and $\mathfrak{R}_A = \mathfrak{H}_2$. In this case we say that *A realizes an isometric mapping of $\mathfrak{H}_1$ onto $\mathfrak{H}_2$*. [The concept of isometry introduced above for normed spaces (see subsection 1, § 4) is equivalent to the concept of isometry defined here for the case of Hilbert spaces because the inner product is expressed in terms of the norm by the formula

$$(x, y) = \tfrac{1}{4}\{|x+y|^2 - |x-y|^2 + i|x+iy|^2 - i|x-iy|^2\}.]$$

XI. *Two Hilbert spaces are isometric if and only if they have the same dimension.*

Proof. The necessity of the condition is obvious. Conversely, suppose $\mathfrak{H}'$, $\mathfrak{H}''$ are two Hilbert spaces of the same dimension and that $\{e'_\nu\}$, $\{e''_\nu\}$ are complete orthonormal systems in $\mathfrak{H}'$ and $\mathfrak{H}''$ respectively; by assumption, their powers are equal, so that the set of indices may be considered to be the same for both systems. To each vector $f' = \sum_\nu \alpha_\nu e'_\nu$ in $\mathfrak{H}'$ we assign the vector $f'' = \sum_\nu \alpha_\nu e''_\nu$ with the same Fourier coefficients. By Proposition II and formulas (8)–(9), this correspondence is an isometric mapping of the space $\mathfrak{H}'$ onto the entire space $\mathfrak{H}''$, so that $\mathfrak{H}'$ and $\mathfrak{H}''$ are isometric.

EXAMPLES. Suppose $\{\nu\}$ is an arbitrary set of power $\mathfrak{a}$. We consider all possible numerical functions $x = \langle x_\nu \rangle$, $\nu \in \{\nu\}$, which possess the following properties:

1) $\langle x_\nu \rangle$ is different from zero only for a denumerable number of elements ν;

2) the series $\Sigma_\nu |x_\nu|^2$ converges.

We denote the set of all such functions $x = \langle x_\nu \rangle$ by $l^2_{\mathfrak{a}}$ and define the operations of scalar multiplication, addition and formation of the inner product in $l^2_{\mathfrak{a}}$ by means of the formulas

$$\alpha x = \langle \alpha x_\nu \rangle, \quad x+y = \langle x_\nu + y_\nu \rangle, \quad (x, y) = \sum_\nu x_\nu \bar{y}_\nu,$$

where $x = \langle x_\nu \rangle \in l^2_{\mathfrak{a}}$, $y = \langle y_\nu \rangle \in l^2_{\mathfrak{a}}$. Repeating the usual reasoning (see the example in subsection 1), one can show that $l^2_{\mathfrak{a}}$ is a Hilbert space. The simplest complete orthonormal system in $l^2_{\mathfrak{a}}$ is obtained by setting $e_{\nu_0} = \{\delta_{\nu\nu_0}\}$, where $\delta_{\nu\nu_0} = 0$ for $\nu \neq \nu_0$ and $\delta_{\nu_0\nu_0} = 1$. Its power equals $\mathfrak{a}$ and, consequently, $l^2_{\mathfrak{a}}$ is a Hilbert space of dimension $\mathfrak{a}$.

This example shows that *there exist Hilbert spaces of arbitrary dimension.*

5. Orthogonal sum of subspaces. Suppose $\mathfrak{M}_\nu$ are closed, mutually orthogonal subspaces in the Hilbert space $\mathfrak{H}$. The minimal closed subspace which contains all the $\mathfrak{M}_\nu$ is called the *orthogonal sum* of the subspaces $\mathfrak{M}_\nu$ and is denoted by $\sum_\nu \oplus \mathfrak{M}_\nu$. If the subspaces $\mathfrak{M}_\nu$ are finite in number, say $\mathfrak{M}_1, \mathfrak{M}_2, \cdots, \mathfrak{M}_n$, or form a denumerable set, say $\mathfrak{M}_1, \mathfrak{M}_2, \mathfrak{M}_3, \cdots$, then we also use the notation

$$\mathfrak{M}_1 \oplus \mathfrak{M}_2 \oplus \cdots \oplus \mathfrak{M}_n$$

and

$$\mathfrak{M}_1 \oplus \mathfrak{M}_2 \oplus \mathfrak{M}_3 \oplus \cdots$$

respectively.

We shall now elucidate of what sort of vectors the orthogonal sum of subspaces consists.

I. *The orthogonal sum* $\mathfrak{M} = \mathfrak{M}_1 \oplus \mathfrak{M}_2 \oplus \cdots \oplus \mathfrak{M}_n$ *of a finite number of subspaces is the set of all vectors*

$$x = x_1 + x_2 + \cdots + x_n, \quad x_k \in \mathfrak{M}_k.$$

Proof. Each such vector belongs to $\mathfrak{M}$. Conversely, if $x \in \mathfrak{M}$, then, denoting the projection of x on $\mathfrak{M}_k$ by x_k, we obtain that the vector $x-(x_1 + \cdots + x_n)$ belongs to $\mathfrak{M}$ on the one hand, and that, on the other hand, it is orthogonal to each $\mathfrak{M}_k$ [since this vector can be written in the form of the sum $(x-x_k)+(-x_1)+ \cdots + (-x_{k-1})+(-x_{k+1})+ \cdots +(-x_n)$, all of whose terms are orthogonal to $\mathfrak{M}_k$] and hence also to $\mathfrak{M}$. Consequently, this vector equals zero and hence $x = x_1 + \cdots + x_n$.

II. *The orthogonal sum* $\mathfrak{M} = \mathfrak{M}_1 \oplus \mathfrak{M}_2 \oplus \mathfrak{M}_3 \oplus \cdots$ *of a denumerable number of subspaces is the set of all vectors*

$$x = x_1 + x_2 + x_3 + \cdots,$$

where $x_k \in \mathfrak{M}_k$ *and the series in the right member converges.*

Proof. Every such vector belongs to $\mathfrak{M}$. Conversely, suppose $x \in \mathfrak{M}$ and let x' be the projection of x on $\mathfrak{M}_1 \oplus \mathfrak{M}_2 \oplus \cdots \oplus \mathfrak{M}_n$; by I, $x' = x_1 + \cdots + x_n$, where $x_k \in \mathfrak{M}_k$. It follows that

$$|x_1 + \cdots + x_n|^2 = |x'|^2 \leqq |x|^2,$$

i.e.

$$|x_1|^2 + \cdots + |x_n|^2 \leqq |x|^2.$$

Therefore the series $|x_1|^2 + |x_2|^2 + \cdots$, and hence the series $x_1 + x_2 + \cdots$ also, converges; the difference $x - (x_1 + x_2 + \cdots)$ belongs to $\mathfrak{M}$ and is orthogonal to every $\mathfrak{M}_k$ and hence to $\mathfrak{M}$ – therefore it equals zero.

III. *The orthogonal sum* $\mathfrak{M} = \sum_\nu \oplus \mathfrak{M}_\nu$ *of an arbitrary number of subspaces is the set of all vectors* $x = \sum_\nu x_\nu$, *where*:

1) $x_\nu \in \mathfrak{M}_\nu$;
2) x_ν *is different from zero only for a finite or denumerable number of indices* ν;
3) *the series* $\sum_\nu x_\nu$ *converges.*

Proof. Obviously, each such vector belongs to $\mathfrak{M}$. Conversely, if $x \in \mathfrak{M}$, then x is the limit of finite linear combinations of vectors in $\mathfrak{M}_\nu$ and hence it belongs to the orthogonal sum of a denumerable number of the $\mathfrak{M}_\nu$. Therefore it only remains to apply II.

If $x = \sum_\nu x_\nu$, $y = \sum_\nu y_\nu$ are vectors in $\mathfrak{M} = \sum_\nu \oplus \mathfrak{M}_\nu$, then, clearly,

$$\alpha x = \sum_\nu \alpha x_\nu, \; x + y = \sum_\nu (x_\nu + y_\nu), \; (x, y) = \sum_\nu (x_\nu, y_\nu).$$

6. Direct sum of Hilbert spaces. We shall first define the direct sum of a finite number of Hilbert spaces. Suppose $\mathfrak{H}_1, \mathfrak{H}_2, \cdots, \mathfrak{H}_n$ are n Hilbert spaces; we denote by $\mathfrak{H} = \mathfrak{H}_1 \oplus \mathfrak{H}_2 \oplus \cdots \oplus \mathfrak{H}_n$ the set of all n-tuples $x = \langle x_1, x_2, \cdots, x_n \rangle$, where $x_k \in \mathfrak{H}_k$, $k = 1, 2, \cdots, n$. We define sum, scalar product and inner product in $\mathfrak{H}$ by means of the formulas

$$x + y = \langle x_1 + y_1, x_2 + y_2, \cdots, x_n + y_n \rangle,$$

$$\alpha x = \langle \alpha x_1, \alpha x_2, \cdots, \alpha x_n \rangle, \; (x, y) = (x_1, y_1) + (x_2, y_2) + \cdots + (x_n, y_n)$$

where $x = \langle x_1, x_2, \cdots, x_n \rangle$, $y = \langle y_1, y_2, \cdots, y_n \rangle$. It is easily verified that then $\mathfrak{H}$ becomes a Hilbert space. This space is called the *direct sum* of the spaces $\mathfrak{H}_1$, $\mathfrak{H}_2, \cdots, \mathfrak{H}_n$.

Now suppose given a denumerable set of Hilbert spaces $\mathfrak{H}_1, \mathfrak{H}_2, \mathfrak{H}_3, \cdots$; we denote by $\mathfrak{H} = \mathfrak{H}_1 \oplus \mathfrak{H}_2 \oplus \mathfrak{H}_3 \oplus \cdots$ the set of all denumerable systems $x = \langle x_1, x_2, x_3, \cdots \rangle$, $x_k \in \mathfrak{H}_k$, such that $\sum_{k=1}^\infty |x_k|^2 < \infty$. We define sum, scalar product and inner product in $\mathfrak{H}$ by means of the formulas

$$x + y = \langle x_1 + y_1, x_2 + y_2, x_3 + y_3, \cdots \rangle,$$

$$\alpha x = \langle \alpha x_1, \alpha x_2, \cdots \rangle, \; (x, y) = \sum_{k=1}^{\infty} (x_k, y_k),$$

where $x = \langle x_1, x_2, \cdots \rangle \in \mathfrak{H}$, $y = \langle y_1, y_2, \cdots \rangle \in \mathfrak{H}$. It is easy to show that $\mathfrak{H}$ is a Hilbert space. It is called the *direct sum* of the spaces $\mathfrak{H}_1, \mathfrak{H}_2, \mathfrak{H}_3, \cdots$.

We now go over to the case of an arbitrary, even nondenumerable, set of Hilbert spaces. Suppose $\mathfrak{N}$ is an arbitrary set of indices ν and suppose a Hilbert space $\mathfrak{H}_\nu$ is assigned to each index ν. If we assign an element x_ν of the space $\mathfrak{H}_\nu$ to each index ν, we obtain the complex $x = \langle x_\nu \rangle$.

We consider the set $\mathfrak{H}$ of all such complexes $\langle x_\nu \rangle$ which satisfy the following conditions:

1° $\langle x_\nu \rangle$ contains at most a denumerable number of elements which are different from zero;

2° the series $\sum |x_\nu|^2$ converges.

For elements $x = \langle x_\nu \rangle$, $y = \langle y_\nu \rangle$ of the set $\mathfrak{H}$ we put

$$x+y = \langle x_\nu + y_\nu \rangle, \quad \alpha x = \langle \alpha x_\nu \rangle, \quad (x, y) = \sum_\nu (x_\nu, y_\nu);$$

with these definitions of the basic operations, $\mathfrak{H}$ forms a Hilbert space which we shall call the *direct sum* of the spaces $\mathfrak{H}_\nu$ and denote by $\sum_\nu \oplus \mathfrak{H}_\nu$.

The elements x_{ν_0} of the space $\mathfrak{H}_{\nu_0}$ can be identified with the complexes $x = \langle x_\nu \rangle$ in which $x_\nu = 0$ for $\nu \neq \nu_0$. Then $\mathfrak{H}_{\nu_0}$ becomes a subspace of the space $\mathfrak{H} = \sum_\nu \oplus \mathfrak{H}_\nu$. If this is done for every $\nu \in \mathfrak{N}$, then the space $\mathfrak{H}$ turns out to be the direct orthogonal sum of its subspaces $\mathfrak{H}_\nu$.

We note that some of the spaces $\mathfrak{H}_\nu$ or even all of them can coincide; thus, for example, $\mathfrak{H} \oplus \mathfrak{H}$ is the set of all pairs $\langle x_1, x_2 \rangle$ where $x_1 \in \mathfrak{H}$ and $x_2 \in \mathfrak{H}$.

7. Graph of an operator. Suppose A is an operator from $\mathfrak{H}_1$ into $\mathfrak{H}_2$ which is not necessarily linear. The set $\mathfrak{B}_A$ of all pairs $\langle x, Ax \rangle$, $x \in \mathfrak{D}_A$ in the direct sum $\mathfrak{H}_1 \oplus \mathfrak{H}_2$ is called the *graph of the operator A*.

The concept of the graph of an operator is a natural generalization of the usual concept of the graph of a function of one real variable, $y = f(x)$. Namely, the usual graph is nothing else than the set of all points $\langle x, f(x) \rangle$ in the plane, which can be considered as the direct sum of two one-dimensional spaces.

Clearly, two operators coincide if and only if their graphs coincide.

A set $S \subset \mathfrak{H}_1 \oplus \mathfrak{H}_2$ *is the graph of an operator if and only if the relations* $\langle x, y \rangle \in S$, $\langle x, y' \rangle \in S$ *imply* $y = y'$.

In fact, every graph satisfies this condition because $y = Ax$; conversely, if this condition is satisfied then the equality $y = Ax$ defines the operator A whose graph is S. It is also easily seen that *the operator A is linear if and only if its graph* $\mathfrak{B}_A$ *is a subspace in* $\mathfrak{H}_1 \oplus \mathfrak{H}_2$.

8. Closed operators; closure of an operator. An operator A from $\mathfrak{H}_1$ into $\mathfrak{H}_2$ is said to be *closed* if its graph $\mathfrak{B}_A$ is closed in $\mathfrak{H}_1 \oplus \mathfrak{H}_2$. Thus, an operator A is closed if the relations

$$x_n \in \mathfrak{D}_A, \langle x_n, Ax_n \rangle \to \langle x, y \rangle$$

imply the relation $\langle x, y\rangle \in \mathfrak{B}_A$, i.e. $x \in \mathfrak{D}_A$ and $y = Ax$. In other words, an operator A is closed means that the relations

$$x_n \in \mathfrak{D}_A,\, x_n \to x,\, Ax_n \to y$$

imply the relations

$$x \in \mathfrak{D}_A \text{ and } y = Ax.$$

I. *Every bounded linear operator defined in the entire space $\mathfrak{H}$ is closed.*

In fact, such an operator is continuous and therefore the relation $x_n \to x$ already implies the relation $Ax_n \to Ax$. (The difference between an operator A being continuous and being closed consists in exactly this. If the operator A is closed then $x_n \to x$, $x_n \in \mathfrak{D}_A$ does not in general imply that $\langle Ax_n\rangle$ converges.) Hence, if $Ax_n \to y$, then $Ax = y$.

The reader can also easily verify the following propositions:

II. *If the operator A is closed, then the operator $A - \lambda 1$ is also closed.*

III. *If the operator A is closed and the operator A^{-1} exists then the operator A^{-1} is also closed.*

If the operator A is not closed, then by definition its graph $\mathfrak{B}_A$ is not closed in $\mathfrak{H}_1 \oplus \mathfrak{H}_2$. It can occur that the closure $\overline{\mathfrak{B}}_A$ of the set $\mathfrak{B}_A$ in $\mathfrak{H}_1 \oplus \mathfrak{H}_2$ is also the graph of some operator. This operator is called the *closure* of the operator A; we shall denote it by $\tilde{A}$. In this case, we also say that the operator A *admits the closure* $\tilde{A}$. Thus, by definition,

$$\mathfrak{B}_{\tilde{A}} = \overline{\mathfrak{B}}_A.$$

Clearly, $\tilde{A}$ is the minimal closed extension of the operator A. It is easy to formulate a condition that the closure exist without using the concept of graph. The set $\overline{\mathfrak{B}}_A$ which must be the graph of some operator consists of elements of the form $\langle x, Ax\rangle$, $x \in \mathfrak{D}_A$ and their limits; therefore

IV. *An operator A admits the closure $\tilde{A}$ if and only if the relations*

$$x_n \in \mathfrak{D}_A,\, x_n' \in \mathfrak{D}_A,\, x_n \to x,\; x_n' \to x,\; Ax_n \to y,\; Ax_n' \to y'$$

imply that $y = y'$. In this case, the domain $\mathfrak{D}_{\tilde{A}}$ of the closure consists of those and only those vectors x for which there exists a sequence $x_n \in \mathfrak{D}_A$ which satisfies the conditions: $x_n \to x$, Ax_n converges. Moreover,

$$\tilde{A}x = \lim_{n\to\infty} Ax_n.$$

If the operator A is linear, then, obviously, the preceding condition can be simplified by requiring that the relations

$$x_n \in \mathfrak{D}_A,\, x_n \to 0,\; Ax_n \to y$$

imply $y = 0$.

9. Adjoint operator. We now consider an arbitrary operator A from $\mathfrak{H}_1$ into $\mathfrak{H}_2$ whose domain is dense in $\mathfrak{H}_1$. It can occur that for some $y \in \mathfrak{H}_2$ a representation of the form

$$(Ax, y) = (x, z)$$

holds for all $x \in \mathfrak{D}_A$. We denote by $\mathfrak{D}^*$ the set of all such vectors y and define the operator A^* from $\mathfrak{H}_2$ into $\mathfrak{H}_1$ with domain $\mathfrak{D}_{A^*} = \mathfrak{D}^*$ by means of the formula

$$A^*y = z.$$

This operator A^* is called the *adjoint* of the operator A. We note that the vector z is defined uniquely by the vector y. In fact, if we also had

$$(Ax, y) = (x, z'),$$

then $(x, z-z') = 0$, i.e. the vector $z-z'$ is orthogonal to the domain $\mathfrak{D}_A$ of the operator A. Since the latter is dense in $\mathfrak{H}_1$, this is possible only if $z-z' = 0$, i.e. $z = z'$.

Here we see that the initial assumption on the density of $\mathfrak{D}_A$ is very essential; in the contrary case, it would be impossible to define the adjoint operator uniquely.

We note that the adjoint operator is always linear.

I. *If the operator A has an inverse A^{-1} and if $\mathfrak{D}_A$ and $\mathfrak{D}_{A^{-1}}$ are dense in $\mathfrak{H}_1$, $\mathfrak{H}_2$ respectively, then*

$$(A^{-1})^* = A^{*-1}. \tag{1}$$

Proof. For $x \in \mathfrak{D}_A$, $y \in \mathfrak{D}_{(A^{-1})^*}$, we have

$$(x, y) = (A^{-1}Ax, y) = (Ax, (A^{-1})^*y);$$

this shows that

$$(A^{-1})^*y \in \mathfrak{D}_{A^*}$$

and

$$A^*(A^{-1})^*y = y. \tag{2}$$

On the other hand, for $x \in \mathfrak{D}_{A^{-1}}$, $y \in \mathfrak{D}_{A^*}$,

$$(x, y) = (AA^{-1}x, y) = (A^{-1}x, A^*y);$$

this shows that

$$A^*y \in \mathfrak{D}_{(A^{-1})^*}$$

and

$$(A^{-1})^*A^*y = y. \tag{3}$$

Equalities (2) and (3) together mean that the operator $(A^{-1})^*$ is the inverse to A^*, i.e. formula (1) is valid.

The reader can also easily verify the following properties of the adjoint operator (it is assumed that the domains of all the operators considered are dense in $\mathfrak{H}$ and that A, B are operators from $\mathfrak{H}_1$ into $\mathfrak{H}_2$):

a) $(\lambda A)^* = \bar{\lambda} A^* (\lambda \neq 0)$;
b) if $A \subset B$, then $A^* \supset B^*$;
c) $(A+B)^* \supset A^* + B^*$;
if A, B are operators from $\mathfrak{H}_2$ into $\mathfrak{H}_3$ and from $\mathfrak{H}_1$ into $\mathfrak{H}_2$ respectively, then
d) $(AB)^* \supset B^* A^*$;
and if A is an operator in $\mathfrak{H}$, then
e) $(A+\lambda 1)^* = A^* + \bar{\lambda} 1$.

The adjoint operator can be described with the aid of a graph. To this end, we define the operator U_1 from $\mathfrak{H}_1 \oplus \mathfrak{H}_2$ into $\mathfrak{H}_2 \oplus \mathfrak{H}_1$ by means of the formula

$$U_1 \langle x, y \rangle = \langle iy, -ix \rangle.$$

It is easy to verify that U_1 maps $\mathfrak{H}_1 \oplus \mathfrak{H}_2$ isometrically onto $\mathfrak{H}_2 \oplus \mathfrak{H}_1$. Analogously, we define the operator U_2 which maps $\mathfrak{H}_2 \oplus \mathfrak{H}_1$ isometrically onto $\mathfrak{H}_1 \oplus \mathfrak{H}_2$ by

$$U_2 \langle y, x \rangle = \langle ix, -iy \rangle.$$

It is easily seen that

$$U_2 U_1 = 1. \tag{4}$$

We apply the operator U_1 to all the vectors of the graph $\mathfrak{B}_A$; we obtain the set of all pairs $\langle iAx, -ix \rangle$, $x \in \mathfrak{D}_A$; we denote it by $\mathfrak{B}'_A$. Then

$$\mathfrak{B}_{A^*} = (\mathfrak{H}_2 \oplus \mathfrak{H}_1) \ominus \mathfrak{B}'_A, \tag{5}$$

i.e. *the graph of the operator A^* is the orthogonal complement of the set $\mathfrak{B}'_A$ in $\mathfrak{H}_2 \oplus \mathfrak{H}_1$*. In fact, this orthogonal complement consists of those and only those pairs $\langle y, z \rangle$ which satisfy the condition

$$(\langle iAx, -ix \rangle, \langle y, z \rangle) = 0$$

for all $x \in \mathfrak{D}_A$. But this condition is equivalent to the condition

$$(Ax, y) - (x, z) = 0,$$

from which it follows that

$$y \in \mathfrak{D}_{A^*}, \; z = A^* y, \; \langle y, z \rangle \in \mathfrak{B}_{A^*}.$$

Since the orthogonal complement is a closed subspace, it follows from (5) that *A^* is always a closed linear operator*. We shall now prove the following important proposition:

II (Von Neumann [2]). *If the linear operator A with dense domain has the closure $\tilde{A}$, then*

$$\tilde{A}^* = A^*, \tag{6}$$

$\mathfrak{D}_{A^*}$ is dense in $\mathfrak{H}_2$ and

$$A^{**} = \tilde{A}.$$

If, in particular, the operator A is closed then

$$A^{**} = A.$$

Proof. Since $\mathfrak{B}_{\tilde{A}} = \overline{\mathfrak{B}}_A$, we have $\mathfrak{B}'_{\tilde{A}} = U_1\overline{\mathfrak{B}}_A = \overline{U_1\mathfrak{B}_A} = \overline{\mathfrak{B}}'_A$, and therefore

$$\mathfrak{B}_{\tilde{A}^*} = (\mathfrak{H}_2 \oplus \mathfrak{H}_1) \ominus \mathfrak{B}'_{\tilde{A}} = (\mathfrak{H}_2 \oplus \mathfrak{H}_1) \ominus \mathfrak{B}'_A = \mathfrak{B}_{A^*}.$$

Equality (6) follows from this. Further, we conclude from (5) that

$$\overline{\mathfrak{B}}'_A = (\mathfrak{H}_2 \oplus \mathfrak{H}_1) \ominus \mathfrak{B}_{A^*}.$$

We now apply the operator U_2 to all the vectors in the left and right members of the last relation. By (4), U_2 maps $\overline{\mathfrak{B}}'_A$ onto $\overline{\mathfrak{B}}_A$; moreover, by definition, U_2 maps $\mathfrak{B}_{A^*}$ onto $\mathfrak{B}'_{A^*}$, and therefore $(\mathfrak{H}_2 \oplus \mathfrak{H}_1) \ominus \mathfrak{B}_{A^*}$ onto $(\mathfrak{H}_1 \oplus \mathfrak{H}_2) \ominus \mathfrak{B}'_{A^*}$. Consequently, we obtain the relation

$$\overline{\mathfrak{B}}_A = (\mathfrak{H}_1 \oplus \mathfrak{H}_2) \ominus \mathfrak{B}'_{A^*}. \tag{7}$$

It follows from (7) that $\mathfrak{D}_{A^*}$ is dense in $\mathfrak{H}_2$. In fact, otherwise there would exist a nonzero vector $z \in \mathfrak{H}_2$ which is orthogonal to $\mathfrak{D}_{A^*}$, hence $\langle z, 0\rangle \perp \mathfrak{D}_{A^*}$, and therefore $\langle 0, -iz\rangle \in (\mathfrak{H}_1 \oplus \mathfrak{H}_2) \ominus \mathfrak{D}'_{A^*} = \overline{\mathfrak{D}}_A$, which is impossible for $z \neq 0$. Moreover, (7) means that $\overline{\mathfrak{B}}_A$ is the graph of the operator A^{**}; on the other hand, $\overline{\mathfrak{B}}_A$, being the closure of the set $\mathfrak{B}_A$, is the graph of the operator $\tilde{A}$. Consequently, $A^{**} = \tilde{A}$. If the operator A is closed, then $\tilde{A} = A$ and $A^{**} = A$.

An operator A (which is not necessarily linear) in a Hilbert space is said to be *Hermitian* if $(Ax, y) = (x, Ay)$ for all $x, y \in \mathfrak{D}_A$.

A Hermitian operator with domain dense in $\mathfrak{H}$ is called a *symmetric* operator. Clearly, an operator A in $\mathfrak{H}$ with domain dense in $\mathfrak{H}$ will be symmetric if and only if

$$A \subset A^*. \tag{8}$$

Since the adjoint operator is linear and closed, we conclude from (8) that *a symmetric operator A admits a closure and is linear if $\mathfrak{D}_A$ is a subspace.*

An operator A in $\mathfrak{H}$ with domain dense in $\mathfrak{H}$ is said to be *self-adjoint* if $A = A^*$.

It follows directly from this definition that *a self-adjoint operator is closed.* Moreover, we conclude from relations a), d) in the remarks following I above that:

III. *If A is a self-adjoint operator then for arbitrary real α and β the operator $\alpha A+\beta 1$ is also self-adjoint.*

IV. *A symmetric operator A in $\mathfrak{H}$ whose range $\mathfrak{R}_A$ coincides with $\mathfrak{H}$ is a self-adjoint operator.*

Proof. It is clear that one must only prove that $\mathfrak{D}_{A^*} = \mathfrak{D}_A$. But if $y \in \mathfrak{D}_{A^*}$ and $z = A^*y$, then in virtue of the condition $\mathfrak{R}_A = \mathfrak{H}$ there exists a vector $y' \in \mathfrak{D}_A$ for which $z = Ay'$. It follows that for arbitrary $x \in \mathfrak{D}_A$ we have

$$(Ax, y) = (x, A^*y) = (x, Ay') = (Ax, y')$$

and $(Ax, y-y') = 0$. Since the set of all vectors Ax coincides with $\mathfrak{H}$, we conclude from this that $y-y' = 0$, i.e. $y = y' \in \mathfrak{D}_A$.

V. *If A is a self-adjoint operator in $\mathfrak{H}$ then $U = (A+i1)(A-i1)^{-1}$ is a unitary operator in $\mathfrak{H}$.*

Proof. Suppose A is a self-adjoint operator in $\mathfrak{H}$. Then for $x \in \mathfrak{D}_A$ we have

$$|Ax \pm ix|^2 = (Ax, Ax) \mp i(Ax, x) \pm i(x, Ax) + (x, x) = |Ax|^2 + |x|^2. \tag{9}$$

Therefore the equality $Ax \pm ix = 0$ is possible only for $x = 0$ and the operator $U = (A+i1)(A-i1)^{-1}$ exists. Further, $\mathfrak{R}_{A \pm i1}$ is dense in $\mathfrak{H}$. In fact, if, for example, $z \perp \mathfrak{R}_{A+i1}$, then $0 = (z, Ax+ix) = (z, Ax)-(iz, x)$, from which it follows that $z \in \mathfrak{D}_{A^*} = \mathfrak{D}_A$ and $Az = iz$, which, as we have just seen, is possible only for $z = 0$. We shall prove that $\mathfrak{R}_{A \pm i1} = \mathfrak{H}$. Suppose $y \in \mathfrak{H}$; since $\mathfrak{R}_{A+i1}$ is dense in $\mathfrak{H}$,

$$y_n = Ax_n + ix_n \to y \tag{10}$$

exists. In virtue of (9),

$$|y_n - y_m|^2 = |A(x_n - x_m) + i(x_n - x_m)|^2 = |A(x_n - x_m)|^2 + |x_n - x_m|^2$$

and therefore $\langle x_n \rangle$ and $\langle Ax_n \rangle$ converge to some vectors x and z; since A is closed we have $x \in \mathfrak{D}_A$ and $z = Ax$. But then, in virtue of (10), $y = Ax+ix \in \mathfrak{R}_{A+i1}$; consequently, $\mathfrak{R}_{A+i1} = \mathfrak{H}$ and analogously $\mathfrak{R}_{A-i1} = \mathfrak{H}$. This means that $\mathfrak{D}_U = \mathfrak{R}_U = \mathfrak{H}$. Moreover, it follows from (9) that U is isometric and therefore U is unitary.

In fact, if $y \in \mathfrak{H}$, then $y = (A-i1)x$ and $Uy = (A+i1)(A-i1)^{-1}(A-i1)x = (A+i1)x$. Therefore (9) means that $|Uy|^2 = |y|^2$.

An operator A is said to be *positive definite* if

$$(Ax, x) \geqq 0 \text{ for all } x \in \mathfrak{D}_A.$$

VI (Von Neumann [2]). *If A is a closed linear operator from $\mathfrak{H}_1$ into $\mathfrak{H}_2$ with domain $\mathfrak{D}_A$, dense in $\mathfrak{H}_1$, then A^*A is a positive definite self-adjoint operator in $\mathfrak{H}_1$.*

Proof. The fact that the operator A^*A is positive definite follows directly from the relations $(A^*Ax, x) = (Ax, Ax) \geqq 0$ for $x \in \mathfrak{D}_{A^*A}$. It remains to prove that A^*A is a self-adjoint operator. To this end, we rewrite (5) in the form

$$\mathfrak{B}'_A \oplus \mathfrak{B}_{A^*} = \mathfrak{H}_2 \oplus \mathfrak{H}_1. \tag{11}$$

In virtue of (11), the vector $\langle 0, -ix \rangle$ in $\mathfrak{H}_2 \oplus \mathfrak{H}_1$ can be represented in the form

$$\langle 0, -ix \rangle = \langle iAy, -iy \rangle + \langle z, A^*z \rangle, \quad y \in \mathfrak{D}_A, \ z \in \mathfrak{D}_{A^*},$$

i.e.

$$0 = iAy + z, \quad -ix = -iy + A^*z = -iy - iA^*Ay.$$

It follows that $x = (1+A^*A)y$, so that the range of the operator $1+A^*A$ coincides with the entire space $\mathfrak{H}_1$. We shall prove that A^*A+1 is a symmetric operator; in virtue of IV, this will also prove that A^*A+1, and hence, in virtue of III, that also A^*A, is a self-adjoint operator.

It is clear that A^*A+1 is Hermitian and consequently one must only prove that $\mathfrak{D}_{A^*A+1}$ is dense in $\mathfrak{H}_1$. Suppose x_0 is a vector in $\mathfrak{H}_1$ which is orthogonal to $\mathfrak{D}_{A^*A+1}$. According to what was proved above, x_0 can be represented in the form $x_0 = (A^*A+1)y_0$, so that $((A^*A+1)y_0, y) = 0$ for all $y \in \mathfrak{D}_{A^*A+1}$. In particular, putting $y = y_0$, we obtain

$$((A^*A+1)y_0, y_0) = |Ay_0|^2 + |y_0|^2 = 0,$$

from which it follows that $y_0 = 0$ and $x_0 = (A^*A+1)y_0 = 0$. Thus, there does not exist a vector $x_0 \neq 0$ in $\mathfrak{H}_1$ which is orthogonal to $\mathfrak{D}_{A^*A+1}$; consequently, $\mathfrak{D}_{A^*A+1}$ is dense in $\mathfrak{H}_1$.

10. The case of a bounded operator.

I. *If A is a bounded operator from $\mathfrak{H}_1$ into $\mathfrak{H}_2$ then A admits a closure $\tilde{A}$; moreover, $\mathfrak{D}_{\tilde{A}} = \overline{\mathfrak{D}}_A$ and $|\tilde{A}| = |A|$.*

Proof. In virtue of II, subsection 4, § 4, the operator A is uniquely extendible to a bounded operator $\tilde{A}$ with domain $\mathfrak{D}_{\tilde{A}} = \overline{\mathfrak{D}}_A$ and norm $|\tilde{A}| = |A|$. It is easily seen that this operator $\tilde{A}$ is the closure of the operator A.

II. *The closure of an isometric operator U is also an isometric operator with domain $\overline{\mathfrak{D}}_U$ and range $\overline{\mathfrak{R}}_U$.*

Proof. Since the isometric operator U is bounded, in virtue of I it admits a closure $\tilde{U}$, where $\mathfrak{D}_{\tilde{U}} = \overline{\mathfrak{D}}_U$. If $x, y \in \mathfrak{D}_{\tilde{U}}$, then there exist x_n, $y_n \in \mathfrak{D}_U$ such that $x_n \to x$, $y_n \to y$. Passing to the limit in the equality $(Ux_n, Uy_n) = (x_n, y_n)$, we obtain that $(\tilde{U}x, \tilde{U}y) = (x, y)$, and consequently $\tilde{U}$ is isometric. The relation $\mathfrak{R}_{\tilde{U}} = \overline{\mathfrak{R}}_U$ is obtained if we apply the formula $\mathfrak{D}_{\tilde{U}} = \overline{\mathfrak{D}}_U$ to the inverse operator U^{-1}.

Since $A^* = \tilde{A}^*$, it is sufficient, in the study of the operator adjoint to the bounded operator A, to consider the case of an operator A with domain $= \mathfrak{H}_1$. Thus, let A be a bounded operator from $\mathfrak{H}_1$ into $\mathfrak{H}_2$ with domain $\mathfrak{D}_A = \mathfrak{H}_1$. Then for arbitrary $y \in \mathfrak{H}_2$ the linear functional

$$f(x) = (Ax, y)$$

is bounded, because

$$|f(x)| = |(Ax, y)| \leqq |Ax|\,|y| \leqq |A|\,|x|\,|y|.$$

On the basis of the Riesz theorem (see subsection 3), there exists a vector $z \in \mathfrak{H}_1$ such that $f(x) = (x, z)$, i.e.

$$(Ax, y) = (x, z);$$

moreover,

$$|z| \leqq |A|\,|y|.$$

This means that $\mathfrak{D}_{A^*} = \mathfrak{H}_2$ and

$$|A^*y| = |z| \leqq |A|\,|y|;$$

consequently, A^* is bounded and

$$|A^*| \leqq |A|. \tag{1}$$

Applying this result to A^*, we obtain that A^{**} is bounded and that

$$|A^{**}| \leqq |A^*|.$$

But, obviously,

$$A^{**} = A,$$

so that $|A| \leqq |A^*|$. Combining this result with (1), we obtain that

$$|A^*| = |A|.$$

Thus,

III. *If A is a bounded operator from $\mathfrak{H}_1$ into $\mathfrak{H}_2$ with domain $= \mathfrak{H}_1$, then A^* is a bounded operator from $\mathfrak{H}_2$ into $\mathfrak{H}_1$ with domain $= \mathfrak{H}_2$ and $|A^*| = |A|$. Moreover,*

a) $A^{**} = A$;

b) $(\alpha A)^* = \bar{\alpha} A^* (\alpha \neq 0)$;

c) *if A, B are bounded operators from $\mathfrak{H}_1$ into $\mathfrak{H}_2$, then*

$$(A+B)^* = A^* + B^*;$$

d) *if A, B are bounded operators in $\mathfrak{H}$, then $(AB)^* = B^*A^*$.*

e) *if T is an operator from $\mathfrak{H}_2$ into $\mathfrak{H}_1$, $\mathfrak{D}_T$ is dense in $\mathfrak{H}_2$ and A is a bounded operator in $\mathfrak{H}_1$ with domain $\mathfrak{H}_1$, then*

$$(AT)^* = T^*A^*.$$

Relations b)–d) are a particular case of relations a), c), d) of subsection 9.

We shall prove assertion c). In virtue of c), subsection 9, $(AT)^* \supset T^*A^*$; therefore, it suffices to prove the reverse inclusion.

Let $\eta \in \mathfrak{D}_{(AT)^*}$. Then for all $\xi \in \mathfrak{D}_{AT} = \mathfrak{D}_T$,

$$(T\xi, A^*\eta) = (AT\xi, \eta) = (\xi, (AT)^*\eta);$$

consequently, $A^*\eta \in \mathfrak{D}_{T^*}$, $\eta \in \mathfrak{D}_{T^*A^*}$ and $T^*A^*\eta = (AT)^*\eta$.

We note further that

$$|A^*A| = |A|^2. \tag{2}$$

In fact,

$$|A|^2 = \sup_{|x|=1} |Ax|^2 = \sup_{|x|=1} (Ax, Ax) = \sup_{|x|=1} (A^*Ax, x) \leqq \sup_{|x|=1} |A^*Ax|\,|x| = |A^*A|;$$

on the other hand,

$$|A^*A| \leqq |A^*|\,|A| = |A|^2.$$

IV. *A bounded operator A in $\mathfrak{H}$ with domain $\mathfrak{D}_A = \mathfrak{H}$ is Hermitian if, and only if, $A^* = A$.*

In fact, since $\mathfrak{D}_A = \mathfrak{H}$, the relation $A^* \supset A$ goes over into $A^* = A$.

V. *An operator U from $\mathfrak{H}_1$ into $\mathfrak{H}_2$ with domain $\mathfrak{D}_U = \mathfrak{H}_1$ is isometric if, and only if, $U^*U = 1$.*

Proof. If U is isometric, then for all $x, y \in \mathfrak{H}_1$, we have

$$(x, y) = (Ux, Uy).$$

Therefore, $Ux \in \mathfrak{D}_{U^*}$ and $U^*Ux = x$, i.e. $U^*U = 1$. Conversely, if $U^*U = 1$, then

$$(x, y) = (U^*Ux, y) = (Ux, Uy),$$

i.e. U is isometric.

VI. *An operator U from $\mathfrak{H}_1$ into $\mathfrak{H}_2$ maps $\mathfrak{H}_1$ isometrically onto $\mathfrak{H}_2$ if, and only if,*

$$U^*U = 1, \; UU^* = 1.$$

Proof. It follows from $U^*U = 1$ that $\mathfrak{D}_U = \mathfrak{H}_1$ and that U maps $\mathfrak{H}_1$ isometrically into $\mathfrak{H}_2$. On the other hand, it follows from $UU^*y = y$ that $\mathfrak{R}_U = \mathfrak{H}_2$.

The following is a particular case of this proposition:

VII. *An operator U in $\mathfrak{H}$ is unitary if, and only if,*

$$U^*U = 1, \; UU^* = 1.$$

VIII. *A bounded linear operator A from $\mathfrak{H}_1$ into $\mathfrak{H}_2$ defined everywhere in $\mathfrak{H}_1$ is compact if, and only if, the operator A^*A is compact.*

Proof. The necessity is a particular case of Proposition II, subsection 6, § 4; we shall prove the sufficiency. Suppose the operator A^*A is compact and let $\langle x_n \rangle$ be a bounded sequence of elements in $\mathfrak{H}_1$, $|x_n| \leqq c$. From $\langle x_n \rangle$ one can select a subsequence $\langle x'_n \rangle$ such that $\langle A^*Ax'_n \rangle$ converges. But then

$$\begin{aligned} |Ax'_n - Ax'_m|^2 &= (A(x'_n - x'_m), \; A(x'_n - x'_m)) \\ &= (A^*Ax'_n - A^*Ax'_m, \; x'_n - x'_m) \leqq |A^*Ax'_n - A^*Ax'_m||x'_n - x'_m| \\ &\leqq 2c|A^*Ax'_n - A^*Ax'_m| \to 0 \text{ for } n, m \to \infty, \end{aligned}$$

so that the sequence $\langle Ax'_n \rangle$ converges. But this means that the operator A is compact.

IX. *If the linear operator A from $\mathfrak{H}_1$ into $\mathfrak{H}_2$, defined everywhere in $\mathfrak{H}_1$, is compact, then the operator A^* is also compact.*

This assertion follows from Proposition VIII because the operator $(A^*)^*A^* = AA^*$ is compact.

11. Generalization to operators in a Banach space. The concept of an adjoint operator extends in a natural way to operators in a Banach space. We shall consider only the case of a bounded linear operator A in the Banach space X with domain $= X$.

In this case, for arbitrary $f \in X'$ the formula $f_1(x) = f(Ax)$ defines a linear functional $f_1 \in X'$ and the correspondence $f \to f_1$ defines a linear operator in X', considered as a Banach space. This operator is called the *adjoint* to A and is denoted by A'; thus, by definition, we have

$$(A'f)(x) = f(Ax). \tag{1}$$

Equality (1) can be written in a more convenient form if we agree to write (x, f) instead of $f(x)$; then (1) goes over into the form

$$(x, A'f) = (Ax, f).$$

It is easily seen that A' is also bounded and that $|A'| = |A|$; it is also easy to verify that the relations

$$(\alpha A)' = \alpha A', (A+B)' = A'+B', (AB)' = B'A'$$

are valid.

Comparing these relations with b)–d) in subsection 11, we see that in going over to a Hilbert space this definition of an adjoint operator does not coincide entirely with the definition of an adjoint operator in a Hilbert space. This is explained by the *anti-linearity* of the correspondence $f \leftrightarrow y$ between functionals and elements of a Hilbert space (see subsection 3).

12. Projection operators. Suppose $\mathfrak{M}$ is a closed subspace in the Hilbert space $\mathfrak{H}$. To each vector $x \in \mathfrak{H}$ we assign its projection x_1 on $\mathfrak{M}$. The correspondence thus obtained is an operator in $\mathfrak{H}$; we denote it by P so that by definition $Px = x_1$. The operator P is called the *projection operator* onto $\mathfrak{M}$. If it is necessary to underscore that P is the projection operator namely onto $\mathfrak{M}$ then we write $P_{\mathfrak{M}}$ instead of P. It follows from the definition of a projection that $|Px| \leqq |x|$; consequently, *the projection operator is bounded.*

I. *The operator P is Hermitian, linear and satisfies the relation $P^2 = P$.*

In fact, setting

$$Px = x_1, Py = y_1, x_2 = x - x_1, y_2 = y - y_1,$$

we have

$$x_2 \perp x_1, y_1; y_2 \perp x_1, y_1.$$

Consequently,

$$(Px, y) = (x_1, y_1 + y_2) = (x_1, y_1)$$

and

$$(x, Py) = (x_1 + x_2, y_1) = (x_1, y_1),$$

so that $(Px, y) = (x, Py)$ and P is Hermitian, and hence also linear (since it is defined everywhere). Since $x_1 \in \mathfrak{M}$, we have $Px_1 = x_1$, i.e. $P^2x = Px$ for arbitrary $x \in \mathfrak{H}$; this means that $P^2 = P$.

Conversely,

II. *Every Hermitian operator P in $\mathfrak{H}$ with domain $= \mathfrak{H}$, satisfying the relation $P^2 = P$, is a projection operator.*

Proof. Since $P^2 = P$, we have

$$|Px|^2 = (Px, Px) = (P^2x, x) = (Px, x) \leqq |Px|\,|x|,$$

from which it follows that $|Px| \leqq |x|$. This means that P is bounded and consequently it is continuous. We denote the set of all vectors x which satisfy the condition $Px = x$ by $\mathfrak{M}$. In virtue of the continuity of the linear operator P, the set $\mathfrak{M}$ is a closed subspace in $\mathfrak{H}$. If x is an arbitrary vector, then, setting $x_1 = Px$, $x_2 = (1-P)x$, we have

$$x = x_1 + x_2 \text{ and } Px_1 = P^2x = Px = x_1;$$

consequently, $x_1 \in \mathfrak{M}$. Moreover, for arbitrary $y \in \mathfrak{M}$ we have $Py = y$, and therefore

$$(y, x_2) = (y, (1-P)x) = ((1-P)y, x) = (y-Py, x) = 0,$$

so that $x_2 \perp \mathfrak{M}$. This means that $x_1 = Px$ is the projection of the vector x on $\mathfrak{M}$, so that P is a projection operator on $\mathfrak{M}$.

III. *The product of two projection operators $P_{\mathfrak{M}_1}$, $P_{\mathfrak{M}_2}$ is also a projection operator if and only if the operators $P_{\mathfrak{M}_1}$, $P_{\mathfrak{M}_2}$ commute. In this case,*

$$P_{\mathfrak{M}_1} P_{\mathfrak{M}_2} = P_{\mathfrak{M}},$$

where $\mathfrak{M} = \mathfrak{M}_1 \cap \mathfrak{M}_2$.

Proof. If $P_{\mathfrak{M}_1} P_{\mathfrak{M}_2}$ is a projection operator, then $P_{\mathfrak{M}_1} P_{\mathfrak{M}_2}$ is Hermitian, i.e.

$$P_{\mathfrak{M}_1} P_{\mathfrak{M}_2} = (P_{\mathfrak{M}_1} P_{\mathfrak{M}_2})^* = P^*_{\mathfrak{M}_2} P^*_{\mathfrak{M}_1} = P_{\mathfrak{M}_2} P_{\mathfrak{M}_1}.$$

This means that $P_{\mathfrak{M}_1}$, $P_{\mathfrak{M}_2}$ commute.

Conversely, if this condition is satisfied, then $P_{\mathfrak{M}_1} P_{\mathfrak{M}_2}$ is Hermitian and

$$(P_{\mathfrak{M}_1} P_{\mathfrak{M}_2})^2 = P^2_{\mathfrak{M}_1} P^2_{\mathfrak{M}_2} = P_{\mathfrak{M}_1} P_{\mathfrak{M}_2};$$

by II, this means that $P_{\mathfrak{M}_1} P_{\mathfrak{M}_2}$ is a projection operator. Suppose $\mathfrak{M}$ is a subspace onto which it projects. Since $P_{\mathfrak{M}_1} P_{\mathfrak{M}_2} x \in \mathfrak{M}_1$ for arbitrary $x \in \mathfrak{H}$, we have $\mathfrak{M} \subset \mathfrak{M}_1$; analogously, $\mathfrak{M} \subset \mathfrak{M}_2$, and therefore $\mathfrak{M} \subset \mathfrak{M}_1 \cap \mathfrak{M}_2$. Conversely, if $x \in \mathfrak{M}_1 \cap \mathfrak{M}_2$, then $P_{\mathfrak{M}_1} P_{\mathfrak{M}_2} x = P_{\mathfrak{M}_1} x = x$; consequently, $x \in \mathfrak{M}$; this means that $\mathfrak{M}_1 \cap \mathfrak{M}_2 \subset \mathfrak{M}$, and therefore $\mathfrak{M}_1 \cap \mathfrak{M}_2 = \mathfrak{M}$.

IV. *The subspaces $\mathfrak{M}_1$, $\mathfrak{M}_2$ are orthogonal if and only if*

$$P_{\mathfrak{M}_1} P_{\mathfrak{M}_2} = O.$$

Proof. This assertion follows directly from the equality

$$(P_{\mathfrak{M}_2} x, P_{\mathfrak{M}_1} y) = (P_{\mathfrak{M}_1} P_{\mathfrak{M}_2} x, y), \; x, y \in \mathfrak{H}.$$

V. *The sum*

$$Q = P_{\mathfrak{M}_1} + P_{\mathfrak{M}_2} + \cdots + P_{\mathfrak{M}_n}$$

of a finite number of projection operators is a projection operator if and only if

$$P_{\mathfrak{M}_j} P_{\mathfrak{M}_k} = O \textit{ for } j \neq k, \tag{1}$$

i.e. if and only if the subspaces $\mathfrak{M}_1, \mathfrak{M}_2, \cdots, \mathfrak{M}_n$ *are mutually orthogonal*; *in this case*

$$P_{\mathfrak{M}_1} + P_{\mathfrak{M}_2} + \cdots + P_{\mathfrak{M}_n} = P_{\mathfrak{M}},$$

where

$$\mathfrak{M} = \mathfrak{M}_1 \oplus \mathfrak{M}_2 \oplus \cdots \oplus \mathfrak{M}_n.$$

Proof. If conditions (1) are satisfied, then, as can easily be verified, $Q^2 = Q$, so that Q is a projection operator. Conversely, suppose Q is a projection operator; we shall prove that conditions (1) are satisfied. If Q is a projection operator, then for arbitrary $x \in \mathfrak{H}$, we have

$$|x|^2 \geqq (Qx, x) = \sum_{i=1}^{n} (P_{\mathfrak{M}_i} x, x) \geqq (P_{\mathfrak{M}_j} x, x) + (P_{\mathfrak{M}_k} x, x),$$

so that

$$|P_{\mathfrak{M}_j} x|^2 + |P_{\mathfrak{M}_k} x|^2 \leqq |x|^2.$$

If we set $x = P_{\mathfrak{M}_k} y$ in this inequality, we obtain

$$|P_{\mathfrak{M}_j} P_{\mathfrak{M}_k} y|^2 + |P_{\mathfrak{M}_k} y|^2 \leqq |P_{\mathfrak{M}_k} y|^2;$$

consequently, $|P_{\mathfrak{M}_j} P_{\mathfrak{M}_k} y| = 0$ and $P_{\mathfrak{M}_j} P_{\mathfrak{M}_k} = O$.

We shall prove the last assertion. Suppose conditions (1) are satisfied and let $\mathfrak{M}$ be a subspace onto which Q projects. Then for arbitrary $x \in \mathfrak{H}$,

$$Qx = P_{\mathfrak{M}_1} x + P_{\mathfrak{M}_2} x + \cdots + P_{\mathfrak{M}_n} x \subset \mathfrak{M}_1 \oplus \mathfrak{M}_2 \oplus \cdots \oplus \mathfrak{M}_n,$$

so that

$$\mathfrak{M} \subset \mathfrak{M}_1 \oplus \mathfrak{M}_2 \oplus \cdots \oplus \mathfrak{M}_n. \tag{2}$$

Conversely, if $x \in \mathfrak{M}_1 \oplus \mathfrak{M}_2 \oplus \cdots \oplus \mathfrak{M}_n$, then $x = x_1 + x_2 + \cdots + x_n$, where $x_k \in \mathfrak{M}_k$. In virtue of (1), we have

$$P_{\mathfrak{M}_j} x_k = \begin{cases} 0 & \text{for } j \neq k, \\ x_k & \text{for } j = k. \end{cases}$$

It follows that

$$Qx = P_{\mathfrak{M}_1} x + P_{\mathfrak{M}_2} x + \cdots + P_{\mathfrak{M}_n} x = x_1 + x_2 + \cdots + x_n = x,$$

and therefore $x \in \mathfrak{M}$. Thus, we also have $\mathfrak{M}_1 \oplus \mathfrak{M}_2 \oplus \cdots \oplus \mathfrak{M}_n \subset \mathfrak{M}$ and comparison with (2) yields

$$\mathfrak{M}_1 \oplus \mathfrak{M}_2 \oplus \cdots \oplus \mathfrak{M}_n = \mathfrak{M}.$$

VI. *The relation* $\mathfrak{M}_1 \supset \mathfrak{M}_2$ *is equivalent to each of the relations*: a) $P_{\mathfrak{M}_1} P_{\mathfrak{M}_2} = P_{\mathfrak{M}_2} P_{\mathfrak{M}_1} = P_{\mathfrak{M}_2}$; *b*) $|P_{\mathfrak{M}} x| \leqq |P_{\mathfrak{M}_1} x|$ *for all* $x \in \mathfrak{H}$.

Proof. Suppose $\mathfrak{M}_1 \supset \mathfrak{M}_2$; then $P_{\mathfrak{M}_2}x \subset \mathfrak{M}_1$, and therefore

$$P_{\mathfrak{M}_1} P_{\mathfrak{M}_2} x = P_{\mathfrak{M}_2} x \text{ for all } x \in \mathfrak{H}.$$

This means that $P_{\mathfrak{M}_1}P_{\mathfrak{M}_2} = P_{\mathfrak{M}_2}$; applying the * operation to both members of this equality, we obtain that also $P_{\mathfrak{M}_2}P_{\mathfrak{M}_1} = P_{\mathfrak{M}_2}$.

Thus, $\mathfrak{M}_1 \supset \mathfrak{M}_2$ implies a). Now suppose a) holds. Then $|P_{\mathfrak{M}_2}x| = |P_{\mathfrak{M}_2}P_{\mathfrak{M}_1}x| \leqq |P_{\mathfrak{M}_1}x|$; consequently, b) follows from a). Finally, suppose b) holds. We note that in virtue of the relation $|x|^2 = |P_{\mathfrak{M}_1}x|^2 + |(1-P_{\mathfrak{M}_1})x|^2$ the subspace $\mathfrak{M}_1$ is the set of all vectors x for which $|x| = |P_{\mathfrak{M}_1}x|$ and $\mathfrak{M}_2$ is the set of all vectors x for which $|x| = |P_{\mathfrak{M}_2}x|$. But, it follows from b) that if $|x| = |P_{\mathfrak{M}_2}x|$, then $|x| = |P_{\mathfrak{M}_1}x|$; this means that if $x \in \mathfrak{M}_2$, then $x \in \mathfrak{M}_1$, i.e. that $\mathfrak{M}_2 \subset \mathfrak{M}_1$.

VII. *The difference $P_{\mathfrak{M}_1} - P_{\mathfrak{M}_2}$ of two projection operators is a projection operator if and only if $\mathfrak{M}_2 \subset \mathfrak{M}_1$. In this case, we have*

$$P_{\mathfrak{M}_1} - P_{\mathfrak{M}_2} = P_{\mathfrak{M}_1 - \mathfrak{M}_2}.$$

Proof. If $\mathfrak{M}_2 \subset \mathfrak{M}_1$, then according to VI, $P_{\mathfrak{M}_1}P_{\mathfrak{M}_2} = P_{\mathfrak{M}_2}P_{\mathfrak{M}_1} = P_{\mathfrak{M}_2}$; it easily follows that $(P_{\mathfrak{M}_1} - P_{\mathfrak{M}_2})^2 = P_{\mathfrak{M}_1} - P_{\mathfrak{M}_2}$, so that $P_{\mathfrak{M}_1} - P_{\mathfrak{M}_2}$ is a projection operator. Conversely, suppose $P_{\mathfrak{M}_1} - P_{\mathfrak{M}_2} = P_{\mathfrak{M}}$ is a projection operator on some subspace $\mathfrak{M}$. Then

$$P_{\mathfrak{M}_1} = P_{\mathfrak{M}_2} + P_{\mathfrak{M}} \tag{3}$$

and, consequently, by V, we have

$$O = P_{\mathfrak{M}_2} P_{\mathfrak{M}} = P_{\mathfrak{M}_2}(P_{\mathfrak{M}_1} - P_{\mathfrak{M}_2}) = P_{\mathfrak{M}_2} P_{\mathfrak{M}_1} - P_{\mathfrak{M}_2};$$

by VI, it follows that $\mathfrak{M}_2 \subset \mathfrak{M}_1$. Moreover, we conclude from (3) that $\mathfrak{M}_1 = \mathfrak{M}_2 \oplus \mathfrak{M}$ and therefore $\mathfrak{M} = \mathfrak{M}_1 \ominus \mathfrak{M}_2$.

VIII. *If $\mathfrak{M}_1, \mathfrak{M}_2, \mathfrak{M}_3, \cdots$ is a denumerable set of mutually orthogonal subspaces, then for arbitrary vector $x \in \mathfrak{H}$ the series*

$$P_{\mathfrak{M}_1} x + P_{\mathfrak{M}_2} x + P_{\mathfrak{M}_3} x + \cdots \tag{4}$$

converges and its sum equals $P_{\mathfrak{M}} x$, where

$$\mathfrak{M} = \mathfrak{M}_1 \oplus \mathfrak{M}_2 \oplus \cdots.$$

Proof. We set $P_{\mathfrak{M}_k} x = x_k$; the vectors x_k are mutually orthogonal and

$$|x_1|^2 + \cdots + |x_n|^2 = (P_{\mathfrak{M}_1} x, x) + \cdots + (P_{\mathfrak{M}_n} x, x) = (P_{\mathfrak{M}_1 \oplus \cdots \oplus \mathfrak{M}_n} x, x) \leqq |x|^2;$$

consequently, the series $\sum_{k=1}^{\infty} |x_k|^2$ and therefore series (4) converges. We denote its sum by Px. Passing to the limit in the relations

$$\begin{aligned}(P_{\mathfrak{M}_1} x + \cdots + P_{\mathfrak{M}_n} x, y) &= (x, P_{\mathfrak{M}_1} y + \cdots + P_{\mathfrak{M}_n} y) \\ &= (P_{\mathfrak{M}_1} x + \cdots + P_{\mathfrak{M}_n} x, P_{\mathfrak{M}_1} y + \cdots + P_{\mathfrak{M}_n} y),\end{aligned}$$

which express the fact that $P_{\mathfrak{M}_1}+\cdots+P_{\mathfrak{M}_n}$ is a projection operator (see I and II), we conclude that

$$(Px, y) = (x, Py) = (Px, Py).$$

Consequently, $P = P^* = P^2$ and P is a projection operator. Suppose $P = P_{\mathfrak{M}}$; clearly, $Px \in \mathfrak{M}_1 \oplus \mathfrak{M}_2 \oplus \cdots$, and therefore

$$\mathfrak{M} \subset \mathfrak{M}_1 \oplus \mathfrak{M}_2 \oplus \cdots. \tag{5}$$

Conversely, if $x \in \mathfrak{M}_1 \oplus \mathfrak{M}_2 \oplus \cdots$, then, setting $x = x_1+x_2+\cdots$, $x_k \in \mathfrak{M}_k$, we easily conclude that $Px = x$; consequently, $x \in \mathfrak{M}$. Thus, $\mathfrak{M}_1 \oplus \mathfrak{M}_2 \oplus \cdots \subset \mathfrak{M}$ and comparison with (5) yields $\mathfrak{M} = \mathfrak{M}_1 \oplus \mathfrak{M}_2 \oplus \cdots$.

Applying similar reasoning, one can prove the following proposition.

IX. *If $\mathfrak{M}_\nu$ is a system of mutually orthogonal subspaces then for arbitrary vector $x \in \mathfrak{H}$ the series*

$$\sum_\nu P_{\mathfrak{M}_\nu} x$$

contains only a denumerable number of terms different from zero and converges. Its sum equals $P_{\mathfrak{M}}x$, where $\mathfrak{M} = \sum_\nu \oplus \mathfrak{M}_\nu$.

13. Reducibility. Suppose $\mathfrak{M}$ is a closed subspace in $\mathfrak{H}$ and that P is a projection operator on $\mathfrak{M}$. We shall say that *the subspace $\mathfrak{M}$ reduces the operator A* (or that *the operator A is reduced by the subspace $\mathfrak{M}$*) if $x \in \mathfrak{D}_A$ implies that also $Px \in \mathfrak{D}_A$ and

$$APx = PAx.$$

I. *A bounded operator A, defined on the entire space $\mathfrak{H}$, is reduced by a closed subspace $\mathfrak{M}$ if and only if it commutes with the projection operator P on this subspace.*

This assertion follows directly from the very definition of reducibility.

II. *A bounded Hermitian operator A, defined on the entire space $\mathfrak{H}$, is reducible by the closed subspace $\mathfrak{M}$ if and only if $\mathfrak{M}$ is invariant with respect to A, i.e. if $x \in \mathfrak{M}$ implies $Ax \in \mathfrak{M}$.*

Proof. The necessity of this condition is obvious. We shall prove its sufficiency. Thus, suppose $\mathfrak{M}$ is invariant with respect to A. Since $P_{\mathfrak{M}}x \in \mathfrak{M}$, we have $AP_{\mathfrak{M}}x \in \mathfrak{M}$; consequently, $P_{\mathfrak{M}}AP_{\mathfrak{M}}x = AP_{\mathfrak{M}}x$ for all $x \in \mathfrak{H}$, i.e.

$$P_{\mathfrak{M}}AP_{\mathfrak{M}} = AP_{\mathfrak{M}}. \tag{1}$$

Going over to the adjoint operator in both members of this equality, we obtain

$$P_{\mathfrak{M}}AP_{\mathfrak{M}} = P_{\mathfrak{M}}A. \tag{2}$$

Comparison of (1) with (2) yields $AP_{\mathfrak{M}} = P_{\mathfrak{M}}A$, so that A and $P_{\mathfrak{M}}$ commute. It follows, in virtue of I, that $\mathfrak{M}$ reduces A.

14. Partially isometric operators. We shall say that an operator U from $\mathfrak{H}_1$ into $\mathfrak{H}_2$ is *partially isometric* if it is isometric on some closed subspace $\mathfrak{M}$ of the space $\mathfrak{H}_1$ and equal to zero on its orthogonal complement $\mathfrak{H}_1 \ominus \mathfrak{M}$. The subspace $\mathfrak{M}$ is called the *initial domain* and its U-image in $\mathfrak{H}_2$ is called the *terminal domain* of the operator U. It is easily verified that $U^*U = P_{\mathfrak{M}}$, $UU^* = P_{\mathfrak{N}}$, where $\mathfrak{N}$ is the terminal domain of the operator U and $P_{\mathfrak{M}}$, $P_{\mathfrak{N}}$ are projection operators in $\mathfrak{H}_1$, $\mathfrak{H}_2$ on $\mathfrak{M}$ and $\mathfrak{N}$ respectively.

In fact, U^*U is a self-adjoint operator in $\mathfrak{H}_1$ which is equal to zero in $\mathfrak{H}_1 \ominus \mathfrak{M}$. Therefore $(U^*Ux, y) = (x, U^*Uy) = 0$ for any $x \in \mathfrak{H}_1$, $y \in \mathfrak{H}_1 \ominus \mathfrak{M}$, and hence $U^*Ux \in \mathfrak{M}$ for all $x \in \mathfrak{H}_1$; therefore, for $x \in \mathfrak{M}$, we also have $x - U^*Ux \in \mathfrak{M}$. On the other hand,

$$(x, y) = (Ux, Uy) = (U^*Ux, y), (x - U^*Ux, y) = 0$$

for all $x \in \mathfrak{M}$, $y \in \mathfrak{M}$. Consequently, the vector $x - U^*Ux$ is orthogonal to $\mathfrak{M}$. But then $x = U^*Ux$, i.e. $U^*U = 1$ on $\mathfrak{M}$, $U^*U = P_{\mathfrak{M}}$. We prove analogously that $UU^* = P_{\mathfrak{N}}$. At the same time we see that U^* is a partially isometric operator from $\mathfrak{H}_2$ into $\mathfrak{H}_1$ with initial domain $\mathfrak{N}$ and terminal domain $\mathfrak{M}$.

The operator U is partially isometric if, and only if, any one of the following conditions is satisfied:

1) *U^*U is a projection operator*;
2) *$UU^*U = U$*;
3) *UU^* is a projection operator*;
4) *$U^*UU^* = U^*$*.

In fact, if U is a partially isometric operator with initial domain $\mathfrak{M}$, then $U^*U = P_{\mathfrak{M}}$ is a projection operator so that condition 1) is satisfied. Conversely, suppose U^*U is a projection operator and that $\mathfrak{M}$ is a subspace in $\mathfrak{H}$ onto which U^*U projects. Then $U^*Ux = 0$ for $x \in \mathfrak{H} \ominus \mathfrak{M}$ and, consequently, also

$$|Ux|^2 = (U^*Ux, x) = 0, Ux = 0.$$

On the other hand, for $x \in \mathfrak{M}$ we have

$$U^*Ux = x, |Ux|^2 = (U^*Ux, x) = (x, x) = |x|^2,$$

so that U is a partially isometric operator. Condition 2) is verified directly; conversely, if 2) holds then, operating on the left by U^*, we obtain that $(U^*U)^2 = U^*U$, i.e. 1) holds. Conditions 3) and 4) are obtained by replacing the operator U by the operator U^*.

15. Matrix representation of an operator. Suppose $\mathfrak{A}$, $\mathfrak{B}$ are two sets; let the spaces $\mathfrak{H}_\alpha$ correspond to the elements α of the set $\mathfrak{A}$ and the spaces $\mathfrak{H}'_\beta$ to the elements β of the set $\mathfrak{B}$. We shall now see how to represent the bounded operators from $\mathfrak{H} = \sum_{\alpha\in\mathfrak{A}} \oplus \mathfrak{H}_\alpha$ into $\mathfrak{H}' = \sum_{\beta\in\mathfrak{B}} \oplus \mathfrak{H}'_\beta$ with the aid of the bounded operators from $\mathfrak{H}_\alpha$ into $\mathfrak{H}'_\beta$.

Suppose A is such an operator; we consider the element $x^{(\alpha_0)} = \{x_\alpha\}$ in which $x_\alpha = 0$ for $\alpha \neq \alpha_0$. Suppose $Ax^{(\alpha_0)} = \{y_\beta\}$; the inequality

$$|Ax|^2 \leqq C^2|x|^2 \tag{1}$$

assumes the form $\sum_{\beta\in\mathfrak{B}}|y_\beta|^2 \leqq C^2|x_{\alpha_0}|^2$ for $x = x^{(\alpha_0)}$. It follows that

$$|y_{\beta_0}| \leqq C|x_{\alpha_0}|. \tag{2}$$

We set

$$A_{\beta_0\alpha_0}x_{\alpha_0} = y_{\beta_0};$$

in virtue of inequality (2) $A_{\beta_0\alpha_0}$ is a bounded operator from $\mathfrak{H}_{\alpha_0}$ into $\mathfrak{H}'_{\beta_0}$ and

$$Ax^{(\alpha_0)} = \{A_{\beta_0\alpha_0}x_{\alpha_0}\}.$$

We thus see that the matrix $||A_{\beta\alpha}||$ of bounded operators from $\mathfrak{H}_\alpha$ into $\mathfrak{H}'_\beta$ corresponds to the operator A.

We shall now consider the properties of this matrix in more detail. An arbitrary element x in $\mathfrak{H}$ can be represented in the form

$$x = \sum_{\alpha\in\mathfrak{A}} x^{(\alpha)},$$

where the sum in the right member contains at most a denumerable number of nonzero elements. It follows, in virtue of the continuity of the operator A, that

$$Ax = \sum_{\alpha\in\mathfrak{A}} Ax^{(\alpha)} = \sum_{\alpha\in\mathfrak{A}} \{A_{\beta\alpha}x_\alpha\} = \{\sum_{\alpha\in\mathfrak{A}} A_{\beta\alpha}x_\alpha\}, \tag{3}$$

and condition (1), that the operator A is bounded, can be rewritten in the form

$$\sum_{\beta\in\mathfrak{B}} |\sum_{\alpha\in\mathfrak{A}} A_{\beta\alpha}x_\alpha|^2 \leqq C^2 \sum_{\alpha\in\mathfrak{A}} |x_\alpha|^2. \tag{4}$$

We shall say that the matrix $A_{\beta\alpha}$, satisfying inequality (4), is *bounded*.

Equality (3) thus establishes a one-to-one correspondence $A \sim ||A_{\beta\alpha}||$ between the set of all bounded operators A from $\mathfrak{H}$ into $\mathfrak{H}'$ and the set of all bounded matrices $||A_{\beta\alpha}||$ of operators from $\mathfrak{H}_\alpha$ into $\mathfrak{H}'_\beta$. Moreover, the norm of the operator A, as follows from its very definition, is the smallest of the numbers C for which inequality (4) is satisfied.

Suppose, in particular, that $||A_{\beta\alpha}||$ is a diagonal matrix, i.e. that $A_{\beta\alpha} = 0$ for $\beta \neq \alpha$. Then inequality (4) assumes the form

$$\sum_\alpha |A_{\alpha\alpha}x_\alpha|^2 \leqq C^2 \sum_\alpha |x_\alpha|^2.$$

We set

$$C' = \sup_\alpha |A_{\alpha\alpha}|.$$

Then

$$|A_{\alpha\alpha}x_\alpha|^2 \leqq |A_{\alpha\alpha}|^2|x_\alpha|^2 \leqq C'^2|x_\alpha|^2$$

and consequently

$$\sum_{\alpha} |A_{\alpha\alpha} x_\alpha|^2 \leqq C'^2 \sum_{\alpha} |x_\alpha|^2.$$

It follows that

$$|A|^2 \leqq C'^2.$$

On the other hand, for given $\varepsilon > 0$ there exists an α_0 such that

$$|A_{\alpha_0\alpha_0}|^2 > C'^2 - \varepsilon;$$

hence there exists a vector $x^0_{\alpha_0} \in \mathfrak{H}_{\alpha_0}$ such that

$$|A_{\alpha_0\alpha_0} x^0_{\alpha_0}|^2 > (C'^2 - \varepsilon)|x^0_{\alpha_0}|^2.$$

But then inequality (4) is not satisfied for $C^2 = C'^2 - \varepsilon$, $x_{\alpha_0} = x^0_{\alpha_0}$, $x_\alpha = 0$ for $\alpha \neq \alpha_0$, and consequently C'^2 is the smallest of the numbers C^2 for which inequality (4) is satisfied. In other words, C coincides with the norm of the operator A, i.e.

$$|A| = \sup_{\alpha} |A_{\alpha\alpha}|.$$

If $A \sim ||A_{\beta\alpha}||$, $B \sim ||B_{\beta\alpha}||$, $C \sim ||C_{\beta\alpha}||$, *then*

$$A + B \sim ||A_{\beta\alpha} + B_{\beta\alpha}||,\ aA \sim ||aA_{\beta\alpha}||,$$
$$A^* \sim ||A^*_{\alpha\beta}||,\ CA \sim ||\sum_{\beta \in \mathfrak{B}} C_{\gamma\beta} A_{\beta\alpha}||,$$

where the series $\sum_{\beta \in \mathfrak{B}} C_{\gamma\beta} A_{\beta\alpha} x_\alpha$ *converges strongly for arbitrary vector* $x_\alpha \in \mathfrak{H}_\alpha$. All these relations are verified directly and we shall leave the proofs for the reader.

We shall further consider the particular case when all the spaces $\mathfrak{H}_\alpha$, $\mathfrak{H}'_\beta$ are one-dimensional. We choose the vectors e_α, e'_β, of unit length, in the spaces $\mathfrak{H}_\alpha$, $\mathfrak{H}'_\beta$. Since $A_{\beta\alpha} e_\alpha \in \mathfrak{H}'_\beta$, the vector $A_{\beta\alpha} e_\alpha$ is a multiple of the vector e'_β; we set

$$A_{\beta\alpha} e_\alpha = a_{\beta\alpha} e'_\beta.$$

The operator $A_{\beta\alpha}$ is completely defined by the numbers $a_{\beta\alpha}$; consequently, the matrix $||A_{\beta\alpha}||$ and hence also the operator A are completely defined by the numerical matrix $||a_{\beta\alpha}||$. Every vector $x \in \mathfrak{H}$ has the form

$$x = \langle \xi_\alpha e_\alpha \rangle,$$

where at most a denumerable set of the numbers ξ_α is different from zero and $\sum_\alpha |\xi_\alpha|^2 < \infty$; therefore formula (3) can be rewritten in the form

$$Ax = \{(\sum_{\alpha \in \mathfrak{A}} a_{\beta\alpha} \xi_\alpha) e'_\beta\}.$$

The properties formulated above of the correspondence $A \sim ||A_{\beta\alpha}||$ now mean that to scalar multiplication, addition, and multiplication of operators A there correspond analogous operations on the corresponding matrices $||a_{\beta\alpha}||$.

§ 6. Integration on locally compact spaces

1. Fundamental concepts; formulation of the problem. Suppose T is a locally compact Hausdorff topological space. We agree to denote by L the set of all complex-valued functions x, continuous on T, and equal to zero in the exterior of some compact set, its own for each function, and further we denote by L^r the set of all real-valued, and by L^+ the set of all nonnegative real-valued, functions in L. If it is necessary to emphasize that L, L^r and L^+ are referred namely to T, then we shall write $L(T)$, $L^r(T)$, $L^+(T)$ instead of L, L^r, L^+.

Clearly, L and L^r are (complex respectively real) linear and L^r contains, together with the function x, also $|x|$ and $x \cap 1 = \min\{x, 1\}$; therefore, together with every two functions x, y, it also contains

$$x \cap y = \min(x, y) = \tfrac{1}{2}(x+y-|x-y|)$$

and

$$x \cup y = \max(x, y) = \tfrac{1}{2}(x+y+|x-y|).$$

Further, it is obvious that if $x, y \in L^+$ and $c \geqq 0$, then also cx, $x+y$, $x \cap y$, $x \cup y$ are $\in L^+$. It is easy to see that $L^r = L^+ - L^+$. In fact, if $x \in L^r$, then, setting

$$x^+ = x \cup 0, \qquad x^- = (-x) \cup 0,$$

we have $x^+ \in L^+$, $x^- \in L^+$ and $x = x^+ - x^-$.

An *integral* on L is any linear functional I which satisfies the condition

$$I(x) \geqq 0 \text{ for } x \in L^+.$$

It is obvious that if $x \in L^r$, then $I(x)$ is real and if $x, y \in L^r$, then $I(x) \leqq I(y)$ for $x \leqq y$.

The fundamental problem consists in the extension of the functional I to a more comprehensive function space than L. If, for example, we take the interval $[0, 1]$ with the usual topology for T, and let the Riemann integral $\int_0^1 x(t)dt$ play the role of $I(x)$, then the method of extension described below leads to the Lebesgue integral on the interval $[0, 1]$.

2. Fundamental properties of the integral. We shall call the smallest compact set $Q_x \subset T$ in the exterior of which $x(t) = 0$ the *carrier* (or *support*) of the function $x \in L$; we set

$$||x|| = \sup_{t \in Q_x} |x(t)| = \sup_{t \in T} |x(t)|.$$

In the propositions discussed below, I denotes an integral on L. We note that the equality $(x_1, x_2) = I(x_1 \bar{x}_2)$ defines a positive definite Hermitian form on L (see subsection 1, § 5).

I. *For arbitrary function $x \in L$, we have*

$$|I(x)| \leqq I(|x|). \tag{1}$$

Proof. If $I(x) = 0$, then (1) is obvious. We shall therefore assume that $I(x) \neq 0$ and set $I(x) = \rho e^{i\varphi}$, $\rho > 0$, $0 \leqq \varphi < 2\pi$. If we put $e^{-i\varphi} x = y_1 + iy_2$, where $y_1, y_2 \in L^r$, then $|I(x)| = \rho = I(e^{-i\varphi}x) = I(y_1) + iI(y_2) = I(y_1) \leqq I(|y_1|) \leqq I(|x|)$. (See the similar argument in the proof of Theorem 2, subsection 9, § 1.)

II. *For arbitrary compact set $Q \subset T$, there exists a constant C_Q which depends only on Q such that*

$$|I(x)| \leqq C_Q \|x\| \tag{2}$$

for all $x \in L$, which satisfy the condition $Q_x \subset Q$.

Proof. On the basis of the Urysohn lemma (see II, subsection 8, § 2) there exists a function $y_Q(t) \in L^+$ nowhere greater than unity and equal to unity on Q. It follows from $|x| \leqq y_Q \|x\|$ and (1) that $|I(x)| \leqq I(|x|) = \|x\| I(y_Q)$, and consequently (2) holds for $C_Q = I(y_Q)$.

III. *If X is some set, directed downward, of functions $x \in L^+$, partially ordered with the aid of the relation $\leqq$, and if*

$$\inf_{x \in X} x = 0, \tag{3}$$

then for every $\varepsilon > 0$ there exists a function $x_\varepsilon \in X$ such that $x(t) < \varepsilon$ for all functions x in X satisfying the condition $x(t) \leqq x_\varepsilon(t)$, and therefore

$$\inf_{x \in X} I(x) = 0. \tag{4}$$

Proof. We set $T_x = \{t: x(t) \geqq \varepsilon\}$ for $x \in X$ and fixed ε; then T_x is compact, $T_{x_2} \subset T_{x_1}$ for $x_2 \leqq x_1$, and in virtue of (3) the intersection of all the T_x is void. On the basis of I, subsection 6, § 2, there exists a function $x_\varepsilon \in X$ such that $T_{x_\varepsilon} = \phi$, and consequently, also $T_x = \phi$ for all $x \leqq x_\varepsilon$, $x \in X$. This means that $x(t) < \varepsilon$ for all $x \leqq x_\varepsilon$. Since $Q_x \subset Q_{x_\varepsilon} \subset Q_{x_1}$ for $\varepsilon < 1$, $x < x_\varepsilon$, (2) now yields $I(x) \leqq C_{Q_{x_1}} \cdot \varepsilon$; this implies (4).

3. Extension of the integral to lower semi-continuous functions. We shall now consider real-valued functions which can also assume the values $+\infty$, $-\infty$. Such a function x is said to be *lower semi-continuous* if for arbitrary t_0 and arbitrary $h < x(t_0)$ there exists a neighborhood $U(t_0)$ such that

$$h < x(t) \text{ for all } t \in U(t_0). \tag{1}$$

Analogously, the function x is said to be *upper semi-continuous* if for arbitrary point t_0 and arbitrary $h > x(t_0)$ there exists a neighborhood $V(t_0)$ such that

$$h > x(t) \text{ for all } t \in V(t_0).$$

Clearly, every continuous real-valued function is both lower and upper semi-continuous.

We shall denote the set of all nonnegative lower semi-continuous functions by M^+ and the set of all nonnegative upper semi-continuous functions by N^+. We shall agree that $a+\infty = +\infty+a = +\infty$ for all $a > -\infty$, that $a\cdot(+\infty) = +\infty$ for arbitrary $a > 0$ and that $0\cdot(+\infty) = 0$. For an arbitrary set $X \subset M^+$ we set

$$\sum_{x\in X} x = \sup\,\{y\}, \tag{1'}$$

where $\{y\}$ is the set of all possible finite sums of functions $x \in X$.

I. 1) *If* $x \in M^+$ *and* $\alpha \geqq 0$, *then also* $\alpha x \in M^+$;

2) *if* $x_1, \cdots, x_n \in M^+$, *then also* $x_1 \cap \cdots \cap x_n \in M^+$;

3) *if* $X \subset M^+$, *then* $\sup_{x\in X} x \in M^+$ *and* $\sum_{x\in X} x \in M^+$.

Proof. Assertion 1) is obvious. Further, if $h < (x_1 \cap \cdots \cap x_n)(t_0)$, then $h < x_k(t_0)$ for $k = 1, \cdots, n$; hence there exist neighborhoods $U_k(t_0)$ such that $h < x_k(t)$ for $t \in U_k(t_0)$ and therefore

$$h < (x_1 \cap \cdots \cap x_n)(t) \text{ for } t \in \bigcap_{k=1}^{n} U_k(t_0);$$

this proves the validity of assertion 2). Now let $h < \sup_{x\in X} x(t_0)$; then $h < x_0(t_0)$ for some $x_0 \in X$; therefore, there exists a neighborhood $U(t_0)$ such that $h < x_0(t)$ for $t \in U(t_0)$ and *a fortiori* $h < \sup_{x\in X} x(t)$ for $t \in U(t_0)$. This proves that $\sup_{x\in X} x(t) \in M^+$. Finally, let $x_1, x_2 \in M^+$. If $h < x_1(t_0)+x_2(t_0)$, then h can be written in the form $h = h_1+h_2$, where $h_1 < x_1(t_0)$, $h_2 < x_2(t_0)$. Then there exist neighborhoods $U_1(t_0)$, $U_2(t_0)$ such that $h_1 < x_1(t)$ for $t \in U_1(t_0)$, $h_2 < x_2(t)$ for $t \in U_2(t_0)$ and therefore $h = h_1+h_2 < x_1(t)+x_2(t)$ for $t \in U_1(t_0) \cap U_2(t_0)$. Hence $x_1+x_2 \in M^+$. By induction, we deduce from this that the sum of an arbitrary finite number of functions in M^+ also belongs to M^+. As was indicated above, by the sum of an arbitrary number of functions in M^+ we understand the least upper bound of the sums of a finite number of such functions; therefore this sum also belongs to M^+.

II. *Every function* x *in* M^+ *is the least upper bound of the set* Y_x *of all functions* $y \in L^+$ *which satisfy the condition* $y \leqq x$.

Proof. The assertion is obvious for points at which $x(t)$ vanishes. But if $x(t_0) > 0$, then for arbitrary h, $0 < h < x(t_0)$ there exists a neighborhood $U(t_0)$ in which $x(t) > h$. On the other hand, in virtue of the Urysohn lemma there exists a function $y_0(t) \in L^+$ with carrier $Q_{y_0} \subset U(t_0)$, satisfying the conditions: $y_0(t_0) = h$, $y_0(t) \leqq h$. Since $0 \leqq y \leqq x$ and h is arbitrary in $(0, x(t_0))$, it then follows that $\sup_{y\in Y_x} y(t_0) = x(t_0)$.

The *upper integral* $\bar{I}(x)$ of the function $x \in M^+$ is defined by

$$\bar{I}(x) = \sup_{y\in Y_x} I(y),$$

where Y_x is the set of all functions $y \in L^+$, satisfying the condition $y \leqq x$. If $x \in L^+$, then Y_x contains x and therefore $\bar{I}(x) = I(x)$. Further, it follows directly from the definition that

$$\bar{I}(x_1) \leqq \bar{I}(x_2) \text{ for } x_1 \leqq x_2, x_1, x_2 \in M^+ \tag{2}$$

and that

$$\bar{I}(cx) = c\bar{I}(x) \text{ for } x \in M^+, c > 0.$$

III. *Suppose X is an arbitrary set in M^+, partially ordered with the aid of the relation $\leqq$ and directed upward. Then*

$$\bar{I}(\sup_{x \in X} x) = \sup_{x \in X} \bar{I}(x).$$

Proof. We set $x_0 = \sup_{x \in X} x$. If $X \subset L^+$ and $x_0 \in L^+$, then the assertion follows from III, subsection 2, because then $\inf_{x \in X} (x_0 - x) = 0$, and consequently $I(x_0) - \sup_{x \in X} I(x) = \inf_{x \in X} I(x_0 - x) = 0$. Now suppose X is an arbitrary set in M^+. In virtue of (2), $\bar{I}(x_0) \geqq \bar{I}(x)$ for all $x \in X$, and therefore $\bar{I}(x_0) \geqq \sup_{x \in X} \bar{I}(x)$. Hence we must prove the reverse inequality. By the definition of $\bar{I}(x_0)$, for this it is sufficient to show that $I(y_0) \leqq \sup_{x \in X} \bar{I}(x)$ for all $y_0 \in Y_{x_0}$, i.e. for all $y_0 \in L^+$, satisfying the inequality $y_0 \leqq x_0$. To this end, we denote by Y the union of all Y_x, $x \in X$. Clearly, $Y \subset Y_{x_0}$ and

$$x_0 = \sup_{x \in X} x = \sup_{x \in X} (\sup_{y \in Y_x} y) = \sup_{y \in Y} y. \tag{3}$$

Suppose $y_0 \in Y_{x_0}$, i.e. $y_0 \leqq x_0$. We conclude from (3) that

$$y_0 = y_0 \cap x_0 = \sup_{y \in Y} (y_0 \cap y);$$

consequently, in virtue of what was stated in the beginning of the proof,

$$I(y_0) = \sup_{y \in Y} I(y_0 \cap y).$$

Thus, taking (3) into consideration, we have

$$I(y_0) = \sup_{y \in Y} I(y_0 \cap y) \leqq \sup_{x \in Y} I(y) = \sup_{x \in X} \sup_{y \in Y_x} I(y) = \sup_{x \in X} \bar{I}(x),$$

and the proposition is completely proved.

IV. *If $x_1, x_2 \in M^+$, then $\bar{I}(x_1 + x_2) = \bar{I}(x_1) + \bar{I}(x_2)$.*

Proof. The functions $y_1 + y_2$, where $y_1 \leqq x_1$, $y_2 \leqq x_2$, $y_1, y_2 \in L^+$, form a set which is directed upward in M^+, is partially ordered with the aid of the relation $\leqq$ and satisfies the condition $\sup (y_1 + y_2) = x_1 + x_2$. It follows, in virtue of III, that

$$\bar{I}(x_1 + x_2) = \sup [I(y_1) + I(y_2)] = \bar{I}(x_1) + \bar{I}(x_2).$$

V. *If $X \subset M^+$, then*

$$\bar{I}(\sum_{x \in X} x) = \sum_{x \in X} \bar{I}(x). \tag{4}$$

Proof. For finite sums, (4) is obtained by induction from IV and, in virtue of III, from this (4) follows for an arbitrary sum because it is the least upper bound of finite sums.

4. Upper integral of an arbitrary nonnegative real-valued function. We shall now consider arbitrary nonnegative real-valued functions which can also assume the value $+\infty$. For any such function x we denote by Z_x the set of all functions z in M^+ satisfying the condition $z \geqq x$; such functions always exist, for example, the function $z(t) = +\infty$ for all $t \in T$. We shall call

$$\bar{I}(x) = \inf_{z \in Z_x} \bar{I}(z)$$

the *upper integral* of the function x. If $x \in M^+$, then $x \in Z_x$, and therefore, for $x \in M^+$ this definition of the upper integral coincides with the former. It is clear from the definition that

$$\bar{I}(x_1) \leqq \bar{I}(x_2) \text{ for } x_1 \leqq x_2 \tag{1}$$

and

$$\bar{I}(cx) = c\bar{I}(x) \text{ for } c \geqq 0. \tag{2}$$

Moreover,

$$\bar{I}(x_1+x_2) \leqq \bar{I}(x_1)+\bar{I}(x_2). \tag{3}$$

In fact, if $z_1 \geqq x_1$, $z_2 \geqq x_2$, then $z_1+z_2 \leqq x_1+x_2$, and therefore $\bar{I}(x_1+x_2) \leqq \bar{I}(z_1+z_2) = \bar{I}(z_1)+\bar{I}(z_2)$. Passing to the greatest lower bounds with respect to z_1, z_2, we obtain (3). We note further the following properties of the upper integral.

I. *For an arbitrary nondecreasing sequence of functions $x_n \geqq 0$, we have*

$$\bar{I}(\sup_n x_n) = \sup_n \bar{I}(x_n). \tag{4}$$

Proof. Since $x_n \leqq \sup_n x_n$, we have $\bar{I}(x_n) \leqq \bar{I}(\sup_n x_n)$, and therefore $\sup_n \bar{I}(x_n) \leqq \bar{I}(\sup_n x_n)$; consequently, it suffices to prove the inequality in the other direction. This inequality is obvious if $\sup_n \bar{I}(x_n) = +\infty$; therefore, we consider the case when $\sup_n \bar{I}(x_n) < +\infty$. Suppose ε is an arbitrary positive number. By the definition of the upper integral there exist functions $z_n \in M^+$ such that $z_n \geqq x_n$ and $\bar{I}(z_n) < \bar{I}(x_n)+\varepsilon/2^n$. We set $u_1 = z_1$, $u_n = z_1 \cup z_2 \cup \cdots \cup z_n (n > 1)$. Then $u_{n+1} \geqq u_n, u_n \geqq x_n$ and $u_{n+1}+u_n \cap z_{n+1} = u_n \cup z_{n+1}+u_n \cap z_{n+1} = u_n+z_{n+1}$ from which it follows that $\bar{I}(u_{n+1})+\bar{I}(u_n \cap z_{n+1}) = \bar{I}(u_n)+\bar{I}(z_{n+1})$. Consequently, $\bar{I}(u_{n+1}) = \bar{I}(u_n)+\bar{I}(z_{n+1})-\bar{I}(u_n \cap z_{n+1}) \leqq \bar{I}(u_n)+\bar{I}(z_{n+1})-\bar{I}(x_n) \leqq \bar{I}(u_n) + \bar{I}(x_{n+1})-\bar{I}(x_n)+\varepsilon/2^{n+1}$. Adding these inequalities termwise for $n = 1, 2, \cdots, m-1$, we obtain that $\bar{I}(u_m) \leqq \bar{I}(x_m)+\varepsilon$, and consequently, in virtue of III, subsection 3,

$$\sup_m \bar{I}(x_m)+\varepsilon \geqq \bar{I}(\sup_m u_m) \geqq \bar{I}(\sup_m x_m)$$

Since the number $\varepsilon > 0$ is arbitrary, this implies formula (4).

Clearly, (4) can also be written in the form

$$\bar{I}(\lim_{n\to\infty} x_n) = \lim_{n\to\infty} \bar{I}(x_n). \tag{5}$$

II. *For an arbitrary finite or denumerable number of functions* $x_n \geqq 0$,

$$\bar{I}(\sum_n x_n) \leqq \sum_n \bar{I}(x_n). \tag{6}$$

Proof. For a finite number of terms, the assertion is obtained by induction from (3); and then for a denumerable number of terms it is obtained if one applies (5) to the finite sums $\sum_{k=1}^{n} x_k$ instead of to x_n.

A function $x(t) \geqq 0$ is called a *zero* function if $\bar{I}(x) = 0$. From (1), (2), (4) and (6), we conclude that

III. *If* $0 \leqq x_1 \leqq x_2$ *and* x_2 *is a zero function, then* x_1 *and* αx_2 *for* $\alpha \geqq 0$ *are also zero functions; the least upper bound of a nondecreasing sequence of zero functions is a zero function; the sum of a finite or a denumerable number of zero functions is a zero function.*

5. Exterior measure of a set. The *characteristic function* of a set $A \subset T$ is the function

$$\xi_A(t) = \begin{cases} 1 & \text{for } t \in A, \\ 0 & \text{for } t \notin A. \end{cases}$$

The number $\bar{\mu}(A) = \bar{I}(\xi_A)$ is called the *exterior measure* of the set A.

Since $\xi_{A_1} \leqq \xi_{A_2}$ for $A_1 \subset A_2$, then in virtue of (1), subsection 4, we have

$$\bar{\mu}(A_1) \leqq \bar{\mu}(A_2) \text{ for } A_1 \subset A_2. \tag{1}$$

Further, it follows from the relation $\xi_{A_1 \cup A_2 \cup \cdots} \leqq \xi_{A_1} + \xi_{A_2} + \cdots$ and II, subsection 4, that for an arbitrary finite or denumerable number of sets $A_1, A_2, \cdots$,

$$\bar{\mu}(A_1 \cup A_2 \cup \cdots) \leqq \bar{\mu}(A_1) + \bar{\mu}(A_2) + \cdots. \tag{2}$$

Analogously, if $A_1 \subset A_2 \subset A_3 \subset \cdots$, then $\xi_{A_1} \leqq \xi_{A_2} \leqq \cdots$ and $\xi_{A_1 \cup A_2 \cup \cdots} = \sup_n \xi_{A_n}$; applying I, subsection 4, we conclude that

$$\bar{\mu}(A_1 \cup A_2 \cup \cdots) = \lim_n \bar{\mu}(A_n) \text{ for } A_1 \subset A_2 \subset \cdots. \tag{3}$$

I. *If* $\mathfrak{B}$ *is an arbitrary family of pairwise disjoint open sets* $V \subset T$, *then*

$$\bar{\mu}(\bigcup_{V \in \mathfrak{B}} V) = \sum_{V \in \mathfrak{B}} \bar{\mu}(V). \tag{4}$$

The assertion follows directly from Proposition V, subsection 3, because the characteristic function of an open set V is lower semi-continuous. [In fact, if

$t_0 \in V$, then some neighborhood $U(t_0) \subset V$, and therefore $1 = \xi_V(t_0) > h$ implies that $\xi_V(t) > h$ for $t \in U(t_0)$; but if $t_0 \notin V$, then, for arbitrary $t \in T$,

$$\xi_V(t) \geqq 0 > h.$$

Analogously, if A is closed then ξ_A is upper semi-continuous.]

Analogously, we conclude from III, subsection 3, that

II. *If $\mathfrak{B}$ is an arbitrary directed family of open sets V, ordered by the inclusion relation $\subset$, then*

$$\bar{\mu}(\bigcup_{V \in \mathfrak{B}} V) = \sup_{V \in \mathfrak{B}} \bar{\mu}(V).$$

III. *The exterior measure of any compact set Q is finite.*

In fact, on the basis of the Urysohn lemma, there exists a function $y \in L^+$, not greater than unity, which is equal to unity on Q. Then $\xi_Q \leqq y$, and therefore

$$\bar{\mu}(Q) = \bar{I}(\xi_Q) \leqq \bar{I}(y) = I(y) < +\infty.$$

IV. *For an arbitrary set A, the exterior measure $\bar{\mu}(A)$ is the greatest lower bound of the exterior measures $\bar{\mu}(U)$ of open sets $U \supset A$.*

Proof. The assertion is obvious if $\bar{\mu}(A) = +\infty$.

Suppose $\bar{\mu}(A) < +\infty$; then for every $\varepsilon > 0$, $\varepsilon < 1$, there exists a function $x \in M^+$ such that $\xi_A(t) \leqq x(t)$ and $\bar{I}(x) < \bar{\mu}(A) + \varepsilon$. We set $U = \{t: x(t) > 1-\varepsilon\}$; since $x \in M^+$, U is evidently open and $U \supset A$. On the other hand, $x \geqq (1-\varepsilon)\xi_U$ and therefore $\bar{\mu}(U) \leqq \{1/(1-\varepsilon)\}\bar{I}(x) < \{1/(1-\varepsilon)\}[\bar{\mu}(A)+\varepsilon]$; the assertion follows since ε is arbitrary.

A set A is called a *zero* set if $\bar{\mu}(A) = 0$. From (1) and (2), we conclude that

V. *Every subset of a zero set is a zero set; the union of a finite or denumerable number of zero sets is a zero set.*

Moreover,

VI. *The function $x(t) \geqq 0$ is a zero function if, and only if, $A = \{t: x(t) > 0\}$ is a zero set.*

Proof. If $x(t)$ is a zero function, then, since $\xi_A \leqq \sup_n (nx)$, in virtue of (4), subsection 4 we have $\bar{\mu}(A) \leqq \sup_n n\bar{I}(x) = \sup 0 = 0$. Conversely, if A is a zero set, then, since $x \leqq \sup_n n\xi_A$, we have $\bar{I}(x) \leqq \sup_n n\bar{I}(\xi_A) = 0$.

VII. *If $\bar{I}(x) < +\infty$, then $A = \{t: x(t) = +\infty\}$ is a zero set.*

Proof. Since $\xi_A \leqq (1/n)x$, we have

$$\bar{\mu}(A) = \bar{I}(\xi_A) \leqq \frac{1}{n}\bar{I}(x);$$

passing to the limit here as $n \to \infty$, we obtain that $\bar{\mu}(A) = 0$.

6. Equivalent functions. We shall say that a certain property holds *almost everywhere on T* if the set of all points $t \in T$ for which this property does not hold is a zero set.

Two complex-valued, finite functions x_1, x_2, defined everywhere, will be called *equivalent* if $x_1(t) = x_2(t)$ almost everywhere on T; we shall write $x_1 \sim x_2$ to indicate that x_1 and x_2 are equivalent. It easily follows from V, subsection 5, that all the properties of equivalence (see subsection 4, § 1) will be satisfied, so that the set of all complex-valued functions decomposes into classes of mutually equivalent functions.

It is easily seen that if $x_1 \sim x_2$, $y_1 \sim y_2$, then $cx_1 \sim cx_2$ (for arbitrary complex c) and $x_1+y_1 \sim x_2+y_2$.

Up to this point, we considered functions which are defined for all $t \in T$; for the sequel, it will be convenient to consider functions which are defined and finite only almost everywhere on T. As before, we shall say that two such functions x_1, x_2 are *equivalent*, and we shall write $x_1 \sim x_2$, if $x_1(t) = x_2(t)$ almost everywhere on T. The set of all complex-valued functions, defined and finite almost everywhere on T, also decomposes into classes of mutually equivalent functions.

Each such class ξ contains functions which are defined everywhere and finite. We define the sum $\xi+\eta$ of two classes ξ, η and the product $c\xi$ of the class ξ with the complex number c as the classes containing the sum $x+y$ and the product cx, where x, y are functions in ξ, η respectively, which are defined everywhere and finite. On the basis of what was stated above, this definition does not depend on the choice of x and y.

It is easily verified that all the linear space axioms will then be satisfied and, consequently, the classes of equivalent almost everywhere defined functions form a linear space.

We shall now extend the upper integral $\bar{I}(x)$ to all functions $x(t) \geqq 0$, which are defined almost everywhere on T, assuming that $\bar{I}(x) = \bar{I}(y)$, where y is an arbitrary nonnegative everywhere defined, finite function which is equivalent to x.

This definition does not depend on the choice of the function $y \sim x$. In fact, if y_1, y_2 are everywhere defined and finite, and $y_1 \sim x$ and $y_2 \sim x$, then $y_1 \sim y_2$ and, consequently, $\bar{I}(|y_1-y_2|) = 0$. Since $y_1 \leqq y_2+|y_1-y_2|$, $\bar{I}(y_1) \leqq \bar{I}(y_2)+\bar{I}(|y_1-y_2|) = \bar{I}(y_2)$ and analogously $\bar{I}(y_2) \leqq \bar{I}(y_1)$; hence $\bar{I}(y_1) = \bar{I}(y_2)$.

It is easily seen that the properties of the upper integral listed in subsection 4 remain valid: for example, if $x_1 \leqq x_2$ almost everywhere, then $\bar{I}(x_1) \leqq \bar{I}(x_2)$, and so forth.

We define $\bar{I}(|\xi|)$ by means of the formula

$$\bar{I}(|\xi|) = \bar{I}(|x|) \text{ for } x \in \xi.$$

This definition is independent of the choice of the function $x \in \xi$, because if $x_1 \sim x_2$, then also $|x_1| \sim |x_2|$; consequently, by what was proved above, we have $\bar{I}(|x_1|) = \bar{I}(|x_2|)$. Moreover, in virtue of (1), (2) and (3), subsection 4,

$$\bar{I}(|c\xi|) = |c|\bar{I}(|\xi|), \bar{I}(|\xi+\eta|) \leqq \bar{I}(|\xi|+|\eta|) \leqq \bar{I}(|\xi|)+\bar{I}(|\eta|), \tag{1}$$

and in virtue of VI, subsection 5,

$$\bar{I}(|\xi|) = 0 \text{ if and only if } \xi = 0. \tag{2}$$

7. The spaces $\mathscr{L}^1$ and L^1. We denote by $\mathscr{L}^1$ the set of all classes ξ of equivalent complex-valued functions x, defined for almost all $t \in T$ and satisfying the condition $\bar{I}(|\xi|) < +\infty$. These classes form a linear space, because in virtue of (1) and (2) of subsection 6, if $\xi_1, \xi_2 \in \mathscr{L}^1$ then

$$\bar{I}(|c\xi_1|) = |c|\bar{I}(|\xi_1|) < +\infty,$$
$$\bar{I}(|\xi_1+\xi_2|) \leqq \bar{I}(|\xi_1|+|\xi_2|) \leqq \bar{I}(|\xi_1|)+\bar{I}(|\xi_2|) < +\infty.$$

We define a norm in $\mathscr{L}^1$ by setting $\|\xi\|_1 = \bar{I}(|\xi|)$; the preceding inequalities show that all the norm axioms are satisfied so that $\mathscr{L}^1$, with this definition of norm, is a normed space.

In the sequel, it will be convenient not to distinguish mutually equivalent functions; in this case, we can replace the class ξ by any function $x \in \xi$ and write $\|x\|_1 = \bar{I}(|x|)$ instead of $\|\xi\|_1$.

We note that if $\langle x_n \rangle$ is an arbitrary sequence of almost everywhere defined functions so that $x_n(t)$ is not defined on a zero set A_n, then all the functions $x_n(t)$ are defined on the set $T - \bigcup_{n=1}^{\infty} A_n$, where $\bigcup_{n=1}^{\infty} A_n$ is a zero set.

I. *If $x_n \in \mathscr{L}^1$ and the series $\sum_{n=1}^{\infty} \|x_n\|_1$ converges, then*

a) *the series $\sum_{n=1}^{\infty} x_n(t)$ converges absolutely almost everywhere;*

b) *the function* $s(t) = \begin{cases} \sum_{n=1}^{\infty} x_n(t) \text{ if all the } x_n(t) \text{ are defined and the series converges,} \\ 0 \text{ otherwise} \end{cases}$

is an element of the space $\mathscr{L}^1$;

c) *the series $\sum_{n=1}^{\infty} x_n$ converges in norm to s.*

Proof. In virtue of (6), subsection 4, we have

$$\bar{I}\left(\sum_{n=1}^{\infty} |x_n|\right) \leqq \sum_{n=1}^{\infty} \bar{I}(|x_n|) = \sum_{n=1}^{\infty} \|x_n\|_1 < +\infty;$$

on the basis of VII, subsection 5, we conclude from this that the series $\sum_{n=1}^{\infty}|x_n(t)|$ converges almost everywhere on T.

Further, $|s| \leqq \sum_{n=1}^{\infty}|x_n|$ almost everywhere on T, and therefore

$$\bar{I}(|s|) \leqq \bar{I}\left(\sum_{n=1}^{\infty} |x_n|\right) \leqq \sum_{n=1}^{\infty} \|x_n\|_1 < +\infty.$$

Consequently, $s \in \mathscr{L}^1$.

Finally, from the inequality $|s - \sum_{k=1}^{n} x_k| \leqq \sum_{k=n+1}^{\infty} |x_k|$, which is valid almost everywhere on T, we conclude that

$$\left\|s - \sum_{k=1}^{n} x_k\right\|_1 \leqq \sum_{k=n+1}^{\infty} \|x_k\|_1 \to 0$$

as $n \to \infty$, so that the series $\sum_{k=1}^{\infty} x_k$ converges in norm to s.

II. *The space $\mathscr{L}^1$ is complete.*

Proof. Suppose $\langle x_n \rangle$ is a fundamental sequence in $\mathscr{L}^1$; then $\langle x_n \rangle$ contains a subsequence $\langle x_{n_k} \rangle$ such that $||x_{n_{k+1}} - x_{n_k}||_1 \leqq 1/2^{k+1}$, and therefore the series $||x_{n_1}||_1 + ||x_{n_2} - x_{n_1}||_1 + \cdots$ converges. In virtue of I, the series $x_{n_1} + (x_{n_2} - x_{n_1}) + (x_{n_3} - x_{n_2}) + \cdots$ then converges in norm to some element $x \in \mathscr{L}^1$, i.e. $||x - x_{n_k}||_1 \to 0$ as $k \to \infty$. Since the initial sequence $\langle x_n \rangle$ is fundamental, $||x - x_n||_1$ also $\to 0$ as $n \to \infty$, and this proves the completeness of $\mathscr{L}^1$.

REMARK. It follows from this proof and Proposition I that *if $x_n \in \mathscr{L}^1$ and $||x - x_n||_1 \to 0$, then there exists a subsequence $\langle x_{n_k} \rangle$ such that $x_{n_k}(t) \to x(t)$ for almost every $t \in T$.*

If $x \in L$, then $\bar{I}(|x|) = I(|x|) < \infty$; consequently, x occurs in one of the classes $\xi \in \mathscr{L}^1$. Identifying x with this class ξ, we can consider L to be a subspace of $\mathscr{L}^1$.

The closure of L in $\mathscr{L}^1$ is a subspace in $\mathscr{L}^1$, complete with respect to the norm $||x||_1$; this closure will be denoted by L^1. In virtue of this definition, L is *dense* in L^1. A function belonging to L^1 will be called a *summable* function.

In virtue of inequality (1), subsection 2, we have

$$|I(x)| \leqq ||x||_1 \tag{1}$$

for all $x \in L$; this means that $I(x)$ is a bounded linear functional in L and, consequently, it is uniquely extendible, with preservation of inequality (1), to a bounded linear functional in L^1 (see II, subsection 4, § 4), which we shall also denote by $I(x)$. This functional $I(x)$ is called the *integral* of the function $x \in L^1$.

Clearly, *every zero function x is summable and $I(|x|) = 0$*, because a zero function can be considered as the limit in the norm $||\cdot||_1$ of the sequence of functions $x_n = 0$, belonging to L.

The set of all real-valued functions $x \in L^1$ forms a real linear subspace in L^1; it is called the *real space L^1*.

It follows from the preceding discussion that *the real space L^1 is complete.*

III. *If $x \in L^1$, then $|x| \in L^1$ and*

$$|I(x)| \leqq I(|x|). \tag{2}$$

Proof. Suppose $x_n \in L$ and $||x - x_n||_1 \to 0$; it follows from the inequality $||x| - |x_n|| \leqq |x - x_n|$ that $||\,|x| - |x_n|\,||_1 \leqq ||x - x_n||_1 \to 0$. Since $|x_n| \in L^+$, it follows that $|x| \in L^1$. Moreover, passing to the limit in the inequality $|I(x_n)| \leqq I(|x_n|)$ and making use of the continuity of the functional $I(x)$ with respect to the norm $||x||_1$, we obtain (2).

IV. *If $x \in L^1$ and $x \geqq 0$, then*

$$I(x) = \bar{I}(x) = ||x||_1. \tag{3}$$

In fact, (3) is valid for $x \in L^+$; since $I(x)$ and $||x||_1$ are continuous with respect to the norm $||x||_1$ and every nonnegative element x in L^1 is the limit with respect to this norm of a sequence $x_n \in L^+$ (see the proof of Proposition III), (3) is valid for all nonnegative x in L^1.

It follows from III that

V. *If the real-valued functions* x, y *belong to* L^1, *then also* $x \cap y = \frac{1}{2}(x+y-|x-y|)$ *and* $x \cup y = \frac{1}{2}(x+y+|x-y|)$ *belong to* L^1.

VI. *If* $x_n \geqq 0$ *almost everywhere*, $x_n \in L^1$, *and the series* $\sum_{n=1}^{\infty} I(x_n)$ *converges, then the series* $\sum_{n=1}^{\infty} x_n$ *converges in norm in* L^1 *to the element* $x \in L^1$ *and* $I(x) = \sum_{n=1}^{\infty} I(x_n)$.

This assertion follows directly from Proposition I and the continuity in norm of the functional $I(x)$ in L^1.

VII. *If* $\langle x_n \rangle$ *is a nondecreasing (nonincreasing) sequence of real-valued functions in* L^1, *and the sequence* $\langle I(x_n) \rangle$ *is bounded above (below), then the function* $x = \lim x_n \in L^1$ *and* $I(x) = \lim I(x_n)$.

Proof. We consider the case of a nondecreasing sequence (the case of a nonincreasing sequence is reducible to it by going over from x_n, x to $-x_n$, $-x$). We have $x_{n+1} - x_n \geqq 0$ for $n = 1, 2, 3, \cdots$. Then $x = x_1 + (x_2 - x_1) + (x_3 - x_2) + \cdots$ almost everywhere, where the series $I(x_2 - x_1) + I(x_3 - x_2) + \cdots = \lim I(x_n) - I(x_1)$ converges. On the basis of VI, $x - x_1 = (x_2 - x_1) + (x_3 - x_2) + \cdots \in L^1$ and $I(x - x_1) = \lim_{n \to \infty} I(x_n) - I(x_1)$. It follows that $x \in L^1$ and $I(x) = \lim_{n \to \infty} I(x_n)$.

VIII. *A nonnegative lower semi-continuous function* x *is summable if, and only if*, $\bar{I}(x) < +\infty$.

Proof. The necessity is obvious; we shall prove the sufficiency. Suppose $x \in M^+$ and $\bar{I}(x) < +\infty$. By the definition of $\bar{I}(x)$, for arbitrary $\varepsilon > 0$ there exists a $y \in L^+$ such that $y \leqq x$ and $\bar{I}(x) < I(y) + \varepsilon$. But $x - y$ is $\geqq 0$ and lower semi-continuous; consequently, in virtue of IV, subsection 3, $\bar{I}(x) = \bar{I}(y + x - y) = I(y) + \bar{I}(x - y)$ and $\bar{I}(x - y) = \bar{I}(x) - I(y) < \varepsilon$. It follows that $x \in L^1$ by the definition of L^1.

IX. *A finite nonnegative upper semi-continuous function* x *is summable if, and only if*, $\bar{I}(x) < +\infty$.

Proof. Clearly, in the proof we need only the sufficiency of the condition. Thus, suppose $x(t) \geqq 0$ is finite, upper semi-continuous and that $\bar{I}(x) < +\infty$. By the definition of $\bar{I}(x)$, for arbitrary $\varepsilon > 0$ there exists a function $y \in M^+$ such that $y \geqq x$, $\bar{I}(y) < \bar{I}(x) + \varepsilon$. Then $y - x \in M^+$ and $y - x \leqq y$. It follows that $\bar{I}(y - x) = \bar{I}(y) < \bar{I}(x) + \varepsilon < \infty$ and hence (see VIII) $y - x \in L^1$, $y \in L^1$; consequently, also $x = y - (y - x) \in L^1$.

X. *If a nonnegative function* x *is summable, then for every* $\varepsilon > 0$ *there exists a finite function* $z \in N^+$ *with compact carrier and a function* $y \in M^+$ *such that* $z \leqq x \leqq y$ *almost everywhere and* $\bar{I}(y - z) < \varepsilon$.

Proof. If $x \geqq 0$ and $x \in L^1$, then there exists a function $u \in L^+$ such that $\bar{I}(|x - u|)$

$< \varepsilon/4$; by the definition of $\bar{I}$, this means there exists a function $v \in M^+$ for which $\bar{I}(v) < \varepsilon/2$ and $|x-u| \leqq v$ almost everywhere. Thus, $-v \leqq x-u \leqq v$ and $(u-v) \cup 0 \leqq x \leqq u+v$ almost everywhere; consequently, the functions $z = (u-v) \cup 0$, $y = u+v$ will satisfy the listed requirements. [For an everywhere defined and finite function x, the relation $z \leqq x \leqq y$ will be satisfied everywhere because in this case there exists a function $v \in M^+$ for which $\bar{I}(v) < \varepsilon/2$ and $|x-u| \leqq v$ everywhere (see subsection 4).]

XI. *For every nonnegative summable function x there exist a nondecreasing sequence of finite $z_n \in N^+$ with compact carriers and a nonincreasing sequence $y_n \in M^+$ such that $\lim_{n\to\infty} z_n \leqq x \leqq \lim_{n\to\infty} y_n$ almost everywhere and*

$$I(\lim_{n\to\infty} z_n) = I(x) = I(\lim_{n\to\infty} y_n) \text{ and } \lim_{n\to\infty} z_n \sim x \sim \lim_{n\to\infty} y_n.$$

Proof. Setting $\varepsilon = 1/n$ in X, we obtain finite functions $z'_n \in N^+$ with compact carriers and functions $y'_n \in M^+$ such that $z'_n \leqq x \leqq y'_n$ and $\bar{I}(y'_n - z'_n) < 1/n$. Then the functions

$$z_n = z'_1 \cup z'_2 \cup \cdots \cup z'_n, \; y_n = y'_1 \cap y'_2 \cap \cdots \cap y'_n$$

will satisfy the listed requirements, because, in virtue of VII,

$$I(x - \lim_{n\to\infty} z_n) = \lim_{n\to\infty} I(x-z_n) \leqq \lim_{n\to\infty} I(y_n - z_n) \leqq \lim_{n\to\infty} \bar{I}(y'_n - z'_n) \leqq \lim_{n\to\infty} \frac{1}{n} = 0$$

and analogously, $I(\lim_{n\to\infty} y_n - x) = 0$.

XII. *If a nonnegative function x is summable, then for every $\varepsilon > 0$ there exists a compact set Q such that $\bar{I}(x\xi_{T-Q}) < \varepsilon$.*

Proof. Let y, z be as in Proposition X and let Q be the carrier of z. Then $\bar{I}(x\xi_{T-Q}) = \bar{I}((x-z)\xi_{T-Q}) \leqq \bar{I}(y-z) < \varepsilon$.

8. Summable sets. A set A is said to be *summable* if its characteristic function $\xi_A(t)$ is summable; in this case, the number $\mu(A) = I(\xi_A)$ is called the *measure* (specifically, the I-measure) of the set A. [In the sequel, the term "measure" will always mean measure defined in this way by some integral I.] In virtue of IV, subsection 7, *the measure of a summable set coincides with its exterior measure*, $\mu(A) = \bar{\mu}(A)$.

I. *Every zero set is summable, and its measure equals zero.*

In fact, if A is a zero set, then ξ_A is a zero function and, consequently, $\xi_A \in L^1$ and $\mu(A) = I(\xi_A) = 0$.

II. *If A_1, A_2 are summable and $A_1 \subset A_2$, then $\mu(A_1) \leqq \mu(A_2)$.*

This assertion follows directly from (1), subsection 5.

III. *The union of a finite number and the intersection of a finite or denumerable number of summable sets are summable, and for a finite number of pairwise disjoint summable sets* $A_1, A_2, \cdots, A_n$,

$$\mu(A_1 \cup A_2 \cup \cdots \cup A_n) = \mu(A_1)+\mu(A_2)+ \cdots +\mu(A_n).$$

The assertion follows directly from the relations

$$\xi_{A_1 \cup A_2 \cup \cdots \cup A_n} = \xi_{A_1} \cup \xi_{A_2} \cup \cdots \cup \xi_{A_n},$$
$$\xi_{A_1 \cap A_2 \cap \cdots \cap A_n} = \xi_{A_1} \cap \xi_{A_2} \cap \cdots \cap \xi_{A_n},$$
$$\xi_{A_1 \cap A_2 \cap \cdots \cap A_n} \searrow \xi_{A_1 \cap A_2 \cap \cdots},$$
$$\xi_{A_1 \cup A_2 \cup \cdots \cup A_n} = \xi_{A_1}+\xi_{A_2}+ \cdots +\xi_{A_n} \text{ if } A_j \cap A_k = \phi \text{ for } j \neq k,$$

Propositions V, VII of subsection 7, and the linearity of the space L^1 and functional $I(x)$. [The symbols $x_n \searrow x (x_n \nearrow x)$ denote that $\langle x_n \rangle$ is a non-increasing (non-decreasing) sequence which converges to x.]

IV. *If* $A_1, A_2, \cdots$ *are summable and the series* $\mu(A_1)+\mu(A_2)+ \cdots$ *converges, then the set* $A_1 \cup A_2 \cup \cdots$ *is summable and*

$$\mu(A_1 \cup A_2 \cup \cdots) \leqq \mu(A_1)+\mu(A_2)+ \cdots.$$

If, furthermore, the sets A_k *are pairwise disjoint, then*

$$\mu(A_1 \cup A_2 \cup \cdots) = \mu(A_1)+\mu(A_2)+ \cdots.$$

The assertions follow from the relations

$$\xi_{A_1 \cup A_2 \cup \cdots \cup A_n} \nearrow \xi_{A_1 \cup A_2 \cup \cdots} \leqq \xi_{A_1}+\xi_{A_2}+ \cdots,$$
$$\xi_{A_1 \cup A_2 \cup \cdots} = \xi_{A_1}+\xi_{A_2}+ \cdots, \text{ if } A_j \cap A_k = \phi \text{ for } j \neq k,$$

and Propositions V, VI, VII, subsection 7.

V. *If* A_1, A_2 *are summable and* $A_1 \supset A_2$, *then* $A_1 - A_2$ *is summable and*

$$\mu(A_1 - A_2) = \mu(A_1) - \mu(A_2).$$

In fact, if $A_1 \supset A_2$, then

$$\xi_{A_1 - A_2} = \xi_{A_1} - \xi_{A_2}$$

and the assertion follows from the linearity of L^1 and $I(x)$.

VI. *If* $A_1, A_2, A_3, \cdots$ *are summable and* $A_1 \supset A_2 \supset A_3 \supset \cdots$, *then*

$$\mu(A_1 \cap A_2 \cap \cdots) = \lim_{n \to \infty} \mu(A_n);$$

but if $A_1 \subset A_2 \subset \cdots$ *and* $\lim_{n \to \infty} \mu(A_n)$ *is finite, then* $A_1 \cup A_2 \cup A_3 \cup \cdots$ *is summable and*

$$\mu(A_1 \cup A_2 \cup A_3 \cup \cdots) = \lim_{n \to \infty} \mu(A_n).$$

Proof. In the first case, $\xi_{A_n} \searrow \xi_{A_1 \cap A_2 \cap \cdots}$, and in the second, $\xi_{A_n} \nearrow \xi_{A_1 \cup A_2 \cup \cdots}$, and the assertion follows from VII, subsection 7.

VII. *An open (closed) set is summable if* $\bar{\mu}(A) < +\infty$; *in particular, an open (closed) set, contained in a compact set, is summable.*

In fact, the characteristic function of an open (closed) set A is lower (upper) semi-continuous and it remains to apply VIII and IX, subsection 7, and III, subsection 5.

VIII. *Every compact set is summable; every open set with compact closure is summable.*

The assertions follow directly from VII.

IX. *A necessary and sufficient condition that a set A be summable is that for every* $\varepsilon > 0$ *there exist a summable open set* $U \supset A$ *and a compact set* $Q \subset A$ *such that*

$$\mu(U-Q) < \varepsilon.$$

Proof. The condition is sufficient, because it means that $\xi_Q \leqq \xi_A \leqq \xi_U, I(\xi_U-\xi_Q) < \varepsilon$, and therefore

$$\|\xi_A-\xi_Q\|_1 \leqq \|\xi_U-\xi_Q\|_1 = I(\xi_U-\xi_Q) < \varepsilon.$$

Since $\xi_Q \in L^1$, we also have $\xi_A \in L^1$.

Conversely, suppose A is summable. In virtue of IV, subsection 5, there exists a set U such that $\mu(U) < \mu(A)+\varepsilon/2$. Further, on the basis of X, subsection 7, there exists a finite function $z \in N^+$ with compact carrier Q_z such that $z \leqq \xi_A$ (see end of proof of X, subsection 7, above) and $I(\xi_A-z) < \varepsilon/4$. We set $Q = \{t: z(t) \geqq \delta\}$, where $\delta > 0$; then Q is a closed subset in Q_z and therefore it is compact. Let $t \in Q$; then $\xi_A(t) \geqq z(t) \geqq \delta > 0$ and hence $\xi_A(t) = 1$ and $t \in A$; consequently $Q \subset A$ and the set $B = A-Q$ is summable. It follows from the inequality $z \leqq \xi_Q+\delta\xi_B$ that $I(z) \leqq \mu(Q)+\delta\mu(B) \leqq \mu(Q)+\delta\mu(A)$, and therefore $\mu(A) < I(z)+\varepsilon/4 < \mu(Q) +\delta\mu(A)+\varepsilon/4$. Taking δ so that $\delta\mu(A) < \varepsilon/4$, we conclude that the sets U and Q will possess the required properties.

It follows from IX that *the measure of any summable set A is the least upper bound of the measures of compact sets* $Q \subset A$.

X. *If A is summable, then there exist a sequence of open sets* $U_n \supset A$ *and a finite or denumerable sequence of disjoint compact sets* $Q_n \subset A$ *such that* $\bigcap_{n=1}^{\infty} U_n - A$ *and* $A-\bigcup_{n=1}^{\infty} Q_n$ *are zero sets.*

Proof. We set $\varepsilon = 1/n$ in IX; we obtain open sets U_n such that $\mu(U_n-A) < 1/n$, and therefore $\mu(\bigcap_{n=1}^{\infty} U_n-A) < 1/n$ for all n, i.e. $\mu(\bigcap_{n=1}^{\infty} U_n-A) = 0$.

We construct the sets Q_n by induction. In virtue of IX (with $\varepsilon = 1$) there exists a $Q_1 \subset A$ such that $\mu(A-Q_1) < 1$; further, in virtue of IX (with $\varepsilon = \frac{1}{2}$) there exists a $Q_2 \subset A-Q_1$ such that $\mu(A-Q_1-Q_2) < \frac{1}{2}$, and so on. At the n-th step, if $A-Q_1- \cdots -Q_{n-1}$ is still not void, we obtain $Q_n \subset A-Q_1- \cdots -Q_{n-1}$ such that $\mu(A-Q_1- \cdots -Q_n) < 1/n$. It follows that $\mu(A-\bigcup_{n=1}^{\infty} Q_n) < 1/n$ for all n, i.e. $\mu(A-\bigcup_{n=1}^{\infty} Q_n) = 0$.

XI. *If* $\bar{\mu}(A) < \infty$, *then* A *is contained in the union of a zero set and a finite or denumerable number of compact sets, which are disjoint.*

For the proof it is sufficient to apply X to a summable open set $U \supset A$ (see IV, subsection 5).

9. Measurable sets. A set A is said to be *measurable* if its intersection with an arbitrary compact set is summable. From this definition and Propositions III and VIII, subsection 8, it follows that *every summable set is measurable and that the space* T *is measurable.*

I. *The union and intersection of a finite or denumerable number of measurable sets are measurable.*

Proof. Suppose A_n are measurable sets and that Q is an arbitrary compact set; then $A_n \cap Q$ are summable sets. It follows from the relation

$$(\bigcap_n A_n) \cap Q = \bigcap_n (A_n \cap Q)$$

and III, subsection 8, that $(\bigcap_n A_n) \cap Q$ is summable and, consequently, $\bigcap_n A_n$ is measurable. Further, from the relations

$$(\bigcup_n A_n) \cap Q = \bigcup_n (A_n \cap Q),$$

$$\mu(\bigcup_{k=1}^{n} (A_n \cap Q)) \leqq \mu(Q)$$

and Proposition VI, subsection 8, we conclude that $(\bigcup_n A_n) \cap Q$ is also summable and consequently $\bigcup_n A_n$ is measurable.

II. *If* A, B *are measurable and* $A \supset B$, *then* $A-B$ *is measurable.*

In fact, for an arbitrary compact Q, $(A-B) \cap Q = (A \cap Q)-(B \cap Q)$, and the assertion follows from V, subsection 8.

In particular, *the complement* $T-A$ *of every measurable set* A *is measurable.*

III. *Every closed set and every open set are measurable.*

Proof. If A is closed and Q is compact, then $A \cap Q$ is compact and hence it is summable; therefore, all closed sets are measurable. But then open sets, which are the complements of closed sets, are also measurable.

IV. *If* A *is measurable and* $\bar{\mu}(A) < +\infty$, *then* A *is summable.*

Proof. In virtue of XI, subsection 8, $A \subset \bigcup Q_n$, where $Q_0, Q_1, Q_2, \cdots$ are pairwise disjoint, Q_0 is a zero set, and the Q_n $(n \geqq 1)$ are compact sets; consequently, $A = \bigcup (A \cap Q_n)$, where $A \cap Q_0, A \cap Q_1, A \cap Q_2, \cdots$ are disjoint summable sets; it follows that for arbitrary n,

$$\mu(A \cap Q_0)+\mu(A \cap Q_1)+ \cdots +\mu(A \cap Q_n)$$
$$= \mu[(A \cap Q_0) \cup (A \cap Q_1) \cup \cdots \cup (A \cap Q_n)] \leqq \bar{\mu}(A),$$

and consequently the series $\mu(A \cap Q_0)+\mu(A \cap Q_1)+\mu(A \cap Q_2)+ \cdots$ converges and A is summable on the basis of IV, subsection 8.

V. *If A measurable, $A \subset B$, and B is summable, then A is summable.*

The assertion follows directly from IV, because $\bar{\mu}(A) \leqq \mu(B) < +\infty$.

A set A is said to be *locally zero* if its intersection with every compact set is a zero set. It follows directly from this definition that

VI. *Every locally zero set is measurable; the union and intersection of a finite or denumerable number of locally zero sets are locally zero sets.*

VII. *If A is a locally zero set and $\bar{\mu}(A) < +\infty$, then A is a zero set.*

The proof is analogous to that of Proposition IV.

VIII. *If A is a locally zero set, then $||x\xi_A||_1 = 0$ and, consequently, $I(x\xi_A) = 0$ for an arbitrary function $x \in L^1$.*

Proof. The assertion is valid for functions $x \in L$, because then $x\xi_A$ is different from zero only on the zero set $Q_x \cap A$, where Q_x is the carrier of x. But for an arbitrary function $x \in L^1$ there exists a sequence $x_n \in L$ such that $||x-x_n||_1 \to 0$. Then $||x\xi_A - x_n\xi_A||_1$ also $\to 0$, and therefore

$$||x\xi_A||_1 \leqq ||x\xi_A - x_n\xi_A||_1 + ||x_n\xi_A||_1 = ||x\xi_A - x_n\xi_A||_1 \to 0,$$

which is possible only for $||x\xi_A||_1 = 0$.

We shall say that a property holds *locally almost everywhere on T* if the set of all points t for which this property does not hold is a locally zero set.

10. Measurable functions. We shall now consider real-valued functions defined and finite locally almost everywhere. Such a function x is said to be *measurable* if for an arbitrary real a the set $A = \{t: x(t) > a\}$ is measurable. (Here, we consider only those $t \in T$ for which $x(t)$ is defined.) Clearly, a constant is a measurable function. Furthermore, it follows from the relations

$$\{t: x(t) \leqq a\} = T - \{t: x(t) > a\},$$

$$\{t: a < x(t) \leqq b\} = \{t: x(t) > a\} - \{t: x(t) > b\},$$

$$\{t: x(t) = a\} = \bigcap_{n=1}^{\infty} \left\{t: a - \frac{1}{n} < x(t) \leqq a\right\},$$

$$\{t: x(t) < a\} = \{t: x(t) \leqq a\} - \{t: x(t) = a\},$$

$$\{t: x(t) \geqq a\} = \{t: x(t) > a\} + \{t: x(t) = a\}$$

and I–II, subsection 9, that in the case of a measurable function x, each of the sets $\{t: x(t) \leqq a\}$, $\{t: a < x(t) \leqq b\}$, $\{t: x(t) = a\}$, $\{t: x(t) < a\}$, $\{t: x(t) \geqq a\}$ is also measurable.

Two functions x_1, x_2 are said to be *locally equivalent* if the set $\{t: x_1(t) \neq x_2(t)\}$ is locally zero.

I. *If x_1, x_2 are locally equivalent and x_1 is measurable, then x_2 is also measurable.*

Proof. We set

$$A_1 = \{t: x_1(t) > a\}, A_2 = \{t: x_2(t) > a\}, B = \{t: x_1(t) \neq x_2(t)\}.$$

In virtue of the relations $A_2-(A_1 \cap A_2) \subset B, A_1-(A_1 \cap A_2) \subset B$, the sets $A_2-(A_1 \cap A_2)$, $A_1-(A_1 \cap A_2)$ are locally zero; from this and the measurability of set A_1 we conclude that the sets $A_1 \cap A_2 = A_1-(A_1-(A_1 \cap A_2))$ and $A_2 = (A_1 \cap A_2) \cup (A_2-(A_1 \cap A_2))$ are measurable.

II. *A function x is measurable if the set $\{t: x(t) > r\}$ is measurable for arbitrary rational r.*

Proof. Suppose a is an arbitrary real number and $\langle r_n \rangle$ is a decreasing sequence of rational numbers which converges to a. Then $\{t: x(t) > a\} = \bigcup_{n=1}^{\infty} \{t: x(t) > r_n\}$; consequently, the set $\{t: x(t) > a\}$ is measurable.

III. *If x_1, x_2 are measurable, then the set $A = \{t: x_1(t) < x_2(t)\}$ is measurable.*

Proof. Suppose $\langle r_n \rangle$ is the sequence of all rational numbers; then the set A is measurable in virtue of the relation

$$A = \bigcup_{n=1}^{\infty} [\{t: x_1(t) < r_n\} \cap \{t: x_2(t) > r_n\}].$$

IV. *If x_1, x_2 are measurable and c is an arbitrary real number, then*: 1) *cx_1 is measurable*; 2) *x_1+c is measurable*; 3) *x_1+x_2 is measurable*; 4) *$|x_1|$ is measurable*; 5) *x_1^2 is measurable*; 6) *$x_1 x_2$ is measurable*; 7) *if, furthermore, x_2 vanishes only on a locally zero set, then x_1/x_2 is measurable.* [At points t where $x_1(t)$, $x_2(t)$ are infinite of different sign we can assign any value to the sum $x_1(t)+x_2(t)$ or assume it to be undefined. Since these points form a locally zero set, the sum x_1+x_2 will be measurable with respect to either definition.]

Proof. Assertion 1) follows from the relations

$$\{t: cx_1 > a\} = \left\{t: x_1 > \frac{a}{c}\right\} \text{ for } c > 0,$$

$$\{t: cx_1 > a\} = \left\{t: x_1 < \frac{a}{c}\right\} \text{ for } c < 0;$$

when $c = 0$, cx_1 is measurable, being a constant function, equal to zero. Assertion 2) follows from the relation

$$\{t: x_1+c > a\} = \{t: x_1 > a-c\},$$

and then assertion 3) follows from the relation

$$\{t: x_1+x_2 > a\} = \{t: x_1 > a+(-1)x_2\}$$

and Proposition III. Further, assertions 4) and 5) follow respectively from the relations

$$\{t: |x_1| > a\} = \{t: x_1 > a\} \cup \{t: x_1 < -a\},$$

$$\{t: x_1^2 > a\} = \begin{cases} \{t: x_1 > \sqrt{a}\} \cup \{t: x_1 < -\sqrt{a}\} & \text{for } a \geqq 0, \\ T & \text{for } a < 0, \end{cases}$$

and then assertion 6) follows from the formula

$$x_1 x_2 = \tfrac{1}{4}[(x_1+x_2)^2-(x_1-x_2)^2].$$

Finally, if x_2 vanishes only on a locally zero set, then $1/x_2$ is measurable in virtue of the relations

$$\left\{t: \frac{1}{x_2} > a\right\} = \left\{t: x_2 < \frac{1}{a}\right\} \text{ for } a > 0,$$

$$\left\{t: \frac{1}{x_2} > a\right\} = \{t: x_2 \geqq 0\} \cup \left\{t: x_2 < \frac{1}{a}\right\} \text{ for } a < 0,$$

$$\left\{t: \frac{1}{x_2} > 0\right\} = \{t: x_2 \geqq 0\};$$

consequently, thc function $x_1/x_2 = x_1 \cdot 1/x_2$ is also measurable.

[At those points t where $x_2(t) = 0$, for definiteness, we set $1/x_2(t) = +\infty$; any other definition does not disturb the measurability of the function $1/x_2(t)$ because the set of such points t is, by assumption, a locally zero set.]

V. *If the nonnegative function x is measurable, then the function x^c is measurable for arbitrary positive c.*

In fact,

$$\{t: [x(t)]^c > a\} = \{t: x(t) > a^{1/c}\} \text{ for } a \geqq 0,$$
$$\{t: [x(t)]^c > a\} = \quad T \quad \text{for } a < 0.$$

VI. *If $\langle x_n \rangle$ is a sequence of measurable functions, bounded locally almost everywhere, then $\sup_n x_n$ and $\inf_n x_n$ are measurable.*

Proof. The measurability of the function $\sup_n x_n$ follows from the relation

$$\{t: \sup_n x_n(t) > a\} = \bigcup_{n=1}^{\infty} \{t: x_n(t) > a\},$$

and of the function $\inf_n x_n$ by passing over to $-x_n$.

VII. *If $\langle x_n \rangle$ is a sequence of measurable functions which converges to a finite limit locally almost everywhere, then the limit function x is measurable.* (Here, we consider only those t for which $x(t)$ is defined.)

The assertion follows directly from the formulas

$$\lim_n x_n = \inf_n \sup_k x_{n+k} = \sup_n \inf_k x_{n+k}$$

and Proposition VI.

A complex-valued function $x = x_1 + ix_2$ is said to be *measurable* if its real and imaginary parts x_1, x_2 are measurable.

From this definition and Propositions IV and V it follows that if x, y are measurable, then $|x|$, cx (with arbitrary complex c), $x+y$, xy, x/y (if $\{t: y = 0\}$ is a locally zero set) are also measurable. Moreover, Proposition VII remains valid for complex-valued functions also.

VIII. *A function x is summable if, and only if, it is measurable and $\bar{I}(|x|) < +\infty$.*

Proof. Suppose $x \in L^1$. Then there exists a sequence $x_n \in L$ such that $\|x - x_n\|_1 \to 0$. It follows from the proof of Propositions II and I, subsection 7, that some subsequence $\langle x_{n_k} \rangle$ converges almost everywhere to x. Since the x_{n_k} are measurable, x is measurable in virtue of VII. Moreover, $\bar{I}(|x|) = I(|x|) < \infty$ because $|x| \in L^1$.

Conversely, suppose x is measurable and $\bar{I}(|x|) < +\infty$. Then $x = y + iz$, where y, z are real-valued measurable functions and also $\bar{I}(|y|) < +\infty$, $\bar{I}(|z|) < +\infty$. Therefore, it suffices to consider the case of a real-valued function. Further, in this case, $x = x \cup 0 - (-x) \cup 0$ and $\bar{I}(x \cup 0) < +\infty$, $\bar{I}((-x) \cup 0) < +\infty$. Therefore, it suffices to consider the case $x \geqq 0$.

We set $A = \{t: x(t) > a\}$, where $a > 0$. The set A is measurable and [in virtue of the inequality $\xi_A \leqq (1/a)x$] we have $\bar{\mu}(A) \leqq (1/a)\bar{I}(x) < +\infty$; consequently, A is summable (IV, subsection 9). Further, it follows from the condition $\bar{I}(x) < +\infty$ that $\{t: x(t) = +\infty\}$ is a zero set (VII, subsection 5); therefore, replacing x by a function equivalent to it, we can assume x to be everywhere defined and finite. We set

$$A_{nm} = \{t: 2^{-n}m < x(t) \leqq 2^{-n}(m+1)\}, \; \xi_{nm} = \xi_{A_{nm}}$$

and

$$x_n = \sum_{m=1}^{\infty} 2^{-n} m \xi_{nm}.$$

By what was proved above, the sets

$$A_{nm} = \{t: x(t) > 2^{-n}m\} - \{t: x(t) > 2^{-n}(m+1)\}$$

are summable and, consequently, $\xi_{nm} \in L^1$. Moreover, $x_n \leqq x$ and therefore $\sum_{m=1}^{\infty} 2^{-n} m I(\xi_{nm}) \leqq \bar{I}(x) < +\infty$; consequently, $x_n \in L^1$ (see VII, subsection 7). But $x_n \nearrow x$, $I(x_n) \leqq \bar{I}(x) < +\infty$; hence, also $x \in L^1$ (VII, subsection 7).

IX. *If x is measurable and $|x| \leqq y$ almost everywhere, where $y \in L^1$, then also $x \in L^1$.*

In fact, $\bar{I}(|x|) \leqq I(y) < +\infty$ and the assertion follows from VIII.

A nonnegative function x is said to be *Lebesgue-summable* if the set $A = \{t: x(t) > a\}$ is summable for arbitrary $a > 0$ and if $\sup \underline{s} = \inf \bar{s} < +\infty$, where $\underline{s} = \sum_{k=-\infty}^{\infty} c_k \mu(A_k)$, $\bar{s} = \sum_{k=-\infty}^{\infty} c_{k+1} \mu(A_k)$, $A_k = \{t: c_k < x(t) \leqq c_{k+1}\}$, $0 < c_k < c_{k+1}$, $c_k \to +\infty$ as $k \to +\infty$, $c_k \to 0$ as $k \to -\infty$ and $\sup \underline{s}$, $\inf \bar{s}$ are taken over all possible such systems of numbers c_k; furthermore, the sums $\underline{s}$ and $\bar{s}$ are assigned the value $+\infty$ if the corresponding series diverge.

In the case of a nonnegative Lebesgue-summable function x, the number $\sup \underline{s} = \inf \bar{s}$ is called the *Lebesgue integral* of the function x and is denoted by $\int x(t)d\mu$ or $\int_T x(t)d\mu$.

An arbitrary real-valued function x is said to be Lebesgue-summable if the functions $x_1 = x \cup 0$, $x_2 = (-x) \cup 0$ are Lebesgue-summable, and in this case we have, by definition, $\int x(t)d\mu = \int x_1(t)d\mu - \int x_2(t)d\mu$. Finally, a complex-valued function $x = x_1 + ix_2$ is said to be Lebesgue-summable if the real-valued functions x_1, x_2 are Lebesgue-summable, and in this case we have, by definition, $\int x(t)d\mu = \int x_1(t)d\mu + i\int x_2(t)d\mu$.

We note that $\underline{s}(\bar{s})$ is a directed nondecreasing (respectively nonincreasing) set, partially ordered by the condition $\underline{s}' \succ \underline{s}(\bar{s}' \succ \bar{s})$, if the set of subdivision points c_k for $\underline{s}(\bar{s})$ is a subset of the set of subdivision points c_k' for $\underline{s}'(\bar{s}')$. Therefore, $\sup \underline{s} = \lim \underline{s}$, $\inf \bar{s} = \lim \bar{s}$ and for a nonnegative Lebesgue-summable function x, we have $\int x(t)d\mu = \lim \underline{s} = \lim \bar{s}$.

X. *If $x \in L^1$, then x is Lebesgue-summable and $I(x) = \int_T x(t)d\mu$. Conversely, if x is Lebesgue-summable then $x \in L^1$ and $I(x) = \int_T x(t)d\mu$.*

Proof. As in the proof of Proposition VIII, it suffices to restrict oneself to the case of functions $x \geqq 0$, which are everywhere defined and finite. Let $x \in L^1$ and suppose A_{nm}, ξ_{nm} are the same as in the proof of Proposition VIII.

Furthermore, we set

$$B_{nm} = \left\{t\colon x(t) > 2^{-n}\frac{1}{m+1}\right\} - \left\{t\colon x(t) > 2^{-n}\frac{1}{m}\right\}, \eta_{nm} = \xi_{B_{nm}};$$

$$x_n = \sum_{m=1}^{\infty} 2^{-n} m\xi_{nm} + \sum_{m=1}^{\infty} 2^{-n}\frac{1}{m+1}\eta_{nm},$$

$$y_n = \sum_{m=1}^{\infty} 2^{-n}(m+1)\xi_{nm} + \sum_{m=1}^{\infty} 2^{-n}\frac{1}{m}\eta_{nm}.$$

Then $x_n \leqq x \leqq y_n$, $x_n \nearrow x$, $y_n \searrow y$; moreover, since

$$y_n - x_n = \sum_{m=1}^{\infty}\left(2^{-n}\xi_{nm} + \frac{1}{m(m+1)}\eta_{nm}\right) \leqq$$

$$\leqq \sum_{m=1}^{\infty}\left(2^{-n}m\xi_{nm} + 2^{-m}\frac{1}{m+1}\eta_{nm}\right) = x_n \leqq x,$$

we then have that $\bar{I}(y_n) \leqq 2I(x) < +\infty$. Therefore $y_n \in L^1$ and

$$I(x_n) \to I(x), I(y_n) \to I(x) \tag{1}$$

(see VII, subsection 7). But

$$\begin{aligned} I(x_n) &= \sum_{m=1}^{\infty}\left(2^{-n}m\mu(A_{nm}) + 2^{-n}\frac{1}{m+1}\mu(B_{nm})\right), \\ I(y_n) &= \sum_{m=1}^{\infty}\left(2^{-n}(m+1)\mu(A_{nm}) + 2^{-n}\frac{1}{m}\mu(B_{nm})\right), \end{aligned} \tag{2}$$

which are sums of the form $\underline{s}$, $\bar{s}$ for x; therefore (1) means that x is Lebesgue-summable and $\int x(t)d\mu = I(x)$.

Conversely, if x is Lebesgue-summable, then the sums (2) converge to $\int_T x(t)d\mu$ (see remark just before Proposition X, above), and therefore $\|x-x_n\|_1 = \bar{I}(x-x_n) \leqq I(y_n-x_n) \to 0$; it follows that $x \in L^1$, and consequently, $I(x) = \int_T x(t)d\mu$.

XI. *If $x \in L^1$ and $\langle A_n \rangle$ is a sequence of summable sets such that $\mu(A_n) \to 0$, then* $I(x\xi_{A_n}) = \int_{A_n} x(t)d\mu \to 0$. [If A is an arbitrarily summable set and x is a summable function, then $\int_A x(t)d\mu$ is understood to mean $\int \xi_A(t)x(t)d\mu$.]

Proof. The function $x\xi_{A_n}$ is measurable and in virtue of the inequality $|x\xi_{A_n}| \leqq |x|$ it is summable. We choose a function $y \in L$ such that $\|x-y\|_1 < \varepsilon/2$. Then also

$$\|x\xi_{A_n} - y\xi_{A_n}\|_1 < \frac{\varepsilon}{2}, \text{ and therefore } |I(x\xi_{A_n}) - I(y\xi_{A_n})| < \frac{\varepsilon}{2}.$$

On the other hand, if $|y| \leqq c$, then $|y\xi_{A_n}| \leqq c\xi_{A_n}$ and consequently $\|y\xi_{A_n}\|_1 \leqq c\mu(A_n)$.

Thus, for sufficiently large n we have

$$|I(x\xi_{A_n})| \leqq |I(x\xi_{A_n}) - I(y\xi_{A_n})| + \|y\xi_{A_n}\|_1 < \frac{\varepsilon}{2} + c\mu(A_n) < \varepsilon.$$

XII. *Every continuous function is measurable.*

In fact, if x is a continuous real-valued function, then the set $\{t: x(t) > a\}$ is open and hence measurable; from this it follows that the assertion is also valid for an arbitrary continuous complex-valued function.

One can show analogously that *all almost everywhere finite functions in M^+ and N^+ are measurable.*

XIII. *If $x \in L^1$ and $x \geqq 0$, then for every $\varepsilon > 0$ there exists a function $u \in L^+$ such that $\|x-u\|_1 < \varepsilon$; if, moreover, $x \leqq C$, then the function u can also be chosen to be $\leqq C$.*

Proof. In virtue of X, subsection 7, there exist $z \in N^+$ and $y \in M^+$ such that $z \leqq x \leqq y$ almost everywhere and $I(y-z) < \varepsilon/2$. Then also

$$\|y-x\|_1 = I(y-x) \leqq I(y-z) < \frac{\varepsilon}{2}. \tag{3}$$

Further, by the definition of the integral of the function $y \in M^+$ (see subsection 3), there exists a function $u \in L^+$ such that $u \leqq y$ and $I(y) < I(u) + \varepsilon/2$; consequently,

$$\|y-u\| = I(y-u) < \frac{\varepsilon}{2}. \tag{4}$$

From (3) and (4) we conclude that $\|x-u\|_1 < \varepsilon$. If, moreover, $x \leqq C$, then replacing y by the function $y_1 = y \cap C$ we do not disturb the inequalities $x \leqq y$ and (3); then $u \leqq y_1 \leqq C$.

11. The real space L^2. We shall denote by L^2 the set of all measurable real-valued functions x for which $x^2 \in L^1$; here also we again do not consider equivalent functions to be distinct. If $x \in L^2$, then, obviously, $cx \in L^2$ for arbitrary real c. Moreover, if $x, y \in L^2$, then also $x+y \in L^2$, because $(x+y)^2 \leqq 2x^2+2y^2$ (see IV and IX, subsection 10). Consequently, L^2 is a linear space.

It now follows from the formula

$$xy = \tfrac{1}{4}(|x+y|^2-|x-y|^2)$$

that if $x, y \in L^2$, then $xy \in L^1$. We define the inner product in L^2 by means of the formula

$$(x, y) = I(xy).$$

It is easily verified that all the inner product axioms are satisfied so that L^2 is then a real pre-Hilbert space. We shall call it the *real space* L^2. We shall denote the norm in this space by $||x||_2$ so that

$$||x||_2 = \sqrt{(x, x)} = \sqrt{I(x^2)}.$$

Then the Cauchy-Bunyakovsky inequality (see subsection 1, § 5) assumes the form

$$|I(xy)|^2 \leqq I(x^2)I(y^2).$$

I. *The real space L^2 is complete.*

Proof. Suppose $\langle x_n \rangle$ is a fundamental sequence in L^2; then $\langle x_n \rangle$ contains a subsequence $\langle x_{n_k} \rangle$ such that

$$||x_{n_{k+1}}-x_{n_k}||_2 < \frac{1}{2^{k+1}},$$

and therefore the series

$$||x_{n_1}||_2+||x_{n_2}-x_{n_1}||_2+||x_{n_3}-x_{n_2}||_2+ \cdots$$

converges. From this it follows that the function

$$z = |x_{n_1}|+|x_{n_2}-x_{n_1}|+|x_{n_3}-x_{n_2}|+ \cdots$$

belongs to L^2, because, denoting the k-th partial sum of this series by z_k, we have: $z_k^2 \nearrow z^2$, $z_k^2 \in L^1$ and

$$||z_k^2||_1 = ||z_k||_2^2 \leqq (||x_{n_1}||_2+||x_{n_2}-x_{n_1}||_2+||x_{n_3}-x_{n_2}||_2+ \cdots)^2$$

(see VII, subsection 7). Analogously, we find that the function

$$u = (|x_{n_1}|-x_{n_1})+[|x_{n_2}-x_{n_1}|-(x_{n_2}-x_{n_1})]+ \cdots$$

belongs to L^2 and therefore also

$$x = z-u = x_{n_1}+(x_{n_2}-x_{n_1})+ \cdots = \lim_{k\to\infty} x_{n_k} \in L^2.$$

Moreover,

$$||x-x_{n_k}||_2 = ||(x_{n_{k+1}}-x_{n_k})+(x_{n_{k+2}}-x_{n_{k+1}})+\cdots||_2$$
$$\leqq ||x_{n_{k+1}}-x_{n_k}||_2+||x_{n_{k+2}}-x_{n_{k+1}}||_2+\cdots \to 0 \text{ as } k\to\infty.$$

Since $\langle x_n\rangle$ is fundamental in L^2, then also $||x-x_n||_2 \to 0$ as $n\to\infty$, and the completeness of L^2 is proved.

Proposition I just proved signifies that the *real space L^2 is a real Hilbert space.*

II. *The set L^r is dense in L^2.*

Proof. Obviously, it suffices to prove that for every nonnegative function $x \in L^2$ and every $\varepsilon > 0$ there exists a function $y \in L^+$ such that $||x-y||_2 < \varepsilon$.

To this end, we set $x_n = x \cap n$; then x_n is measurable, $0 \leqq x_n^2 < x^2$, $x_n \nearrow x$. On the basis of IX, subsection 10, we conclude from this that $x_n^2 \in L^1$ and, consequently,

$$x_n \in L^2,\ x-x_n \in L^2,\ (x-x_n)^2 \in L^1 \text{ and } (x-x_n)^2 \searrow 0.$$

But then (see VII, subsection 7)

$$||x-x_n||_2^2 = I((x-x_n)^2) \to 0,$$

so that, choosing a sufficiently large n, we have:

$$||x-x_n||_2 \leqq \frac{\varepsilon}{3}. \tag{1}$$

We set

$$A_m = \left\{t\colon x_n > \frac{1}{m}\right\} = \left\{t\colon x_n^2 > \frac{1}{m^2}\right\},\quad y_m = x_n\xi_{A_m}.$$

Since $x_n^2 \in L^1$, A_m is a summable set. From this and the inequalities

$$0 \leqq y_m \leqq n\xi_{A_m} \tag{2}$$

it follows that also $y_m \in L^1$. Moreover, $0 \leqq (x_n-y_m)^2 \searrow 0$ as $m\to\infty$ and $(x_n-y_m)^2 \in L^1$; consequently (see VII, subsection 7), $||x_n-y_m||_2^2 = I((x_n-y_m)^2) \to 0$ as $m\to\infty$. Therefore, choosing m sufficiently large, we have:

$$||x_n-y_m||_2 < \frac{\varepsilon}{3}. \tag{3}$$

Since $y_m \in L^1$ and satisfies condition (2), there exists in virtue of XIII, subsection 10 a function $y \in L^+$ such that

$$||y_m-y||_1 < \frac{\varepsilon^2}{18n},\quad 0 \leqq y \leqq n. \tag{4}$$

But then:

$$||y_m-y||_2^2 = I((y_m-y)^2) \leqq I(2n|y_m-y|) = 2n||y_m-y||_1 < 2n\frac{\varepsilon^2}{18n} = \frac{\varepsilon^2}{9},$$

which in combination with (1) and (3) yields $||x-y||_2 < \varepsilon$.

12. The complex space L^2. The *complex space* L^2 is the set of all complex-valued measurable functions $z = x+iy$, for which the function $|z|^2$ is summable. As before, two equivalent functions will not be considered distinct.

If, in L^2, we define the operations of addition and multiplication by a complex number as usual, we transform it into a complex linear space. Further, setting

$$(z_1, z_2) = I(z_1\bar{z}_2) \text{ for } z_1, z_2 \in L^2,$$

we transform L^2 into a pre-Hilbert space. This space is complete. In fact, it follows from the relations

$$||x+iy||_2^2 = ||x||_2^2 + ||y||_2^2 \text{ for } x, y \in L^2$$

that $\langle x_n + iy_n \rangle$ is a fundamental sequence in the complex space L^2 if and only if $\langle x_n \rangle$, $\langle y_n \rangle$ are fundamental sequences in the real space L^2. Thus, *the complex space L^2 is a complex Hilbert space.*

13. The space L^∞. A measurable function $x = x(t)$ will be said to be *essentially bounded* if it is locally equivalent to a bounded function. The set of all essentially bounded real-valued (respectively complex-valued) functions will be called the *real* (respectively *complex*) *space* L^∞; in this connection two locally equivalent essentially bounded functions will not be considered distinct, so that actually the elements of the space L^∞ will be classes of mutually locally equivalent essentially bounded functions.

If we define the operations of addition and multiplication by a number in the usual way, we transform L^∞ into a linear space. We shall now denote by $||x||_\infty$ the greatest lower bound of the set of all numbers which are least upper bounds of the moduli of bounded functions which are locally equivalent to the functions $x \in L^\infty$; it is easily verified that $||x||_\infty$ will then satisfy the norm axioms. (The notation *ess sup* $|x|$ is also used for this norm.)

The space L^∞ is complete. In fact, this space can be considered as the quotient space of the space of all bounded measurable functions x with norm $||x|| = \sup |x(t)|$, which is complete in virtue of VII, subsection 10, with respect to the closed subspace of all those of these functions which are locally almost everywhere equal to zero (see subsection 3, § 4).

14. The positive and negative parts of a linear functional. Suppose p denotes one of the numbers 1, 2, ∞.

THEOREM 1. *Every bounded linear functional f in the real space L^p can be represented in the form $f = f^+ - f^-$, where f^+, f^- are bounded linear functionals in L^p, with norms not surpassing $|f|$, which are nonnegative on all nonnegative functions $x \in L^p$.*

Proof. We shall denote, for brevity, the norm of x in L^p by $||x||$ (so that $||x|| =$

$||x||_p$, $p = 1, 2, \infty$) and for $x \in L^p$, $x \geqq 0$ we set:

$$f^+(x) = \sup\{f(y): 0 \leqq y \leqq x, y \in L^p\};$$

since $f(y) = 0$ (for $y = 0$) is among the numbers $f(y)$ and since $|f(y)| \leqq |f|\,||y|| \leqq |f|\,||x||$, we have

$$f^+(x) \geqq 0, f^+(x) \leqq |f|\,||x||. \tag{1}$$

Moreover, it is obvious that $f^+(cx) = cf^+(x)$ for $c > 0$. We shall prove that also

$$f^+(x_1+x_2) = f^+(x_1)+f^+(x_2) \text{ for } x_1,\ x_2 \in L^p, x_1 \geqq 0, x_2 \geqq 0.$$

If $0 \leqq y_1 \leqq x_1, 0 \leqq y_2 \leqq x_2$, then $0 \leqq y_1+y_2 \leqq x_1+x_2$, and therefore $f^+(x_1+x_2) \geqq \sup f(y_1+y_2) = \sup f(y_1)+\sup f(y_2) = f^+(x_1)+f^+(x_2)$. On the other hand, if $0 \leqq y \leqq x_1+x_2$, then $0 \leqq x_1 \cap y \leqq x_1$, and since $y \leqq x_1 \cap y+x_2$, moreover $0 \leqq y-x_1 \cap y \leqq x_2$; it follows that

$$f^+(x_1+x_2) = \sup f(y) \leqq \sup f(x_1 \cap y)+\sup f(y-x_1 \cap y) \leqq f^+(x_1)+f^+(x_2).$$

Thus, f^+ is additive on nonnegative functions; consequently, f^+ can be extended to a linear functional on all of L^p by setting $f^+(x_1-x_2) = f^+(x_1)-f^+(x_2)$, where x_1, x_2 are nonnegative. In fact, if $x = x_1-x_2 = y_1-y_2$, where x_1, x_2, y_1, $y_2 \geqq 0$ and $\in L^p$, then $x_1+y_2 = y_1+x_2$, from which it follows that $f^+(x_1)+f^+(y_2) = f^+(y_1)+f^+(x_2)$ and consequently $f^+(x_1)-f^+(x_2) = f^+(y_1)-f^+(y_2)$ so that f^+ is uniquely defined.

f^+ is bounded because in virtue of (1)

$$|f^+(x)| \leqq f^+(|x|) \leqq |f|\,||x||.$$

Hence f^+ is a nonnegative bounded linear functional.

We set $f^- = f^+ - f$. Since $f^+(x) \geqq f(x)$ for $x \geqq 0$, we have $f^-(x) \geqq 0$ for $x \geqq 0$. We shall prove that

$$|f^-(x)| \leqq f^-(|x|) \leqq |f|\,||x||. \tag{2}$$

Suppose $x \geqq 0$; by the definition of f^+, for every $\varepsilon > 0$ there exists a function y such that $0 \leqq y \leqq x$ and $f^+(x) < f(y)+\varepsilon$. It follows that $f^-(x) = f^+(x)-f(x) < f(y)-f(x)+\varepsilon = -f(x-y)+\varepsilon \leqq |f|\,||x-y|| + \varepsilon \leqq |f|\,||x|| + \varepsilon$; consequently, $f^-(x) \leqq |f|\,||x||$ in view of the fact that the number ε is arbitrary. Therefore, for an arbitrary real-valued function x in L^p we have

$$|f^-(x)| \leqq f^-(|x|) \leqq |f|\,||x||.$$

Thus, f^- is also a nonnegative bounded linear functional and the equation $f = f^+-f^-$ is the required representation of the functional f.

The functionals f^+ and f^- are called the *positive and negative parts* of the functional f.

15. The Radon-Nikodým theorem. An integral I is said to be *bounded* if the entire set T is summable; clearly, this means that the function $x(t) \equiv 1$ is summable. Further, if I and J are two given integrals on L, then J is said to be *absolutely continuous with respect to I* if every function which is zero with respect to I is also a zero function with respect to J.

In this case, the J-measure μ_J is said to be *dominated* by the I-measure μ_I and we write $\mu_J < \mu_I$; if also $\mu_I < \mu_J$, then μ_I and μ_J are said to be *equivalent* and we write $\mu_I \sim \mu_J$.

THEOREM 2 (RADON-NIKODÝM). *If the bounded integral J is absolutely continuous with respect to the bounded integral I, then there exists an I-summable function x_0, unique to within equivalence, such that for arbitrary function $x \in L^1(J)$ the function xx_0 is also I-summable and $J(x) = I(xx_0)$.*

Proof. We consider the integral $K = I+J$ and the real Hilbert space $L^2(K)$. Suppose $x \in L^2(K)$; since K is bounded, $1 \in L^2(K)$ and, consequently, $x = x\cdot 1 \in L^1(K)$ and

$$|J(x)| \leqq J(|x|) \leqq K(|x|) \leqq ||x||_2 \cdot ||1||_2,$$

where $||x||_2$, $||1||_2$ are norms in $L^2(K)$. [$\bar{J}(x) \leqq \bar{K}(x)$ and $\bar{I}(x) \leqq \bar{K}(x)$ for $x \geqq 0$; consequently, if $x \in L^1(K)$, then $x \in L^1(J)$ and $x \in L^1(I)$.] Thus, J is a linear functional in $L^2(K)$. On the basis of th Riesz theorem (see subsection 3, § 5) there exists a function $y \in L^2(K)$ such that

$$J(x) = (x, y) = K(xy), \tag{1}$$

i.e.

$$J(x) = I(xy)+J(xy), \; J(x(1-y)) = I(xy). \tag{2}$$

Since J is an integral, we conclude from (1) that y is nonnegative with the exception of some set A of K-measure zero; replacing y by the function $(1-\xi_A)y$, we may assume that $y \geqq 0$ for all $t \in T$.

Further, we may also assume that $y < 1$ for all $t \in T$. In fact, setting $B = \{t: y(t) \geqq 1\}$ and replacing x by $x\xi_B$ in (2), we obtain:

$$J(x\xi_B(1-y)) = I(x\xi_B y). \tag{3}$$

But for $x \geqq 0$ the left member in (3) will be $\leqq 0$ (because $x\xi_B(1-y) \leqq 0$), and the right member will be $\geqq 0$ (because $x\xi_B y \geqq 0$); consequently, $J(x\xi_B(1-y)) = I(x\xi_B y) = 0$, and this means that also $J(x\xi_B y) = 0$ for all $x \geqq 0$ in $L^2(K)$ (because J is absolutely continuous with respect to I), and hence also for all $x \in L^2(K)$. It follows also that

$$J(x(1-(1-\xi_B)y)) = I(xy(1-\xi_B)). \tag{4}$$

In fact,

$$J(x(1-(1-\xi_B)y) = J(x(1-y))+J(x\xi_B y) = J(x(1-y)),$$
$$I(xy(1-\xi_B)) = I(xy)-I(xy\xi_B) = I(xy),$$

so that (4) follows from (2).

Therefore, replacing y by $(1-\xi_B)y$, we do not disturb the validity of relation (2) and obtain a function $y < 1$, $y \in L^2(K)$. Consequently, replacing the function $x \in L^2(K)$, $x \geqq 0$ by the function $x(1+y+\cdots+y^{n-1})$ in (2), we obtain that

$$I(x(y(1+y+\cdots+y^{n-1}))) = J(x(1-y^n)),$$

and therefore

$$I(xy)+I(xy^2)+\cdots+I(xy^n) = J(x(1-y^n)) \leqq J(x) < +\infty$$

for $x \in L^2(K)$, $x \geqq 0$. On the basis of VI, subsection 7, we conclude from this that $xy(1-y)^{-1} = xy+xy^2+\cdots \in L^1(I)$ and

$$I(xy(1-y)^{-1}) = \lim_{n\to\infty} J(x(1-y^n)) = J(x),$$

because $x(1-y^n) \nearrow x$, $x(1-y^n) \in L^1(J)$ (see VII, subsection 7). Thus, setting $x_0 = y(1-y)^{-1}$, we have

$$J(x) = I(xx_0) \text{ for all } x \in L^2(K),\ x \geqq 0. \tag{5}$$

Suppose now that $x \in M^+$ and that $J(x) < +\infty$ and put $x_n = x \cap n$; then $x_n \in L^2(K)$, $x_n \geqq 0$, $x_n \nearrow x$, and consequently $x_n x_0 \nearrow xx_0$. Substituting x_n in place of x in (5) and passing to the limit, we obtain that (5) also holds for all $x \in M^+ \cap L^1(J)$ (see VII, subsection 7). Suppose, finally, that x is an arbitrary nonnegative function in $L^1(J)$. There exists a sequence $x_n \in M^+$ such that $x_n \searrow x$ almost everywhere with respect to J. Then $x_n x_0 \searrow xx_0$ almost everywhere with respect to I. In fact, if z is a nonnegative zero function with respect to J and $y \in M^+ \cap L^1(J)$, $y \geqq z$ (such a function y exists by the definition of $\bar{J}(z)$; see subsection 4), then $J(y) = I(yx_0) \geqq I(zx_0)$, and therefore $0 = J(z) = \inf J(y) \geqq I(zx_0)$; consequently, zx_0 is a zero function with respect to I. Substituting x_n for x in (5) and passing to the limit, we obtain (see VII, subsection 7) that (5) holds for all nonnegative, and hence in general for all, $x \in L^1(J)$.

Setting $x = 1$ in (5), we obtain that $x_0 \in L^1(I)$.

If also $J(x) = I(xx_0')$, where $x_0' \in L^1(I)$, then $I(x(x_0-x_0')) = 0$; setting $x = \operatorname{sgn}(x_0-x_0')$ in the last equality (it is easily seen that this function is measurable with respect to I), we obtain that $I(|x_0-x_0'|) = 0$; consequently, x_0 and x_0' are equivalent and the theorem is proved.

16. The space conjugate to L^1.

THEOREM 3. *If I is a bounded integral and f is a bounded linear functional in $L^1(I)$, then there exists a unique function $y_0 \in L^\infty(I)$, such that*

$$f(x) = I(xy_0) \text{ for all } x \in L^1(I), \tag{1}$$

where

$$|f| = ||y_0||_\infty . \tag{2}$$

Conversely, for an arbitrary function $y_0 \in L^\infty(I)$, formula (1) *defines a bounded linear functional in $L^1(I)$.*

Proof. The positive and negative parts f^+, f^- of the functional f are integrals on L; moreover, if $I(|x|) = ||x||_1 = 0$, then also $f^+(x) = 0$ and $f^-(x) = 0$; consequently, f^+ and f^- are absolutely continuous with respect to I.

On the basis of the Radon-Nikodým theorem, there exist functions $y^+, y^- \in L^1(I)$ such that

$$f^+(x) = I(xy^+), f^-(x) = I(xy^-)$$

for all functions which are simultaneously f^+- and f^--summable, in particular, for all $x \in L^1(I)$. In fact, for $x \in L^1(I)$,

$$f^+(|x|) \leqq |f|\, ||x||_1 < \infty, f^-(|x|) \leqq |f|\, ||x||_1 < \infty$$

(see (1) and (2), subsection 14). Setting $y_0 = y^+ - y^-$, we see that

$$f(x) = I(xy_0) \text{ for all } x \in L^1(I). \tag{1}$$

We shall prove that $y_0 \in L^\infty(I)$ and that $||y_0||_\infty \leqq |f|$. To this end, we set $x = \xi_A \operatorname{sgn} y_0$, where $A = \{t: |y_0(t)| \geqq |f| + \varepsilon\}$, $\varepsilon > 0$ in (1); we then obtain

$$I(\xi_A \operatorname{sgn} y_0 \cdot y_0) = f(\xi_A \operatorname{sgn} y_0) \leqq |f|\, ||\xi_A||_1 = |f|\mu(A). \tag{3}$$

On the other hand,

$$(|f|+\varepsilon)\mu(A) = I((|f|+\varepsilon)\xi_A) \leqq I(|y_0|\xi_A) = I(\xi_A \operatorname{sgn} y_0 \cdot y_0),$$

which, for $\mu(A) > 0$, contradicts inequality (3).

Consequently, $\mu(A) = 0$, and $|y_0(t)| \leqq |f| + \varepsilon$ in the exterior of this set A of measure zero. But this means that $y_0 \in L^\infty(I)$ and $||y_0||_\infty \leqq |f| + \varepsilon$; it follows, since $\varepsilon > 0$ is arbitrary, that

$$||y_0||_\infty \leqq |f|. \tag{4}$$

On the other hand, it is obvious that for an arbitrary function $y_0 \in L^\infty(I)$, formula (1) defines a linear functional in $L^1(I)$, bounded in virtue of the inequality

$$|f(x)| = |I(xy_0)| \leqq ||y_0||_\infty ||x||_1 .$$

From this it follows that $|f| \leqq ||y||_\infty$, which in combination with (4) yields $||y_0||_\infty = |f|$; this completes the proof of the theorem.

This theorem signifies that the correspondence $f \to y_0$ is an isometric mapping of the space $[L^1(I)]'$ onto the space $L^\infty(I)$; therefore, we may put

$$[L^1(I)]' = L^\infty(I).$$

REMARKS. 1) *The assertion of Theorem 3 is valid for arbitrary* (*not necessarily bounded*) *integral I on L, if T is the union* (*of perhaps a nondenumerable number*) *of a disjoint locally zero set T_0 and compact sets T_α af positive measure such that every summable set intersects at most a denumerable number of sets T_α.*

In fact the assertion of Theorem 3 is valid for an arbitrary (not necessarily bounded) integral I without any additional conditions (see, e.g., N. BOURBAKI [1], Ch. 5, § 5, subsection 8); however, we do not need this type of general result.

Proof. We set $\xi_0 = \xi_{T_0}$, $\xi_{T_\alpha} = \xi_\alpha$,

$$I_\alpha(x) = I(x\xi_\alpha), f_\alpha(x) = f(x\xi_\alpha) \text{ for } x \in L^1(I).$$

Then I_α is a bounded integral, $f_\alpha(x)$ is a functional which is bounded with respect to I_α, with $|f_\alpha| \leqq |f|$; consequently, $f_\alpha(x) = I_\alpha(xy_\alpha)$, where $y_\alpha \in L^\infty(I_\alpha)$ and

$$||y_\alpha||_\infty = |f_\alpha| \leqq |f|, \tag{5}$$

where $||y_\alpha||_\infty$ is the norm in $L^\infty(I_\alpha)$.

Thus,

$$f(x\xi_\alpha) = I(x\xi_\alpha y_\alpha) \text{ for all } x \in L^1(I). \tag{6}$$

We set $y = y_\alpha$ on T_α; in virtue of (5), $y \in L^\infty$. The set $T - T_\alpha$ is locally zero for I_α, because for arbitrary compact set Q, $I_\alpha(\xi_{Q\cap(T-T_\alpha)}) = I(\xi_Q \xi_\alpha(1-\xi_\alpha)) = 0$. Consequently,

$$||y||_\infty = \sup_\alpha ||y_\alpha||_\infty \leqq |f|. \tag{7}$$

Clearly, $\xi_\alpha y = y_\alpha$; therefore, (6) can be rewritten in the form

$$f(x\xi_\alpha) = I(xy\xi_\alpha) \text{ for all } x \in L^1(I). \tag{8}$$

Suppose $x \in L^1$; then each of the sets $A_n = \{t: |x(t)| > 1/n\}$ is summable (see the proof of Proposition VIII, subsection 10) and therefore it intersects at most a denumerable number of the sets T_α; consequently, the set $A = \{t: |x(t)| > 0\} = \bigcup_{n=1}^\infty A_n$ possesses the same property.

Suppose $T_{\alpha_1}, T_{\alpha_2}, \cdots$ are sets which are intersected by A. Then $|x| = |x\xi_0| + |x\xi_{\alpha_1}| + |x\xi_{\alpha_2}| + \cdots$ and consequently, in virtue of VIII, subsection 9, and VI, subsection 7, $x = x\xi_{\alpha_1} + x\xi_{\alpha_2} + \cdots$, where the series converges in norm in L^1. Therefore, the series $xy = xy\xi_{\alpha_1} + xy\xi_{\alpha_2} + \cdots$ also converges in norm in L^1 and in virtue of (6) and (8), we have

$$f(x) = f(x\xi_{\alpha_1}) + f(x\xi_{\alpha_2}) + \cdots = I(xy_{\alpha_1}\xi_{\alpha_1}) + I(xy_{\alpha_2}\xi_{\alpha_2}) + \cdots$$
$$= I(xy\xi_{\alpha_1}) + I(xy\xi_{\alpha_2}) + \cdots = I(xy).$$

It follows that

$$|f(x)| \leqq ||y_{\alpha_1}||_\infty ||x\xi_{\alpha_1}||_1 + ||y_{\alpha_2}||_\infty ||x\xi_{\alpha_2}||_1 + \cdots$$
$$\leqq ||y||_\infty (||x\xi_{\alpha_1}||_1 + ||x\xi_{\alpha_2}||_1 + \cdots) = ||y||_\infty \cdot ||x||_1,$$

and consequently $|f| \leqq ||y||_\infty$. In combination with (7), this yields $||y||_\infty = |f|$.

2) Theorem 3 can be extended also to the complex space L^1; in this case, the conjugate space $(L^1)'$ is isometric to the complex space L^∞.

In fact, suppose f is a bounded linear functional in the complex space L^1; its real and imaginary parts, considered only on the real space L^1, are bounded linear functionals in the real space L^1. Applying Theorem 3 to each of them, we obtain that

$$f(x) = I(xy_0) \tag{1}$$

for all real-valued $x \in L^1$, where y_0 is a complex-valued function in L^∞. In view of the complex homogeneity of the functional f, equality (1) remains valid for all complex-valued $x \in L^1$. It follows, as above, that $||y_0||_\infty = |f|$.

17. Complex measures. Above (X, subsection 10), we saw that every integral I on L is representable in the form

$$I(x) = \int x(t)d\mu \text{ for all } x \in L, \tag{1}$$

where μ is the measure defined by the functional I.

Equality (1) is easily extendible to arbitrary real linear functionals f in L, which are bounded with respect to the norm $||x||_\infty$. In fact, suppose $f = f^+ - f^-$ is the decomposition of f into positive and negative parts according to the theorem of subsection 14 and let μ^+, μ^- be the measures defined by the nonnegative functionals f^+, f^-; setting $\mu = \mu^+ - \mu^-$, we obtain

$$f(x) = f^+(x) - f^-(x) = \int x(t)d\mu^+ - \int x(t)d\mu^- = \int x(t)d\mu, \tag{2}$$

so that (1) will be valid for $f(x)$ instead of $I(x)$. We shall, as before, call this μ the measure (which is now of arbitrary sign) corresponding to the functional f. [The last equality in (2) is simply the definition of the integral $\int x(t)d\mu$.]

We now consider the complex linear functional f in L, bounded with respect to the norm $||x||_\infty$. For its real and imaginary parts f_1, f_2, considered only on L^r, formula (2) holds; suppose μ_1, μ_2 are the corresponding measures. Setting $\mu = \mu_1 + i\mu_2$, we obtain

$$f(x) = f_1(x) + if_2(x) = \int x(t)d\mu_1 + i\int x(t)d\mu_2 = \int x(t)d(\mu_1 + i\mu_2),$$

i.e.

$$f(x) = \int x(t)d\mu \tag{3}$$

for all $x \in L^r$. [For $x(t) \geqq 0$, the integrals in formulas (2) and (3) can also be defined as the common limits of sums $\underline{s}$, $\bar{s}$, analogous to the sums in subsection 10 but constructed for $\mu = \mu^+ - \mu^-$ and $\mu = \mu_1 + i\mu_2$ respectively. These limits exist. In fact, for $\mu = \mu^+ - \mu^-$, we have $\underline{s} = \underline{s}^+ - \underline{s}^-$, $\bar{s} = \bar{s}^+ - \bar{s}^-$, where $\underline{s}^+$, $\underline{s}^-$, $\bar{s}^+$, $\bar{s}^-$ are the sums $\underline{s}$, $\bar{s}$ constructed for μ^+, μ^- respectively; the limits of the sums $\underline{s}^+$, $\bar{s}^+$ and $\underline{s}^-$, $\bar{s}^-$ exist according to X, subsection 10. The case $\mu = \mu_1 + i\mu_2$ is considered analogously. For arbitrary x with finite norm $||x||_\infty$, the integral $\int x(t)d\mu$ is now defined in the usual way as a linear combination of integrals with $x(t) \geqq 0$.] In view of the complex homogeneity of the functional f, formula (3) will also be valid for all $x \in L$. The function μ is called the *complex measure* defined by the functional f.

We arrive at the following result.

THEOREM 4. *Every complex linear functional f on L, which is bounded in the norm $||x||_\infty$, is representable in the form*

$$f(x) = \int x(t)d\mu,$$

where μ is some (in general, complex) measure on T.

18. Integrals on the direct product of spaces.

THEOREM 5. *Suppose T_1 and T_2 are locally compact spaces and that I and J are integrals on $L(T_1)$ and $L(T_2)$ respectively. Then*

$$I_t(J_s x(t, s)) = J_s(I_t x(t, s)) \tag{1}$$

for an arbitrary function $x(t, s) \in L(T_1 \times T_2)$ and the functional $K(x) = I_t J_s x = J_s I_t x$ is an integral on $L(T_1 \times T_2)$.

Proof. Suppose $x(t, s) \in L(T_1 \times T_2)$ and let A_1, A_2 be compact sets in T_1, T_2 respectively such that x vanishes in the exterior of $A_1 \times A_2$. We denote the norms of I and J on $L_{A_1}(T_1)$ and $L_{A_2}(T_2)$ by C_1 and C_2 respectively (see II, subsection 2), where in general $L_A(T)$ denotes the set of all functions $x \in L(T)$, equal to zero in the exterior of A. According to Stone's theorem (see Theorem 1, subsection 10, § 2), one can approximate an arbitrary function in $L_{A_1 \times A_2}(T_1 \times T_2)$ uniformly on $A_1 \times A_2$ by means of functions of the form

$$u(t, s) = \sum_{k=1}^{n} y_k(t) z_k(s),\ y_k \in L_{A_1}(T_1),\ z_k \in L_{A_2}(T_2).$$

Consequently, for given $\varepsilon > 0$, there exists a function $u(t, s)$ of this sort such that

$$|x(t, s) - u(t, s)| < \varepsilon \text{ on } A_1 \times A_2.$$

It follows that

$$|J_s x(t, s) - \sum_k J(z_k) y_k(t)| = |J_s(x-u)| < \varepsilon C_2, \tag{2}$$

so that $J_s x(t, s)$ is the uniform limit of functions, continuous on A_1, and therefore it is also a continuous function, equal to zero in the exterior of A_1. Moreover, it follows from (2) that

$$|I_t J_s x - \sum_k I(y_k)J(z_k)| < \varepsilon C_1 C_2.$$

From this and the analogous inequality for $J_s I_t x$ we conclude that $|I_t J_s x - J_s I_t x| < 2\varepsilon C_1 C_2$, so that $I_t J_s x = J_s I_t x$ since $\varepsilon > 0$ is arbitrary. Clearly, $I_t J_s x$ is a nonnegative linear functional, i.e. it is an integral.

The measure μ, defined by means of the integral K, is called the *product of the measures* μ_1, μ_2, defined by the integrals I, J, and is denoted by $\mu_1 \times \mu_2$.

I. *For an arbitrary function* $x \in M^+(T_1 \times T_2)$,

$$\bar{K}(x) = \bar{I}_t \bar{J}_s(x) = \bar{J}_s \bar{I}_t(x). \tag{3}$$

Proof. In virtue of II, subsection 3, we have $x = \sup y$ over the set $Y_x = \{y \in L^+(T_1 \times T_2): y \leqq x\}$. But in virtue of (1), $K(y) = I_t J_s(y)$; consequently, applying III, subsection 3, we obtain

$$\bar{K}(x) = \sup_{y \in Y_x} I_t[J_s(y)] = \bar{I}_t[\sup_{y \in Y_x} J_s(y)] = \bar{I}_t \bar{J}_s(\sup_{y \in Y_x} y) = \bar{I}_t \bar{J}_s(x).$$

[We note that $J_s(y) \in L^+(T_1)$(see the proof of Theorem 5), and therefore $\bar{J}_s(x) = \sup_{y \in Y_x} J_s(y) \in M^+(T_1)$ (see I, subsection 3).] Analogously, $\bar{K}(x) = \bar{J}_s \bar{I}_t(x)$.

II. *For an arbitrary nonnegative everywhere defined function* $x(t, s)$,

$$\bar{K}(x) \geqq \bar{I}_t \bar{J}_s(x). \tag{4}$$

Proof. Suppose $y \in M^+(T_1 \times T_2)$, $y \geqq x$. In virtue of (3), then $\bar{K}(y) = \bar{I}_t \bar{J}_s(y) \geqq \bar{I}_t \bar{J}_s(x)$; going over to the greatest lower bound with respect to y, we obtain (4).

III. *If the nonnegative everywhere defined function* $x(t, s)$, *is a zero function with respect to* K, *then for almost every (with respect to* I) $t \in T_1$ *it is a zero function with respect to* J.

Proof. If $\bar{K}(x) = 0$, then it follows from (4) that also $\bar{I}_t \bar{J}_s(x) = 0$. In virtue of VI, subsection 5, $\bar{J}_s(x) = 0$ for almost every (with respect to I) $t \in T_1$, i.e. x is a zero function with respect to J for almost every $t \in T_1$ (with respect to I).

IV. *If* $A \subset T_1 \times T_2$ *is a zero set with respect to* K, *then for almost every* $t \in T_1$ *(with respect to* I), $A_t = \{s: t \times s \in A\}$ *is a zero set with respect to* J.

For the proof it suffices to apply III to the function $x(t, s) = \xi_A(t, s)$.

We shall say that the *iterated integral* $I_t J_s$ of the function $x(t, s)$ *exists* if $x(t, s) \in L^1(J)$ for almost every $t \in T_1$ (with respect to I) and $J_s(x) \in L^1(I)$; one defines the iterated integral $J_s I_t$ analogously.

THEOREM 6 (FUBINI). *If* $x(t, s) \in L^1(K)$, *then both iterated integrals* $I_t J_s$, $J_s I_t$ *exist and*

$$K(x) = I_t J_s(x) = J_s I_t(x). \tag{5}$$

Proof. Suppose first of all that $x(t, s)$ is defined everywhere and that $x(t, s) \in L^1(K)$. Then there exists a sequence $x_n(t, s) \in L(K)$ such that $\|x-x_n\|_1 = K(|x-x_n|) \to 0$ as $n \to \infty$. But in virtue of (4), $K(|x-x_n|) \geqq \bar{I}_t \bar{J}_s(|x-x_n|)$; therefore we also have that $\bar{I}_t \bar{J}_s(|x-x_n|) \to 0$. On the basis of the remark to the proof of II, subsection 7, $\langle x_n \rangle$ contains a subsequence, which we shall again denote by $\langle x_n \rangle$, such that $\bar{J}_s(|x-x_n|) \to 0$ for all $t \in T_1 - A_1$, where A_1 is a set of I-measure zero. Since x_n is a continuous function of s, then, by the very definition of $L^1(J)$ (see subsection 7), it follows from this that $x \in L^1(J)$ for all $t \in T_1 - A_1$ and $|J_s x - J_s x_n| = |J_s(x-x_n)| \leqq J_s(|x-x_n|)$ for all $t \in T_1 - A_1$. This implies that

$$\bar{I}_t(|J_s(x) - J_s(x_n)|) \leqq \bar{I}_t J_s(|x-x_n|) \to 0 \text{ as } n \to \infty.$$

Since the $J_s(x_n)$ are continuous functions of t (see the proof of Theorem 5), this means that $J_s(x) \in L^1(I)$ and $I_t(|J_s(x) - J_s(x_n)|) \to 0$. But then also

$$|I_t J_s(x) - I_t J_s(x_n)| = |I_t(J_s(x) - J_s(x_n))| \leqq I_t(|J_s(x) - J_s(x_n)|) \to 0;$$

therefore, passing to the limit in the equality

$$K(x_n) = I_t J_s(x_n),$$

which is valid in virtue of Theorem 5, we obtain $K(x) = I_t J_s(x)$. If we interchange the roles of I and J in this argument, we obtain that $J_s I_t$ also exists and that $J_s I_t(x) = K(x)$. This proves the theorem for an everywhere defined function $x(t, s) \in L^1(K)$.

Now let $x(t, s) \in L^1(K)$ and suppose $x(t, s)$ is defined on the set $T_1 \times T_2 - A$ where A is a zero set with respect to K. Then there exists an everywhere defined function $y(t, s)$ in $L^1(K)$, which is equal to $x(t, s)$ on $T_1 \times T_2 - A$. By what was proved above, $y \in L^1(J)$ for $t \in T_1 - B_1$, where B_1 is a zero set with respect to I, $J_s(y) \in L^1(I)$, and

$$K(x) = K(y) = I_t J_s(y). \tag{6}$$

But in virtue of IV, $A_t = \{s\colon t \times s \in A\}$ is a zero set with respect to J for $t \in T_1 - C_1$, where C_1 is a zero set with respect to I; on the other hand, by the definition of the set A, $x(t, s) = y(t, s)$ for $s \in T_2 - A_t$ for every $t \in T_1 - C_1$. Therefore $x(t, s)$, as a function of s, is equivalent to the summable function $y(t, s)$ for every $t \in T_1 - (B_1 \cup C_1)$. Consequently, $J_s(x) \in L^1(I)$ and $J_s(x) = J_s(y)$ for every $t \in T_1 - (B_1 \cup C_1)$; since $B_1 \cup C_1$ is a zero set with respect to I and $J_s(y) \in L^1(I)$, we also have $J_s(x) \in L^1(I)$ and $I_t J_s(x) = I_t J_s(y)$. From this and (6) we conclude that $I_t J_s(x) = K(x)$. The existence of the iterated integral $J_s I_t$ and the equality $J_s I_t(x) = K(x)$ are proved analogously.

With certain supplementary restrictions, the converse assertion can be proved, i.e. that the existence of one of the iterated integrals $I_t J_s$, $J_s I_t$ implies the relation $x \in L^1(K)$.

We note only those particular cases which we shall require in the sequel.

V. *If $x(t, s) = x_1(t)x_2(s)$, where $x_1(t), x_2(s)$ are nonnegative everywhere defined functions on T_1, T_2, then*

$$\bar{K}(x) = \bar{I}(x_1)\bar{J}(x_2), \tag{7}$$

with the exclusion of the case when $\bar{I}(x_1) = 0, \bar{J}(x_2) = +\infty$, or $\bar{I}(x_1) = +\infty$, $\bar{J}(x_2) = 0$.

Proof. Suppose $y_1 \in M^+(T_1)$, $y_2 \in M^+(T_2)$ and $y_1 \geqq x_1$, $y_2 \geqq x_2$; setting $y(t, s) = y_1(t)y_2(s)$, we have $y \in M^+(T_1 \times T_2)$, $y \geqq x$; but in virtue of I,

$$\bar{K}(y) = \bar{I}_t[\bar{J}_s(y_1 y_2)] = \bar{I}[y_1 \bar{J}(y_2)] = \bar{I}(y_1)\bar{J}(y_2),$$

and therefore

$$\bar{K}(x) = \inf_{y \in Z_x} \bar{K}(y) \leqq \inf_{y_1 \in Z_{x_1}, y_2 \in Z_{x_2}} [\bar{I}(y_1)\bar{J}(y_2)] = \bar{I}(x_1)\bar{J}(x_2).$$

On the other hand, in virtue of (4),

$$\bar{K}(x) \geqq \bar{I}_t[\bar{J}_s(x_1 x_2)] = \bar{I}[x_1 \bar{J}(x_2)] = \bar{I}(x_1)\bar{J}(x_2);$$

consequently, $\bar{K}(x) = \bar{I}(x_1)\bar{J}(x_2)$.

VI. *If one of the sets $A_1 \subset T_1$, $A_2 \subset T_2$ is a zero set with respect to I (respectively J), and the other set has finite exterior measure with respect to J (respectively I), then the set $A_1 \times A_2$ is a zero set with respect to K.*

The assertion follows directly from (7) and the relation $\xi_{A_1 \times A_2}(t, s) = \xi_{A_1}(t)\xi_{A_2}(s)$.

VII. *If one of the sets $A_1 \subset T_1$, $A_2 \subset T_2$ is a zero set with respect to I (respectively J), and the other is the union of a denumerable number of sets Q_ν, $\nu = 1, 2, \cdots$, of finite exterior measure with respect to J (respectively I), then $A_1 \times A_2$ is a zero set with respect to K.*

Proof. Suppose, for example, that A_1 is a zero set with respect to I, $A_2 = \bigcup_{j=1}^{\infty} Q_j$, where $\bar{\mu}_2(Q_j) < +\infty$. Then $A_1 \times A_2 = \bigcup_{j=1}^{\infty} A_1 \times Q_j$, where the $A_1 \times Q_j$ are zero sets with respect to K in virtue of VI; consequently, $A_1 \times A_2$ is a zero set with respect to K.

VIII. *If $x_1(t) \in L^1(I)$, $x_2(s) \in L^1(J)$, then $x = x_1(t)x_2(s) \in L^1(K)$ and*

$$K(x) = I(x_1)J(x_2). \tag{8}$$

[We assume here and in the sequel that $x_1(t)x_2(s) = 0$ at all points $t \times s$ for which one of the factors vanishes independently of whether or not the second factor is defined.]

Proof. Suppose $x_1(t)$, $x_2(s)$ are defined and finite everywhere on T_1, T_2 respectively and that $x_1(t) \in L^1(I)$, $x_2(s) \in L^1(J)$. Then there exist sequences $x_{1n}(t) \in L(T_1)$, $x_{2n}(s) \in L(T_2)$ such that $I(|x_1 - x_{1n}|) \to 0$, $J(|x_2 - x_{2n}|) \to 0$. Since

$$\begin{aligned}\bar{K}(|x_1(t)x_2(s)-x_{1n}(t)x_{2n}(s)|)&\\ \leqq \bar{K}(|x_1(t)|\,|x_2(s)-x_{2n}(s)|+|x_{2n}(s)|\,|x_1(t)-x_{1n}(t)|)&\\ \leqq \bar{K}(|x_1(t)|\,|x_2(s)-x_{2n}(s)|)+\bar{K}(|x_{2n}(s)|\,|x_1(t)-x_{1n}(t)|)&\\ = I(|x_1|)J(|x_2-x_{2n}|)+J(|x_{2n}|)I(|x_1-x_{1n}|)\to 0,&\end{aligned}$$

and

$$x_{1n}(t)x_{2n}(s)\in L(T_1\times T_2),\text{ we have } x = x_1(t)x_2(s)\in L^1(K)$$

and

$$K(x) = \lim_{n\to\infty} K(x_{1n}(t)x_{2n}(s)) = \lim_{n\to\infty} I(x_{1n})J(x_{2n}) = I(x)J(x).$$

Thus, the proposition is proved for functions $x_1(t)\in L^1(I)$, $x_2(s)\in L^1(J)$ which are defined and finite everywhere.

Now suppose $x_1(t)$, $x_2(s)$ are arbitrary functions in $L^1(I)$, $L^1(J)$ and that A_1, A_2 are zero sets with respect to I respectively J in T_1, T_2 on which $x_1(t), x_2(s)$ are not defined or infinite. Then $x_1(t) = y_1(t)$ on T_1-A_1, $x_2(s) = y_2(s)$ on T_2-A_2, where $y_1(t), y_2(s)$ are functions from $L^1(I)$, $L^1(J)$, which are defined and finite everywhere. We set $B_1 = \{t\colon |y_1(t)| > 0\}$, $B_2 = \{s\colon |y_2(s)| > 0\}$; then the function $x_1(t)x_2(s)$ may not be defined or different from $y_1(t)y_2(s)$ only on the set $(A_1\times B_2)\cup(B_1\times A_2)\cup(A_1\times A_2)$, which is a zero set with respect to K in virtue of VII, because B_1, B_2 are the unions of a denumerable number of summable sets $\{t\colon |y_1(t)| > 1/n\}$, $\{s\colon |y_2(s)| > 1/n\}$, $n = 1, 2, 3, \cdots$ (see the proof of Proposition VIII, subsection 10). Consequently, $K(x_1(t)x_2(s)) = K(y_1(t)y_2(s)) = I(y_1)J(y_2) = I(x_1)J(x_2)$, and the proposition is proved.

IX. *If $x(t, s)\in M^+(T_1\times T_2)$, then it follows from the existence of either of the iterated integrals $I_tJ_s(x)$, $J_sI_t(x)$ that $x\in L^1(K)$.*

Proof. Suppose $x\in M^+(T_1\times T_2)$ and that $I_tJ_s(x)$ exists; then in virtue of (3), $\bar{K}(x) = I_tJ_s(x) < +\infty$ and the assertion follows from VIII, subsection 7.

X. *If $T_1\times T_2$ has a denumerable neighborhood basis with compact closures and the function $x(t, s)$ is measurable with respect to K, then the existence of either of the iterated integrals $I_tJ_s(|x|)$, $J_sI_t(|x|)$ implies that $x\in L^1(K)$.*

Proof. Obviously, it suffices to prove the assertion for nonnegative measurable functions $x(t, s)$ and for one of the iterated integrals $I_tJ_s(x)$, $J_sI_t(x)$. We denote by $\mathfrak{X}$ the set of all nonnegative functions $x(t, s)$, measurable with respect to K, and satisfying the following conditions:

1) $x(t, s)$ is a J-measurable function of s for I-almost all $t\in T_1$;

2) if $y(t) = J_s(x)$ exists for I-almost all $t\in T_1$, then $y(t)$ is I-measurable;

3) if $I_tJ_s(x)$ exists, then $x\in L^1(K)$.

We must prove that $\mathfrak{X}$ contains all nonnegative functions, measurable with respect to K. We shall carry out the proof in a number of steps:

a) *If $x_n\in\mathfrak{X}$ and $x_n\nearrow x$, then $x\in\mathfrak{X}$.*

Conditions 1) and 2) are satisfied for x in virtue of VII, subsection 10. We shall now prove that x also satisfies condition 3). Suppose $I_t J_s(x)$ exists; then $x(t, s) \in L^1(J)$ for $t \in T_1 - A_1$, where A_1 is a zero set (with respect to I), and $J(x) \in L^1(I)$. Since $x_n \leqq x$, then also $x_n(t, s) \in L^1(I)$ for $t \in T_1 - A_1$ and $J(x_n) \in L^1(I)$. Since, moreover, $x_n \nearrow x$, then $J_s(x_n) \nearrow J_s(x)$, and therefore $I_t J_s(x_n) \nearrow I_t J_s(x)$ (see VII, subsection 7). But $x_n \in \mathfrak{X}$ and $I_t J_s(x_n)$ exists; consequently, $x_n \in L^1(K)$ and $K(x_n) = I_t J_s(x_n) \leqq I_t J_s(x) < +\infty$. On the basis of Proposition VII, subsection 7, we conclude from this that $x \in L^1(K)$, $x \in \mathfrak{X}$.

b) *$\xi_A \in \mathfrak{X}$ for every set A which is measurable with respect to K.*

Suppose U_n, $n = 1, 2, 3, \cdots$, is a denumerable neighborhood basis in $T_1 \times T_2$ with compact closures, so that $T_1 \times T_2$ is the union of sets $\overline{U}_n$ which are summable with respect to K, and therefore A is the union of the summable sets $A \cap \overline{U}_n$. We set $B_n = (A \cap \overline{U}_1) \cup \cdots \cup (A \cap \overline{U}_n)$; B_n is summable with respect to K and therefore, according to Theorem 6, $\xi_{B_n} \in \mathfrak{X}$. Since $\xi_{B_n} \nearrow \xi_A$, the assertion follows from step a) of the proof.

c) *$\sum_{i=1}^n c_i \xi_{A_i} \in \mathfrak{X}$ for every finite number n of measurable sets A_i and nonnegative numbers c_i.*

The assertion follows from step b).

d) *$\sum_{i=1}^\infty c_i \xi_{A_i} \in \mathfrak{X}$ for an arbitrary denumerable number of measurable sets A_i and nonnegative numbers c_i.*

The assertion follows from steps a) and c) because $\sum_{i=1}^n c_i \xi_{A_i} \nearrow \sum_{i=1}^\infty c_i \xi_{A_i}$.

Now suppose x is an arbitrary nonnegative function which is measurable with respect to K. We set

$$A_{nm} = \{t: 2^{-n}m < x(t) \leqq 2^{-n}(m+1)\}, \ \xi_{nm} = \xi_{A_{nm}}$$

and

$$x_n = \sum_{m=1}^{\infty} 2^{-n} m \xi_{nm}.$$

Then $x_n \nearrow x$, $x_n \in \mathfrak{X}$ (in virtue of step d)); consequently, $x \in \mathfrak{X}$ on the basis of step a), and the proposition is proved.

19. The integration of vector-valued and operator-valued functions. Suppose x is a vector-valued function on the locally compact space T with values in some Hilbert space $\mathfrak{H}$; let I be an integral on $L(T)$ and let μ be the measure defined by this integral. The function x is said to be *measurable* (with respect to I) if the inner product $(y, x(t))$ is a measurable numerical function for arbitrary $y \in \mathfrak{H}$. Let x be a measurable vector-valued function and suppose $|x(t)| \leqq \alpha(t)$, where $\alpha \in L^1(I)$.

In virtue of the inequalities $|(y, x(t))| \leqq |y| \, |x(t)| \leqq \alpha(t)|y|$, we also have that $(y, x(t)) \in L^1(I)$ and $| \int (y, x(t)) d\mu| \leqq |y| \int \alpha(t) d\mu$. The latter inequality shows that $\int (y, x(t)) d\mu$ is a bounded linear functional with respect to y in $\mathfrak{H}$. On the basis of the Riesz theorem (subsection 3, § 5) there exists a unique element $z \in \mathfrak{H}$ such that $\int (y, x(t))d\mu = (y, z)$, where $|z| \leqq \int \alpha(t)d\mu$. This element z is called the *integral of the function x with respect to the measure* μ and is denoted by $\int x(t)d\mu$. Thus, by the very definition,

$$\left(y, \int x(t)d\mu\right) = \int (y, x(t))d\mu \tag{1}$$

and

$$\left|\int x(t)d\mu\right| \leqq \int \alpha(t)d\mu \text{ for } |x(t)| \leqq \alpha(t). \tag{2}$$

Clearly, the usual properties of the integral are preserved:

$$\int cx(t)d\mu = c\int x(t)d\mu,$$

$$\int [x_1(t)+x_2(t)]d\mu = \int x_1(t)d\mu + \int x_2(t)d\mu.$$

Moreover,

$$\int Ax(t)d\mu = A\int x(t)d\mu \tag{3}$$

for an arbitrary bounded linear operator A in $\mathfrak{H}$. In fact, in virtue of the relations $(y, Ax(t)) = (A^*y, x(t))$, $|Ax(t)| \leqq |A|\,|x(t)| \leqq |A|\alpha(t)$ the function $Ax(t)$ is also measurable and its integral also exists. Moreover, in virtue of (1),

$$\left(y, A\int x(t)d\mu\right) = \left(A^*y, \int x(t)d\mu\right) = \int (A^*y, x(t))d\mu$$

$$= \int (y, Ax(t))d\mu = \left(y, \int Ax(t)d\mu\right);$$

(3) follows from this.

Now suppose $A(t)$ is an arbitrary operator function on T, every value of which is a bounded linear operator in $\mathfrak{H}$. This function is said to be *measurable with respect to I* if $A(t)x$ is a measurable vector-valued function for arbitrary $x \in \mathfrak{H}$. Suppose $|A(t)| \leqq \alpha(t)$, where $\alpha(t) \in L^1(I)$. In virtue of the inequality $|A(t)x| \leqq \alpha(t)|x|$, we conclude from this that $\int A(t)xd\mu$ exists, where, in virtue of (2),

$$\left|\int A(t)x\,d\mu\right| \leqq \int \alpha(t)d\mu|x|. \tag{4}$$

From this it follows that the formula $y = \int A(t)xd\mu$ defines a bounded linear operator in $\mathfrak{H}$ which is called the *integral of the operator function* $A(t)$ *with respect to the measure* μ and is denoted by $\int A(t)d\mu$. Thus, by the very definition,

$$\left(\int A(t)d\mu\right) x = \int A(t)x\, d\mu$$

and in virtue of (4),

$$\left|\int A(t)d\mu\right| \leqq \int \alpha(t)d\mu \text{ for } |A(t)| \leqq \alpha(t).$$

It is also easily verified that

$$\int cA(t)d\mu = c\int A(t)d\mu,$$

$$\int [A_1(t)+A_2(t)]d\mu = \int A_1(t)d\mu + \int A_2(t)d\mu$$

and that

$$B\int A(t)d\mu = \int BA(t)d\mu, \int A(t)B\, d\mu = \int A(t)d\mu \cdot B.$$

For other expositions of integration theory, see, for instance, N. Bourbaki [1], P. Halmos [1] and S. Saks [1]. The exposition here is in many respects near to that of N. Bourbaki [1].

CHAPTER II

FUNDAMENTAL CONCEPTS AND PROPOSITIONS IN THE THEORY OF NORMED ALGEBRAS

§ 7. Fundamental algebraic concepts

1. Definition of a linear algebra. We shall say that R is a *linear algebra* if:

1) R is a linear space;

2) an operation of multiplication (which in general is not commutative) is defined in R satisfying the following conditions: a) $\alpha(xy) = (\alpha x)y = x(\alpha y)$, b) $(xy)z = x(yz)$, c) $(x+y)z = xz+yz$, d) $x(y+z) = xy+xz$ for arbitrary $x, y, z \in R$ and any number α.

In the sequel we shall consider only linear algebras and the term "algebra" will denote a linear algebra.

Two elements x, y in the algebra R are said to *commute* if $xy = yx$. An algebra is said to be *commutative* if any two of its elements commute. In the sequel we shall not assume in general that the algebras under consideration are commutative.

A subset $R_1 \subset R$ is called a *subalgebra* of the algebra R if the application of the operations of addition, scalar multiplication, and multiplication to elements of R_1 yields again elements in R_1.

Clearly, the intersection of any set of subalgebras of the algebra R is again a subalgebra of R. In particular, the intersection of the set of all subalgebras containing a given set $S \subset R$ is the minimal subalgebra containing S. We denote it by $R_a(S)$.

Obviously, $R_a(S)$ is the set of all finite sums of the form $s = \sum_k \lambda_k a_k$ where the a_k are products of a finite number of elements from S. In fact, $R_a(S)$ must contain all such elements s, and their linear combinations form an algebra.

It follows from this that if all the elements in the set S are pairwise commutative, then $R_a(S)$ is a commutative algebra. If, in particular, S consists of a single element, x, then $R_a(x)$, the minimal subalgebra containing the element x, is commutative.

A commutative subalgebra is said to be *maximal* if it is not contained in any other commutative subalgebra. It follows from the preceding discussion that

I. *Every commutative subalgebra is contained in a maximal commutative subalgebra.*

Proof. The set $\sum$ of all commutative subalgebras of the algebra R, which contain a given commutative subalgebra $\mathfrak{A}$, is a partially ordered set, ordered by inclusion, which satisfies the condition of Zorn's lemma; namely, the least upper bound of any linearly ordered set of these subalgebras is simply their union. On the basis of the Zorn lemma, $\sum$ contains a maximal element which will then be the maximal commutative subalgebra containing $\mathfrak{A}$.

Since every element x is contained in the commutative subalgebra $R_a(x)$, it follows from Proposition I that

II. *Every element x is contained in a maximal commutative subalgebra.*

EXAMPLES. 1. We denote the set of all continuous complex-valued functions on the topological space X by $C(X)$; in $C(X)$ we define operations of addition, scalar multiplication, and multiplication respectively as the addition of functions, the multiplication of functions by a number, and the multiplication of functions. Clearly, $C(X)$ will then be an algebra; this algebra is commutative.

2. Suppose X is an arbitrary linear space; we denote the set of all linear operators in X with domain X by $A(X)$. In $A(X)$ we define operations of addition, scalar multiplication, and multiplication as the corresponding operations on operators (see subsection 6, § 1). Then $A(X)$ is an algebra; $A(X)$ is commutative only in the case when X is one-dimensional.

If X is finite-dimensional, namely n-dimensional, then by assigning to each operator in X its matrix with respect to a fixed basis, we can represent $A(X)$ as the set of all matrices of order n; then the algebra operations are represented as the corresponding operations on matrices.

3. Let X be a Banach space; we denote the set of all bounded linear operators in X with domain X (see subsection 4, § 4) by $\mathfrak{B}(X)$. We again define the operations of addition, scalar multiplication, and multiplication as the corresponding operations on operators. Then $\mathfrak{B}(X)$ is an algebra. Clearly, $\mathfrak{B}(X)$ is a subalgebra of the algebra $A(X)$; $\mathfrak{B}(X)$ is commutative only if X is one-dimensional, as is the case for $A(X)$.

4. We denote the set of all functions e^{int}, $n = 0 \pm 1, \pm 2, \cdots$ by W. Clearly, W is a subset of the algebra $C(-\infty, \infty)$(see Example 1); $R_a(W)$ is the set of all finite trigonometric sums $\Sigma\, c_k e^{ikt}$.

5. We denote the set of all compact linear operators in the Banach space X by $\mathfrak{C}(X)$ (see subsection 6, § 4); the operations of addition, scalar multiplication, and multiplication are defined as the corresponding operations on operators. Then $\mathfrak{C}(X)$ is an algebra which evidently is a subalgebra of the algebra $\mathfrak{B}(X)$ (see Example 3).

2. Algebras with identity. An algebra R is called an *algebra with identity* if R contains an element e which satisfies the condition

$$ex = xe = x \text{ for all } x \in R. \tag{1}$$

The element e itself which satisfies condition (1) is called an *identity* of the algebra R.

We note that *an algebra cannot have more than one identity*. In fact, if e' is also an identity in R, then

$$e'x = xe' = x \text{ for all } x \in R. \tag{1'}$$

Setting $x = e'$ in (1) and $x = e$ in (1′), we obtain

$$ee' = e'e = e', e'e = ee' = e,$$

and consequently, $e' = e$.

I. *Every algebra R without identity can be considered as a subalgebra of an algebra R' with identity.*

Proof. The algebra sought must contain all sums $x' = \alpha e + x$, $x \in R$; on the other hand, the set of all such sums forms an algebra R' in which the basic operations are defined by the formulas

$$\begin{aligned} &\beta(\alpha e+x) = \beta\alpha e+\beta x,\ (\alpha_1 e+x_1)+(\alpha_2 e+x_2) = (\alpha_1+\alpha_2)e+(x_1+x_2), \\ &(\alpha_1 e+x_1)(\alpha_2 e+x_2) = \alpha_1\alpha_2 e+(\alpha_1 x_2+\alpha_2 x_1+x_1 x_2). \end{aligned} \tag{2}$$

[We note that every element x' in R' is uniquely representable in the form $x' = \alpha e+x$, $x \in R$, because by assumption R does not contain the identity.] Therefore, the algebra R' can be realized as the set of all formal sums $x' = \alpha e+x$, $x \in R$ in which the basic operations are defined by formulas (2); R itself is obtained when $\alpha = 0$.

The algebra R' can also be realized as the set of all pairs $\langle\alpha, x\rangle$, $x \in R$ in which the basic operations are defined by the formulas

$$\begin{aligned} &\beta\langle\alpha, x\rangle = \langle\beta\alpha, \beta x\rangle,\ \langle\alpha_1, x_1\rangle+\langle\alpha_2, x_2\rangle = \langle\alpha_1+\alpha_2, x_1+x_2\rangle, \\ &\langle\alpha_1, x_1\rangle\langle\alpha_2, x_2\rangle = \langle\alpha_1\alpha_2, \alpha_1 x_2+\alpha_2 x_1+x_1 x_2\rangle, \end{aligned} \tag{2'}$$

which is analogous to the way one defines complex numbers. Then the algebra R itself can be considered as the set of all ordered pairs $\langle 0, x\rangle$, $x \in R$ and considering x and $\langle 0, x\rangle$ to be identical. If we set $e = \langle 1, 0\rangle$, we obtain

$$\langle\alpha, x\rangle = \alpha\langle 1, 0\rangle+\langle 0, x\rangle = \alpha e+x,$$

so that the second realization of the algebra R' is equivalent to the first.

Clearly, R' is the minimal algebra with identity which contains R. The transition from R to R' is called *adjunction of the identity*.

Proposition I shows that the study of algebras without identity can be reduced to the study of algebras with identity. However, for certain applications it is useful to have propositions directly on algebras without identity. Therefore, together with propositions on algebras with identity, we shall frequently discuss the corresponding propositions for algebras without identity.

If R is an algebra with identity and S is an arbitrary set in R, then the intersection of all subalgebras with identity of the algebra R containing S is the *minimal subalgebra with identity* containing S; we denote it by $R'_a(S)$. Clearly, $R'_a(S)$ is obtained from $R_a(S)$ by adjoining the identity and therefore it is the set of all sums $\alpha e+\sum_k \alpha_k a_k$ where the a_k are products of a finite number of elements from S.

II. *A maximal commutative subalgebra R_1 of the algebra R with identity is also an algebra with identity.*

In fact, if this were not the case, we could adjoin the identity to R_1 and obtain a commutative subalgebra R'_1 containing R_1 as a proper subset, which contradicts the fact that R_1 is maximal.

An element y is called a *left inverse* of x if $yx = e$. A left inverse of the element x will be denoted by x_l^{-1}. One can define a right inverse x_r^{-1} in an analogous manner. If the element x has a left and a right inverse then all left and right inverses of the element x coincide. In fact, if we multiply both members of the equality $x_l^{-1}x = e$ on the right by x_r^{-1}, we obtain

$$x_r^{-1} = (x_l^{-1}x)x_r^{-1} = x_l^{-1}(xx_r^{-1}) = x_l^{-1}e = x_l^{-1}.$$

In this case we shall say that the inverse x^{-1} of the element x exists.

III. *If x^{-1} exists and if x, y commute, then x^{-1} and y also commute.*

In fact, multiplying both members of the equality $xy = yx$ on the left and right by x^{-1}, we obtain $yx^{-1} = x^{-1}y$.

IV. *If $\mathfrak{A}$ is the maximal commutative subalgebra which contains x and if x^{-1} exists, then $x^{-1} \in \mathfrak{A}$.*

In fact, by III, x^{-1} commutes with all elements in $\mathfrak{A}$; since $\mathfrak{A}$ is maximal, we have $x^{-1} \in \mathfrak{A}$.

An algebra R with identity is called a *division algebra* if every element $x \neq 0$ in R has an inverse.

V. *If every element $x \neq 0$ in the algebra R with identity has a left inverse, then R is a division algebra.*

Proof. Let $x \in R$, $x \neq 0$. By assumption, x has a left inverse a in R so that

$$ax = e. \tag{3}$$

Set $xa = b$; then $b \neq 0$ since otherwise we would obtain $a = axa = ab = 0$, i.e. $a = 0$, and this contradicts (3). Therefore b has a left inverse y in R so that $yb = e$, i.e.

$$yxa = e. \tag{4}$$

But then a has a right inverse x and a left inverse yx in R; consequently, $x = yx$ (see the discussion after II, above) and it follows from (4) that $xa = e$. This means that a is simultaneously a left and a right inverse, i.e. a is an inverse for x.

Clearly, an analogous proposition holds for algebras in which every element $x \neq 0$ has a right inverse.

The concept of an inverse element can be modified in such a way that it is also applicable to algebras without identity.

An element $y \in R$ is called a *left quasi-inverse* of the element $x \in R$ if $e+y$ is a left inverse of the element $e+x$ in R', i.e. if

$$(e+y)(e+x) = e. \tag{5}$$

Multiplying out the left member in (5) and then cancelling e in both members of the resultant equality, we obtain

$$x+y+yx = 0,$$

so that the identity does not actually occur in the definition of a left quasi-inverse element. A right quasi-inverse element and a quasi-inverse element are defined analogously.

It is easily seen that propositions analogous to III and IV are valid for a quasi-inverse element.

EXAMPLES. 1. The algebra $C(X)$ (see Example 1, subsection 1) is an algebra with identity; namely, the identity of this algebra is the function which is identically equal to unity on X.

2. The algebras $A(X)$ and $\mathfrak{B}(X)$ (see Examples 2 and 3, subsection 1) are algebras with identity, which is the identity operator.

3. If X is infinite-dimensional, the algebra $\mathfrak{C}(X)$(see Example 5, subsection 1) is an algebra without identity. In fact, if E is an operator in $\mathfrak{C}(X)$ satisfying the condition $EA = AE = A$ for all $A \in \mathfrak{C}(X)$ then, as can easily be seen, E is the identity operator. But in case X is infinite-dimensional the identity operator cannot be compact (see subsection 6, § 4).

3. Center. The *center* of an algebra R is the set of those elements $a \in R$ which commute with all the elements of R.

The center is a commutative subalgebra of the algebra R.

In fact, suppose Z is the center of the algebra R. Its elements form a subalgebra because if the elements $a, b \in R$ commute with all elements in the algebra R, then the elements $\lambda a + \mu b$, ab also have this property. The elements of the subalgebra Z, being commutative with all the elements of the algebra, are mutually commutative. Consequently, Z is a commutative subalgebra.

Clearly, the center contains all elements of the form λe (if R is an algebra with identity). If the algebra R itself is commutative, then the center coincides with R.

4. Ideals. A set I_l of elements of the algebra R is called a *left ideal* of R if:

α) $I_l \neq R$;

β) I_l is a subspace of the linear space R;

γ) if $x \in I_l$, $a \in R$ then $ax \in I_l$.

[In abstract algebra the entire algebra R is also considered to be a (two-sided and, consequently, also a left and right) ideal; in contrast to the ideals introduced in the text, R is then called an *improper ideal.*]

A *right ideal* I_r is defined analogously.

We note that neither a right nor a left ideal can contain the identity of the algebra R (if R is an algebra with identity). In fact, if $e \in I_l$ then it would follow from condition γ) that $a = ae \in I_l$ for arbitrary $a \in R$, i.e. that $I_l = R$.

I. *An element x of an algebra with identity has a left (right) inverse if, and only if, it is not contained in any left (right) ideal.*

In fact, suppose the element x does not have a left inverse. The set I_l of all elements of the form yx cannot coincide with the entire algebra because then we should have $yx = e$ for some y, i.e. the element x would have a left inverse y. Therefore I_l is a left ideal containing the element x (for $y = e$).

Conversely, suppose x belongs to the left ideal I_l. If x has a left inverse then $e = x_l^{-1} x \in I_l$, which is impossible. Consequently x_l^{-1} does not exist.

A left (right) ideal of the algebra R is said to be *maximal* if it is not contained in any other left (right) ideal of R.

II. *Every left (right) ideal of the algebra R with identity is contained in a maximal left (right) ideal.*

Proof. The set of all left ideals $I_l' \supset I_l$ of the algebra R is a partially ordered set, ordered by inclusion, which satisfies the condition of Zorn's lemma. In fact, an upper bound of any (linearly ordered) set J of ideals $I_l' \supset I_l$ is the union of all ideals I_l' of the set J; this union satisfies conditions β), γ) and does not coincide with R, because it does not contain the identity of the algebra; consequently, this union is also a left ideal. On the basis of Zorn's lemma, there is at least one maximal ideal among the ideals $I_l' \supset I_l$.

It follows from Propositions I and II that

III. *An element x of an algebra with identity has a left (right) inverse if, and only if, it is not contained in any maximal left (right) ideal.*

A set I of elements of the algebra R is called a *two-sided ideal* in R if I is simultaneously a right and a left ideal in R. An algebra R is called *simple* if it contains no two-sided ideals different from (0).

Suppose I is a two-sided ideal in R. Two elements x_1, x_2 in R are said to be *equivalent modulo I* if $x_1 - x_2 \in I$. Then the entire algebra R decomposes into classes ξ, η, $\cdots$ of mutually equivalent elements, which are called the *residue classes* of the algebra R with respect to the ideal (or modulo the ideal) I. We denote the set of all these classes by R_1. We introduce operations of addition, scalar multiplication, and multiplication in R_1 by carrying out these operations on representatives of classes. Since I is a two-sided ideal, the result of the operations does not depend on the choice of these representatives.

Consequently, R_1 is an algebra. This algebra is called the *quotient algebra* (or *residue class algebra* or *factor algebra*) *of the algebra R modulo the ideal I* and it is denoted by R/I.

A two-sided ideal is said to be *maximal* if it is not contained in any other two-sided ideal.

IV. *Every two-sided ideal of an algebra with identity is contained in a maximal two-sided ideal.*

The proof of this proposition is analogous to the proof of Proposition II.

Propositions II – IV can be generalized to algebras without identity in the following manner.

A left ideal I_l of the algebra R is said to be *regular* if there exists an element u in R such that $xu - x \in I_l$ for all $x \in R$; the element u itself is called an *identity modulo the ideal* I_l. A regular right ideal I_r and an identity element modulo a regular right ideal are defined analogously. Finally, a two-sided ideal I is said to be *regular* if there exists an element u such that

$$ux - x \in I \text{ and } xu - x \in I$$

for all $x \in R$.

If R is an algebra with identity e, then e can be taken as the element u; it is clear from this that *in an algebra with identity every ideal is regular.*

V. *Every regular (right, left, two-sided) ideal can be extended to a maximal (right, left, respectively two-sided) ideal (which is obviously regular also).*

Proof. Suppose, for instance, that I_l is a regular left ideal and that u is an identity element modulo the ideal I_l; then u is not contained in any left ideal $I_l' \supset I_l$. In fact, if $u \in I_l'$, then we also have $xu \in I_l'$; but, furthermore, $x - xu \in I_l \subset I_l'$, and therefore every element

$$x = (x - xu) + xu \in I_l',$$

i.e. $I_l' = R$.

Therefore the reasoning used in the proof of Proposition II is applicable to the ideals $I_l' \supset I_l$.

The cases of regular right and two-sided ideals are treated analogously.

VI. *Suppose R is an algebra without identity and that R' is the algebra obtained from R by adjunction of the identity. Then the correspondence $I_r' \to I_r = I_r' \cap R$ is a mapping of the set of all right ideals I_r' of the algebra R' which are not contained entirely in R onto the set of all regular right ideals of the algebra R. This correspondence is a one-to-one mapping:*

a) *of the set of all two-sided ideals of the algebra R', not contained entirely in R, onto the set of all regular two-sided ideals of the algebra R;*

b) *of the set of all maximal right ideals of the algebra R', not contained entirely in R, onto the set of all maximal right regular ideals of the algebra R.*

Analogous assertions hold for left ideals.

Proof. 1) Suppose I_r' is a right ideal in the algebra R' which is not contained entirely in R; we set $I_r = I_r' \cap R$. Clearly, I_r is a right ideal in R. We shall prove that I_r is regular. Since I_r' is not contained entirely in R, I_r' contains an element $y = -e + u$, $u \in R$; then u is an identity element modulo the ideal I_r because

$$-x + ux = (-e + u)x \in I_r' \cap R = I_r \text{ for all } x \in R.$$

Consequently, I_r is regular.

Conversely, suppose I_r is a regular right ideal in R and let u be an identity element modulo the ideal I_r. Then, as can easily be verified, the set I_r' of all elements $y \in R'$, for which $uy \in I_r$, is a right ideal in R' which contains I_r. Moreover, I_r' is not contained entirely in R because the relation

$$u(u-e) = u^2 - u \in I_r$$

implies that $u - e \in I_r'$ where $u - e \notin R$. Now suppose $y \in R$; since $uy - y \in I_r$ for all $y \in R$, we have that $y \in I_r$ if and only if $y \in I_r'$. Consequently, $I_r = I_r' \cap R$ and this completes the proof of the first proposition.

2) We prove now the one-to-oneness of the mapping $I' \to I = I' \cap R$ for two-sided

ideals. We must establish that if I', J' are two-sided ideals in R', which are not contained entirely in R, and $I = I' \cap R = J' \cap R = J$, then $I' = J'$. Since I', J' are not contained entirely in R, there exist $u, v \in R$ such that $-e+u \in I'$, $-e+v \in J'$; inasmuch as I', J' are two-sided ideals, we conclude from this that $-v+vu = v(-e+u) \in I$, $-u+vu = (-e+v)u \in I$ and hence $u-v \in I$. Suppose $\lambda e+y \in I'$; then $\lambda u + uy = u(\lambda e+y) \in I$, $-y+uy = y(-e+u) \in I$; consequently, also $\lambda u+y = \lambda u+uy-(-y+uy) \in I$. Since $I = J \subset J'$, it follows that $\lambda e+y = \lambda(e-v)+\lambda(v-u)+\lambda u+y \in J'$, consequently $I' \subset J'$. Analogously, $J' \subset I'$, and hence $I' = J'$.

We carry out the proof of the last assertion in several steps:

3) *If M_r is a maximal regular right ideal of the algebra R and u, v are elements of R which are identities modulo the ideal M_r, then $v-u \in M_r$.*

We set $w_0 = v-u$. It follows from the relations $-x+ux \in M_r$, $-x+vx \in M_r$ that $w_0 x \in M_r$ for all $x \in R$. Further, we set $W = \{w: w \in R, wx \in M_r \text{ for all } x \in R\}$ and prove that W is a regular right ideal in R which contains M_r. In virtue of the maximality of the ideal M_r it will follow from this that $W = M_r$ and hence $v-u = w_0 \in W = M_r$. Clearly, W is linear, and if $w \in M$, then $wx \in W$ for all $x \in R$. Moreover $u \notin W$. In fact, if $u \in W$, then $ux \in M_r$ and, consequently, $x = x - ux+ux \in M_r$ for all $x \in R$, which is impossible. Therefore $W \neq R$ and W is a right ideal in R. If $x_0 \in M_r$, then $x_0 x \in M_r$ for all $x \in R$; consequently, $x_0 \in W$. This means that $M_r \subset W$. Finally, $x-ux \in M_r \subset W$ for all $x \in R$, and hence W is regular. This proves 3).

4) *Let M_r be a maximal regular right ideal of the algebra R, u an element of R which is an identity modulo the ideal M_r, and I'_r a right ideal in R such that $I'_r \cap R \subset M_r$ and $J'_r = \{y: y \in R', uy \in M_r\}$. Then $I'_r \subset J'_r$ and J'_r does not depend on the choice of u.*

If v also is an identity modulo M_r, then, in virtue of 3), $w_0 = v-u \in M_r$. Therefore, for $y = \lambda e+x$, $x \in R$, we have $vy-uy = w_0 y = \lambda w_0+w_0 x \in M_r$, so that uy and vy simultaneously belong or do not belong to M_r. This means that J'_r does not depend on the choice of u. Now let $y \in I'_r$. We shall prove that $y \in J'_r$. Since $y \in R'$, we have $y = \lambda e+x$, $x \in R$.

If $\lambda = 0$, then $y = x \in I'_r \cap R \subset M_r$. It follows that $uy = ux = ux-x+x \in M_r$ and hence $y \in J'_r$. But if $\lambda \neq 0$, then, multiplying by $-\lambda^{-1}$ if necessary, we can assume that $y = -e+v$, $v \in R$. Then v is an identity modulo $I'_r \cap R$ (see 1)), and hence also modulo M_r. It follows, in virtue of 3), that $v = u+w_0$, where $w_0 \in M_r$ and hence

$$uy = u(-e+u+w_0) = (-u+u^2)+(uw_0-w_0)+w_0 \in M_r.$$

Consequently, $y \in J'_r$.

5) *Let M_r, J'_r be the same as in 4). Then J'_r is a maximal right ideal in R', which is not contained entirely in R, and $J'_r \cap R = M_r$. Every maximal right ideal M'_r in*

R', not contained entirely in R, for which $M_r' \cap R = M_r$, coincides with J_r'; consequently, it is uniquely determined by the condition $M_r' \cap R = M_r$ for given M_r.

Let M_r' be a maximal right ideal containing J_r'. Then $M_r' \cap R \supset J_r' \cap R = M_r$ (see 1)); in virtue of the maximality of the ideal M_r, we conclude that $M_r' \cap R = M_r$ or $M_r' \cap R = R$. But the second case is impossible. In fact, then $M_r' = R$ or $M_r' = R'$, which is impossible. Thus, $M_r' \cap R = M_r$, so that by 4), $M_r' \subset J_r'$. Consequently, $M_r' = J_r'$ in virtue of the maximality of the ideal M_r'. We see at the same time that J_r' is maximal and that every maximal right ideal M_r', not contained entirely in R and satisfying the condition $M_r' \cap R = M_r$, coincides with J_r'.

6) *Let M_r' be a maximal right ideal in R', not contained entirely in R. Then $I_r = M_r' \cap R$ is a maximal regular right ideal in R.*

The regularity of the ideal I_r was proved in 1). Let M_r be a maximal regular right ideal in R, containing I_r. By 5), there exists a unique maximal right ideal J_r', not contained entirely in R and such that $J_r' \cap R = M_r$. By assumption, $M_r' \cap R = I_r \subset M_r$; consequently, by 5), $M_r' \subset J_r'$. It follows, by virtue of the maximality, that $M_r' = J_r'$ and hence $I_r = M_r \cap R = J_r' \cap R = M_r$.

Assertion b) follows from propositions 5) and 6).

For non-maximal right regular ideals the one-to-oneness of the mapping $I_r' \to I_r' \cap R$ is disturbed. A simple example is the algebra R of all matrices of the form $\begin{pmatrix} \alpha & \beta \\ 0 & 0 \end{pmatrix}$, where α, β are arbitrary complex numbers. Then R' is the algebra of all matrices $\begin{pmatrix} \alpha & \beta \\ 0 & \gamma \end{pmatrix}$, where α, β, γ are arbitrary complex numbers.

Every set $I_r' = \left\{\begin{pmatrix} 0 & \lambda\beta_0 \\ 0 & \lambda\gamma_0 \end{pmatrix}\right\}$, where $\gamma_0 \neq 0$ and β_0 is fixed, whereas λ runs through all complex numbers, is a right ideal in R', not contained entirely in R, and $I' \cap R = (0)$. Consequently, all these ideals I_r' map into the same ideal (0) under the mapping $I_r' \to I_r' \cap R$.

The proposition proved above reduces the study of regular ideals in R to the study of ideals in R'. In particular, it follows from this proposition that *the maximal regular (right, left, two-sided) ideals in R are the intersections with R of maximal (right, left, respectively two-sided) ideals in R' which do not coincide with R.*

VII. *An element x in the algebra R does not have a left quasi-inverse if and only if*

$$I_l = \{z + zx\}, z \in R$$

is a left ideal. In this case, I_l is a regular left ideal which does not contain x.

Proof. Clearly, I_l satisfies conditions β), γ) in the definition of a left ideal. Therefore, if I_l is not a left ideal, we have $I_l = R$; consequently, there exists an element $z \in R$ such that

$$z + zx = -x, \text{ i.e. } x + z + zx = 0.$$

This means that z is a left quasi-inverse of x. Conversely, if x has a left quasi-inverse, we have $x \in I_l$ by what was just proved. But then we also have $-zx \in I_l$, and hence also

$$z = (z + zx) - zx \in I_l$$

for arbitrary $z \in R$. This means that $I_l = R$. It also follows from this reasoning that if I_l is an ideal then $x \notin I_l$. Finally, the element $u = -x$ is an identity modulo the ideal I_l, so that I_l is regular.

The analogous proposition can be proved for elements which do not have a right quasi-inverse.

VIII. *An element x in the algebra R has a left quasi-inverse if, and only if, for arbitrary maximal regular left ideal M_l there exists an element y such that*

$$x+y+yx \in M_l.$$

Proof. The condition is necessary because it suffices to take the left quasi-inverse of the element x for y. Suppose now that this condition is satisfied and assume x does not have a left quasi-inverse. Then $I_l = \{z+zx\}$ is a regular left ideal in R and therefore there exists a maximal regular left ideal $M_l \supset I_l$. By assumption, there exists an element y such that

$$x+y+yx \in M_l.$$

But $y+yx \in I_l \subset M_l$ and consequently $x \in M_l$. It follows that $-zx \in M_l$ and therefore

$$z = -zx+(z+zx) \in M_l$$

for arbitrary $z \in R$; hence $M_l = R$, which is impossible.

5. The (Jacobson) radical. An element x_0 in the algebra R with identity is said to be *generalized nilpotent* if $(e+yx_0)_l^{-1}$ exists for an arbitrary element $y \in R$. The set of all generalized nilpotent elements in the algebra R is called its *(Jacobson) radical.*

I. *The radical of an algebra with identity coincides with the intersection of all its maximal left ideals.*

Proof. Suppose the element x_0 belongs to all maximal left ideals. If $(e+yx_0)_l^{-1}$ does not exist for some $y \in R$, then the element $z = e+yx_0$ belongs to at least one left, and hence also to a maximal left, ideal I_l. By assumption, the element yx_0 belongs to this ideal and, consequently, also

$$e = z-yx_0 \in I_l,$$

which is impossible. Thus, $(e+yx_0)_l^{-1}$ exists for an arbitrary $y \in R$, i.e. x_0 belongs to the radical. Suppose, conversely, that x_0 belongs to the radical. We shall prove that x_0 is contained in all maximal left ideals of the algebra R. Let us assume the contrary. Suppose $x_0 \notin I_l$. We then consider the set of all elements z of the form $z = a-yx_0$ where $a \in I_l$, $y \in R$. This set coincides with the entire algebra because otherwise it would be a left ideal containing the maximal left ideal I_l as a proper subset. In particular, the identity e can be represented in the form $e = a-yx_0$. It follows that the element $a = e+yx_0$, as an element in the ideal I_l, does not have a left inverse, which contradicts our assumption. This also completes the proof of Proposition I.

It follows from Proposition I that the radical is a left ideal.

II. *An element* x_0 *belongs to the radical of an algebra with identity if, and only if, a two-sided inverse* $(e+ax_0)^{-1}$ *exists for every element a of the algebra.*

Proof. If $(e+ax_0)^{-1}$ exists for all $a \in R$, then *a fortiori* $(e+ax_0)_l^{-1}$ exists; consequently, x_0 belongs to the radical. Conversely, suppose x_0 belongs to the radical and that $a \in R$. By the definition of the radical, a left inverse $(e+ax_0)_l^{-1}$ exists. We denote one of these left inverses by $e+y$ so that

$$(e+y)(e+ax_0) = e. \tag{1}$$

This means that $e+y$ has $e+ax_0$ for its right inverse. Furthermore, if we expand the left member of (1), we obtain that

$$y = -yax_0 - ax_0. \tag{2}$$

We shall denote the radical of the algebra R by I. Since $x_0 \in I$ and I is a left ideal, it follows from (2) that $y \in I$; consequently, a left inverse $(e+by)_l^{-1}$ exists for arbitrary $b \in R$. Setting $b = e$, we obtain that $(e+y)_l^{-1}$ exists. On the other hand, as we saw, $(e+y)_r^{-1} = e+ax_0$ exists; consequently, the two-sided inverse $(e+y)^{-1} = e+ax_0$ exists. But then the element $e+y$ is a two-sided inverse of the element $e+ax_0$.

Above, we defined the radical in terms of left inverses. One can define it with the aid of right inverses analogously. Namely, we denote the set of elements x such that $(e+xa)_r^{-1}$ exists for all elements a of the algebra by I'. Reasoning as in the proof of Proposition I, we obtain that I' is the intersection of all maximal right ideals of the algebra and consequently it is a right ideal. Further, as in the proof of Proposition II, we obtain that I' consists of those and only those elements x_0 for which the two-sided inverse $(e+x_0a)^{-1}$ exists for arbitrary a in the algebra.

We shall prove that the two-sided inverses $(e+x_0a)^{-1}$ and $(e+ax_0)^{-1}$ exist or do not exist simultaneously. This will also prove that I' coincides with the radical. Suppose, for example, that

$$(e+ax_0)^{-1} = e+y$$

exists. Then

$$e-x_0a-x_0ya = (e+x_0a)^{-1}.$$

In fact,

$$(e+x_0a)(e-x_0a-x_0ya)-e = -x_0[(e+ax_0)(e+y)-e]a = 0$$

and

$$(e-x_0a-x_0ya)(e+x_0a)-e = -x_0[(e+y)(e+ax_0)-e]a = 0.$$

This also proves that

III. *The intersection of all maximal left ideals coincides with the intersection of all maximal right ideals and is the radical of the algebra.*

It follows, in particular, that *the radical is a two-sided ideal.*

An algebra is said to be *semisimple* if its radical consists of only the zero element.

Suppose now that R is an algebra without identity and that R' is the algebra obtained from R by adjoining the identity. The existence of a left inverse of the element $e+yx_0$ in R' is equivalent to the existence of a left quasi-inverse of the element yx_0 for all $y \in R'$. Therefore, if we set $y = \alpha e+z$, $z \in R$, we can give the following definition of a generalized nilpotent element which is equivalent to the orginal definition in the case of an algebra with identity. An element x_0 is said to be *generalized nilpotent* if αx_0+zx_0 has a left quasi-inverse for arbitrary $z \in R$ and arbitrary numbers α; in this definition, R is no longer necessarily an algebra with identity. The set of all generalized nilpotent elements in the algebra R (not necessarily with identity) is called its *radical*. It follows directly from this definition that

$$I = R \cap I', \tag{3}$$

where I and I' are the radicals of the algebras R and R', respectively.

An algebra R is called a *radical algebra* if it coincides with its radical; otherwise, R is called a *non-radical algebra*. It follows from VII, subsection 4, that *in a non-radical algebra there exist regular, and hence also maximal regular, left and right ideals*. Combining this result with VI, subsection 4, and formula (3), we conclude that:

III′. *In a non-radical algebra, the radical is the intersection of all maximal regular left ideals and also the intersection of all maximal regular right ideals and therefore it is a two-sided ideal.*

IV. *The quotient algebra modulo the radical is a semisimple algebra.*

Proof. Suppose I is the radical of the algebra R and that J is the radical of the algebra R/I. Suppose the element ξ of the algebra R/I belongs to J. This means that the element $\zeta = \alpha\xi+\eta\xi$ has a left quasi-inverse in R/I for arbitrary $\eta \in R/I$ and arbitrary α. Let ζ' be this quasi-inverse, so that

$$\zeta'+\zeta+\zeta'\zeta = 0, \tag{4}$$

and let x, y, z, z' be any representatives of the classes ξ, η, ζ, ζ', respectively. Then (4) means that

$$z'+z+z'z = p, \text{ with } p \in I,$$

where we may assume that

$$z = \alpha x+yx.$$

Since $p \in I$, p has a left quasi-inverse; we denote it by p'. We can verify directly that $p'+z'+p'z'$ is a left quasi-inverse of z. Consequently, the element $z = \alpha x+yx$ has a quasi-inverse for arbitrary $y \in R$ and arbitrary α. This means that $x \in I$, i.e. that $\xi = 0$. Thus, every element $\xi \in J$ is equal to zero, $J = (0)$; consequently, R/I is a semisimple algebra.

We note the following important example of a semisimple algebra. The algebra R of linear operators in the vector space X is said to be *irreducible* if X does not contain subspaces, distinct from (0) and all of X, which are invariant with respect to all the operators in R.

V. *Every irreducible algebra R, different from* (0), *of linear operators in the vector space X is a semisimple algebra.*

Proof. For fixed $x \in X$, $x \neq 0$, the set Rx of all vectors Ax, $A \in R$, is a subspace in X which is invariant with respect to all operators in R; consequently, either $Rx = (0)$ or $Rx = X$. The set $\mathfrak{N}$ of all vectors x for which $Rx = (0)$ is a subspace in X which is invariant with respect to all operators $A \in R$; therefore, either $\mathfrak{N} = (0)$ or $\mathfrak{N} = X$. But the second case would mean that $R = (0)$, contrary to assumption; consequently, $\mathfrak{N} = (0)$ and $Rx = X$ for every $x \in X$, $x \neq 0$. This reasoning is also applicable to an arbitrary two-sided ideal $I \neq (0)$ in R, because for an arbitrary two-sided ideal I in R the sets Ix and $\{x \in X; Ix = 0\}$ are subspaces in X, which are invariant with respect to all the operators $A \in R$; consequently, $Ix = X$ for any such ideal I and arbitrary $x \in X$, $x \neq 0$.

Suppose now that I is the radical in R and that $A \in I$; if $A \neq 0$, then there exists an element $x \in X$ such that $Ax \neq 0$, and this means that $RAx = X$. Therefore, there exists an operator $B \in R$ for which $BAx = -x$. But BA has a left quasi-inverse because $A \in I$; suppose C is this quasi-inverse. Then $C+BA+CBA = 0$, and therefore $Cx+BAx+CBAx = 0$. But this implies that $x = -BAx = Cx+CBAx = Cx-Cx = 0$, which contradicts our assumption. Thus, $U = (0)$, i.e. R is a semisimple algebra.

6. Homomorphism and isomorphism of algebras. A mapping $x \to x'$ of the algebra R into an arbitrary algebra R' is called a *homomorphism of the algebra R into R'* if $x \to x'$, $y \to y'$ imply that $\lambda x \to \lambda x'$, $x+y \to x'+y'$, $xy \to x'y'$. If R' is the image of the algebra R, then the homomorphism is called a *homomorphism of R onto R'*.

Under a homomorphism of R onto R' all concepts, defined in the preceding subsections, including the notion of an identity, remain invariant. In other words, the identity, (right, left, or two-sided) ideal, subalgebra of the algebra R go over into the identity, ideal and subalgebra, respectively, of the algebra R'.

If the homomorphism is a one-to-one mapping, then it is called an *isomorphism*. Two algebras R and R' are said to be *isomorphic* if there exists an isomorphism of R onto R'.

A homomorphism may be described with the aid of two-sided ideals of the algebra R.

I. *Under a homomorphism of the algebra R into the algebra R', the inverse image I of the zero $0'$ of R' is a two-sided ideal in R.*

In fact, suppose $x \in I$, $y \in I$; this means that $x \to 0'$, $y \to 0'$; consequently, for arbitrary $a \in R$ and arbitrary numbers α, β,

$$ax \to a'0' = 0', \; xa \to 0'a' = 0', \; \alpha x + \beta y \to 0' + 0' = 0'.$$

These relations imply that $ax \in I$, $xa \in I$, $x+y \in I$, i.e. I is a two-sided ideal.

This ideal is called the *kernel of the homomorphism* of the algebra R into the algebra R'.

Two elements x, y of the algebra R go over into the same element of the algebra R' if and only if $x-y \in I$. This shows that the inverse images of elements of the algebra R' are residue classes modulo the ideal I, i.e. they are elements of the algebra R/I. Thus,

II. *Under a homomorphic mapping of the algebra R, the inverse image I of the zero element is a two-sided ideal of this algebra, and the homomorphic image itself is isomorphic to the quotient algebra R modulo I. Conversely, every two-sided ideal I of the algebra R induces a homomorphism of the algebra R onto the quotient algebra R/I.*

This homomorphism of the algebra R onto R/I, consisting in this that every element $x \in R$ is assigned the class ξ which contains x, is called the *natural* (or *canonical*) *homomorphism* of the algebra R onto the algebra R/I.

Proposition II signifies that if we identify the images of the elements of the algebra R under the given homomorphism with the elements of the algebra R/I which correspond to them, then this homomorphism becomes the natural homomorphism of the algebra R onto the algebra R/I.

III. *The quotient algebra R/I is simple if and only if I is a maximal two-sided ideal in R.*

Proof. Suppose I_1 is a two-sided ideal in the algebra R/I. The inverse image I' of I_1 under the homomorphism $R \to R/I$ is a two-sided ideal in the algebra R, containing I. Therefore, if I is a maximal ideal in R, then I' coincides with I; consequently, every ideal I_1 in R/I consists only of the zero; R/I is a simple algebra. Conversely, suppose I is not a maximal two-sided ideal in R. Then there exists a two-sided ideal $I_1' \supset I$, $I_1' \neq I$.

The image of the ideal I_1' under the homomorphism $R \to R/I$ is a two-sided ideal in R/I which is different from (0). Consequently, R/I is not a simple algebra.

In certain cases it is convenient to consider the *anti-homomorphism* of algebras instead of their homomorphism; an anti-homomorphism is a mapping $x \to x'$ satisfying the conditions

$$\lambda x \to \lambda x', \; x+y \to x'+y', \; xy \to y'x'$$

for $x \to x'$, $y \to y'$. Clearly, Proposition I remains valid for an anti-homomorphism of algebras.

7. Regular representations of algebras. An example of a homomorphism of an algebra is the so-called left regular representation of the algebra. Each element $a \in R$ is assigned the operator A_a of left multiplication by a:

$$A_a x = ax;$$

clearly, A_a is a linear operator in R, considered as a vector space.

It is also easily seen that the correspondence $a \to A_a$ is a homomorphism of the algebra R into the algebra of all linear operators in R. This homomorphism is called a *left regular representation* of the algebra R. The kernel of this homomorphism consists of all those elements a of the algebra R which satisfy the condition

$$ax = 0 \text{ for all } x \in R.$$

It follows, in particular, that *if R is an algebra with identity, then a left regular representation is an isomorphism.*

Obviously, a left ideal I_l of the algebra R can be defined as a subspace in R, which is distinct from R and invariant with respect to all the operators A_a of the left regular representation. Hence, the operator A_a induces a linear operator $A_a^{\wedge}$ in the quotient space R/I_l (see subsection 7, § 1) and, clearly, the correspondence $a \to A_a^{\wedge}$ is a homomorphism of the algebra R into the algebra of linear operators in the space R/I_l.

An algebra R is said to be *primitive* if it contains a maximal left ideal I_l for which this homomorphism $a \to A_a^{\wedge}$ is an isomorphism. In this case, the set of all operators $A_a^{\wedge}$ is an irreducible algebra of operators in R/I_l. In fact, suppose $\mathfrak{M}$ is a subspace in R/I_l which is invariant with respect to all the operators $A_a^{\wedge}$, $a \in R$; then the set J of all representatives x of classes $\xi \in \mathfrak{M}$ is a (proper or improper) left ideal in R, containing I_l, and therefore either $J = I_l$ or $J = R$. In the first case, $\mathfrak{M} = (0)$, and in the second, $\mathfrak{M} = R/I_l$.

Thus,

I. *Every primitive algebra is isomorphic to an irreducible algebra of linear operators in some vector space.* (We note that the converse assertion is also true; its proof will be left to the reader.)

Combining this result with Proposition V, subsection 5, we conclude that:

II. *Every primitive algebra is semisimple.*

III. *If $I \neq (0)$ is a two-sided ideal in the primitive algebra R and if a is an arbitrary nonzero element of the algebra R, then $Ia \neq (0)$.*

Proof. In virtue of Proposition I, we may assume that R is an irreducible algebra of linear operators in some vector space X; then $Ix = X$ for $x \in X$, $x \neq 0$. But if the operator $a \neq 0$, then there exists a vector $x \in X$ such that $ax \neq 0$, and this means that $Iax = X$. From this it follows that $Ia \neq (0)$.

A two-sided ideal I in the algebra R is said to be *primitive* if the quotient algebra R/I is primitive.

We have already defined a left regular representation; a right regular representation is defined analogously. Namely, each element $a \in R$ is assigned the operator B_a of right multiplication by a:

$$B_a x = xa.$$

The correspondence $a \to B_a$ is called a *right regular representation* of the algebra R; it is easily seen that this correspondence is an anti-homomorphism of the algebra R into the algebra of linear operators in the space R. The preceding discussion carries over completely to right regular representations; one must only replace the left ideals by right ideals.

§ 8. Topological algebras

1. Definition of a topological algebra. R is called a *topological algebra* if:

1) R is an algebra;

2) R is a locally convex topological linear space;

3) the product xy is a continuous function of each of the factors x, y provided the other factor is fixed.

[We note that many results of this section remain valid also for the case when the algebra R is an arbitrary topological linear space (not necessarily locally convex), satisfying condition 3). In particular, this applies to all the results of subsections 1 and 2.]

A mapping $x \to x'$ of the topological algebra R into the topological algebra R' is called a *continuous homomorphism* if:

1) $x \to x'$ is a homomorphism of the algebra R into the algebra R';

2) $x \to x'$ is a continuous mapping of the topological space R into the topological space R'.

In particular, if the mapping $x \to x'$ is an isomorphism, then it is called a *continuous isomorphism*.

A continuous anti-homomorphism and a continuous anti-isomorphism are defined analogously.

Two topological algebras R and R' are said to be *topologically isomorphic* if there exists a continuous isomorphism of R onto R' whose inverse is also continuous.

A subset $R_1 \subset R$ is said to be a *closed subalgebra* of the algebra R if:

α') R_1 is a subalgebra of the algebra R;

β') R_1 is a closed subspace of the topological space R.

I. *If R_1 is a subalgebra of the algebra R, then its closure $\bar{R}_1$ is a closed subalgebra of R.*

Proof. Clearly, it will suffice to show that $\bar{R}_1$ is a subalgebra of the algebra R. In virtue of II, subsection 2, § 3, $\bar{R}_1$ is a linear subspace in R, and hence it will be sufficient to establish that if $x, y \in \bar{R}_1$ then also $xy \in \bar{R}_1$. But, for fixed $y \in R_1$, the continuous function $f(x) = xy$ maps R_1 into R_1 and hence $\bar{R}_1$ into $\bar{R}_1$; this means that $xy \in \bar{R}_1$ for $x \in \bar{R}_1$, $y \in R_1$. But then, for fixed $x \in \bar{R}_1$, the continuous function $\varphi(y) = xy$ maps R_1 into $\bar{R}_1$ and therefore also $\bar{R}_1$ into $\bar{R}_1$; hence $x, y \in \bar{R}_1$ implies that $xy \in \bar{R}_1$.

Clearly, the intersection of closed subalgebras of the algebra R is also a closed subalgebra of the algebra R. In particular, the intersection of all closed subalgebras of the algebra R which contain a given set $S \subset R$ is the *minimal closed subalgebra containing* S. We denote it by $R_t(S)$.

II. *The algebra* $R_t(S)$ *is the closure of the algebra* $R_a(S)$:

$$R_t(S) = \overline{R_a(S)}.$$

In fact, every closed subalgebra which contains S must also contain $\overline{R_a(S)}$; on the other hand, in virtue of I, $\overline{R_a(S)}$ itself is a closed subalgebra; therefore, $\overline{R_a(S)}$ is the minimal closed subalgebra containing S.

III. *The closure of a commutative subalgebra of a topological algebra is commutative.*

Proof. Suppose $\mathfrak{A}$ is a commutative subalgebra of the topological algebra R; we shall prove that $\bar{\mathfrak{A}}$ is also commutative. For fixed $y \in \mathfrak{A}$ the continuous mappings $f(x) = xy$ and $\varphi(x) = yx$ coincide on $\mathfrak{A}$, and hence also on $\bar{\mathfrak{A}}$ (see IX, subsection 7, § 2). Consequently, $xy = yx$ for all $x \in \bar{\mathfrak{A}}$. But then for fixed $x \in \bar{\mathfrak{A}}$ the continuous mappings $f_1(y) = xy$ and $\varphi_1(y) = yx$ coincide on $\mathfrak{A}$, and hence also on $\bar{\mathfrak{A}}$; consequently, $xy = yx$ for all $x, y \in \bar{\mathfrak{A}}$.

IV. *A maximal commutative subalgebra of a topological algebra is closed.*

Proof. The closure $\bar{\mathfrak{A}}$ of a maximal commutative subalgebra $\mathfrak{A}$ is a commutative subalgebra which contains $\mathfrak{A}$. In virtue of the maximality of $\mathfrak{A}$, this is possible only when $\bar{\mathfrak{A}} = \mathfrak{A}$.

V. *The set* R_S *of all elements* x *of a topological algebra* R, *which commute with all elements of some set* $S \subset R$, *is a closed subalgebra of the algebra* R.

Proof. If x, y commute with all the elements in S, then clearly, $\alpha x, x+y$ and xy possess the same property; consequently, R_S is a subalgebra. We shall prove that R_S is closed. To this end, we denote by R_y the set of all elements $x \in R$ for which $xy = yx$. Since R_y is the set of all elements x on which the continuous mappings $f_1(x) = xy$ and $f_2(x) = yx$ coincide, R_y is closed. Therefore R_S, being the intersection of all R_y, $y \in S$, is also closed.

The algebra R_S is called the *commutant* of the set S; we denote it by S'. The following relations are easily verified:

1) if R contains the identity e, then $e \in S'$;

2) if $S_1 \subset S_2$, then $S_1' \supset S_2'$;

3) $S'' \supset S$.

VI. *The center* Z *of a topological algebra* R *is a closed commutative subalgebra in* R.

In fact, since $Z = R'$ the assertion follows directly from V.

VII. *The closure of a (left, right, two-sided) ideal in a topological algebra, which does not coincide with the entire algebra, is also a (left, right, respectively two-sided) ideal in this algebra.*

Proof. Suppose, for instance, that I_l is a left ideal in the topological algebra R and that $\bar{I}_l \neq R$. Since I_l is a subspace in R, its closure $\bar{I}_l$ is also a subspace in R. Therefore it will be sufficient to show that if $x \in \bar{I}_l$, $y \in R$, then $yx \in \bar{I}_l$. But this follows from the fact that for fixed $y \in R$, the continuous function $f(x) = yx$ maps I_l into I_l, and hence $\bar{I}_l$ into $\bar{I}_l$.

2. Topological adjunction of the identity. Suppose R is a topological algebra without identity and that R' is the algebra obtained from R upon adjunction of the identity. The algebra R' may be considered as the set of all pairs $\langle \alpha, x \rangle$, where $x \in R$ and the α's are numbers; therefore, a topology can be introduced in R' by considering R' to be the topological product of the topological spaces $\mathfrak{C}$ (set of all complex numbers) and R. It is easily seen that then R' becomes a topological algebra. The transition from the topological algebra R to the topological algebra R' will be called the *topological adjunction of the identity*.

3. Algebras with continuous inverse. A topological algebra R with identity is called an *algebra with continuous inverse* if there exists a neighborhood $U_0(e)$ possessing the following properties:

1) every element $x \in U_0(e)$ has an inverse x^{-1};

2) x^{-1} is a continuous function of x at the point $x = e$.

In the remainder of this subsection, R will denote an algebra with continuous inverse.

I. *If x_0 is an element of the algebra R having an inverse x_0^{-1}, then there exists a neighborhood $U(x_0)$ such that:*

a) *every element $x \in U(x_0)$ has an inverse x^{-1};*

b) *x^{-1} is a continuous function of x at the point $x = x_0$.*

Proof. The element $e+x_0^{-1}y$ is a continuous function of y in the entire algebra, in particular, at the point $y = 0$. Consequently, a neighborhood $U(0)$ can be found such that $e+x_0^{-1}y \in U_0(e)$ for $y \in U(0)$, and hence $(e+x_0^{-1}y)^{-1}$ exists. Suppose $U(x_0)$ is the neighborhood of the point x_0, consisting of all the elements x_0+y, $y \in U(0)$. If $x_0+y \in U(x_0)$, then the element

$$x = x_0+y = x_0(e+x_0^{-1}y)$$

has an inverse, because x_0^{-1} and $(e+x_0^{-1}y)^{-1}$ exist; moreover,

$$x^{-1} = (x_0+y)^{-1} = (e+x_0^{-1}y)^{-1}x_0^{-1}. \tag{1}$$

Now $(e+x_0^{-1}y)^{-1}$ is a continuous function of $u = e+x_0^{-1}y$ at the point $u = e$; but u is a continuous function of y at the point $y = 0$ and hence also of $x = x_0+y$

at the point $x = x_0$. Consequently, $(e+x_0^{-1}y)^{-1}$ and hence also x^{-1} are continuous functions of x at the point $x = x_0$.

The following proposition is proved analogously:

II. *If x_0 is an element of the algebra R having a left (right) inverse, then there exists a neighborhood $U(x_0)$ such that:*

1) *every element $x \in U(x_0)$ has a left (right) inverse;*

2) *the left (right) inverse of the element x is a continuous function of x at the point x_0.*

In this connection, formula (1) *must be replaced by the formulas*

$$x_l^{-1} = (x_0+y)_l^{-1} = x_{0,l}^{-1}(e+yx_{0,l}^{-1})^{-1}, \tag{2a}$$

$$x_r^{-1} = (x_0+y)_r^{-1} = (e+x_{0,r}^{-1}y)^{-1}x_{0,r}^{-1}. \tag{2b}$$

From Propositions I and II, we conclude that

III. *The set of all elements of the algebra R which have an inverse, and also the set of all elements of the algebra R which have a left (right) inverse, are open sets in R. The set of all elements of the algebra R which do not have an inverse and also the set of all elements of the algebra R which do not have a left (right) inverse are closed sets in R.*

Using Proposition III, it is easily proved that

IV. *The closure of a (left, right, two-sided) ideal in R is also a (left, right, respectively two-sided) ideal in R.*

Proof. Suppose I_l is a left ideal in R, and that S_l is the set of all elements of the algebra R, which do not have a left inverse. Then $I_l \subset S_l$ (see I, subsection 4, § 7); consequently, its closure $\bar{I}_l$ is also contained in S_l and hence it does not coincide with R. On the basis of VII, subsection 1, $\bar{I}_l$ is a left ideal in R. The assertion for a right ideal is proved analogously. A two-sided ideal I is also a left ideal; consequently, its closure $\bar{I}$ does not coincide with R; in virtue of VII, subsection 1, it follows from this that $\bar{I}$ is a two-sided ideal.

V. *A maximal (left, right, two-sided) ideal in the algebra R is closed.*

Proof. The closure $\bar{I}_l$ of a maximal left ideal I_l is an ideal containing I_l; in virtue of the fact that I_l is maximal, this is possible only when $\bar{I}_l = I_l$. Consequently, I_l is closed. That a maximal right ideal and a maximal two-sided ideal are closed is proved analogously.

EXAMPLES. 1. Suppose G is an open set in the complex plane, and that R_G is the set of all functions $f(z)$ which are holomorphic in G. With the usual definitions of the operations, R_G forms an algebra, where the function $f_0(z) \equiv 1$ will be the identity in R_G. We define a topology in R_G by taking as the neighborhood basis all possible finite intersections of sets $U(f_0; K; \varepsilon)$ of functions $f(z) \in R_G$ which satisfy the condition $|f(z)-f_0(z)| < \varepsilon$ for all $z \in K$, where $f_0 \in R_G$, and K is a closed bounded set in G. It is easily verified that with the topology defined in this way, R_G becomes a topological algebra.

Suppose now that S is an arbitrary closed set in the complex plane. We shall agree to call any open set $U \supset S$ consisting of a finite number of connected parts, each intersecting S, a neighborhood $U(S)$ of the set S. A function $f(z)$ is said to be *analytic on S* if it belongs to some $R_{U(S)}$. We shall denote by $R(S)$ the set of all functions $f(z)$ which are analytic on S; in this connection the functions $f_1(z) \in R_{U_1(S)}, f_2(z) \in R_{U_2(S)}$, which coincide on some $U(S) \subset U_1(S) \cap U_2(S)$, will not be considered distinct. Defining again the operations in the usual way, we transform $R(S)$ into an algebra. Here, as before, the function $f(z) \equiv 1$ will be the identity in $R(S)$. We shall now define a topology in $R(S)$ by considering a neighborhood in $R(S)$ to be any convex set whose intersection with every $R_{U(S)}$ is a neighborhood in $R_{U(S)}$. [This means that $R(S)$ is the *inductive limit* of the algebras $R_{U(S)}$ (in this connection, see Chapter III, § 11, subsection 7).] It is easily verified that with this definition of a topology, $R(S)$ becomes a topological algebra.

It can also be shown (see WAELBROECK [1]) that *$R(S)$ is an algebra with continuous inverse.*

The algebra $R(S)$ and also its generalization to functions of several complex variables were investigated by WAELBROECK [1].

For other examples of algebras with continuous inverse, see below in § 9, subsections 1 and 3.

2. Consider the algebra $\mathfrak{B}(\mathfrak{H})$ of all bounded linear operators in the Hilbert space $\mathfrak{H}$ (see Example 3, subsection 1, § 7). We define a locally convex topology in $\mathfrak{B}(\mathfrak{H})$ by taking the set of all functionals

$$p_{x,y}(A) = |(Ax, y)|, x, y \in \mathfrak{H}$$

as a sufficient family of seminorms. Then all possible sets $U(A_0; x_1, \cdots, x_n; y_1, \cdots, y_n; \varepsilon)$ of the operators $A \in \mathfrak{B}(\mathfrak{H})$ which satisfy the inequalities

$$|((A-A_0)x_k, y_k)| < \varepsilon, k = 1, 2, \cdots, n$$

for all possible fixed $x_1, \cdots, x_n; y_1, \cdots, y_n \in \mathfrak{H}$ and $\varepsilon > 0$, form a neighborhood basis in $\mathfrak{B}(\mathfrak{H})$. The topology defined in this manner is called the *weak operator topology* in $\mathfrak{B}(\mathfrak{H})$.

In virtue of III, subsection 4, § 3, $\mathfrak{B}(\mathfrak{H})$ is a locally convex space in this topology. It is also easily verified that the product AB is continuous in the weak topology with respect to either of the factors, the other factor fixed; consequently, $\mathfrak{B}(\mathfrak{H})$ is a topological algebra in the weak topology. In this topology, $\mathfrak{B}(\mathfrak{H})$ is not an algebra with continuous inverse. In fact, suppose given a neighborhood $U(A_0; x_1, \cdots, x_n; y_1, \cdots, y_n; \varepsilon)$, $\mathfrak{M}$ the linear hull of the elements $x_1, \cdots, x_n$ defining a neighborhood of the operator $A_0 = 1$ and $P = P_{\mathfrak{M}}$ the projection operator on $\mathfrak{M}$. Then P belongs to this neighborhood and has no inverse in $\mathfrak{B}(\mathfrak{H})$.

4. Resolvents in an algebra with continuous inverse. Suppose R denotes, as before, a topological algebra with continuous inverse. The complex number λ is called a *regular point* of the element $x \in R$ if the inverse $(x-\lambda e)^{-1}$ exists; all non-regular points of the element x are called *points of its spectrum*, and the set of all non-regular points is called the *spectrum* of the element x.

We call $(x-\lambda e)^{-1}$ the *resolvent* of the element x; thus, the resolvent is a vector-valued function

$$x_\lambda = (x-\lambda e)^{-1},$$

which is defined for all regular points λ of the element x.

I. *The resolvent x_λ satisfies the relation*

$$x_{\lambda_2} - x_{\lambda_1} = (\lambda_2 - \lambda_1) x_{\lambda_1} x_{\lambda_2} \tag{1}$$

for any two regular points λ_1, λ_2 of the element x.

In fact, multiplying both members of the obvious identity

$$(x-\lambda_1 e)-(x-\lambda_2 e) = (\lambda_2-\lambda_1)e \tag{2}$$

on the left by x_{λ_1} and then on the right by x_{λ_2} we obtain (1).

II. *The set $\mathfrak{R}_x$ of all regular points of the element x is open and the resolvent x_λ is an analytic vector-valued function on $\mathfrak{R}_x$.*

Proof. Suppose $x_{\lambda_0} = (x-\lambda_0 e)^{-1}$ exists; then there is a neighborhood $U(x-\lambda_0 e)$ such that every element $y \in U(x-\lambda_0 e)$ has an inverse. Since $y(\lambda) = x-\lambda e$ is a continuous function of λ, to this neighborhood there corresponds a neighborhood $U(\lambda_0)$ such that $y(\lambda) \in U(x-\lambda_0 e)$ provided $\lambda \in U(\lambda_0)$; consequently, $x_\lambda = [y(\lambda)]^{-1}$ will exist. By the same token we have proved that $\mathfrak{R}_x$ is open. At the same time we see that x_λ is a continuous function at each point $\lambda_0 \in \mathfrak{R}_x$.

Suppose now that $\lambda \in U(\lambda_0)$. According to (1),

$$x_\lambda - x_{\lambda_0} = (\lambda-\lambda_0)x_{\lambda_0}x_\lambda;$$

consequently,

$$\frac{x_\lambda - x_{\lambda_0}}{\lambda-\lambda_0} = x_{\lambda_0}x_\lambda .$$

Passing to the limit as $\lambda \to \lambda_0$ in the last equality and using the continuity of x_λ at the point λ_0, we conclude that the

$$\lim_{\lambda\to\lambda_0} \frac{x_\lambda - x_{\lambda_0}}{\lambda-\lambda_0} = x_{\lambda_0}^2$$

exists, so that x_λ is analytic in $\mathfrak{R}_x$.

III. *The set $\mathfrak{R}_x$ contains a neighborhood of the point at infinity and the resolvent x_λ is regular at infinity.*

Proof. Suppose $U_0(e)$ is a neighborhood of the identity, each element of which has an inverse. We set $\mu = 1/\lambda$. Since the element $e-\mu x$ is a continuous function of μ, there exists a neighborhood $|\mu| < \varepsilon$ such that $e-\mu x \in U_0(e)$, and therefore $(e-\mu x)^{-1}$ exists for $|\mu| < \varepsilon$. Furthermore,

$$\lim_{\mu\to 0}(e-\mu x)^{-1} = e. \tag{3}$$

This means that $\{e-(1/\lambda)x\}^{-1}$ exists in the neighborhood $|\lambda| > 1/\varepsilon$ of the point at infinity. But then

$$x_\lambda = (x-\lambda e)^{-1} = \left[-\lambda\left(e-\frac{1}{\lambda}x\right)\right]^{-1} = -\frac{1}{\lambda}\left(e-\frac{1}{\lambda}x\right)^{-1} = -\mu(e-\mu x)^{-1} \tag{4}$$

also exists for these values of λ. It remains to show that x_λ is an analytic function of μ in the entire neighborhood $|\mu| < \varepsilon$. But the existence of the derivative $(x_\lambda)'_\mu$

for $\mu \neq 0$, $|\mu| < \varepsilon$, follows from II, and at the point $\mu = 0$ in virtue of (3) and (4) this derivative equals

$$\lim_{\mu \to 0} \frac{x_\lambda}{\mu} = -\lim_{\mu \to 0} (e - \mu x)^{-1} = -e.$$

This proves that x_λ is analytic at infinity.

THEOREM 1. *The spectrum of every element x of an algebra with continuous inverse is a nonvoid set.*

In fact, if all points in the plane were regular points of the element $x \in R$, then the resolvent x_λ of the element x would be an analytic function on the entire plane and on the basis of the Liouville theorem (see II, subsection 12, § 3), x_λ would be a constant element: $x_\lambda = c$; then

$$e = (x - \lambda e)c. \tag{5}$$

Setting $\lambda = 0$, we would then obtain

$$e = xc, \tag{6}$$

and (5) would yield $\lambda c = 0$, from which it would follow that $c = 0$ inasmuch as λ is arbitrary; the last equality contradicts equality (6).

5. Topological division algebras with continuous inverse. We recall (see subsection 3, § 7) that an algebra R is called a *division algebra* if every nonzero element in R has an inverse. If, in particular, a topological algebra with continuous inverse is a division algebra then it is called a *topological division algebra with continuous inverse.*

THEOREM 2. *Every topological division algebra R with continuous inverse is isomorphic to the field of complex numbers.*

Proof. Clearly, it will suffice to show that every element $x \in R$ has the form $x = \lambda e$. The correspondence $\lambda e \to \lambda$ will then be an isomorphism of R onto the field of complex numbers. But if $x - \lambda e$ does not vanish for any λ, then $(x - \lambda e)^{-1}$ exists for all λ (because R is a division algebra), i.e. the spectrum of x is the void set, which is impossible in virtue of Theorem 1 in subsection 4.

For the particular case of normed algebras (see below, § 9), Theorem 2 was first published by MAZUR [1] who arrived at it in an entirely different way. The proof given here is due essentially to GELFAND [1, 4] who also considered the particular case of normed algebras. Later, vari ous generalizations of this theorem were given (see, for example, ARENS [3] and STONE [8]).

6. Algebras with continuous quasi-inverse. A topological algebra R is called an *algebra with continuous quasi-inverse* if there exists a neighborhood $U(0)$ in R which has the following properties:

1) every element $x \in U(0)$ has a quasi-inverse x';

2) the quasi-inverse x' is a continuous function of x at the point $x = 0$.

If an algebra R with continuous quasi-inverse has an identity, then, clearly, R is an algebra with continuous inverse. From this we conclude that

I. *If R is an algebra with continuous quasi-inverse, not containing the identity, then the algebra R', obtained from R by the topological adjunction of the identity, is an algebra with continuous inverse.*

Therefore the study of algebras with continuous quasi-inverse reduces to the study of algebras with continuous inverse,

II. *Every two-sided ideal I in the algebra R with continuous quasi-inverse is also an algebra with continuous quasi-inverse.*

Proof. Suppose $U(0)$ is a neighborhood of zero in R which satisfies conditions 1) and 2). Then $U(0) \cap I$ is a neighborhood of zero in I, possessing the analogous property; to prove this, it suffices to note that if $x \in I$, then also $x' \in I$ in virtue of the relation $x' = -x-xx'$.

Various generalizations of algebras with continuous quasi-inverse under additional restrictions have been made in a number of papers (see, for example, KAPLANSKY [1–14]).

In the papers of a number of authors there are investigated generalizations of the Gelfand theory to topological algebras which are not locally convex spaces. Among such algebras, of greatest interest are the following two classes:

1) Algebras in which there exists a bounded neighborhood of zero; they are said to be *locally bounded.*

2) Algebras in which the topology is defined by a metric; they are called *F-algebras.*

The reader will find a detailed discussion of the theory of such algebras in the monograph by ŻELAZKO [6] (also see ŻELAZKO [1, 2], MITYAGIN, ROLEWICZ and ŻELAZKO [1], ROLEWICZ and ŻELAZKO [1, 2], ROLEWICZ [2] and TURPIN [1, 2]).

§ 9. Normed algebras

1. Definition of a normed algebra. R is called a *normed algebra* if:

1) R is an algebra;

2) R is a normed space;

3) for any two elements $x, y \in R$,

$$|xy| \leqq |x|\,|y|; \tag{1}$$

4) if R contains an identity e, then $|e| = 1$.

The norm in a normed algebra R defines a topology in R in a natural manner (see subsection 1, §4); recall that in this topology, the open balls $|x-x_0| < r$ with center at x_0 form a neighborhood basis of the element $x_0 \in R$.

I. *In the norm topology, the product xy is a continuous function of the variables x, y simultaneously.*

In fact, in virtue of (1),

$$\begin{aligned} |xy-x_0y_0| &= |(x-x_0)(y-y_0)+(x-x_0)y_0+x_0(y-y_0)| \\ &\leqq |x-x_0||y-y_0|+|x-x_0||y_0|+|x_0||y-y_0|; \end{aligned} \tag{2}$$

now the assertion follows directly from this.

Since a normed space R is a topological linear space, we conclude from I that

II. *In the topology defined by the norm, a normed algebra is a topological algebra.*

A normed algebra R is said to be *complete* if R is a complete normed space. A complete normed algebra will also be called a *Banach* algebra.

III. *Every non-complete normed algebra can be embedded in a complete normed algebra.*

Proof. Suppose $\tilde{R}$ is the completion of the normed space R (see II, subsection 1, § 4). Now define multiplication in $\tilde{R}$. Suppose $\tilde{x}, \tilde{y} \in \tilde{R}$ and let $\langle x_n \rangle$, $\langle y_n \rangle$ be fundamental sequences in R, which define $\tilde{x}, \tilde{y}$ respectively. It follows from inequality (2) with x_n, x_m in place of x, x_0 and y_n, y_m in place of y, y_0 that $\langle x_n y_n \rangle$ also is a fundamental sequence. The element in $\tilde{R}$ which it defines will be considered to be the product $\tilde{x}\tilde{y}$ of the elements $\tilde{x}, \tilde{y}$. Again applying inequality (2), it can also be easily verified that $\tilde{x}\tilde{y}$ does not depend on the choice of the fundamental sequences $\langle x_n \rangle$, $\langle y_n \rangle$ which define $\tilde{x}, \tilde{y}$. If, in particular, $\tilde{x} = x \in R$, $\tilde{y} = y \in R$, then, setting $x_n = x$, $y_n = y$, we conclude that in this case the product coincides with the product in R. Passing to the limit in the relations for the elements in the algebra R, it is easily shown that $\tilde{R}$ is an algebra and that the inequality $|\tilde{x}\tilde{y}| \leqq |\tilde{x}||\tilde{y}|$ is satisfied for elements of the ring $\tilde{R}$. Consequently, $\tilde{R}$ is a complete normed algebra which contains R as a subalgebra.

The algebra $\tilde{R}$ is called the *completion* of the algebra R.

EXAMPLES. 1. *The algebra $C(T)$.* Suppose T is a topological space. The set $C(T)$ of all bounded continuous functions $x(t)$ on T forms a Banach space (see Example 2, subsection 1, § 4). Recall that the norm $|x|$ in $C(T)$ is defined by the formula

$$|x| = \sup_{t \in T} |x(t)|.$$

Multiplication in $C(T)$ can be defined as the multiplication of functions, i.e. $(xy)(t) = x(t)y(t)$. It is easily seen that the condition $|xy| \leqq |x|\,|y|$ will be satisfied so that $C(T)$ becomes a Banach algebra. If T is compact, then the boundedness condition on the functions $x(t)$ is redundant in virtue of VII, subsection 7, § 2.

2. *The algebra $\mathfrak{B}(X)$.* Recall that $\mathfrak{B}(X)$ denotes the set of all bounded linear operators in the Banach space X (see subsection 4, § 4). We saw, above, that $\mathfrak{B}(X)$ is a Banach space where the norm in $\mathfrak{B}(X)$ is the operator norm. But multiplication in $\mathfrak{B}(X)$ is also defined as the multiplication of operators with

$$|AB| \leqq |A||B|$$

according to what we proved in subsection 4, § 4. Consequently, $\mathfrak{B}(X)$ is a Banach algebra.

3. *The algebra W.* We denote by W the set of all absolutely convergent series $x(t) = \sum_{n=-\infty}^{\infty} c_n e^{int}$ with norm

$$|x| = \sum_{n=-\infty}^{\infty} |c_n|.$$

We obtain a Banach algebra by defining addition, scalar multiplication, and multiplication as the corresponding operations on functions $x(t)$.

2. Adjunction of the identity. Suppose R is a normed algebra without identity, and let R' be the algebra obtained from R upon adjunction of the identity. We may introduce a norm in R' by setting

$$|\alpha e + x| = |\alpha| + |x|;$$

it is easily verified that R' then becomes a normed algebra.

If R is a complete algebra without identity, then R′ is also a complete algebra.

The proof is simple and so we shall omit it.

3. The radical in a normed algebra.

I. *For every element x of the normed algebra R,* $\lim_{n\to\infty} \sqrt[n]{|x^n|} < \infty$ *exists.*

Proof. For $n \geqq m$, $n = km + l$, in virtue of inequality (1), subsection 1,

$$|x^n| = |x^{km+l}| \leqq |x^k|^m |x^l| \qquad (0 \leqq l < m),$$

from which it follows that

$$|x^n|^{1/n} \leqq |x^k|^{1/k - 1/kn} |x^l|^{1/n}.$$

Fixing k, we let n tend to infinity; then

$$\overline{\lim_{n\to\infty}} \sqrt[n]{|x^n|} \leqq \sqrt[k]{|x^k|} \leqq |x| < \infty. \tag{1}$$

Now we let k tend to infinity. We find that

$$\overline{\lim_{n\to\infty}} \sqrt[n]{|x^n|} \leqq \underline{\lim_{n\to\infty}} \sqrt[n]{|x^n|};$$

thus, $\lim_{n\to\infty} \sqrt[n]{|x^n|}$ exists. In virtue of (1), we see at the same time that this limit is finite.

II. *If the element x of the normed algebra R belongs to the radical of the algebra, then*

$$\lim_{n\to\infty} \sqrt[n]{|x^n|} = 0. \tag{2}$$

Proof. It obviously suffices to consider the case of an algebra with identity (see (3), subsection 5, § 7). If x belongs to the radical, then for an arbitrary $y \in R$, $(e+yx)^{-1}$ exists. In particular, for $y = -\lambda e$, for all λ (in the algebra R and hence also in its completion $\tilde{R}$) the element $(e-\lambda x)^{-1}$ exists, which therefore is an entire function of λ (see II, subsection 4, § 8). But then the series $(e-\lambda x)^{-1} = e+\lambda x +\lambda^2 x^2 + \cdots$ converges absolutely in the norm in $\tilde{R}$ for all values of λ (see III, subsection 7, § 4).

By virtue of formula (5), subsection 7, § 4 for the radius of convergence, we conclude from this that

$$\lim_{n\to\infty} \sqrt[n]{|x^n|} = 0.$$

4. Banach algebras with identity.

I. *Every Banach algebra with identity is an algebra with continuous inverse; moreover, every element x, satisfying the inequality $|x-e| < 1$, is invertible.*

Proof. Consider the neighborhood $U_0(e)$ of the identity e, defined by the inequality

$$|x-e| < 1.$$

Every element $x \in U_0(e)$ is representable in the form $x = e+y$, where $|y| < 1$. In virtue of the completeness of R, the series

$$e-y+y^2-y^3+\cdots \tag{1}$$

converges absolutely in R, because its terms are at most equal in norm to the corresponding terms of the convergent series

$$1+|y|+|y|^2+|y|^3+\cdots.$$

Denoting the sum of series (1) by z, we have

$$z = e-y(e-y+y^2-y^3+\cdots) = e-yz, \tag{2}$$

from which it follows that

$$z+yz = e,\ (e+y)z = e.$$

Analogously, $z(e+y) = e$, so that z is the inverse of the element $x = e+y$.

Thus, every element $x \in U(e)$ has an inverse $x^{-1} = z$. We shall now prove that x^{-1} is continuous at the point $x = e$. It follows from (2) that

$$|x^{-1}-e| = |z-e| \leqq |y||z|;$$

on the other hand,

$$|z| \leqq 1+|y|+|y|^2+\cdots = \frac{1}{1-|y|}.$$

Therefore, for $|y| < \varepsilon$,

$$|x^{-1}-e| \leqq |y|\,|z| \leqq \frac{|y|}{1-|y|} < \frac{\varepsilon}{1-\varepsilon}$$

and so the function x^{-1} is continuous at the point $x = e$.

Thus, all results proved for algebras with continuous inverse (see subsections 3–5, § 8) carry over to Banach algebras with identity. In particular,

II. *In a Banach algebra R with identity*:

1) *the set of all elements x having a (left, right, two-sided) inverse is an open set*;

2) *the inverse* x^{-1} *is a continuous function of x at all points for which* x^{-1} *exists*;

3) *the closure of a (left, right, two-sided) ideal is a (left, right, respectively two-sided) ideal*;

4) *a maximal (left, right, two-sided) ideal is closed*;

5) *the set* $\mathfrak{R}_x$ *of all regular points of the element* $x \in R$ *is open and the resolvent* $x_\lambda = (x-\lambda e)^{-1}$ *is an analytic function of* λ;

6) *the spectrum of every element* $x \in R$ *is a nonvoid set.*

III (Gelfand [1] and Mazur [1]). *Every complete normed division algebra is isomorphic to the field of complex numbers.*

Furthermore,

IV. *In a Banach algebra R, the quotient algebra R/I modulo a closed two-sided ideal I is a Banach algebra.*

Proof. Since I is closed, the quotient space R/I is a Banach space (see subsection 3, § 4); on the other hand, since I is a two-sided ideal, R/I is an algebra (see subsection 4, § 7). It remains to show that the following conditions are satisfied in R/I: 1) $|\tilde{x}\tilde{y}| \leqq |\tilde{x}||\tilde{y}|$; 2) if R/I contains the identity $\tilde{e}$, then $|\tilde{e}| = 1$.

Since

$$|\tilde{x}| = \inf_{x\in\tilde{x}} |x|, \ |\tilde{y}| = \inf_{y\in\tilde{y}} |y|,$$

for arbitrary $\varepsilon > 0$ there exist $x_0 \in \tilde{x}$, $y_0 \in \tilde{y}$ such that

$$|x_0| < |\tilde{x}|+\varepsilon, \ |y_0| < |\tilde{y}|+\varepsilon.$$

Then

$$|\tilde{x}\tilde{y}| = \inf_{a\in I} |x_0 y_0+a| \leqq |x_0 y_0| \leqq |x_0|\,|y_0| < (|\tilde{x}|+\varepsilon)(|\tilde{y}|+\varepsilon),$$

from which it follows that

$$|\tilde{x}\tilde{y}| \leqq |\tilde{x}||\tilde{y}|$$

inasmuch as the number $\varepsilon > 0$ is arbitrary.

We shall now prove that $|\tilde{e}| = 1$. Since $e \in \tilde{e}$, $|\tilde{e}| \leqq |e| = 1$. On the other hand, if $y \in \tilde{e}$, then $y = e-x$, where $x \in I$. Moreover, $|y| \geqq 1$ since otherwise, by virtue of I, the element $x = e-y$ is invertible, which is impossible. Therefore $|\tilde{e}| = \inf_{y\in\tilde{e}} |y| \geqq 1$.

Proposition III admits the following strengthening.

V. *If in the Banach algebra R with identity every element* $x \neq 0$ *has a left inverse, then R is isomorphic to the field of complex numbers.*

In fact, in virtue of V, subsection 2, § 7, R is a division algebra, and the assertion follows from Proposition III.

Clearly, the analogous proposition also holds for algebras all of whose elements with the exception of $x = 0$ have a right inverse.

5. Resolvent in a Banach algebra with identity. Proposition III, subsection 4, § 8 can be strengthened in Banach algebras with identity.

I. *For* $|\lambda| > \lim_{n\to\infty} \sqrt[n]{|x^n|}$, *the resolvent* x_λ *can be expanded in an absolutely convergent Laurent series*

$$x_\lambda = \sum_{n=0}^{\infty} \lambda^{-(n+1)} x^n; \tag{1}$$

series (1) *diverges for* $|\lambda| < \lim_{n\to\infty} \sqrt[n]{|x^n|}$.

Proof. It follows from III, subsection 7, § 4 that the series (1) converges absolutely for $|\lambda| > \lim_{n\to\infty} \sqrt[n]{|x^n|}$ and diverges for $|\lambda| < \lim_{n\to\infty} \sqrt[n]{|x^n|}$. Multiplying this series on the right or on the left by $x - \lambda e$, we obtain e. Consequently, for $|\lambda| > \lim_{n\to\infty} \sqrt[n]{|x^n|}$ the series (1) represents the resolvent, which is what was required.

Thus, to the set $\mathfrak{R}_x$ of all regular points of the element x there belongs the region exterior to the circle $|\lambda| = \lim_{n\to\infty} \sqrt[n]{|x^n|}$. This leads to the following definition: the number

$$r(x) = \sup_{\lambda \in S_x} |\lambda|,$$

where S_x is the spectrum of the element x, is called the *spectral radius of the element* x.

II. *For an arbitrary* $x \in R$,

$$r(x) = \lim_{n\to\infty} \sqrt[n]{|x^n|}. \tag{2}$$

Proof. In virtue of I, every number λ satisfying the inequality $|\lambda| > \lim_{n\to\infty} \sqrt[n]{|x^n|}$ belongs to $\mathfrak{R}_x$, i.e., $r(x) \leqq \lim_{n\to\infty} \sqrt[n]{|x^n|}$. On the other hand, for $|\lambda| > r(x)$, the resolvent x_λ is regular; thus, for $|\lambda| > r(x)$, the Laurent series (1) converges, and again in virtue of I it follows from this that

$$r(x) \geqq \lim_{n\to\infty} \sqrt[n]{|x^n|}.$$

III. *The spectral radius possesses the following properties*:
1) $r(x^k) = [r(x)]^k$, 2) $r(\alpha x) = |\alpha| r(\alpha)$, 3) $r(x) \leqq |x|$.

6. Continuous homomorphisms of normed algebras.

I. *Every continuous homomorphism* $x \to x'$ *of the normed algebra* R *into the normed algebra* R' *satisfies the inequality*

$$|x'| \leqq C|x|, \tag{1}$$

where C *is a constant.*

In fact, such a homomorphism is, in particular, a continuous linear operator from R into R'; in virtue of I, subsection 4, § 4, such an operator is bounded.

II. *Every continuous homomorphism* $x \to x'$ *of a normed algebra* R *into a normed* *algebra R' is uniquely extendible to a continuous homomorphism of the completion $\tilde{R}$ of the algebra R into the completion $\tilde{R}'$ of the algebra R'.*

Proof. Since the homomorphism $x \to x'$ is, in particular, a bounded linear operator from R into R', it is uniquely extendible to a bounded linear operator from $\tilde{R}$ into $\tilde{R}'$. This operator is also a homomorphism. In fact, if $x, y \in \tilde{R}$, $x_n, y_n \in R$ and $x_n \to x$, $y_n \to y$ as $n \to \infty$, then $x_n' \to x'$, $y_n' \to y'$ as $n \to \infty$; hence, passing to the limit in the relation $(x_n y_n)' = x_n' y_n'$, we obtain that $(xy)' = x'y'$.

III. *Every continuous isomorphism of a Banach algebra R onto a Banach algebra R' is a topological isomorphism.*

The assertion follows directly from the Banach theorem (see VI, subsection 4, § 4) because such an isomorphism is, in particular, a one-to-one bounded linear operator which maps the Banach space R onto the Banach space R'.

IV. *Under a continuous homomorphism of the Banach algebra R onto the Banach algebra R', the kernel I of the homomorphism is a closed two-sided ideal in R, and the algebra R' itself is topologically isomorphic to the quotient algebra R/I. Conversely, every closed two-sided ideal I of the Banach algebra R induces a continuous homomorphism (the so-called natural homomorphism) of the algebra R onto the algebra R/I.*

Proof. In virtue of II, subsection 6, § 7, the kernel I of the homomorphism of R onto R' is an ideal and R' is isomorphic to R/I; as the inverse image of a closed set, consisting only of the zero of the algebra R', the ideal I is closed. Consequently, R/I is a Banach algebra. Taking the greatest lower bound in (1) over all x in the class $\xi \in R/I$ which corresponds to a given x', we obtain that $|x'| \leqq C|\xi|$; consequently, our isomorphism of R/I onto R' is continuous and hence topological. The converse assertion is obvious because $|\xi| \leqq |x|$ for $x \in \xi \in R/I$.

The mapping $x \to x'$ of the normed algebra R onto the normed algebra R' is said to be an *isometric isomorphism* if :

1) $x \to x'$ is an isomorphism of the algebra R onto the algebra R';

2) $x \to x'$ is an isometric mapping of the normed space R onto the normed space R'.

An *isometric anti-isomorphism* is defined analogously.

Clearly, an isometric isomorphism is also a topological isomorphism.

Two normed algebras R and R' are said to be *isometrically isomorphic* if there exists an isometric isomorphism of R onto R'.

7. Regular representations of a normed algebra. Recall that the left and right regular representations $a \to A_a$ and $a \to B_a$ of the algebra R are defined by means of the formulas

$$A_a x = ax, \; B_a x = xa$$

(see subsection 7, § 7).

I. *A left (right) regular representation of a normed algebra R is a continuous* *homomorphism (anti-homomorphism) of the algebra R into the algebra* $\mathfrak{B}(R)$ *of all bounded linear operators in the space R.*

In fact, the inequalities

$$|A_a x| \leqq |a||x|, \; |B_a x| \leqq |a||x| \tag{1}$$

imply that

$$|A_a| \leqq |a|, \; |B_a| \leqq |a|. \tag{2}$$

II. *If R is a normed algebra with identity, then a left (right) regular representation of the algebra R is an isometric isomorphism (anti-isomorphism) of the algebra R into the algebra* $\mathfrak{B}(R)$.

In fact, for $x = e$, inequalities (1) go over into equalities, and hence $|A_a| = |a|$, $|B_a| = |a|$.

We now consider, further, representations which are adjoint to regular representations. We denote by R' the Banach space conjugate to the Banach space R and by A'_a, B'_a the operators in R' which are adjoint to A_a, B_a respectively. From Propositions I, II and the properties of an adjoint operator (see subsection 11, § 5), we conclude that

III. *The correspondence* $a \to B'_a$ *is a continuous homomorphism and the correspondence* $a \to A'_a$ *is a continuous anti-homomorphism of the normed algebra R into the algebra* $\mathfrak{B}(R')$. *But if R contains the identity then* $a \to B'_a$ *is an isometric isomorphism and* $a \to A'_a$ *is an isometric anti-isomorphism.*

IV. *The annihilator of a closed right ideal of a Banach algebra R is a closed subspace in R′ which is distinct from* (0) *and invariant with respect to all operators* B'_a. (Everywhere in this subsection, closure in R' is understood in the sense of the weak topology in R'.) *Conversely, every such closed invariant subspace is an annihilator of some closed right ideal in R.*

Proof. Suppose $\mathfrak{N}$ is the annihilator of the closed right ideal I_r; in virtue of I, subsection 11, § 3, $\mathfrak{N}$ is a closed subspace in R'; we shall prove that $B'_a f \in \mathfrak{N}$ for all $a \in R$ and $f \in \mathfrak{N}$. Suppose $f \in \mathfrak{N}$; by the definition of the annihilator (see subsection 11, § 5),

$$f(x) = 0 \text{ for all } x \in I_r.$$

Since $x \in I_r$ implies that $xa \in I_r$, we also have (see subsection 11, § 5)

$$(x, B'_a f) = (B_a x, f) = f(xa) = 0 \text{ for all } x \in I_r.$$

But this means that $B_a f \in \mathfrak{N}$.

Conversely, suppose $\mathfrak{N}$ is a closed subspace in R' which is distinct from (0) and invariant with respect to all the operators B'_a. According to II, subsection 11, § 3, $\mathfrak{N}$ is the annihilator of some closed subspace $S \subset R$. We shall prove that S is a right ideal in R. Since $\mathfrak{N} \neq (0)$, $S \neq R$. Suppose, furthermore, that $x \in S$, $a \in R$; we shall prove that $xa \subset S$. By the same token, this will prove that S is a right ideal in R. According to III, subsection 11, § 3, S consists of those, and only those, elements x which satisfy the condition

$$f(x) = 0 \text{ for all } f \in \mathfrak{N}.$$

Since $\mathfrak{N}$ is an invariant subspace, it follows from $f \in \mathfrak{N}$ that $B'_a f \in \mathfrak{N}$. Consequently, we also have

$$f(xa) = (x, B'_a f) = 0 \text{ for all } f \in \mathfrak{N}.$$

According to what we proved above, this means that $xa \in S$.

A closed invariant subspace is said to be *minimal* if it is $\neq (0)$ and does not contain any other closed invariant subspace. (Everywhere in the remainder of this subsection, "invariant subspace" means a subspace in R' which is invariant with respect to all the operators B'_a.) A functional in a minimal invariant subspace is called *elementary*.

Combining Propositions IV, subsection 11, § 3, and IV of this subsection, we obtain:

V. *$\mathfrak{N}$ is a minimal invariant subspace in R' if and only if it is the annihilator of a maximal right ideal in R.*

According to II, subsection 4, § 7, every right ideal of an algebra with identity is contained in a maximal right ideal. From this and IV, V, we conclude that:

VI. *If R is an algebra with identity, then every closed invariant subspace in R' which is distinct from* (0) *contains a minimal invariant subspace and, consequently, it contains elementary functionals.*

Suppose, again, that R is a Banach algebra with identity and let $f \neq 0$ be an arbitrary functional in R'. We shall consider the set of all functionals of the form $f_a(x) = f(xa)$, $a \in R$. Clearly it is linear. Its closure in R' forms a closed invariant subspace $\mathfrak{N}$. Since $f \neq 0$, $\mathfrak{N} \neq (0)$. Consequently, $\mathfrak{N}$ contains elementary functionals. It follows from the construction of $\mathfrak{N}$ that every functional $f_0 \in \mathfrak{N}$ is a contact point (in the sense of the weak topology in R') of functionals $f_a(x) = f(xa)$. Hence the following proposition is proved:

VII. *Suppose R is an algebra with identity and let $f \neq 0$ be a functional in R'; there exists an elementary functional which is a weak contact point of functionals of the form $f_a(x) = f(xa)$.*

Regular representations of an algebra can also be used in the proof of the following proposition.

VIII. *Suppose R is a complete topological algebra with identity in which the topology is defined by the norm $|x|$; then R is topologically isomorphic to a Banach algebra.*

Proof. Consider the left regular representation $x \to A_x$ of the algebra R. Since R contains the identity, this representation is an isomorphism with

$$|A_x| = \sup_{|y|=1} |xy| \geqq \left| x \frac{e}{|e|} \right| = \frac{|x|}{|e|},$$

or

$$|x| \leqq |e||A_x|;$$

consequently, the isomorphism $x \to A_x$ is continuous.

We shall prove that the image of the algebra R under this isomorphism is closed in $\mathfrak{B}(R)$ and hence it is a complete algebra. From this and III, subsection 4, it will then follow that $x \to A_x$ is a topological isomorphism, i.e. that R is topologically isomorphic to the Banach algebra of all operators A_x.

To prove this, let us note that the operators A_x satisfy the relation

$$A_x(yz) = A_x y \cdot z;$$

in fact, $A_x(yz) = x(yz) = (xy)z = A_x y \cdot z$. Conversely, every operator A in R which satisfies the relation

$$A(yz) = Ay \cdot z \text{ for all } y, z \in R \tag{3}$$

is an operator of the form A_x, where $x = Ae$. In order to verify this, it is sufficient to set $y = e$ in (3); we then obtain

$$Az = A(ez) = Ae \cdot z = xz.$$

Suppose now that the sequence $\langle A_{x_n} \rangle$ converges to some operator $A \in \mathfrak{B}(R)$ in the sense of the norm in $\mathfrak{B}(R)$. Then, passing to the limit as $n \to \infty$ in the equality $A_{x_n}(yz) = A_{x_n} y \cdot z$, we obtain that A

satisfies condition (3) and hence it is an operator of the form A_x. Consequently, the set of all operators A_x is closed and Proposition VIII is proved.

It follows from this proposition that if R is a complete topological algebra with identity in which the topology is defined by the norm, then the product xy of elements of the algebra is a continuous function with respect to both variables x, y simultaneously.

§ 10. Symmetric algebras

1. Definition and simplest properties of a symmetric algebra. R is called a *symmetric algebra* if:

1) R is an algebra;

2) an operation is defined in R which assigns to each element x in R the element x^* in R in such a way that the following conditions are satisfied:

a) $(\lambda x+\mu y)^* = \bar{\lambda}x^*+\bar{\mu}y^*$;

b) $x^{**} = x$;

c) $(xy)^* = y^*x^*$.

[In the literature, the expression "algebra with involution" is frequently used instead of "symmetric algebra", and the expression "symmetric algebra" is used in another sense (in this connection, see Chapter III, § 14, subsection 1).]

The operation $x \to x^*$ will be called *involution* and the elements x^* and x will be said to be *adjoint* to each other.

An element x is said to be *Hermitian* if $x^* = x$.

I. *Every element x of a symmetric algebra can be uniquely represented in the form $x = x_1+ix_2$, where x_1, x_2 are Hermitian elements.*

In fact, if such a representation holds, then $x^* = x_1-ix_2$; consequently,

$$x_1 = \frac{x+x^*}{2}, \quad x_2 = \frac{x-x^*}{2i}. \tag{1}$$

Thus, this representation is unique. Conversely, the elements x_1, x_2 defined by equalities (1) are Hermitian and $x = x_1+ix_2$.

These elements x_1, x_2 will be called the *Hermitian components of the element x*.

An element x is called *normal* if $x^*x = xx^*$.

If x is a normal element then it follows from formula (1) that x_1 and x_2 commute; conversely, if x_1 and x_2 commute, then the elements $x = x_1+ix_2$ and $x^* = x_1-ix_2$ also commute and, consequently, x is normal. In particular, every Hermitian element is normal.

II. *Every element of the form x^*x is Hermitian.*

In fact, in virtue of c) and b), $(x^*x)^* = x^*x^{**} = x^*x$.

III. *The identity e is a Hermitian element.*

In fact, $e^* = e^*e$ is a Hermitian element; consequently, $e^* = e$.

If R is a symmetric algebra without identity and R' is the algebra obtained from R by adjunction of the identity, then, setting $(\lambda e+x)^* = \bar{\lambda}e+x^*$ for $x \in R$, we define an involution in R' which satisfies conditions a), b) and c), so that R' becomes a symmetric algebra. We shall say that R' is the symmetric algebra obtained from R upon adjunction of the identity.

IV. *If x^{-1} exists, then $(x^*)^{-1}$ also exists and*

$$(x^*)^{-1} = (x^{-1})^*.$$

Proof. Applying the * operation to both parts of the relation

$$x^{-1}x = xx^{-1} = e,$$

we obtain

$$x^*(x^{-1})^* = (x^{-1})^*x^* = e.$$

But this means that $(x^{-1})^*$ is the inverse of x^*.

A subalgebra R_1 of the symmetric algebra R is said to be *symmetric* if $x \in R_1$ implies that $x^* \in R_1$.

The intersection of symmetric subalgebras is also a symmetric subalgebra. In particular, the intersection of all symmetric subalgebras containing a given set $S \subset R$ is the minimal symmetric subalgebra containing S. We shall denote it by $R_{a^*}(S)$. Clearly,

$$R_{a^*}(S) = R_a(S, S^*),$$

where S^* denotes the set of all elements x^*, $x \in S$. Hence, if all the elements of the set $S \cup S^*$ commute pairwise then the algebra $R_{a^*}(S)$ is commutative. In particular, the algebra $R_{a^*}(x)$ is commutative if and only if $x^*x = xx^*$, i.e. when x is normal.

A commutative symmetric subalgebra is said to be *maximal* if it is not contained in any other commutative symmetric subalgebra. As in subsection 1, § 7, it can be shown that *every commutative symmetric subalgebra is contained in some maximal commutative symmetric subalgebra.*

V. *If $\mathfrak{A}$ is a maximal commutative symmetric subalgebra containing a normal element x, and if x^{-1} exists, then $x^{-1} \in \mathfrak{A}$.*

In fact, since x and x^* commute with all elements in $\mathfrak{A}$, x^{-1} and $(x^*)^{-1} = (x^{-1})^*$ possess the same property (see III, subsection 2, § 7). Since $\mathfrak{A}$ is maximal it follows that $x^{-1} \in \mathfrak{A}$.

The mapping $x \to x'$ of a symmetric algebra R into the symmetric algebra R' is called a *symmetric homomorphism* if:

α) $x \to x'$ is a homomorphism;

β) $x \to x'$ implies that $x^* \to x'^*$.

If, furthermore, the mapping $x \to x'$ is an isomorphism, then it is called a *symmetric isomorphism.*

A symmetric homomorphism of algebras is described with the aid of so-called *symmetric two-sided ideals.*

A (left, right, or two-sided) ideal I is said to be *symmetric* if $x \in I$ implies that $x^* \in I$.

A symmetric ideal is automatically a two-sided ideal. In fact, the mapping $x \to x^*$ carries a left ideal into a right ideal and a right ideal into a left ideal; therefore, if the mapping $x \to x^*$ carries I into I, then I is simultaneously a left and a right ideal.

In the quotient algebra R/I modulo a symmetric two-sided ideal I, involution can be defined in the following way. If $x_1 - x_2 \in I$, then $x_1^* - x_2^* \in I$. Hence, in passing from x to x^* every residue class $\tilde{x}$ modulo the ideal I goes over into some other residue class modulo I, which we shall denote by $\tilde{x}^*$. Clearly, conditions a), b), c) will be satisfied; consequently, R/I is a symmetric algebra.

If $x \to x'$ is a symmetric homomorphism of R onto R', then the inverse image I of the zero is a symmetric two-sided ideal in R. The quotient algebra R/I is symmetrically isomorphic to the algebra R'.

Conversely, the mapping $x \to \tilde{x}$ of every element $x \in R$ into the residue class modulo I which contains it is a symmetric homomorphism of the algebra R onto R/I.

VI. *The radical of a symmetric algebra is a symmetric two-sided ideal.*

In fact, under the mapping $x \to x^*$ all maximal regular left ideals go over into maximal regular right ideals. Consequently, the radical, being the intersection of all maximal regular left ideals, as well as of all maximal regular right ideals, maps onto itself under the mapping $x \to x^*$.

EXAMPLES. 1. The algebra $C(T)$ (see Example, subsection 1, § 9) is a symmetric algebra if we set $x^* = \overline{x(t)}$ for $x = x(t)$(where the vinculum denotes the conjugate complex number).

2. Suppose $\mathfrak{H}$ is a Hilbert space. The algebra $\mathfrak{B}(\mathfrak{H})$, i.e. $\mathfrak{B}(X)$ with $X = \mathfrak{H}$ (see Example 2, subsection 1, § 9), is a symmetric algebra if involution is understood to be passage over to the adjoint operator (see subsection 10, § 5).

3. The algebra W (see Example, 3 subsection 1, § 9) is a symmetric algebra if we set

$$x^* = \sum_{n=-\infty}^{\infty} \bar{c}_{-n} e^{int} \text{ for } x = \sum_{n=-\infty}^{\infty} c_n e^{int}.$$

2. Positive functionals. A linear functional f in the symmetric algebra R is said to be *real-valued* if f assumes real values on all Hermitian elements of the algebra R.

I. *Every linear functional in a symmetric algebra can be represented in the form $f = f_1 + if_2$, where f_1, f_2 are real-valued functionals.*

Namely, it suffices to set

$$f_1(x) = \tfrac{1}{2}[f(x) + \overline{f(x^*)}], f_2(x) = \frac{1}{2i}[f(x) - \overline{f(x^*)}];$$

then f_1, f_2 are real-valued functionals and

$$f(x) = f_1(x) + if_2(x).$$

These functionals f_1, f_2 are called the *real components* of the functional f.

I'. *If f is a real-valued functional, then $f(x^*) = \overline{f(x)}$ for an arbitrary $x \in R$.*

In fact, setting $x = x_1 + ix_2$, where x_1, x_2 are Hermitian, we have

$$f(x^*) = f(x_1 - ix_2) = f(x_1) - if(x_2) = \overline{f(x_1) + if(x_2)} = \overline{f(x)},$$

inasmuch as $f(x_1), f(x_2)$ are real-valued by assumption.

A linear functional f is said to be *positive* if $f(x^*x) \geqq 0$ for an arbitrary element x of the algebra R.

II. *For every positive functional f in the symmetric algebra R,*

$$f(y^*x) = \overline{f(x^*y)}, \tag{1}$$

$$|f(y^*x)|^2 \leqq f(y^*y)f(x^*x). \tag{2}$$

In fact, for arbitrary complex λ and μ,

$$f((\lambda x + \mu y)^*(\lambda x + \mu y)) > 0,$$

i.e.

$$|\lambda|^2 f(x^*x) + \lambda\bar{\mu} f(y^*x) + \bar{\lambda}\mu f(x^*y) + |\mu|^2 f(y^*y) \geqq 0. \tag{3}$$

First setting $\lambda = \mu = 1$ and then $\lambda = 1$, $\mu = i$ in (3), we obtain (1). Further, by (1), the left member in (3) is a nonnegative Hermitian form with respect to λ, μ. And inequality (2) is a necessary condition that this form be nonnegative (see Remark in subsection 1, § 5). We shall call this inequality the *Cauchy-Bunyakovsky inequality.*

III. *Every positive functional f in a symmetric algebra R with identity is real and*

$$|f(x)|^2 \leqq f(e)f(x^*x). \tag{4}$$

The assertion follows from II with $y = e$.

Obviously, in a symmetric algebra with identity every linear combination, with real coefficients, of positive functionals is a real functional. We shall see later on that the converse is not true in general [see Example a), subsection 3, § 20].

IV. *Suppose R is a symmetric algebra without identity and that R' is the symmetric algebra obtained from R by adjunction of the identity. A positive functional f in R can be extended to a positive functional in R' if and only if f is real and satisfies the inequality*

$$|f(x)|^2 \leqq Cf(x^*x) \text{ for all } x \in R, \tag{5}$$

where C is some constant.

Proof. If such an extension is possible, then, in virtue of III, f is real and (5) holds for all x, $y \in R'$ with $C = f(e)$. Conversely, if (5) is satisfied and f is real, then, setting

$$f(\lambda e + x) = \lambda C + f(x),$$

we obtain a linear functional in R'. This functional is positive because, in virtue of (5) (see Remark on p. 84),

$$f((\lambda e+x)^*(\lambda e+x)) = f(\lambda\bar{\lambda}e+\lambda x^*+\bar{\lambda}x+x^*x)$$
$$= \lambda\bar{\lambda}C+\lambda f(x^*)+\bar{\lambda}f(x)+f(x^*x) \geqq 0.$$

We note further the following important particular case. The set $\{e_\alpha\}$ in the normed algebra R is an *approximating identity* if

1) $|e_\alpha| \leqq 1$;

2) for every $\varepsilon > 0$ and every $x \in R$ there exists an element e_α such that $|x-xe_\alpha| < \varepsilon$.

V. *If in a symmetric normed algebra* R: *a*) $|x^*| = |x|$, *b*) *there exists a set* $\{e_\alpha\}$, *approximating the identity, then every continuous positive functional in* R *can be extended to a positive functional in* R'.

Proof. According to the Cauchy-Bunyakovsky inequality,

$$|f(e_\alpha x)|^2 \leqq f(e_\alpha e_\alpha^*)f(x^*x),$$

and consequently, in virtue of condition 1), $|f(e_\alpha x)|^2 \leqq |f|f(x^*x)$. From this, taking condition 2) and the continuity of the functional f into consideration, we conclude that also $|f(x)|^2 \leqq |f|f(x^*x)$, so that condition (5) is satisfied. Furthermore (see (1)), $f(e_\alpha^* x^*) = \overline{f(xe_\alpha)}$; from this, taking 2), a) and the continuity of f into consideration, we conclude that $f(x^*) = \overline{f(x)}$, and the assertion follows from Proposition IV.

We shall say that a Banach algebra R *approximates the identity strongly* if for every finite set of elements $x_j \in R$, $j = 1, \cdots, n$ and every $\varepsilon > 0$ there exists an element $e \in R$ such that $|e| \leqq A$ and $|ex_j - x_j| < \varepsilon$, $j = 1, \cdots, n$, where A is a fixed constant for R.

Varopoulos [1] proved that *in a symmetric Banach algebra which strongly approximates the identity, every positive functional is continuous*. Therefore, for such an algebra we can omit the condition of continuity for the functionals in Proposition V.

3. Normed symmetric algebras. R is called a *normed symmetric algebra* if:

a) R is a normed algebra;

b) R is a symmetric algebra;

c) $|x^*| = |x|$.

It follows from condition c) that the involution operation is continuous.

If R is a non-complete normed algebra, then, as we saw in subsection 1, § 9, it can be uniquely completed to a complete normed algebra $\tilde{R}$. If $x_0 \in \tilde{R}$ and $|x_n - x_0| \to 0$, $x_n \in R$, then $|x_n - x_m| \to 0$ as $n, m \to \infty$; consequently, $|x_n^* - x_m^*| \to 0$ also. Therefore, $\langle x_n^* \rangle$ is a fundamental sequence in R. The element in $\tilde{R}$ defined by this sequence will be denoted by x_0^*.

It is easily seen that the operation $x_0 \to x_0^*$ defined in this way satisfies conditions a), b), c) of subsection 1 and condition c) of the present subsection. Consequently with this definition of involution, $\tilde{R}$ is a complete normed symmetric algebra.

Thus, *every normed symmetric algebra R can be extended to a symmetric Banach algebra $\tilde{R}$ in which R is dense.* This algebra $\tilde{R}$ is called the *completion* of the normed symmetric algebra R.

In a symmetric normed algebra one may consider closed symmetric subalgebras. We shall denote the minimal closed symmetric subalgebra, containing a given set S, by $R(S)$.

For normed symmetric algebras, it is natural to consider continuous symmetric homomorphisms and continuous symmetric isomorphisms. In the case of a continuous symmetric homomorphism of complete algebras, Proposition IV, subsection 6, § 9, remains valid if only closed symmetric ideals of the algebra being mapped are considered. The mapping $x \to x'$ of a normed symmetric algebra R onto the normed symmetric algebra R' is called a *complete isomorphism* if:

1) $x \to x'$ is an isometric mapping of the space R onto the space R';

2) $x \to x'$ is a symmetric isomorphism of the symmetric algebra R onto the symmetric algebra R'.

Two normed symmetric algebras R, R' are said to be *completely isomorphic* if there exists a complete isomorphism of the algebra R onto the algebra R'.

If an element x of a normed symmetric algebra belongs to the radical, then

$$r(x^*x) = \lim_{n\to\infty} \sqrt[n]{|(x^*x)^n|} = 0. \tag{1}$$

In fact, if x belongs to the radical of a symmetric normed algebra, then x^*x also belongs to the radical because the radical is an ideal.

Therefore, applying II, subsection 3, § 9 to the element x^*x, we obtain the desired result.

4. Positive functionals in a symmetric Banach algebra.

I. *Every positive functional f in a symmetric Banach algebra with identity is bounded, namely, it satisfies the inequality*

$$|f(x)| \leqq f(e)|x|. \tag{1}$$

Proof. Suppose first of all that x is a Hermitian element and that $|x| \leqq 1$. Consider the binomial series

$$(1-\lambda)^{\frac{1}{2}} = 1-\tfrac{1}{2}\lambda - \frac{1}{2!}\tfrac{1}{2}\cdot\tfrac{1}{2}\lambda^2 - \cdots;$$

it converges absolutely for $|\lambda| \leqq 1$. Replace λ by x and 1 by e in this series; since R is a complete space and $|x| \leqq 1$, the series thus obtained

$$e-\tfrac{1}{2}x - \frac{1}{2!}\tfrac{1}{2}\cdot\tfrac{1}{2}x^2 - \cdots$$

converges absolutely in R. In virtue of the continuity of involution, its sum, which we shall denote by y, is a Hermitian element; furthermore,

$$y^*y = y^2 = e-x.$$

It follows that

$$f(e-x) = f(y^*y) \geqq 0;$$

consequently,

$$f(x) \leqq f(e).$$

Replacing x everywhere by $-x$, we obtain that also $-f(x) \leqq f(e)$; consequently,

$$|f(x)| \leqq f(e).$$

Suppose now that x is an arbitrary Hermitian element and that $x \neq 0$ (for $x = 0$, inequality (1) is obviously satisfied). We set $x_1 = (1/|x|)x$; then x_1 is also Hermitian and $|x_1| = 1$. Consequently, by what we have already proved, $|f(x_1)| \leqq f(e)$, i.e. $|f(x)|/|x| \leqq f(e)$ and thus inequality (1) has been proved for an arbitrary Hermitian element.

Now suppose that x is an arbitrary element of the algebra R. Then x^*x is a Hermitian element; consequently,

$$f(x^*x) \leqq f(e)|x^*x| \leqq f(e)|x^*||x| = f(e)|x|^2. \tag{2}$$

On the other hand, (see (4), subsection 2),

$$|f(x)|^2 \leqq f(e)f(x^*x).$$

It follows from this and inequality (2) that

$$|f(x)|^2 \leqq |f(e)|^2|x|^2,$$

and by the same token inequality (1) has been proved for arbitrary element $x \in R$.

II. *The norm* $|f|$ *of a positive functional* f *in a symmetric Banach algebra with identity* e *equals* $f(e)$.

In fact, by the definition of the norm of a functional,

$$|f| = \sup_{|x|=1} |f(x)|.$$

It follows from inequality (1) that $\sup_{|x|=1} |f(x)| \leqq f(e)$. On the other hand, this equality is attained for $x = e$.

III. *For an arbitrary positive functional* f *in a symmetric Banach algebra with identity* e,

$$f(x^*x) \leqq f(e) \lim_{n\to\infty} \sqrt[n]{|(x^*x)^n|} = f(e)r(x^*x). \tag{3}$$

Proof. A repeated application of the inequality (4), subsection 2, and then the use of inequality (1) and Proposition II yield:

$$\begin{aligned}|f(x)| &\leqq [f(e)]^{\frac{1}{2}}[f(x^*x)]^{\frac{1}{2}} \leqq [f(e)]^{\frac{1}{2}+\frac{1}{4}}[f((x^*x)^2)]^{\frac{1}{4}} \leqq \cdots \\ &\leqq [f(e)]^{\frac{1}{2}+\frac{1}{4}+\cdots+1/2^m}[f((x^*x)^{2^{m-1}}]^{1/2^m} \\ &\leqq [f(e)]^{1-1/2^m}|f|^{1/2^m}|(x^*x)^{2^{m-1}}|^{1/2^m} = f(e)|(x^*x)^{2^{m-1}}|^{1/2^m}.\end{aligned}$$

It follows that

$$|f(x)| \leqq f(e) \lim_{n\to\infty} \sqrt[2n]{|(x^*x)^n|} \tag{4}$$

as $m \to \infty$. Replacing x by x^*x in (4), we obtain inequality (3).

REMARK. *If x is a normal element, then*

$$|f(x)| \leqq f(e) \lim_{n\to\infty} \sqrt[n]{|x^n|} = f(e)r(x).$$

The assertion follows from (4) and the inequality

$$|(x^*x)^n| = |(x^*)^n x^n| \leqq |x^n|^2,$$

which holds for a normal x.

IV. *If the element x belongs to the radical of a symmetric Banach algebra R with identity, then $f(x) = 0$ and consequently $f(x^*x) = 0$ for an arbitrary positive functional f in R.*

In fact, if x belongs to the radical, then $\lim_{n\to\infty} \sqrt[n]{|(x^*x)^n|} = \lim_{n\to\infty} \sqrt[2n]{|(x^*x)^n|} = 0$, and the assertion follows directly from inequalities (3) and (4).

The uniqueness of a complete norm and the continuity of derivations in a semisimple Banach algebra were recently proved by JOHNSON [2] and JOHNSON and SINCLAIR [1]. A short review of results is given in BONSALL [1].

CHAPTER III

COMMUTATIVE NORMED ALGEBRAS

§ 11. Realization of a commutative normed algebra in the form of an algebra of functions

One of the basic results in the theory of commutative normed algebras consists in this that under certain conditions every such algebra is isomorphic to an algebra of functions. (See below, subsection 3, for a precise formulation of this result.) This result is obtained by investigating the quotient algebra of a given algebra modulo a maximal ideal.

1. Quotient algebra modulo a maximal ideal. Suppose R is a commutative Banach algebra with identity; since R is commutative, every left or right ideal is two-sided. We shall consider an arbitrary maximal ideal M in R. Since M is closed, the quotient algebra R/M is a Banach algebra which also contains an identity (see II and IV, subsection 3, § 9). Since M is a maximal ideal, the algebra R/M is simple, i.e. R/M contains no ideals different from (0); therefore, an element $\tilde{x} \neq 0$ of the algebra R/M is not contained in any ideal of this algebra, and this means it has an inverse (see III, subsection 4 and III, subsection 6, § 7). In other words, R/M is a field and moreover it is a complete normed field. On the basis of the Gelfand-Mazur theorem (III, subsection 3, § 9), R/M is isomorphic to the field of complex numbers, i.e. every element $\tilde{x}$ in the algebra R/M has the form $\tilde{x} = \lambda\tilde{e}$, where $\tilde{e}$ is the identity in R/M. We have thus proved the following theorem.

THEOREM 1. *The quotient algebra of a commutative Banach algebra with identity modulo a maximal ideal is isomorphic to the field of complex numbers.*

We conclude from this theorem that the natural homomorphism of the algebra R onto R/M is essentially a homomorphism of the algebra R onto the field of complex numbers. Consequently:

I. *Every maximal ideal in a commutative Banach algebra R with identity induces a homomorphism of the algebra R onto the field of complex numbers.*

Conversely,

II. *Every homomorphism of a commutative Banach algebra R with identity onto the field of complex numbers is the natural homomorphism generated by some maximal ideal M in R.*

In fact, the inverse image of zero under a homomorphism of R onto the field of complex numbers is an ideal M in R and the given homomorphism is the natural homomorphism of the algebra R onto R/M (see II, subsection 6, § 7). Moreover, M is a maximal ideal: for otherwise M would be contained in some maximal ideal M_1 whose image under the given homomorphism would be an ideal, different from the ideal (0), in the field of complex numbers, which is impossible.

2. Functions on maximal ideals generated by elements of an algebra. We denote the set of all maximal ideals M of a given commutative Banach algebra R with identity e by $\mathfrak{M}$. By I, subsection 1, every ideal $M \in \mathfrak{M}$ generates a homomorphism of the algebra R onto the field of complex numbers; we denote the number corresponding to the element $x \in R$ under this homomorphism by $x(M)$. For fixed $x \in R$ we obtain in this way a function $x(M)$ on the set $\mathfrak{M}$. Consequently, we obtain a correspondence $x \to x(M)$ between the elements x of the algebra R and functions $x(M)$ on the set $\mathfrak{M}$.

I. *The correspondence* $x \to x(M)$ *has the following properties*:

1) *if* $x = x_1+x_2$, *then* $x(M) = x_1(M)+x_2(M)$;

2) *if* $x = \alpha x_1$, *then* $x(M) = \alpha x_1(M)$;

3) *if* $x = x_1 x_2$, *then* $x(M) = x_1(M)x_2(M)$;

4) $e(M) = 1$;

5) $x(M_0) = 0$ *if and only if* $x \in M_0$;

6) *if* $M_1 \neq M_2$, *then there exists an element* $x \in R$ *such that* $x(M_1) \neq x(M_2)$;

7) $|x(M)| \leqq |x|$.

Proof. Assertions 1)–4) follow from the fact that for every fixed M_0, the mapping $x \to x(M_0)$ is a homomorphism. The equation $x(M_0) = 0$ means that x maps into zero under this homomorphism and, consequently, that $x \in M_0$; this also proves assertion 5). Further, if $M_1 \neq M_2$, then there exists an element x in R such that $x \in M_1$, $x \notin M_2$. By 5), this means that $x(M_1) = 0$, $x(M_2) \neq 0$; therefore $x(M_1) \neq x(M_2)$ and 6) is also proved.

In order to prove assertion 7), we note that $|x(M)|$ is the norm of the image $x(M)\tilde{e}$ of the element x under the homomorphism of R onto R/M. By the definition of the norm in R/M (see subsection 3, § 9), it follows from this that

$$|x(M)| = \inf |x'|,$$

where x' runs through the residue class modulo M containing x; therefore $|x(M)| \leqq |x|$.

II. *An element* x *in the algebra* R *has an inverse if, and only if, the function* $x(M)$ *vanishes nowhere in* $\mathfrak{M}$.

In fact, x has an inverse if, and only if, x does not belong to any maximal ideal; by I, 5) this condition is equivalent to the condition that $x(M) \neq 0$ for all $M \in \mathfrak{M}$.

III. *For fixed* $x \in R$ *the set of all values assumed by the function* $x(M)$ *coincides with the spectrum of the element* x.

Proof. If $x(M_0) = \lambda_0$, then $(x-\lambda_0 e)(M_0) = 0$ and therefore $(x-\lambda_0 e)^{-1}$ does not exist, i.e. λ_0 belongs to the spectrum of the element x. Conversely, if λ_0 belongs to the spectrum of the element x, then $(x-\lambda_0 e)^{-1}$ does not exist, and therefore $x-\lambda_0 e$ vanishes on some maximal ideal M_0. But this means that $x(M_0) = \lambda_0$.

IV. *For an arbitrary* $x \in R$,

$$\sup_{M \in \mathfrak{M}} |x(M)| = \lim_{n \to \infty} \sqrt[n]{|x^n|}. \tag{1}$$

This follows directly from III and equality (2), subsection 5, § 9.

We denote the set of all functions on $\mathfrak{M}$ by $\mathscr{F}(\mathfrak{M})$; $\mathscr{F}(\mathfrak{M})$ forms an algebra if we define the operations of scalar multiplication, addition, and multiplication in $\mathscr{F}(\mathfrak{M})$ as the multiplication of a function by a number, addition of functions, and multiplication of functions, respectively. Assertions 1)–4) of Proposition I mean that

V. *The correspondence $x \to x(M)$ is a homomorphism of the algebra R into the algebra $\mathscr{F}(\mathfrak{M})$.*

VI. *The inverse image of zero under the homomorphism $x \to x(M)$ of the algebra R into $\mathscr{F}(\mathfrak{M})$ is the radical of R.*

In fact, this inverse image consists of those, and only those, elements x of the algebra R for which $x(M) = 0$ on all of $\mathfrak{M}$, i.e. which belong to the intersection of all maximal ideals in R. On the other hand, this intersection is the radical of the algebra R (see III, subsection 5, § 7).

COROLLARY. *In a commutative Banach algebra R with identity the element x belongs to the radical* (i.e., *x is a generalized nilpotent; see subsection* 5, § 7) *if and only if* $\lim_{n\to\infty} \sqrt[n]{|x^n|} = 0$.

[We note that in algebra an element x is called *nilpotent* if $x^n = 0$ for some natural number n.]

This assertion follows directly from VI and IV.

VII. *If R is a semisimple algebra, then the correspondence $x \to \{x(M)\}$ is an isomorphism of R into the algebra $\mathscr{F}(\mathfrak{M})$.*

Proof. By VI, in this case the inverse image of zero under the homomorphism $x \to x(M)$ is (0), and therefore this homomorphism is an isomorphism.

Proposition VII means that *every semisimple commutative Banach algebra R with identity is isomorphic with some algebra of functions on the set of all maximal ideals of the algebra R.*

EXAMPLES. 1. We shall consider the algebra W of all functions $x(t) = \sum_{n=-\infty}^{\infty} c_n e^{int}$ where $\sum_{n=-\infty}^{\infty} |c_n| < \infty$ (Example 3, subsection 1, § 9). We shall find all the maximal ideals of the algebra W. Suppose the element $x_0 = e^{it}$ goes over into the number a and, consequently, that the element $x_0^{-1} = e^{-it}$ goes over into the number a^{-1} under the homomorphism of the algebra W with respect to its maximal ideal M. The norm of the element $x_0 = e^{it}$ equals unity; consequently, by I, 7), $|a| \leqq |x_0| = 1$; analogously, $|a^{-1}| \leqq |x_0^{-1}| = 1$ because the norm of the element $x_0^{-1} = e^{-it}$ also equals unity. Therefore $|a| = 1$ and $a = e^{it_0}$ for some t_0, $0 \leqq t_0 \leqq 2\pi$. But then e^{int} goes over into e^{int_0} and therefore $\sum_{n=-\infty}^{\infty} c_n e^{int}$ goes over into $\sum_{n=-\infty}^{\infty} c_n e^{int_0}$. Thus, every maximal ideal

M of the algebra W is defined by some number t_0 in the interval $0 \leqq t \leqq 2\pi$; namely, M consists of those and only those functions $x(t) = \sum_{n=-\infty}^{\infty} c_n e^{int}$ of the ring W which vanish at $t = t_0$.

Suppose the function $x(t) = \sum_{n=-\infty}^{\infty} c_n e^{int}$ of the algebra W does not vanish for any value of t. This means that $x(t)$ does not belong to any maximal ideal and therefore it has an inverse in W. This inverse must coincide with the function $1/x(t)$, which, consequently, must also belong to the algebra W. We thus arrive at the following theorem, due to WIENER [1, 2].

If the sum of an absolutely convergent trigonometric series $\sum_{n=-\infty}^{\infty} c_n e^{int}$ vanishes nowhere, then the function $1/\sum_{n=-\infty}^{\infty} c_n e^{int}$ can also be expanded in an absolutely convergent trigonometric series.

2. Consider the algebra $C(T)$ of all functions $x = x(t)$ which are continuous on the compact space T (Example 1, subsection 1, § 9); we shall find all the maximal ideals of this algebra.

If t_0 is a fixed point in T, then the correspondence $x(t) \to x(t_0)$ is a homomorphism of the algebra $C(T)$ into the field of complex numbers and therefore it is generated by some maximal ideal; let us denote it by M_{t_0}. This ideal is the set of all functions x of the algebra $C(T)$, which vanish at the point t_0.

Conversely, *every maximal ideal M_0 in $C(T)$ is a set M_{t_0} of all those functions x in the algebra $C(T)$ which vanish at some fixed point $t_0 \in T$.*

In fact, suppose M_0 is a maximal ideal in $C(T)$; we shall prove that there exists a point t_0 at which all functions in M_0 vanish. Let us assume the contrary. Then there corresponds to every point $\tau \in T$ a function $x_\tau \in M_0$, different from zero at the point τ; since x_τ is continuous, it is also different from zero in some neighborhood $U(\tau)$ of the point τ. Since T is compact, a finite number of neighborhoods $U(\tau_1), \cdots, U(\tau_n)$ covering all of T can be chosen from these neighborhoods; suppose $x_{\tau_1}, \cdots, x_{\tau_n}$ are the corresponding functions $x_\tau \in M_0$. Then the function

$$x(t) = x_{\tau_1}(t)\overline{x_{\tau_1}(t)} + \cdots + x_{\tau_n}(t)\overline{x_{\tau_n}(t)} = |x_{\tau_1}(t)|^2 + \cdots + |x_{\tau_n}(t)|^2$$

also belongs to the ideal M_0 and does not vanish anywhere in T. Therefore $1/x(t)$ is a continuous function on T and, consequently, it also belongs to the algebra $C(T)$. But this means that the element $x = x(t)$ has an inverse, and hence that $x = x(t)$ cannot belong to any maximal ideal, in particular, it cannot belong to the ideal M_0. The contradiction thus arrived at shows that there exists a point $t_0 \in T$ at which all functions of the ideal M_0 vanish. But then $M_0 \subset M_{t_0}$; consequently, $M_0 = M_{t_0}$ in virtue of the fact that M_0 is a maximal ideal. Clearly, the correspondence $t_0 \to M_{t_0}$ between the set of all points $t_0 \in T$ and all maximal ideals of the algebra $C(T)$ is one-to-one.

Since $x(t) \to x(t_0) \in M_0$, we have $x(M_{t_0}) = x(t_0)$; consequently, under the correspondence $t_0 \to M_{t_0}$ the functions $x(M)$ go over into the initial functions $x(t)$.

In the preceding discussion we actually used only the following properties of the algebra $R = C(T)$: 1) all functions $x(t) \in R$ are continuous; 2) if $x(t) \in R$, then $\overline{x(t)} \in R$ also; 3) if the function $x(t)$ of the algebra R does not vanish anywhere on T, then $1/x(t) \in R$ also. Therefore, we proved the following more general proposition.

VIII. *If R is the algebra of functions on the compact space T satisfying the indicated conditions* 1), 2), 3), *then every maximal ideal M in R coincides with some ideal M_{t_0}, $t_0 \in T$.*

3. We denote by $D_n(a, b)$ the set of all complex-valued functions $x = x(t)$ which are defined and have continuous n-th order derivatives in the closed interval $[a, b]$. We define the operations of addition, scalar multiplication, and multiplication in $D_n(a, b)$ as the addition of functions, the multiplication of a function by a number, and the multiplication of functions, respectively. Furthermore, we define the norm in $D_n(a, b)$ by the formula

$$|x| = \sum_{k=0}^{n} \frac{1}{k!} \max_{a \leqq t \leqq b} |x^{(k)}(t)|.$$

Then $D_n(a, b)$ is a commutative Banach algebra with identity. Clearly, this algebra satisfies all the conditions of Proposition VIII in the preceding example and, consequently, every maximal ideal in $D_n(a, b)$ is the set M_{t_0} of all functions $x(t) \in D_n(a, b)$ which vanish at some fixed point $t_0 \in [a, b]$.

3. Topologization of the set of all maximal ideals. We shall now define the topology in the set $\mathfrak{M}$ of all maximal ideals of the algebra R as the weak topology defined by the family of all functions $x(M)$, $x \in R$ (see subsection 11, § 2). Thus, a neighborhood basis in this topology is formed by all possible sets $U(M_0; x_1, \cdots, x_n; \varepsilon)$ of ideals M, defined by inequalities of the form

$$|x_k(M) - x_k(M_0)| < \varepsilon, \quad k = 1, 2, \cdots, n,$$

for all possible fixed $\varepsilon > 0$ and $x_1, x_2, \cdots, x_n \in R$. Then, by the very definition of the topology in $\mathfrak{M}$, all functions $x(M)$, $x \in R$, are continuous on $\mathfrak{M}$. Furthermore, $\mathfrak{M}$ is a Hausdorff space because by I, subsection 2, the functions $x(M)$, $x \in R$, separate points in $\mathfrak{M}$.

THEOREM 2. *The space $\mathfrak{M}$ of maximal ideals in a commutative Banach algebra R with identity is compact.*

Proof. We assign to each element x in the algebra R a circle Q_x in the complex plane with center at the point O and radius $|x|$. Suppose Q is the topological product of all these circles. Thus, Q is a space whose points are all possible sets $\Lambda = \{\lambda_x\}$ of numbers $\lambda_x \in Q_x$, where x runs through the entire algebra R; a neighborhood basis of the point $\Lambda^0 = \{\lambda_x^0\}$ in Q is given by all possible finite systems of elements $x_1, x_2, \cdots, x_n \in R$ and all possible positive numbers ε as a system of sets in Q defined by inequalities of the form

$$|\lambda_{x_1} - \lambda_{x_1}^0| < \varepsilon, \cdots, |\lambda_{x_n} - \lambda_{x_n}^0| < \varepsilon \tag{1}$$

(see subsection 12, § 2); we denote the neighborhood defined by inequalities (1) by $U(\Lambda^0; x_1, \cdots, x_n; \varepsilon)$. According to Tikhonov's theorem (II, subsection 12, § 2) the space Q is compact.

Since $|x(M)| \leqq |x|$, there corresponds to every maximal ideal M a point $\{\lambda_x\} \in Q$, where $\lambda_x = x(M)$ and to distinct maximal ideals there correspond distinct points. In fact, if $M \neq M_1$, then there exists an element $x_0 \in R$ for which $x_0(M_1) \neq x_0(M)$. This means that the corresponding points $\{\lambda_x\}$, $\{\lambda'_x\}$ differ by at least one coordinate, namely the one with index x_0, and therefore they are distinct. Thus, the correspondence $M \to \{\lambda_x\}$ established by the equality $\lambda_x = x(M)$ is one-to-one. Upon comparison of the topologies in $\mathfrak{M}$ and Q, it follows that this correspondence is a homeomorphism. We shall prove that the image Q_1 of the space $\mathfrak{M}$ in Q under this homeomorphism is closed. It will then follow that Q_1, as a closed subset of a compact space, is compact (see II, subsection 6, § 2) and therefore its homeomorphic inverse image $\mathfrak{M}$ is also compact.

Suppose $\Lambda^0 = \{\lambda_x^0\}$ is an arbitrary contact point of the set Q_1; we shall prove that there exists a maximal ideal M_0 such that $x(M_0) = \lambda_x^0$; this will mean that $\{\lambda_x^0\} \in Q_1$ and hence that Q_1 is closed in Q will be proved. Clearly, it will suffice to show that the correspondence $x \to \lambda_x^0$ is a homomorphism of the algebra R onto the field of complex numbers, i.e. that

$$\lambda_x^0 \not\equiv 0,\ \lambda_{\alpha x}^0 = \alpha\lambda_x^0,\ \lambda_{x+y}^0 = \lambda_x^0+\lambda_y^0,\ \lambda_{xy}^0 = \lambda_x^0\lambda_y^0; \tag{2}$$

by II, subsection 1, this homomorphism is then generated by some maximal ideal M_0, i.e. $\lambda_x^0 = x(M_0)$ for some $M_0 \in \mathfrak{M}$. We shall restrict ourselves to the proof of the last of relations (2); the remaining ones are proved analogously. To this end, we consider a neighborhood $U(\Lambda^0; x, y, xy; \varepsilon)$ of the point Λ^0. Since Λ^0 is a contact point of the set Q_1, there exists a maximal ideal M such that the corresponding point $\Lambda = \{\lambda_x\}$, $\lambda_x = x(M)$ belongs to this neighborhood. This means that

$$|\lambda_x^0 - x(M)| < \varepsilon,\ |\lambda_y^0 - y(M)| < \varepsilon,$$
$$|\lambda_{xy}^0 - (xy)(M)| = |\lambda_{xy}^0 - x(M)y(M)| < \varepsilon.$$

But then

$$\begin{aligned}|\lambda_{xy}^0 - \lambda_x^0\lambda_y^0| &\leqq |\lambda_{xy}^0 - x(M)y(M)| + |x(M)[y(M)-\lambda_y^0]|\\ &+ |\lambda_y^0[x(M)-\lambda_x^0]| \leqq \varepsilon + |x|\,|y(M)-\lambda_y^0| + |\lambda_y^0|\,|x(M)-\lambda_x^0|\\ &< \varepsilon(1+|x|+|\lambda_y^0|);\end{aligned}$$

consequently,

$$\lambda_{xy}^0 = \lambda_x^0\lambda_y^0$$

since the number $\varepsilon > 0$ is arbitrary.

Thus, the correspondence $x \to \lambda_x^0$ is a homomorphism and the theorem is proved.

The topology in the set of all maximal ideals is unique in the following sense.

THEOREM 3. *Suppose $\mathfrak{M}'$ is the set of all maximal ideals of the algebra R topologized in such a way that*

1) *$\mathfrak{M}'$ is compact;*

2) *all functions $x(M)$ corresponding to elements in the algebra R are continuous on $\mathfrak{M}'$.*

Then $\mathfrak{M}'$ is homeomorphic to the space $\mathfrak{M}$ defined above.

The assertion follows directly from VI, subsection 7, § 2 (see also II, subsection 11, § 2).

EXAMPLE. Let us consider the space $\mathfrak{M}$ of all maximal ideals in the algebra $C(T)$ (Example 2, subsection 2). There exists a one-to-one correspondence $t_0 \to M_{t_0}$ between points $t \in T$ and ideals $M \in \mathfrak{M}$; we define a topology in the set of all maximal ideals of the algebra $C(T)$ by considering neighborhoods of ideals to be the images of neighborhoods of the space T under this correspondence. We obtain a topological space $\mathfrak{M}'$ which is homeomorphic to the space T. Then the functions $x(t)$ go over into the functions $x(M)$ which are continuous on $\mathfrak{M}'$; on the basis of Theorem 3, the spaces $\mathfrak{M}'$ and $\mathfrak{M}$, and hence also T and $\mathfrak{M}$, are homeomorphic.

Thus, *the space $\mathfrak{M}$ of maximal ideals of the algebra $C(T)$ of all continuous functions on the compact space T is homeomorphic to the space T.*

Combining Theorem 2 with the results of subsection 2, we arrive at the following theorem.

THEOREM 4. *For every commutative Banach algebra R with identity the correspondence $x \to x(M)$ is a homomorphism of the algebra R onto some algebra of continuous functions on a compact space, where the kernel of this homomorphism is the radical of the algebra. If, in particular, the algebra R is semisimple, then R is isomorphic to an algebra of continuous functions on a compacı space.*

If the algebra R itself is already an algebra of (in general not necessarily all) continuous functions $x(t)$ on some topological space T, then every point $t_0 \in T$ defines a maximal ideal M_{t_0} in R, namely, the set of all functions x in R which vanish at the point t_0. In this connection, it might turn out that different points t_0 determine the same maximal ideal; consequently, the transition to the maximal ideals can in the general case lead to the "pasting together" of points of the space T.

Furthermore, it might turn out that not every maximal ideal of the algebra R is defined by a point of the space T. Therefore, the transition to the maximal ideals can also mean the extension of thc initial space.

We shall clarify this by means of the following examples:

a) In the algebra W of all absolutely convergent trigonometric series $x(t) = \sum_{n=-\infty}^{\infty} c_n e^{int}$, defined on the interval $(-\infty, \infty)$, every real number t_0 generates a maximal ideal, namely, the set of all $x(t)$ which vanish at the point t_0. Two distinct numbers t_1, t_2 define the same maximal ideal if and only if $t_1 - t_2$ is a multiple of 2π. Therefore the transition to the maximal ideals means the "pasting together" of points which differ by a multiple of 2π, and the space of maximal ideals of the algebra W is homeomorphic to the circle which is the natural dom ainof definition of the functions of the algebra W.

b) Suppose A is the set of all functions $x = x(\zeta)$, regular in the open disc $|\zeta| < 1$ and continuous in the closed disc $|\zeta| \leqq 1$, with norm

$$|x| = \sup_{|\zeta| \leqq 1} |x(\zeta)|,$$

and with the usual definitions of scalar multiplication, addition, and multiplication. We shall find all the maximal ideals of the algebra A. Every point ζ_0 of the disc $|\zeta| \leqq 1$ generates a maximal ideal in A, namely the set of all functions $x(\zeta)$ of the algebra A which vanish at the point ζ_0. Conversely, every maximal ideal of the algebra A is generated in this manner by some point of the disc $|\zeta| \leqq 1$. In fact, suppose M_0 is a maximal ideal in A and let ζ_0 be the number into which the function $x_0(\zeta) = \zeta$ maps under the homomorphism $x \to x(M_0)$; since $|x_0| = 1$, we have $|\zeta_0| \leqq 1$. For an arbitrary polynomial $x(\zeta) = c_0 + c_1\zeta + \cdots + c_n\zeta^n$, we have $x(M_0) = c_0 + c_1\zeta_0 + \cdots + c_n\zeta_0^n = x(\zeta_0)$. But all thc functions in A are the limits of sequences of such polynomials which are uniformly convergent in the disc $|\zeta| \leqq 1$ (i.e. convergent in the sense of the norm in A); therefore, $x(M_0) = x(\zeta_0)$ for an arbitrary function $x(\zeta) \in A$ and M_0 is generated by the point ζ_0.

In fact, the function $x(\zeta) \in A$ is the uniform limit in the disc $|\zeta| \leqq 1$ of functions $x(\zeta/(1+\varepsilon))$ as $\varepsilon \to +0$, each of which, as an analytic function in the disc $|\zeta| \leqq 1+\varepsilon$, is approximated uniformly by polynomials (e.g. the partial sums of its Taylor series) in the disc $|\zeta| \leqq 1$.

The equality $x(M_0) = x(\zeta_0)$ shows that the maximal ideals can be identified with the corresponding points in the disc $|\zeta| \leqq 1$.

We shall now denote the function $x(\zeta)$ of the algebra A considered only on the circle $|\zeta| = 1$ by $x^\wedge(\zeta)$. We obtain the algebra $A^\wedge$ of functions on the circle, isomorphic to the algebra A. In fact, the correspondence $x \to x^\wedge$ is, clearly, a homomorphism; its kernel consists of analytic functions $x(\zeta)$ which vanish on the circle $|\zeta| = 1$ and are therefore identically equal to zero. Consequently, the correspondence $x \to x^\wedge$ is an isomorphism.

We define the norm in $A^\wedge$ by means of the formula

$$|x^\wedge| = \sup_{|\zeta|=1} |x^\wedge(\zeta)|;$$

by the maximum modulus principle, this norm coincides with

$$|x| = \sup_{|\zeta|\leqq 1} |x(\zeta)|$$

so that the isomorphism $x \to x^\wedge$ also preserves norm. Therefore the maximal ideals of the algebra $A^\wedge$ are also described by points of the disc $|\zeta| \leqq 1$, although $A^\wedge$ consists of functions defined only on the circle $|\zeta| = 1$. Thus, the closed disc $|\zeta| \leqq 1$, and not the circle $|\zeta| = 1$, is the natural domain of definition of the algebra $A^\wedge$.

Summarizing, we can say that the transition from the space T to the space $\mathfrak{M}$ of maximal ideals can lead to the following operations:

1) "pasting together" of points of the space T;
2) the adjunction of new elements to T.

4. The case of an algebra without identity. Suppose R is a commutative Banach algebra without identity and that R' is the algebra obtained from R by the adjunction of the identity. We denote the set of all maximal regular ideals M of the algebra R by $\mathfrak{M}$ and the set of all maximal ideals M' of the algebra R' by $\mathfrak{M}'$; suppose $\mathfrak{M}_0'$ is the set obtained from $\mathfrak{M}'$ upon exclusion of the one ideal $M_0' = R$ of the algebra R'. Since $\mathfrak{M}'$ is compact, $\mathfrak{M}_0'$ is locally compact. On the other hand, there exists a one-to-one correspondence between the ideals $M \in \mathfrak{M}$ and $M' \in \mathfrak{M}_0'$ (see VI, subsection 4, § 7). In virtue of this correspondence, the topology in $\mathfrak{M}_0'$ can be carried over to $\mathfrak{M}$; then $\mathfrak{M}$ will be a locally compact space of maximal regular ideals of the algebra R. Moreover, if $x \in R$, then $x(M_0') = 0$. This means that the function $x(M')$, $x \in R$, can be considered as the function $x(M)$ which satisfies the condition

$$\lim_{M\to\infty} x(M) = 0. \tag{1}$$

From this it is clear that the preceding results carry over to commutative Banach algebras without identity with the following modifications. The role of the space of maximal ideals is assumed by the locally compact space $\mathfrak{M}$ of maximal regular ideals, and instead of arbitrary continuous functions we must consider only continuous functions on $\mathfrak{M}$, satisfying condition (1). In particular, we obtain the following result.

Every semisimple commutative Banach algebra without identity is isomorphic to some algebra of functions $x(M)$ on the locally compact space $\mathfrak{M}$ which satisfy condition (1).

We shall leave it to the reader to carry over in more detail the preceding and also further results of this chapter to algebras without identity (in this connection, see Loomis [2]).

5. System of generators of an algebra. A set K of elements in a Banach algebra R with identity is called a *system of generators* of this algebra if the smallest closed subalgebra with identity, containing K, coincides with R. Here, the identity of the algebra is not included in the set of generators.

Theorem 5. *The set of all neighborhoods $U(M; y_1, \cdots, y_n; \varepsilon)$, where the y_k run through all elements of the system K of generators in R, is a neighborhood basis in the space $\mathfrak{M}$ of maximal ideals of the algebra R.*

Proof. We must prove that every neighborhood $U(M_0; x_1, \cdots, x_m; \varepsilon)$ contains some neighborhood $U(M_0; y_1, \cdots, y_n; \delta)$. To this end, we note that by the definition of the system of generators there exist polynomials

$$p_k = p_k(y_1, \cdots, y_n), k = 1, 2, \cdots, m,$$

of a finite number of elements $y_j \in K$, which differ in norm from x_k by less than $\varepsilon/3$:

$$|x_k - p_k| < \frac{\varepsilon}{3}, \quad k = 1, 2, \cdots, m.$$

Since $p_k(M) = p_k(y_1(M), \cdots, y_n(M))$ is a continuous function of $y_1(M), \cdots, y_n(M)$, there exists a number $\delta > 0$ such that

$$U(M_0; y_1, \cdots, y_n; \delta) \subset U\left(M_0; p_1, \cdots, p_m; \frac{\varepsilon}{3}\right),$$

so that for all $M \in U(M_0; y_1, \cdots, y_n; \delta)$,

$$|p_k(M) - p_k(M_0)| < \frac{\varepsilon}{3}.$$

The neighborhood $U(M_0; y_1, \cdots, y_n; \delta)$ is contained in $U(M_0; x_1, \cdots, x_m; \varepsilon)$ because for all $M \in U(M_0; y_1, \cdots, y_n; \delta)$ and all $k = 1, 2, \cdots, m$, we have

$$|x_k(M) - x_k(M_0)| \leqq |x_k(M) - p_k(M)| + |p_k(M) - p_k(M_0)| + |p_k(M_0) - x_k(M_0)| \leqq |x_k - p_k| + \frac{\varepsilon}{3} + |x_k - p_k| < 3 \cdot \frac{\varepsilon}{3} = \varepsilon.$$

This completes the proof of the theorem.

Corollary. *If R is an algebra with a finite or countable number of generators, then the space $\mathfrak{M}$ of its maximal ideals has a countable neighborhood basis.*

In fact, if $y_1, y_2, \cdots$ are the generators of the algebra R, then $U(y_1, \cdots, y_n; 1/m)$, $n, m = 1, 2, 3, \cdots$, is a countable neighborhood basis in $\mathfrak{M}$.

We shall now consider the particular case when the algebra R has a finite number of generators.

THEOREM 6. *If the algebra R has a finite number of generators $y_1, y_2, \cdots, y_n$, then the space $\mathfrak{M}$ of its maximal ideals is homeomorphic to a closed bounded set in complex n-dimensional space.*

Proof. The single-valued functions $y_1(M), y_2(M), \cdots, y_n(M)$ map $\mathfrak{M}$ continuously onto some subset $\mathfrak{M}'$ of the complex n-dimensional space C^n. We shall prove that this mapping is one-to-one. In view of the compactness of the space $\mathfrak{M}$, the theorem will follow from this (see III and V, subsection 7, § 2).

Suppose two points $M_1, M_2 \in \mathfrak{M}$ map into the same point of the space $\mathfrak{M}'$; this means that

$$y_k(M_1) = y_k(M_2), k = 1, 2, \cdots, n.$$

But then for arbitrary polynomial $p = p(y_1, \cdots, y_n)$, we have $p(M_1) = p(M_2)$, and since the polynomials $p(y_1, \cdots, y_n)$ form a dense set in R, $x(M_1) = x(M_2)$ for an arbitrary element $x \in R$. It follows, in virtue of the separation property (see I, 6), subsection 2), that $M_1 = M_2$.

It follows from the theorem just proved, in particular, that *if R is an algebra with one generator, then $\mathfrak{M}$ is homeomorphic to a closed bounded set in the complex plane.*

A bounded closed set $F \subset C^n$ is homeomorphic to the space $\mathfrak{M}$ of maximal ideals of a commutative Banach algebra with n generators if, and only if, for each point $\zeta_0 \in C^n - F$ there exists a polynomial, equal to 1 at the point ζ_0 and in absolute value < 1 on F. In the case $n = 1$, the sets F which are homeomorphic to $\mathfrak{M}$ admit the following characterization: they are precisely those sets which do not decompose the plane; for $n > 1$ there does not exist a purely topological characterization of such sets F (for details, see GELFAND, RAIKOV and SHILOV [1], pp. 44–46).

6. Analytic functions of algebra elements. A Banach algebra R with identity contains, together with the element x, every polynomial $p(x) = c_0 e + c_1 x + \cdots + c_n x^n$ and in general every function $f(x) = \sum_{n=0}^{\infty} c_n x^n$, where $f(\zeta) = \sum_{n=0}^{\infty} c_n \zeta^n$ is an entire function of ζ. In fact, in the case of an entire function $f(\zeta)$, the series $\sum_{n=0}^{\infty} |c_n x^n|$ is majorized by the convergent series $\sum_{n=0}^{\infty} |c_n| |x|^n$; consequently, the series $\sum_{n=0}^{\infty} c_n x^n$ converges absolutely to some element of the algebra R; this element is denoted by $f(x)$ and is called an *entire analytic function f of the element* $x \in R$.

On the other hand, the function $f(\zeta) = 1/\zeta$, having the single singular point $\zeta = 0$, is assigned the element $f(x) = x^{-1}$ of the algebra R if, and only if, $x(M)$ does not vanish for any M, i.e. if the singular point $\zeta = 0$ does not belong to the spectrum of the element x.

We shall show that an analogous construction can also be carried out for other analytic functions. Suppose x is an arbitrary element of the algebra R and that

$f(\zeta)$ is a function which is regular in the bounded region D which contains the spectrum S_x of the element x entirely in its interior. We set

$$f(x) = \frac{1}{2\pi i} \oint_{\Gamma} (\zeta e - x)^{-1} f(\zeta) d\zeta, \tag{1}$$

where Γ is a closed rectifiable Jordan curve in D, which contains S_x in its interior (consequently, it is at a positive distance from S_x). The element $f(x)$ of the algebra R, so defined, does not depend on the choice of the curve Γ, satisfying these conditions, because the vector-valued function $(1/2\pi i)(\zeta e - x)^{-1} f(\zeta)$ is regular in $D - S_x$, and the Cauchy theorem is valid for regular vector-valued functions. We shall agree to say that *the analytic function $f(\zeta)$ is applicable to the element $x \in R$*, if $f(\zeta)$ is regular on S_x, and hence also in some bounded region containing S_x. In virtue of what was stated above, formula (1) defines the element $f(x)$ of the algebra for an arbitrary analytic function $f(\zeta)$, which is applicable to the element x.

THEOREM 7. *Suppose F_x is the set of all analytic functions $f(\zeta)$ applicable to the element $x \in R$. The correspondence $f(\zeta) \to f(x)$ between the functions $f \in F_x$ and elements $f(x)$ in R possesses the following properties*:

1) *if* $f(\zeta) = 1$, *then* $f(x) = e$;
2) *if* $f(\zeta) = \zeta$, *then* $f(x) = x$;
3) *if* $f(\zeta) = \alpha_1 f_1(\zeta) + \alpha_2 f_2(\zeta)$, *then* $f(x) = \alpha_1 f_1(x) + \alpha_2 f_2(x)$;
4) *if* $f(\zeta) = f_1(\zeta) f_2(\zeta)$, *then* $f(x) = f_1(x) f_2(x)$;
5) *if the sequence $\langle f_n(\zeta) \rangle$ converges uniformly to $f(\zeta)$ in some region D, containing S_x in its interior, then the sequence $\langle f_n(x) \rangle$ converges in norm to $f(x)$.*

Proof. The functions $f(\zeta) = 1$ and $f(\zeta) = \zeta$ are regular in the entire complex plane; consequently, for $f(\zeta) = 1$ and $f(\zeta) = \zeta$ we can take any circle with center at the origin of the coordinate system and with sufficiently large radius for the curve Γ. But on such a circle,

$$(\lambda e - x)^{-1} = \lambda^{-1} e + \lambda^{-2} x + \lambda^{-3} x^2 + \cdots.$$

Therefore, for $f(\zeta) \equiv 1$,

$$f(x) = \frac{1}{2\pi i} \oint_{\Gamma} (\lambda^{-1} e + \lambda^{-2} x + \lambda^{-3} x^2 + \cdots) d\lambda = \frac{1}{2\pi i} \oint_{\Gamma} \lambda^{-1} e \, d\lambda = e,$$

and for $f(\zeta) = \zeta$,

$$f(x) = \frac{1}{2\pi i} \oint_{\Gamma} \lambda(\lambda^{-1} e + \lambda^{-2} x + \lambda^{-3} x + \cdots) d\lambda = \frac{1}{2\pi i} \oint_{\Gamma} \lambda^{-1} x \, d\lambda = x.$$

Further, it is obvious that condition 3) is satisfied. We shall now prove that condition 4) is also satisfied. This condition means that for any two functions $f_1(\lambda)$, $f_2(\lambda)$, which are regular in $D \supset S_x$,

$$\frac{1}{2\pi i}\oint_\Gamma (\lambda e-x)^{-1}f_1(\lambda)f_2(\lambda)d\lambda$$
$$= \frac{1}{2\pi i}\oint_\Gamma (\lambda e-x)^{-1}f_1(\lambda)d\lambda \cdot \frac{1}{2\pi i}\oint_\Gamma (\lambda e-x)^{-1}f_2(\lambda)d\lambda. \tag{2}$$

In the second integral in the right member, we replace the curve Γ by another curve Γ_1, which is contained in D and which contains Γ is its interior. Then the right member in (2) can be written in the form

$$\frac{1}{2\pi i}\oint_\Gamma (\lambda e-x)^{-1}f_1(\lambda)d\lambda \cdot \frac{1}{2\pi i}\oint_{\Gamma_1} (\mu e-x)^{-1}f_2(\mu)d\mu$$
$$= -\frac{1}{4\pi^2}\oint_\Gamma\oint_{\Gamma_1} (\lambda e-x)^{-1}(\mu e-x)^{-1}f_1(\lambda)f_2(\mu)d\lambda\, d\mu$$
$$= -\frac{1}{4\pi^2}\oint_\Gamma\oint_{\Gamma_1} \frac{f_1(\lambda)f_2(\mu)}{\mu-\lambda}[(\lambda e-x)^{-1}-(\mu e-x)^{-1}]d\lambda\, d\mu$$
$$= \frac{1}{2\pi i}\oint_\Gamma (\lambda e-x)^{-1}f_1(\lambda)\left(\frac{1}{2\pi i}\oint_{\Gamma_1}\frac{f_2(\mu)d\mu}{\mu-\lambda}\right)d\lambda$$
$$-\frac{1}{4\pi^2}\oint_{\Gamma_1} (\mu e-x)^{-1}f_2(\mu)\left(\oint_\Gamma \frac{f_1(\lambda)d\lambda}{\mu-\lambda}\right)d\mu.$$

Since the function $f_1(\lambda)/(\mu-\lambda)$ is regular on and in the interior of Γ, the second term equals zero; on the other hand, by the Cauchy formula, the inside integral in the first term equals $f_2(\lambda)$. This completes the proof of formula (2).

We shall now prove that condition 5) is satisfied. Suppose D is a bounded region which contains S_x in its interior, and suppose $\langle f_n(\zeta)\rangle$ is a sequence of functions which are regular in D and converge to $f(\zeta)$ uniformly in D, so that

$$\max_{\lambda\in D}|f(\lambda)-f_n(\lambda)| \to 0 \text{ as } n\to\infty.$$

Then, taking in D the closed curve Γ, where Γ contains S_x in its interior, we have

$$|f(x)-f_n(x)| = \left|\frac{1}{2\pi i}\oint_\Gamma [f(\lambda)-f_n(\lambda)](\lambda e-x)^{-1}d\lambda\right|$$
$$\leq \max_{\lambda\in D}|f(\lambda)-f_n(\lambda)|\cdot\frac{1}{2\pi}\oint_\Gamma |(\lambda e-x)^{-1}|\,|d\lambda| \to 0 \text{ as } n\to\infty.$$

REMARK. It can be shown (see GELFAND, RAIKOV and SHILOV [1]) that every correspondence $f(\zeta)\to f(x)$, $f\in F_x$, satisfying conditions 1)–5) of Theorem 7, is necessarily given by formula (1).

COROLLARY. *If the function $f(\zeta)$ is regular in the open region D which contains all the values of the function $x(M)$, then there exists an element y in the algebra such that $y(M) = f(x(M))$ for all maximal ideals M.*

Proof. By hypothesis, $f \in F_x$, because S_x is at the same time the set of all values of the function $x(M)$ (see III, subsection 2); therefore, the element $y = f(x)$ exists in the algebra R; moreover,

$$y = \frac{1}{2\pi i} \oint_{\Gamma} f(\lambda)(\lambda e - x)^{-1} d\lambda, \tag{3}$$

where Γ is a closed rectifiable Jordan curve which is contained entirely in the region of regularity of the function $f(\zeta)$ and containing S_x in its interior.

Since the correspondence $x \to x(M)$ is a continuous homomorphism and the integral in the right member of (3) converges in norm, it follows from (3) that

$$y(M) = \frac{1}{2\pi i} \oint_{\Gamma} \frac{f(\lambda)}{\lambda - x(M)} d\lambda = f(x(M)).$$

From this we obtain as a particular case the following theorem due to Lévy [1] which generalizes the Wiener theorem (see § 11, subsection 2, above).

If the Fourier series of the function $x(t)$ converges absolutely and all the values of this function are contained in the disc $|\zeta - \zeta_0| \leqq \rho$, then for an arbitrary function $f(\zeta)$, regular at all the points of this disc, the Fourier series of the function $f(x(t))$ also converges absolutely.

In fact, $x(t)$ is an element of the algebra W, with $x(M) = x(t)$, where M is the ideal generated by the point t. Consequently, there is an element $y = f(x)$ in the algebra W; this means that the function $y(t) = y(M) = f(x(M)) = f(x(t))$ also is expandable in an absolutely convergent trigonometric series.

Katznelson [1] proved the following result: If a function $F(\tau)$, $\tau \in (a, b)$, $a \geqq -\infty$, $b \leqq +\infty$ has the property that $F(x(t)) \in W$ for every real-valued $x(t) \in W$ with values in (a, b), then $F(\tau)$ is analytic in (a, b). For other results in this direction, see Katznelson [1–3] and Kahane [1].

Another application of the corollary to Theorem 7 is obtained if we take for the algebra R the algebra of all functions $x(\zeta) = \sum_{n=0}^{\infty} c_n \zeta^n$ in the disc $|\zeta| \leqq 1$, such that $\sum_{n=0}^{\infty} |c_n| < \infty$; the operations of scalar multiplication, addition, and multiplication in R are defined in the usual way as the corresponding operations with functions, and the norm $|x|$ is defined by the formula $|x| = \sum_{n=0}^{\infty} |c_n|$. Repeating the line of reasoning followed in the proof of Theorem 4, in § 11, subsection 3, above, we verify that the formula $x(M) = x(\zeta_0)$ establishes a one-to-one correspondence between the maximal ideals M of the algebra R and the points ζ_0 of the disc $|\zeta| \leqq 1$. Therefore, applying the corollary to Theorem 7, we arrive at the following result.

If the Maclaurin series of the function $x(\zeta)$ converges absolutely in the disc $|\zeta| \leqq 1$ and all the values of this function are contained in the interior of a bounded region D, then for an arbitrary function $f(z)$, regular in D, the Maclaurin series of the function $f(x(\zeta))$ also converges absolutely in the disc $|\zeta| \leqq 1$.

In fact, there exists an element y in R such that

$$y(\zeta) = y(M) = f(x(M)) = f(x(\zeta)).$$

7. **Wiener pairs of algebras.** Let R, R_1 ($R \supset R_1$) be (not necessarily commutative) Banach algebras with common identity e and norms $|x|$ and $|x|_1$ respectively. We shall say that the algebras R, R_1 form a *Wiener pair* if every element $x \in R_1$ which is invertible in R is also invertible in R_1.

If we take for R_1 the algebra of absolutely convergent Fourier series and for R the algebra of all continuous functions on the real line with the usual norms, then the preceding condition is the well-known Wiener theorem (see Example 1, subsection 2).

I. *The following assertions are equivalent*:

a) *the algebras* R *and* R_1 *form a Wiener pair*;

b) *the closure* $\overline{N}_1$ *with respect to the norm* $|x|$ *of an arbitrary set* $N_1 \subset R_1$, *not containing elements which are invertible in* R_1, *does not contain elements which are invertible in* R.

c) *the closure* $\overline{M}_1$ *with respect to the norm* $|x|$ *of every one-sided maximal ideal* M_1 *of the algebra* R_1 *does not contain elements which are invertible in* R;

d) $S_x = S_x^1$ *for every* $x \in R_1$ (S_x, S_x^1 are the spectra of the element x in the algebras R, R_1 respectively).

Proof. a) $\to$ b). We assume that for some set $N_1 \subset R_1$ which does not contain elements which are invertible in R_1, its closure $\overline{N}_1$ in R contains an element x which is invertible in R. Since the set of all invertible elements in R is an open set (see II, subsection 4, § 9), there exists a neighborhood $U(x)$ of the point x consisting solely of invertible elements in R. Therefore, if the sequence $\langle x_n \rangle \subset N_1$ converges in R to x, then, beginning with some index n_0, all the elements x_n ($n > n_0$) lie in $U(x)$ and hence will be invertible in R, and by condition a) also invertible in R_1, which is impossible.

b) $\to$ c). This is obvious since a maximal ideal does not contain invertible elements.

c) $\to$ d). The inclusion $S_x \subset S_x^1$ for $x \in R_1$ is satisfied in arbitrary algebras R, R_1 ($R \supset R_1$) with common identity, for if $x - \lambda_0 e$ does not have an inverse in R then also it does not have an inverse in R_1.

We shall prove that $S_x^1 \subset S_x$. Let $\lambda_1 \in S_x^1$, i.e. the element $x - \lambda_1 e$ belongs to some one-sided maximal ideal $M_1 \subset R_1$; since M_1, together with its closure $\overline{M}_1$ in the norm $|x|$, does not contain invertible elements in R, then $x - \lambda_1 e$ is not invertible in R, i.e. $\lambda_1 \in S_x$.

d) $\to$ a). Let $x \in R_1$ and suppose x^{-1} exists in R; then $0 \notin S_x = S_x^1$, i.e. the element x is invertible in R_1.

Corollary. *If* R, R_1 ($R \supset R_1$) *is a Wiener pair, then, for every* $x \in R_1$, $r(x) = r_1(x)$, *where* $r(x)$, $r_1(x)$ *are the spectral radii of the element* x *relative to the algebras* R, R_1.

II. *Let* R, R_1 ($R \supset R_1$) *be a Wiener pair of commutative Banach algebras with common identity* e *and norms* $|x|$, $|x|_1$. *Assume* R_1 *is dense in* R *with respect to the norm* $|x|$. *Then the closure of every proper ideal of the algebra* R_1 *with respect to the norm* $|x|$ *is a proper ideal in* R; *the closure of a maximal ideal in* R_1 *is a maximal ideal in* R.

Proof. Since R_1 is dense in R it is obvious that the closure $\overline{I}_1$ of a proper ideal $I_1 \subset R_1$ is an ideal in R. By I b), the ideal $\overline{I}_1$ is proper.

Now let M_1 be a maximal ideal of the algebra R_1. It suffices to prove that every element x in R is representable in the form $x = \lambda e + y$, where $y \in \overline{M}_1$. Inasmuch as R_1 is dense in R, there exists a sequence of elements $\langle x_n \rangle \subset R_1$ which converges in R to x, and since M_1 is maximal in R_1, we have that $x_n = \lambda_n e + y_n$, where $y_n \in M_1$. The sequence $\langle \lambda_n \rangle$ is bounded, for otherwise it would contain a subsequence $\lambda_{n_m} \to \infty$ and, since the sequence $\langle x_n \rangle$ is bounded,

$$e + \frac{y_{n_m}}{\lambda_{n_m}} = \frac{\lambda_{n_m} e + y_{n_m}}{\lambda_{n_m}} = \frac{x_{n_m}}{\lambda_{n_m}} \to 0,$$

i.e. $e \in \overline{M}_1$, which is impossible. From the sequence $\langle \lambda_n \rangle$ we choose a convergent subsequence $\lambda_{n_k} \to \lambda$; x_{n_k} converges in R to x and therefore $y_{n_k} = x_{n_k} - \lambda_{n_k} e$ converges in R to $y = x - \lambda e \in \overline{M}_1$, which is what was required.

We make the following simple remark.

III. *Let* R, R_1 ($R \supset R_1$) *be commutative Banach algebras with common identity* e. *If* M *is maximal in* R, *then* $M \cap R_1$ *is maximal in* R_1.

This follows from the fact that if $x \in R_1$ and $x = \lambda e + y$, where $y \in M$, then $y = x - \lambda e \in R_1$ and therefore $y \in M \cap R_1$. Since $x \in R_1$ is arbitrary, it follows from this that $M \cap R_1$ is maximal in R_1.

IV. *Let* R, R_1 ($R \supset R_1$) *be a Wiener pair of commutative algebras with common identity* e *and assume that* R_1 *is dense in* R. *Then, for any maximal ideals* M_1, M *of the algebras* R_1, R,

$$M_1 = \overline{M}_1 \cap R_1, M = \overline{M \cap R_1},$$

where the vinculum denotes closure with respect to the norm $|x|$.

Proof. By II, the ideal $\overline{M}_1 \cap R_1$ is proper (for otherwise $e \in R_1 = \overline{M}_1 \cap R_1$) and contains a maximal ideal M_1, i.e. coincides with it.

In virtue of III, the ideal $M \cap R$ is maximal in R_1 and by II its closure $\overline{M \cap R_1}$ is a maximal ideal in R, contained in M, i.e. coinciding with M.

V. *If* R, R_1 ($R \supset R_1$) *is a Wiener pair of commutative algebras with common identity* e *and* R_1 *is dense in* R, *then the spaces of maximal ideals* $\mathfrak{M}(R)$ *and* $\mathfrak{M}(R_1)$ *are homeomorphic.*

Proof. Consider the mapping F: $\mathfrak{M}(R_1) \to \mathfrak{M}(R)$ such that $F(M_1) = \overline{M}_1 = M$. This mapping is one-to-one, for if $F(M_1) = F(M_1')$, then

$$M_1' = \overline{M}_1' \cap R_1 = F(M_1') \cap R_1 = F(M_1) \cap R_1 = \overline{M}_1 \cap R_1 = M_1 .$$

We shall show that the mapping F is continuous; then, by V, subsection 7, § 2, F will be a homeomorphism and $\mathfrak{M}(R_1)$, $\mathfrak{M}(R)$ homeomorphic.

Since R_1 is dense in R, then $R_1 - \{e\}$ can be considered to be a system of generators of the algebra R. In virtue of Theorem 5, subsection 4, § 11, the neighborhoods of the form $U(M; y_1, \cdots, y_n; \varepsilon)$, where $y_k \in R_1$, form a basis in $\mathfrak{M}(R)$. Consider one of them and let V be a neighborhood of the point M_1 in $\mathfrak{M}(R_1)$, defined by the same collection of elements $y_1, \cdots, y_n$ and the same ε: $V = V(M_1; y_1, \cdots, y_n; \varepsilon)$. To each $M_1' \in V$ there corresponds a unique $M' = \overline{M}_1'$ and since $M' \supset M_1'$, then $y_k(M_1') = y_k(M')$, since $y_k - y_k(M_1')e \in M_1' \subset M'$. Consequenctly, $M' \in U$, i.e. $F(V) \subset U$.

8. **Functions of several algebra elements; locally analytic functions.** Theorem 7, subsection 6 admits a generalization to functions of several algebra elements and to certain multi-valued analytic functions.

Let R be a commutative Banach algebra with identity and let $\mathfrak{M}$ be the space of its maximal ideals. The set of all points $\langle x_1(M), \cdots, x_p(M) \rangle$ ($M \in \mathfrak{M}$) of the space C^n is called the *joint spectrum* of the elements $x_1, \cdots, x_p \in R$.

The following theorem holds:

THEOREM 8. *For every (single-valued) function* $f(\zeta) = f(\zeta_1, \cdots, \zeta_p)$ *which is analytic on the joint spectrum of the elements* $x_1, \cdots, x_p \in R$ *there exists an element* $x \in R$ *such that* $x(M) = f(x_1(M), \cdots, x_p(M))$ *for all* $M \in \mathfrak{M}$.

This theorem enables us to construct an operator calculus on the elements of the algebra R (for more details, see, e.g., GELFAND, RAIKOV and SHILOV [1], § 13).

Now let $f(M)$ be a single-valued continuous function on $\mathfrak{M}$ and let $z \in R$. The function $f(M)$ is called a *locally analytic function* of z if, for every point $M_0 \in \mathfrak{M}$, there exists a neighborhood $V(M_0)$ of M_0, in which

$$f(M) = \sum_{k=1}^{\infty} a_k(M_0)[z(M) - z(M_0)]^k. \tag{1}$$

In this connection, $f(M)$ can also be a non-single-valued function of z, for there can exist different points M_1 and M_2 at which $z(M_1) = z(M_2)$, but $f(M_1) \neq f(M_2)$ inasmuch as the coefficients $a_k(M_1)$ and $a_k(M_2)$ in the expansion (1) can be distinct. A locally analytic function $f(M)$ is called a *normally multi-valued function* of z if every set $\{M: z(M) = C\}$ is the union of a finite number of disjoint closed sets, on each of which the coefficients $a_k(M_0)$ in (1) do not depend on M_0. For example, the function $f(M) = \ln z = \ln z(M)$, if it is single-valued and continuous on $\mathfrak{M}$, is a normally multi-valued function of z. In fact,

$$\{M\colon z(M) = C\} = \bigcup_k \{M\colon \ln z(M) = \ln |C| + i \arg C + 2\pi ik\}, \quad 0 \leqq \arg C < 2\pi,$$

where, in virtue of the continuity of the function $f(M)$, the union is over only a finite number of values of k and for a given k the coefficients $a_k(M_0)$ of an expansion of the form (1) of the function $\ln |z(M)| + i \arg z(M) + 2\pi ik$ depend only on $z(M_0)$; consequently, for fixed $z(M_0)$ and k, these coefficients do not depend on M_0.

[The union is over only a finite number of values of k, for otherwise $f(M) = \ln |C| + i \arg C + 2\pi ik \to \infty$ as $k \to \infty$ which is impossible in virtue of the continuity of the function $f(M)$ and the compactness of the space $\mathfrak{M}$.]

Clearly, any *normally multi-valued* function $f(z)$ of the complex variable z possesses this same property. This is what we call a (multi-valued) function, all values $f(z_0)$ of which at each point z_0, not a branch point, are distinct and isolated. For example, $f(z) = \sqrt[m]{z}$ is normally multi-valued, and $f(z) = (1+z)\sqrt{z}$ is not normally multi-valued, since, for $z = -1$, its values on both sheets of the Riemann surface coincide.

THEOREM 9. *If $f(M)$ is a locally analytic and normally multi-valued function of $z \in R$, then there exists an element $x \in R$ such that $x(M) = f(M)$ for all $M \in \mathfrak{M}$.*

In the works of KREIN [11] and GOKHBERG [3], Theorem 9 is applied to the investigation of analytic functions of elements in new classes of Banach algebras constructed in these works. We note also the application of theorem 9 in the theory of singular integral equations (see, e.g., GOKHBERG [2]).

The concept of analytic function was carried over by LORCH [2] to functions $f(x)$ with domains and ranges in an algebra R. Such functions were studied in the works of LORCH [2] and BLUM [1].

The case of the algebra generated by all the bounded rational functions of a bounded linear operator in a Banach space was considered by TOMITA [4].

The concept of an analytic function of several algebra elements was later carried over by WAELBROECK [1] to complete topological algebras with continuous inverse.

Waelbroeck uses this concept for the study of the structure of topological algebras with continuous inverse in the following manner. The *spectrum* $S(x_1, x_2, \cdots, x_n)$ *of the system of elements* $x_1, x_2, \cdots, x_n \in R$, in Waelbroeck's terminology, is the set of all systems of complex numbers $(\lambda_1, \lambda_2, \cdots, \lambda_n)$ such that $P(\lambda_1, \lambda_2, \cdots, \lambda_n)$ lies in the spectrum of the element $P(x_1, x_2, \cdots, x_n)$ for all polynomials P in n variables. Further, he calls the set $\{x_\alpha\}$, $\alpha \in \mathfrak{A}$, in R the *system of analytic generators* of the algebra R if every element $x \in R$ is an analytic function of a finite number of elements x_α.

Suppose now that $\mathscr{H}(S(x_{\alpha_1}, \cdots, x_{\alpha_n}))$ is the algebra of all analytic functions on $S(x_{\alpha_1}, \cdots, x_{\alpha_n})$ with the topology defined in a way analogous to how it was done in Example 1, subsection 3, § 8, and suppose H is the inductive limit of the algebras $\mathscr{H}(S(x_{\alpha_1}, \cdots, x_{\alpha_n}))$. [Suppose $\{F_\alpha\}$, $\alpha \in \mathfrak{A}$ is a directed set of locally convex spaces and that E is a vector space, initially without a topology, and assume that for every α, g_α is a given linear mapping of the space F_α into E, where $g_\alpha(F_\alpha) \subset g_\beta(F_\beta)$ for $\alpha < \beta$ and $\bigcup g_\alpha(F_\alpha) = E$. The space E with the strongest locally convex topology for which all the functions g_α are continuous is called the *inductive limit* of the spaces F_α.] If, in this connection, the separation axiom is not satisfied in H, then H is replaced by the quotient algebra, whose completion, provided this quotient algebra is not complete, is called by Waelbroeck the *algebra of analytic functions* on the space $\mathfrak{M}$ of maximal ideals of the algebra R, topologized as in the case of a normed algebra. Waelbroeck calls this space the *spectrum* of the algebra R.

One of the fundamental results due to Waelbroeck consists of the following.

Every complete commutative algebra R with continuous inverse is isomorphic to the quotient *algebra of the algebra of analytic functions on the spectrum $\mathfrak{M}$ of the algebra R with respect to the closed ideal of these functions which vanish on $\mathfrak{M}$.*

9. **Decomposition of an algebra into the direct sum of ideals.** Suppose R is a commutative Banach algebra with identity and that $\mathfrak{M}$ is the space of its maximal ideals. The algebra R is said to be the *direct sum of its ideals* I_1, I_2 and this fact is denoted by $R = I_1 \dot{+} I_2$, if 1) $I_1 \cap I_2 = (0)$; 2) every element $x \in R$ can be represented in the form $x = x_1 + x_2$, where $x_1 \in I_1$, $x_2 \in I_2$; it follows from condition 1) that such a representation is unique.

Suppose $R = I_1 \dot{+} I_2$ is the direct sum of its ideals I_1, I_2. Applying condition 2) to the identity e of the algebra R, we obtain $e = e_1 + e_2$, $e_1 \in I_1$, $e_2 \in I_2$; moreover, $e_1 e_2 \in I_1 \cap I_2 = (0)$, so that $e_1 e_2 = (0)$, and therefore $e_1 + e_2 = e = e^2 = e_1^2 + e_2^2$; $e_1^2 = e_1$, $e_2^2 = e_2$. Moreover, $e_1 I_2 \subset I_1 \cap I_2 = (0)$, i.e. $e_1 I_2 = (0)$, and analogously $e_2 I_1 = (0)$.

From this we easily conclude that

$$I_1 = \{x \in R \colon e_1 x = x\}, I_2 = \{x \in R \colon e_2 x = x\}; \tag{1}$$

consequently, I_1, I_2 are closed ideals and e_1, e_2 are the identities in the algebras I_1, I_2, respectively.

Conversely, if R contains nonzero elements e_1, e_2, satisfying the conditions $e = e_1 + e_2$, $e_1^2 = e_1$, $e_2^2 = e_2$, $e_1 e_2 = 0$, then, as is easily seen, equalities (1) define the closed ideals I_1, I_2 in R and $R = I_1 \dot{+} I_2$; the last assertion follows immediately from the equalities $x = xe = xe_1 + xe_2$.

Since $e_1^2 = e_1$, $e_2^2 = e_2$, the functions $e_1(M)$, $e_2(M)$ assume only the values 0 and 1, where

$$e_1(M) + e_2(M) = 1, \; e_1(M) e_2(M) = 0.$$

Consequently, setting

$$F_1 = \{M \colon e_1(M) = 1\}, F_2 = \{M \colon e_2(M) = 1\},$$

we obtain that $\mathfrak{M}$ is the union of two nonvoid disjoint closed sets F_1, F_2 and therefore it is not connected.

A topological space is said to be *not connected* if it is the union of two disjoint closed sets, and *connected* otherwise. The sets F_1, F_2 defined in the text are nonvoid, since, e.g.,

$$\lim_{n \to \infty} \sqrt[n]{|e_1^n|} = \lim_{n \to \infty} \sqrt[n]{|e_1|} = 1;$$

consequently, e_1, e_2 do not belong to the radical of the algebra.

Assigning to each $x \in I_1$ the number $x(M)$, $M \in F_1$, we obtain a homomorphism of the algebra I_1 onto the field of complex numbers and, consequently, onto the ideal M_1 of the algebra I_1. It is easily seen that the correspondence $M \leftrightarrow M_1$ is a homeomorphism of F_1 onto the space $\mathfrak{M}(I_1)$ of maximal ideals of the algebra I_1; analogously, F_2 is homeomorphic to the space $\mathfrak{M}(I_2)$. Thus, *if $R = I_1 \dot{+} I_2$, then $\mathfrak{M}$ is the union of two closed sets which are the spaces of maximal ideals of the algebras I_1, I_2.*

Shilov [18], utilizing his theory of analytic functions of several algebra elements (see subsection 8), proved that the converse assertion is also true:

If the space $\mathfrak{M}$ is not connected, then R is decomposable into the direct sum of its ideals which are uniquely defined by the closed sets F_1, F_2 for which $\mathfrak{M} = F_1 \cup F_2$, $F_1 \cap F_2 = \emptyset$.

Żelazko [1] carried this result over to algebras R in which $|\alpha x| = |\alpha|^p |x|$ (instead of $|\alpha x| = |\alpha| \, |x|$) for arbitrary complex α and arbitrary $x \in R$, where $0 < p \leqq 1$, p fixed.

10. **Algebras with radical.** The structure of non-semisimple commutative Banach algebras R has been investigated relatively recently and the basic direction of the investigation is the study of

the conditions under which the well-known Wedderburn theorem (on the decomposition of a finite-dimensional complex algebra, as a linear space, into the direct sum of the radical and a subalgebra; see, e.g., JACOBSON [1]) carries over to Banach algebras. Usually the problem is stated so that the subalgebra $\hat{R}$ and the radical I are given in advance and $\hat{R}$ is extended to an algebra R so that I is the radical of the algebra R and R/I is isomorphic to $\hat{R}$ (it is assumed that R contains an identity). Apparently these questions first appeared in the article by FELDMAN [1] who showed by examples that there exist decomposable, but not strongly decomposable algebras. [An algebra R is called *strongly decomposable* if there exists a decomposition $R = I \dotplus \hat{R}$, where the subalgebra $\hat{R}$ is closed.] Among other works in this direction (see, e.g. ZELINSKY [1], GLAESER [1]), we mention the works of BADE and CURTIS [1] and KHELEMSKY [1–4]. In the article by BADE and CURTIS [1], among other results, it is proved that *if the radical I of the algebra R is finite-dimensional and the functions $x(M)$, corresponding to elements of the quotient algebra R/I, form the entire algebra $C(\mathfrak{M})$ ($\mathfrak{M}$ is the space of maximal ideals of R and $R(I)$), then R decomposes, as a linear space, into the direct sum of the radical I and the subalgebra $\hat{R}$.*

KHELEMSKY [1] obtained this same result independently in another way; he succeeded in generalizing this result to non-commutative symmetric algebras R, for which R/I is completely symmetric (see §§ 14, 23). In the works of KHELEMSKY [1–4] there are investigated conditions for the decomposability and the strong decomposability of an algebra into a subalgebra and the radical and the structure of certain algebras for which such a decomposition is valid.

We now give some of these results.

I. *Suppose R decomposes into the direct sum of its radical I and the subalgebra $\hat{R}$. Then there exists a finite number of primary ideals $\mathfrak{J}_l$, $l = 1, \cdots, k$, of the algebra $\hat{R}$, having a finite co-dimension and contained in distinct maximal ideals. Moreover, I decomposes into the direct sum of subalgebras I_s $(s = 1, \cdots, k)$ such that $I_s I_{s'} = (0)$ for $s \neq s'$.* [An ideal is called *primary* if it is contained in only one maximal ideal. For details concerning primary ideals, see subsection 7, § 15.]

It is described how the operators of left multiplication by $x \in R$ act in the quotient spaces $\hat{R}/\mathfrak{J}_l$.

II. *Let*

$$\lim_{n\to\infty} \sqrt[n^2]{\sup_{y \in I,\, |y|=1} |y^n|} = 0$$

and $R/I = C(Q)$, where Q is a totally disconnected compact space, or $R/I = l_1$; then R is strongly decomposable.

Also given are necessary and sufficient conditions, in terms of $\hat{R}$, for the strong decomposability of an arbitrary algebra R, for which I is finite-dimensional and $\hat{R} = R/I$. In the article by KAMOWITZ [1], connections between this problem and the cohomology groups of the algebra are established.

§ 12. Homomorphism and isomorphism of commutative algebras

1. Uniqueness of the norm in a semisimple algebra. One of the important results in the theory of commutative algebras consists in the fact that the norm in a semisimple algebra is uniquely defined, to within equivalence, by its supply of elements.

We shall first of all prove the following lemma.

LEMMA. *Suppose R_1 is a commutative algebra, R is a subalgebra of R_1, and that $|x|_1$, $|x|$ are given norms in R_1, R respectively, with respect to which R_1 and R are Banach algebras. If $\mathfrak{M}_1$ and $\mathfrak{M}$ are the spaces of maximal regular ideals of the algebras R_1 and R, then for arbitrary element $x \in R$,*

$$\max_{M_1 \in \mathfrak{M}_1} |x(M_1)| \leqq \max_{M \in \mathfrak{M}} |x(M)|. \tag{1}$$

Proof. Suppose S_x, S_x^1 are the spectra of the element $x \in R$ in the algebras R, R_1 respectively. (If R is an algebra without identity, then the spectrum of the element $x \in R$ is its spectrum in the algebra obtained from R upon adjunction of the identity.) Clearly,

$$S_x \supset S_x^1. \tag{2}$$

On the other hand, S_x is the set of all values of the function $x(M)$, $M \in \mathfrak{M}$, and S_x^1 is the set of all values of the function $x(M_1)$, $M_1 \in \mathfrak{M}_1$. From this and from (2) it follows immediately that (1) is valid.

THEOREM 1. *Suppose a second norm $|x|_1$ is given in the commutative Banach algebra R with norm $|x|$ and that the completion R_1 of the algebra R with respect to the norm $|x|_1$ is a semisimple algebra. Then $|x|_1 \leqq C|x|$ for all $x \in R$, where C is some constant.*

Proof. We define a new norm $|x|_2$ in R by setting

$$|x|_2 = \max \{|x|, |x|_1\}. \tag{3}$$

It is easily verified that $|x|_2$ satisfies all the norm axioms in the algebra, so that R is also a normed algebra with respect to the norm $|x|_2$.

We shall prove that R is complete with respect to the norm $|x|_2$.

Let $\langle x_n \rangle$ be a fundamental sequence in R with respect to the norm $|x|_2$. In virtue of (3), it is also fundamental with respect to each of the norms $|x|$, $|x|_1$. Suppose x_0 and x_0' are the limits of this sequence relative to the norms $|x|$ and $|x|_1$ respectively.

We shall prove that $x_0' = x_0$; this also will prove that the algebra R is complete with respect to the norm $|x|_2$. Applying the preceding lemma, we have that

$$\max_{M_1 \in \mathfrak{M}_1} |x_0(M_1) - x_n(M_1)| \leqq \max_{M \in \mathfrak{M}} |x_0(M) - x_n(M)| \leqq |x_0 - x_n| \to 0,$$

$$\max_{M_1 \in \mathfrak{M}_1} |x_0'(M_1) - x_n(M_1)| \leqq |x_0' - x_n|_1 \to 0$$

as $n \to \infty$ [see I, 7), subsection 2, § 11] and therefore

$$x_0'(M_1) - x_0(M_1) = 0 \text{ for all } M_1 \in \mathfrak{M}_1.$$

This means that $x_0' - x_0$ belongs to the radical of the algebra R_1; since the algebra R_1 is semisimple, we conclude from this that $x_0' - x_0 = 0$.

Thus, R is complete with respect to the norm $|x|$ as well as with respect to the norm $|x|_2 \geqq |x|$. On the basis of the corollary to the Banach theorem (see VII, subsection 4, § 4), there exists a constant C such that $|x|_2 \leqq C|x|$ for all $x \in R$. But then in virtue of (3), we also have $|x|_1 \leqq C|x|$ for all $x \in R$, and this proves the theorem. (The theorem discussed above is a generalization of a theorem due to GELFAND [4]; this generalization is due to SHILOV [5] and RICKART [3].)

COROLLARY 1. *Every isomorphism of a commutative Banach algebra R onto a dense subset R_1' of a semisimple commutative Banach algebra R_1 is continuous.*

Proof. Carrying over to R_1' the norm from R, we obtain two norms in R_1' to which Theorem 1 is applicable. Consequently, if x' and x are elements of the algebras R_1' and R, corresponding to each other under a given isomorphism, then $|x'|_1 \leqq C|x|$ where C is some constant.

COROLLARY 2. *If two semisimple commutative Banach algebras are isomorphic then they are also topologically isomorphic.*

Corollary 2 means that the norm in a semisimple commutative Banach algebra is defined uniquely, to within equivalence, by the supply of elements of this algebra and by its algebraic properties.

COROLLARY 3. *Every automorphism (i.e. isomorphism onto itself) of a semisimple commutative Banach algebra is continuous.*

Corollary 2 was generalized by RICKART [3] to certain noncommutative algebras. In particular, he showed that *in a simple Banach algebra with identity, the norm is uniquely defined, to within equivalence.*

For other generalizations of the results of this subsection, see YOOD [5] and IX, subsection 2, § 18.

2. The case of symmetric algebras.

THEOREM 2. *Every involution in a semisimple commutative Banach algebra is continuous.*

Proof. We denote by $R^\wedge$ the algebra consisting of the same elements as the given algebra R, with the same definitions of addition, multiplication of elements, and norm, but in which the operation of scalar multiplication (we shall denote it by a dot) is defined by the formula $\lambda \cdot x = \bar{\lambda}x$. Clearly, $R^\wedge$ is a semisimple commutative Banach algebra and the correspondence $x \to x^*$ is an isomorphism of the algebras R and $R^\wedge$; therefore, it suffices to apply Corollary 2, subsection 1.

§ 13. Algebra (or Shilov) boundary

1. Definition and fundamental properties of the algebra boundary. Suppose R is a commutative Banach algebra with identity and that $\mathfrak{M}$ is the space of maximal ideals of the algebra R. A closed set $F \subset \mathfrak{M}$ is said to be a *determining set* if the absolute value of every function $x(M)$, $x \in R$, attains its maximum at some point of the set F. Evidently at least one determining set exists: namely $\mathfrak{M}$ itself. A determining set is said to be *minimal* if none of its proper closed subsets is a determining set. A minimal determining set is called an *algebra boundary* (or *Shilov boundary*) of the space $\mathfrak{M}$.

THEOREM 1. *In the space of maximal ideals of a commutative Banach algebra with identity, an algebra boundary exists and, furthermore, it is unique.*

Proof. We shall first prove the existence of an algebra boundary.

The set $\{F\}$ of all determining sets, ordered in a natural way with the aid of the inclusion relation $\supset$, is a partially ordered set which satisfies the condition of Zorn's lemma. Namely, a lower bound of a linearly ordered family $\{F_\alpha\}$ of determining sets (ordered by inclusion) will be simply the intersection of all sets of this family (which is nonvoid in view of the fact that $\mathfrak{M}$ is compact), which is also a determining set. In fact, if $x(M)$ is a fixed function, $x \in R$, and F_x is the set of all points $M \in \mathfrak{M}$, at which its absolute value attains a maximum, then the compact sets $F_x \cap F_\alpha$ have nonvoid intersection and, consequently, the intersection of F_x with $\bigcap_\alpha F_\alpha$ is also nonvoid.

Therefore, $\{F\}$ contains a minimal element, which is then an algebra boundary.

We shall now prove the uniqueness of the algebra boundary. Suppose Γ_1 and Γ_2 are two algebra boundaries. We shall prove that every neighborhood of an arbitrary point $M_1 \in \Gamma_1$ contains points of Γ_2. Inasmuch as the set Γ_2 is closed, it will follow that $\Gamma_1 \subset \Gamma_2$, and hence $\Gamma_1 = \Gamma_2$, because we also have $\Gamma_2 \subset \Gamma_1$.

Thus, suppose $U = U(M_1; x_1, \cdots, x_n; \varepsilon)$ is an arbitrary neighborhood of the neighborhood basis of the point $M_1 \in \Gamma_1$, defined by the inequalities

$$|x_k(M) - x_k(M_1)| < \varepsilon, \quad k = 1, 2, \cdots, n.$$

Replacing the elements x_k by the elements $x_k - x_k(M_1)e$, we may assume that $x_k \in M_1$ and that the neighborhood U is defined by the inequalities

$$|x_k(M)| < \varepsilon, \quad k = 1, 2, \cdots, n. \tag{1}$$

There exists a function $y = y(M)$ in the algebra R which assumes its maximum modulus

$$m = \max_{M \in \mathfrak{M}} |y(M)|$$

in $\Gamma_1 \cap U$ and remains in modulus less than m on $\Gamma_1 - \Gamma_1 \cap U$; in fact, in the contrary case, the set $\Gamma_1 - \Gamma_1 \cap U$ would be a determining set in contradiction to the fact that Γ_1 is minimal. Replacing y by $(1/m)y$, we may, without loss of generality, assume that $m = 1$. Further, replacing y by some one of its powers y^n if necessary, we may assume that

$$|y(M)| < \frac{\varepsilon}{\max\limits_{1 \leq k \leq n} |x_k|} \quad \text{for } M \in \Gamma_1 - U \cap \Gamma_1.$$

In virtue of (1), the functions $(x_k y)(M) = x_k(M)y(M)$, $k = 1, 2, \cdots, n$, are then less in absolute value than ε everywhere on Γ_1, and hence also on all of $\mathfrak{M}$:

$$|x_k(M)y(M)| < \varepsilon \text{ for all } M \in \mathfrak{M} \text{ and } k = 1, 2, \cdots, n. \tag{2}$$

Since Γ_2 is an algebra boundary, there exists a point $M_2 \in \Gamma_2$ on which $|y(M)|$ attains its maximum $m = 1$:

$$|y(M_2)| = 1.$$

Setting $M = M_2$ in (2), we obtain

$$|x_k(M_2)| < \varepsilon, \quad k = 1, 2, \cdots, n;$$

consequently, $M_2 \subset U$. In virtue of what was stated above, we conclude from this that $\Gamma_1 = \Gamma_2$; this completes the proof of the theorem.

THEOREM 2. *A maximal ideal M_0 of a commutative Banach algebra R with identity belongs to the algebra boundary Γ if, and only if, for an arbitrary neighborhood $U(M_0)$ of the point M_0 there exists a function $y(M)$, $y \in R$, which attains its maximal absolute value in $U(M_0)$ and whose absolute value is less than this maximal value everywhere in the exterior of $U(M_0)$.*

Proof. If such a function $y(M)$ does not exist, then the modulus of every function $y(M)$, $y \in R$, attains its largest value in $\mathfrak{M} - U(M_0)$ for some $U(M_0)$ and therefore $\Gamma \subset \mathfrak{M} - U(M_0)$, $M_0 \notin \Gamma$; hence the condition of the theorem is necessary. Conversely, if this condition is satisfied, then an arbitrary neighborhood of the point M_0 contains points of the boundary and therefore M_0 belongs to the boundary.

EXAMPLE. Suppose A is the algebra of all functions $x = x(\zeta)$ which are regular in the open disc $|\zeta| < 1$ and continuous in the closed disc $|\zeta| \leqq 1$ (Example b), subsection 3, § 11). The space $\mathfrak{M}$ of maximal ideals of the algebra A is the disc $|\zeta| \leqq 1$. By the maximum modulus principle, the algebra boundary is the circle $|\zeta| = 1$.

REMARK. In the general case the algebra boundary of an algebra with a finite number n of generators may not coincide with the topological boundary of a set in C^n, which is homeomorphic to the space of maximal ideals (see Theorem 6, subsection 5, § 11). An example is the algebra obtained from polynomials $P(z_1, z_2)$ of two complex variables by means of closure with respect to the norm

$$|P| = \max_{|z_1| \leqq 1,\, |z_2| \leqq 1} |P(z_1, z_2)|.$$

(For details, see GELFAND, RAIKOV and SHILOV [1], pp. 78–79.)

2. Extension of maximal ideals. Suppose R_1 is a commutative Banach algebra with identity and that R is an arbitrary commutative Banach algebra containing R_1 as a subalgebra, with the same norm. We shall denote the space of maximal ideals of the algebra R by $\mathfrak{M}$ and the space of maximal ideals of the algebra R_1 by $\mathfrak{M}_1$. For an element x of the algebra R_1 we can simultaneously consider the function $x(M)$ on $\mathfrak{M}$ as well as the function $x(M_1)$ on $\mathfrak{M}_1$. In this connection, the following lemma holds.

LEMMA. *For arbitrary element $x \in R_1$, we have*

$$\max_{M \in \mathfrak{M}} |x(M)| = \max_{M_1 \in \mathfrak{M}_1} |x(M_1)|. \tag{1}$$

The assertion of the lemma follows directly from the formula

$$\max_{M \in \mathfrak{M}} |x(M)| = \lim_{n \to \infty} \sqrt[n]{|x^n|}$$

(see IV, subsection 2, § 11), because the powers x^n are contained simultaneously in R_1 and in R, and their norms in R_1 and R coincide.

We can now prove the following theorem.

THEOREM 3. *Every maximal ideal of the algebra boundary Γ_1 of the algebra R_1 is extendible to a maximal ideal of an arbitrary algebra R with identity which contains R_1 as a closed subalgebra.*

Proof. Let $M_0 \in \Gamma_1$ and suppose M_0 is not contained in any maximal (and hence in no) ideal of the algebra $R \supset R_1$. Then the set of all sums of the form $\sum_{k=1}^{n} x_k z_k$, $x_k \in M_0$, $z_k \in R$, must coincide with the entire algebra R; otherwise, this set would be an ideal I in R, containing M_0.

In particular, one of the sums of this form must be equal to the identity e of the algebra R:

$$e = \sum_{k=1}^{n} x_k z_k. \tag{2}$$

Replacing x_k, z_k by $\alpha_k x_k$, $(1/\alpha_k) z_k$ if necessary, we may assume that

$$\max_{M \in \mathfrak{M}} |x_k(M)| \leqq 1. \tag{3}$$

We denote the largest of the numbers

$$\mu_k = \max_{M \in \mathfrak{M}} |z_k(M)|, \quad k = 1, 2, \cdots, n \tag{4}$$

by μ and consider the neighborhood $U(M_0)$ of the point M_0 in $\mathfrak{M}_1$, defined by the inequalities

$$|x_k(M_1)| < \frac{1}{2n\mu}, \quad k = 1, 2, \cdots, n \, (M_1 \in \mathfrak{M}_1). \tag{5}$$

According to Theorem 2, subsection 1, there exists a function $y(M_1)$, $y \in R_1$, whose modulus attains its largest value 1 in $U(M_0)$ and remains less than unity in the exterior of $U(M_0)$. Replacing the element y by a power of y if necessary, we may assume that

$$|y(M_1)| < \frac{1}{2n\mu} \text{ in } \mathfrak{M}_1 - U(M_0). \tag{6}$$

Then applying the preceding lemma and taking (2)–(6) into consideration, we have

$$\max_{M\in\mathfrak{M}} |y(M)| = \max_{M\in\mathfrak{M}} |\sum_{k=1}^{n} x_k(M)y(M)z_k(M)| \leqq \mu \sum_{k=1}^{n} \max_{M\in\mathfrak{M}} |x_k(M)y(M)|$$
$$= \mu \sum_{k=1}^{n} \max_{M_1\in\mathfrak{M}_1} |x_k(M_1)y(M_1)| < \mu \cdot n \frac{1}{2n\mu} = \frac{1}{2};$$

on the other hand, in virtue of the same lemma,

$$\max_{M\in\mathfrak{M}} |y(M)| = \max_{M_1\in\mathfrak{M}_1} |y(M_1)| = 1.$$

Now the validity of the theorem follows from the contradiction thus obtained.

THEOREM 3'. *Let R, R_1 ($R \supset R_1$) be commutative Banach algebras with common identity e and norms $|x|$, $|x|_1$ respectively. A necessary and sufficient condition for every maximal ideal of the algebra boundary Γ_1 of the algebra R_1 to be extendible to a maximal ideal of the algebra R is that $r(x) = r_1(x)$ for every $x \in R_1$.*

The proof of Theorem 3 means precisely the establishing of the sufficiency of the condition of Theorem 3'. We shall prove the necessity. Let us assume the contrary: suppose there exists an element $x \in R_1$ such that $r_1(x) \neq r(x)$. By the lemma of § 12, subsection 1, $r_1(x) > r(x)$. Consider the maximal ideal $M_1 \in \Gamma_1$ such that the absolute value of the function $x(M_1)$, considered on $\mathfrak{M}(R_1)$, takes on at the point M_1 its largest value $|x(M_1)| = r_1(x)$. Let $M \in \mathfrak{M}(R)$ be a maximal ideal of the algebra R which is an extension of M_1. The element $x - x(M_1)e \in M_1 \subset M$ and therefore $x(M) - x(M_1)e(M) = 0$, i.e. $x(M) = x(M_1)$, which is impossible since $|x(M)| \leqq r(x) < r_1(x) = |x(M_1)|$.

COROLLARY 1. *For every maximal ideal of the algebra R_1 to be extendible to a maximal ideal of the algebra R, it is necessary, and if the space of maximal ideals $\mathfrak{M}(R_1)$ coincides with the algebra boundary Γ_1 of the algebra R_1, it is also sufficient that the algebras R, R_1 form a Wiener pair.*

Proof. Necessity. Consider an arbitrary maximal ideal $M_1 \in R_1$ and its extension M. Let $\overline{M}_1$ be the closure of M_1 in R. Since $\overline{M}_1 \subset M$, then by I, subsection 7, § 11 the algebras R and R_1 form a Wiener pair.

Sufficiency. If the algebras R and R_1 form a Wiener pair, then by the corollary to I, subsection 7, § 11, $r(x) = r_1(x)$ for every $x \in R_1$. By Theorem 3' this is sufficient for the extendibility of maximal ideals of Γ_1. But by assumption $\Gamma_1 = \mathfrak{M}(R_1)$.

COROLLARY 2. *If the closed subalgebra R_1 contains the identity e of the algebra R and $\Gamma_1 = \mathfrak{M}(R_1)$, then the algebras R and R_1 form a Wiener pair.*

COROLLARY 3. *If $R \supset R_1$, R_1 is complete in its norm and contains the identity e of the algebra R, and $\Gamma_1 = \mathfrak{M}(R_1)$, then the equality $S_x = S_x^1$ for all $x \in R_1$ is equivalent to the equality $r(x) = r_1(x)$ for all $x \in R_1$.*

THEOREM 4. *Let R, R_1 ($R \supset R_1$) be commutative Banach algebras with common identity and suppose that only maximal ideals of the algebra boundary Γ_1 of the algebra R_1 extend to maximal ideals of the algebra R. A necessary and sufficient condition for the algebras R and R_1 to form a Wiener pair is that every non-invertible element in R_1 belongs to some maximal ideal of the algebra boundary Γ_1.*

Proof. Necessity. Let $x \in R_1$ and suppose that x is not invertible in R_1. Then x is not invertible in R and hence belongs to some maximal ideal $M \subset R$, i.e. $x \in M \cap R_1$. By III, subsection 7,

§ 11, the ideal $M \cap R_1$ is maximal in R_1 and since it can be extended to M, then $M \cap R_1$ belongs to the algebra boundary Γ_1.

Sufficiency. Let us assume the contrary: suppose there exists an element x in R_1 which is not invertible in R_1 but is invertible in R. Since $x \in M_1$, where $M_1 \in \Gamma_1$, and hence it is extendible to a maximal ideal $M \subset R$, then $x \in M$ in contradiction to our assumption that the element x is invertible in R.

§ 14. Completely symmetric commutative algebras

1. Definition of a completely symmetric algebra. If R is a commutative Banach symmetric algebra with identity and M is a maximal ideal in R, then for arbitrary element $x \in R$,

$$x = x(M)e + m, \; m \in M.$$

It follows that

$$x^* = \overline{x(M)}e + m^*. \tag{1}$$

Obviously, M^* (i.e. the set of all m^*, $m \in M$) is also a maximal ideal in R; therefore (1) means that

$$x^*(M^*) = \overline{x(M)}. \tag{2}$$

A commutative Banach symmetric algebra R with identity is said to be *completely symmetric* if $x^*(M) = \overline{x(M)}$ for all $x \in R$ and all $M \in \mathfrak{M}$. (In the literature, completely symmetric algebras are frequently called symmetric algebras and symmetric algebras are called algebras with involution.)

A commutative Banach symmetric algebra R with identity is completely symmetric if, and only if, all its maximal ideals are symmetric.

In fact, if all the maximal ideals M are symmetric ($M^* = M$), then we conclude from (2) that $x^*(M) = \overline{x(M)}$; consequently, R is completely symmetric. Conversely, if R is completely symmetric, then it follows from $x(M) = 0$ that $x^*(M) = 0$, i.e. $x \in M$ implies $x^* \in M$. This means that every maximal ideal M is symmetric.

EXAMPLES 1. The algebra $C(T)$ of all continuous functions on a compact space T is clearly completely symmetric.

2. The algebra W of all absolutely convergent trigonometric series $x = x(t) = \sum_{n=-\infty}^{\infty} c_n e^{int}$ is also completely symmetric.

3. The algebra A of all functions $x(\zeta)$, regular inside the open disc $|\zeta| < 1$ and continuous in the closed disc $|\zeta| \leq 1$, is not completely symmetric. In fact, the functions $x(\zeta)$ and $\overline{x(\bar{\zeta})}$ can simultaneously be regular in the closed disc $|\zeta| \leq 1$ only for $x(\zeta) =$ constant.

We note that one can introduce an involution in the algebra A by setting $x^*(\zeta) = \overline{x(\bar{\zeta})}$; obviously, all the involution axioms (see subsection 1, § 10) will then be satisfied. A is thus an example of a symmetric algebra which is not completely symmetric.

2. Criterion for complete symmetry.

THEOREM 1. *Suppose R is a commutative Banach algebra with identity and that an involution $x \to x^*$ is defined in R. This involution satisfies the condition*

$$x^*(M) = \overline{x(M)} \textit{ for all } M \in \mathfrak{M}$$

if, and only if, every element $e+x^*x$ *has an inverse in* R.

Proof. The necessity of the condition is obvious. In fact, if R is completely symmetric, then

$$(e+x^*x)(M) = 1+\overline{x(M)}x(M) \geqq 1,$$

so that the element $e+x^*x$ does not vanish on any maximal ideal and therefore it has an inverse.

We shall now prove the sufficiency of the condition. Suppose every element $e+x^*x$ has an inverse. We first prove that $\overline{x(M)} = x(M)$ for every Hermitian element $x \in R$. To this end, it suffices to prove that the spectrum of the Hermitian element x is real, i.e. that $(x-(\lambda+i\mu)e)^{-1}$, where λ, μ are real numbers, exists for $\mu \neq 0$. We set $y = (1/\mu)(x-\lambda e)$; then $y^* = (1/\mu)(x-\lambda e)$ and

$$(x-(\lambda+i\mu)e)(x-(\lambda-i\mu)e) = (x-\lambda e)^2+\mu^2 e = \mu^2(e+y^*y).$$

By hypothesis, the right, and hence also the left, member has an inverse; consequently

$$(x-(\lambda+i\mu)e)^{-1} = (x-(\lambda-i\mu)e)[(x-\lambda e)^2+\mu^2 e]^{-1}$$

exists. Thus, for a Hermitian element x, we have proved that $x(M)$ is real. But an arbitrary element x can be represented in the form $x = x_1+ix_2$, where x_1, x_2 are Hermitian. Therefore,

$$x^*(M) = (x_1-ix_2)(M) = x_1(M)-ix_2(M) = \overline{x_1(M)+ix_2(M)} = \overline{x(M)},$$

which completes the proof of the theorem.

We may now give the following more general definition of a completely symmetric algebra: a symmetric Banach algebra R with identity is said to be *completely symmetric* if every element in R of the form $e+x^*x$ has an inverse.

Theorem 1 means that in the case of a *commutative* algebra, this second definition is equivalent to the original definition.

3. Application of Stone's theorem.

THEOREM 2. *If* R *is a completely symmetric commutative algebra and* $\mathfrak{M}$ *is the space of its maximal ideals then every continuous function* $f(M)$ *on* $\mathfrak{M}$ *is the limit function of a uniformly convergent sequence of functions* $x(M)$, $x \in R$.

Proof. We denote the set of all real-valued functions $x(M)$, $x \in R$ by K; then K is the algebra of continuous real-valued functions on the compact space $\mathfrak{M}$. If $M_1 \neq M_2$, then there exists a function $x(M) = x_1(M)+ix_2(M)$, $x, x_1, x_2 \in R$ (see I, subsection 1, § 10 and I, subsection 2, § 11), such that $x(M_1) \neq x(M_2)$; consequently, at least one of the real-valued functions $x_1(M)$, $x_2(M) \in K$ assumes distinct values at the points M_1 and M_2. But then K satisfies all the conditions of Stone's theorem (see subsection 10, § 2); therefore every continuous real-valued function on $\mathfrak{M}$ is the limit function of a uniformly convergent sequence of real-valued functions $x(M) \in K$. Applying this result to the continuous real-valued functions $\operatorname{Re} f(M)$ and $\operatorname{Im} f(M)$, we find that the theorem is valid.

Corollary 1. *If in the completely symmetric algebra R,*

$$|x| = \max_{M \in \mathfrak{M}} |x(M)|,$$

then R is isometrically isomorphic to the algebra $C(\mathfrak{M})$. [We recall that $C(\mathfrak{M})$ is the algebra of all continuous functions on $\mathfrak{M}$ with the usual definitions of the operations and with norm $\max_{M \in \mathfrak{M}} |x(M)|$ (see Example 1, subsection 1, § 9).]

In fact, in this case the functions $x(M)$ form a complete algebra in the sense of the norm $\max_{M \in \mathfrak{M}} |x(M)|$, because R is complete in the sense of the norm $|x|$. Therefore, every function in $C(\mathfrak{M})$, being the limit function of a sequence of functions $x(M)$, $x \in R$, must itself be a function of this form.

Corollary 2. *If in the completely symmetric algebra R,*

$$|x^2| = |x|^2$$

for all $x \in R$, *then R is isometrically isomorphic to the algebra* $C(\mathfrak{M})$.

In fact, in this case

$$\max_{M \in \mathfrak{M}} |x(M)| = \lim_{n \to \infty} \sqrt[2^n]{|x^{2^n}|} = \lim_{n \to \infty} \sqrt[2^n]{|x|^{2^n}} = |x|$$

and it remains to apply Corollary 1.

A commutative symmetric Banach algebra R without identity is said to be *completely symmetric* if the algebra R_1, obtained from R by the adjunction of the identity e, is completely symmetric.

Corollary 3. *If R is a completely symmetric algebra without identity and* $\mathfrak{M}$ *is the space of maximal regular ideals of the algebra R, then every continuous function* $f(M)$ *on* $\mathfrak{M}$, *which equals zero at infinity, is the limit function of a sequence of functions* $\langle x_n(M) \rangle$, $x_n \in R$, *which is uniformly convergent on* $\mathfrak{M}$.

Proof. The space $\mathfrak{M}$ may be considered as the locally compact space which is obtained from the space $\mathfrak{M}_1$ of maximal ideals of the algebra R_1 upon discarding the ideal $M_0 = R$, which plays the role of the infinitely distant point in $\mathfrak{M}_1$ (see subsection 4, § 11), and the elements $x \in R$ as the functions $x(M)$ on $\mathfrak{M}_1$ which vanish at the point M_0.

Suppose $f(M)$ is an arbitrary continuous function on $\mathfrak{M}_1$ which vanishes at the point M_0.

According to Theorem 2, there exists a sequence $x_n \in R_1$ such that $\sup_{\mathfrak{M}_1} |f(M) - x_n(M)| \to 0$ as $n \to \infty$.

Since $f(M_0) = 0$, it follows that

$$|x_n(M_0)| = |x_n(M_0) - f(M_0)| \to 0 \text{ as } n \to \infty;$$

therefore, setting $y_n = x_n - x_n(M_0)e$, we obtain a sequence $y_n \in R$ such that $\sup_{\mathfrak{M}} |y_n(M) - f(M)| \to 0$ as $n \to \infty$.

4. The algebra boundary of a completely symmetric algebra.

THEOREM 3. *The algebra boundary of a completely symmetric commutative algebra coincides with its space of maximal ideals.*

Proof. Suppose R is a completely symmetric commutative algebra, $\mathfrak{M}$ is the space of its maximal ideals, M_0 is an arbitrary maximal ideal of the algebra R, and $f(M)$ is a continuous function which vanishes in the exterior of a given neighborhood $U(M_0)$ of the point M_0 and is equal to unity at the point M_0. According to Theorem 2, subsection 3, there exists a function $x(M)$, $x \in R$, which satisfies the inequality $|f(M) - x(M)| < 1/3$ on all of $\mathfrak{M}$. But then $|x(M)|$ assumes its maximal value only in the neighborhood $U(M_0)$; consequently, the condition of Theorem 2, subsection 1, § 13, is satisfied for the point M_0 and M_0 belongs to the algebra boundary. Inasmuch as the point $M_0 \in \mathfrak{M}$ is arbitrary, we conclude from this that the algebra boundary coincides with $\mathfrak{M}$.

COROLLARY. *Every maximal ideal of a completely symmetric commutative algebra R_1 is extendible to a maximal ideal of an arbitrary commutative Banach algebra with identity which contains R_1 as a closed subalgebra.*

To prove this, it suffices to apply Theorem 3, subsection 2, § 13 to the present case.

§ 15. Regular algebras

1. Definition of a regular algebra. A family $\mathscr{F}$ of functions on a topological space T is said to be *regular* if for an arbitrary closed set $F \subset T$ and any point $t_0 \notin F$ there exists a function $x \in \mathscr{F}$ which satisfies the conditions:

$$x(t_0) \neq 0, \ x(t) = 0 \text{ on } F.$$

The family $\mathscr{F}$ is said to be *normal* if for any two disjoint closed sets $F_1, F_2 \subset T$ there exists a function $x \in \mathscr{F}$, satisfying the conditions:

$$x(t) = 0 \text{ on } F_1, \ x(t) = 1 \text{ on } F_2.$$

A Banach algebra R with identity is said to be *regular* if the family of all functions $x(M)$, corresponding to all possible elements x of the algebra R, is regular, and *normal* if this family of functions is normal.

Clearly, every normal family is regular and therefore every normal algebra is regular. We shall see later (see subsection 4) that the converse proposition is also true.

EXAMPLES. 1. The algebra $C(T)$ of all continuous functions on the compact Hausdorff space T is normal in virtue of the Urysohn lemma (see II, subsection 8, § 2) and consequently it is also regular.

2. The algebra $D_n(a, b)$ (Example 3, subsection 2, § 11) is also regular.

3. The algebra A (Example b), subsection 3, § 11) is not regular. In fact, every function $x(\zeta)$, regular in the open disc $|\zeta| < 1$ and equal to zero on some open set of this disc, vanishes identically.

4. The algebra W (see Example 3, subsection 1, § 9) is regular since it contains all 2π-periodic continuously differentiable functions. For the same reason the algebra W_m of all functions

$$x(t_1, \cdots, t_m) = \sum_{k_1=-\infty}^{\infty} \cdots \sum_{k_m=-\infty}^{\infty} c_{k_1 \ldots k_m} \exp i(k_1 t_1 + \cdots + k_m t_m),$$

for which

$$\sum_{k_1=-\infty}^{\infty} \cdots \sum_{k_m=-\infty}^{\infty} |c_{k_1 \ldots k_m}| < \infty$$

with the usual definition of the operations and with the norm

$$|x| = \sum_{k_1=-\infty}^{\infty} \cdots \sum_{k_m=-\infty}^{\infty} |c_{k_1 \ldots k_m}|$$

(m-dimensional analogue of the algebra W) is regular.

5. McKissick [1] constructed an example of a normal uniformly closed subalgebra of the algebra $C(T)$, where T is compact, which does not coincide with $C(T)$.

2. Normal algebras of functions.

I. *Suppose $\mathscr{F}$ is a normal family of functions defined on the normal topological space T, forming an algebra with identity with respect to the usual operations, and suppose $\{U_1, U_2, \cdots, U_n\}$ is a finite covering of the space T with open sets $U_1, \cdots, U_n$. Then there exist functions $x_1, \cdots, x_n$ in $\mathscr{F}$ possessing the following properties*:

1) *$x_k(t) = 0$ in the exterior of U_k, $k = 1, 2, \cdots, n$;*

2) *$x_1(t)+x_2(t)+ \cdots +x_n(t) = 1$.*

Proof. We shall prove the proposition by induction on n. Suppose first of all that $n = 2$. Then the closed sets $F_1 = T-U_1$, $F_2 = T-U_2$ are disjoint; consequently, in the normal family $\mathscr{F}$ there exists a function x_1, which is equal to zero on F_1 and to unity on F_2. Setting $x_2 = 1-x_1$, we obtain the functions x_1, x_2, which satisfy conditions 1)–2) for $n = 2$.

We now assume that the proposition has already been proved for coverings by $n-1$ open sets. We consider a covering with n open sets $U_1, U_2, \cdots, U_{n-1}$, U_n. We set $F = T-U_n$; then F is closed and $F \subset U_1 \cup \cdots \cup U_{n-1}$. Hence (since T is normal), there exists an open set $U \subset T$ such that $F \subset U \subset \overline{U} \subset U_1 \cup \cdots \cup U_{n-1}$. We further set $V_k = \overline{U} \cap U_k$; then the V_k are open in $\overline{U}$ and $\{V_1, \cdots, V_{n-1}\}$ forms a covering of F. The functions of the family $\mathscr{F}$, considered only on $\overline{U}$, also form a normal algebra; consequently, in virtue of the induction assumption, there exist functions $y_1, \cdots, y_{n-1}$ in $\mathscr{F}$ satisfying the conditions:

a) $y_k(t) = 0$ in $\bar{U} - V_k$, $k = 1, 2, \cdots, n-1$;
b) $y_1(t) + \cdots + y_{n-1}(t) = 1$ on $\bar{U}$.

On the other hand, the open sets U, U_n form a covering of T; hence, there exist functions x, x_n in $\mathscr{F}$ such that

a') $x(t) = 0$ on $T - U$, $x_n(t) = 0$ on $T - U_n$;
b') $x(t) + x_n(t) = 1$ on T.

Setting $x_k = y_k x$, $k = 1, \cdots, n-1$, and taking conditions a), b) and a'), b') into consideration, we obtain the functions $x_1, \cdots, x_n \in \mathscr{F}$, satisfying the conditions 1), 2).

The functions $x_1, \cdots, x_n$ of the family $\mathscr{F}$, satisfying conditions 1), 2), will be called a *partition of unity* corresponding to the covering $\{U_1, \cdots, U_n\}$.

In particular, if $\mathscr{F}$ is the algebra $C(T)$ of all continuous functions on the compact Hausdorff space T, then the conditions of Proposition I are satisfied (see Example 1, subsection 1). Hence the following proposition holds.

II. *If $\{U_1, \cdots, U_n\}$ is a finite covering of the compact Hausdorff space T by open sets $U_1, \cdots, U_n$, then there exist continuous functions $x_1, x_2, \cdots, x_n$, satisfying the conditions*:

1) $x_k(t) = 0$ *in the exterior of* U_k, $k = 1, 2, \cdots, n$;
2) $x_1(t) + \cdots + x_n(t) = 1$.

Suppose $\mathscr{F}$ is a family of functions defined on the topological space T. A function y, defined on T, is said to *locally belong to the family $\mathscr{F}$ at the point $\tau \in T$* if there exist a function $x_\tau \in \mathscr{F}$ and a neighborhood $U(\tau)$ of the point τ such that $y(t) = x_\tau(t)$ for all $t \in U(\tau)$. A function y is said to *locally belong to the family $\mathscr{F}$* if it locally belongs to $\mathscr{F}$ at every point $\tau \in T$.

Now let R be a commutative Banach algebra with identity, $\mathfrak{M}$ its space of maximal ideals, and I an ideal of the algebra R. We shall say that the function $f(M)$, defined on $\mathfrak{M}$, belongs (locally belongs) to the algebra R or to its ideal I, if $f(M)$ belongs (locally belongs) to the family of functions $\{x(M): x \in R\}$ ($\{x(M): x \in I\}$).

THEOREM 1. *Suppose $\mathscr{F}$ is a normal family of functions on the compact Hausdorff space T, forming an algebra with identity with respect to the usual operations. If a function y, defined on T, locally belongs to $\mathscr{F}$, then y is contained in $\mathscr{F}$; moreover, y belongs to the ideal generated by the functions x_τ.*

Proof. By assumption, to every point $\tau \in T$ there correspond $x_\tau \in \mathscr{F}$ and $U(\tau)$ such that $y(t) = x_\tau(t)$ in $U(\tau)$. The neighborhoods $U(\tau)$ form a covering of the space T, which contains a finite subcovering $U(\tau_1), \cdots, U(\tau_n)$ in virtue of the fact that T is compact. Suppose $\{y_1, \cdots, y_n\}$ is the partition of unity corresponding to this covering and consisting of the functions $y_1, \cdots, y_n \in \mathscr{F}$. Then

$$y y_k = x_{\tau_k} y_k, \tag{1}$$

because $y(t) = x_{\tau_k}(t)$ in $U(\tau_k)$ and $y_k(t) = 0$ in the exterior of $U(\tau_k)$; therefore, we also have $y = yy_1 + \cdots + yy_n \in \mathscr{F}$. Moreover, in virtue of (1), y belongs to the ideal generated by the functions x_τ.

3. Structure space of an algebera. Suppose Δ is an arbitrary set of primitive (see subsection 7, § 7) closed two-sided ideals of an, in general, non-commutative Banach algebra R with identity. We shall agree to call the intersection of all ideals of the set Δ the *kernel* of the set Δ and to denote it by $k(\Delta)$; clearly, $k(\Delta)$ is a two-sided ideal in R. Further, suppose I is a two-sided ideal in R; we shall agree to call the set of all primitive closed two-sided ideals containing I the *hull* (or *envelope*) of the ideal I and to denote it by $h(I)$. Clearly,

$$\Delta \subset hk(\Delta), I \subset kh(I). \tag{1}$$

We denote the set of all primitive closed two-sided ideals of the algebra R by $\mathfrak{M}$; we define a closure operation in $\mathfrak{M}$ (see beginning of subsection 3, § 2, Chapter I) by setting $\bar{\Delta} = hk(\Delta)$ if $\Delta \neq \phi$ and $\bar{\phi} = \phi$. The closure axioms (see subsection 3, § 2) will then be satisfied. In fact, it is clear from the very definition of the closure operation that $\Delta \subset \bar{\Delta}$, $\bar{\bar{\Delta}} = \bar{\Delta}$, $\bar{\phi} = \phi$; therefore we need only prove the relation $\overline{\Delta_1 \cup \Delta_2} = \bar{\Delta}_1 \cup \bar{\Delta}_2$.

Suppose $I \in \bar{\Delta}_1 \cup \bar{\Delta}_2$, say, $I \in \bar{\Delta}_1$. Then $I \supset k(\Delta_1)$, and consequently *a fortiori* $I \supset k(\Delta_1 \cup \Delta_2)$; but this means that $I \in hk(\Delta_1 \cup \Delta_2) = \overline{\Delta_1 \cup \Delta_2}$. Thus, $\bar{\Delta}_1 \cup \bar{\Delta}_2 \subset \overline{\Delta_1 \cup \Delta_2}$ and we have only to prove the reverse inclusion.

Suppose I is a primitive closed two-sided ideal and that $I \notin \bar{\Delta}_1 \cup \bar{\Delta}_2$; then $I \not\supset k(\Delta_1)$, $I \not\supset k(\Delta_2)$. The quotient algebra R/I is primitive and, consequently, $J_1 J_2 \neq (0)$ for any two two-sided ideals $J_1 \neq (0)$, $J_2 \neq (0)$ in R/I (see III, subsection 7, § 7). This is true, in particular, for the images J_1, J_2 of the ideals $k(\Delta_1)+I$, $k(\Delta_2)+I$ under the natural homomorphism $R \to R/I$, because in virtue of the relations $k(\Delta_1)+I \supsetneq I$, $k(\Delta_2)+I \supsetneq I$, we have $J_1 \neq (0)$, $J_2 \neq (0)$. But the relation $J_1 J_2 \neq (0)$ means that $k(\Delta_1)k(\Delta_2) \not\subset I$, so that *a fortiori* $k(\Delta_1 \cup \Delta_2) \not\subset I$ [because $k(\Delta_1)k(\Delta_2) \subset k(\Delta_1) \cap k(\Delta_2)$]. But this means that $I \notin \overline{\Delta_1 \cup \Delta_2}$.

Thus, the closure operation $\bar{\Delta} = hk(\Delta)$ satisfies the axiom requirements; in the topology determined by it, $\mathfrak{M}$ becomes a topological space (which in general is not Hausdorff); it is called the *structure space* of the algebra R and is denoted by $\mathfrak{M}_s$.

It is easily seen that *in the case of a commutative Banach algebra R with identity a closed ideal in R is primitive if, and only if, it is maximal.* In fact, in this case the maximal left ideal M_l coincides with the two-sided maximal ideal M and the quotient algebra R/M is isomorphic to the field of complex numbers (Theorem 1, subsection 1, § 11). On the other hand, if R is primitive then the mapping $x \to A_x^\wedge$ of the algebra R onto R/M must be an isomorphism for some M (see subsection 7, § 7); consequently, R must be isomorphic to the field of complex numbers. Therefore, if I is a closed primitive ideal in an algebra R which is already non-primitive, then, being primitive, R/I is, by what we have just proved, isomorphic to the field of complex numbers; in virtue of II, subsection 1, § 11, this is possible if, and only if, I is a maximal ideal. The converse assertion is obvious. Thus, in the case of a *commutative* algebra R, the space $\mathfrak{M}_s$ consists of the same elements as the space $\mathfrak{M}$ of maximal ideals; but, in general, the topology in $\mathfrak{M}_s$ does not coincide with the topology in $\mathfrak{M}$.

I. *Let R be a commutative Banach algebra with identity. Then the topology in $\mathfrak{M}_s$ is majorized by the topology in $\mathfrak{M}$. These topologies coincide if, and only if, R is regular.*

Proof. Suppose I is an ideal of the algebra R and that $\mathfrak{M}_x$ is the set of all maximal ideals of the algebra R, which contain a given element x; $\mathfrak{M}_x$, as the set of all points M on which the continuous function $x(M)$ vanishes, is closed in $\mathfrak{M}$. Therefore, $h(I) = \bigcap_{x \in I} \mathfrak{M}_x$ is also closed in $\mathfrak{M}$. In particular, $hk(\Delta)$ is closed in $\mathfrak{M}$, and hence it contains the closure $\bar{\Delta}$ of the set Δ in $\mathfrak{M}$. Consequently, the topology in $\mathfrak{M}_s$ is majorized by the topology in $\mathfrak{M}$. The topologies in $\mathfrak{M}_s$ and $\mathfrak{M}$ coincide if, and only if,

$$\Delta = hk(\Delta) = \bigcap_{x \in k(\Delta)} \mathfrak{M}_x$$

for arbitrary closed set Δ in $\mathfrak{M}$. In virtue of (1) this holds if, and only if, for $M_0 \notin \Delta$ we also have $M_0 \notin \bigcap_{x \in k(\Delta)} \mathfrak{M}_x$, i.e. $M_0 \notin \mathfrak{M}_{x_0}$ for at least one $x_0 \in k(\Delta)$.

But this means that $x_0(M_0) \neq 0$ and $x_0(M) = 0$ on Δ, i.e. that R is regular.

The preceding line of reasoning is also applicable to an arbitrary algebra R (in general, not of all) continuous functions on the topological space T. In this case, T is in general a subset of the set $\mathfrak{M}$ of all maximal ideals of the algebra R. Denoting the set T, considered as a topological subspace in $\mathfrak{M}_s$ by T_s, and repeating the line of reasoning followed in the proof of Proposition I, we arrive at the following result.

II. *Suppose R is the algebra of continuous functions on the topological space T, where R contains an identity. Then the topology in T_s is majorized by the topology in T. These topologies coincide if, and only if, R is a regular family of functions.*

4. Properties of regular algebras.

I. *The algebra boundary of a regular algebra R coincides with the space $\mathfrak{M}$ of all its maximal ideals.*

Proof. Suppose $M_0 \in \mathfrak{M}$ and that $U(M_0)$ is an arbitrary neighborhood of the ideal M_0. By the definition of a regular algebra, there exists a function $x(M)$ in R which vanishes in the exterior of $U(M_0)$ and is equal to unity in M_0 and, consequently, assumes its maximum modulus only in $U(M_0)$. According to Theorem 2, subsection 1, § 13, it follows from this that M_0 belongs to the algebra boundary. Consequently, the algebra boundary coincides with $\mathfrak{M}$.

II. *Suppose R is a regular algebra and that F is a closed set in the space $\mathfrak{M}$ of its* *maximal ideals. Then F is the space of maximal ideals of the quotient algebra* $R_F = R/k(F)$.

Proof. Suppose $x^\wedge \in R_F$ and that x is a representative of the class $x^\wedge$; we set $x^\wedge(M) = x(M)$ for $M \in F$. This definition does not depend on the choice of the representative $x \in x^\wedge$. In fact, if also $x' \in x^\wedge$, then $x' - x \in k(F)$, and hence $x'(M) - x(M) = 0$ for $M \in F$. Thus, for $M \in F$, the correspondence $x^\wedge \to x^\wedge(M)$ defines a homomorphism of the algebra R_F onto the field of complex numbers, and consequently, it defines some maximal ideal $M^\wedge$ of the algebra R_F. This maximal ideal $M^\wedge$ will be set into correspondence with the ideal M. Evidently, we then have

$$x^\wedge(M^\wedge) = x^\wedge(M) = x(M).$$

The correspondence $M \to M^\wedge$, established in this way, is one-to-one. In fact, if $M_1 \neq M_2$, $M_1, M_2 \in F$, then there exists a function $x(M)$ in R such that $x(M_1) \neq x(M_2)$; if $x^\wedge$ is the residue class modulo $k(F)$ which contains x, then $x^\wedge(M_1^\wedge) = x(M_1)$, $x^\wedge(M_2^\wedge) = x(M_2)$, and therefore also $x^\wedge(M_1^\wedge) \neq x^\wedge(M_2^\wedge)$; consequently, $M_1^\wedge \neq M_2^\wedge$.

We shall now prove that the correspondence $M \to M^\wedge$ maps F onto the space $\mathfrak{M}^\wedge$ of all maximal ideals of the algebra $R/k(F)$. Every ideal $M_0^\wedge$ defines a homomorphism of the algebra $R/k(F)$, and hence also of the algebra R, onto the field of complex numbers; consequently, $M_0^\wedge$ defines an ideal $M_0 \in \mathfrak{M}$ such that

$$x^\wedge(M_0^\wedge) = x(M_0) \text{ for } x \in x^\wedge.$$

Moreover, $M_0 \in F$; in fact, otherwise there would exist a function $x(M)$ in the algebra R (in virtue of its regularity) satisfying the conditions

$$x(M_0) = 1, x(M) = 0 \text{ on } F;$$

but then we would have $x \in k(F)$ and $x^\wedge = 0$, and this would contradict the equality

$$x^\wedge(M_0^\wedge) = x(M_0) = 1.$$

Thus the correspondence $M \to M^\wedge$ is a one-to-one mapping to F onto $\mathfrak{M}^\wedge$ which maps the continuous functions $x(M)$, $M \in F$, into the continuous functions $x^\wedge(M^\wedge)$, $x^\wedge \in R/k(F)$. Since F is compact, the correspondence $M \to M^\wedge$ is, on the basis of Theorem 3, subsection 3, § 11, a homeomorphism and hence we may identify $\mathfrak{M}^\wedge$ with F.

THEOREM 2. *Suppose I is an ideal of the regular algebra R. Then for arbitrary closed set F in* $\mathfrak{M}$ *which does not intersect* $h(I)$, *there exists a function* $x(M)$ *belonging to the ideal I which is equal to unity on F.*

Proof. We consider the quotient algebra $R/k(F)$. In virtue of II, F is the space of maximal ideals of the algebra $R/k(F)$. Suppose I' is the image of the ideal I under the natural homomorphism $R \to R/k(F)$. We shall prove that $I' = R/k(F)$.

In fact, in the contrary case, I' is an ideal in $R/k(F)$ and hence it is contained in some maximal ideal $M^\wedge$ of the algebra $R/k(F)$. But then I is contained in some ideal $M \in F$, which is impossible because by assumption $h(I)$ and F are disjoint. Consequently, $I' = R/k(F)$; in particular, I' contains the identity $e^\wedge$ of the algebra $R/k(F)$. This means that the ideal I contains an element x_0, which maps into $e^\wedge$ under the homomorphism $R \to R/k(F)$. But e also maps into $e^\wedge$ under the homomorphism $R \to R/k(F)$. Therefore, $x_0 - e \in k(F)$. This means that $x_0(M) - 1 = 0$ on F and, consequently, $x_0(M) = 1$ on F.

COROLLARY. *Every regular algebra is normal.*

Proof. Suppose F_1, F_2 are disjoint closed sets in $\mathfrak{M}$. Applying the preceding theorem to the ideal $I = k(F_1)$ and to the set $F = F_2$, we obtain the function $x_0(M)$, which vanishes on F_1 and equals unity on F_2.

Now combining this corollary with Theorem 1, subsection 2, we arrive at the following result.

THEOREM 3. *Suppose R is a regular algebra and that $y(M)$ is a function defined on the space $\mathfrak{M}$ of maximal ideals of the algebra R. Then: if $y(M)$ locally belongs to the algebra of all functions $x(M)$, $x \in R$, then there exists an element $x \in R$ such that $x(M) = y(M)$ for all $M \in \mathfrak{M}$. Moreover, $y(M)$ belongs to the ideal generated by the functions $x_\tau(M)$, $\tau \in \mathfrak{M}$, with which $y(M)$ coincides locally.*

At this date it is not known whether this theorem is true without the condition that the algebra R is regular.

The following result due to WIENER [1] follows from Theorem 3.

Let x be a 2π-periodic function. If in some neighborhood of each point t_0, $-\infty < t_0 < \infty$, the function x coincides with the function y_{t_0}, which can be expanded as an absolutely convergent Fourier series, then x also can be expanded as an absolutely convergent Fourier series.

In fact, the condition of the theorem means that x locally belongs to the regular algebra W at every point of its space of maximal ideals; therefore, $x \in W$ on the basis of Theorem 3.

Another consequence of Theorem 3 is the following theorem.

THEOREM 3'. *On the space $\mathfrak{M}$ of maximal ideals of the regular algebra R let there be defined a function $f(M)$ which in some neighborhood U of each point $M_0 \in \mathfrak{M}$ coincides with an analytic function of an arbitrary element $x_0 \in R$ (which, in general, depends on M_0 and U). Then there exists an element $x \in R$ such thot $x(M) = f(M)$ for all $M \in \mathfrak{M}$.*

Proof. Let V be a neighborhood of the point M_0 such that $\overline{V} \subset U$ and let $I = k(\overline{V})$. Then $\overline{V}$ is the space of maximal ideals of the algebra R/I and by Theorem 7, subsection 6, § 11, the restriction of the function $f(M)$ to $\overline{V}$ is an element of the algebra R/I. Inasmuch as this is true for some neighborhood of an arbitrary point $M_0 \in \mathfrak{M}$, $f(M)$ locally belongs to the algebra R and the assertion of the theorem follows from Theorem 3.

Clearly, the assertion of Theorem 3 remains valid if we replace the given regular algebra by some ideal in it; consequently, the following theorem holds.

THEOREM 3''. *Suppose R is a regular algebra, I is an ideal in R, and $y(M)$ is a function defined on the space $\mathfrak{M}$ of maximal ideals of R. Then if $y(M)$ locally belongs* *to the ideal I, there exists an element $x \in I$ such that $x(M) = y(M)$ for all $M \in \mathfrak{M}$.*

The condition of Theorem 3″ can be weakened if we make use of the following proposition:

III. *If I is an ideal in the regular algebra R and y is an element in R, then the function $y(M)$ locally belongs to I at each point of $\mathfrak{M}-h(I)$ and at each interior point of the set $h(y)$.*

Proof. If $M_0 \notin h(I)$, then there exists a neighborhood $U(M_0)$ whose closure does not intersect $h(I)$; consequently, by Theorem 2, there exists a function $x_0(M)$ belonging to the ideal I which equals unity in a neighborhood $U(M_0)$ of the point M_0. Then $y(M)x_0(M) = y(M)$ in $U(M_0)$ and $yx_0 \in I$; consequently, setting $x_{M_0}(M) = y(M)x_0(M)$, we see that $y(M)$ locally belongs to I at the point M_0. But if $U(M_0) \subset h(y)$, then $y(M) = 0$ in $U(M_0)$; setting $x_{M_0}(M) \equiv 0$, we again see that $y(M)$ locally belongs to I at the point M_0.

Thus, *the assertion of Theorem 3″ remains valid if $y(M)$ is an element of the algebra R, which locally belongs to the ideal I at every point of the set $h(I)$ which is not an interior point of the set $h(y)$.*

Suppose Δ is any set of ideals of the algebra R. An ideal $I \in \Delta$ is said to be *minimal in Δ* if there is no ideal $\neq I$ in Δ which is contained in I.

THEOREM 4. *Suppose F is a closed set in the space $\mathfrak{M}$ of maximal ideals of the semisimple regular algebra R and that $I_0(F)$ is the set of all $x \in R$ for which the function $x(M)$ equals zero on an open set containing F (each on its own open set). Then $I_0(F)$ is the minimal ideal I of the algebra R satisfying the condition $h(I) = F$.*

Proof. Clearly, $hI_0(F) \supset F$. On the other hand, if $M_0 \notin F$, then there exist an open set $U \supset F$, not containing M_0, and a function $x_0(M)$, $x_0 \in R$, equal to unity in M_0 and to zero on U; consequently, we also have $M_0 \notin hI_0(F)$. Therefore,

$$hI_0(F) = F. \tag{1}$$

Now suppose I is an arbitrary ideal satisfying the condition $h(I) = F$; we shall prove that $I \supset I_0(F)$. Suppose $y \in I_0(F)$ and that $y(M) = 0$ on an open set $U \supset F$. The closed sets F and $\mathfrak{M}-U$ do not intersect, and hence (by Theorem 2) there exists a function $z(M)$, $z \in R$, belonging to I, which is equal to unity on $\mathfrak{M}-U$. Then $y(M)-y(M)z(M) \equiv 0$ and since R is a semisimple algebra, $y = yz \in I$. Consequently, $I \supset I_0(F)$ and the theorem is proved.

IV. *The ideal $\overline{I_0(F)}$ is the minimal closed ideal I of the semisimple regular algebra R, satisfying the condition $h(I) = F$. It consists of all $x \in R$, for which there exists a sequence $x_n \in R$ satisfying the conditions*:

1) $\lim_{n\to\infty} |x_n| = 0$; 2) $x_n(M) = x(M)$ *on an open set* $U_n \supset F$.

Proof. In virtue of (1), $hI_0(F) = F$. On the other hand, suppose I is a closed ideal, satisfying the condition $h(I) = F$. Then $I_0(F) \subset I$, and therefore $\overline{I_0(F)} \subset I$. Consequently, $\overline{I_0(F)}$ is the minimal closed ideal I for which $h(I) = F$.

Now suppose that $x \in \overline{I_0(F)}$, i.e. $x = \lim_{n\to\infty} y_n$, where $y_n \in I_0(F)$. Then $y_n(M) = 0$ on some open set $U_n \supset F$; consequently, the function $x_n(M) = x(M) - y_n(M)$ coincides with $x(M)$ in U_n and $\lim_{n\to\infty} |x_n| = 0$, so that conditions 1), 2) are satisfied. Conversely, if these conditions are satisfied for some sequence $x_n \in R$, then $y_n(M) = x(M) - x_n(M) = 0$ on U_n, and therefore $y_n \in I_0(F)$. But then $x = \lim_{n\to\infty} y_n \in \overline{I_0(F)}$, and Proposition IV is proved.

Let us recall (see subsection 10, § 11) that an ideal I in the Banach algebra R with identity is said to be *primary* if it is contained in only one maximal ideal of this algebra. Setting $F = \{M_0\}$ in Proposition IV, we obtain the minimal closed ideal $\overline{I_0(M_0)}$, satisfying the condition $h\overline{I_0(M_0)} = M_0$. In other words:

V. *For every maximal ideal of a semisimple regular algebra there exists a minimal closed primary ideal contained in it.*

We shall say that the commutative Banach algebra R with identity *satisfies condition* (D) if for every $M_0 \subset \mathfrak{M}$ and every $x \in \mathfrak{M}_0$ there exists a sequence $x_n \in R$ such that:

1) $x_n(M) = 0$ in some neighborhood $U_n(M_0)$;

2) $xx_n \to x$ as $n \to \infty$.

THEOREM 5 (SHILOV [5]). *Suppose R is a semisimple regular algebra satisfying condition (D) and that I is a closed ideal of the algebra R. Then I contains every element x in $kh(I)$ for which the intersection of the boundary of $h(x)$ with $h(I)$ does not contain a nonvoid perfect set.* (This theorem is a generalization of a lemma due to DITKIN [1].)

Proof. We denote the set of all maximal ideals M in which $x(M)$ locally does not belong to I by Δ. We shall prove that Δ is a perfect set. Evidently, Δ is closed and in virtue of III, $\Delta \subset h(I)$. We shall assume that it has an isolated point M_0; suppose U is a neighborhood of the point M_0 at every point $M \neq M_0$ of whose closure $\overline{U}$ $x(M)$ locally belongs to I. Since $M_0 \in \Delta \subset h(I)$ and $x \in kh(I)$, $x \in M_0$. But then according to condition (D), there exists a sequence $y_n \in R$ such that $y_n x \to x$ as $n \to \infty$ and $y_n(M) = 0$ in some neighborhood $U_n(M_0)$. Suppose z is an element of the algebra R such that $z(M) = 0$ in $\mathfrak{M} - U$ and $z(M) = 1$ in some neighborhood $V(M_0) \subset U$. Then $y_n xz$ locally belongs to the ideal I at every point $M \in \mathfrak{M}$, and hence $y_n xz \in I$ (see Theorem 3″). Since I is closed and $y_n x \to x$, we also have $xz \in I$ and, consequently, $x(M)$ locally belongs to I at the point M_0. The contradiction thus obtained shows that Δ is perfect. In virtue of III, Δ is contained in the intersection of $h(I)$ with the boundary of the set $h(x)$; in virtue of the condition of the theorem we conclude from this that Δ is the void set, i.e. x locally belongs to the ideal I, and hence $x \in I$.

5. The case of an algebra without identity. The preceding results are easily carried over to an algebra without identity. Then the role of $\mathfrak{M}$ is played by the locally compact space of maximal regular ideals of the algebra R with the infinitely distant point M_∞ adjoined to R, and all functions $x(M)$, $x \in R$ vanish at the point M_∞. Therefore, in all the preceding definitions and theorems, the point M_∞ must be excluded every time we are dealing with values of $x(M) \neq 0$, in particular, the value $x(M) = 1$. We leave to the reader the detailed formulations and deductions (in this connection, see LOOMIS [2]). We shall note only the following corollary to Theorem 4, subsection 4:

COROLLARY. *Let R be a regular semisimple algebra such that the set R' of all elements x for which $x(M)$ vanishes in the exterior of some compact set in $\mathfrak{M} - M_\infty$ (in general depending on $x(M)$) is dense in R. Then every closed ideal I in R is contained in some maximal regular ideal.*

Proof. Let us assume the contrary: suppose some closed ideal I is not contained in any maximal regular ideal. Then $h(I)$ is the void set. But, besides this, $h(R')$ is also a void set, because R' is dense in R; in virtue of Theorem 4, subsection 4, R' is the minimal ideal I' for which $h(I')$ is void. Consequently, $I \supset R'$; since R' is dense in R and I is closed, it follows that $I = R$ contrary to the assumption that I is an ideal.

6. **Sufficient condition that an algebra be regular.** We note yet the following condition for regularity (proof given in SHILOV [5]): *if R is a Banach algebra with identity and with real generators and if the condition*

$$\int_{-\infty}^{\infty} \ln |e^{itz}| \frac{dt}{1+t^2} < \infty$$

is satisfied for each of these generators z, then R is regular. (An element x of the algebra R is said to be *real* if the corresponding function $x(M)$ is real-valued for all $M \in \mathfrak{M}$.)

In this subsection, we did not touch upon the applications of the theory of regular algebras discussed above. It will be convenient for us to consider these applications further on, in subsection 8, § 31.

7. **Primary ideals.** Recall (see subsection 10, § 11) that an ideal I of a Banach algebra R with identity is called *primary* if it is contained in only one maximal ideal of R. The algebra R itself is called primary if it possesses only one maximal ideal. In the algebra of all functions $\{x(M): x \in R\}$ the primary ideals I are characterized by the situation that the set $\{M: x(M) = 0 \text{ for all } x \in I\}$ consists of one point. Therefore, *if I is a closed primary ideal, then R/I is primary.*

Now let R be regular. According to IV, subsection 4, for every maximal ideal M_0 (in the given case $F = \{M_0\}$) there exists a smallest closed primary ideal $\overline{I(M_0)}$. Moreover, $x \in \overline{I(M_0)}$ if, and only if, there exists a sequence $x_n \in R$ such that $x_n \to 0$ and $x(M) = x_n(M)$ in some neighborhood of the point M_0.

More general conditions for the existence of minimal primary ideals in a not necessarily regular algebra were obtained by GELFAND [5] (see also KORENBLYUM [3]).

Example. We shall find the minimal closed primary ideals of the algebra $D_m(a, b)$ (see Example 3, subsection 2, § 11). Let J be the set of all $x \in D_m(a, b)$ which equal zero for $t = t_0$, $t_0 \in [a, b]$, together with all their derivatives to order m inclusive. Clearly, J is a closed primary ideal in $D_m(a, b)$. We shall prove that J is minimal. Let $x \in J$ and suppose the function $h \in D_m(a, b)$

equals unity in some neighborhood of the point t_0 and equals zero exterior to some other neighborhood $|t-t_0| \leqq c$ of the point t_0. Set

$$h_\nu(t) = \begin{cases} h(\nu(t-t_0)) & \text{for } t \in [a, b] \text{ and } \nu(t-t_0) \in [a, b], \\ 0 & \text{for } t \in [a, b] \text{ and } \nu(t-t_0) \notin [a, b] \end{cases}$$

and

$$A = \sup_{\substack{0 \leqq t \leqq 1 \\ 0 \leqq k \leqq m}} |h^{(k)}(t)|.$$

Then $|h_\nu^{(k)}(t)| \leqq \nu^k A$; moreover, by successive integration, we find that on the segment $|t-t_0| \leqq \rho$, $|x^{(r)}(t)| = o(\rho^{m-r})$. From this and Leibniz's formula it follows that $|(h_\nu x)^{(q)}| = o(1/\nu^{m-q})$ and hence $\lim_{\nu\to\infty} ||h_\nu x|| = 0$. [Since $h_\nu(t) = 0$ for $|t-t_0| \geqq c$, we can set $\rho = c/\nu$ in the estimates for $x^{(r)}(t)$.] By IV, subsection 4, x belongs to the minimal closed primary ideal corresponding to the point t_0; consequently, I coincides with this minimal primary ideal. It is also easy to show that D_m/I is finite-dimensional and isomorphic to the algebra of polynomials of degree $\leqq m$ (discarding terms of degree $> m$ if they are obtained in multiplication). An analogous result holds for the algebras $D_m(G)$ of functions $x(t_1, \cdots, t_m)$, $(t_1, \cdots, t_m) \in G$ having partial derivatives to the m-th order inclusive in the closed region $G \subset R^m$.

§ 16. Completely regular commutative algebras

1. Definition and simplest properties of a completely regular algebra. Suppose R is a symmetric algebra (which is not necessarily commutative). The norm $|x|$ in the algebra R is said to be *completely regular* if

$$|x^*x| = |x|^2 \text{ for all } x \in R. \tag{1}$$

If $|x|$ is a completely regular norm, then we conclude from the relations

$$|x|^2 = |x^*x| \leqq |x|\,|x^*|$$

that $|x| \leqq |x^*|$, and replacing x by x^* that also $|x^*| \leqq |x|$. Therefore, $|x^*| = |x|$, i.e. *every algebra with a completely regular norm is a normed symmetric algebra* (see subsection 3, § 10). A normed symmetric algebra is said to be *completely regular* if its norm is completely regular. Evidently, every symmetric subalgebra of a completely regular algebra is also completely regular.

I. *The completion of a completely regular algebra is completely regular.*

A complete completely regular algebra will be called a *C*-algebra.*

Proof. Suppose $\tilde{R}$ is the completion of the completely regular algebra R. If $x \in \tilde{R}$, then there exists a sequence $x_n \in R$ such that $|x-x_n| \to 0$ and, consequently, $|x^*-x_n^*| \to 0$ as $n \to \infty$. Hence, passing to the limit in the equality $|x_n^*x_n| = |x_n|^2$ we obtain $|x^*x| = |x|^2$.

II. *If R is a completely regular algebra, then*

$$|x| = \sup_{y \neq 0} \frac{|xy|}{|y|}.$$

In fact, it follows from the inequality $|xy| \leqq |x|\,|y|$ that

$$\frac{|xy|}{|y|} \leqq |x|,$$

where, in virtue of (1), equality is attained for $y = x^*$.

III. *Suppose R is a completely regular algebra without identity (which is not necessarily commutative), and that R_1 is the algebra obtained from R by the adjunction of the identity. Then a completely regular norm in R can be extended to a completely regular norm in R_1 in such a way that if R is a complete algebra then R_1 will also be a complete algebra.*

Proof. We shall define the norm in R_1 by the formula

$$|\lambda e + x| = \sup_{y \neq 0} \frac{|\lambda y + xy|}{|y|}. \tag{2}$$

Then $|\lambda e + x|$ will be the norm of an operator in R, namely, the norm of the operator of left multiplication by $\lambda e + x$. Hence $|\lambda e + x|$ satisfies all the norm axioms in the algebra.

We shall prove that this norm is completely regular.

Suppose c is an arbitrary positive number, less than 1. In virtue of (2), there exists an element $y \in R$ such that

$$|y| = 1 \text{ and } c|\lambda e + x| < |\lambda y + xy|.$$

Then

$$\begin{aligned} c^2|\lambda e + x|^2 < |\lambda y + xy|^2 &= |(\lambda y + xy)^*(\lambda y + xy)| \\ &= |y^*(\lambda e + x)^*(\lambda e + x)y| \leqq |y^*||(\lambda e + x)^*(\lambda e + x)y| \\ &= |((\lambda e + x)^*(\lambda e + x))y| \leqq |(\lambda e + x)^*(\lambda e + x)|, \end{aligned}$$

so that

$$c^2|\lambda e + x|^2 \leqq |(\lambda e + x)^*(\lambda e + x)|.$$

In view of the fact that the number c, $0 < c < 1$, is arbitrary, it follows that

$$|\lambda e + x|^2 \leqq |(\lambda e + x)^*(\lambda e + x)|. \tag{3}$$

On the other hand, according to a property of the algebra norm, we have

$$|(\lambda e + x)^*(\lambda e + x)| \leq |(\lambda e + x)^*||\lambda e + x|.$$

Consequently, $|\lambda e + x| \leqq |(\lambda e + x)^*|$. If we replace $\lambda e + x$ by $(\lambda e + x)^*$ in this inequality, we obtain that $|(\lambda e + x)^*| = |\lambda e + x|$. Consequently,

$$|(\lambda e + x)^*(\lambda e + x)| \leqq |(\lambda e + x)^*||\lambda e + x| = |\lambda e + x|^2$$

and comparison with (3) yields

$$|\lambda e + x|^2 = |(\lambda e + x)^*(\lambda e + x)|.$$

Let us now assume that R is complete. We shall prove that R_1 is also complete in the norm introduced above. Suppose $\langle\lambda_n e+x_n\rangle$ is a fundamental sequence in R_1. Then the sequence $\langle\lambda_n\rangle$ is bounded. In fact, if this were not the case, there would exist a subsequence $\lambda_{k_n}\to\infty$ and hence $e+(1/\lambda_{k_n})x_{k_n} = (1/\lambda_{k_n})(\lambda_{k_n}e+x_{k_n}) \to 0$ as $n\to\infty$.

But then e is the limit of the sequence $\langle -(1/\lambda_{k_n})x_{k_n}\rangle$ of elements of the complete space R and therefore, contrary to assumption, it belongs to R. Thus, the sequence $\langle\lambda_n\rangle$ is bounded, and hence there exists a subsequence $\langle\lambda_{k_n}\rangle$ which has a finite limit λ. But then $\langle\lambda_{k_n}e\rangle$ is a fundamental sequence, and therefore

$$\langle x_{k_n}\rangle = \langle(\lambda_{k_n}e+x_{k_n})-\lambda_{k_n}e\rangle$$

is also a fundamental sequence. In view of the completeness of the space R, the sequence $\langle x_{k_n}\rangle$ has some limit, say x, in R. It follows that $\lambda_{k_n}e+x_{k_n}\to\lambda e+x\in R_1$; since $\langle\lambda_n e+x_n\rangle$ is a fundamental sequence, we also have that $\lambda_n e+x_n\to\lambda e+x$, and the completeness of R_1 is proved.

EXAMPLES. 1. The algebra $C(T)$ of all bounded continuous functions on the topological space T (see Example 1, subsection 1, § 10) is completely regular. In fact,

$$|x^*x| = \sup_{t\in T}|\overline{x(t)}x(t)| = \sup_{t\in T}|x(t)|^2 = (\sup_{t\in T}|x(t)|)^2 = |x|^2.$$

2. The algebra $\mathfrak{B}(\mathfrak{H})$ of all bounded operators in the Hilbert space $\mathfrak{H}$ is completely regular. In fact, for an operator A in Hilbert space, we have

$$|A^*A| = |A|^2$$

(see (2), subsection 10, § 5).

2. Realization of completely regular commutative algebras.

THEOREM 1 (GELFAND and NAIMARK [1]). *Suppose R is a commutative Banach algebra with identity and that an involution is defined in R satisfying the usual algebraic conditions*

$$(\lambda x+\mu y)^* = \bar{\lambda}x^*+\bar{\mu}y^*,$$
$$x^{**} = x,$$
$$(xy)^* = y^*x^*$$

and, furthermore, the condition

$$|x^*x| = |x^*||x|. \tag{1}$$

(We note it is not assumed that the condition $|x^*| = |x|$ is satisfied. It is a consequence of the remaining conditions of the theorem.) *Then the algebra R is completely isomorphic to the algebra $C(\mathfrak{M})$ of all continuous functions $x(M)$ on the space $\mathfrak{M}$ of maximal ideals of the algebra R.* [We recall that in the algebra $C(\mathfrak{M})$, the norm and involution are defined by means of the formulas $|x| = \sup_{M\in\mathfrak{M}}|x(M)|$, $x^*(M) = \overline{x(M)}$, as was the situation in the more general case of the algebra $C(T)$.]

Proof. We shall first prove that

$$|x^2| = |x|^2$$

in the algebra R. To this end, we note that in virtue of (1),

$$|x^{*2}||x^2| = |x^{*2}x^2| = |(x^*x)^*(x^*x)| = |(x^*x)^*||x^*x| = |x^*x||x^*x| = |x^*x|^2 = |x^*|^2|x|^2; \qquad (2)$$

on the other hand,

$$|x^{*2}| \leqq |x^*|^2, \; |x^2| \leqq |x|^2.$$

Consequently, (2) holds if, and only if,

$$|x^{*2}| = |x^*|^2, \; |x^2| = |x|^2.$$

It follows that

$$\sup_{M\in\mathfrak{M}} |x(M)| = \lim_{n\to\infty} \sqrt[2^n]{|x^{2^n}|} = \lim_{n\to\infty} \sqrt[2^n]{|x|^{2^n}} = |x| \qquad (3)$$

(see IV, subsection 2, § 11). Therefore [see (2), subsection 1, § 24]

$$|x^*| = \sup_{M\in\mathfrak{M}} |x^*(M)| = \sup_{M^*\in\mathfrak{M}} |x(M^*)| = \sup_{M\in\mathfrak{M}} |x(M)| = |x|.$$

We shall prove that $x^*(M) = \overline{x(M)}$, i.e. that R is completely symmetric. From this and (3) the assertion of the theorem will follow (see Corollary 1 to Theorem 2, subsection 3, § 14).

It suffices to show that if x is Hermitian then $x(M)$ is real on all $M \in \mathfrak{M}$.

[The proof of the relation $x^*(M) = \overline{x(M)}$ discussed here is due to FUKAMIYA [2]. The original proof of GELFAND and NAIMARK [1] (given in the first edition of this book) is based on the use of the concept of algebra boundary. Another proof, not using the concept of algebra boundary, is given by ARENS [1]; it is discussed in the book of GELFAND, RAIKOV and SHILOV [1].]

In fact, then for an arbitrary $x = x_1 + ix_2 \in R$, where x_1, x_2 are Hermitian,

$$x^*(M) = (x_1 - ix_2)(M) = x_1(M) - ix_2(M) = \overline{x_1(M)+ix_2(M)} = \overline{x(M)}.$$

Thus, let x be Hermitian. Set

$$u = \exp(ix) = e + \frac{1}{1!}(ix) + \frac{1}{2!}(ix)^2 + \cdots;$$

in virtue of the relation $|x^*| = |x|$, the mapping $x \to x^*$ is continuous and hence

$$u^* = \left(e + \frac{1}{1!}(ix) + \frac{1}{2!}(ix)^2 + \cdots\right)^* = e + \frac{1}{1!}(-ix) + \frac{1}{2!}(-ix)^2 + \cdots = \exp(-ix) = u^{-1}.$$

Consequently, $u^*u = uu^* = e$. From this, in virtue of (1), $1 = |e| = |u^*u| = |u^*||u| = |u|^2$ and hence $|u| = 1, |u^{-1}| = |u^*| = |u| = 1$. But then $\sup_{M\in\mathfrak{M}} |u(M)|$

$= |u| = 1$, $|u(M)| \leqq 1$, and analogously $|u^{-1}(M)| = |u(M)|^{-1} \leqq 1$; consequently,

$$|u(M)| = 1 \text{ for all } M \in \mathfrak{M}. \tag{4}$$

On the other hand, $u(M) = \exp\,(ix(M))$(see subsection 6, § 11); therefore (4) is possible only when $x(M)$ is real, and the theorem is proved.

THEOREM 2. *Every commutative C^*-algebra with identity is completely isomorphic to the algebra $C(\mathfrak{M})$ of all continuous functions on the space of all maximal ideals of this algebra.* (The difference between this theorem and Theorem 1 consists in this that in Theorem 1, instead of the condition $|x^*x| = |x|^2$, we put on the weaker condition $|x^*x| = |x^*||x|$.) In fact, a completely regular algebra satisfies the condition $|x^*x| = |x^*||x|$, and therefore it only remains to apply Theorem 1.

COROLLARY 1. *Suppose R is a commutative Banach algebra without identity and that an involution is defined in R satisfying the usual algebraic conditions and the condition*

$$|x^*x| = |x^*||x|. \tag{5}$$

Then R is completely isomorphic to the algebra $C_0(T)$ of all continuous functions $x(t)$ on some locally compact space T, which satisfy the condition

$$\lim_{t\to\infty} x(t) = 0.$$

Proof. Suppose R_1 is the symmetric algebra obtained from R by adjunction of the identity. We define the norm in R_1 by means of formula (2) of subsection 1. It is clear from the reasoning pursued in the proof of Proposition III, subsection 1, that this norm also satisfies condition (5) and that the algebra R_1 is complete in this norm. Therefore R_1 is completely isomorphic to the algebra $C(\mathfrak{M})$, where $\mathfrak{M}$ is the space of maximal ideals of the algebra R_1. One such maximal ideal is the algebra R. We set $M_0 = R$ and denote by T the space obtained from $\mathfrak{M}$ by removing the point M_0. Then T is locally compact. The function $x(M)$ belongs to $R = M_0$ if and only if $x(M_0) = 0$, i.e. if, and only if, $\lim_{M\to M_0} x(M) = 0$. Since M_0 is the infinitely distant point in T, this proves our corollary.

COROLLARY 2. *Every commutative C^*-algebra without identity is completely isomorphic to the algebra $C_0(T)$ of all continuous functions on some locally compact space T, which satisfy the condition $\lim_{t\to\infty} x(t) = 0$.*

The assertion follows directly from Corollary 1 because a completely regular algebra satisfies the condition $|x^*x| = |x^*||x|$.

COROLLARY 3. *Every completely regular commutative algebra is completely isomorphic to some symmetric subalgebra of the algebra $C(T)$, where T is a compact space.*

Proof. Adjoining the identity to R if necessary and completing the algebra thus obtained, we arrive at the algebra R_1 which is also completely regular and contains R as one of its symmetric subalgebras. Suppose T is the space of maximal ideals of the algebra R_1; then R_1 is completely isomorphic to the algebra $C(T)$ and therefore R is completely isomorphic to some symmetric subalgebra of the algebra $C(T)$.

COROLLARY 4. *Every algebra R satisfying the conditions of Theorem* 1 (*and consequently every commutative C^*-algebra with identity*) *is completely symmetric.*

In fact, every such algebra R is completely isomorphic to the algebra $C(\mathfrak{M})$, which is completely symmetric.

COROLLARY 5. *Every C^*-algebra with identity is regular.*

In fact, such an algebra is completely isomorphic to the algebra $C(\mathfrak{M})$, which is regular.

We note that the converse assertion is not true; thus, the algebra $D_n(a, b)$ (see Example 3, subsection 2, § 11) is regular, but it is not completely regular.

Everywhere in the following corollaries, R denotes a commutative C^*-algebra.

COROLLARY 6. *If x is an element in the algebra R with identity and $f(\lambda)$ is a function which is continuous on the spectrum of the element x, then there exists a unique element y in R such that $y(M) = f(x(M))$ on all the maximal ideals M of the algebra R. The spectrum of the element y consists of all the numbers $f(\lambda)$, where λ runs through the spectrum S_x of the element x, and $|y| = \sup_{\lambda \in S_x} |f(\lambda)|$.*

Proof. The function $f(x(M))$ is continuous and hence belongs to the algebra $C(\mathfrak{M})$ which is completely isomorphic to the algebra R. Consequently, there exists a unique element y in R for which $y(M) = f(x(M))$, $M \in \mathfrak{M}$. The spectrum of the element y is the set of all values $y(M) = f(\lambda)$, where $\lambda = x(M)$; therefore

$$|y| = \sup_{M \in \mathfrak{M}} |y(M)| = \sup_{M \in \mathfrak{M}} |f(x(M))| = \sup_{\lambda \in S_x} |f(\lambda)|. \tag{6}$$

The element y is denoted by $f(x)$ and is called a *continuous function of the element* x. In virtue of (6),

$$|f(x)| = \sup_{\lambda \in S_x} |f(\lambda)|.$$

Corollary 6 is easily generalized to continuous functions of several variables. We recall (see subsection 7, § 11) that the set of all systems $\{\lambda_1, \cdots, \lambda_n\}$, $\lambda_k = x_k(M)$ is called the *joint spectrum of the elements* $x_1, \cdots, x_n$ of the algebra R. Repeating the preceding arguments, we arrive at the following result, which is a generalization of Corollary 6.

COROLLARY 7. *If $x_1, \cdots, x_n$ are elements of the algebra R with identity, then for an arbitrary function $f(\lambda_1, \cdots, \lambda_n)$, continuous on the joint spectrum of the elements $x_1, \cdots, x_n$, there exists a unique element $y = f(x_1, \cdots, x_n)$ in the algebra R*

such that $y(M) = f(x_1(M), \cdots, x_n(M))$ *on all maximal ideals* M *of the algebra* R. *The spectrum of the element* $f(x_1, \cdots, x_n)$ *consists of all the numbers* $f(\lambda_1, \cdots, \lambda_n)$, *where* $\{\lambda_1, \cdots, \lambda_n\}$ *runs through the joint spectrum of the elements* $x_1, \cdots, x_n$; *furthermore,*

$$|f(x_1, \cdots, x_n)| = \sup |f(\lambda_1, \cdots, \lambda_n)|, \tag{7}$$

where the least upper bound is taken over the joint spectrum of the elements $x_1, \cdots, x_n$.

The assertions of Corollaries 6 and 7 carry over to C^*-algebras R without identity. In this case, $y = f(x_1, \cdots, x_n)$ exists for an arbitrary function $f(\lambda_1, \cdots, \lambda_n)$, continuous on the joint spectrum of the elements $x_1, \cdots, x_n$ and equal to zero for $\lambda_1 = 0, \cdots, \lambda_n = 0$. In order to verify this, it suffices to adjoin the identity to R; the element $y = f(x_1, \cdots, x_n)$ exists in the algebra R' obtained after this adjunction and it belongs to the algebra R, because $x_1(R) = \cdots = x_n(R) = 0$ on the maximal ideal R of the algebra R' and therefore $y(R) = f(x_1(R), \cdots, x_n(R)) = f(0, \cdots, 0) = 0$.

COROLLARY 8. *Every Hermitian element* $x \in R$ *can be represented in the form* $x = u - v$, *where* u, v *are Hermitian elements in* R *with nonnegative spectrum and* $uv = 0$.

Proof. Since the spectrum of a Hermitian element is real, it suffices to set $u = f_1(x)$, $v = f_2(x)$, where $f_1(\lambda) = \max\{\lambda, 0\}$, $f_2(\lambda) = \max\{-\lambda, 0\}$. Since $\lambda = f_1(\lambda) - f_2(\lambda)$ and $f_1(\lambda)f_2(\lambda) = 0$, we have $x = u - v$ and $uv = 0$.

COROLLARY 9. *The element* $y = f(x)$ *is contained in every closed symmetric subalgebra* R_1 *with identity of the algebra* R, *where* R_1 *contains* x, *and if* $f(0) = 0$, *then* y *is also contained in every closed symmetric subalgebra without identity which contains* x *of the algebra* R. *In particular, for* $f(0) = 0$, *the element* $f(x)$ *is contained in every closed symmetric ideal, which contains* x, *of the algebra* R.

Proof. We set $x = x_1 + ix_2$, $\lambda = \lambda_1 + i\lambda_2$, $f(\lambda) = f_1(\lambda_1, \lambda_2) + if_2(\lambda_1, \lambda_2)$, where x_1, x_2 are Hermitian elements, and $\lambda_1, \lambda_2, f_1, f_2$ are real. Then, as is easily seen, $f(x) = f_1(x_1, x_2) + if_2(x_1, x_2)$. Suppose S is the joint spectrum of the elements x_1, x_2. According to the Weierstrass theorem (see subsection 10, § 2) there exists a sequence of polynomials $p_n(\lambda_1, \lambda_2)$, which converges uniformly on S to $f_1(\lambda_1, \lambda_2)$. In virtue of (7),

$$|f_1(x_1, x_2) - p_n(x_1, x_2)| = \sup_S |f_1(\lambda_1, \lambda_2) - p_n(\lambda_1, \lambda_2)| \to 0 \text{ as } n \to \infty.$$

But if R_1 is a closed symmetric subalgebra with identity of the algebra R, containing x, then $p_n(x_1, x_2) \in R_1$, and therefore also $f_1(x_1, x_2) \in R_1$. Analogously, $f_2(x_1, x_2) \in R_1$, and hence $f(x) = f_1(x_1, x_2) + if_2(x_1, x_2) \in R_1$. But if R_1 is an algebra without identity and $f(0) = 0$, then $p_n(0, 0) \to 0$, and therefore, setting $\tilde{p}_n(\lambda_1, \lambda_2) = p_n(\lambda_1, \lambda_2) - p_n(0, 0)$, we obtain that $\sup_S |f_1(\lambda_1, \lambda_2) - \tilde{p}_n(\lambda_1, \lambda_2)| \to 0$

and $\tilde{p}_n(x_1, x_2) \in R_1$. From this, we conclude, the same as above, that $f(x) \in R_1$.

The assertion with respect to an ideal is a particular case of the proposition already proved, because an ideal may be considered as a subalgebra without identity.

We note the requirement that the ideal $I \subset R$ be symmetric is in reality superfluous in virtue of the following theorem. (For algebras of real-valued functions, this theorem was first proved by STONE [3], and in the complex case by Gelfand and Shilov (see GELFAND and SHILOV [1] and SHILOV [6])).

THEOREM 3. *Every closed ideal I in the algebra $C(\mathfrak{M})$ is the set of all functions $x(M)$ of the algebra $C(\mathfrak{M})$ which vanish on the closed set $F = h(I)$, i.e.*

$$I = kh(I).$$

Proof. Consider the closed ideals I' of the algebra $C(\mathfrak{M})$ for which $h(I') = F$. $C(\mathfrak{M})$ is regular and therefore $\overline{I_0(F)}$ is minimal among such ideals I' (IV, subsection 4, § 15). Consequently, it will suffice to show that $\overline{I_0(F)}$ coincides with $k(F) = kh(I)$ which is maximal among the ideals I'. Let $x(M) \in k(F)$. For an arbitrary natural number n the closed sets

$$F_{1n} = \left\{M \in \mathfrak{M}; |x(M)| \leqq \frac{1}{n}\right\}, \quad F_{2n} = \left\{M \in \mathfrak{M}; |x(M)| \geqq \frac{2}{n}\right\}$$

are disjoint; consequently, by Urysohn's lemma, there exists a function $y_n(M) \in C(\mathfrak{M})$ which satisfies the conditions $0 \leqq y_n(M) \leqq 1, y_n(M) = 1$ on $F_{1n}, y_n(M) = 0$ on F_{2n}. Then $(xy_n)(M) = x(M)$ on Int $F_{1n} \supset F$, $|xy_n| = \sup_{M \in \mathfrak{M}} |(xy_n)(M)| \leqq 2/n$ and $x \in \overline{I_0(F)}$ in virtue of IV, subsection 4, § 15.

Theorem 3 can also be formulated in the following way.

Every closed ideal in a commutative C^*-algebra with identity is the intersection of all maximal ideals which contain it and hence it is symmetric.

There are still several types of algebras other than the completely regular ones for which all closed ideals are known. Thus, in the algebra $D_n(a, b)$, every closed ideal is the intersection of primary ideals (SHILOV [1]); this same assertion is also valid for the multi-dimensional analogue of the algebra $D_n(a, b)$ (WHITNEY [1]).

However, as yet we do not have a description of all closed ideals, for example, in the algebra W_m, in particular in W.

MALLIAVIN [2] showed that in each W_m there are closed ideals which are not the intersection of maximal ideals, despite the fact that there are no primary ideals different from maximal ideals in W_m; for the case of the algebra W_3 the simplest example of such an ideal was pointed out earlier by L. SCHWARTZ [1]. (This example is given in the book by GELFAND, RAIKOV and SHILOV [1], pp. 244–247.) The question in what other commutative algebras, besides the completely regular algebras, is every closed ideal the intersection of maximal ideals is still open (concerning this, see SHILOV [6]).

We note still another corollary to Theorem 3.

COROLLARY 10. *If I is a closed ideal in the commutative C^*-algebra R with identity, then the quotient algebra R/I is a C^*-algebra with identity which is completely isomorphic to the algebra $C(F)$, where $F = h(I)$.*

Proof. We assign to every function $x = x(M)$ of the algebra $R = C(\mathfrak{M})$ the function $x\hat{}(M) = x(M)$, but considered only on $F = h(I)$. The correspondence $x(M) \rightarrow x\hat{}(M)$ thus obtained is a symmetric homomorphism of the algebra $C(\mathfrak{M})$ onto the algebra $C(F)$; the kernel of this homomorphism is $kh(I) = I$, and therefore $C(F)$ is symmetrically isomorphic to the algebra R/I. We shall prove that this correspondence is also isometric.

Suppose $\xi \in R/I$; then ξ consists of functions $x(M) = x\hat{}(M)$ on F. Moreover,

$$|x| = \sup_{M \in \mathfrak{M}} |x(M)| \geqq \sup_{M \in F} |x(M)| = |x\hat{}|,$$

and therefore also

$$|\xi| = \inf_{x \in \xi} |x| \geqq |x\hat{}|.$$

We shall now prove the reverse inequality. Suppose $x(M) = x\hat{}(M)$ on F and that $\varepsilon > 0$; we denote by U the set of all points $M \in \mathfrak{M}$ which are such that $|x(M)-x(M')| < \varepsilon$ for some $M' \in F$. Then U is an open set which contains F and on this set $|x(M)| < |x\hat{}|+\varepsilon$. But, on the basis of the Urysohn lemma, there exists a continuous function $y(M)$, $0 \leqq y(M) \leqq 1$, equal to unity on F and to zero in the exterior of U. Since $x(M)y(M) = x(M) = x\hat{}(M)$ on F, then also $xy \in \xi$, and

$$|\xi| \leqq |xy| = \sup_{M \in \mathfrak{M}} |x(M)y(M)| \leqq \sup_{M \in U} |x(M)| \leqq |x\hat{}|+\varepsilon.$$

Since the number $\varepsilon > 0$ is arbitrary, we conclude from this that $|\xi| \leqq |x\hat{}|$, which also completes the proof.

REMARK. Corollary 10 can be generalized to algebras without identity, namely:

If I is a closed ideal in the C^*-algebra R then the quotient algebra R/I is also a C^*-algebra and, consequently, it is completely isomorphic to some algebra $C_0(T)$. [We recall that $C_0(T)$ denotes the set of all continuous functions on the locally compact space T, which vanish at infinity (see Corollary 1 to Theorem 2, above).]

In fact, if R is an algebra without identity, then the algebra R' obtained from R by adjoining the identity, is completely regular, and hence R'/I is completely isomorphic to some $C(T)$; consequently, R/I is completely isomorphic to some $C_0(T)$.

3. **Generalization to multinormed algebras.** A seminorm p in the algebra R is called a *multiplicative seminorm* if $p(x) \not\equiv 0$ in R and

$$p(xy) \leqq p(x)p(y) \text{ for all } x, y \in R.$$

A topological algebra R is said to be *multinormed* if there exists a sufficient set P of multiplicative seminorms which define the topology in R.

An example of such an algebra is the algebra $C^{\wedge}(T)$ of all continuous complex-valued functions on a locally compact topological space T with the usual definition of the operations of addition, multiplication by a number and multiplication.

We obtain a sufficient set of seminorms $\{p_K\}$ which are multiplicative in $C^{\wedge}(T)$ by setting

$$p_K(x) = \max_{t \in K} |x(t)|$$

for arbitrary compact set $K \subset T$. The algebra $C^{\wedge}(T)$ can be considered as a symmetric algebra if we set

$$x^*(t) = \overline{x(t)}.$$

The question arises, is it possible to characterize all commutative multinormed algebras which are topologically and symmetrically isomorphic to this algebra $C^{\wedge}(T)$? ARENS [5] gave the answer to this question under the following additional restrictions.

We shall agree to call every function x_p, $p \in P$, with values in R, satisfying the following conditions *a locally finite partition of unity*:

1) for arbitrary $p \in P$, the function $p'(x_p)$ is different from zero only for a finite set S_p of functionals $p' \in P$;

2) $p'(x_p) \leqq 1$ for all $p' \in S_p$;

3) for arbitrary $p \in P$, the function $p(x_{p'})$ can be different from zero only for $p' \in S_p$;

4) for arbitrary $p' \in P$ and $y \in R$,

$$p'(y - \sum_{p \in S_{p'}} y x_p) = 0.$$

Thus, in the algebra $C^{\wedge}(T)$ considered above there exists a locally finite partition of unity if the space T is such that for each of its open covers $\{G\}$ there exists an open cover $\{G'\}$ possessing the following properties:

a) for every G' there exists a $G \supset G'$;

b) for every point $t \in T$ there exists a neighborhood $U(t)$ which intersects only a finite number of sets G'.

Such a space T is called *paracompact* (DIEUDONNÉ [1], BOURBAKI [3]).

The fundamental result of Arens consists in the following. *Every commutative complete multinormed symmetric algebra R having a locally finite partition of unity and satisfying the condition*

$$p(xx^*) \geqq C_p p(x) p(x^*)$$

for all multiplicative seminorms p in R, is topologically and symmetrically isomorphic to some algebra $C^{\wedge}(T)$, *where T is a locally compact paracompact topological space.*

A more general result was later obtained by TAO-HSING HSIA [1, 2]. He proved that *if a complete multinormed symmetric algebra R with identity satisfies the condition* $\sup_p p(x^*x) = \sup_p p(x) \sup_p p(x^*)$ $(\leqq \infty)$, *then* $p(x^*x) = p(x)p(x^*)$ *holds for all p and R is topologically and symmetrically isomorphic to some algebra* $C^{\wedge}(T)$.

4. **Symmetric subalgebras of the algebra $C(T)$ and compact extensions of the space T.** A compact Hausdorff space Q containing a subset which is homeomorphic to a Hausdorff topological space T as an everywhere dense subset is called a *compact extension* of T. Since Q is normal, then for every set $A \subset Q$ and point $t_0 \notin \bar{A}$ there must exist a continuous function on Q which equals zero on A and is different from zero at the point t_0 (see I and II, subsection 8, § 2). A Hausdorff topological space possessing this property is called *completely regular* so that Q, and hence also T, are completely regular.

Let $C(T)$ be the algebra of all bounded continuous functions on a completely regular space with norm $|x| = \sup_T |x(t)|$ and let $\mathfrak{M}$ be the space of maximal ideals of the algebra $C(T)$. Every point $t_0 \in T$ generates a maximal ideal in $C(T)$ consisting of all functions $x(t) \in C(T)$, vanishing at the point t_0, and to distinct points t_0 there correspond distinct maximal ideals.

Therefore we can assume that $T \subset \mathfrak{M}$. It turns out that *the initial topology of the space T coincides with the topology induced in T from* $\mathfrak{M}$ *and* $\mathfrak{M}$ *is a compact extension of the space T which is maximal in the sense that every compact extension of the space T is obtained from* $\mathfrak{M}$ *by a continuous mapping which leaves all points of T fixed.* [The existence of a compact extension of a completely regular space was proved by TIKHONOV [1], and of a maximal compact extension by ČECH [1].]

Now let Q be an arbitrary compact extension of the space T. Every function in $C(Q)$ is continuous and bounded on T, and hence its restriction to T is a function in $C(T)$; consequently, $C(Q)$ can be considered as a closed subalgebra of the algebra $C(T)$. *A closed subalgebra R with identity of the algebra C(T) coincides with some C(Q), where Q is a compact extension of the space T, if, and only if,* 1) $x(t) \in R$ *implies* $\overline{x(t)} \in R$, 2) *for every set* $A \subset T$ *and point* $t_0 \notin \bar{A}$ *there exists a function* $x \in R$ *which equals zero on A and is different from zero at the point* t_0.

For the proof of these statements, see GELFAND, RAIKOV and SHILOV [1], § 43.

5. **Antisymmetric subalgebras of the algebra $C(T)$.** A closed subalgebra R with identity of the algebra $C(T)$ is called *antisymmetric* if $x(t) \in R$, $\overline{x(t)} \in R$ imply that $x(t)$ is a constant. An example of such an algebra is the algebra A of functions which are analytic in the interior of the unit disc and continuous in the closed unit disc [see Example b), subsection 3, § 11]. Clearly, the algebra $A(G)$ of all functions which are analytic in the interior of a plane region G and continuous in a closed region $\bar{G}$ of the Riemann sphere, and also every closed subalgebra with identity of the algebra $A(G)$ with the same space of maximal ideals, are also antisymmetric algebras. HOFFMAN and SINGER [1] pointed out the following method of constructing antisymmetric algebras with spaces of maximal ideals of arbitrary dimension $\geqq 2$. Let G be a plane region, Q a compact space and R the algebra of all functions $f(z, q)$ which are continuous on $\bar{G} \times Q$, analytic in G with respect to z for every fixed $q \in Q$ and constant with respect to q for some fixed $z = z_0 \in G$. It is not difficult to verify that R is antisymmetric. This example shows that there exist antisymmetric algebras of functions which are not analytic with respect to some variable. The problem of the structure of arbitrary antisymmetric algebras or even of antisymmetric algebras with a finite number of generators is still open. On the other hand, the solution of this problem is of interest in connection with the following theorem of BISHOP [12]:

Let T be a compact Hausdorff space and R a closed subalgebra with identity of the algebra C(T), which separates points of T. Then there exists a decomposition $P = \{F\}$ *of the space T into pairwise disjoint closed sets F such that:* 1) *the restriction* R_F *of the algebra R to each F is antisymmetric*; 2) *if* $f \in C(T)$ *and its restriction to each F belongs to* R_F, *then* $f \in R$; 3) *for every* $F \in P$, *every closed set* $A \subset T - F$ *and every* $\varepsilon > 0$ *there exists a function* $f \in R$, *satisfying the following conditions*:

$$\sup_{t \in T} |f(t)| \leqq 1, \ |f(t)-1| < \varepsilon \text{ for } t \in F \text{ and } |f(t)| < \varepsilon \quad \text{for } t \in A.$$

This theorem is an answer to a question posed by SHILOV [20].

The sets F in Bishop's theorem are maximal sets of antisymmetry for the algebra R. This means that every F is a maximal set (with respect to inclusion) on which R_F is antisymmetric.

Recently, ARENSON [1] showed that a result close to Bishop's theorem holds if maximal sets of antisymmetry of the algebra are replaced by some finer sets of weak analyticity. A closed set F is called *a set of weak analyticity* for the algebra R if for every function f on F, which is the uniform limit on F of the restrictions to F of functions in R, the condition $|f(x)| \leqq 1$ for all $x \in F$ implies that $f^{-1}(1)$ either coincides with F or it is nowhere dense in F.

A positive solution of the corona problem is due to CARLESON [4]: the open unit disc forms an everywhere dense set in the space of maximal ideals of the algebra of bounded analytic functions in this disc with the natural norm.

6. **Subalgebras of the algebra $C(T)$ and certain problems in approximation theory.** Let T be an arbitrary compact set in the complex space C^n of the complex variables $z_1, \cdots, z_n$ (see Example 2, subsection 2, § 2). We denote by $P(T)$ the algebra of all functions on T which are the uniform limits on T of sequences of polynomials in $z_1, \cdots, z_n$, and by $R(T)$ the algebra of all functions on T which are the uniform limits of rational functions with singularities outside of T. Clearly, $P(T) \subset R(T) \subset C(T)$. One of the basic problems in approximation theory consists in the following:

1) Under what conditions is $P(T) = C(T)$ or $R(T) = C(T)$?

2) If $P(T) \neq C(T)$ or $R(T) \neq C(T)$, then give a description of all functions occurring in $P(T)$ or in $R(T)$, respectively.

For $n = 1$ and the algebra $P(T)$, the first result in this direction was obtained by LAVRENTEV [1]. He showed that for $n = 1$, $P(T) = C(T)$ if, and only if,

a) T does not have interior points;

b) the complement of T is connected.

Further, KELDYSH [3] proved that if a closed bounded region T in C^1 has a connected complement, then $P(T)$ coincides with the algebra $A(T)$ of functions which are continuous on T and analytic at all interior points of T. [It is not difficult to see that this condition is also necessary for the equality $P(T) = A(T)$.] A complete solution of 2) for the algebra $P(T)$ over an arbitrary compact $T \subset C^1$ was given by MERGELYAN [1]. He showed that in this case $P(T) = A(T)$ if, and only if, the complement of T is connected.

Recently, Bishop, Wermer, Rudin and others developed a new approach to these problems, based on the theory of normed algebras. In particular, BISHOP [8] (see also WERMER [13]) obtained a new proof of the Mergelyan theorem stated above. In the work of BISHOP [8], there is given a necessary and sufficient condition for the equality $R(T) = C(T)$ in terms of the *minimal boundary*. Let T be a compact space and B a closed subalgebra in $C(T)$, which separates points of T and contains an identity. For functions $x \in C(T)$, we set

$$S(x) = \{t : t \in T, |x(t)| = \sup_T |x(t)|\}.$$

If among the sets $N \subset T$ satisfying the condition $N \cap S(x) \neq \emptyset$ for all $x \in B$ there exists a smallest one N_0, then it is called the *minimal boundary* of the algebra B. Clearly, in this case the closure of N_0 coincides with the algebra boundary Γ of the algebra B (see § 13). Generally speaking, N_0 does not always exist, but *if T is metrizable, then N_0 exists and consists of all points $t \in T$ at which $S(x) = \{t\}$ for some function $x \in B$.* [A point t in which $S(x) = \{t\}$ is called a *peak* point of the function $x(t)$.] *If, in particular, T is a compact subset of C^1, not having interior points, then $R(T) = C(T)$ if, and only if, the minimal boundary N_0 of the algebra $R(T)$ coincides with T. Moreover, if $T - N_0$ has zero measure, then $N_0 = T$.*

Necessary and sufficient conditions for the equality $R(T) = C(T)$ with $T \subset C^1$ are given in other terms by MERGELYAN [3] and VITUSHKIN [1].

In the most general situation, the condition $N_0 = \mathfrak{M}_B = T$, where $\mathfrak{M}_B$ is the space of maximal ideals of the algebra B, is necessary for the equality $B = C(T)$ (WERMER [13]); the question when is it sufficient remains open (with the exception of the case $T \subset C^1$ considered above). Also open is the problem of describing the functions in $P(T)$ with $T \subset C^n$, $n > 1$ and $P(T) \neq C(T)$. For the case of a polynomially convex T, SHILOV [18] proved that *if $x(t)$ is an analytic function in some neighborhood of the set T, then the restriction of $x(t)$ to T belongs to $P(T)$.* [T is called *polynomially* *convex* if for every point $t_0 \notin T$ there exists a polynomial $P(t)$ such that $|P(t_0)| > \max_{t \in T} |P(t)|$.]

The problem of describing the maximal subalgebras in $C(T)$ is closely related to the problems discussed above. Let B be a closed subalgebra of the algebra $C(T)$ which does not coincide with $C(T)$ and contains an identity; B is called *maximal* if the relation $B \subset B_1 \subset C(T)$, where B_1 is also a closed subalgebra of the algebra $C(T)$, implies $B_1 = B$ or $B_1 = C(T)$. WERMER [13] proved that if T is the unit disc in C^1, then the algebra A (see subsection 5) is maximal in $C(T)$; an analogous result no longer holds in C^n with $n > 1$, but if the maximal algebra B in $C(T)$ (where T is an arbitrary compact space) is antisymmetric, then the functions $x(t) \in B$ possess many of the properties of analytic functions.

WERMER [13] pointed out a general method for constructing Dirichlet algebras certain of which are, and others are not, maximal subalgebras in $C(T)$. [A subalgebra B of the algebra $C(T)$ is called a *Dirichlet algebra* if the real parts of functions in B form a dense set in $C^r(T)$. For more details on Dirichlet algebras and related problems, see the book by HOFFMAN [4].] [Concerning maximal subalgebras, see the article by HOFFMAN and WERMER [1].] On the other hand, GORIN [1] showed that if T is the n-dimensional torus $|z_j| = 1$ $(j = 1, \cdots, n)$ in C^n and A is the algebra of all continuous functions on T which admit a holomorphic continuation in the interior of a polycylinder $|z_j| < 1$ $(j = 1, \cdots, n)$, then A is a maximal subalgebra in $C(T)$ among the closed subalgebras with identity which are invariant relative to some semigroup of automorphisms of the torus T. GORIN [1] obtained a general theorem which contains this result of his andalso various generalizations of Wermer's theorem as special cases (see, e.g., HOFFMAN and SINGER [1]); in another article, GORIN [3] considered maximal subalgebras of a symmetric Banach algebra.

The following problem is related to this same family of problems. Let A be a closed subalgebra of the algebra $C(T)$ which separates points of T and contains all constants. Under what conditions is $A = C(T)$?

We note here only the following result of HOFFMAN and WERMER [1]: $A = C(T)$ if, and only if, the set of all real parts of the functions in A is closed in the norm of $C^r(T)$. [A simple proof of this result was recently given by BROWDER [2].]

In conclusion, we note that in certain cases (e.g., if T is finite or countable), $C(T)$ does not have proper closed subalgebras B, which separate points on T and contain an identity; but if T is null-dimensional and $\mathfrak{M}_B = T$, then $B = C(T)$.

The reader will find a detailed discussion of the problems considered here and a further literature list in the collection: *Some problems in approximation theory*, Moscow, 1963; see also the survey articles of BISHOP [18], SEMADENI [1], BRELOT [1], HOFFMAN [3], the monographs of WERMER [18] and PHELPS [1], and the journal *Function Algebras*.

The results of § 11, with the exception of those discussed in the text and in subsection 7, are due to GELFAND [1–4]; those of subsection 7, § 11 are due to SONIS [1–2]. For algebras with a finite number of generators, Theorem 8, subsection 8, § 11 was obtained by SHILOV [18], and for an arbitrary commutative Banach algebra with identity by ARENS and CALDERÓN [1]; Theorem 9, subsection 8, § 11 is due to SHILOV [22] (in the original formulation of SHILOV [21] the condition of normal multi-valuedness was absent; however, as a counter-example of KALLIN [1] showed, in this original formulation the assertion of the theorem was false); § 12 is due to Gelfand and Shilov (see GELFAND and SHILOV [1] and SHILOV [5]); § 13, subsection 4, § 14 and § 15 is due to SHILOV [3, 5] (see also GELFAND, RAIKOV and SHILOV [1]); Theorems 3′ and 4, subsection 2, § 13 are due to SONIS [1, 2]; subsections 1, 2, § 14 are basically due to Raikov (see GELFAND, RAIKOV and SHILOV [1]). To Shilov belong also a large number of works devoted to regular algebras and a number of other special problems in the theory of commutative normed algebras (see SHILOV [1–22]).

The concept of an algebra boundary (see § 13) received further generalization in the works of MILMAN [3] and ARENS and SINGER [1].

The concept of a minimal boundary was first introduced in the work of MILMAN [3].

Theorem 1, § 15, in a somewhat different formulation, was first obtained by KREIN [4, 5]. The concept of a structure space is due to JACOBSON [3]; it is a generalization of the results of GELFAND and SHILOV [1].

The results of § 16 are basically due to GELFAND and NAIMARK [1]; Proposition III, subsection 1, § 16 for the case of a commutative algebra and the result of subsection 3, § 16 are due to ARENS [2, 4]; Theorem 3, subsection 2, § 16 is due to GELFAND and SHILOV [1].

A survey of some recent works on Banach algebra theory is given in BONSALL [1].

CHAPTER IV

REPRESENTATIONS OF SYMMETRIC ALGEBRAS

§ 17. Fundamental concepts and propositions in the theory of representations

1. Definition and simplest properties of a representation. A *representation of the algebra R* is any homomorphism of R into the algebra of linear operators on some vector space; this vector space is called a *representation space.*

In the sequel, we shall restrict ourselves to the study of symmetric representations of symmetric algebras. One example of a symmetric algebra is the set $\mathfrak{B}(\mathfrak{H})$ of all bounded linear operators in the given Hilbert space $\mathfrak{H}$ (see Example 2, subsection 1, § 10). The algebra $\mathfrak{B}(\mathfrak{H})$ will play an essential role in the sequel. A *symmetric representation* of a symmetric algebra R is a symmetric homomorphism $x \to A_x$ of R into the algebra $\mathfrak{B}(\mathfrak{H})$. Everywhere in the following discussion, the term "representation" means a symmetric representation. A representation is said to be *continuous* if this homomorphism is continuous. A representation is said to be *cyclic* if there exists a vector ξ_0 in the space $\mathfrak{H}$ such that the set of all vectors $A_x\xi_0$ is dense in $\mathfrak{H}$. The vector ξ_0 itself and also the space $\mathfrak{H}$ are then said to be *cyclic* for the representation $x \to A_x$.

Two representations $x \to A_x$, $x \to B_x$ in the spaces $\mathfrak{H}$ and $\mathfrak{H}'$, respectively, are said to be *equivalent* if there exists an isometric mapping of $\mathfrak{H}$ onto $\mathfrak{H}'$ under which the operator A_x maps into B_x for all $x \in R$. In other words, if U is this isometric mapping and if $\xi' = U\xi$, then $B_x\xi' = UA_x\xi$. Consequently, $B_xU\xi = UA_x\xi$ for all elements ξ in $\mathfrak{H}$, i.e. $B_xU = UA_x$.

Thus, the equivalence of the representations $x \to A_x$ and $x \to B_x$ in the spaces $\mathfrak{H}$, $\mathfrak{H}'$ respectively means that there exists an operator U which maps $\mathfrak{H}$ isometrically onto $\mathfrak{H}'$, and is such that $B_xU = UA_x$.

A subspace $\mathfrak{H}_1 \subset \mathfrak{H}$ is said to be *invariant* if every vector in $\mathfrak{H}_1$ is mapped by any operator A_x again into a vector in $\mathfrak{H}_1$.

If $\mathfrak{H}_1$ is a closed invariant subspace, then all the operators A_x can be considered as operators in $\mathfrak{H}_1$. We then obtain a representation in the space $\mathfrak{H}_1$. This representation is called a *part* of the initial representation.

I. *If $\mathfrak{H}_1 \subset \mathfrak{H}$ is invariant, then its orthogonal complement is also invariant.*

Proof. Suppose ξ is orthogonal to $\mathfrak{H}_1$, i.e. $(\xi, \eta) = 0$ for all $\eta \in \mathfrak{H}_1$. Then $(A_x\xi, \eta) = (\xi, A_x^*\eta) = (\xi, A_{x^*}\eta) = 0$, because $A_{x^*}\eta \in \mathfrak{H}_1$ also. Consequently, the vector $A_x\xi$ is also orthogonal to $\mathfrak{H}_1$.

We denote the projection operator in $\mathfrak{H}$ onto the closed subspace $\mathfrak{H}_1$ by P_1.

II. *$\mathfrak{H}_1$ is a closed invariant subspace if, and only if, all the representation operators commute with the projection operator P_1 onto $\mathfrak{H}_1$.*

If fact, if $\mathfrak{H}_1$ is a closed invariant subspace and $\xi \in \mathfrak{H}_1$, then also $A_x \xi \in \mathfrak{H}_1$. From this it follows that for an arbitrary vector $\xi \in \mathfrak{H}$,

$$A_x P_1 \xi \in \mathfrak{H}_1;$$

consequently,

$$P_1 A_x P_1 \xi = A_x P_1 \xi,$$

i.e.

$$P_1 A_x P_1 = A_x P_1.$$

Applying the involution operation to both members of this equality and then replacing x by x^*, we obtain that also

$$P_1 A_x P_1 = P_1 A_x.$$

Consequently, $P_1 A_x = A_x P_1$; the operators P_1 and A_x commute.

Conversely, if these operators commute, then for $\xi \in \mathfrak{H}_1$ we have

$$P_1 A_x \xi = A_x P_1 \xi = A_x \xi;$$

consequently, also $A_x \xi \in \mathfrak{H}_1$. This means that $\mathfrak{H}_1$ is an invariant subspace.

III. *The closed linear hull $\mathfrak{M}$ of invariant subspaces is also an invariant subspace.*

In fact, every element ξ in $\mathfrak{M}$ is the limit of finite sums of the form $\xi' = \xi_1 + \cdots + \xi_n$, where $\xi_1, \cdots, \xi_n$ are vectors in the initial subspaces. On the other hand, $A_x \xi' = A_x \xi_1 + \cdots + A_x \xi_n$ is a sum of the same form and has $A_x \xi$ as its limit.

2. Direct sum of representations. Suppose $\mathfrak{A}$ is an arbitrary set and let $x \to A_x^{(\alpha)}$, $\alpha \in \mathfrak{A}$ be the set of representations of the symmetric algebra R in the spaces $\mathfrak{H}_\alpha$. Suppose

$$|A_x^{(\alpha)}| \leqq C_x, \tag{1}$$

where C_x does not depend on α. We denote by $\mathfrak{H}$ the direct sum of the spaces $\mathfrak{H}_\alpha$, i.e. the set of all elements $\xi = \langle \xi_\alpha \rangle$ such that

$$\sum_\alpha |\xi_\alpha|^2 < +\infty$$

(see subsection 6, § 5). We set

$$A_x \xi = \langle A_x^{(\alpha)} \xi_\alpha \rangle.$$

By (1), A_x is a bounded operator in $\mathfrak{H}$ and the mapping $x \to A_x$ is a representation in the space $\mathfrak{H}$. This representation will be called the *direct sum* of the initial representations $x \to A_x^{(\alpha)}$. Every element $\xi = \langle \xi_\alpha \rangle$ which is such that $\xi_\alpha = 0$ for $\alpha \neq \alpha_0$ will be identified with ξ_{α_0}. Then every space $\mathfrak{H}_{\alpha_0}$ will be a closed invariant subspace of the space $\mathfrak{H}$, and the part of the representation $x \to A_x$ in this subspace will coincide with the representation $x \to A_x^{(\alpha_0)}$. The space $\mathfrak{H}$ is then the direct sum of its invariant subspaces.

Every representation is the direct sum of cyclic representations.

Proof. Suppose $\xi_0 \neq 0$ is any vector in $\mathfrak{H}$. We consider the set of all vectors $A_x\xi_0$ where x runs through the entire algebra R. The closure of this set will be denoted by $\mathfrak{H}_1$; clearly, $\mathfrak{H}_1$ is a closed invariant subspace in which ξ_0 is a cyclic vector. In other words, $\mathfrak{H}_1$ is a cyclic subspace of the representation $x \to A_x$.

If $\mathfrak{H}_1 = \mathfrak{H}$, then the proposition is proved; otherwise, $\mathfrak{H} \ominus \mathfrak{H}_1$ is an invariant subspace which is distinct from (0). Applying the same method to it, we can select a cyclic subspace $\mathfrak{H}_2$ such that $\mathfrak{H}_2 \perp \mathfrak{H}_1$.

We shall now denote by M the set of all systems $\{\mathfrak{H}_\alpha\}$ consisting of mutually orthogonal cyclic subspaces of the representation $x \to A_x$; one of these systems is the system $\{\mathfrak{H}_1, \mathfrak{H}_2\}$ constructed above. The set M, ordered by the inclusion relation $\subset$, forms a partially ordered set, satisfying the condition of Zorn's lemma; namely, an upper bound of a linearly ordered set of systems $\{\mathfrak{H}_\alpha\} \in M$ will be the union of these systems. Therefore, M contains a maximal system $\{\mathfrak{H}_\alpha\}$. But then $\mathfrak{H} = \sum_\alpha \oplus \mathfrak{H}_\alpha$; for otherwise there would exist a cyclic subspace $\mathfrak{H}_0$ distinct from (0) in the invariant subspace $\mathfrak{H} \ominus \sum_\alpha \oplus \mathfrak{H}_\alpha$ and we would obtain a system $\{\mathfrak{H}_\alpha\} \bigcup \mathfrak{H}_0 \in M$ containing the maximal system $\{\mathfrak{H}_\alpha\}$, which is impossible.

3. Description of representations in terms of positive functionals. Suppose $x \to A_x$ is a representation of a symmetric algebra in the space $\mathfrak{H}$ and that $\xi_0 \neq 0$ is a vector $\mathfrak{H}$. We set

$$f(x) = (A_x\xi_0, \xi_0);$$

then f is a positive functional. In fact,

$$f(x^*x) = (A_{x^*x}\xi_0, \xi_0) = (A_{x^*}A_x\xi_0, \xi_0) = (A_x^*A_x\xi_0, \xi_0) = (A_x\xi_0, A_x\xi_0) \geqq 0.$$

THEOREM 1. *Every symmetric representation of a symmetric Banach algebra R is continuous and* $|A_x| \leqq |x|$.

Proof. Without loss of generality, we may assume that R contains an identity. In fact, in the contrary case, by setting $A_{\lambda e+x} = \lambda 1 + A_x$, we could extend the representation $x \to A_x$ of the algebra R to a representation of the symmetric Banach algebra obtained from R by adjunction of the identity. But if R contains an identity, then, applying inequality (1), subsection 4, § 10, to the positive functional $f(x) = (A_x\xi_0, \xi_0)$, we obtain

$$|(A_x\xi_0, \xi_0)| \leqq |x|(\xi_0, \xi_0).$$

If we replace x by x^*x everywhere in the last inequality, we conclude that

$$(A_{x^*x}\xi_0, \xi_0) \leqq |x^*x|(\xi_0, \xi_0) \leqq |x|^2(\xi_0, \xi_0),$$

i.e.

$$|A_x\xi_0|^2 \leqq |x|^2|\xi_0|^2.$$

Since ξ_0 is an arbitrary nonzero vector in $\mathfrak{H}$, it follows from the last inequality that

$$|A_x| \leqq |x|.$$

Our immediate goal is to give a description of a representation in terms of positive functionals. This is best done for cyclic representations.

Suppose $x \to A_x$ and $x \to B_x$ are two cyclic representations in the spaces $\mathfrak{H}$ and $\mathfrak{H}'$, respectively, and let ξ_0 and ξ'_0 be cyclic vectors of these representations. We set

$$f(x) = (A_x\xi_0, \xi_0), f'(x) = (B_x\xi'_0, \xi'_0).$$

I. *If $f(x) = f'(x)$ for all x in R, then the representations $x \to A_x$ and $x \to B_x$ are equivalent.*

Proof. We assign to each vector $\xi = A_x\xi_0$ of the space $\mathfrak{H}$ the vector $\xi' = B_x\xi'_0$ of the space $\mathfrak{H}'$. This mapping is isometric. In fact, suppose $\xi_1 = A_{x_1}\xi_0, \xi_2 = A_{x_2}\xi_0$; consequently, $\xi'_1 = B_{x_1}\xi'_0$, $\xi'_2 = B_{x_2}\xi'_0$. Then

$$(\xi_1, \xi_2) = (A_{x_1}\xi_0, A_{x_2}\xi_0) = (A_{x_2{}^*x_1}\xi_0, \xi_0) = f(x_2^*x_1),$$

$$(\xi'_1, \xi'_2) = (B_{x_1}\xi'_0, B_{x_2}\xi'_0) = (B_{x_2{}^*x_1}\xi'_0, \xi'_0) = f'(x_2^*x_1).$$

By assumption, $f(x_2^*x_1) = f'(x_2^*x_1)$; consequently, $(\xi_1, \xi_2) = (\xi'_1, \xi'_2)$ and this proves that the mapping is isometric. Since ξ_0, ξ'_0 are cyclic elements, the elements of the form $A_x\xi_0$, $B_x\xi'_0$ constitute sets dense in $\mathfrak{H}$, $\mathfrak{H}'$ respectively. Thus, our isometric mapping $\xi \to \xi'$ maps a set dense in $\mathfrak{H}$ into a set dense in $\mathfrak{H}'$. Consequently, this mapping can be uniquely extended to an isometric mapping of the space $\mathfrak{H}$ onto $\mathfrak{H}'$ (see II, subsection 10, § 5).

If $\xi = A_x\xi_0$, $\xi' = B_x\xi'_0$ are corresponding vectors in $\mathfrak{H}$ and $\mathfrak{H}'$, then $A_{x_0}\xi = A_{x_0x}\xi_0$, $B_{x_0}\xi' = B_{x_0x}\xi'_0$ also correspond. This means that our mapping maps the operator A_{x_0} into the operator B_{x_0}; consequently, the representations are equivalent.

The proposition just proved shows that *a cyclic representation $x \to A_x$ with cyclic vector ξ_0 is uniquely defined to within equivalence by the positive functional $f(x) = (A_x\xi_0, \xi_0)$.*

The question arises whether a cyclic representation $x \to A_x$ exists for every positive functional f, such that $f(x) = (A_x\xi_0, \xi_0)$. The answer to this question is in the affirmative in the case of a symmetric Banach algebra R with identity.

Namely, suppose f is a positive functional in R. We introduce a Hermitian form (see subsection 1, § 5) $(x, y) = f(y^*x)$ in R. It may turn out that $(x, x) = f(x^*x) = 0$ for some elements x. Any element x of this sort will be said to be *equivalent to zero* and we shall write this symbolically as $x \sim 0$.

We denote the set of all elements x which are equivalent to zero by $I_l^{(f)}$.

$I_l^{(f)}$ is a left ideal of the algebra R.

Proof. If $x \sim 0$ and y is an arbitrary element of the algebra R then $f(yx) = 0$. In fact, by the Cauchy-Bunyakovsky inequality,

$$|f(yx)|^2 \leqq f(yy^*)f(x^*x) = 0.$$

Now suppose $x_1, x_2 \in I_l^{(f)}$. Then, by the above remark,

$$f((x_1+x_2)^*(x_1+x_2)) = f(x_1^*x_1)+f(x_1^*x_2)+f(x_2^*x_1)+f(x_2^*x_2) = 0,$$

i.e. $x_1+x_2 \in I_l^{(f)}$. Further, it is clear that $x \in I_l^{(f)}$ implies $\alpha x \in I_l^{(f)}$. Finally, suppose $x \sim 0$ and that y is an arbitrary element in the algebra R; we set $z = x^*y^*y$. Then, by the Cauchy-Bunyakovsky inequality,

$$f((yx)^*yx) = f(x^*y^*yx) = f(zx) = 0,$$

i.e. $yx \in I_l^{(f)}$.

Thus, $I_l^{(f)}$ is a left ideal.

We denote the quotient space $R/I_l^{(f)}$ by $\mathfrak{H}'$ and the elements of this space, i.e. the residue classes with respect to $I_l^{(f)}$, by $\xi, \eta, \cdots$. If ξ, η are two such classes, and x, y are their representatives, then we set

$$(\xi, \eta) = (x, y) = f(y^*x).$$

The expression (ξ, η) does not depend on the choice of the representatives x, y. In fact, if, for example, x', x'' are two elements in the class ξ, then $x'-x'' \sim 0$, and consequently, $(x', y)-(x'', y) = (x'-x'', y) = f(y^*(x'-x'')) = 0$.

It is easily verified that (ξ, η) has all the properties of an inner product, i.e.

1° $(\xi, \eta) = \overline{(\eta, \xi)}$;

2° $(\lambda\xi+\mu\eta, \zeta) = \lambda(\xi, \zeta)+\mu(\eta, \zeta)$;

3° $(\xi, \xi) > 0$ for $\xi \neq 0$.

Namely, 2° is obvious; 1° follows from the fact that f is a real-valued functional. We shall go into more detail concerning property 3°.

Suppose x is an element in the class ξ. Since f is a positive functional, we have

$$(\xi, \xi) = f(x^*x) \geqq 0;$$

moreover, if $(\xi, \xi) = f(x^*x) = 0$, then $x \sim 0$, i.e. $\xi = 0$.

An inner product has thus been constructed in $\mathfrak{H}'$. The completion of $\mathfrak{H}'$ with respect to this inner product will be denoted by $\mathfrak{H}$; $\mathfrak{H}$ is a Hilbert space.

We now construct a representation in the space $\mathfrak{H}$. Suppose $\xi = \{x\}$ is any class, with representative x; we denote the class $\{x_0x\}$ with representative x_0x by η; since $I_l^{(f)}$ is a left ideal, η does not depend on the choice of the representative x of the class ξ. Setting

$$A_{x_0}\xi = \{x_0 x\},$$

we obtain a linear operator in the space $\mathfrak{H}'$. We shall prove that it is bounded. By the definition of the inner product in $\mathfrak{H}'$,

$$|A_{x_0}\xi|^2 = (A_{x_0}\xi, A_{x_0}\xi) = (x_0 x, x_0 x) = f(x^* x_0^* x_0 x). \tag{2}$$

We set

$$f_1(y) = f(x^* y x);$$

$f_1(y)$ is a positive functional. In fact,

$$f_1(y^* y) = f(x^* y^* y x) = f((yx)^* yx) \geqq 0.$$

Applying inequality (1), subsection 4, § 10, to $f_1(y)$, we obtain

$$|f_1(y)| \leqq f_1(e)|y|;$$

in particular, for $y = x_0^* x_0$,

$$|f_1(x_0^* x_0)| \leqq f_1(e)|x_0^* x_0| \leqq f_1(e)|x_0|^2,$$

i.e.

$$|f(x^* x_0^* x_0 x)| \leqq f(x^* x)|x_0|^2 = (\xi, \xi)|x_0|^2.$$

In virtue of (2), this means that

$$|A_{x_0}\xi|^2 \leqq |x_0|^2 |\xi|^2.$$

Thus, A_{x_0} is a bounded operator in $\mathfrak{H}'$ and its norm is less than or equal to $|x_0|$:

$$|A_{x_0}| \leqq |x_0|. \tag{3}$$

Consequently, this operator can be uniquely extended to a bounded operator in $\mathfrak{H}$. This extension does not change the norm $|A_{x_0}|$ and, consequently, inequality (3) also remains valid for the norm of the operator A_{x_0} in the space $\mathfrak{H}$. The mapping $x_0 \to A_{x_0}$ is a representation of the algebra R. In fact,

$$A_{\lambda_1 x_1 + \lambda_2 x_2}\xi = \{(\lambda_1 x_1 + \lambda_2 x_2)x\} = \lambda_1\{x_1 x\} + \lambda_2\{x_2 x\} = \lambda_1 A_{x_1}\xi + \lambda_2 A_{x_2}\xi$$

and

$$A_{x_1 x_2}\xi = \{x_1 x_2 x\} = A_{x_1}\{x_2 x\} = A_{x_1} A_{x_2}\xi.$$

It remains to prove that $(A_{x_0})^* = A_{x_0^*}$, i.e. that $(A_{x_0}\xi, \eta) = (\xi, A_{x_0^*}\eta)$. Suppose x, y are elements in the classes ξ, η respectively. Then

$$(A_{x_0}\xi, \eta) = (x_0 x, y) = f(y^* x_0 x),$$

$$(\xi, A_{x_0^*}\eta) = (x, x_0^* y) = f((x_0^* y)^* x) = f(y^* x_0 x);$$

consequently, $(A_{x_0}\xi, \eta) = (\xi, A_{x_0^*}\eta)$.

We shall show that the representation thus obtained is cyclic. Suppose ξ_0 is the class containing the identity e of the algebra R and that x_0 is an arbitrary element of R. Then $A_{x_0}\xi_0$ is the class containing the element x_0. Consequently, the set of all vectors of the form $A_x\xi_0$, $x \in R$, coincides with the set $\mathfrak{H}'$ of all classes; since $\mathfrak{H}'$ is dense in $\mathfrak{H}$, it follows from this that ξ_0 is a cyclic vector.

Finally, we find $(A_x\xi_0, \xi_0)$. The class ξ_0 contains e and, therefore,

$$(A_x\xi_0, \xi_0) = (xe, e) = f(e^*xe) = f(x).$$

We have thus proved the following theorem.

THEOREM 2. *To every cyclic representation $x \to A_x$ of the symmetric algebra R with cyclic vector ξ_0, there corresponds a positive functional*

$$f(x) = (A_x\xi_0, \xi_0).$$

The representation $x \to A_x$ is defined uniquely to within equivalence by the functio nal $f(x)$. Conversely, to every positive functional in the symmetric Banach algebra R with identity there corresponds a cyclic representation $x \to A_x$ such that

$$f(x) = (A_x\xi_0, \xi_0).$$

4. Representations of completely regular commutative algebras; spectral theorem.

I. *Every cyclic representation of a commutative C^*–algebra R with identity is equivalent to its representation in some space $L^2(f)$, defined by the formula*

$$A_x\xi(M) = x(M)\xi(M),$$

where f is an integral in the algebra $C(\mathfrak{M})$ of all continuous functions on the space $\mathfrak{M}$ of maximal ideals of the algebra R. [This representation in $L^2(f)$ will be called the *spectral* representation.]

Proof. The algebra R is completely isomorphic to the algebra $C(\mathfrak{M})$ (see Theo-rem 2, subsection 2, § 16), and therefore the positive functional $f(x)$, which defines the given cyclic representation, may be considered as a positive functional in the algebra $C(\mathfrak{M})$. Now if $x(M) \in C(\mathfrak{M})$ and $x(M) \geqq 0$ then, setting $y(M) = \sqrt{x(M)}$, we have $y(M) \in C(\mathfrak{M})$ and $f(x) = f(y^2) = f(y^*y) \geqq 0$. Consequently, $f(x)$ is an integral on $C(\mathfrak{M})$ and hence (see X, subsection 10, § 6) it is representable in the form

$$f(x) = \int_{\mathfrak{M}} x(M)d\mu(M),$$

where μ is the measure on $\mathfrak{M}$ defined by the integral $f(x)$. We now assign to each class $\xi \in \mathfrak{H}'$ (see subsection 3) the function $x = x(M) \in \xi$, considered as an element in the space $L^2(f)$. [In this connection, to the cyclic vector ξ_0 there corresponds the class containing the identity of the algebra and hence the function $x_0(M) = 1$.] The correspondence thus obtained will be isometric because for $\xi, \eta \in \mathfrak{H}'$ and $x \in \xi$, $y \in \eta$ we have

$$(\xi, \eta) = f(y^*x) = \int_{\mathfrak{M}} x(M)\overline{y(M)}d\mu(M).$$

Since $\mathfrak{H}'$ is dense in $\mathfrak{H}$ and $C(\mathfrak{M})$ is dense in $L^2(f)$, this correspondence is uniquely extendible to an isometric mapping $\xi \to x$ of the space $\mathfrak{H}$ onto the space $L^2(f)$.

For $\xi = \{x\}, \xi \in \mathfrak{H}'$,

$$A_{x_0}\xi = \{x_0 x\} \to x_0(M)x(M);$$

consequently, under the mapping $\xi \to x$ the operators A_x map into the operators of multiplication by $x(M)$.

A *spectral measure* on the locally compact space T is any operator-valued function $P(\Delta)$ defined on all Borel sets $\Delta \subset T$ and possessing the following properties:

1) $P(\Delta)$ is a projection operator in the fixed Hilbert space $\mathfrak{H}$;

2) for arbitrary $\xi \in \mathfrak{H}$ the function $(P(\Delta)\xi, \xi)$ is the restriction to Borel sets of a measure defined by some integral on $L(T)$.

It follows from property 2) that $P(\Delta_1 \bigcup \Delta_2) = P(\Delta_1)+P(\Delta_2)$, where $\Delta_1 \bigcap \Delta_2 = \phi$. Consequently,

$$P(\Delta_1)P(\Delta_2) = 0 \text{ for } \Delta_1 \bigcap \Delta_2 = \phi. \tag{1}$$

[The *Borel family of sets* in T is the minimal system $\sum$ of sets in T such that 1) $\sum$ contains all open sets $\Delta \subset T$; 2) if $\Delta \in \sum$, then $T-\Delta \in \sum$; 3) the union of a finite or denumerable number of sets in $\sum$ also belongs to $\sum$. Each set $\Delta \in \sum$ is called a *Borel* set. In virtue of I–III, subsection 9, § 6, all Borel sets are measurable with respect to an arbitrary integral on $L(T)$: in virtue of IX, subsection 8, § 6, the measure corresponding to an integral on $L(T)$ is completely determined when it is defined on Borel sets.]

II. *Every representation of a commutative C^*-algebra R with identity is given by the formula*

$$A_x = \int_{\mathfrak{M}} x(M)dP(M), \tag{2}$$

where $\mathfrak{M}$ is the space of maximal ideals of the algebra R, $P(\Delta)$ is a spectral measure on $\mathfrak{M}$ which commutes with all the operators A_x and with all the bounded operators in $\mathfrak{H}$ which commute with all the operators A_x, and the integral converges in the sense of the operator norm.

Equality (2) *defines the spectral measure uniquely.*

Proof. We first consider the case of a cyclic representation $x \to A_x$. In virtue of I, the representation $x \to A_x$ is then equivalent to the spectral representation in some space $L^2(f)$ and we may obviously assume that it simply coincides with this spectral representation. We define the operator $P(\Delta)$ in $\mathfrak{H} = L^2(f)$ by the formula

$$P(\Delta)\xi(M) = P_\Delta(M)\xi(M),$$

where $P_\Delta(M)$ is the characteristic function of the Borel set Δ. Clearly, $P(\Delta)$ is then a projection operator in $L^2(f)$ and

$$(P(\Delta)\xi, \xi) = \int_{\mathfrak{M}} P_{\Delta}(M)|\xi(M)|^2 d\mu(M) = \int_{\Delta} |\xi(M)|^2 d\mu(M) \tag{3}$$

is a measure in $\mathfrak{M}$, majorized by the measure μ.

We now prove formula (2). Decomposing $\mathfrak{M}$ into a finite number of Borel sets $\Delta_1, \cdots, \Delta_n$, on each of which $|x(M')-x(M'')| < \varepsilon$, and choosing $M_k \in \Delta_k$, we have

$$\begin{aligned}|A_x\xi - \sum_{k=1}^{n} x(M_k)P(\Delta_k)\xi|^2 &= \int |x(M)\xi(M) - \sum_{k=1}^{n} x(M_k)P_{\Delta_k}(M)\xi(M)|^2 d\mu(M) \\ &= \int |\sum_{k=1}^{n} [x(M)-x(M_k)]P_{\Delta_k}(M)\xi(M)|^2 d\mu(M) \\ &= \int \sum_{k=1}^{n} |x(M)-x(M_k)|^2 P_{\Delta_k}(M)|\xi(M)|^2 d\mu(M) \\ &< \varepsilon^2 \int \sum_{k=1}^{n} P_{\Delta_k}(M)|\xi(M)|^2 d\mu(M) = \varepsilon^2|\xi|^2.\end{aligned}$$

Therefore

$$|A_x - \sum_{k=1}^{n} x(M_k)P(\Delta_k)| < \varepsilon, \tag{3'}$$

and formula (2) is proved.

Clearly, the operator $P(\Delta)$ commutes with all the operators of multiplication by the functions $x(M)$, i.e. with all the operators A_x.

Now if $x \to A_x$ is an arbitrary representation of the algebra R in the space $\mathfrak{H}$, it can be decomposed into the direct sum of cyclic representations $x \to A_x^{(\alpha)}$ in certain spaces $\mathfrak{H}_\alpha$. Suppose $P^{(\alpha)}(\Delta)$ are the corresponding spectral measures. Setting $P(\Delta)\xi = \{P^{(\alpha)}(\Delta)\xi_\alpha\}$ for $\xi = \langle\xi_\alpha\rangle \in \mathfrak{H}$, we obtain a spectral measure in $\mathfrak{H}$, which commutes with all the operators A_x and satisfies condition (2).

We shall now prove the uniqueness of this spectral measure.

If $P(\Delta)$ and $P'(\Delta)$ are two spectral measures satisfying condition (2), then for arbitrary $\xi, \eta \in \mathfrak{H}$ and any continuous function $x(M)$

$$\int_{\mathfrak{M}} x(M)d(P'(M)\xi, \eta) = \int_{\mathfrak{M}} x(M)d(P(M)\xi, \eta),$$

which is possible if, and only if, $(P'(\Delta)\xi, \eta) = (P(\Delta)\,\xi, \eta)$, and consequently, $P'(\Delta) = P(\Delta)$.

Suppose now that B is a bounded linear operator in $\mathfrak{H}$ which commutes with all the operators A_x; we must prove that B commutes with all the operators $P(\Delta)$. To this end, we note that for arbitrary $\xi, \eta \in \mathfrak{H}$ and arbitrary $x \in R$,

$$(A_x B\xi, \eta) = (BA_x\xi, \eta) = (A_x\xi, B^*\eta),$$

i.e. in virtue of (2),

$$\int_{\mathfrak{M}} x(M)d(P(M)B\xi, \eta) = \int_{\mathfrak{M}} x(M)d(P(M)\xi, B^*\eta)$$

for an arbitrary continuous function $x(M)$. This is possible if, and only if,

$$(P(\Delta)B\xi, \eta) = (P(\Delta)\xi, B^*\eta) = (BP(\Delta)\xi, \eta),$$

and consequently, if, and only if, $P(\Delta)B = BP(\Delta)$.

III. *For every commutative symmetric algebra $R \subset \mathfrak{B}(\mathfrak{H})$ there exists a unique spectral measure $P(\Delta)$, which commutes with all the operators of the algebra R and with all the operators in $\mathfrak{B}(\mathfrak{H})$ which commute with all the operators in R, such that*

$$A = \int_{\mathfrak{M}} A(M)dP(M) \tag{4}$$

for all $A \in R$.

Proof. Clearly, we may assume that R is complete and contains an identity, because it suffices to prove formula (4) for the algebra obtained from R by adjunction of an identity and completion. But a symmetric subalgebra of the algebra $\mathfrak{B}(\mathfrak{H})$ is completely regular (see subsection 1, § 16) and hence it suffices to apply Proposition II to the representation $A \to A$, which assigns to every operator $A \in R$ the operator A itself.

IV (Spectral Theorem). *For every Hermitian operator $H \in \mathfrak{B}(\mathfrak{H})$ there exists a unique operator-valued function $P(\lambda)$, $-\infty < \lambda < \infty$, possessing the following properties*:

1) *$P(\lambda)$ is a projection operator*:
2) *$P(\lambda)P(\mu) = P(\lambda)$ for $\lambda \leqq \mu$*;
3) *$P(\lambda)$ commutes with every operator A in $\mathfrak{B}(\mathfrak{H})$ which commutes with H*;
4) *$P(\lambda)\xi$ is a left-continuous function for arbitrary $\xi \in \mathfrak{H}$*;
5) *if $[a, b]$ is a closed interval containing in its interior the entire spectrum of the operator H and $f(\lambda)$ is an arbitrary continuous function on $[a, b]$, then $P(\lambda) = 0$ for $\lambda < a$, $P(\lambda) = 1$ for $\lambda > b$ and*

$$f(H) = \int_a^b f(\lambda)dP(\lambda); \tag{5}$$

in particular,

$$H = \int_a^b \lambda dP(\lambda). \tag{6}$$

Proof. Suppose R is the minimal closed symmetric commutative subalgebra with identity in $\mathfrak{B}(\mathfrak{H})$, containing H, and that $P(\Delta)$ is the spectral measure corresponding to R in virtue of III. We set $\Delta_\lambda = \{M: H(M) < \lambda\}$; since Δ_λ is an open set, $P(\Delta_\lambda)$ has meaning. We set $P(\lambda) = P(\Delta_\lambda)$. Conditions 1), 2), 4) will evidently be satisfied, where condition 2) is satisfied in virtue of (1). Further, if the operator A in $\mathfrak{B}(\mathfrak{H})$ commutes with H, then it commutes with all the operators in R (since R consists of polynomials in H and their limits with respect to the operator norm), and hence also with $P(\Delta_\lambda) = P(\lambda)$. Furthermore, $\Delta_\lambda = \phi$ for $\lambda < a$ and $\Delta_\lambda = \mathfrak{M}$ for $> b$; from this we conclude that $P(\lambda) = 0$ for $\lambda < a$ and $P(\lambda) = 1$ for $\lambda > b$.

We shall prove formula (5). To this end, we decompose the closed interval $I = [a, b]$ into a finite number of subintervals I_k and set

$$\Delta_k = \{M: H(M) \in I_k\}, \ \lambda_k = H(M_k), \text{ where } M_k \in \Delta_k.$$

Then, denoting the increment of the function $P(\lambda)$ in the interval I_k by P_k, we have $P_k = P(\Delta_k)$. Consequently, applying inequality (3′) to the operator $A_x = f(H)$, we obtain that for a sufficiently fine decomposition of the interval I,

$$|f(H) - \sum_{k=1}^{n} f(\lambda_k)P_k| = |f(H) - \sum_{k=1}^{n} f[H(M_k)]P(\Delta_k)| < \varepsilon$$

and formula (5) is proved. Formula (6) is the particular case when $f(\lambda) = \lambda$.

We now assume that there exist two functions $P(\lambda)$ and $P'(\lambda)$ satisfying conditions 1)–5). Then for an arbitrary continuous function $f(\lambda)$ on $[a, b]$ and any vectors $\xi, \eta \in \mathfrak{H}$ we have

$$\int_a^a f(\lambda)d(P(\lambda)\xi, \eta) = \int_a^b f(\lambda)d(P'(\lambda)\xi, \eta) = (f(H)\xi, \eta);$$

but this is possible if, and only if, $(P'(\lambda)\xi, \eta) = (P(\lambda)\xi, \eta)$ for all $\xi, \eta \in \mathfrak{H}$, and consequently if, and only if, $P'(\lambda) = P(\lambda)$. We note that the uniqueness of the function $P(\lambda)$ also follows easily from the uniqueness of the spectral measure for the algebra R.

A function $P(\lambda)$, satisfying the conditions of Proposition IV, is called the *spectral function of the operator* H, and formula (6) is called the *spectral decomposition of* H.

Remark 1. Condition 4) in IV (and also 5) in VII, below) is inessential; it is a sort of normalization of the spectral function $P(\lambda)$ and it can be replaced, for instance, by right continuity if we set $\Delta_\lambda = \{M: H(M) \leqq \lambda\}$. This second normalization will be used in Chapter VII.

Remark 2. We note that actually formula (5) follows from formula (6) and all the remaining conditions of Proposition IV. In fact, it follows from (6) that for arbitrary natural number m,

$$H^m = \lim \left(\sum_{k=1}^{n} \lambda_k P_k\right)^m = \lim \sum_{k=1}^{n} \lambda_k^m P_k = \int_a^b \lambda^m dP(\lambda)$$

and, consequently, for an arbitrary polynomial $p(\lambda)$,

$$p(H) = \int_a^b p(\lambda)dP(\lambda).$$

Now if $f(\lambda)$ is an arbitrary continuous function on $[a, b]$ and $\langle p_n(\lambda)\rangle$ is a sequence of polynomials which converges uniformly to $f(\lambda)$ on $[a, b]$ then, passing to the limit in the equality

$$(p_n(H)\xi, \eta) = \int_a^b p_n(\lambda)d(P(\lambda)\xi, \eta),$$

we obtain that

$$(f(H)\xi, \eta) = \int_a^b f(\lambda)d(P(\lambda)\xi, \eta).$$

But this implies (5).

V. *Every bounded Hermitian operator H can be represented in the form $H = H_+ - H_-$, where H_+, H_- are bounded positive definite operators.*

In fact, it suffices to set $H_+ = \int_0^b \lambda dP(\lambda)$ and $H_- = -\int_a^0 \lambda dP(\lambda)$; furthermore, one of these operators may be equal to zero (for instance, if $0 \notin [a, b]$).

[Proposition V also follows from Corollary 8, subsection 2, § 16.]

We note further the case of a compact operator.

VI. *A bounded Hermitian operator H is compact if, and only if, it has the form*

$$H = \sum_{k=1}^{\infty} \lambda_k P_k, \tag{7}$$

where the P_k are mutually orthogonal projection operators onto finite-dimensional subspaces, and $\langle \lambda_k \rangle$ is a sequence of real numbers which tends to zero as $k \to \infty$.

Proof. Suppose the operator H is compact, $P(\lambda)$ is its spectral function, $\mathfrak{M}_\varepsilon$ is a subspace which is invariant with respect to H such that the operator $P_\varepsilon = [P(b)-P(\varepsilon)]+[P(-\varepsilon)-P(a)]$ projects onto $\mathfrak{M}_\varepsilon$, and H_ε is the restriction of the operator H to $\mathfrak{M}_\varepsilon$. Then H_ε is compact and has the inverse

$$H_\varepsilon^{-1} = \int_a^{-\varepsilon} \frac{1}{\lambda} dP(\lambda) + \int_\varepsilon^b \frac{1}{\lambda} dP(\lambda).$$

Consequently, the identity operator $1 = H_\varepsilon^{-1} H_\varepsilon$ in $\mathfrak{M}_\varepsilon$ is compact, which is possible if, and only if, $\mathfrak{M}_\varepsilon$ is finite-dimensional. But then $P(\lambda)$ can vary in the intervals $[a, -\varepsilon]$ and $[\varepsilon, b]$ only by jumps at a finite number of points and, consequently, a formula of the form $\int_a^{-\varepsilon} \lambda dP(\lambda)+\int_\varepsilon^b \lambda dP(\lambda) = \sum_{k=1}^n \lambda_k P_k$ holds, where the P_k are mutually orthogonal projection operators onto finite-dimensional subspaces, and λ_k, $k = 1, 2, \cdots, n$ are saltus (i.e. jump) points of the function $P(\lambda)$ in the intervals $[a, -\varepsilon]$ and $[\varepsilon, b]$. Passing to the limit as $\varepsilon \to 0$ and taking formula (6) into consideration, we obtain (7). Since there are only finitely many of the numbers λ_k outside the interval $(-\varepsilon, \varepsilon)$, we have that $\lambda_k \to 0$ as $k \to \infty$.

Conversely, if the conditions of Proposition VI are satisfied, then

$$|(H-\sum_{k=1}^{n}\lambda_k P_k)\xi|^2 = |\sum_{k=n+1}^{\infty}\lambda_k P_k\xi|^2 = \sum_{k=n+1}^{\infty}\lambda_k^2|P_k\xi|^2$$

$$\leqq \max_{k>n}\lambda_k^2 \sum_{k=n+1}^{\infty}|P_k\xi|^2 \leqq \max_{k>n}\lambda_k^2|\xi|^2$$

for every $\xi \in \mathfrak{H}$; it follows that $|H-\sum_{k=1}^{n}\lambda_k P_k| \leqq \max_{k>n}|\lambda_k| \to 0$ as $n \to \infty$; since the operators $\sum_{k=1}^{n}\lambda_k P_k$ are compact, H is also compact (see IV, subsection 6, § 4).

It follows from formula (7) that: 1) the spectrum of the operator H in the algebra $\mathfrak{B}(\mathfrak{H})$ consists of only the characteristic values λ_k (see subsection 6, § 4) and perhaps of the number 0 which may or may not be a characteristic value; moreover, if $\mathfrak{H}$ is infinite-dimensional, then the point 0 belongs to the spectrum of the operator H; 2) the subspace $\mathfrak{M}_k = P_k\mathfrak{H}$ is the characteristic subspace of the operator H, corresponding to the characteristic value λ_k.

The spectral theorem can be generalized to unbounded self-adjoint operators in the following way. An operator $A \in \mathfrak{B}(\mathfrak{H})$ is said to *commute with the operator* T if $AT \subset TA$. It is easily verified that the operator $A \in \mathfrak{B}(\mathfrak{H})$ commutes with the self-adjoint operator H if, and only if, it commutes with the unitary operator $U = (H+i1)(H-i1)^{-1}$ (see V, subsection 9, § 5).

VII (Spectral Theorem for an Arbitrary Self-Adjoint Operator). *For every self-adjoint operator H in the Hilbert space $\mathfrak{H}$ there exists a unique operator-valued function $P(\lambda)$, $-\infty < \lambda < \infty$, which possesses the following properties:*

1) *$P(\lambda)$ is a projection operator;*
2) *$P(\lambda)P(\mu) = P(\lambda)$ for $\lambda \leqq \mu$;*
3) *$P(\lambda)$ commutes with every operator A in $\mathfrak{B}(\mathfrak{H})$ which commutes with H;*
4) *$\lim_{\lambda\to-\infty} P(\lambda)\xi = 0$, $\lim_{\lambda\to+\infty} P(\lambda)\xi = \xi$ for arbitrary $\xi \in \mathfrak{H}$;*
5) *$P(\lambda)\xi$ is a function which is left-continuous for arbitrary $\xi \in \mathfrak{H}$;*
6) *$\xi \in \mathfrak{D}_H$ if, and only if,*

$$\int_{-\infty}^{\infty}|\lambda|^2 d|P(\lambda)\xi|^2 < \infty \tag{8}$$

and in this case

$$H\xi = \int_{-\infty}^{\infty}\lambda dP(\lambda)\xi. \tag{9}$$

Proof. Since $U = (H+i1)(H-i1)^{-1}$ is unitary and hence normal, there exists a minimal symmetric commutative algebra R which contains U. Suppose $P(\Delta)$ is the spectral measure corresponding to this algebra so that

$$U = \int_{\mathfrak{M}} U(M)dP(M). \tag{10}$$

Then $|U(M)| = 1$ in virtue of the fact that the operator U is unitary, and hence $i[U(M)+1][U(M)-1]^{-1}$ is a real number, defined for $U(M) \neq 1$. Setting $\Delta_\lambda = \{M: i[U(M)+1] \cdot [U(M)-1]^{-1} < \lambda\}$ and $P(\lambda) = P(\Delta_\lambda)$, we obtain an operator-valued function satisfying conditions 1), 2), 4) and 5). If the operator $A \in \mathfrak{B}(\mathfrak{H})$ commutes with H, then it also commutes with U, and hence with all the operators of the algebra R; consequently, it also commutes with $P(\lambda)$ and condition 3) is also satisfied.

The uniqueness of $P(\lambda)$ follows from the uniqueness of the spectral measure $P(\Delta)$.

It remains to verify that condition 6) is satisfied.

Setting $\eta = (1/i)(H-i1)\xi$ so that consequently $U\eta = (1/i)(H+i1)\xi$, we see that $\mathfrak{D}_H$ consists of all vectors ξ of the form $\xi = U\eta-\eta$ where $H\xi = i(U\eta+\eta)$. But if $\xi = U\eta-\eta$, then in virtue of (10), $P(\Delta)\xi = \int_\Delta [U(M)-1]dP(M)\eta$, and therefore $|P(\Delta)\xi|^2 = \int_\Delta |U(M)-1|^2 d|P(M)\eta|^2$. From this we conclude that

$$\int_{\mathfrak{M}} \frac{1}{|U(M)-1|^2} d|P(M)\xi|^2 = \int_{\mathfrak{M}} \frac{1}{|U(M)-1|^2} d\int_{\Delta_M} |U(M')-1|^2 d|P(M')\eta|^2$$

$$= \int_{\mathfrak{M}} d|P(M)\eta|^2 = |\eta|^2,$$

so that

$$\int_{\mathfrak{M}} \frac{1}{|U(M)-1|^2} d|P(M)\xi|^2 < \infty. \tag{11}$$

Conversely, if condition (11) is satisfied, then the integral in the formula

$$\eta = \int_{\mathfrak{M}} \frac{1}{U(M)-1} dP(M)\xi$$

converges and

$$U\eta-\eta = \int_{\mathfrak{M}} (U(M)-1)dP(\Delta_M)\int_{\mathfrak{M}} \frac{1}{U(M')-1} dP(M')\xi$$

$$= \int_{\mathfrak{M}} (U(M)-1)d\int_{\Delta_M} \frac{1}{U(M')-1} dP(M')\xi = \int_{\mathfrak{M}} dP(M)\xi = \xi.$$

But the convergence of the integral in (11) is equivalent to the convergence of the integral

$$\int_{\mathfrak{M}} \left| i\frac{U(M)+1}{U(M)-1}\right|^2 d|P(M)\xi|^2; \tag{12}$$

therefore $\xi \in \mathfrak{D}_H$ if, and only if, the integral (12) converges, and in this case,

$$H\xi = i(U\eta+\eta) = \int_{\mathfrak{M}} i[U(M)+1]dP(M)\eta,$$

i.e.

$$H\xi = \int_{\mathfrak{M}} i\,\frac{U(M)+1}{U(M)-1}\,dP(M)\xi. \tag{13}$$

If we now take into consideration the definition of the function $P(\lambda)$ and repeat the arguments followed in the proof of Proposition IV, we see that the integrals in (12) and (13) coincide with the integrals in (8) and (9), respectively, which completes the proof.

A function $P(\lambda)$ satisfying conditions 1)–6) of Proposition VII is called the *spectral function* of the operator H, and formula (9) is called the *spectral decomposition* of H.

The spectral decomposition enables one to define the operator $f(H)$ for an arbitrary continuous bounded function $f(\lambda)$, $-\infty < \lambda < \infty$, by the formula

$$f(H) = \int_{-\infty}^{\infty} f(\lambda)dP(\lambda).$$

It is easily seen that then

1) if $f = \lambda_1 f_1 + \lambda_2 f_2$, then $f(H) = \lambda_1 f_1(H) + \lambda_2 f_2(H)$;
2) if $f = f_1 f_2$, then $f(H) = f_1(H) f_2(H)$;
3) $|f(H)| \leqq |f|_\infty$.

Moreover, if H is a bounded Hermitian operator and $f(\lambda)$ is continuous on the spectrum of H, then this definition of the operator $f(H)$ coincides with the definition in subsection 2, § 16. We leave the proof of these simple propositions for the reader.

[For a detailed discussion of the spectral theory of operators, see, e.g., the books of AKHIEZER and GLAZMAN [1], LYUSTERNIK and SOBOLEV [1], PLESNER [2], DUNFORD and SCHWARTZ [1]; also see the articles by PLESNER [1] and PLESNER and ROKHLIN [1], and Volume V of the book by V. SMIRNOV [1].]

5. **Spectral operators.** The spectral decomposition of an operator can be carried over to certain classes of linear operators in a Banach space in the following manner. A bounded linear operator P in the Banach space X is said to be a *projection* if $P^2 = P$.

We denote the complex plane by Π and the family of all Borel sets $\Delta \subset \Pi$ by B; an operator-valued function $P(\Delta)$, $\Delta \in B$, is called a *spectral measure* if for all $\Delta_1, \Delta_2, \Delta \in B$:

1) $P(\Delta_1 \cap \Delta_2) = P(\Delta_1)P(\Delta_2)$; 2) $P(\Delta_1 \cup \Delta_2) = P(\Delta_1) + P(\Delta_2)$, where by definition $P_1 \cup P_2 = P_1 + P_2 - P_1 P_2$; 3) $P(\Pi - \Delta) = 1 - P(\Delta)$; 4) $|P(\Delta)| \leqq K$, where K is some constant.

It follows from condition 1) for $\Delta_1 = \Delta_2$ that $P(\Delta)$ is a projection operator for arbitrary $\Delta \in B$; moreover, it also follows from condition 1) that $P(\Delta_1)$, $P(\Delta_2)$ commute for arbitrary $\Delta_1, \Delta_2 \in B$. Further, we conclude from conditions 1) and 3) that $P(\phi) = 0$. In fact, $P(\phi) = P((\Pi - \Delta) \cap \Delta) = P(\Pi - \Delta)P(\Delta) = (1 - P(\Delta))P(\Delta) = 0$.

A bounded linear operator A in X is called a *spectral operator* if there exists a spectral measure $P(\Delta)$ such that:

a) A commutes with $P(\Delta)$ for arbitrary $\Delta \in B$;

b) for arbitrary $\Delta \in B$ the spectrum $S(T, P(\Delta)X)$ of the operator T on the subspace $P(\Delta)X$ is contained in the closure $\bar{\Delta}$ of the set Δ: $S(T, P(\Delta)X) \subset \bar{\Delta}$;

c) there exists a dense set Γ in the conjugate space X' such that

$$f(P(\Delta_1 \cup \Delta_2 \cup \cdots)) = f(P(\Delta_1)) + f(P(\Delta_2)) + \cdots$$

for arbitrary $f \in \Gamma$ and arbitrary denumerable number of pairwise disjoint sets $\Delta_k \in B$.

If these conditions are satisfied, $P(\Delta)$ is called the *spectral measure* of the operator A; it can be shown that conditions a), b), c) define the spectral measure of a spectral operator uniquely. [For detailed proofs of results discussed here, see FAGE [1], DUNFORD [2] and DUNFORD and SCHWARTZ [1].] Furthermore, it can also be shown that the spectral measure $P(\Delta)$ of the operator A commutes with every bounded linear operator in X which commutes with A.

The spectral operator Λ is said to be an *operator of scalar type* if it is representable in the form

$$\Lambda = \int_S \lambda dP(\lambda), \tag{1}$$

where S is the spectrum of the operator Λ and $P(\lambda)$ is its spectral measure. One of the basic results of the theory under discussion consists in the following:

THEOREM 3. *A is a spectral operator if, and only if, it is representable in the form*

$$A = \Lambda + N, \tag{2}$$

where Λ is an operator of scalar type and N is a generalized nilpotent element which commutes with Λ. [N is a generalized nilpotent element means that

$$\lim_{n\to\infty} \sqrt[n]{|N^n|} = 0$$

(see subsection 2, § 11). Clearly, in the finite-dimensional case simply some power of the operator N will equal zero. Thus, every linear operator in a finite-dimensional space is spectral; but in the infinite-dimensional case, there also exist non-spectral operators.] *In this case, representation* (2) *is unique, where A and Λ have the same spectrum and the same spectral measure.*

The idea of the proof consists in the following. Suppose A is a spectral operator, $P(\Delta)$ its spectral function, and that S is its spectrum. We denote by R the minimal algebra, closed with respect to the operator norm, of bounded linear operators in X which contains A and all the $P(\Delta)$, and possessing the following property:

$$\text{if } B \in R \text{ and } B^{-1} \text{ exists and is bounded, then } B^{-1} \in R. \tag{3}$$

We set $\Lambda = \int_S \lambda dP(\lambda)$, $N = A - \Lambda$. It easily follows from the definition of the spectral measure that $A(M) = \Lambda(M)$ and, consequently, $N(M) = A(M) - \Lambda(M) = 0$ on all maximal ideals M of the algebra R. From this we conclude (see IV and VI, subsection 2, § 11), that N is a generalized nilpotent element. Moreover, it follows from the very formula $\Lambda = \int_S \lambda dP(\lambda)$ that Λ is an operator of scalar type and that $P(\Delta)$ is its spectral measure. This also proves decomposition (2); its uniqueness follows easily from the uniqueness of the spectral measure of the given spectral operator.

The converse assertion is obtained by considering the minimal complete (with respect to the operator norm) commutative algebra R of bounded linear operators in X, containing Λ, N and all the $P(\Delta)$ (where $P(\Delta)$ is the spectral measure of the operator Λ) and also possessing property (3). Namely, suppose $A = \Lambda + N$, where Λ and N satisfy the conditions of the theorem, and suppose $\mathfrak{M}_\Delta$ is the space of maximal ideals M of the restriction of this algebra to the subspace $P(\Delta)X$; then

$$\begin{aligned} S(A, P(\Delta)X) &= [\lambda\colon \lambda = \Lambda(M) + N(M),\ M \in \mathfrak{M}_\Delta] \\ &= [\lambda\colon \lambda = \Lambda(M),\ M \in \mathfrak{M}_\Delta] = S(\Lambda, P(\Delta)X) \subset \bar{\Delta}. \end{aligned}$$

From this we conclude that A is a spectral operator that has the same spectrum and the same spectral measure as A.

In conclusion, we note that the structure of algebras of operators, generated by the spectral operator and the values of its spectral measure, has also been investigated; we, however, shall not go into the details of this problem here (in this connection, see DUNFORD [2]).

For a discussion of further results in the theory of spectral operators, see the survey article by DUNFORD [3] and also the book of DUNFORD and SCHWARTZ [1]. LYANTSE [1] generalized this theory in an essential way, freeing himself from the condition $|P(\Delta)| \leqq K$ which is quite restrictive for applications.

6. Irreducible representations. A representation is said to be *irreducible* if there does not exist a closed invariant subspace in the space $\mathfrak{H}$ which is different from (0) and all of $\mathfrak{H}$.

According to Proposition II, subsection 1, this means that every projection operator, which commutes with all the operators of the representation, equals 0 or 1.

Clearly, every representation in one-dimensional space is irreducible; therefore, we shall consider representations which are not one-dimensional.

I. *The representation $x \to A_x$ in the space $\mathfrak{H}$, of dimension $\neq 1$, is irreducible if, and only if, every nonzero vector in the space $\mathfrak{H}$ is a cyclic vector of this representation.*

Proof. Suppose the representation $x \to A_x$ is irreducible. For $\xi \in \mathfrak{H}$, $\xi \neq 0$, the closed subspace enveloping the vectors $A_x\xi$, $x \in R$, is an invariant subspace; in virtue of the irreducibility of the representation, it coincides with (0) or $\mathfrak{H}$. But the first case is impossible because then the one-dimensional space $\{\alpha\xi\}$ is invariant and hence coincides with $\mathfrak{H}$, i.e. $A_x = 0$ in $\mathfrak{H}$. And in the second case, ξ is a cyclic vector.

Conversely, if the representation $x \to A_x$ is reducible and $\mathfrak{M}$ is a closed invariant subspace, different from (0) and $\mathfrak{H}$, in $\mathfrak{H}$, then no vector ξ in $\mathfrak{M}$ will be cyclic for the representation $x \to A_x$ in $\mathfrak{H}$.

II. *The representation $x \to A_x$ is irreducible if, and only if, every bounded linear operator which commutes with all the operators A_x of the representation is a multiple of the identity (i.e. it is a multiple of the identity operator).*

In fact, suppose the representation is irreducible and that the bounded operator B commutes with all the operators A_x. We shall first assume that B is a Hermitian operator; we denote by $P(\lambda)$ the spectral function of the operator B. Then, for arbitrary λ, the operator $P(\lambda)$ commutes with all the operators A_x; in view of the irreducibility of the representation, $P(\lambda) = 0$ or $P(\lambda) = 1$. Since $(P(\lambda)\xi, \xi)$ does not decrease as λ increases, it follows from this that there exists a λ_0 such that $P(\lambda) = 0$ for $\lambda < \lambda_0$ and $P(\lambda) = 1$ for $\lambda > \lambda_0$. It follows that

$$B = \int \lambda dP(\lambda) = \lambda_0 1.$$

Now suppose that B is an arbitrary bounded operator which commutes with all the operators A_x. Then B^* also commutes with all the operators A_x. In fact,

$$B^*A_x = (A_{x^*}B)^* = (BA_{x^*})^* = A_xB^*.$$

Therefore, the Hermitian operators $B_1 = (B+B^*)/2$, $B_2 = (B-B^*)/2i$ also commute with all the operators A_x and, consequently, they are multiples of the identity. But then the operator $B = B_1+iB_2$ is also a multiple of the identity.

Conversely, suppose every bounded operator which commutes with all the operators A_x is a multiple of the identity. Then, in particular, every projection operator, which commutes with all the operators A_x, is a multiple of the identity. But a projection operator can be a multiple of the identity only when it equals 0 or 1. Consequently, the representation is irreducible.

REMARK. It also follows from the above discussion that *every* (a priori *unbounded*) *self-adjoint operator H which commutes with all the operators A_x of an irreducible representation $x \to A_x$ is a multiple of the identity.*

7. Connection between vectors and positive functionals. We saw above that every vector ξ in $\mathfrak{H}$ generates a positive functional $f(x) = (A_x\xi, \xi)$. If $\xi_1 = \lambda\xi_2$, where $|\lambda| = 1$, then the corresponding functionals coincide. In general, non-proportional vectors can also yield one and the same positive functional. However, if the representation is irreducible, then the following theorem holds.

THEOREM 4. *Suppose $x \to A_x$ is an irreducible representation of the symmetric algebra R. If $(A_x\xi_1, \xi_1) = (A_x\xi_2, \xi_2)$ for all $x \in R$, then $\xi_1 = \lambda\xi_2$, where $|\lambda| = 1$.*

Proof. Clearly, it suffices to consider the case of a non-one-dimensional representation of the algebra R with identity. Suppose $\xi_1 \neq 0$; in virtue of I, subsection 6, the set of all vectors $A_x\xi_1$ is dense in $\mathfrak{H}$. We specify an operator U in $\mathfrak{H}$ in the following way: if $\xi = A_x\xi_1$, then $U\xi = A_x\xi_2$; in particular, for $x = e$, $U\xi_1 = \xi_2$. U is an isometric operator. In fact,

$$(U\xi, U\xi) = (A_x\xi_2, A_x\xi_2) = (A_{x^*x}\xi_2, \xi_2),$$
$$(\xi, \xi) = (A_x\xi_1, A_x\xi_1) = (A_{x^*x}\xi_1, \xi_1);$$

but by assumption $(A_{x^*x}\xi_2, \xi_2) = (A_{x^*x}\xi_1, \xi_1)$. It follows that the operator U is uniquely defined; namely, if $A_{x_1}\xi_1 = A_{x_2}\xi_1$, then $A_{x_1-x_2}\xi_1 = 0$; consequently, also $A_{x_1-x_2}\xi_2 = 0$, i.e. $A_{x_1}\xi_2 = A_{x_2}\xi_2$.

The operator U is defined on the set of all elements of the form $A_x\xi_1$; this set is dense in $\mathfrak{H}$; consequently, U can be uniquely extended to a bounded operator in $\mathfrak{H}$. This operator commutes with all the operators A_x. In fact, if $\xi = A_x\xi_1$ then

$$A_{x_0}U\xi = A_{x_0}A_x\xi_2 = A_{x_0x}\xi_2 = UA_{x_0x}\xi_1 = UA_{x_0}\xi.$$

In other words, $A_{x_0}U\xi = UA_{x_0}\xi$ on vectors of the form $\xi = A_x\xi_1$. Since these vectors form a dense set in $\mathfrak{H}$ and the operators A_{x_0}, U are continuous, the equality $A_{x_0} U\xi = UA_{x_0}\xi$ holds for all vectors ξ in $\mathfrak{H}$.

Thus, the operator U commutes with all the operators A_x of an irreducible representation. Consequently (see II, subsection 6), it is a multiple of the identity: $U = \lambda 1$. Hence $\xi_2 = U\xi_1 = \lambda\xi_1$. Since U is an isometric operator, $|\lambda| = 1$.

Now suppose $\xi_1 = 0$; then $\xi_2 = 0$ also. In fact, if $\xi_2 \neq 0$, then the equality

$$0 = (A_x\xi_1, \xi_1) = (A_x\xi_2, \xi_2)$$

means that the vector $\xi_2 \neq 0$ is orthogonal to all the vectors $A_x\xi_2$; consequently, this set is not dense in $\mathfrak{H}$. But the latter contradicts the irreducibility of the representation. Thus, for $\xi_1 = 0$ we also have $\xi_2 = 0$, so that the theorem is trivial in this case.

§ 18. Embedding of a symmetric algebra in an algebra of operators

1. Regular norm. Suppose R is a symmetric algebra. We denote the set of all positive functionals in R by F_R. Clearly, F_R remains unchanged if a norm $|x|$ is arbitrarily introduced in the algebra R so that R becomes a normed symmetric algebra. In fact, the definition of a positive functional does not depend on the norm in the algebra R. However, in passing from the algebra R to its completion $\tilde{R}$ with respect to this norm, the set F_R, generally speaking, may change. In other words, not every positive functional in the algebra R can be extended to a positive functional in the algebra $\tilde{R}$.

Suppose, for instance, that R is the algebra of all polynomials $p(t)$ with complex coefficients and that involution is introduced in R by means of the equality $(p(t))^* = \overline{p(\bar{t})}$. For arbitrary real t_0, the functional $f_{t_0}(p) = p(t_0)$ is a positive functional in R. We introduce a norm in the algebra R by the formula

$$|p| = \max_{0 \leqq t \leqq 1} |p(t)|.$$

With this definition of a norm, R is a normed algebra (which is not complete). From the Weierstrass theorem (see subsection 10, § 2) it follows that the completion $\tilde{R}$ of the algebra R with respect to this norm is the algebra of all continuous functions $x(t)$ on the interval $0 \leqq t \leqq 1$ with norm $|x| = \max_{0 \leqq t \leqq 1} |x(t)|$.

For $|t_0| > 1$, none of the functionals $f_{t_0}(p) = p(t_0)$ can be extended to a positive functional in the algebra $\tilde{R}$. In fact, let us assume that such an extension is possible. Then, according to Proposition I, subsection 4, § 10,

$$|f_{t_0}(p)| \leqq f_{t_0}(e)|p|,$$

i.e.

$$|p(t_0)| \leqq \max_{0 \leqq t \leqq 1} |p(t)|$$

for every polynomial $p(t)$, which is impossible.

The norm $|x|$ in the symmetric algebra R with identity will be called *regular* if every positive functional f in the algebra R can be extended to a positive functional

$\tilde{f}$ in the completion $\tilde{R}$ of the algebra R with respect to the norm $|x|$. In this case, $\tilde{f}$ is a positive functional in the complete algebra $\tilde{R}$ with identity; it follows, in virtue of (1), subsection 4, § 10, that

$$|\tilde{f}(\tilde{x})| \leqq \tilde{f}(e)|\tilde{x}| = f(e)|\tilde{x}|;$$

in particular, for $\tilde{x} = x \in R$,

$$|f(x)| \leqq f(e)|x|. \tag{1}$$

Consequently, the functional f is continuous in R and hence it is uniquely extendible to $\tilde{f}$ in $\tilde{R}$.

Thus, (1) *is a necessary and sufficient condition for a positive functional in the symmetric algebra R with identity to be extendible to a positive functional in its completion $\tilde{R}$ relative to the norm $|x|$.*

An element x in the symmetric algebra R with identity is said to be *bounded* if for all positive functionals f in R,

$$f(x^*x) \leqq Cf(e),$$

where C is a constant which depends only on x.

I. *In a symmetric Banach algebra R with identity, all elements are bounded.*

In fact, according to inequality (1), $f(x^*x) \leqq |x^*x|f(e)$ for all positive functionals in the algebra R.

II. *If the symmetric algebra R with identity has a regular norm, then all elements in R are bounded.*

This proposition is also an immediate consequence of inequality (1).

2. Reduced algebras. Suppose R is a symmetric algebra with identity. In all of this subsection we shall assume that R contains at least one positive functional which is not identically equal to zero. An element x in R will be called a *generalized zero* (denoted by $x \approx 0$) if $f(x^*x) = 0$ for all positive functionals f on R.

I. *An element x is a generalized zero if, and only if, $f(yx) = 0$ for all positive functionals f and for all elements y in the algebra R.*

The sufficiency of the condition is evident because for $y = x^*$ we obtain $f(x^*x) = 0$. And the necessity follows from the Cauchy-Bunyakovsky inequality:

$$|f(yx)| \leqq f(yy^*)f(x^*x).$$

COROLLARY 1. *e is not a generalized zero.*

In fact, if we had $e \approx 0$, then from Proposition I, with $x = e$, we would obtain that $f(y) = 0$ for all $y \in R$ and all positive functionals f, i.e. that R contains no positive functionals which are distinct from the identically zero functional; the latter contradicts our assumption relative to the algebra R.

II. *An element x_0 is a generalized zero if, and only if, $f(x_0) = 0$ for all positive functionals f.*

The necessity follows from Proposition I with $y = e$. We shall prove the sufficiency. Suppose $f(x_0) = 0$ for all positive functionals f. For arbitrary $y \in R$ and any λ we have that $\varphi(x) = f((\lambda e+y)x(\bar{\lambda}e+y^*))$ is also a positive functional; therefore $f((\lambda e+y)x_0(\bar{\lambda}e+y^*)) = 0$, i.e. $\bar{\lambda}f(yx_0)+\lambda(x_0y^*)+f(yx_0y^*) = 0$ for all numbers λ and for all $y \in R$. Setting $\lambda = 1, -1, i, -i$, multiplying the equations obtained by $1, -1, i, -i$ respectively, and then adding them, we arrive at the equality $4f(yx_0) = 0$. Consequently, $f(yx_0) = 0$ for all $y \in R$ and therefore x_0 is a generalized zero in virtue of I.

If $x \approx 0$, then we also have $x_1 x \approx 0$ for an arbitrary element x_1 in R. In fact, if $f(yx) = 0$ for all $y \in R$, then also $f(yx_1 x) = 0$ for all $y \in R$. Analogously, if $x_1 \approx 0$ and $x_2 \approx 0$, then also $\alpha_1 x_1 + \alpha_2 x_2 \approx 0$.

III. *If $x \approx 0$, then also $x^* \approx 0$.*

In fact, suppose $f(x^*x) = 0$ for all positive functionals f. We set $f_1(y) = f(xyx^*)$; f_1 is also a positive functional in R. Therefore $f_1(x^*x) = 0$, i.e. $f(xx^*xx^*) = 0$. But in virtue of the Cauchy-Bunyakovsky inequality, $|f(xx^*)|^2 \leqq f(xx^*xx^*)f(e)$; consequently, $f(xx^*) = 0$ for arbitrary positive functional f. Thus, we have

IV. *The set I of all generalized zeros is a symmetric two-sided ideal.*

[See Proposition I.]

This ideal will be called a *reducing ideal* of the algebra R, and the quotient algebra $R' = R/I$ will be called a *reduced algebra.*

If, in particular, $I = (0)$, then the algebra R itself is reduced.

V. *The algebras R' and R have the same set of positive functionals; specifically, every positive functional f on the algebra R defines a positive functional f' on the algebra R' by the formula*

$$f'(x') = f(x) \text{ for all } x \in x'. \tag{1}$$

Conversely, every positive functional f' on the algebra R' uniquely defines a positive functional f on the algebra R by formula (1).

In fact, if x_1 and x_2 belong to the same class x' modulo I, then $x_1 - x_2 \in I$ and, consequently, $f(x_1 - x_2) = 0$, $f(x_1) = f(x_2)$. Thus, the functional f remains constant on each class; hence, formula (1) defines a linear functional f' on R'. It will, clearly, be a positive functional on R'. Conversely, suppose f' is a given positive functional on R'; we set $f(x) = f'(x')$ for $x \in x'$. Then f is a positive functional on R.

In the sequel we shall not distinguish between f and f'.

VI. *In a reduced algebra R', for every element $x' \neq 0$ there exists a positive functional f such that $f(x'^*x') \neq 0$.*

In fact, suppose $f(x'^*x') = 0$ for all positive functionals on R' and that $x \in x'$. Then $f(x^*x) = 0$ for all positive functionals on R, i.e. $x \approx 0$. But this means that $x' = 0$.

COROLLARY 2. *The algebra R is reduced if, and only if, its reducing ideal is $I = (0)$.*

We now consider the case of a Banach symmetric algebra with identity.

VII. *The reducing ideal of a Banach symmetric algebra with identity contains the radical of this algebra.*

Proof. If the element x belongs to the radical, then, in virtue of IV, subsection 4, § 10, $f(x^*x) = 0$ for all positive functionals on R, and this means that x belongs to the reducing ideal.

VIII. *A reduced Banach algebra is semisimple.*

This follows directly from Proposition VII.

As also in the commutative case (see subsection 1, § 12), the norm in a reduced symmetric algebra with identity is defined uniquely to within equivalence by the collection of elements of the algebra.

IX. *In the symmetric Banach algebra R with identity e and norm $|x|$ let there be given a second norm $|x|_1$ such that $|x^*|_1 = |x^*|$. Consider the completion R_1 of the algebra R with respect to this norm* (see subsection 3, § 10). *If R_1 is a reduced Banach algebra with identity e_1, then $|x|_1 \leqq C|x|$ for all $x \in R$, where C is some constant.*

Proof. We introduce the new norm

$$|x|_2 = \max\{|x|, |x|_1\} \tag{2}$$

in R and prove that R is complete with respect to this norm; then, in virtue of VII, subsection 4, § 4, there exists a constant C such that $|x|_2 \leqq C|x|$ for all $x \in R$ and hence, in virtue of (2), $|x|_1 \leqq |x|_2 \leqq C|x|$ for all $x \in R$.

Let $\langle x_n \rangle \subset R$ be a fundamental sequence with respect to the norm $|x|_2$. In virtue of (2), it is fundamental with respect to the norms $|x|$ and $|x|_1$; therefore, in virtue of the completeness of R and R_1, it has the limits x_0 and x_0^1 in R and R_1 respectively. We shall prove that $x_0 = x_0^1$; this will prove that the algebra R is complete with respect to the norm $|x|_2$.

Let f_1 be an arbitrary positive functional on R_1. Denote its restriction to R by f. Clearly, f is a positive functional on R.

By virtue of inequality (1), subsection 4, § 10,

$$\begin{aligned}|f_1(x_0 - x_0^1)| \leqq |f_1(x_0 - x_n)| + |f_1(x_n - x_0^1)| \\ = |f(x_0 - x_n)| + |f_1(x_n - x_0^1)| \\ \leqq f(e)|x_0 - x_n| + f_1(e_1)|x_n - x_0^1|_1 \to 0\end{aligned}$$

as $n \to \infty$. Therefore, $f_1(x_0 - x_0^1) = 0$ for every positive functional f_1 and since R_1 is a reduced algebra, $x_0 - x_0^1 = 0$, which is what was required.

COROLLARY 3. *Every symmetric isomorphism of a symmetric Banach algebra R with identity onto a dense subset R' of a reduced symmetric Banach algebra R_1 with identity is continuous.*

COROLLARY 4. *If two reduced symmetric Banach algebras with identity are symmetrically isomorphic, then they are also topologically isomorphic.*

COROLLARY 5. *Every automorphism of a reduced symmetric Banach algebra with identity is continuous.*

3. Minimal regular norm.

THEOREM 1. *If there exists a regular norm in the reduced algebra R, then R also has a minimal regular norm. The completion of the algebra R with respect to this minimal norm is completely isomorphic to some algebra of operators in a Hilbert space.*

Proof. Suppose $|x|_1$ is any regular norm in the algebra R, and let R_1 be the completion of the algebra R with respect to this norm. Every positive functional f in the algebra R can be extended to a positive functional (we denote it also by f) on the algebra R_1; to it corresponds a cyclic representation of the algebra R_1.

We denote this representation by $x \to A_x^{(f)}$. Further, suppose $x \to A_x$ is the direct sum of all these representations. According tc Theorem 1, subsection 3, § 17,

$$|A_x| \leqq |x|_1, \tag{1}$$

where $|A_x|$ is the norm of the operator A_x. We note that for $x \in R$, $|A_x|$ is completely determined by the supply of positive functionals in R and therefore does not depend on the choice of the regular norm $|x|_1$.

The mapping $x \to A_x$ is an isomorphism of the algebra R. In fact, suppose $A_x = 0$. Then also $A_{x^*x} = A_{x^*}A_x = 0$. It follows that also $A_{x^*x}^{(f)} = 0$, i.e. $(A_{x^*x}^{(f)}\xi, \eta) = 0$ for arbitrary vectors ξ, η in the corresponding space $\mathfrak{H}^f$. We set, in particular, $\xi = \eta = \xi_0$, where ξ_0 is the class containing the identity of the algebra R. We obtain that $f(x^*x) = 0$ for all positive functionals on R. Sirce R is a reduced algebra, $x = 0$.

Thus, $x \to A_x$ is an isomorphism of the algebra R. We introduce a norm in R by setting $|x| = |A_x|$. Clearly, R with the norm $|x| = |A_x|$, and hence also its completion $\tilde{R}$ with respect to this norm, are completely isomorphic to some algebra of operators in Hilbert space (namely, to the algebra of all operators A_x).

We shall prove that $|x|$ is a regular norm. Suppose f is a positive functional in the algebra R. Then, for arbitrary element $x \in R$,

$$|f(x)| = |(A_x^{(f)}\xi_0, \xi_0)| \leqq |A_x^{(f)}|\,|\xi_0|^2 \leqq |A_x|\,|\xi_0|^2 = |x| f(e).$$

This inequality shows that the functional f is continuous with respect to the norm $|x|$. Consequently, it is uniquely extendible to a positive functional in the completion $\tilde{R}$ of the algebra R with respect to the norm $|x|$. And this means that $|x|$ is a regular norm.

Inequality (1) shows that $|x|$ is a minimal regular norm.

COROLLARY 1. *The minimal regular norm is completely regular.*

In fact, in this norm the algebra R is completely isomorphic to some subalgebra in $\mathfrak{B}(\mathfrak{H})$ and for operators $A \in \mathfrak{B}(\mathfrak{H})$ we have $|A^*A| = |A|^2$ (see (2), subsection 10, § 5 and subsection 1, § 16).

COROLLARY 2. *Let R be a symmetric algebra with identity; let R_1 be its reduced algebra and suppose there exists a regular norm, and hence also a minimal regular norm $|x'|$, in R_1. To each representation $x \to A_x$ of the algebra R there corresponds a representation $x' \to A_{x'}$ of the algebra R_1 by the formula $A_{x'} = A_x$ with $x \in x'$, $x' \in R_1$. This representation is uniquely extendible to the representation $\tilde{x} \to A_x$ of the completion $\tilde{R}_1$ of the algebra R_1 with respect to the norm $|x|_1$. Conversely, to each representation $\tilde{x} \to A_{\tilde{x}}$ of the algebra $\tilde{R}_1$ there corresponds a representation $x \to A_x$ of the algebra R by the formula $A_x = A_{x'} = A_{\tilde{x}}$ with $x \in x' = \tilde{x} \in \tilde{R}_1$.*

Proof. Clearly, it suffices to prove the assertion for cyclic representations. Let ξ_0 be a cyclic vector of the representation $x \to A_x$ and let I be a reducing ideal of the algebra R. If $x \in I$, then $f(x) = (A_x\xi_0, \xi_0) = 0$. From this it follows that $A_x = 0$.

[In fact, for arbitrary $x \in I$, $y, z \in R$, we also have $z^*xy \in I$ because I is a two-sided ideal. It follows that $0 = (A_{z^*xy}\xi_0, \xi_0) = (A_x A_y \xi_0, A_z\xi_0)$; since ξ_0 is cyclic, we have that $A_y\xi_0$ and $A_z\xi_0$ form dense sets in the representation space and hence $A_x = 0$.]

Consequently, $x \to A_x$ can be considered as the representation $x' \to A_{x'} = A_{x'}^{(f)}$ of the algebra $R_1 = R/I$ by setting $A_{x'} = A_x$ for $x \in x' \in R_1$. It follows directly from the definition of $|x'|$ (see the proof of Theorem 1) that $|A_{x'}| = |A_{x'}^{(f)}| \leqq |x'|$, and hence the representation $x' \to A_{x'}$ is uniquely extendible to the representation $\tilde{x} \to A_x$ of the algebra $\tilde{R}_1$. The converse assertion is obvious.

THEOREM 2. *The equality*

$$|x| = \sqrt{\sup f(x^*x)}$$

holds for the minimal regular norm where the least upper bound is taken over all positive functionals f on the algebra R for which $f(e) = 1$.

Proof. Suppose x is an element of the algebra R and that ξ is the corresponding element of the space $\mathfrak{H}^f$. Then

$$|A_{x_0}^{(f)}|^2 = \sup_{|\xi|=1} |A_{x_0}^{(f)}\xi|^2 = \sup_{f(x^*x)=1} f(x^*x_0^*x_0x),$$

where the least upper bound is taken over all elements x in the algebra R for which $f(x^*x) = 1$. We set $f_x(y) = f(x^*yx)$; f_x is a positive functional in R and $f_x(e) = f(x^*x) = 1$. Therefore

$$f(x^*x_0^*x_0x) = f_x(x_0^*x_0) \leqq \sup_{f(e)=1} f(x_0^*x_0).$$

It follows that

$$|A_{x_0}^{(f)}|^2 \leqq \sup_{f(e)=1} f(x_0^* x_0);$$

consequently, we also have

$$|x_0|^2 = |A_{x_0}|^2 = \sup_f |A_{x_0}^{(f)}|^2 \leqq \sup_{f(e)=1} f(x_0^* x_0).$$

On the other hand, for $f(e) = 1$,

$$f(x_0^* x_0) = |A_{x_0}^{(f)} \xi_0|^2 \leqq |A_{x_0}^{(f)}|^2 \leqq |A_{x_0}|^2 = |x_0|^2.$$

Therefore $\sup_{f(e)=1} f(x_0^* x_0) = |x_0|^2$ and the theorem is proved.

THEOREM 3. *A necessary and sufficient condition for a regular norm to exist in a reduced algebra R is that all elements of R be bounded.*

Proof. The necessity was proved in subsection 1; we shall prove the sufficiency. Suppose all elements of the algebra R are bounded. Then

$$f(x_0^* x_0) \leqq C_{x_0} f(e)$$

for all positive functionals f in the algebra R. In particular, again setting $f_x(y) = f(x^*yx)$, we have that

$$f_x(x_0^* x_0) \leqq C_{x_0} f_x(e),$$

i.e.

$$f(x^* x_0^* x_0 x) \leqq C_{x_0} f(x^* x).$$

This inequality may be rewritten in the form

$$|A_{x_0}^{(f)} \xi|^2 \leqq C_{x_0} |\xi|^2;$$

consequently, it means that $A_{x_0}^{(f)}$ is a bounded operator and that

$$|A_{x_0}^{(f)}|^2 \leqq C_{x_0}.$$

Thus, for fixed x_0 the norms of the operators $A_{x_0}^{(f)}$ are bounded by the same number. Therefore, one can construct the direct sum $x \to A_x$ of all the representations $x \to A_x^{(f)}$. As was seen in the proof of Theorem 1, the norm $|x| = |A_x|$ will be regular.

COROLLARY. *In a reduced symmetric Banach algebra with identity there exists a regular* (*and hence also a minimal regular*) *norm.*

This assertion follows directly from Theorems 1, 3 and Proposition I, subsection 1.

§ 19. Indecomposable functionals and irreducible representations

1. Positive functionals, dominated by a given positive functional. Suppose R is a symmetric algebra and that f, f_1 ($f \neq 0$) are positive functionals in R. We shall say that f_1 is dominated by f, and write $f_1 \prec f$, if there exists a number λ such that

$\lambda f - f_1$ is a positive functional in R. Clearly, in this case λ is real and $\geqq 0$, because for $\lambda < 0$ and $f(x^*x) > 0$ we should have $\lambda f(x^*x) - f_1(x^*x) < 0$.

THEOREM 1. *Suppose $x \to A_x$ is a cyclic representation of the symmetric algebra R with identity in the space $\mathfrak{H}$, defined by a given positive functional f, and that ξ_0 is the corresponding cyclic vector. Then every positive functional f_1, dominated by f, has the form*

$$f_1(x) = (A_x B\xi_0, \xi_0), \tag{1}$$

where B is a bounded positive definite operator in the space $\mathfrak{H}$, which commutes with all the operators A_x, $x \in R$. Conversely, every such operator B defines, by formula (1), *a positive functional f_1 which is dominated by f.*

The correspondence $f_1 \sim B$ which has thus been established is one-to-one.

Proof. Suppose B is a bounded positive definite operator in $\mathfrak{H}$, which commutes with all the operators A_x. Then $f_1(x) = (A_x B\xi_0, \xi_0)$ is a positive functional. In fact,

$$f_1(x^*x) = (A_{x^*x} B\xi_0, \xi_0) = (A_x B\xi_0, A_x\xi_0) = (BA_x\xi_0, A_x\xi_0) \geqq 0$$

because B is a positive definite operator. The functional f_1 is dominated by the functional f. In fact, the operator B is bounded and, consequently, there exists a number λ such that $(B\xi, \xi) \leqq \lambda(\xi, \xi)$, i.e. $\lambda(\xi, \xi) - (B\xi, \xi) \geqq 0$ for all vectors ξ in $\mathfrak{H}$. Setting $\xi = A_x\xi_0$ in this inequality, we obtain

$$\lambda(A_{x^*x}\xi_0, \xi_0) - (A_{x^*x}B\xi_0, \xi_0) \geqq 0,$$

i.e.

$$\lambda f(x^*x) - f_1(x^*x) \geqq 0.$$

But this means that $\lambda f - f_1$ is a positive functional. Consequently, the functional f_1 is dominated by the functional f.

Conversely, suppose f_1 is a positive functional dominated by the functional f; namely, suppose $\lambda f - f_1$ is a positive functional and hence that $\lambda f(x^*x) - f_1(x^*x) \geqq 0$, i.e.

$$0 \leqq f_1(x^*x) \leqq \lambda f(x^*x). \tag{2}$$

For

$$\xi = A_x\xi_0, \eta = A_y\xi_0$$

we set

$$(\xi, \eta)_1 = f_1(y^*x).$$

This expression depends only on the elements ξ and η. In fact, suppose, for instance, that $\xi = A_x\xi_0$ and $\xi = A_{x'}\xi_0$; then $A_{x-x'}\xi_0 = 0$. Consequently

$$f((x-x')^*(x-x')) = (A_{x-x'}\xi_0, A_{x-x'}\xi_0) = 0.$$

In virtue of the inequality (2), it follows from this that also $f_1((x-x')^*(x-x')) = 0$;

consequently, by the Cauchy-Bunyakovsky inequality,

$$f_1(y^*(x-x')) = 0, f_1(y^*x) = f_1(y^*x').$$

This proves that the expression $(\xi, \eta)_1$ is independent of the method of representing the element ξ in the form $\xi = A_x\xi_0$. Clearly, the analogous situation is also true for the element η.

We denote the set of all vectors ξ of the form $A_x\xi_0$ by $\mathfrak{H}'$; since ξ_0 is a cyclic element, $\mathfrak{H}'$ is dense in $\mathfrak{H}$. In virtue of what we just proved, the expression $(\xi, \eta)_1$ is uniquely defined in $\mathfrak{H}'$; clearly, it is a bilinear form with respect to ξ, η. Inequalities (2) may be rewritten in the form

$$0 \leqq (\xi, \xi)_1 \leqq \lambda(\xi, \xi); \tag{2'}$$

consequently, $(\xi, \eta)_1$ is a bounded bilinear form in $\mathfrak{H}'$. It can therefore be uniquely extended to a bounded bilinear form in the entire space $\mathfrak{H}$. Then inequalities (2') will hold in the entire space $\mathfrak{H}$. But to a bounded bilinear form $(\xi, \eta)_1$ in $\mathfrak{H}$ there corresponds a bounded operator B such that $(\xi, \eta)_1 = (B\xi, \eta)$ (see subsection 3, § 5). Since $(B\xi, \xi) = (\xi, \xi)_1 = f_1(x^*x) \geqq 0$, B is a positive definite operator. We shall prove that B commutes with all the operators A_x. To this end, we note that for $\xi = A_x\xi_0$, $\eta = A_y\xi_0$,

$$\begin{aligned}(BA_{x_0}\xi, \eta) &= (BA_{x_0x}\xi_0, A_y\xi_0) = (A_{x_0x}\xi_0, A_y\xi_0)_1 = f_1(y^*x_0x),\\ (A_{x_0}B\xi, \eta) &= (B\xi, A_{x_0^*}\eta) = (BA_x\xi_0, A_{x_0^*y}\xi_0)\\ &= (A_x\xi_0, A_{x_0^*y}\xi_0)_1 = f_1((x_0^*y)^*x) = f_1(y^*x_0x),\end{aligned}$$

and hence

$$(A_{x_0}B\xi, \eta) = (BA_{x_0}\xi, \eta) \tag{3}$$

for all vectors ξ and η in $\mathfrak{H}'$. Since both members of equality (3) are continuous functions of ξ and η with respect to the norm in $\mathfrak{H}$ and since $\mathfrak{H}'$ is dense in $\mathfrak{H}$, this equality holds for all vectors ξ, η in $\mathfrak{H}$. But the latter means that $A_{x_0}B = BA_{x_0}$, i.e. B commutes with A_{x_0}. Finally, $f_1(x) = f_1(e^*x) = (A_x\xi_0, A_e\xi_0)_1 = (BA_x\xi_0, \xi_0) = (A_xB\xi_0, \xi_0)$, i.e. B satisfies condition (1).

Clearly, for given fixed operator B, which commutes with all the operators A_x, equality (1) uniquely defines the functional f_1. Conversely, for given positive functional f_1, dominated by the functional f, this equality uniquely defines a bounded positive definite operator B which commutes with all the operators A_x.

In fact, setting y^*x in place of x in equality (1) and rewriting the equality thus obtained in the form

$$f_1(y^*x) = (BA_x\xi_0, A_y\xi_0),$$

we see that the inner product $(B\xi, \eta)$ is uniquely defined for all vectors ξ, η in $\mathfrak{H}'$. Since $\mathfrak{H}'$ is dense in $\mathfrak{H}$, this also defines the bounded operator B uniquely.

2. The algebra C_f. We shall say that a functional f_1 in the symmetric algebra R is *dominated by a given positive functional* f in R if f_1 is a linear combination with complex coefficients of positive functionals dominated by the functional f.

It follows from Theorem 1 that every functional f' which is dominated by the functional $f(x) = (A_x\xi_0, \xi_0)$ has the form

$$f'(x) = (BA_x\xi_0, \xi_0),$$

where B is a bounded operator which commutes with all the operators A_x. In fact, suppose $f'(x) = \sum_{k=1}^{n} \lambda_k f_k(x)$, where the f_k are positive functionals dominated by the functional f. Then $f_k(x) = (B_k A_x \xi_0, \xi_0)$, where B_k is a positive definite operator which commutes with all the operators A_x. We set $B = \sum_{k=1}^{n} \lambda_k B_k$. Clearly, B is a bounded operator which commutes with all the operators A_x and, furthermore,

$$f'(x) = \sum_{k=1}^{n} \lambda_k (B_k A_x \xi_0, \xi_0) = (BA_x\xi_0, \xi_0).$$

Conversely, every bounded operator B which commutes with all the operators A_x can be represented in the form of a linear combination of positive definite operators which commute with all the operators A_x (see V, subsection 4, § 17). Therefore, the functional $f'(x) = (BA_x\xi_0, \xi_0)$ is a linear combination of positive functionals which are dominated by the functional f. Thus, *there exists a one-to-one correspondence between the functionals f', dominated by a given positive functional f, and all bounded operators B which commute with all the operators A_x. This correspondence is established by the formula*

$$f'(x) = (BA_x\xi_0, \xi_0).$$

We denote the set of all functionals dominated by the positive functional f by C_f. In virtue of what was just stated, this set can be considered as a symmetric algebra. Namely, if

$$f_1(x) = (B_1 A_x \xi_0, \xi_0), f_2(x) = (B_2 A_x \xi_0, \xi_0),$$

then we define the product $f_1 \cdot f_2$ of the functionals f_1 and f_2 by the equality

$$(f_1 \cdot f_2)(x) = (B_1 B_2 A_x \xi_0, \xi_0),$$

and involution by the equality

$$f_1^*(x) = (B_1^* A_x \xi_0, \xi_0),$$

from which it follows that

$$f_1^*(x) = \overline{f_1(x^*)}.$$

With these definitions of multiplication and involution and with the usual definitions of addition and scalar multiplication, the set C_f becomes an algebra which is symmetrically isomorphic to the algebra R_f' of all bounded operators B which commute with every operator A_x. In this correspondence, the functional f itself corresponds to the identity operator.

3. Indecomposable positive functionals. A positive functional f is said to be *indecomposable* if every functional f_1, dominated by the functional f, is a multiple of f, i.e. $f_1 = \lambda f$. In other words, f is indecomposable if the algebra C_f consists of only the functionals λf.

THEOREM 2. *A cyclic representation $x \to A_x$ is irreducible if, and only if, the positive functional $f(x) = (A_x \xi_0, \xi_0)$ defining it is indecomposable.*

Proof. According to II, subsection 6, § 17, the irreducibility of the representation $x \to A_x$ is equivalent to the fact that the algebra R_f (see subsection 2) consists only of operators which are multiples of the identity. Since the algebras R_f and C_f are isomorphic, it follows that the algebra C_f consists of the multiples λf. But the latter means that f is an indecomposable functional.

Let R be a closed symmetric subalgebra of the algebra $\mathfrak{B}(\mathfrak{H})$, i.e. the algebra – in general, not of all – bounded linear operators in a Hilbert space $\mathfrak{H}$, which contains together with A also A^*, and is closed with respect to the operator norm. KADISON [8] proved that:

1) *If R is irreducible, then it is also algebraically irreducible, i.e. every* (*and not only a closed*) *subspace which is invariant with respect to all $A \in R$ is either* (0) *or all of $\mathfrak{H}$.*

2) *If f is a nonzero indecomposable positive functional on R and $I_l = \{A: A \subset R, f(A^*A) = 0\}$, then R/I_l is complete with respect to the inner product $(\xi, \eta) = f(B^*A)$, where $A \in \xi$, $B \in \eta$, $\xi, \eta \in R/I_l$.*

3) *Every closed left ideal in R is the intersection of the left maximal ideals containing it.*

4. Completeness and approximation theorems. We shall now apply the results of subsections 7 and 9, § 3.

Suppose R is a Banach symmetric algebra with identity. We denote by H the set of all Hermitian elements of the algebra R. Clearly, H is a real Banach space. According to I, subsection 4, § 10, every positive functional f can be considered as a bounded linear functional on H, where $|f| = f(e)$. Therefore, f can be considered as an element of the conjugate space H'. We denote the set of all positive functionals which satisfy the condition $f(e) = 1$ by K. (A positive functional f which satisfies the condition $f(e) = 1$ is said to be *normalized.*) Then also $|f| = 1$ and, consequently, K is a subset of the unit sphere Q in H'. In virtue of III, subsection 7, § 3, Q is compact in the weak topology in H' because Q is the set of all linear functionals $f \in H'$, satisfying the inequality

$$|f(x)| \leqq p(x), \text{ where } p(x) = |x|.$$

Since the function $F(f) = f(e)$ is continuous and the set of all positive functionals is closed in H', K is a closed subset of the compact set Q and, consequently, K is compact.

Finally, K is convex. In fact, if $f_1, f_2 \in K$, i.e. $f_1(e) = f_2(e) = 1$, then

$$f = tf_1 + (1-t)f_2, \quad 0 \leqq t \leqq 1,$$

is a positive functional and $f(e) = 1$ and, consequently, $f \in K$.

I. *A positive functional f satisfying the condition $f(e) = 1$ is indecomposable if, and only if, it is an extreme point of the set K.*

Proof. Suppose the indecomposable positive functional f belongs to the segment $[f_1, f_2]$ in K, i.e. that it has the form

$$f = tf_1 + (1-t)f_2, \quad 0 \leqq t \leqq 1.$$

Then $f - tf_1 = (1-t)f_2$ is a positive functional, i.e. if $t > 0$, then the functional f_1 is dominated by the functional f. Since f is indecomposable, $f_1 = t_1 f$. In particular, $f_1(e) = t_1 f(e)$, from which it follows that $t_1 = 1$ and $f_1 = f$. Therefore f is not an interior point of any segment $[f_1, f_2] \subset K$, i.e. f is an extreme point. Conversely, suppose f is an extreme point of the set K and that f_1 is a functional in K which is dominated by the functional f. This means that for some $\lambda > 0$, $f' = f - \lambda f_1$ is a positive functional. If $f'(e) = 0$ then $f - \lambda f_1 = 0$, $f = \lambda f_1$ and $1 = f(e) = \lambda f_1(e) = \lambda$; therefore, in this case, $f_1 = f$.

We shall now prove that the case $f'(e) > 0$ is impossible. Let $f'(e) > 0$. Put $f_2 = f'/\mu$, where $\mu = f'(e)$. Then $f_2(e) = 1$; consequently, $f_2 \in K$ and

$$f = \lambda f_1 + \mu f_2, \tag{1}$$

where $\lambda > 0$, $\mu > 0$, $\lambda + \mu = 1$ (we obtain the last relation by setting $x = e$ in (1)), i.e., f is an interior point of the segment $[f_1, f_2]$ in K, which contradicts the fact that f is an extreme point. Thus, $f_1 = f$; in order words, every functional in K which is dominated by f coincides with f. Consequently, the functional f is indecomposable and Proposition I is proved.

We now consider a set of irreducible representations of the symmetric algebra R. In this connection, two equivalent irreducible representations will not be considered distinct. This set of representations is said to be *complete* if for every element $x_0 \neq 0$ of the algebra R there exists in this set an irreducible representation $x \to A_x$ such that $A_{x_0} \neq 0$.

THEOREM 3. *The set of all irreducible representations of a Banach reduced algebra forms a complete system.*

Proof. Suppose R is a Banach reduced algebra (see subsection 2, § 18) and that $x_0 \neq 0$ is an element in R. By the definition of a reduced algebra, there exists a positive functional f_1 such that

$$f_1(e) = 1 \text{ and } f_1(x_0^* x_0) > 0.$$

Suppose, further, that $P_{x_0^* x_0}$ is the hyperplane of support to K which is defined by the element $x_0^* x_0$ (see subsection 9, § 3). According to Proposition VII, subsection 9, § 3, $P_{x_0^* x_0}$ contains an extreme point f_0.

We have:

$$f_0(x_0^* x_0) = \max_{f \in K} f(x_0^* x_0) \geqq f_1(x_0^* x_0) > 0. \tag{2}$$

On the other hand, according to Proposition I, every extreme point f_0 of the set K is an indecomposable positive functional. This functional defines an irreducible representation $x \to A_x$ of R (see Theorem 2, subsection 3), with

$$(A_x \xi_0, \xi_0) = f_0(x),$$

where ξ_0 is some vector in the representation space. In particular, in virtue of (2),

$$|A_{x_0} \xi_0|^2 = (A_{x_0^* x_0} \xi_0, \xi_0) = f_0(x_0^* x_0) > 0;$$

consequently, $A_{x_0} \neq 0$, and the theorem is proved.

We shall now apply the Krein-Milman theorem (see subsection 9, § 3) to the set K. The set K contains all finite linear combinations of the form

$$\sum_{i=1}^{n} \lambda_i f_i,$$

where the f_i are indecomposable positive functionals satisfying the condition $f_i(e) = 1$ and the λ_i are nonnegative real numbers such that $\sum_{i=1}^{n} \lambda_i = 1$; we denote the set of all such linear combinations by K_1. Further, K contains all weak limit points of the set K_1; we denote by K_2 the closure of the set K_1 in the weak topology in H'. The set K_2 is the minimal compact convex set which contains all the extreme points of the set K and hence $K_2 = K$. This proves the following theorem.

THEOREM 4. *Suppose f is a nonzero positive functional in the symmetric Banach algebra R with identity; then for every $x_0 \in R$ and $\varepsilon > 0$ there exists a linear combination $\sum_{i=1}^{n} f_i$ of indecomposable positive functionals on R satisfying the following conditions*:

$$\sum_{i=1}^{n} f_i(e) = f(e), \quad |f(x_0) - \sum_{i=1}^{n} f_i(x_0)| < \varepsilon.$$

[Here, the functionals f, f_i are no longer normalized by the condition $f(e) = 1$, and the number λ_i is incorporated in the functional f_i.]

REMARK. It follows from Theorem 4 that in the definition of a reducing ideal and of a minimal regular norm (see subsections 2, 3, § 18) we can restrict ourselves to indecomposable positive functionals.

Theorem 1 can also be formulated in terms of representations of the algebra R. To this end, it suffices to use Theorem 2, subsection 3, § 17 and Theorem 2, subsection 3. We then obtain the following proposition.

II. *Let $x \to A_x$ be a representation in the space $\mathfrak{H}$ of the symmetric Banach algebra R with identity. For all $x_0 \in R$, $\xi^0 \in \mathfrak{H}$ and $\varepsilon > 0$ there exist irreducible representations $x \to A_{1,x}, \cdots, x \to A_{n,x}$ and vectors $\xi_1^0, \cdots, \xi_n^0$ in the space of these representations such that*

$$\|\xi_1^0\|^2 + \cdots + \|\xi_n^0\|^2 = \|\xi^0\|^2$$

and

$$|(A_x \xi^0, \xi^0) - \sum_{i=1}^{n} (A_{i,x} \xi_i^0, \xi_i^0)| < \varepsilon.$$

Sometimes, instead of the weak topology in H', it is convenient to make use of another topology which we shall call the *uniform-on-compacta* topology in which the neighborhood basis is defined in the following way.

Let Q be an arbitrary set in H, precompact in the topology defined by the norm, and let ε be an arbitrary positive number. The neighborhood $U(f_0; Q; \varepsilon)$ of the functional f_0 is the set of all functionals $f \in H'$ satisfying the inequality

$$|f(x) - f_0(x)| < \varepsilon \text{ for all } x \in Q.$$

III. *If $\mathfrak{S}$ is a set in H' which is uniformly bounded in norm in H', then every weak contact point of this set is also a contact point of it in the sense of the uniform-on-compacta topology.* [From the proof below it is also clear that on such sets $\mathfrak{S}$ the weak topology and uniform-on-compacta topology coincide.]

Proof. Suppose $|f| \leqq C$ for all $f \in \mathfrak{S}$ and that f_0 is a weak contact point of the set $\mathfrak{S}$. We are required to prove that every neighborhood $U(f_0; Q; \varepsilon)$ contains elements $f \in \mathfrak{S}$.

We set $\varepsilon_1 = \varepsilon/(2C+1)$; there exists a finite set of elements $x_1, \cdots, x_n \in Q$ such that every other element $x \in Q$ is at a distance less than ε_1 from one of the x_p, $p = 1, \cdots, n$ (see I, subsection 14, § 2). The weak neighborhood $U(f_0; x_1, \cdots, x_n; \varepsilon_1)$ contains some element $f \in \mathfrak{S}$; we shall prove that $f \in U(f_0; Q; \varepsilon)$. In fact, $|f(x_p) - f_0(x_p)| < \varepsilon_1$, $p = 1, \cdots, n$, and for given $x \in Q$ there exists an x_p such that $|x_p - x| < \varepsilon_1$. Therefore

$$\begin{aligned} |f(x) - f_0(x)| &\leqq |f(x) - f(x_p)| + |f(x_p) - f_0(x_p)| + |f_0(x_p) - f_0(x)| \\ &< |f|\,|x - x_p| + \varepsilon_1 + |f_0|\,|x - x_p| \\ &< C\varepsilon_1 + \varepsilon_1 + C\varepsilon_1 = \varepsilon. \end{aligned}$$

If the space H is separable, then the weak neighborhoods on $\mathfrak{S}$ have a countable basis. Every weak contact point f_0 of the set $\mathfrak{S}$ is then a weak limit of some sequence of functionals $\varphi_n \in \mathfrak{S}$. Since $\mathfrak{S}$ is assumed to be bounded in norm, this sequence is uniformly bounded in norm and, consequently, in virtue of III, it converges to f_0 uniformly on every compactum. Theorem 4 therefore signifies that *every positive functional f on R is a contact point in the sense of the uniform-on-compacta topology of the functionals $\sum_{i=1}^{n} f_i$, where the f_i are indecomposable functionals and*

$\sum_{i=1}^{n} f_i(e) = f(e)$. *But if the algebra R is separable, then f is the limit of a sequence of functionals* $\langle \sum_{i=1}^{n} f_i^{(n)} \rangle$ *which is uniformly convergent on every precompact set.*

§ 20. Application to commutative symmetric algebras

1. Minimal regular norm in a commutative symmetric algebra

THEOREM 1. *Suppose R is a Banach commutative symmetric algebra with identity. The norm in this algebra coincides with the minimal regular norm if, and only if, it is completely regular and, consequently, if, and only if, it is completely isomorphic to some algebra* $C(\mathfrak{M})$.

Proof. Suppose the norm in the algebra R coincides with the minimal regular norm. On the basis of Corollary 1, subsection 3, § 18, this norm is completely regular and hence R is completely isomorphic to the algebra $C(\mathfrak{M})$ (see Theorem 2, subsection 2, § 16).

Conversely, suppose the algebra R is completely isomorphic to the algebra $C(\mathfrak{M})$. Every positive functional f in $C(\mathfrak{M})$ is also a bounded functional and hence it is a bounded integral (see subsection 15, § 6). In fact, if $x(M) \geqq 0$, then $y(M) = \sqrt{x(M)} \in C(\mathfrak{M})$, $x = y^2 = y^*y$, and hence $f(x) \geqq 0$. Therefore a positive functional in $C(\mathfrak{M})$ has the form

$$f(x) = \int_{\mathfrak{M}} x(M) d\mu(M),$$

where μ is the measure corresponding to this integral. Moreover,

$$f(e) = \int_{\mathfrak{M}} d\mu(M) = \mu(\mathfrak{M}).$$

We denote the minimal regular norm in the algebra R by $|x|_0$. Then

$$|x|_0^2 = \sup_{f(e)=1} f(x^*x) = \sup_{\mu(\mathfrak{M})=1} \int_{\mathfrak{M}} |x(M)|^2 d\mu(M) = \max_{\mathfrak{M}} |x(M)|^2 = |x|^2,$$

consequently, the norm in the algebra R coincides with the minimal regular norm.

COROLLARY 1. *Suppose* $\mathfrak{M}$ *and* $\mathfrak{M}'$ *are compact spaces; then every symmetric isomorphism of the algebra* $C(\mathfrak{M})$ *onto a subalgebra* R' *of the algebra* $C(\mathfrak{M}')$, *which is dense in* $C(\mathfrak{M}')$, *preserves norm and hence is an isomorphism onto the entire algebra* $C(\mathfrak{M}')$.

Proof. According to Theorem 1, the norm in the algebras $C(\mathfrak{M})$ and $C(\mathfrak{M}')$ is the minimal regular norm. The algebra $C(\mathfrak{M}')$ is the closure of the algebra R' relative to such a norm; consequently, according to the definition of a regular norm, the algebras R' and $C(\mathfrak{M}')$ have the same supply of positive functionals. On the other hand, the algebras $C(\mathfrak{M})$ and R' also have the same supply of positive functionals, because these algebras are symmetrically isomorphic. Moreover, a correspondence between the positive functionals f and f' in the algebras $C(\mathfrak{M})$ and R' is established by the equality $f(s) = f'(x')$, where $x \in C(\mathfrak{M})$ and $x' \in R'$ are corresponding elements under the given isomorphism. It follows that

$$|x| = \sqrt{\sup_{f(e)=1} f(x^*x)} = \sqrt{\sup_{f'(e')=1} f'(x'^*x')} = |x'|.$$

COROLLARY 2. *Every symmetric isomorphism of a commutative C^*-algebra R into a commutative C^*-algebra R' preserves norm.*

Proof. Without loss of generality, we may assume that R and R' contain the identity. In fact, otherwise, adjoining the identity to them, we obtain the C*-algebras R_1, R_1' with identity; the initial isomorphism $x \to x'$ of the algebras R, R' can then be extended to an isomorphism of the algebras R_1 and R_1' by setting

$$\lambda e + x \to \lambda e' + x'.$$

Thus, suppose R contains the identity and that R_0' is the image of the algebra R under the given isomorphism $x \to x'$; we may assume that R_0' is dense in R' because otherwise we could replace R' by the closure $\bar{R}_0'$ of the algebra R_0' in R'. Then R and R' are completely isomorphic to certain algebras $C(\mathfrak{M})$ and $C(\mathfrak{M}')$ and our isomorphism is a symmetric isomorphism of the algebra $C(\mathfrak{M})$ into the dense subalgebra R_0' of the algebra $C(\mathfrak{M}')$. Therefore the assertion follows directly from Corollary 1.

2. Positive functionals in a commutative symmetric algebra. We now go over to the study of positive functionals in an arbitrary Banach commutative symmetric algebra with identity.

We denote the compact space of all maximal ideals of the algebra R by $\mathfrak{M}$. It easily follows from the relation $x^*(M^*) = \overline{x(M)}$ (see (2), subsection 1, § 14) that the mapping $M \to M^*$ is a homeomorphism of the space $\mathfrak{M}$ onto itself (see subsection 2, § 16); therefore, the set $\mathfrak{M}_0$ of all maximal ideals M satisfying the condition $M^* = M$, i.e. of all symmetric maximal ideals, is a closed subset of the space $\mathfrak{M}$.

We denote the reduced algebra, constructed relative to the algebra R, by R_1. We introduce the minimal regular norm as the norm in R_1. The completion of the algebra R_1 relative to this norm will be called the *completely regular completion of the algebra R* and it will be denoted by $R^\wedge$.

THEOREM 2. *Suppose R is a Banach commutative symmetric algebra with identity. Then the algebra $R^\wedge$ is completely isomorphic to the algebra $C(\mathfrak{M}_0)$, where $\mathfrak{M}_0$ is the space of all symmetric maximal ideals of R.*

Proof. Let $\mathfrak{M}^\wedge$ be the space of all maximal ideals of the algebra $R^\wedge$. According to Theorem 1, it will suffice to prove that $\mathfrak{M}^\wedge$ and $\mathfrak{M}_0$ are homeomorphic.

Let $M^\wedge$ be a maximal ideal of the algebra $R^\wedge$. $M^\wedge$ determines a homomorphism of the algebra $R^\wedge$ and, consequently, also of the algebra R_1 into the field of complex numbers. But R_1 is the quotient algebra of the algebra R modulo some two-sided symmetric ideal I. Therefore the homomorphism of the algebra R_1 is at the same time a homomorphism of the algebra R into the field of complex numbers. Namely, it suffices to map all elements x of the class $\tilde{x} \in R_1$ into the same number – the image of the class $\tilde{x}$. This homomorphism defines the maximal ideal M_0 of the algebra R; we correspond the ideal M_0 with the ideal $M^\wedge$.

The ideal M_0 is symmetric. In fact, suppose $x \in M_0$. This means that the element x maps into the zero under the homomorphism of R considered. But then the corresponding element $\tilde{x}$ of the algebra $R_1 \subset R^\wedge$ also maps into zero. In other words, $\tilde{x}(M^\wedge) = 0$. In virtue of Theorem 1, the involution operation in $R^\wedge$ is the passage from the function $\tilde{x}(M^\wedge)$ to the complex conjugate function $\overline{\tilde{x}(M^\wedge)}$, i.e. $\tilde{x}^*(M^\wedge) = \overline{\tilde{x}(M^\wedge)}$; consequently, also $\tilde{x}^*(M^\wedge) = 0$. In other words, $\tilde{x}^*$ also maps into zero. But then x^* also maps into zero and, consequently, $x^* \in M_0$. By the same token, $M_0^* = M_0$.

Conversely, suppose M_0 is a symmetric maximal ideal of the algebra R. Let us consider the functional $f(x) = x(M_0)$; f is a positive functional. In fact,

$$f(x^*x) = x^*(M_0)x(M_0) = |x(M_0)|^2 \geqq 0.$$

Moreover,

$$f(e) = e(M_0) = 1.$$

This functional can therefore be considered as a positive functional on $R^\wedge$ (see subsections 1, 2, § 18). The functional f defines a homomorphism of the algebra R, and hence also of the algebra R_1, into the field of complex numbers. Since the functional f is continuous in the algebra R_1 and since the algebra R_1 is dense in $R^\wedge$, this functional also defines a homomorphism of the algebra $R^\wedge$ into the field of complex numbers. The latter homomorphism defines a maximal ideal $M^\wedge$ in the algebra $R^\wedge$.

If $M_1^\wedge \neq M_2^\wedge$, then there exists an element $x \in R^\wedge$ such that $x^\wedge(M_1^\wedge) \neq x^\wedge(M_2^\wedge)$ (see I, subsection 2, § 11); since R_1 is dense in $R^\wedge$, there exists an element $\tilde{x} \in R_1$ with the same property. Then for $x \in \tilde{x}$ and the corresponding ideals M_{10}, M_{20} we will have $x(M_{10}) = \tilde{x}(M_1^\wedge) \neq \tilde{x}(M_2^\wedge) = x(M_{20})$ and hence $M_{10} \neq M_{20}$. This means that the mapping $M^\wedge \to M_0$ constructed above is a one-to-one mapping of the space $\mathfrak{M}^\wedge$ onto the space $\mathfrak{M}_0$. It follows from the definition of the topologies in $\mathfrak{M}^\wedge$ and $\mathfrak{M}_0$ that the inverse image of every neighborhood in $\mathfrak{M}_0$ is a neighborhood in $\mathfrak{M}^\wedge$. Consequently, the mapping is continuous. Since both of these spaces are compact Hausdorff, this mapping is a homeomorphism. This completes the proof of the theorem.

COROLLARY. *Let R be commutative completely symmetric algebra with identity and $\mathfrak{M}$ its space of maximal ideals. Then:*

1) *the minimal regular norm in R is* $|x|_0 = \sup_{M \in \mathfrak{M}} |x(M)|$;

2) *the space* $\mathfrak{M}^\wedge$ *of maximal ideals of the algebra* $R^\wedge$ *is homeomorphic to* $\mathfrak{M}$.

Proof. Since R is completely symmetric, $\mathfrak{M}_0 = \mathfrak{M}$ (see subsection 1, § 14). On the other hand, $\mathfrak{M}^\wedge$ and $\mathfrak{M}_0 = \mathfrak{M}$ are homeomorphic (see the proof of Theorem 2). Assertion 1) follows from Theorem 1, subsection 1.

THEOREM 3. *Every positive functional f in a Banach commutative symmetric algebra R with identity can be uniquely represented in the form*

$$f(x) = \int_{\mathfrak{M}_0} x(M)d\mu(M), \tag{1}$$

where μ is some measure in the space $\mathfrak{M}_0$ of all symmetric maximal ideals of the algebra R.

Proof. Every positive functional in the algebra R can be extended to a positive functional in the algebra $R^\wedge$. But according to Theorem 2, the algebra $R^\wedge$ is completely isomorphic to the algebra $C(\mathfrak{M}_0)$, and thus f can also be considered as a positive functional, and hence as an integral, on $C(\mathfrak{M}_0)$ (see the proof of Theorem 1). Formula (1) and the uniqueness of the measure μ for given f follow directly from this.

REMARK. It follows from Theorem 3 that *a positive functional f in the algebra R is indecomposable if, and only if, $f(x) = x(M)$ for some $M \in \mathfrak{M}_0$*. Consequently, this theorem means that every positive functional in a Banach commutative symmetric algebra with identity decomposes uniquely into positive functionals which cannot be decomposed further. The following theorem shows that the commutativity condition is essential here.

THEOREM 4. *Suppose every positive functional in the Banach symmetric algebra R with identity decomposes uniquely into positive functionals which cannot be decomposed further. Then the reduced algebra R_1 is commutative.*

Proof. We shall prove that every irreducible representation of the algebra R is one-dimensional. Let us assume the contrary. Suppose $x \to A_x$ is an irreducible representation of the algebra R in the space $\mathfrak{H}$ and that $\mathfrak{H}$ is not one-dimensional. Then in $\mathfrak{H}$ there exist at least two linearly independent vectors ξ_1 and ξ_2. We set

$$f_1(x) = (A_x\xi_1, \xi_1), \quad f_2(x) = (A_x\xi_2, \xi_2);$$
$$f_1'(x) = \tfrac{1}{2}(A_x(\xi_1+\xi_2), \xi_1+\xi_2), \quad f_2'(x) = \tfrac{1}{2}(A_x(\xi_1-\xi_2), \xi_1-\xi_2).$$

Then

$$f_1(x)+f_2(x) = f_1'(x)+f_2'(x).$$

Suppose

$$f(x) = f_1(x)+f_2(x) = f_1'(x)+f_2'(x); \tag{2}$$

f is a positive functional. According to Theorem 2, subsection 3, § 19, the functionals

f_1, f_2, f_1', f_2' are indecomposable. Since, furthermore, no two of the vectors ξ_1, ξ_2, $\xi_1+\xi_2$, $\xi_1-\xi_2$ are proportional, all these functionals are distinct (see Theorem 4, subsection 7, § 17). Therefore, (2) exhibits two distinct decompositions of the functional f into indecomposable functionals, in contradiction with the assumption.

Thus, every irreducible representation of the algebra R is one-dimensional, and hence commutative. Therefore, every indecomposable positive functional vanishes on all elements of the form $xy-yx$. But, by assumption, every positive functional decomposes into indecomposable positive functionals. Therefore, every positive functional also vanishes on all elements of the form $xy-yx$. This means that every such element is a generalized zero (see II, subsection 2, § 18). Consequently, in the reduced algebra R_1, we have $\tilde{x}\tilde{y}-\tilde{y}\tilde{x} = 0$, i.e. R_1 is commutative.

For other sufficient conditions that an algebra be commutative, the reader is referred to TURUMARU [1], FUKAMIYA, MISONOU and TAKEDA [1] and OGASAWARA [2].

3. **Examples.** a) Suppose A is the algebra of all functions

$$x(z) = \sum_{n=0}^{\infty} c_n z^n,$$

which are analytic in the open unit disc and continuous on the closed unit disc (see Example b), subsection 3, § 11). We define involution in A by the formula

$$x^*(z) = \sum_{n=0}^{\infty} \bar{c}_n z^n.$$

We shall now find the symmetric maximal ideals of the algebra A.

Every maximal ideal M of the algebra A is defined by some point z_0 of the closed unit disc $|z| \leqq 1$ by the formula

$$x(M) = x(z_0),$$

where the correspondence $M \to z$ between $\mathfrak{M}$ and the points of the closed unit disc $|z| \leqq 1$, thus established, is one-to-one (see Example b), subsection 3, § 11).

Suppose M is a symmetric maximal ideal. Since $x_1 = z$ is a Hermitian element, the number $z_0 = x_1(M)$ must be real. Thus, to each symmetric maximal ideal of the algebra R there corresponds a real number t in the interval $[-1, 1]$. Conversely, to each such number t there corresponds the symmetric maximal ideal of all functions $x(z)$ in A which are such that $x(t) = 0$.

According to Theorem 2, the completely regular completion $A^\wedge$ of the algebra A is the algebra of all continuous functions $x(t)$ in the interval $[-1, 1]$ with norm

$$|x| = \max_{-1 \leqq t \leqq 1} |x(t)|.$$

If $x(z)$ is a polynomial and $y(t)$ is the corresponding element of the algebra $A^\wedge$, then $x(t) = y(t)$ for all $t \in [-1, 1]$; in virtue of the Weierstrass theorem on the approximation by polynomials (see subsection 10, § 2) it follows from this that $x(t) = y(t)$ for an arbitrary function $x(z) \in A$.

Thus, the algebra $A^\wedge$ is obtained if we consider the functions $x(z)$ only on the interval $[-1, 1]$ and then complete the algebra thus obtained relative to the norm $|x| = \max_{-1 \leqq t \leqq 1} |x(t)|$.

Applying Theorem 3 to the algebra A, we obtain that every positive functional on A has the form

$$f(x) = \int_{-1}^{1} x(t)d\sigma(t), \tag{1}$$

where $\sigma(t)$ is a nondecreasing function in the interval $[-1, 1]$.

We note that not every real-valued bounded linear functional in the algebra A can be represented in the form of a linear combination of positive functionals. In fact, we consider as an example the functional

$$f_0(x) = \tfrac{1}{2}[x(z_0)+x(\bar{z}_0)],$$

where z_0 is an arbitrary imaginary number in the closed unit disc $|z| \leqq 1$. Clearly, f_0 is a real bounded linear functional in the algebra A. We shall prove that it is not a linear combination of positive functionals. Let us suppose the contrary. Then, using the general expression (1) for a positive functional, we shall have

$$f_0(x) = \int_{-1}^{1} x(t)d\sigma_1(t),$$

where $\sigma_1(t)$ is a function of finite variation in the interval $[-1, 1]$. Thus,

$$\frac{x(z_0)+x(\bar{z}_0)}{2} = \int_{-1}^{1} x(t)d\sigma_1(t)$$

for an arbitrary function $x(z) \in A$. Applying this equality to the functions $x_n(z) = e^{inz}/n$, we obtain

$$\frac{e^{inz_0}+e^{in\bar{z}_0}}{2n} = \frac{1}{n}\int_{-1}^{1} e^{int}d\sigma_1(t).$$

But as $n \to \infty$ the absolute value of the left member of this equality tends to ∞, and that of the right member to zero. The contradiction thus obtained shows that the functional f_0 cannot be represented in the form of a linear combination of positive functionals.

b) We denote the set of all measurable functions $x(u)$ on the half-line $0 \leqq u < +\infty$ which are such that

$$|x| = \int_0^{\infty} |x(u)| \sinh 2u \, du < \infty \tag{2}$$

by R'_0. We define the operations of addition and scalar multiplication in R'_0 in the usual way; moreover, we define the multiplication operation $x = x_1 \cdot x_2$ by the equality

$$x(u) = (x_1 \cdot x_2)(u) = \int_0^{\infty} \int_{|u-t|}^{u+t} x_1(s)x_2(t)ds\,dt.$$

This definition is legitimate because

$$\begin{aligned}\int_0^{\infty} |x(u)| \sinh 2u\,du &\leqq \int_0^{\infty} \left| \int_0^{\infty} dt \int_{|u-t|}^{u+t} x_1(s)x_2(t)ds \right| \sinh 2u\,du \\ &\leqq \int_0^{\infty} \sinh 2u\,du \int_0^{\infty} dt \int_{|u-t|}^{u+t} |x_1(s)|\,|x_2(t)|ds \\ &= \int_0^{\infty} dt \int_0^{\infty} |x_1(s)|\,|x_2(t)|ds \int_{|t-s|}^{t+s} \sinh 2u\,du \\ &= \int_0^{\infty}\int_0^{\infty} |x_1(s)|\,|x_2(t)| \sinh 2s \sinh 2t\,ds\,dt = |x_1|\,|x_2|;\end{aligned}$$

consequently, if $x_1, x_2 \in R'_0$, then also $x_1 \cdot x_2 \in R'_0$. At the same time we see that the condition $|x_1 \cdot x_2| \leqq |x_1|\,|x_2|$ is satisfied. It is also easily verified that all the other algebra axioms are satisfied and that $x_1 \cdot x_2 = x_2 \cdot x_1$; consequently, R'_0 is a commutative normed algebra which, evidently, is complete.

Adjoining the identity e to R'_0, we obtain a Banach algebra with norm

$$|x+\lambda e| = |x|+|\lambda|,$$

where $|x|$ is defined by equality (2). We denote this algebra by R_0. If we set $x^*(u) = \overline{x(u)}$, $(\lambda e+x)^* = \bar{\lambda} e+x^*$, then R_0 becomes a symmetric algebra.

We shall now find all the maximal ideals in the algebra R_0. First, R'_0 is a maximal ideal on R_0. Suppose M is a maximal ideal in R_0 which is distinct from R'_0. Then

$$f(x) = x(M)$$

is a linear functional on R_0. It can also be considered as a linear functional in the space of all summable functions $x(u)$ sinh $2u$. Therefore,

$$x(M) = f(x) = \int_0^\infty x(u)\omega(u)du,$$

where $\omega(u)/\sinh 2u$ is an essentially bounded function and $\omega(u) \not\equiv 0$ (see Theorem 3, subsection 16, § 6). Since the mapping $x \to x(M)$ is a homomorphism, we must have $f(x_1 \cdot x_2) = f(x_1)f(x_2)$, i.e.

$$\int_0^\infty x_1(s)\omega(s)ds \int_0^\infty x_2(t)\omega(t)dt$$

$$= \int_0^\infty \left[\int_0^\infty dt \int_{|u-t|}^{u+t} x_1(s)x_2(t)ds\right] \omega(u)du$$

$$= \int_0^\infty \int_0^\infty x_1(s)x_2(t)ds\,dt \int_{|t-s|}^{t+s} \omega(u)du.$$

We conclude from this that the function $\omega(u)$ satisfies the functional equation

$$\omega(s)\omega(t) = \int_{|t-s|}^{t+s} \omega(u)du \tag{3}$$

for almost all positive s and t.

The general solution of this equation is $\omega(u) = (2/\rho)\sin \rho u$, where ρ is an arbitrary complex number. In fact, it follows from (3) that $\omega(s)$ is equivalent to a continuous function, so that $\omega(s)$ may be considered continuous. Setting $s = t = 0$ in (3), we obtain $\omega(0) = 0$. Differentiating both members of (3) with respect to s for $s \leqq t$, we obtain

$$\omega'(s)\omega(t) = \omega(t+s)+\omega(t-s). \tag{4}$$

It follows that for $s = 0$, $\omega'(0)\omega(t) = 2\omega(t)$, $\omega'(0) = 2$. Differentiating (4) twice with respect to s and twice with respect to t, subtracting the resultant equations termwise and setting $s = 0$, we conclude that $\omega''(t)-\rho^2\omega(t) = 0$, where $\rho^2 = \omega'''(0)/\omega'(0) = \omega'''(0)/2$ is some number, in general complex. The solution of this equation which satisfies the conditions $\omega(0) = 0$, $\omega'(0) = 2$ is $\omega(t) = (2/\rho)\sin \rho t$.

The function $\omega(u)/\sinh 2u = 2 \sin \rho u/\rho \sinh 2u$ is bounded for $-\infty < u < \infty$ if, and only if,

$|\text{Im}\,\rho| \leqq 2$. Consequently, the maximal ideals of the algebra R_0 which are different from R_0' are defined by the numbers ρ in the strip $|\text{Im}\,\rho| \leqq 2$, and the corresponding homomorphisms are defined by the formula

$$x \to \int_0^\infty x(u) \frac{2 \sin \rho u}{\rho} du.$$

Clearly, the numbers ρ_1 and ρ_2 define the same maximal ideal if, and only if, either $\rho_1 = \rho_2$ or $\rho_1 = -\rho_2$.

We shall now find all the symmetric maximal ideals of the algebra R_0. To this end, we note that we always have $x(M^*) = \overline{x^*(M)}$. Therefore, it follows from the equality

$$x(M) = \int_0^\infty x(u) \frac{2 \sin \rho u}{\rho} du$$

that

$$x(M^*) = \overline{x^*(M)} = \overline{\int_0^\infty x^*(u) \frac{2 \sin \rho u}{\rho} du} = \int_0^\infty x(u) \frac{2 \sin \bar{\rho} u}{\bar{\rho}} du.$$

Thus, if the ideal M is defined by the number ρ then the ideal M^* is defined by the number $\bar{\rho}$. Therefore $M = M^*$ if, and only if, either $\rho = \bar{\rho}$ or $\rho = -\bar{\rho}$, i.e. either the number ρ is real or it is pure imaginary.

Consequently, the symmetric maximal ideals of the algebra R_0 are defined biuniquely by all the numbers on the real half-line $0 \leqq \rho < \infty$ and the segment $\rho = \rho_1 i$, $0 < \rho_1 \leqq 2$, of the imaginary axis. The corresponding homomorphisms have the form

$$x \to \int_0^\infty x(u) \frac{2 \sin \rho u}{\rho} du \text{ for } \rho \text{ real},$$

$$x \to \int_0^\infty x(u) \frac{2 \sinh \rho_1 u}{\rho_1} du \text{ for } \rho = \rho_1 i.$$

Applying Theorem 3 to the algebra R_0, we obtain that every positive functional in the algebra R_0 has the form

$$f(x) = \int_0^\infty \left[\int_0^\infty x(u) \frac{2 \sin \rho u}{\rho} du \right] d\sigma(\rho) + \int_0^2 \left[\int_0^\infty x(u) \frac{2 \sinh \rho_1 u}{\rho_1} du \right] d\sigma_1(\rho_1),$$

where $\sigma(\rho)$, $\sigma_1(\rho_1)$ are nondecreasing functions in the intervals $0 \leqq \rho < \infty$, $0 < \rho_1 \leqq 2$, respectively.

4. The case of a completely symmetric algebra. If the algebra R is completely symmetric, then all its maximal ideals are symmetric so that $\mathfrak{M}_0 = \mathfrak{M}$. Therefore the following corollary holds.

COROLLARY 1. *Every positive functional f in a commutative completely symmetric algebra R with identity can be uniquely represented in the form*

$$f(x) = \int_{\mathfrak{M}} x(M) d\mu(M),$$

where $\mathfrak{M}$ is the space of maximal ideals of the algebra R and μ is some measure on $\mathfrak{M}$.

KOSTYUCHENKO and MITYAGIN [1, 2] obtained an integral representation for positive functionals in a topological algebra with continuous involution which is a so-called nuclear space (see, e.g., GELFAND, VILENKIN and GRAEV [1]), and they applied this result to the multi-dimensional moment problem. Related results were obtained independently by BEREZANSKY [3].

In the applications, it frequently occurs that one must consider positive functionals in an algebra without identity. We recall that the algebra R without identiy is said to be *completely symmetric* if the algebra R_1 obtained from R by adjunction of the identity is completely symmetric (see subsection 3, § 14).

Suppose $\mathfrak{M}_1$ is the space of maximal ideals of the algebra R_1; if the positive functional f in R is defined by the formula

$$f(x) = \int_{\mathfrak{M}_1} x(M) d\mu(M) \text{ for all } x \in R, \tag{1}$$

then, by setting

$$F(x) = \int_{\mathfrak{M}_1} x(M) d\mu(M) \text{ for all } x \in R_1, \tag{2}$$

we extend f to a positive functional F on all of R_1. Conversely, if such an extension F is possible, then in virtue of Corollary 1 formula (2) holds for the functional F and, consequently, formula (1) holds for the functional f.

From this and from IV, subsection 2, § 10, we conclude the following.

COROLLARY 2. *A positive functional f in a commutative completely symmetric algebra R without identity can be represented in the form* (1) *if, and only if, f can be extended to a positive functional on the algebra R_1, obtained from R by adjunction of the identity, i.e. if f is real and satisfies the condition*

$$|f(x)|^2 \leqq Cf(x^*x) \text{ for all } x \in R, \tag{3}$$

where C is some constant.

We note that in virtue of formula (1), such a functional f can be uniquely extended to a positive functional $\tilde{f}$ with the same norm on the entire algebra $C(\mathfrak{M}_1)$.

Up to this point, we considered positive functionals defined on the entire algebra R. However, in some cases one must formulate the problem of the decomposition of a positive functional which is not defined on the entire algebra and which is, in general, not extendible to a positive functional on the entire algebra (see below, subsection 4, § 31). To study such functionals, it turns out that we must first consider some properties of integrals on a locally compact space.

Suppose T is a locally compact Hausdorff space, I is an integral on $L(T)$ (see subsection 1, § 6), and that U is an arbitrary open set in T. Then U, considered as a subspace in T, is also a locally compact space and every function $x \in L(U)$ can be extended to a function $x \in L(T)$, assuming $x(t) = 0$ in the exterior of U.

Consequently, $L(U)$ may be considered as a subspace in $L(T)$ and the initial integral I on $L(T)$, considered only on $L(U)$, will also be an integral on $L(U)$; we shall call it the *restriction* of the initial integral to the set U.

I. *Suppose* $\{U_\alpha\}$ *is a covering of the space* T *by open sets* U_α *and that on every* $L(U_\alpha)$ *there is defined an integral* I_α *such that for arbitrary* α_1, α_2 *the restrictions of the integrals* $I_{\alpha_1}, I_{\alpha_2}$ *to* $U_{\alpha_1} \cap U_{\alpha_2}$ *coincide. Then there exists one, and only one, integral on* $L(T)$ *whose restriction to each* U_α *coincides with* I_α.

Proof. Suppose $x \in L(T)$ and that Q_x is the carrier of x. [The carrier of x is the minimal compact set in the exterior of which $x(t) = 0$ (see subsection 2, § 6).] The intersections of the sets U_α with Q_x form a covering of the compact set Q_x; consequently, there exists a finite number $U_{\alpha_1}, \cdots, U_{\alpha_n}$ of such U_α which also form a covering of Q_x. The functions in $L(T)$, considered only on Q_x, form a normal algebra with identity (in virtue of the Urysohn lemma; see II, subsection 8, § 2 and subsection 1, § 15); consequently (see subsection 2, § 15), there exist functions $y_k \in L(U_{\alpha_k})$, $k = 1, \cdots, n$ such that $y_1(t) + \cdots + y_n(t) = 1$ on Q_x. It follows that $x = xy_1 + \cdots + xy_n$, where $xy_k \in L(U_{\alpha_k})$, and hence if the integral I exists, then we must have

$$I(x) = I_{\alpha_1}(xy_1) + \cdots + I_{\alpha_n}(xy_n). \tag{4}$$

We shall show that formula (4) does indeed define the integral I on $L(T)$; by the same token, our proposition will be proved. But to this end it suffices to show that if $z_k \in L(U_{\alpha'_k})$, $k = 1, \cdots, m$, are any functions satisfying the condition $z_1(t) + \cdots + z_m(t) = 1$ on Q_x, then

$$I_{\alpha_1}(xy_1) + \cdots + I_{\alpha_n}(xy_n) = I_{\alpha'_1}(xz_1) + \cdots + I_{\alpha'_m}(xz_m).$$

But since $xy_k z_l \in U_{\alpha_k} \cap U_{\alpha'_l}$, then by assumption $I_{\alpha_k}(xy_k z_l) = I_{\alpha'_l}(xy_k z_l)$ and hence

$$\sum_{k=1}^{n} I_{\alpha_k}(xy_k) = \sum_{k=1}^{n} \sum_{l=1}^{m} I_{\alpha_k}(xy_k z_l) = \sum_{k=1}^{n} \sum_{l=1}^{m} I_{\alpha'_l}(xy_k z_l) = \sum_{l=1}^{m} I_{\alpha'_l}(xz_l),$$

which completes the proof.

II. *Suppose* T *and* $\{U_\alpha\}$ *are the same as in Proposition* I; *if* I, J *are two integrals on* $L(T)$ *whose restrictions on every* U_α *coincide, then* I *and* J *coincide on all of* $L(T)$.

The assertion follows directly from I.

We now consider all possible open sets U (if there are any) on which the restriction of a given integral equals zero. Let V be the union of all these sets U. Since the sets U form a covering of the space V, the restriction of the integral I to V is also equal to zero, in virtue of II. Clearly, V is the maximal open set on which the restriction of the integral I equals zero; this means that V is the maximal open set which is zero with respect to I (see subsections 5 and 8, § 6). The complement $T - V$ of this set is called the *carrier* of the integral I and also of the measure μ defined by it.

Suppose now that R is a completely symmetric commutative algebra without identity and that f is a real positive functional defined on the symmetric ideal I which is dense in R. Then $px \in I$ for all $p \in I$, $x \in R$, and hence the equality

$$\varphi_p(x) = f(px)$$

defines the linear functional φ_p on the entire algebra R for arbitrary $p \in I$. The element $p \in I$ is said to be *positive relative to the functional* f if φ_p is a real positive functional on R, satisfying condition (3).

We denote the set of all elements $p \in I$ which are positive relative to f by P. Such elements p exist; namely, an arbitrary element of the form

$$p = y_1^* y_1 + y_2^* y_2 + \cdots + y_n^* y_n, \quad y_k \in I \tag{5}$$

will always be an element of this sort. In fact, for an arbitrary $x \in R$,

$$\begin{aligned}\varphi_p(x^*x) = f(px^*x) &= f((y_1^* y_1 + \cdots + y_n^* y_n)x^*x) \\ &= f((y_1 x)^*(y_1 x)) + \cdots + f((y_n x)^*(y_n x)) \geqq 0,\end{aligned}$$

so that φ_p is positive on R. This positive functional satisfies condition (3) because

$$\begin{aligned}|\varphi_p(x)|^2 = |\sum_{k=1}^{n} f(y_k^* y_k x)|^2 &\leqq n^2 \sum_{k=1}^{n} |f(y_k^*(y_k x))|^2 \\ &\leqq n^2 \sum_{k=1}^{n} f(y_k^* y_k) f((y_k x)^*(y_k x)) \leqq n^2 \max_{1 \leqq k \leqq n} f(y_k^* y_k) \cdot f(px^*x) \\ &= n^2 \max_{1 \leqq k \leqq n} f(y_k^* y_k) \cdot \varphi_p(x^*x).\end{aligned}$$

Moreover, since p is Hermitian and f is a real functional, we have, for $x \in R$, that

$$\varphi_p(x^*) = f(px^*) = f((px)^*) = \overline{f(px)} = \overline{\varphi_p(x)},$$

so that φ_p is a real functional.

We denote the set of all elements p of the form (5) by P'; in virtue of what we just proved, $P' \subset P$.

According to the remark immediately following Corollary 2, for $p \in P$ the functional φ_p can be uniquely extended to a positive functional $\tilde{\varphi}_p$ on the entire algebra $C(\mathfrak{M}_1)$, where $\mathfrak{M}_1$ is the space of maximal ideals of the algebra R_1.

By the definition of an integral (see subsection 1, § 6), $\tilde{\varphi}_p$ is an integral on $C(\mathfrak{M}_1)$ and hence it can be uniquely extended to all functions $x(M) \in L^1(\tilde{\varphi}_p)$; this extension will also be denoted by $\tilde{\varphi}_p$.

We set $M_0 = R$ and $\mathfrak{M}' = \mathfrak{M}_1 - M_0$; then $\mathfrak{M}'$ is a locally compact space. Every element $x \in R$ can be considered as a continuous function $x(M)$ on $\mathfrak{M}'$ which is equal to zero at the infinitely distant point M_0. We denote by $L(\mathfrak{M}')$ the set of all continuous functions $x(M)$ on $\mathfrak{M}'$ with compact carrier Q_x (one for each x). Clearly, $\tilde{\varphi}_p(x)$ will be defined, in particular, for all functions $x(M) \in L(\mathfrak{M}')$ and it will be an integral on $L(\mathfrak{M}')$.

III. *For an arbitrary function* $x(M) \in L(\mathfrak{M}')$ *there exist elements* $p \in P'$ *such that* $p(M) > 0$ *on* Q_x.

Proof. Since the ideal I is dense in R, for an arbitrary $M \in \mathfrak{M}'$, in particular for an arbitrary $M \in Q_x$, there exists an element $y \in I$ such that $y(M) \neq 0$. By continuity, this condition is also satisfied in some neighborhood $U(M)$ of the ideal M. In virtue of the compactness of Q_x, we can select a finite number $U(M_1), \cdots, U(M_n)$ of these neighborhoods which also cover Q_x. If $y_1, \cdots, y_n$ are the corresponding elements $y \in I$, then the element $p = y_1^* y_1 + \cdots + y_n^* y_n$ will have the required property.

We denote the set of all elements $p \in P$ for which $p(M) \neq 0$ on Q_x by P_x, and the set of all elements $p \in P$ for which $p(M) > 0$ on Q_x by P_x^+. Clearly, $P_x^+ \subset P_x$. Proposition III means that P_x^+, and hence also P_x, are non-void.

If $x \in L(\mathfrak{M}')$ and $p \in P_x$, then also $x/p \in L(\mathfrak{M}')$ where x/p is the function equal to $x(M)/p(M)$ for $M \in Q_x$ and to zero for $M \notin Q_x$; from this we conclude that $\tilde{\varphi}_p(x/p)$ is meaningful.

We set

$$J(x) = \tilde{\varphi}_p\left(\frac{x}{p}\right) \text{ for } x \in L(\mathfrak{M}'),\ p \in P_x. \tag{6}$$

This definition does not depend on the choice of the element $p \in P_x$. In fact, if $p, q \in P_x$, then, for an arbitrary $z \in R$,

$$\varphi_q(pz) = \varphi_p(qz) = f(pqz).$$

From this it follows that

$$\tilde{\varphi}_q(pz) = \tilde{\varphi}_p(qz) \text{ for all } z \in C(\mathfrak{M}_1). \tag{7}$$

In fact, $\tilde{\varphi}_q$, $\tilde{\varphi}_p$ are integrals on $C(\mathfrak{M}_1)$, and hence they are continuous in norm in $C(\mathfrak{M}_1)$; on the other hand, equality (7) holds for all functions $z(M)$ of the algebra R, and these $z(M)$ form a dense set in the algebra $C_0(\mathfrak{M}_1)$ of all functions in $C(\mathfrak{M}_1)$ which vanish at the point M_0 (see Corollary 3, subsection 3, § 14). Consequently, (7) holds for all $z(M) \in C_0(\mathfrak{M}_1)$. But

$$\tilde{\varphi}_q(p) = \varphi_q(p) = f(pq) = \varphi_p(q) = \tilde{\varphi}_p(q);$$

therefore,

$$\tilde{\varphi}_q(p(\lambda e + z)) = \tilde{\varphi}_p(q(\lambda e + z))$$

for all numbers λ and all $z \in C_0(\mathfrak{M}_1)$. Since the functions $\lambda e + z$ form the entire algebra $C(\mathfrak{M}_1)$, (7) is proved by the same token for all functions in $C(\mathfrak{M}_1)$.

Setting $z = x/pq$ in (7) (which is legitimate because $x/pq \in L(\mathfrak{M}') \subset C(\mathfrak{M}')$), we obtain

$$\tilde{\varphi}_q\left(\frac{x}{q}\right) = \tilde{\varphi}_p\left(\frac{x}{p}\right).$$

IV. *J is an integral on* $L(\mathfrak{M}')$.

Proof. If $x \in L(\mathfrak{M}')$ and $x(M) \geqq 0$, then, choosing $p \in P_x^+$, we have $x(M)/p(M) \geqq 0$; consequently, $J(x) = \tilde{\varphi}_p(x/p) \geqq 0$. This means that J is an integral on $L(\mathfrak{M}')$.

It follows from IV that there exists a measure μ on $\mathfrak{M}'$ such that

$$J(x) = \int_{\mathfrak{M}'} x(M)d\mu \text{ for all } x \in L(\mathfrak{M}') \tag{8}$$

(see X, subsection 10, § 6).

We now denote by $\mathfrak{M}_f$ the set of all points $M \subset \mathfrak{M}'$ on each neighborhood of which the restriction of at least one of the functionals $\tilde{\varphi}_p$, $p \in P$, does not vanish; clearly, $\mathfrak{M}_f$ is a closed subset in $\mathfrak{M}'$, and $\mathfrak{M}' - \mathfrak{M}_f$ is the maximal open set on which the restrictions of all the $\tilde{\varphi}_p$, $p \in P$, equal zero. Clearly, $\mathfrak{M}_f \supset \mathfrak{M}_J$, where $\mathfrak{M}_J$ is the carrier of the integral J.

V. *If* $p \in P$, *then* $p(M) \geqq 0$ *on* $\mathfrak{M}_f$.

Proof. Let us assume that $p(M_0) < 0$ at the point $M_0 \in \mathfrak{M}_f$; then $p(M) < \frac{1}{2}p(M_0) < 0$ in some neighborhood $U = U(M_0)$ of the point M_0. Consequently, for $x \in L^+(U)$ we have $0 \leqq J(x) = \tilde{\varphi}_p(x/p) \leqq 0$, i.e. $J(x) = 0$ on $L^+(U)$ and hence on $L(U)$; from this we conclude that $\tilde{\varphi}_p(x) = J(xp) = 0$ for $x \in L(U)$. But then, for an arbitrary $q \in P$, $x \in L^+(U)$,

$$0 = \tilde{\varphi}_p(qx) = \tilde{\varphi}_q(px) \leqq \tfrac{1}{2}p(M_0)\tilde{\varphi}_q(x) \leqq 0,$$

i.e. $\tilde{\varphi}_q(x) = 0$, which contradicts the definition of $\mathfrak{M}_f$.

We shall now prove that

$$J(py) = \tilde{\varphi}_p(y) \tag{9}$$

for an arbitrary function $y(M) \in C(\mathfrak{M}_1)$, i.e. that for any such function

$$\int_{\mathfrak{M}'} p(M)y(M)d\mu = \int_{\mathfrak{M}'} y(M)d\mu_p, \tag{10}$$

where μ_p is the measure corresponding to the integral $\tilde{\varphi}_p$.

We note first of all that it will suffice to prove the equality

$$\int_{\mathfrak{M}_f} p(M)y(M)d\mu = \int_{\mathfrak{M}_f} y(M)d\mu_p, \tag{11}$$

because, by the very definition of the set $\mathfrak{M}_f$, the restrictions of both integrals to $\mathfrak{M}' - \mathfrak{M}_f$ equal zero.

We set

$$N = \{M: M \in \mathfrak{M}_f, p(M) = 0\},$$

$$A_n = \left\{M: M \in \mathfrak{M}_f,\ p(M) \geqq \frac{1}{n}\right\},\quad B_n = \left\{M: M \in \mathfrak{M}_f,\ p(M) \leqq \frac{1}{n+1}\right\};$$

since $p(M) \to 0$ as $M \to M_0$, each of the A_n is a compact set. We shall prove that N is a zero set with respect to $\tilde{\varphi}_p$. To this end, we note that on the basis of the Urysohn lemma, there exists a function $z_n'(M) \in C(\mathfrak{M}_1)$, satisfying the conditions $z_n'(M) = 0$ on A_n, $z_n'(M) = 1$ on B_n, $0 \leqq z_n'(M) \leqq 1$ on all of $\mathfrak{M}_1$. Put $z_n = z_1' \cap \cdots \cap z_n'$. Clearly, $z_n \searrow \xi_N$ on $\mathfrak{M}_f$; consequently, by V, $pz_n \searrow 0$ and $qz_n \searrow q\xi_N$ on $\mathfrak{M}_f$ for an arbitrary function $q \in P$. [Recall that ξ_A denotes the characteristic function of the set A (see subsection 5, § 6).] Therefore, setting $z = z_n$ in (7) and passing to the limit as $n \to \infty$, we obtain (see VII, subsection 7, § 6) that $\tilde{\varphi}_p(q\xi_N) = 0$ for an arbitrary function $q \in P$. In particular, $\tilde{\varphi}_p(x^*x\xi_N) = 0$ for all $x \in I$; if we substitute $x \pm y$, $x \pm iy$ for x everywhere in the last equality and take the appropriate linear combinations, we obtain that $\tilde{\varphi}_p(y^*x\xi_N) = 0$ for all $x, y \in I$; since I is dense in R, we also have $\tilde{\varphi}_p(y^*x\xi_N) = 0$ for all $y, x \in R$. But R is dense in $C_0(\mathfrak{M}_1)$, and $\tilde{\varphi}_p$ is continuous in the norm $|x| = \sup_{M \in \mathfrak{M}_1} |x(M)|$; therefore, also $\tilde{\varphi}_p(y^*x\xi_N) = 0$ for all $y, x \in C_0(\mathfrak{M}_1)$. Since the functions y^*x form a dense set in $C_0(\mathfrak{M}_1)$, $\tilde{\varphi}_p(z\xi_N) = 0$ for all $z \in C_0(\mathfrak{M}_1)$; consequently, N is a zero set with respect to $\tilde{\varphi}_p$. [In fact, if $0 \leqq y \leqq 1$, $y(M) = 1$ on $F = \{M: |x(M)| \geqq \varepsilon\}$ and $y(M) = 0$ outside an arbitrary neighborhood of the set F, then $|x - yx| < \varepsilon$.]

It follows that $\int_N x(M) d\mu_p = 0$; moreover, it is obvious that $\int_N p(M)x(M) d\mu = 0$.

Therefore, relation (11) will be proved if we show that $\int_W p(M)y(M)d\mu = \int_W y(M)d\mu_p$ where $W = \mathfrak{M}_f - N$, i.e. that

$$J(py\xi_W) = \tilde{\varphi}_p(y\xi_W).$$

Clearly, it suffices to restrict ourselves to the case of functions $y(M) \geqq 0$. We set $u_n = 1 - z_n$; then $pyu_n = 0$ on B_n, and since $p(M) > 1/(n+1) > 0$ on $\mathfrak{M}_f - B_n$, formula (6) is applicable to the functions $x = pyu_n$, i.e.

$$J(pyu_n) = \tilde{\varphi}_p(yu_n).$$

But $pyu_n \nearrow py\xi_W$, $yu_n \nearrow y\xi_W$ on $\mathfrak{M}_f$; therefore, passing to the limit, we obtain that $J(py\xi_W) = \tilde{\varphi}_p(y\xi_W)$. By the same token, formulas (9) and (10) are proved.

Setting $y(M) \equiv 1$ in (10) and taking into consideration that $p(M) = |p(M)|$ on $\mathfrak{M}_f$, we obtain $\int_{\mathfrak{M}'} |p(M)| d\mu = \int_{\mathfrak{M}'} p(M) d\mu = \tilde{\varphi}_p(1) < \infty$, i.e. $p(M) \in L^1(\mu)$.

We denote by H_f the Hilbert space generated by the elements in I with inner product $(x, y) = f(y^*x)$, and by H_f' its closed subspace generated by linear combinations of elements $p \in P$. Setting $y = q^* \in P$ in (10) we obtain

$$(p, q) = f(pq^*) = \varphi_p(q^*) = \int p(M)\overline{q(M)} d\mu;$$

consequently, the correspondence $x \to x(M)$ is an isometric mapping of the space H_f' into the space $L^2(\mu)$.

We shall now assume that I^2 is dense in I relative to the inner product in H_f.

This means that the elements of the form y^*x, where $y, x \in I$, form a dense set in H_f. But in virtue of the identity

$$y^*x = \tfrac{1}{4}[(x+y)^*(x+y)-(x-y)^*(x-y)+i(x+iy)^*(x+iy)-i(x-iy)^*(x-iy)],$$

$y^*x \in H_f'$. Therefore, $H_f' = H_f$.

We have thus proved the following theorem.

THEOREM 5. *Suppose R is a completely symmetric commutative algebra without identity. Then for an arbitrary positive real functional f, defined on the ideal I, dense in R, there exists a unique measure μ on $\mathfrak{M}'$ which possesses the following properties:*

1) *for every element $p \in P$,*

$$p(M) \in L^1(\mu) \text{ and } f(px) = \int x(M)p(M)d\mu;$$

2) *the mapping $p \to p(M)$ is extendible to an isometric mapping of the Hilbert space H_f' into the space $L^2(\mu)$. If I^2 is dense in I relative to the inner product in H_f, then $p \to p(M)$ is extendible to an isometric mapping of all of H_f into $L^2(\mu)$. Here, P denotes the set of all elements $p \in I$ which are positive with respect to f, H_f is the Hilbert space generated by the elements $x \in I$ with inner product $(x, y) = f(xy^*)$, and H_f' is the subspace in H_f generated by the elements $x \in P$.*

§ 21. Generalized Schur lemma

1. Canonical decomposition of an operator.

I. *If A is a positive definite self-adjoint operator, then there exists a unique positive definite self-adjoint operator H such that $H^2 = A$.*

Proof. Suppose $P(\lambda)$ is the spectral function of the operator A. Since A is a positive definite operator, $P(\lambda) = 0$ for $\lambda < 0$. We denote the set of all vectors ξ for which $\int_0^\infty \lambda d|P(\lambda)\xi|^2 < \infty$ by $\mathfrak{D}_H$ and set

$$H\xi = \int_0^\infty \sqrt{\lambda} dP(\lambda)\xi$$

for vectors ξ of this sort. By direct computation, we can verify that H is a positive definite self-adjoint operator and that $H^2 = A$. The uniqueness of the operator H follows from the uniqueness of the spectral function $P(\lambda)$.

We use the notation $H = \sqrt{A}$ for the operator H. It follows from the construction of the operator H that it commutes with every operator in $\mathfrak{B}(\mathfrak{H})$ which commutes with A.

II (VON NEUMANN [2]). *Every closed linear operator A from $\mathfrak{H}_1$ into $\mathfrak{H}_2$ with domain $\mathfrak{D}_A$, dense in $\mathfrak{H}_1$, is representable in the form*

$$A = UH, \tag{1}$$

where H is a self-adjoint operator in $\mathfrak{H}_1$ *with domain* $\mathfrak{D}_H = \mathfrak{D}_A$, and kernel Ker H = Ker A, *and U is a partially isometric operator with initial domain* $\overline{\mathfrak{R}}_{A^*}$ *and terminal domain* $\overline{\mathfrak{R}}_A$. *The operators A, U and H are defined uniquely by these conditions.*

Proof. In virtue of VI, subsection 9, § 5, A^*A is a positive definite self-adjoint operator; consequently, in virtue of I, $H = \sqrt{A^*A}$ exists. We denote by H_1 and A_1 the restrictions of the operators H and A, respectively, to $\mathfrak{D}_{H^2} = \mathfrak{D}_{A^*A}$. Then for $\xi \in \mathfrak{D}_{H^2}$

$$|H_1\xi|^2 = (H_1\xi, H_1\xi) = (H_1^2\xi, \xi) = (H^2\xi, \xi) = (A^*A\xi, \xi) = (A\xi, A\xi) = |A\xi|^2 = |A_1\xi|^2.$$

Consequently,

$$|H_1\xi| = |A_1\xi| \text{ for all } \xi \in \mathfrak{D}_{H_1} = \mathfrak{D}_{A_1}.$$

From this we conclude that also $\mathfrak{D}_{\tilde{H}_1} = \mathfrak{D}_{\tilde{A}_1}$ and

$$|\tilde{H}_1\xi| = |\tilde{A}_1\xi| \text{ for all } \xi \in \mathfrak{D}_{\tilde{H}_1} = \mathfrak{D}_{\tilde{A}_1}. \tag{2}$$

[The operators A_1 and H_1 admit closures $\tilde{A}_1$, $\tilde{H}_1$ respectively since they are restrictions of the closed operators A and H.] We shall prove that $\tilde{H}_1 = H$ and $\tilde{A}_1 = A$, i.e. that $\mathfrak{B}_{\tilde{A}_1} = \mathfrak{B}_A$ and $\mathfrak{B}_{\tilde{H}_1} = \mathfrak{B}_H$, where $\mathfrak{B}_A$ denotes the graph of the operator A (see subsection 7, § 5). Clearly, $\mathfrak{B}_{\tilde{A}_1} \subset \mathfrak{B}_A$; if $\mathfrak{B}_{\tilde{A}_1} \neq \mathfrak{B}_A$, then there exists a nonzero vector $\langle \xi, A\xi \rangle \in \mathfrak{B}_A$, orthogonal to $\mathfrak{B}_{A_1}$, i.e.

$$(\langle \xi, A\xi \rangle, \langle \eta, A_1\eta \rangle) = 0 \text{ for all } \eta \in \mathfrak{D}_{A_1}.$$

This means that

$$0 = (\xi, \eta) + (A\xi, A_1\eta) = (\xi, \eta) + (A\xi, A\eta) = (\xi, \eta) + (\xi, A^*A\eta) = (\xi, (1 + A^*A)\eta).$$

Since the domain of the operator $1 + A^*A$ is all of $\mathfrak{H}_1$ (see VI, subsection 9, § 5), $\xi \perp \mathfrak{H}_1$, and hence $\xi = 0$; consequently, $\tilde{A}_1 = A$, and analogously $\tilde{H}_1 = H$. Therefore, (2) assumes the form

$$|H\xi| = |A\xi| \text{ for all } \xi \in \mathfrak{D}_H = \mathfrak{D}_A. \tag{3}$$

We now set

$$\begin{cases} U'H\xi = A\xi \text{ for all } \xi \in \mathfrak{D}_H, \text{ and} \\ U'\eta = 0 \text{ for } \eta \perp \mathfrak{R}_H. \end{cases} \tag{4}$$

In virtue of (3), U' is a partially isometric operator with initial domain $\mathfrak{R}_H$ and terminal domain $\mathfrak{R}_A$; setting $U = \tilde{U}'$, we obtain a partially isometric operator U with initial domain $\overline{\mathfrak{R}}_H$ and terminal domain $\overline{\mathfrak{R}}_A$. It follows directly from (4) that $A = UH$.

We shall prove that $\overline{\mathfrak{R}}_H = \overline{\mathfrak{R}}_{A^*}$ or, what amounts to the same thing, that

$\mathfrak{H}_1 \ominus \mathfrak{R}_H = \mathfrak{H}_1 \ominus \mathfrak{R}_{A^*}$. As is easily seen, $\mathfrak{H}_1 \ominus \mathfrak{R}_H$ and $\mathfrak{H}_1 \ominus \mathfrak{R}_{A^*}$ consist of those and only those vectors $\xi \in \mathfrak{H}$ on which $H_1 \xi = 0$ and $A\xi = 0$, respectively; and in virtue of (3), these sets coincide.

It remains to prove the uniqueness of the operators U and H. But if we are given a decomposition of the form (1), then

$$A^* = HU^* \text{ and } A^*A = HU^*UH = HP_{\mathfrak{R}_H}H = H^2;$$

consequently, H, and hence also U, are defined uniquely.

Formula (1) is called the *canonical decomposition of the operator A.*

An operator A from $\mathfrak{H}_1$ into $\mathfrak{H}_2$ is said to be *non-singular* if 1) A is closed; 2) $\mathfrak{D}_A$ and $\mathfrak{R}_{A^*}$ are dense in $\mathfrak{H}_1$; 3) $\mathfrak{R}_A$ is dense in $\mathfrak{H}_2$.

Thus, in the case of a non-singular operator A, $\overline{\mathfrak{R}}_A = \mathfrak{H}_2$, $\overline{\mathfrak{R}}_{A^*} = \mathfrak{H}_1$, and hence the operator U in (1) maps $\mathfrak{H}_1$ isometrically onto $\mathfrak{H}_2$.

Consequently,

III. *Every non-singular operator A from $\mathfrak{H}_1$ into $\mathfrak{H}_2$ can be uniquely represented in the form*

$$A = UH,$$

where H is a positive definite self-adjoint operator in $\mathfrak{H}_1$, and U maps $\mathfrak{H}_1$ isometrically onto $\mathfrak{H}_2$.

2. Fundamental theorem. In the representation theory of algebras it happens that we must use the following theorem, which is a generalization of the well-known Schur lemma on finite-dimensional representations (see, for example, PONTRYAGIN [4]). (Everywhere in this subsection we shall be dealing with representations of the same symmetric algebra R.)

THEOREM 1. *Suppose $x \to A_x$, $x \to B_x$ are representations in the spaces $\mathfrak{H}$ and $\mathfrak{H}'$ and that there exists a non-singular operator T from $\mathfrak{H}$ into $\mathfrak{H}'$ such that*

$$B_x T \subset TA_x \tag{1}$$

for all x in R. Then these representations are equivalent.

Proof. Applying involution to both members of relation (1) and taking relations c), subsection 9, § 5 and III, b), subsection 10, § 5 into account, we obtain

$$T^*B_{x^*} = (B_x T)^* \supset (TA_x)^* \supset A_{x^*}T^*;$$

consequently,

$$T^*B_{x^*} \supset A_{x^*}T^*. \tag{2}$$

We substitute x^* everywhere for x; then (2) assumes the form

$$T^*B_x \supset A_x T^*. \tag{3}$$

Operating on the left in both members of relation (1) by T^* and using relation (3), we obtain

$$T^*TA_x \supset T^*B_xT \supset A_xT^*T;$$

consequently,

$$T^*TA_x \supset A_xT^*T,$$

i.e. the operator T^*T commutes with A_x.

On the other hand, according to the Von Neumann theorem (see III, subsection 1), $T = UH$, where $H^2 = T^*T$, and the operator U maps $\mathfrak{H}$ isometrically onto $\mathfrak{H}'$. We now substitute the expression UH for T in relation (1); then (1) assumes the form

$$B_xUH \subset UHA_x.$$

We apply both members of this last relation to the element ξ from the domain of the operator H. We then obtain

$$B_xUH\xi = UHA_x\xi. \tag{4}$$

The operator A_x commutes with T^*T and, consequently, also with $H = \sqrt{T^*T}$. Therefore $HA_x\xi = A_xH\xi$ and equality (4) can be rewritten in the form

$$B_xUH\xi = UA_xH\xi.$$

Thus, $B_xU = UA_x$ on the domain of the operator H. The latter is dense in $\mathfrak{H}$ because T, by assumption, is a non-singular operator. In virtue of the continuity of the operators A_x, B_x, U, it follows from this that the equality $B_xU = UA_x$ holds in the entire space $\mathfrak{H}$, i.e. the representations A_x and B_x are equivalent.

COROLLARY 1. *Suppose $x \to A_x$, $x \to B_x$ are irreducible representations in the spaces $\mathfrak{H}$ and $\mathfrak{H}'$ respectively and that T is a closed linear operator from $\mathfrak{H}$ into $\mathfrak{H}'$, satisfying the condition*

$$B_xT \subset TA_x. \tag{5}$$

Then either $T = 0$, or the representations are equivalent and $T = \rho U$, where ρ is a positive number.

Proof. Suppose $T \neq 0$; consequently, in the domain of the operator T there exists a vector $\xi_0 \neq 0$ such that $T\xi_0 \neq 0$. It follows from condition (5) that every element of the form $A_x\xi_0$ also appears in the domain of the operator T. Since the representation $x \to A_x$ is irreducible, the set of all vectors of the form $A_x\xi_0$ is dense in $\mathfrak{H}$. Consequently, the domain of the operator T is dense in $\mathfrak{H}$.

Further, $TA_x\xi_0 = B_xT\xi_0$; consequently, the range of the operator T contains all the vectors $B_xT\xi_0$. Since $T\xi_0 \neq 0$ and the representation $x \to B_x$ is irreducible, these vectors form a dense set in $\mathfrak{H}'$. Consequently, the range of the operator T is dense in $\mathfrak{H}'$.

We now apply involution to both members of the relation (5) and substitute x^* for x. We then obtain the relation

$$A_x T^* \subset T^* B_x,$$

analogous to (5), in which the role of T is played by the operator T^*. Since $T \neq 0$, also $T^* \neq 0$. Therefore the preceding discussion is applicable to T^*. Thus, the domain and range of the operator T^* are dense in $\mathfrak{H}'$ and $\mathfrak{H}$ respectively. Therefore T is a non-singular operator.

According to Theorem 1, it follows from this that the representations $x \to A_x$ and $x \to B_x$ are equivalent. Then the operator $H = \sqrt{T^*T}$ commutes with all the operators A_x of the irreducible representation $x \to A_x$; consequently, it is a multiple of the identity operator. Suppose $H = \rho 1$; then $\rho > 0$ and $T = UH = \rho U$.

3. Application to direct sums of pairwise non-equivalent representations. We now apply these results to the study of direct sums of representations.

THEOREM 2. *Suppose the representation $x \to A_x$ in the space $\mathfrak{H}$ is a direct sum of irreducible and pairwise non-equivalent representations $x \to A_x^{(\alpha)}$, $\alpha \in \mathfrak{A}$, in the spaces $\mathfrak{H}_\alpha$ respectively. Then every bounded operator B in the space $\mathfrak{H}$ which commutes with all the operators A_x has the form*

$$B\{\xi_\alpha\} = \{\lambda_\alpha \xi_\alpha\}, \tag{1}$$

where λ_α is a scalar.

Proof. According to subsection 15, § 5, every bounded operator B in the space $\mathfrak{H}$ is defined by a matrix $||B_{\alpha\alpha_1}||$, where the $B_{\alpha\alpha_1}$ are operators from $\mathfrak{H}_{\alpha_1}$ into $\mathfrak{H}_\alpha$. In particular, the operator A_x is defined by the diagonal matrix $A_x = ||\delta_{\alpha\alpha_1} A_x^{(\alpha)}||$, where

$$\delta_{\alpha\alpha_1} = \begin{cases} 1 \text{ for } \alpha = \alpha_1, \\ 0 \text{ for } \alpha \neq \alpha_1. \end{cases}$$

Therefore, the condition that the operators B and A_x commute yields

$$A_x^{(\alpha)} B_{\alpha\alpha_1} = B_{\alpha\alpha_1} A_x^{(\alpha_1)}. \tag{2}$$

By assumption, for $\alpha \neq \alpha_1$ the representations $x \to A_x^{(\alpha)}$ and $x \to A_x^{(\alpha_1)}$ are not equivalent. On the other hand, equality (2) means that the operator $T = B_{\alpha\alpha_1}$ satisfies the conditions of Corollary 1, subsection 2. Therefore, $B_{\alpha\alpha_1} = 0$ for $\alpha \neq \alpha_1$. For $\alpha = \alpha_1$, equality (2) means that the operator $B_{\alpha\alpha}$ commutes with all the operators $A_x^{(\alpha)}$ of the irreducible representation $x \to A_x^{(\alpha)}$. Consequently, the operator $B_{\alpha\alpha}$ is a multiple of the identity, i.e. $B_{\alpha\alpha} = \lambda_\alpha 1_\alpha$, where 1_α is the identity operator in the space $\mathfrak{H}_\alpha$. It follows that

$$B\{\xi_\alpha\} = \{\lambda_\alpha \xi_\alpha\}.$$

Below, in Corollaries 2 and 3, the representation $x \to A_x$ in the space $\mathfrak{H}$ is the direct sum of representations $x \to A_x^{(\alpha)}$, $\alpha \in \mathfrak{A}$, in the spaces $\mathfrak{H}_\alpha$.

COROLLARY 2. *Every closed invariant subspace $\mathfrak{M}$ in $\mathfrak{H}$ is the set of all vectors $\xi = \{\xi_\alpha\}$ in $\mathfrak{H}$ which satisfy the condition*

$$\xi_\alpha = 0 \text{ for all } \alpha \in \mathfrak{A}_1,$$

where $\mathfrak{A}_1$ is some subset of the set $\mathfrak{A}$.

In fact, suppose P is a projection operator on $\mathfrak{M}$. Since $\mathfrak{M}$ is an invariant subspace, P commutes with all the operators A_x. According to Theorem 2, $P\{\xi_\alpha\} = \{\lambda_\alpha \xi_\alpha\}$. Since $P^2 = P$, $\lambda_\alpha^2 = \lambda_\alpha$, from which it follows that either $\lambda_\alpha = 0$ or $\lambda_\alpha = 1$. Suppose $\mathfrak{A}_1$ is the set of all indices α for which $\lambda_\alpha = 0$; $\mathfrak{M}$ is the set of those and only those vectors $\{\xi_\alpha\}$ in $\mathfrak{H}$ for which $P\{\xi_\alpha\} = \{\xi_\alpha\}$, i.e. for which

$$\lambda_\alpha \xi_\alpha = \xi_\alpha. \tag{3}$$

If $\alpha \in \mathfrak{A}_1$, then $\lambda_\alpha = 0$, so that condition (3) yields $\xi_\alpha = 0$. But if $\alpha \notin \mathfrak{A}_1$, then $\lambda_\alpha = 1$, and no condition is placed on the corresponding component ξ_α.

COROLLARY 3. *If $\mathfrak{A}$ is denumerable, then every vector $\xi = \{\xi_\alpha\}$ in $\mathfrak{H}$ which is such that $\xi_\alpha \neq 0$ for all $\alpha \in \mathfrak{A}$ is a cyclic vector in $\mathfrak{H}$.*

In fact, suppose $\mathfrak{M}$ is the closed hull of the set of all vectors $A_x\xi$, $x \in R$; $\mathfrak{M}$ is an invariant subspace in $\mathfrak{H}$. According to Corollary 2, it coincides with $\mathfrak{H}$ because $\mathfrak{A}_1$ is the void set.

4. Application to representations which are multiples of a given irreducible representation. Up to this point, we considered direct sums of pairwise non-equivalent representations. We shall now consider another limiting case.

The representation $x \to A_x$ is said to be *a multiple of a given irreducible representation $x \to A_x^{(0)}$* if $x \to A_x$ is the direct sum of representations which are equivalent to the same representation $x \to A_x^{(0)}$.

THEOREM 3. *Every irreducible part of the representation $x \to A_x$ which is a multiple of a given irreducible representation $x \to A_x^{(0)}$ is equivalent to this irreducible representation.*

Proof. Suppose $x \to A'_x$ is the irreducible part of the representation $x \to A_x$ and that $\mathfrak{M}$ is the closed invariant subspace of the space $\mathfrak{H}$ in which this irreducible part is being considered. If $\mathfrak{M} \neq (0)$, then in $\mathfrak{M}$ there exists a vector $\xi^0 = \{\xi_\alpha^0\} \neq 0$; consequently, at least one of the components ξ_α^0 of this vector is nonzero. Suppose this component is $\xi_{\alpha_0}^0$. For brevity, we set $\mathfrak{H}_0 = \mathfrak{H}_{\alpha_0}$, $H = \sum_{\alpha \neq \alpha_0} \oplus \mathfrak{H}_\alpha$. Then the space $\mathfrak{H} = \sum_\alpha \oplus \mathfrak{H}_\alpha$ can be considered as the direct sum of the spaces $\mathfrak{H}_0$ and H; consequently, every element $\xi \in \mathfrak{H}$ can be considered as the pair $\xi = \langle \eta, \zeta \rangle$, where $\eta \in \mathfrak{H}_0$ and $\zeta \in H$. In virtue of our choice of index α_0 there exists at least one pair

$\xi = \langle \eta, \zeta \rangle$, belonging to the space $\mathfrak{M}$, for which $\eta \neq 0$. We set for brevity $B_x = A_x^{(\alpha_0)}$ and denote by $x \to C_x$ the direct sum of all the representations $x \to A_x^{(\alpha)}$, $\alpha \neq \alpha_0$. Then the representation $x \to A_x$ can be considered as the direct sum of the representations $x \to B_x$ and $x \to C_x$. In other words,

$$A_x \langle \eta, \zeta \rangle = \langle B_x \eta, C_x \zeta \rangle.$$

We denote by H' the set of all elements ζ such that $\langle \eta, \zeta \rangle \in \mathfrak{M}$ for some η, and its closure by $\tilde{H}'$. $\tilde{H}'$ is invariant with respect to all the operators C_x. In fact, since $\mathfrak{M}$ is an invariant subspace in $\mathfrak{H}$, it follows from $\langle \eta, \zeta \rangle \in \mathfrak{M}$ that also $\langle B_x \eta, C_x \zeta \rangle \in \mathfrak{M}$. In other words, it follows from $\zeta \in H'$ that also $C_x \zeta \in H'$. Therefore $\tilde{H}'$ is an invariant subspace in H.

We now consider the elements of the subspace $\mathfrak{M}$ which are of the form $\langle \eta, 0 \rangle$. They form an invariant subspace in $\mathfrak{M}$ with respect to the representation $x \to A'_x$. Since the representation $x \to A'_x$ is irreducible, this invariant subspace is either $\mathfrak{M}$ or (0). In the first case, $\mathfrak{M}$ consists of elements of the form $\xi = \langle \eta, 0 \rangle = \langle \xi_{\alpha_0}, 0 \rangle$ and, consequently, coincides with $\mathfrak{H}_{\alpha_0}$. Therefore, the representation $x \to A'_x$ coincides with the representation $x \to A_x^{(\alpha_0)}$, and the theorem is proved in this case.

In the second case, the element $\langle \eta, 0 \rangle$ can belong to the subspace $\mathfrak{M}$ only when $\eta = 0$. Thus,

$$\langle \eta, \zeta \rangle \in \mathfrak{M}, \ \zeta = 0 \text{ implies } \eta = 0. \tag{1}$$

Analogously, the set of elements of the subspace $\mathfrak{M}$ of the form $\langle 0, \zeta \rangle$ forms an invariant subspace in $\mathfrak{M}$; consequently, it also either coincides with $\mathfrak{M}$ or $= (0)$. The first possibility is excluded in view of our choice of α_0. Therefore, the second possibility remains, and

$$\langle \eta, \zeta \rangle \in \mathfrak{M}, \ \eta = 0 \text{ implies } \zeta = 0. \tag{2}$$

Condition (2) means that the subspace $\mathfrak{M}$ can be considered as the graph of a linear operator T from $\mathfrak{H}_0$ into $\tilde{H}'$ (see subsection 7, § 5). In other words, for $\langle \eta, \zeta \rangle \in \mathfrak{M}$ we set $\zeta = T\eta$. Since $\mathfrak{M}$ is closed, the operator T is closed.

Suppose $\langle \eta_0, \zeta_0 \rangle$ is an element in $\mathfrak{M}$ such that $\eta_0 \neq 0$. Then also $\langle B_x \eta_0, C_x \zeta_0 \rangle \in \mathfrak{M}$; consequently, all the vectors $B_x \eta_0$ belong to the domain of the operator T. Since the representation $x \to B_x$ is irreducible, the set of all vectors $B_x \eta_0$ is dense in $\mathfrak{H}_0$. Consequently, the domain of the operator T is dense in $\mathfrak{H}_0$. Further, by the definition of the space H', the range of the operator T coincides with H' and, consequently, it is dense in $\tilde{H}'$.

Finally, condition (1) means that $T\eta = 0$ only for $\eta = 0$ and hence $\tilde{\mathfrak{R}}_{T^*} = \mathfrak{H}_0$. Thus, T is a non-singular operator from $\mathfrak{H}_0$ into $\tilde{H}'$.

This operator satisfies the condition $C_x T \subset TB_x$. In fact, the elements of the subspace $\mathfrak{M}$ have the form $\langle \eta, T\eta \rangle$. Since $\mathfrak{M}$ is invariant, also $\langle B_x \eta, C_x T\eta \rangle \in \mathfrak{M}$ and, consequently, $TB_x \eta = C_x T\eta$ for an arbitrary element η in the domain of the operator T. And this means that

$$C_x T \subset T B_x .$$

According to Theorem 1, it follows from this that the representations $x \to B_x$ and $x \to C_x$ are equivalent. Moreover, the operator $H = \sqrt{T^*T}$ commutes with all the operators of the irreducible representation $x \to B_x$; consequently, $H = \rho 1$, $\rho > 0$. It follows that $T = \rho W$, where W is an isometric mapping of $\mathfrak{H}_0$ onto $\tilde{H}'$. Thus, every element of the subspace $\mathfrak{M}$ has the form $\xi = \langle \eta, \rho W\eta \rangle$, where η runs through $\mathfrak{H}_0$ and $A_x \xi = A'_x \langle \eta, \rho W\eta \rangle = \langle B_x \eta, \rho W B_x \eta \rangle$. To each such element $\xi = \langle \eta, \rho W\eta \rangle$ we assign the element $\eta' = \sqrt{1+\rho^2}\eta$. The correspondence $\xi \to \eta'$ is an isometric mapping of the subspace $\mathfrak{M}$ onto $\mathfrak{H}_0 = \mathfrak{H}_{\alpha_0}$. Moreover, the element $A'_x \xi = \langle B_x \eta, \rho W B_x \eta \rangle$ maps into $\sqrt{1+\rho^2} B_x \eta = B_x \eta'$; consequently, the operator A'_x maps into B_x. This means that the representations $x \to A'_x$ and $x \to B_x = A_x^{(0)}$ are equivalent, and the theorem is proved.

COROLLARY 4. *Suppose the representations $x \to A_x$ and $x \to B_x$ are multiples of the irreducible representations $x \to A_x^{(0)}$ and $x \to B_x^{(0)}$, respectively. If the multiple representations $x \to A_x$ and $x \to B_x$ are equivalent, then the irreducible representations $x \to A_x^{(0)}$ and $x \to B_x^{(0)}$ are also equivalent.*

Proof. Suppose $\mathfrak{H}$, $\mathfrak{H}_1$, $\mathfrak{H}^{(0)}$, $\mathfrak{H}_1^{(0)}$ are the spaces of the representations $x \to A_x$, $x \to B_x$, $x \to A_x^{(0)}$, $x \to B_x^{(0)}$, respectively. The space $\mathfrak{H}^{(0)}$ may be considered as a closed invariant subspace of the space $\mathfrak{H}$, and the representation $x \to A_x^{(0)}$ as the part of the representation $x \to A_x$ in this space. By assumption, there exists an isometric mapping of the space $\mathfrak{H}$ onto $\mathfrak{H}_1$, under which the operator A_x maps into B_x. This mapping maps $\mathfrak{H}^{(0)}$ into an invariant subspace $\mathfrak{N}$ of the space $\mathfrak{H}_1$. Suppose $x \to C_x$ is the part of the representation $x \to B_x$ in this invariant subspace. The representation $x \to C_x$, being equivalent to the irreducible representation $x \to A_x^{(0)}$, is irreducible.

Thus, $x \to C_x$ is an irreducible part of the representation $x \to B_x^{(0)}$ which is a multiple of the irreducible representation $x \to B_x^{(0)}$. According to Theorem 3, the representations $x \to B_x^{(0)}$ and $x \to C_x$ are equivalent. Consequently, the representations $x \to B_x^{(0)}$ and $x \to A_x^{(0)}$ are also equivalent.

§ 22. Some representations of the algebra $\mathfrak{B}(\mathfrak{H})$

One of the important problems in the theory of symmetric algebras is the description, to within equivalence, of all the representations of a given algebra. This problem is completely solved only for certain particular cases (see, e.g., GELFAND and NAIMARK [2–5, 7, 8]). This problem has not yet been solved for the algebra $\mathfrak{B}(\mathfrak{H})$ of all bounded operators in the Hilbert space $\mathfrak{H}$; there are only results of a particular nature which will be discussed in this section.

1. Ideals in the algebra $\mathfrak{B}(\mathfrak{H})$. Every representation of the algebra $\mathfrak{B}(\mathfrak{H})$ is a symmetric homomorphism of $\mathfrak{B}(\mathfrak{H})$ and, consequently, it is a symmetric isomorphism of the quotient algebra $\mathfrak{B}(\mathfrak{H})/I$, where I is the kernel of this homomorphism (see subsection 1, § 10). Since every representation of the algebra $\mathfrak{B}(\mathfrak{H})$ is continuous (Theorem 1, subsection 3, § 17), I is a closed two-sided symmetric ideal in $\mathfrak{B}(\mathfrak{H})$.

One such ideal is the set of all compact operators in $\mathfrak{H}$. In fact: a) the sum of two compact operators is a compact operator; b) the product of a compact operator with a bounded operator is a compact operator; c) the limit, relative to the norm in $\mathfrak{B}(\mathfrak{H})$, of a sequence of compact operators is a compact operator.

We denote this ideal by I_0; it is symmetric because if A is a compact operator then A^* is also a compact operator.

THEOREM 1. *If $\mathfrak{H}$ is separable, then I_0 is the only closed two-sided ideal in $\mathfrak{B}(\mathfrak{H})$ which is different from* (0).

The proof is based on the following propositions. Suppose S is an arbitrary subset in $\mathfrak{B}(\mathfrak{H})$; we denote the set of all projection operators contained in S by S^P. Then:

I. *If I_l is a left ideal in $\mathfrak{B}(\mathfrak{H})$ and $I_l^P = (0)$, then $I_l = (0)$.*

Proof. Suppose $A \in I_l$, $A \neq 0$; we set $H = A^*A$; then also $H \in I_l$ and $H \neq 0$. Suppose $P(\lambda)$ is the spectral function of the operator H. Since this operator is positive definite, there exists an interval $\Delta = (\alpha, \beta)$ $(0 < \alpha < \beta)$ such that $P(\Delta) \neq 0$. Suppose $\mathfrak{M}_\Delta$ is the subspace onto which $P(\Delta)$ projects. The operator H, considered as an operator in the space $\mathfrak{M}_\Delta$, will be denoted by H_Δ; H_Δ has a bounded inverse in $\mathfrak{M}_\Delta$. We set $B = H_\Delta^{-1}P(\Delta)$; then B is a bounded operator. Consequently, $BH \in I_l$. On the other hand, $BH = P(\Delta)$ since $BH = H_\Delta^{-1}P(\Delta)H = H_\Delta^{-1}P(\Delta)HP(\Delta) = H_\Delta^{-1}H_\Delta P(\Delta) = P(\Delta)$; consequently, the ideal I_l contains the projection operator $P(\Delta) \neq 0$. Hence, if $I_l^P = (0)$, then I_l contains no operators $A \neq 0$.

The *dimension of the projection operator P* is the dimension of the subspace on which it projects.

II. *If I is a two-sided ideal in $\mathfrak{B}(\mathfrak{H})$ and P is a projection operator contained in I, then every projection operator Q having the same dimension as P also belongs to I.*

In fact, if the dimensions of the spaces $P\mathfrak{H}$ and $Q\mathfrak{H}$ coincide, then there exists a partially isometric operator U with initial domain $P\mathfrak{H}$ and terminal domain $Q\mathfrak{H}$. This means that $U^*U = P$, $UU^* = Q$ (see subsection 14, § 5). Since I is a two-sided ideal,

$$Q = Q^2 = UU^*UU^* = UPU^* \in I.$$

III. *If $\mathfrak{H}$ is separable, then a two-sided ideal in $\mathfrak{B}(\mathfrak{H})$ does not contain infinite-dimensional projection operators.*

In fact, an infinite-dimensional projection operator P and the identity operator 1 have the same dimension in this case. Since the latter cannot belong to the ideal, the operator P also cannot belong to it, in virtue of II.

IV. *If $P \in I$ and $Q < P$, where P and Q are projection operators, then $Q \in I$.*

Namely, the condition $Q < P$ means that $Q = QP = PQ$. Consequently, $Q \in I$, because I is an ideal.

It follows from Propositions I and IV that the two-sided ideal $I \neq (0)$ contains at least one one-dimensional projection operator. In virtue of II, it then contains all one-dimensional, and hence also all finite-dimensional, projection operators.

Taking Proposition III into consideration, we see that for $I \neq (0)$ and separable $\mathfrak{H}$, I^P consists of all finite-dimensional projection operators.

We shall now prove Theorem 1. Suppose $\mathfrak{H}$ is separable and that I is a closed two-sided ideal, which is different from (0), in $\mathfrak{B}(\mathfrak{H})$; suppose A is a compact operator. We shall prove that $A \in I$. The operator A can be represented in the form $A = H_1 + iH_2$, where H_1 and H_2 are compact Hermitian operators. Therefore it suffices to prove the assertion for the case when A is a Hermitian operator. But in this case the spectral decomposition of the operator A has the form

$$A = \sum_{k=1}^{\infty} \lambda_k P_k, \tag{1}$$

where the P_k are finite-dimensional projection operators and $\lambda_k \to 0$ as $k \to \infty$ (see VI, subsection 4, § 17). Moreover, the series in the right member of (1) converges in the norm in $\mathfrak{B}(\mathfrak{H})$. Since every finite sum $\sum_{k=1}^{n} \lambda_k P_k \in I$, then also $A \in I$. Thus, $I_0 \subset I$.

We shall now prove the reverse inclusion. Suppose $A \in I$. We shall prove that $A \in I_0$. Suppose $A = UH$ is the canonical decomposition of the operator A. Then $A^* = HU^* = U^*AU^*$; consequently, $A^* \in I$ also. Therefore $H_1 = (A + A^*)/2$ and $H_2 = (A - A^*)/2i$ are also elements in I, and we can thus assume that A is a Hermitian operator.

We denote the spectral function of the operator A by $P(\lambda)$. Repeating the argument carried out in the proof of Propositon I, we can verify that $P(\Delta) \in I$ for an arbitrary closed interval which does not contain zero. In virtue of III, the operator $P(\Delta)$ is finite-dimensional and, consequently, $P(\Delta) \in I_0$. Since

$$A = \int \lambda dP(\lambda) = \lim \sum_{k=1}^{n} \lambda_k P(\Delta_k),\ 0 \notin \Delta_k$$

in the sense of convergence in norm in $\mathfrak{B}(\mathfrak{H})$, we also have $A \in I_0$.

Thus $I_0 = I$, and the theorem is proved.

Applying these same arguments, we can prove the following proposition. (Here we shall no longer assume that $\mathfrak{H}$ is separable.)

V. *The algebra I_0 does not contain closed two-sided ideals which are different from* (0).

In fact, if I is such an ideal then Proposition I is applicable to I because $B \in I_0$ for $H \in I_0$. Proposition II is also applicable because $U \in I_0$ for $P \in I$. Finally, Proposition IV can also be applied because if $P \in I_0$ and $Q < P$, then $Q \in I_0$. Therefore, reasoning as above, we conclude that if $I \neq (0)$, then I contains all the finite-dimensional projection operators and hence coincides with I_0.

COROLLARY. *If $\mathfrak{H}$ is separable, then every representation of the algebra $\mathfrak{B}(\mathfrak{H})$ is a symmetric isomorphism of the algebra $\mathfrak{B}(\mathfrak{H})$ itself or of the quotient algebra $\mathfrak{B}(\mathfrak{H})/I_0$.*

In fact, this follows from the fact that the only closed two-sided ideals in $\mathfrak{B}(\mathfrak{H})$ are (0) and I_0.

Among the two-sided ideals of the algebra $\mathfrak{B}(\mathfrak{H})$ an especially important role is played by the symmetrically normable ideals. A two-sided ideal I of the algebra $\mathfrak{B}(\mathfrak{H})$ is said to be *symmetrically normable* if there exists a norm $|X|_s$ on it with respect to which it is a Banach space satisfying the following conditions:

a) $|AXB|_s \leqq |A|\,|X|_s|B|$ for $X \in I$, $A, B \in \mathfrak{B}(\mathfrak{H})$,
b) $|X|_s = |X|$ for a one-dimensional $X \in I$.

A norm $|X|_s$ satisfying these conditions is called a *symmetric* norm.

An example of a symmetric norm is $|A|_p = (\sum_{j=1}^{\infty} s_j^p(A))^{1/p}$, where the $s_j(A)$ are the characteristic values of the operator $\sqrt{A^*A}$, $A \in I_0$; the corresponding symmetrically normable ideal is denoted by $\mathfrak{S}_p$; it consists of all $A \in I_0$ for which $|A|_p < \infty$. The theory of symmetrically normable ideals was first discussed in detail in the monograph by SCHATTEN [1]; it arose in connection with the theory of tensor products of normed spaces, developed by VON NEUMANN and SCHATTEN [1]. In the works of Gokhberg, Krein and Mityagin, the theory of symmetrically normable ideals received further development; for a detailed discussion of this theory, see Chapter III of the monograph by GOKHBERG and KREIN [1].

2. The algebra I_0 and its representations. The set I_0 of all compact operators in the space $\mathfrak{H}$ is, obviously, a Banach symmetric subalgebra (without identity) of the algebra $\mathfrak{B}(\mathfrak{H})$. We shall find all representations of the algebra I_0.

Suppose $\{\varphi_\alpha\}$ is a complete orthonormal system in $\mathfrak{H}$ and that P_α is the projection operator on the one-dimensional subspace $\mathfrak{M}_\alpha$ defined by the vector φ_α, i.e.

$$P_\alpha \xi = (\xi, \varphi_\alpha)\varphi_\alpha. \tag{1}$$

Clearly,

$$P_{\alpha_1} P_{\alpha_2} = 0 \text{ for } \alpha_1 \neq \alpha_2. \tag{2}$$

We denote the operator defined by the equalities

$$U_{\alpha\beta}(\lambda\varphi_\beta) = \lambda\varphi_\alpha, \; U_{\alpha\beta}\xi = 0 \text{ for } \xi \perp \varphi_\beta$$

by $U_{\alpha\beta}$. Clearly, $U_{\alpha\beta}$ is a partially isometric operator with initial domain $\mathfrak{M}_\beta$ and terminal domain $\mathfrak{M}_\alpha$. It follows from the definition of $U_{\alpha\beta}$ that

$$U_{\beta\alpha}^* = U_{\alpha\beta}, \; U_{\alpha\alpha} = P_\alpha, \; U_{\alpha\beta} U_{\beta\gamma} = U_{\alpha\gamma}. \tag{3}$$

Moreover, by (1), for an arbitrary bounded operator A,

$$P_\alpha A P_\beta \xi = P_\alpha A(\xi, \varphi_\beta)\varphi_\beta = (\xi, \varphi_\beta)(A\varphi_\beta, \varphi_\alpha)\varphi_\alpha;$$

consequently,

$$P_\alpha A P_\beta(\lambda\varphi_\beta) = \lambda(A\varphi_\beta, \varphi_\alpha)\varphi_\alpha,$$
$$P_\alpha A P_\beta \xi = 0 \text{ for } \xi \perp \varphi_\beta.$$

Comparison of these equalities with the definition of the operator $U_{\alpha\beta}$ shows that

$$P_\alpha A P_\beta = a_{\alpha\beta} U_{\alpha\beta}, \tag{4}$$

where

$$a_{\alpha\beta} = (A\varphi_\beta, \varphi_\alpha)$$

is an element of the matrix of the operator A in the orthonormal system $\{\varphi_\alpha\}$.

Suppose now that $A \to \bar{A}$ is a representation of the algebra I_0 and that $\bar{\mathfrak{H}}$ is the corresponding representation space. Since the operators P_α, $U_{\alpha\beta}$ belong to the algebra I_0, there correspond to them certain operators $\bar{P}_\alpha$, $\bar{U}_{\alpha\beta}$ of the representation, which in virtue of (2) and (3) satisfy the relations

$$\bar{P}_{\alpha_1}\bar{P}_{\alpha_2} = 0 \text{ for } \alpha_1 \neq \alpha_2, \tag{5}$$

$$\bar{U}^*_{\alpha\beta} = \bar{U}_{\beta\alpha}, \bar{U}_{\alpha\alpha} = \bar{P}_\alpha, \bar{U}_{\alpha\beta}\bar{U}_{\beta\gamma} = \bar{U}_{\alpha\gamma}. \tag{6}$$

Furthermore, if $A \in I_0$ and $\bar{A}$ is the corresponding operator of the representation, then it follows from (4) that

$$\bar{P}_\alpha \bar{A} \bar{P}_\beta = a_{\alpha\beta}\bar{U}_{\alpha\beta}. \tag{7}$$

Clearly, $\bar{P}_\alpha$ is a projection operator in $\bar{\mathfrak{H}}$; suppose $\bar{\mathfrak{M}}_\alpha$ is the subspace in $\bar{\mathfrak{H}}$ on which it projects. By relation (5),

$$\bar{\mathfrak{M}}_{\alpha_1} \perp \bar{\mathfrak{M}}_{\alpha_2} \text{ for } \alpha_1 \neq \alpha_2.$$

Furthermore, it follows from relation (6) that $\bar{U}_{\alpha\beta}$ is a partially isometric operator with initial domain $\bar{\mathfrak{M}}_\beta$ and terminal domain $\bar{\mathfrak{M}}_\alpha$. Therefore all the subspaces $\bar{\mathfrak{M}}_\alpha$ have the same dimension. If this dimension equals zero, then all the operators $\bar{P}_\alpha$ and hence also $\bar{P}_\alpha \bar{A} \bar{P}_\beta$ equal zero. On the other hand, every compact operator A is the limit, in the sense of the norm in $\mathfrak{B}(\mathfrak{H})$, of finite sums $\sum a_{\alpha\beta} U_{\alpha\beta}$. Since every representation $A \to \bar{A}$ of the algebra I_0 is continuous (see Theorem 1, subsection 3, § 17), the image $\bar{A}$ of the compact operator A is the limit in the sense of the norm in $\mathfrak{B}(\bar{\mathfrak{H}})$ of finite sums $\sum a_{\alpha\beta}\bar{U}_{\alpha\beta}$. Therefore, also the image $\bar{A}$ of every element $A \in I_0$ equals zero. In other words, in this case the representation carries every operator $A \in I_0$ into the zero operator.

Discarding this uninteresting case, we can therefore assume that the spaces $\bar{\mathfrak{M}}_\alpha$ have nonzero dimension.

Suppose $\bar{\varphi}_{\alpha_0}$ is a normalized element in the fixed space $\overline{\mathfrak{M}}_{\alpha_0}$. We set

$$\bar{\varphi}_\alpha = \bar{U}_{\alpha\alpha_0} \bar{\varphi}_{\alpha_0}.$$

Then the elements $\bar{\varphi}_\alpha$ form an orthonormal system in $\bar{\mathfrak{H}}$. We denote the subspace constructed on the elements $\{\varphi_\alpha\}$ by $\overline{\mathfrak{N}}$. It follows from relations (5), (6) and (7) that $\overline{\mathfrak{N}}$ is invariant with respect to all the operators A. In fact,

$$\bar{U}_{\alpha\beta} \bar{\varphi}_\gamma = \bar{U}_{\alpha\beta} \bar{U}_{\gamma\alpha_0} \bar{\varphi}_{\alpha_0} = 0 \text{ for } \beta \neq \gamma,$$
$$\bar{U}_{\alpha\beta} \bar{\varphi}_\beta = \bar{U}_{\alpha\beta} \bar{U}_{\beta\alpha_0} \bar{\varphi}_{\alpha_0} = \bar{U}_{\alpha\alpha_0} \bar{\varphi}_{\alpha_0} = \bar{\varphi}_\alpha;$$

consequently

$$\bar{P}_\alpha \bar{A} \bar{P}_\beta \bar{\varphi}_\gamma = 0 \text{ for } \gamma \neq \beta,\ \bar{P}_\alpha \bar{A} \bar{P}_\beta \bar{\varphi}_\beta = a_{\alpha\beta} \bar{\varphi}_\alpha. \tag{8}$$

Thus, the operators $\bar{P}_\alpha \bar{A} \bar{P}_\beta$ and hence also the limit $\bar{A}$ of their finite sums are reduced by the space $\overline{\mathfrak{N}}$. In this space, $\bar{P}_\alpha$ is a projection operator on the one-dimensional subspace generated by the element $\bar{\varphi}_\alpha$. Therefore equality (8) means that $||a_{\alpha\beta}||$ is the matrix of the operator $\bar{A}$ in the space $\overline{\mathfrak{N}}$ with respect to the orthonormal system $\{\bar{\varphi}_\alpha\}$. The correspondence $\varphi_\alpha \to \bar{\varphi}_\alpha$ maps $\mathfrak{H}$ isometrically onto $\overline{\mathfrak{N}}$ and hence it maps the operator A into the operator $\bar{A}$. In other words, the part of the representation $A \to \bar{A}$ in the subspace $\overline{\mathfrak{N}}$ is generated by the isometric mapping of $\mathfrak{H}$ onto $\overline{\mathfrak{N}}$, i.e. it is unitarily equivalent to the identity representation $A \to A$.

We now consider the representation in the orthogonal complement $\bar{\mathfrak{H}} \ominus \overline{\mathfrak{N}}$. If it is different from the identically zero representation in this complement, then we can again select an invariant subspace $\overline{\mathfrak{N}}'$ in which the representation $A \to \bar{A}$ is equivalent to the identity representation $A \to A$. Repeating this reasoning, we arrive at the following theorem.

THEOREM 2. *Every representation $A \to \bar{A}$ of the algebra I_0 is equivalent to the direct sum of the identity representation $A \to A$ and the zero representation $A \to 0$.*

In particular, it follows from Theorem 2 that every irreducible representation $A \to \bar{A}$ of the algebra I_0 is equivalent to the identity representation or to the zero representation. In other words, the algebra I_0 has only one irreducible representation (to within equivalence) which is different from the zero representation.

This property turns out to be characteristic for the algebra I_0; namely, if $\mathfrak{H}$ is separable and a closed irreducible symmetric subalgebra R of the algebra $\mathfrak{B}(\mathfrak{H})$ has only one (two within equivalence) irreducible representation, distinct from the zero representation, then R coincides with I_0 (see NAIMARK [4] and ROSENBERG [1]).

3. Representations of the algebra $\mathfrak{B}(\mathfrak{H})$. We now consider any representation $A \to \bar{A}$ of the algebra $\mathfrak{B}(\mathfrak{H})$. This representation is at the same time a representation of the algebra $I_0 \subset \mathfrak{B}(\mathfrak{H})$. According to Theorem 2, the latter representation is equivalent to the direct sum of identity representations $A \to A$ and the zero representation $A \to 0$ of the algebra I_0. In other words, the space $\bar{\mathfrak{H}}$ of the representation $A \to \bar{A}$ can be represented, to within a unitary equivalence, in the form of the direct sum of copies $\mathfrak{H}_\alpha$ of the space $\mathfrak{H}$ and some space $\mathfrak{H}_0$, where in each of the spaces $\mathfrak{H}_\alpha = \mathfrak{H}$ the representation $A \to \bar{A}$ of the algebra I_0 reduces to the identity representation $A \to A$, and in the space $\mathfrak{H}_0$ to the zero representation $A \to 0$. Moreover, it may of course occur that the spaces $\mathfrak{H}_\alpha$ or the space $\mathfrak{H}_0$ are absent.

We shall prove that *each of the spaces* $\mathfrak{H}_\alpha = \mathfrak{H}$ *and hence also the space* $\mathfrak{H}_0$ *are invariant not only with respect to the images* $\bar{A}$ *of the elements of the algebra* I_0 *but also with respect to the entire algebra* $\mathfrak{B}(\mathfrak{H})$.

Suppose, for example, that $\bar{P}'$ is the projection operator in $\bar{\mathfrak{H}}$ on the space $\mathfrak{H}_{\alpha_0}$. We set

$$\bar{A}'\xi = \bar{P}'\bar{A}\xi$$

for $\xi \in \mathfrak{H}_{\alpha_0}$. Then $\bar{A}'$ is a bounded operator in the space $\mathfrak{H}_{\alpha_0}$. The space $\mathfrak{H}_{\alpha_0}$ can be considered to coincide with the space $\mathfrak{H}$, i.e. we can assume that $\mathfrak{H}$ is embedded in $\bar{\mathfrak{H}}$. Then $\bar{A}'$ can be considered as an operator in the space $\mathfrak{H}$. We shall prove that $A = \bar{A}'$. To this end, it suffices to prove that the matrices of these operators coincide in the orthonormal system $\{\varphi_\alpha\}$. To this end, we consider the operator $P_\alpha A P_\beta = a_{\alpha\beta} U_{\alpha\beta}$; then $||a_{\alpha\beta}||$ is the matrix of the operator A in the system $\{\varphi_\alpha\}$. Our representation $A \to \bar{A}$ in the space $\mathfrak{H}_{\alpha_0} = \mathfrak{H}$ maps this operator into itself, because this operator is an element of the ideal I_0; the same is true of the operators P_α, $U_{\alpha\beta} \in I_0$. On the other hand, the operator $P_\alpha A P_\beta$ goes over into $\bar{P}_\alpha \bar{A} \bar{P}_\beta$. Therefore, $\bar{P}_\alpha \bar{A} \bar{P}_\beta = a_{\alpha\beta} U_{\alpha\beta}$ in the space $\mathfrak{H}_{\alpha_0} = \mathfrak{H}$. It follows that

$$\begin{aligned}(\bar{A}'\varphi_\beta, \varphi_\alpha) = (\bar{P}'\bar{A}\varphi_\beta, \varphi_\alpha) &= (\bar{A}\varphi_\beta, \bar{P}'\varphi_\alpha) = (\bar{A}\varphi_\beta, \varphi_\alpha)\\ &= (\bar{A}P_\beta\varphi_\beta, P_\alpha\varphi_\alpha) = (\bar{A}\bar{P}_\beta\varphi_\beta, \bar{P}_\alpha\varphi_\alpha) = (\bar{P}_\alpha\bar{A}\bar{P}_\beta\varphi_\beta, \varphi_\alpha)\\ &= a_{\alpha\beta}(U_{\alpha\beta}\varphi_\beta, \varphi_\alpha) = a_{\alpha\beta}(\varphi_\alpha, \varphi_\alpha) = a_{\alpha\beta}.\end{aligned}$$

Since $(\bar{A}'\varphi_\beta, \varphi_\alpha)$ is an element $a'_{\alpha\beta}$ of the matrix of the operator $\bar{A}'$ in the system $\{\varphi_\alpha\}$, we have by the same token proved that $A = \bar{A}'$. From this it follows that

$$\bar{P}'(\bar{A}\bar{B})\xi = (AB)\xi, \; \bar{P}'\bar{A}\bar{P}'\bar{B}\xi = A(B\xi),$$

so that $\bar{P}'\bar{A}(\bar{B}\xi) = \bar{P}'\bar{A}\bar{P}'(\bar{B}\xi)$.

In other words, $\bar{P}'\bar{A} = \bar{P}'\bar{A}\bar{P}'$ on all elements of the form $\bar{B}\xi$, where the vector ξ runs through the entire space $\mathfrak{H}_{\alpha_0} = \mathfrak{H}$ and the operator B runs through the entire algebra $\mathfrak{B}(\mathfrak{H})$. We denote the closed linear hull of the set of all these elements $\bar{B}\xi$ by $\bar{\mathfrak{H}}'$. Then the equality

$$\bar{P}'\bar{A} = \bar{P}'\bar{A}\bar{P}' \tag{1}$$

holds in $\bar{\mathfrak{H}}'$. Since $\mathfrak{H}_{\alpha_0} \subset \bar{\mathfrak{H}}'$, $\bar{P}' = 0$ in the complement $\bar{\mathfrak{H}} \ominus \bar{\mathfrak{H}}'$. Moreover, $\bar{\mathfrak{H}}'$ and hence also $\bar{\mathfrak{H}} \ominus \bar{\mathfrak{H}}'$ are invariant with respect to all the operators $\bar{A}$; from this it follows that also $\bar{P}'\bar{A} = 0$ in $\bar{\mathfrak{H}} \ominus \bar{\mathfrak{H}}'$ and therefore equality (1) also holds in

$\overline{\mathfrak{H}} \ominus \overline{\mathfrak{H}}'$. Consequently, equality (1) holds in the entire space $\overline{\mathfrak{H}}$. Applying involution to both members of (1) and then substituting A^* for A, we obtain that also

$$\bar{A}\bar{P}' = \bar{P}'\bar{A}\bar{P}'.$$

It follows that $\bar{P}'\bar{A} = \bar{A}\bar{P}'$, i.e. the space $\mathfrak{H}_{\alpha_0} = \mathfrak{H}$ reduces all the operators in $\bar{A}$.

The equality $A = \bar{A}'$ now assumes the form

$$A\xi = \bar{P}'\bar{A}\xi = \bar{A}\bar{P}'\xi = \bar{A}\xi \text{ for } \xi \in \mathfrak{H},$$

i.e. in the space $\mathfrak{H} = \mathfrak{H}_{\alpha_0}$, the representation $A \to \bar{A}$ of the entire algebra $\mathfrak{B}(\mathfrak{H})$ also reduces to the identity representation $A \to A$.

In the space $\mathfrak{H}_0$ all operators of the algebra I_0 go over into zero. Consequently, in this space our representation is at the same time a representation of $\mathfrak{B}(\mathfrak{H})/I_0$.

We have thus proved the following theorem.

THEOREM 3. *Every representation of the algebra $\mathfrak{B}(\mathfrak{H})$ is the direct sum of identity representations $A \to A$ and a representation of the quotient algebra $\mathfrak{B}(\mathfrak{H})/I_0$.*

If $\mathfrak{H}$ is separable, then I_0 is a maximal two-sided ideal in $\mathfrak{B}(\mathfrak{H})$ and $\mathfrak{B}(\mathfrak{H})/I_0$ is a simple algebra. Therefore every representation of the algebra $\mathfrak{B}(\mathfrak{H})/I_0$, different from the zero representation, is an isomorphic mapping of this algebra into the algebra of operators in a Hilbert space.

The problem of finding *all* representations of the algebra $\mathfrak{B}(\mathfrak{H})/I_0$ has not yet been solved. Only one concrete representation of this algebra, constructed by CALKIN [1], is known.

The results of §§ 17–21 and subsection 2, § 22, except those credited in the text, are due basically to GELFAND and NAIMARK [6] (also see NAIMARK [2]); of subsection 3, § 22 to NAIMARK [2]; subsection 4, § 17, except, of course, the spectral theorem, to DUNFORD [1]; subsection 5, § 17 to FAGE [1] and DUNFORD [2]; certain of these results were also obtained independently by SEGAL [1–4]; the uniqueness of the norm in a reduced symmetric algebra with identity (IX, subsection 2, § 18) was noted by Sonis.

In a recent work of PORTA and SCHWARTZ [1], Theorems 2 and 3 of subsections 2 and 3, § 22, were generalized to arbitrary (not necessarily symmetric) continuous representations of the algebras I_0 and $\mathfrak{B}(\mathfrak{H})$.

Theorem 5, subsection 4, § 20, for the case of the group algebra of a locally compact commutative group (see below, subsection 4, § 31) was first established by WEIL [1] and independently by M. KREIN [7], and for algebras by GODEMENT [1]; in the formulation given here it can be found in the book of LOOMIS [2] (however, the proof in this book contained an essential gap which was filled by Raikov by a remark in the Russian edition). We must note that in all proofs of this theorem the basic idea is the same as that in M. KREIN [7]; the proof of WEIL [1] uses the structure properties of a locally compact commutative group. Theorem 1, subsection 1, § 22, is due to CALKIN [1].

CHAPTER V

SOME SPECIAL ALGEBRAS

§ 23. Completely symmetric algebras

1. Definition and examples of completely symmetric algebras. We recall (see subsection 2, § 14) that a Banach symmetric algebra R with identity is said to be *completely symmetric* if the inverse $(e+x^*x)^{-1}$ exists in the algebra R for an arbitrary element $x \in R$.

An example of a completely symmetric algebra is any algebra $C(\mathfrak{M})$ where $\mathfrak{M}$ is a compact space. In fact, in this case

$$(e+x^*x)^{-1} = \frac{1}{1+|x(M)|^2}$$

is an element of the algebra $C(\mathfrak{M})$.

According to Theorem 2, subsection 2, § 16, every commutative C^*-algebra R with identity is completely isomorphic to some algebra $C(\mathfrak{M})$ and consequently it is a completely symmetric algebra.

Further,

I. *Every closed symmetric subalgebra R of the algebra $\mathfrak{B}(\mathfrak{H})$ with identity is completely symmetric.*

In fact, if we set $|A| = c$ and

$$B = \frac{1}{c^2+1}(c^2 1 - A^*A),$$

then, since $0 \leqq (A^*A\xi, \xi) \leqq c^2(\xi, \xi)$, we have $0 \leqq (B\xi, \xi) \leqq \{c^2/(c^2+1)\}(\xi, \xi)$. From this we conclude that $|B| \leqq c^2/(c^2+1) < 1$, and therefore $1-B$ has an inverse in R (see I, subsection 3, § 9). But then it follows from the relation

$$A^*A+1 = (c^2+1)\left(1-\frac{1}{c^2+1}(c^2 1 - A^*A)\right) = (c^2+1)(1-B)$$

that A^*A+1 also has an inverse in R.

Every Banach reduced algebra with minimal regular norm is completely isomorphic to a closed subalgebra of the algebra $\mathfrak{B}(\mathfrak{H})$(see Theorem 1, subsection 3, § 18). Therefore

II. *Every Banach reduced algebra with minimal regular norm is a completely symmetric algebra.*

Also,

III. *Every maximal symmetric commutative subalgebra of a completely symmetric algebra is completely symmetric.*

Proof. Suppose K is a maximal symmetric commutative subalgebra of a completely symmetric algebra R and that $x \in K$. Then $y = (e+x^*x)^{-1}$ exists in R. But the element y commutes with all the elements in K and, consequently, $y \in K$ in virtue of the fact that the algebra K is maximal. This means that K is completely symmetric.

FELL and THOMA [1] showed that every closed symmetric subalgebra R_1 of the completely symmetric algebra R with identity is completely symmetric and the spectrum of every element $x \in R_1$ on R_1 coincides with its spectrum on R.

2. Spectrum. Let us recall (see subsection 4, § 8) that the *spectrum* of the element x in the algebra R is the set of all complex numbers λ such that $(x-\lambda e)^{-1}$ does not exist in R.

Suppose R is a Banach algebra with identity and that x is an element in R.

I. *If* $|\lambda| > |x|$, *then the number* λ *does not belong to the spectrum of the element* x.

In fact, in this case $x-\lambda e = -\lambda\{e-(1/\lambda)x\}$, where $|(1/\lambda)x| < 1$; therefore, $(x-\lambda e)^{-1} = -\lambda^{-1}\{e-(1/\lambda)x\}^{-1}$ exists in R.

II. *If* x^{-1} *exists in* R *and* $|\lambda|^{-1} > |x^{-1}|$, *then the number* λ *does not belong to the spectrum of the element* x.

This proposition is obtained from the preceding one because

$$(x-\lambda e)^{-1} = -x^{-1}(x^{-1}-\lambda^{-1}e)^{-1}\lambda^{-1}.$$

III. *If the spectrum of the Hermitian element* x *of the symmetric Banach algebra* R *with identity consists of only the zero, then* x *belongs to the reducing ideal of this algebra.*

Proof. Suppose R_1 is the maximal commutative symmetric subalgebra of the algebra R which contains x; the spectrum of the element x with respect to R_1 coincides with the spectrum of x with respect to R (see V, subsection 1, § 10) and therefore it consists of only the zero. This means that x belongs to all maximal ideals of the algebra R_1 and hence to the radical of this algebra. But then

$$\lim_{n\to\infty} \sqrt[n]{|(x^*x)|^n} = 0$$

(see subsection 3, § 10) and inequality (3) subsection 4, § 10 shows that x belongs to the reducing ideal of the algebra R.

IV. *If the spectrum of the Hermitian element* x *of the reduced Banach algebra* R *consists of only the zero then* $x = 0$.

In fact, in this case the reducing ideal $= (0)$, and therefore $x = 0$.

We shall now assume that R is a completely symmetric algebra.

V. *Every Hermitian element of a completely symmetric algebra has a real spectrum.*

Proof. Suppose x is a Hermitian element. We set $y = (1/\tau)x$. By hypothesis,

$$(e+y^*y)^{-1} = \left(e+\frac{1}{|\tau|^2}x^2\right)^{-1} = |\tau|^2(x^2+|\tau|^2e)^{-1}$$

exists; consequently, $z = (x^2+|\tau|^2e)^{-1}$ exists. Then

$$e = z(x^2+|\tau|^2e) = z(x+i|\tau|e)(x-i|\tau|e);$$

hence

$$(x+i|\tau|e)^{-1} \text{ and } (x-i|\tau|e)^{-1}$$

also exist for arbitrary τ.

Now suppose that $\lambda = \sigma+i\tau$ is an arbitrary imaginary number ($\tau \neq 0$). If x is a Hermitian element then $x-\sigma e$ is also a Hermitian element; consequently,

$$(x-\sigma e-i\tau e)^{-1} = (x-\lambda e)^{-1}$$

exists.

Thus, no imaginary λ belongs to the spectrum of the Hermitian element x. Consequently, its spectrum is real.

VI. *In a completely symmetric algebra, every element of the form x^*x has a real nonnegative spectrum.*

Proof. Since x^*x is a Hermitian element, its spectrum is real. Suppose $\lambda > 0$. We set $y = (1/\sqrt{\lambda})x$; then

$$(y^*y+e)^{-1} = \left(\frac{1}{\lambda}x^*x+e\right)^{-1} = \lambda(x^*x+\lambda e)^{-1}$$

exists. Consequently, the negative number $-\lambda$ cannot belong to the spectrum of the element x^*x.

We denote the set of all Hermitian elements of the completely symmetric algebra R whose spectrum is nonnegative by P, and the set of all its Hermitian elements with positive spectrum by P^+.

VII. *Every element $x \in P^+$ can be represented in the form $x = y^2$, where $y \in P^+$.*

Proof. We denote the maximal commutative subalgebra of the algebra R which contains x by K. The spectrum of the element x with respect to K is also positive so that it suffices to apply Theorem 7, subsection 6, § 11, and its corollary to the branch of the analytic function $f(\zeta) = \sqrt{\zeta}$, which is positive on the positive half-axis, and to the element $x \in K$.

VIII. *If $x \in P$ and $\lambda \geqq 0$, then $\lambda x \in P$. But if $x \in P$, $x \neq 0$, $\lambda < 0$ and R is a reduced algebra, then $\lambda x \notin P$.*

In fact, upon multiplication of the element x by the number λ all the numbers of the spectrum are also multiplied by λ. But if $x \in P$ and $\lambda x \in P$ for some $\lambda < 0$, then the spectrum of the element x consists of only the zero and, consequently, if R is a reduced algebra, $x = 0$ in virtue of IV.

IX. *If* $x_1, x_2 \in P$, *then* $x_1+x_2 \in P$.

We must prove that $(x_1+x_2+\lambda e)^{-1}$ exists for $\lambda > 0$.

Suppose $0 < \lambda_1 < \lambda$. We set $\lambda_2 = \lambda-\lambda_1$; then also $0 < \lambda_2 < \lambda$. Since $x_1+\lambda_1 e$, $x_2+\lambda_2 e \in P^+$, then, according to Proposition VII, the elements $x_1+\lambda_1 e$, $x_2+\lambda_2 e$ can be represented in the form

$$x_1+\lambda_1 e = y_1^2,\ y_1 \in P^+,\ x_2+\lambda_2 e = y_2^2,\ y_2 \in P^+.$$

Since $x_1 \in P$,

$$z = (x_1+\lambda_1 e)^{-1}$$

exists, so that

$$y_1(y_1 z) = (x_1+\lambda_1 e)z = e;$$

consequently, since y_1 and z commute, y_1^{-1} exists.

We now rewrite $x_1+x_2+\lambda e$ in the form

$$x_1+x_2+\lambda e = x_1+\lambda_1 e+x_2+\lambda_2 e = y_1^2+y_2^2 = y_1^2(e+y_1^{-2}y_2^2).$$

Since y_1^{-2} exists, it will suffice to prove that $(e+y_1^{-2}y_2^2)^{-1}$ exists. To this end, it suffices to show that the spectrum of the element $y_1^{-2}y_2^2$ is nonnegative. But this follows from the relation

$$y_1^{-2}y_2^2 = y_1^{-1}((y_2 y_1^{-1})^*(y_2 y_1^{-1}))y_1,$$

because by VI the spectrum of the element $(y_2y_1^{-1})^*(y_2y_1^{-1})$ is nonnegative, and the mapping $x \to y_1^{-1}xy_1$ leaves the spectrum of the element x invariant.

This proves Proposition IX.

3. Theorems on extensions. We denote the set of all Hermitian elements of the completely symmetric algebra R by H. H is a real Banach space.

Propositions VIII and IX, subsection 2 mean that *P is a wedge in the space H* and, moreover, *a cone if R is a reduced algebra* (see subsection 10, § 3).

Clearly, P contains the identity e of the algebra R. We shall prove that e is an interior element of the wedge P.

In fact, suppose $x = e+y$, where $y \in H$ and $|y| < 1$. It follows from the proof of Proposition I, subsection 4, § 10, that x can be represented in the form $x = x_1^2$, where $x_1 \in H$. Therefore, $x \in P$. Thus, all the elements x in H which satisfy the condition $|e-x| < 1$ belong to P; consequently, e is an interior element of P.

I. *For every closed left ideal I_l of a completely symmetric algebra R there exists a normalized positive functional f such that*

$$f(x^*x) = 0 \textit{ for all } x \in I_l.$$

[Recall that a positive functional f is said to be normalized if $f(e) = 1$ (see subsection 4, § 19).]

Proof. Suppose $I_l^{(H)}$ is the set of all Hermitian elements of the ideal I_l. Clearly, $I_l^{(H)}$ is a closed subspace of the space H. We assumed that R is a reduced algebra; therefore, P is a cone in R with interior point e. We denote by H' the set of all elements of the form $\lambda e+y$, where λ is a real number and $y \in I_l^{(H)}$; H' is a subspace of the space H. We set

$$f(\lambda e+y) = \lambda$$

on H'; in particular, $f(e) = 1$ and $f(y) = 0$ for $y \in I_l^{(H)}$. The functional obtained in this way is nonnegative on elements $x \in H' \cap P$.

In fact, if $x \in H' \cap P$, then

$$x-\lambda e = y \in I_l^{(H)} \subset I_l,$$

and consequently $(x-\lambda e)^{-1}$ does not exist. Since $x \in P$, this is impossible for $\lambda < 0$. Thus, $\lambda \geqq 0$, i.e. $f(x) = f(y+\lambda e) = \lambda \geqq 0$.

Since P is a wedge in R with interior point e, then, according to M. Krein's theorem (see Theorem 2, subsection 10, § 3 and the remark after it), the functional f can be extended to a real-valued functional on the entire space H in such a way that it assumes nonnegative values on the cone P.

Every element x of the algebra R can be represented in the form

$$x = x_1+ix_2, x_1, x_2 \in H.$$

We set

$$f(x) = f(x_1)+if(x_2);$$

f is a positive functional. In fact, $x^*x \in P$, and consequently, $f(x^*x) \geqq 0$. Further, if $x \in I_l$, then $x^*x \in I_l^{(H)}$ and consequently $f(x^*x) = 0$. It then follows from the Cauchy-Bunyakovsky inequality, $|f(x)|^2 \leqq f(e)f(x^*x)$, that $f(x) = 0$ also.

Corollary 1. *For every maximal left ideal M_l of the completely symmetric algebra R, there exists an indecomposable normalized positive functional f_0 such that*

$$M_l = \{x \in R : f_0(x^*x) = 0\} = I_l^{(f_0)}.$$

Proof. By I, there exists a normalized positive functional f such that

$$M_l \subset \{x \in R : f(x^*x) = 0\} = I_l^{(f)}.$$

Since the set $I_l^{(f)}$ is a proper left ideal (see subsection 3, § 17), then, in virtue of its maximality,

$$M_l = I_l^{(f)}. \tag{1}$$

We denote the set of all normalized positive functionals satisfying condition (1) by K. Clearly, K is a weakly closed convex set in the space H', conjugate to H. According to the Krein-Milman theorem (subsection 9, § 3), it contains at least one extreme point; denote it by f_0.

We shall prove that f_0 is an indecomposable functional in R. Suppose it is represented in the form

$$f_0 = \lambda f_1 + (1-\lambda) f_2,$$

where $0 < \lambda < 1$ and f_1, f_2 are normalized positive functionals in R. Clearly, $M_l = I_l^{(f_0)} \subset I_l^{(f_1)}, I_l^{(f_2)}$ and in virtue of the maximality of M_l, $f_1, f_2 \in K$. Since f_0 is an extreme point of K, then $f_1 = f_2 = f_0$; this proves that the functional f_0 is indecomposable.

COROLLARY 2. *A necessary and sufficient condition for an element x of a completely symmetric algebra R to be left invertible is that for every indecomposable normalized positive functional f the condition $f(x^*x) > 0$ is satisfied.*

Proof. If $f(x^*x) = 0$ for some normalized f, then x belongs to the proper ideal $I_l^{(f)}$ and hence it is left invertible. Conversely, if x is not left invertible, then x belongs to some maximal left ideal M_l. But in virtue of Corollary 1, the latter coincides with $I_l^{(f_0)}$ for some indecomposable normalized positive functional f_0 and hence $f_0(x^*x) = 0$.

In an algebra with identity, we shall call the set of all complex numbers λ for which $x - \lambda e$ does not have a left inverse the *left spectrum* of the element x.

COROLLARY 3. *The left spectrum of the element x in a completely symmetric algebra coincides with the set of numbers λ for which there exists an indecomposable normalized positive functional f such that $\lambda = f(x)$ and $f(x^*x) = f(x)f(x^*) = |\lambda|^2$.*

Proof. Let $\lambda = f(x)$, $f(x^*x) = f(x^*)f(x)$ and $f(e) = 1$. We set $z = x - f(x)e$; then $f(z^*z) = f((x^* - \overline{f(x)}e)(x - f(x)e) = 0$ so that, by Corollary 2, z does not have a left inverse.

Conversely, if λ belongs to the left spectrum of x, then there exists a maximal left ideal M_l such that $z = x - \lambda e \in M_l$. In virtue of Corollary 1, M_l has the form

$$M_1 = \{x \in R : f(x^*x) = 0\} = I_l^{(f)} \quad (f(e) = 1).$$

Therefore $f(z^*z) = 0$, i.e.

$$0 = f(z^*z) = f(x^*x) - \bar{\lambda} f(x) - \lambda f(x^*) + \lambda\bar{\lambda} f(e) = |\lambda - f(x)|^2 + f(x^*x) - |f(x)|^2.$$

But by inequality (4), subsection 2, § 10, $|f(x)|^2 \leqq f(x^*x)$. Therefore the sum of the nonnegative numbers $|\lambda - f(x)|^2$ and $f(x^*x) - |f(x)|^2$ equals zero, i.e.

$$\lambda = f(x) \text{ and } f(x^*x) = f(x^*)f(x).$$

REMARK. Corollaries, analogous to Corollaries 1, 2, 3, can be formulated for maximal right ideals, for right invertible elements, and for the right spectrum.

COROLLARY 4. *In a completely symmetric algebra R, for every normal element $x \in R$,*

$$\sup |f(x)| = r(x),$$

where the supremum in the left member is taken over all indecomposable normalized positive functionals f.

Proof. In virtue of II, subsection 5, § 9 and Corollary 3,

$$\lim_{n \to \infty} \sqrt[n]{|x^n|} = r(x) = \sup_{\lambda \in S_x} |\lambda| \leqq \sup |f(x)|.$$

The reverse inequality follows from the remark to III, subsection 4, § 10.

II. *The radical of a completely symmetric algebra coincides with the reducing ideal.*

Proof. According to VII, subsection 2, § 18, the radical is always contained in the reducing ideal. Therefore, we must prove that every element of the reducing ideal belongs to the radical. Suppose x_0 is an element such that

$$f(x_0^* x_0) = 0$$

for all positive functionals f. We shall prove that x_0 belongs to the radical. In virtue of Corollary 1 to I, x_0 belongs to all maximal left ideals and thus to the radical inasmuch as the radical is the intersection of all maximal left ideals (see I, subsection 5, § 7).

It follows from Proposition II that *completely symmetric semisimple algebras coincide with completely symmetric reduced algebras.*

III. *Suppose R_1 is a closed symmetric subalgebra of the completely symmetric algebra R with identity. Then every positive functional f_0 in R_1 can be extended to a positive functional on R. If, furthermore, the functional f_0 on the algebra R_1 is indecomposable, then it can be extended to an indecomposable functional on the algebra R.*

Proof. We denote by H_1 and H the sets of all Hermitian elements of the algebras R_1 and R, respectively. Then H_1 is a closed subspace in H. A positive functional f in R_1 can be considered as a linear functional in H_1 which assumes nonnegative values on all elements of $P \cap H_1$. According to the Krein theorem, cited above, f can then be extended to a linear functional on H which assumes nonnegative values on all elements in P; hence, it only remains to repeat the reasoning at the end of the proof of Proposition I.

Suppose now that the functional f_0 on the algebra R_1 is indecomposable. Without loss of generality, we can assume that $f_0(e) = 1$. We denote by K the set of all normalized positive functionals f on R, which coincide with f_0 on R_1. Clearly, K is a weakly closed convex set in the space H', conjugate to H. According to the Krein-Milman theorem (subsection 9, § 3), it contains at least one extreme point; we denote this point by F_0.

We shall prove that F_0 is an indecomposable functional on R. Suppose the functional F_0 is represented in the form

$$F_0 = \lambda F_1 + (1 - \lambda) F_2, \tag{1}$$

where $0 < \lambda < 1$ and F_1, F_2 are normalized positive functionals on R. Considering the equality (1) only on R_1, we obtain, in particular, that

$$f_0 = \lambda f_1 + (1-\lambda) f_2,$$

where f_1, f_2 are the functionals F_1, F_2, respectively, considered only on R_1. But since the functional f_0 is indecomposable, it follows from this that the functionals f_1 and f_2 are multiples of f_0. Furthermore, they are normalized; consequently, $f_1 = f_2 = f_0$ on R_1. Other words, $F_1 = F_2 = f_0$ on R_1, i.e. $F_1, F_2 \in K$.

But then it follows from equality (1) that F_1 and F_2 are multiples of the functional F_0, because F_0 is an extreme point of the set K. This proves that the functional F_0 is indecomposable.

The following more general proposition holds.

IV. *Let R and R_1 ($R \supset R_1$) be completely symmetric algebras with the same involution and identity and complete in the norms $|x|$, $|x|_1$ respectively. A necessary and sufficient condition for every positive functional f_1 on R_1 to be extendible (with preservation of indecomposability if f_1 is indecomposable) to a positive functional f on R is that $r_1(x) = r(x)$ for every element $x \in R_1$.*

Proof. Necessity. Let $x \in R_1$ and suppose that x^{-1} exists in R; then, for every indecomposable normalized positive functional f on R the inequalities $f(x^*x) > 0, f(xx^*) > 0$ are satisfied. If furthermore x is not invertible in R_1, then, in virtue of Corollary 1 to I, there exists an indecomposable normalized positive functional f_1 on R_1 such that $f_1(x^*x) = 0$ or $f_1(xx^*) = 0$. Then an indecomposable positive functional on R which is an extension of f_1 vanishes on the element x^*x or xx^*, which is impossible. Thus, the pair of algebras R, R_1 is a Wiener pair and, in virtue of the corollary in subsection 7, § 11, $r(x) = r_1(x)$ for $x \in R_1$.

The proof of *sufficiency* is easily reduced to III in the following way. In virtue of V, subsection 2, § 18, we can assume that the algebras R and R_1 are reduced. In virtue of Corollary 4 to I and Theorem 2, subsection 3, § 18, $\sqrt{r(x^*x)}$ serves as the minimal regular norm in the algebra R.

Let R' be the closure of the algebra R with respect to the norm $\sqrt{r(x^*x)}$. By assumption, the closure R_1' of the algebra R_1 with respect to the norm $\sqrt{r(x^*x)}$ is the closure of the algebra R_1 with respect to its minimal regular norm $\sqrt{r_1(x^*x)}$. Since these norms are regular, then every positive functional of the algebra $R(R_1)$ can be extended to a positive functional of the algebra $R'(R_1')$. Further, it remains to apply III to the pair of algebras R', R_1'.

The proof of the fact that an indecomposable functional can be extended to an indecomposable one can be adapted from III.

COROLLARY 5. *In the completely symmetric algebras R and R_1 ($R \supset R_1$) with common involution and identity, the equality $S_x = S_x^1$ for $x \in R_1$ is equivalent to the equality $r(x) = r_1(x)$.*

In fact, let $r(x) = r_1(x)$ for $x \in R_1$. It follows from IV that it is possible to extend positive functionals with preservation of indecomposability. Consider an arbitrary maximal left ideal $M_l \subset R_1$. According to Corollary 1 to I, $M_l = I_l^{(f_1)}$, where f_1 is an indecomposable normalized positive functional on R_1.

We extend f_1 to an indecomposable positive functional f on R. Then $M_l \subset I_l^{(f)}$ and consequently $\overline{M_l} \subset I_l^{(f)}$, where $\overline{M_l}$ is the closure of M_l in the norm $|x|$. Therefore, $\overline{M_l}$ does not contain elements which are invertible in R. Analogously, $\overline{M_r}$, where M_r is a maximal right ideal in R_1, does not contain elements which are invertible in R.

By I, subsection 7, § 11, the algebras R and R_1 form a Wiener pair. By I d), subsection 7, § 11, it follows from this that $S_x = S_x^1$ for $x \in R_1$.

Suppose R_1 is a subalgebra of the algebra R and let $x \to A_x$, $x \to B_x$ be representations of the algebras R_1 and R in the spaces $\mathfrak{H}_1$ and $\mathfrak{H}$, respectively. The representation $x \to B_x$ is said to be an *extension* of the representation $x \to A_x$ if $\mathfrak{H}_1 \subset \mathfrak{H}$ and for all $x \in R_1$ and $\xi \in \mathfrak{H}_1$,

$$A_x \xi = B_x \xi.$$

THEOREM 1. *Suppose R_1 is a closed symmetric subalgebra of the completely symmetric algebra R, which contains an identity. Then every representation of R_1 can be extended to a representation of the entire algebra R, and every irreducible representation of R_1 can be extended to an irreducible representation of the algebra R.*

Proof. Let us first suppose $x \to A_x$ is a cyclic representation of the algebra R_1 in $\mathfrak{H}_1$ and that ξ_0 is a cyclic vector in $\mathfrak{H}_1$. Then $f_0(x) = (A_x\xi_0, \xi_0)$ is a positive functional on R_1. According to III, it can be extended to a positive functional F_0 on the algebra R. The functional F_0 defines a cyclic representation $x \to B_x$ of the algebra R in some space $\mathfrak{H}$; moreover, $F_0(x) = (B_x\eta_0, \eta_0)$, where η_0 is a cyclic element in $\mathfrak{H}$ (see Theorem 2, subsection 3, § 17).

The space $\mathfrak{H}_1$ is the closure of the set of all elements $A_x\xi_0$, $x \in R_1$, because ξ_0 is a cyclic element. We assign to each such element $A_x\xi_0$ the element $B_x\eta_0$. This correspondence maps $\mathfrak{H}_1$ isometrically onto a subspace in $\mathfrak{H}$. In fact, since $F_0(x) = f_0(x)$ for $x \in R_1$, we have $(A_x\xi_0, A_x\xi_0) = f_0(x^*x) = F_0(x^*x) = (B_x\eta_0, B_x\eta_0)$ for $x \in R_1$.

Therefore, if we identify $\mathfrak{H}_1$ with its image in $\mathfrak{H}$, then $\mathfrak{H}_1$ is a subspace in $\mathfrak{H}$, ξ_0 coincides with η_0, and $A_x\xi$ coincides with $B_x\xi$ for all $\xi \in \mathfrak{H}$, $x \in R_1$. Thus, the first part of the theorem is proved for a cyclic representation $x \to A_x$. But if the representation $x \to A_x$ is not cyclic, then it is the direct sum of cyclic representations. Extending each of them and forming the direct sum of the extensions thus obtained, we obtain, by the same token, an extension of the initial representation $x \to A_x$.

If the representation $x \to A_x$ of the algebra R_1 is irreducible, then f_0 is an indecomposable functional. According to III, the functional F_0 can also be chosen to be indecomposable. Then the corresponding representation $x \to B_x$ of the algebra R will be irreducible.

The following theorem can easily be proved.

THEOREM 1′. *Let R and R_1 ($R \supset R_1$) be completely symmetric algebras with the same involution and identity. A necessary and sufficient condition for every representation of R_1 to be extendible to a representation of R (with preservation of irreducibility) is that $r_1(x) = r(x)$ for every $x \in R_1$.*

V. *If λ_0 is a point in the left spectrum of the element x_0 of a completely symmetric algebra R, then there exists an irreducible representation $x \to A_x$ of the algebra R and a vector ξ_0 in the space of this representation such that $A_{x_0}\xi_0 = \lambda_0\xi_0$.*

Proof. In virtue of Corollary 3 to I, there exists an indecomposable normalized positive functional f on R such that

$$\lambda_0 = f(x_0) \text{ and } f(x_0^* x_0) = f(x_0^*) f(x_0).$$

Consider the irreducible cyclic representation $x \to A_x$ constructed with the aid of the functional f (see Theorem 2, subsection 3, § 17), i.e. such that $f(x) = (A_x \xi_0, \xi_0)$, where ξ_0 is a cyclic vector. Moreover,

$$1 = f(e) = |\xi_0|^2.$$

The equality $f(x_0{}^* x_0) = f(x_0{}^*) f(x_0)$ means that

$$(A_{x_0{}^* x_0} \xi_0, \xi_0) = (A_{x_0{}^*} \xi_0, \xi_0)(A_{x_0} \xi_0, \xi_0),$$

so that

$$|(A_{x_0} \xi_0, \xi_0)|^2 = |A_{x_0} \xi_0|^2 |\xi_0|^2;$$

thus, we obtain equality in the Cauchy-Bunyakovsky inequality, which is possible if, and only if, $A_{x_0} \xi_0 = \lambda_0 \xi_0$. We note that, moreover, $\lambda_0 = (\lambda_0 \xi_0, \xi_0) = (A_{x_0} \xi_0, \xi_0) = f(x_0)$.

Thus, to each point of the left spectrum there corresponds a characteristic vector in the space of some irreducible representation.

COROLLARY. *There exists a representation $x \to A_x$ of the completely symmetric algebra R such that all points of the left spectrum of all elements x map into characteristic values of the operators A_x.*

The direct sum of all the irreducible cyclic representations of the algebra R which correspond to indecomposable functionals, considered in the proof of Theorem 1, subsection 3, § 18, can be taken as such a representation.

Proposition V can be given a simple interpretation in terms of quantum mechanics.

Every quantum-mechanical system S defines, in a natural way, a Banach symmetric algebra, namely the algebra of all bounded operators, which define quantities in this system. As a closed subalgebra of the algebra $\mathfrak{B}(\mathfrak{H})$, it is a completely symmetric algebra (see I, subsection 1). A positive functional f on this algebra is defined by some vector ξ_0: $f(x) = (A_x \xi_0, \xi_0)$, i.e. by means of some state of the system S, where $f(x)$ is the mathematical expectation of the quantity x in this state.

If f is an indecomposable functional in the subalgebra R_1, then Proposition V asserts that in the state ξ_0 of the space $\mathfrak{H}$ of some irreducible representation of the algebra R all quantities x of the subalgebra R_1 have value exactly equal to $f(x)$.

4. Criterion for complete symmetry.

THEOREM 2 (RAIKOV [7]). *A Banach symmetric algebra R with identity is completely symmetric if, and only if, for every element $x \in R$ the equality*

$$\sup f(x^* x) = \lim_{n \to \infty} \sqrt[n]{|(x^* x)^n|} = r(x^* x) \tag{1}$$

is satisfied for all elements $x \in R$, where the supremum in the left member is taken over all indecomposable normalized positive functionals f.

Proof. Suppose R is a completely symmetric ring and let $x \in R$. Since the element $x^* x$ is normal, (1) follows from Corollary 4, subsection 3.

Conversely, suppose equality (1) holds in the algebra R; we shall prove that R is completely symmetric. According to Theorem 2, subsection 3, § 18, the left member of equality (1) is the square of the minimal regular norm of the element x; we can denote this norm by $|x|_0$. (We recall that $|x|_0$ can $= 0$ also for $x \neq 0$, so that $|x|_0$ is in fact the norm in the reduced algebra.) Then, for arbitrary positive functional φ,

$$\varphi(x^*x) \leqq \varphi(e)|x^*x|_0. \tag{2}$$

We shall prove that for arbitrary positive functional f,

$$f(x^*xx^*x) \leqq f(x^*x)|x^*x|_0. \tag{3}$$

If $f(x^*xx^*x) = 0$, then the assertion is trivial. Suppose $f(x^*xx^*x) > 0$. Then

$$f(x^*xx^*x) = f((xx^*x)^*x) \leqq \sqrt{f(x^*xx^*xx^*x)f(x^*x)}. \tag{4}$$

On the other hand, if we apply inequality (2) to the positive functional

$$\varphi(y) = f(x^*xyx^*x),$$

we obtain that

$$f(x^*xx^*xx^*x) \leqq f(x^*xx^*x)|x^*x|_0.$$

If we substitute the last inequality into inequality (4), we obtain

$$f(x^*xx^*x) \leqq \sqrt{f(x^*xx^*x)|x^*x|_0 f(x^*x)},$$

from which (3) follows.

We must prove that $(e+x^*x)^{-1}$ exists for arbitrary $x \in R$. We write the formal equation

$$\begin{aligned}(e+x^*x)^{-1} &= [(|x^*x|_0+1)e-(|x^*x|_0 e-x^*x)]^{-1}\\ &= \frac{1}{|x^*x|_0+1}\left(e-\frac{|x^*x|_0 e-x^*x}{|x^*x|_0+1}\right)^{-1} = \frac{1}{|x^*x|_0+1}\sum_{n=1}^{\infty}\left(\frac{|x^*x|_0 e-x^*x}{|x^*x|_0+1}\right)^n. \end{aligned}\tag{5}$$

The existence of $(e+x^*x)^{-1}$ will be proved if we establish that the series in (5) converges absolutely in the norm in R.

We set $y = |x^*x|_0 e-x^*x$. Then

$$|y|_0 = \sqrt{\sup_{f(e)=1} f(y^*y)} \leqq |x^*x|_0. \tag{6}$$

In fact, in virtue of inequality (3),

$$\begin{aligned} f(y^*y) &= f(|x^*x|_0^2 e-2|x^*x|_0 x^*x+x^*xx^*x)\\ &= |x^*x|_0^2-2|x^*x|_0 f(x^*x)+f(x^*xx^*x)\\ &\leqq |x^*x|_0^2-2|x^*x|_0 f(x^*x)+|x^*x|_0 f(x^*x)\\ &= |x^*x|_0^2-|x^*x|_0 f(x^*x) \leqq |x^*x|_0^2. \end{aligned}$$

Inequality (6) follows from this. Since, by assumption, $|y|_0 = \lim_{n\to\infty} \sqrt[n]{|y^n|}$, (6) can be rewritten in the form

$$\lim_{n\to\infty} \sqrt[n]{|y^n|} \leqq |x^*x|_0,$$

i.e.

$$\lim_{n\to\infty} \sqrt[n]{|(|x^*x|_0 e - x^*x)^n|} \leqq |x^*x|_0.$$

It follows that

$$|(|x^*x|_0 e - x^*x)^n| < (|x^*x|_0 + \tfrac{1}{2})^n$$

for $n \geqq n_0$; consequently,

$$\sum_{n=n_0}^{\infty} \left| \left(\frac{|x^*x|_0 e - x^*x}{|x^*x|_0 + 1} \right)^n \right| \leqq \sum_{n=n_0}^{\infty} \left(\frac{|x^*x|_0 + \frac{1}{2}}{|x^*x|_0 + 1} \right)^n.$$

Therefore, the series in (5) does indeed converge absolutely, and this completes the proof that the algebra R is completely symmetric.

REMARK. It can be shown that the properties of an algebra, formulated in each of the Corollaries 1–3, subsection 3, are also criteria for complete symmetry.

§ 24. Completely regular algebras

1. Fundamental properties of completely regular algebras. We recall that a symmetric normed algebra is said to be *completely regular* if

$$|x^*x| = |x|^2 \tag{1}$$

for all $x \in R$ (see subsection 1, § 16). Examples of a completely regular algebra are the algebra $C(\mathfrak{M})$ of all continuous functions on a compact space $\mathfrak{M}$, and also the algebra $\mathfrak{B}(\mathfrak{H})$ of all bounded linear operators in the Hilbert space $\mathfrak{H}$.

I. *Every completely regular algebra is semisimple.*

In fact, if x is an element of the radical of such an algebra, then (see subsection 3, § 10)

$$0 = \lim_{n\to\infty} \sqrt[2^n]{|(x^*x)^{2^n}|} = \lim_{n\to\infty} \sqrt[2^n]{|x^*x|^{2^n}} = |x^*x| = |x|^2;$$

consequently, $x = 0$.

In particular,

II. *Every symmetric subalgebra in $\mathfrak{B}(\mathfrak{H})$ is semisimple.*

Suppose h is a Hermitian element in a C^*-algebra R and that $\mathfrak{A}$ is a closed commutative symmetric subalgebra in R, which contains h; then $\mathfrak{A}$ is a commutative C^*-algebra and therefore it is completely isomorphic to some algebra $C_0(T)$, where T is a locally compact space (see subsection 2, § 16). The function $h(t)$ corresponding to the element h under this isomorphism will be called the *functional* *representation* of the element h and we shall write $h \underset{\mathfrak{A}}{\sim} h(t)$; of course, this functional representation depends on the choice of the algebra $\mathfrak{A}$, containing h, but in every case

$$|h| = \sup_{t \in T} |h(t)|$$

and $h(t)$ is a real-valued function.

LEMMA 1. *If h is a Hermitian element in the C^*-algebra R, then h^2 has a quasi-inverse $(h^2)'$ in R, and*

$$|(h^2)'| = \frac{|h|^2}{1+|h|^2}. \tag{2}$$

Proof. Suppose $h(t)$ is the functional representation of the element h which corresponds to some closed commutative symmetric subalgebra $\mathfrak{A} \subset R$. Then the function $g(t) = -h^2(t)/\{1+h^2(t)\}$ is continuous and vanishes at infinity; hence, $g(t) \in C_0(T)$. This means that $\mathfrak{A}$ contains an element g such that $g \underset{\mathfrak{A}}{\sim} g(t)$. It follows from the relation $h^2(t)+g(t)+h^2(t)g(t) = 0$, that g is a quasi-inverse of h^2. Since the function $x/(1+x)$, $0 \leqq x \leqq |h|^2$, assumes its largest value in the interval $[0, |h|^2]$ when $x = |h|^2$,

$$|(h^2)'| = |g| = \sup_{t \in T} |g(t)| = \frac{|h|^2}{1+|h|^2}.$$

LEMMA 2. *Suppose x is an element in a commutative, in general not complete, normed algebra R. If for an arbitrary closed maximal regular ideal M in R the image of the element x under the natural homomorphism $R \to R/M$ has a quasi-inverse in R/M, then x has a quasi-inverse in the completion $\tilde{R}$ of R.*

Proof. If x does not have a quasi-inverse in $\tilde{R}$, then there exists in $\tilde{R}$ a maximal regular ideal M', with respect to which $-x$ is the identity element (see VII and V, subsection 4, § 7). Then $M = M' \cap R$ is a closed maximal regular ideal in R and, contrary to assumption, the image $\tilde{x}$ of the element x under the homomorphism $R \to R/M$ does not have a quasi-inverse in R/M.

THEOREM 1. *Suppose $C_0(T)$ is the algebra of all continuous functions $x = x(t)$ on a locally compact space T which vanish at infinity. Then every norm $|x|_1$ in $C_0(T)$, with respect to which $C_0(T)$ is a normed algebra (which may or may not be complete), is not less than the norm $|x| = \sup_{t \in T} |x(t)|$.*

Proof. The maximal regular ideals in $C_0(T)$, which are closed relative to the norm $|x|_1$, form a subspace S in T (see VIII, subsection 2, § 11 and subsection 4, § 11). It suffices to prove that S is dense in T; in fact, then

$$|x|_1 \geqq \sup_{t \in S} |x(t)| = \sup_{t \in T} |x(t)| = |x|.$$

We shall assume that S is not dense in T. Then there exists an open set U in T

which does not intersect S and therefore there exists a function $x(t)$ in $C_0(T)$ which equals -1 on an open set $V \subset U$ and is not equal to -1 anywhere on S. According to the preceding lemma, x has a quasi-inverse z in the completion $\widetilde{C_0(T)}$ of the algebra $C_0(T)$ with respect to the norm $|x|_1$. On the other hand, one can construct a continuous function $y(t) \not\equiv 0$ such that $y+xy = 0$, for which it suffices to take $y \neq 0$ on some set contained in V, and $y = 0$ on the complement of the set V. Multiplying both sides of the equation $x+z+xz = 0$ by y we obtain that $xy+z(y+xy) = 0$, and therefore $xy = 0$. Combining this with the equality $y+xy = 0$, we arrive at the result that $y = 0$, contrary to hypothesis. Consequently, S is dense in T and the theorem is proved.

Theorem 1 shows that the set of functions in the algebra $C_0(T)$ uniquely defines the minimal norm in it.

COROLLARY. *Suppose R is a C^*-algebra, I is a closed two-sided ideal in R, and $\mathfrak{A}$ is a closed subalgebra in R, generated by a Hermitian element, and suppose $J = I \cap \mathfrak{A}$. Then the natural mapping of $\mathfrak{A}/J$ into R/I is isometric.*

Proof. According to the remark at the end of subsection 2, § 16, $\mathfrak{A}/J$ is completely isomorphic to some $C_0(T)$. The natural mapping of $\mathfrak{A}/J$ into R/I is obtained if each class $\xi = a+J$, $a \in \mathfrak{A}$, is assigned the class $\xi' = a+I$; it is easy to see that this mapping is an isomorphism. Since $J \subset I$,

$$|\xi| = \inf_{x \in J} |a+x| \geqq \inf_{x \in I} |a+x| = |\xi'|.$$

Setting $|\xi|_1 = |\xi'|$, we obtain a second norm in $\mathfrak{A}/J$ so that, according to Theorem 1, $|\xi'| \geqq |\xi|$. Therefore, $|\xi'| = |\xi|$.

THEOREM 2. *If another norm $|x|_1$ is given in $C_0(T)$ with respect to which the completion of the algebra $C_0(T)$ is a semisimple algebra, then this norm $|x|_1$ is topologically equivalent to the norm $|x| = \sup_T |x(t)|$.*

Proof. On the basis of Theorem 1, subsection 1, § 12, $|x|_1 \leqq C|x|$; on the other hand, in virtue of the preceding theorem, $|x| \leqq |x|_1$.

THEOREM 3. *Every symmetric isomorphism $x \to x'$ of a C^*-algebra R into a C^*-algebra R' preserves norm.*

Proof. Suppose $x_0 \in R$ and that $\mathfrak{A}$ is the closed subalgebra in R generated by the element $x_0^* x_0$, and let $\mathfrak{A}'$ be its image in R' under the isomorphism $x \to x'$. On the basis of Corollary 2, subsection 1, § 20, the isomorphism $x \to x'$ preserves the norm in $\mathfrak{A}$; in particular, $|x_0^* x_0| = |x_0'^* x_0'|$, i.e. $|x_0|^2 = |x_0'|^2$, and therefore $|x_0| = |x_0'|$.

2. Realization of a completely regular algebra as an algebra of operators. In this subsection, we shall show that every completely regular algebra with identity is completely isomorphic to the algebra of bounded operators in Hilbert space. We shall first prove the following lemmas in which R denotes a C^*-algebra with identity.

I. *The set Q of all squares of Hermitian elements of the algebra R coincides with the set of all Hermitian elements $x \in R$ which satisfy the condition*

$$\left| e - \frac{1}{c} x \right| \leqq 1 \tag{1}$$

with certain c's > 0, in particular, with the set of all Hermitian elements $x \in R$ which satisfy the condition

$$\big| |x|e - x \big| \leqq |x|. \tag{2}$$

Proof. Suppose $x \in Q$; applying to the element $z = (1/c)x$, where $c \geqq |x|$, $c > 0$, any functional representation (see subsection 1), we have $0 \leqq z(t) \leqq 1$, and therefore $|e-(1/c)x| = |e-z| \leqq 1$. Conversely, if $|e-(1/c)x| \leqq 1$, where $c > 0$, then in the functional representation we have $|1-(1/c)x(t)| \leqq 1$, and hence $(1/c)x(t) \geqq 0$, $x(t) \geqq 0$. From this it follows that there exists a Hermitian element $y = y(t) \in R$ such that $y^2 = x$, $x \in Q$.

Clearly, for $x \neq 0$ one can set $c = |x|$; therefore, Q is the set of all Hermitian elements $x \in R$ which satisfy condition (2).

II. *Q is a closed cone in R.*

Proof. It follows directly from condition (2) that Q is closed.

Further, if $x = y^2$, where y is Hermitian and $\lambda \geqq 0$, then $\lambda x = (\sqrt{\lambda} y)^2$, where $\sqrt{\lambda} y$ is Hermitian; consequently, it follows from $x \in Q$, $\lambda \geqq 0$, that $\lambda x \in Q$.

But if $x \in Q$ and also $\lambda x \in Q$ for some $\lambda < 0$, then the functional representation shows that $x(t) \geqq 0$ and $\lambda x(t) \geqq 0$, from which it follows that $x(t) \leqq 0$ and, consequently, $x(t) \equiv 0$, $x = 0$.

Finally, suppose $x, y \in Q$; choosing $c > \max\{|x|, |y|\}$, we have in virtue of I that

$$\left| e - \frac{1}{c} x \right| \leqq 1, \quad \left| e - \frac{1}{c} y \right| \leqq 1,$$

and therefore

$$\left| e - \frac{1}{2c}(x+y) \right| = \tfrac{1}{2} \left| \left(e - \frac{1}{c} x\right) + \left(e - \frac{1}{c} y\right) \right| \leqq 1.$$

Consequently, we also have $x+y \in Q$ in virtue of condition (1).

III. *Q is the set of all Hermitian elements in the algebra R with nonnegative spectrum.*

Proof. The spectrum of a Hermitian element x is the set of all values of the function $x(t)$ in the functional representation. If $x = y^2$ where y is Hermitian, then $x(t) = [y(t)]^2 \geqq 0$. Conversely, if $x(t) \geqq 0$, then, setting $y(t) = \sqrt{x(t)}$, we obtain $x = y^2$, where y is a Hermitian element.

IV. *The elements xy and yx have the same nonzero part of the spectrum.*

Proof. If $(xy-\lambda e)^{-1}$ exists and $\lambda \neq 0$, then

$$(yx-\lambda e)^{-1} = \frac{1}{\lambda}[y(xy-\lambda e)^{-1}x-e]$$

also exists. In fact,

$$(yx-\lambda e)\frac{1}{\lambda}[y(xy-\lambda e)^{-1}x-e] = \frac{1}{\lambda}[y(xy-\lambda e)(xy-\lambda e)^{-1}x-(yx-\lambda e)] = e$$

and, analogously, $(1/\lambda)[y(xy-\lambda e)^{-1}x-e](yx-\lambda e) = e$.

V. *For an arbitrary element $x \in R$ the relation $-x^*x \in Q$ is possible only for $x^*x = 0$.*

Proof. We set $x = u+iv$; then $x^*x+xx^* = 2u^2+2v^2 \in Q$ and

$$x^*x = 2u^2+2v^2+(-xx^*). \tag{3}$$

By assumption, $-x^*x \in Q$ and, consequently, by III, $-x^*x$ has a nonnegative real spectrum; then, by IV, the Hermitian element $-xx^*$ also has a nonnegative spectrum and this means, again in virtue of III, that also $-xx^* \in Q$; then it follows from (3) that $x^*x \in Q$. But since Q is a cone, both relations $-x^*x \in Q$, $x^*x \in Q$ are possible if, and only if, $x^*x = 0$.

THEOREM 4. *Every C^*-algebra with identity is completely symmetric.*

Proof. We must prove that $x^*x \in Q$ for all $x \in R$. The element x^*x is Hermitian and therefore it can be represented in the form $x^*x = u - v$, where $u, v \in Q$, $uv = 0$ (see Corollary 8, subsection 2, § 16). But then

$$(xv)^*(xv) = vx^*xv^* = v(u-v)v = -v^3;$$

consequently, $-(xv)^*(xv) \in Q$. By V, this is possible if, and only if, $(xv)^*(xv) = 0$, i.e. if, and only if, $v^3 = 0$. It follows that $v = 0$ and $x^*x = u \in Q$, i.e. $(e+x^*x)^{-1}$ exists for all $x \in R$, and the theorem is proved.

THEOREM 5. *Every completely regular algebra is completely isomorphic to some algebra of bounded operators in Hilbert space.*

Proof. Without loss of generality, we can assume that R is a C^*-algebra with identity; for, otherwise, R can be embedded in a C^*-algebra R_1 with identity (see I and III, subsection 1, § 16). But in virtue of Theorem 4, the algebra R_1 is also completely symmetric, and therefore (see (1), subsection 4, § 23)

$$\sup_{f(e)=1} f(x^*x) = \lim_{n\to\infty} \sqrt[n]{|(x^*x)^n|}. \tag{4}$$

On the other hand, if we apply (1), subsection 1, to x^*x instead of to x, we obtain that $|(x^*x)^2| = |x^*x|^2$; repeating this reasoning, we find that in general $|(x^*x)^{2^n}| = |x^*x|^{2^n}$. Combining this result with (4), we obtain

$$\sup_{f(e)=1} f(x^*x) = \lim_{n\to\infty} \sqrt[2^n]{|(x^*x)^{2^n}|} = \lim_{n\to\infty} \sqrt[2^n]{|x^*x|^{2^n}} = |x^*x| = |x|^2.$$

Consequently, the norm in the algebra R_1 coincides with the minimal regular norm.

On the basis of Theorem 1, subsection 3, § 18, the algebra R_1, and hence also the initial algebra R, is completely isomorphic to the algebra of bounded operators in a Hilbert space.

3. Quotient algebra of a completely regular algebra.

THEOREM 6. *Every closed two-sided ideal I in the C^*-algebra R is symmetric and the quotient algebra R/I is also completely regular.*

Proof. We set

$$y = x[c(x^*x)^2]' + x \tag{1}$$

for an arbitrary $x \in R$ and arbitrary $c > 0$, where $[c(x^*x)^2]'$ is a quasi-inverse element of $c(x^*x)^2$; in virtue of the complete regularity of the complete algebra R, the element $[c(x^*x)^2]'$ exists (see Lemma 1, subsection 1). Moreover, setting $z = x^*x$, we have

$$y^*y = [(cz^2)' + e]x^*x[(cz^2)' + e] = [(cz^2)' + e]^2 z. \tag{2}$$

But for positive c, the function $\lambda/(1+c\lambda^2)^2$ attains its maximal value $9/(16\sqrt{3c})$ at the point $\lambda = 1/\sqrt{3c}$; consequently, applying the functional representation to (2), we conclude that $|y^*y| \leqq 9/(16\sqrt{3c})$, and hence

$$|y| \leqq 3/(4\sqrt[4]{3c}). \tag{3}$$

Applying the functional representation to (2) yields

$$(y^*y)(t) = \left[-\frac{cz^2(t)}{1+cz^2(t)} + 1\right]^2 z(t) = \frac{z(t)}{[1+cz^2(t)]^2}.$$

Moreover, in virtue of (2), subsection 1,

$$|(cz^2)'| = c|z|^2/(1+c|z|^2). \tag{4}$$

Now suppose I is a closed two-sided ideal in R and that $x \in I$; then also $z = x^*x \in I$ and $(cz^2)' \in I$. But, in virtue of (1),

$$x^* = y^* - (cz^2)'x^*. \tag{5}$$

Since $-(cz^2)'x^* \in I$ and, according to (3), $|y^*| \to 0$ as $c \to \infty$, we conclude from (5) that $x^* \in I$. Consequently, the ideal I is symmetric.

Consider the quotient algebra R/I. Since the natural homomorphism $R \to R/I$ can only decrease the norm, inequality (3) also remains valid in R/I. Suppose $\mathfrak{A}$ is a closed commutative subalgebra in R, generated by the element z; we set $J = I \cap \mathfrak{A}$. Relation (4) holds in $\mathfrak{A}/J$, because $\mathfrak{A}/J$ admits a representation in the form of an algebra of functions. According to the corollary to Theorem 1, subsection 1, relation (4) also remains valid in R/I. Combining (1), (3) and (4), we conclude that

$$|x| \leqq |y|+|x(cz^2)'| \leqq 3/(4\sqrt[4]{3c})+c|x|\,|z|^2/(1+c(|z|^2)$$

in R/I. Setting $c = 1/(3|z|^2)$ in the last inequality, we obtain

$$|x| \leqq \tfrac{3}{4}\sqrt{|z|}+\tfrac{1}{4}|x|.$$

It follows that

$$\tfrac{3}{4}|x| \leqq \tfrac{3}{4}\sqrt{|z|},\ |x|^2 \leqq |z| = |x^*x|.$$

But, on the other hand, $|x^*x| \leqq |x^*|\,|x| = |x|^2$ (because we also have $|x^*| = |x|$ in R/I). Consequently, $|x^*x| = |x|^2$ in R/I and the theorem is proved.

COROLLARY. *Under a symmetric homomorphism $x \to x'$ of a C^*-algebra R into a C^*-algebra R', the image of R is complete.*

Proof. Suppose I is the kernel of the homomorphism $x \to x'$; then I is a closed symmetric ideal in R and, in virtue of Theorem 6, R/I is a C^*-algebra. The homomorphism $x \to x'$ of the algebra R into the algebra R' generates an isomorphism $\tilde{x} \to x'$ of the algebra R/I into the algebra R', where the image of the algebra R under the homomorphism $x \to x'$ coincides with the image of the algebra R/I under the isomorphism $\tilde{x} \to x'$. But according to Theorem 3, subsection 1, the isomorphism $\tilde{x} \to x'$ is isometric, and therefore this image is a complete algebra.

[In virtue of Theorem 5, subsection 2, R' can be considered to be a symmetric subalgebra of the algebra $\mathfrak{B}(\mathfrak{H})$ and hence the homomorphism $x \to x'$ to be a representation of the algebra R; on the basis of Theorem 1, subsection 3, § 17, this homomorphism is continuous.]

EXAMPLE. Suppose R is the algebra of all bounded operators in the Hilbert space $\mathfrak{H}$ ($R = \mathfrak{B}(\mathfrak{H})$), and that I is the set of all compact operators in $\mathfrak{H}$. Then I is a closed two-sided ideal in R; since R is completely regular, then, on the basis of Theorem 6, the quotient algebra R/I is also completely regular. Consequently, applying Theorem 5, subsection 2, to R/I we arrive at the following result.

THEOREM 7. *The quotient algebra $\mathfrak{B}(\mathfrak{H})$ modulo the ideal I of all compact operators in $\mathfrak{H}$* [i.e., $\mathfrak{B}(\mathfrak{H})/I$] *is completely isomorphic to an algebra of bounded operators is a Hilbert space $\mathscr{H}$.*

This theorem was first established by CALKIN [1] who constructed directly a realization of the algebra $\mathfrak{B}(\mathfrak{H})/I$ in the form of an algebra of operators in a Hilbert space.

§ 25. Dual algebras

1. Annihilator algebras and dual algebras. Suppose R is a topological algebra and that S is an arbitrary set in R. The *left annihilator* $\mathfrak{L}(S)$ of the set S is the set of all elements $x \in R$ for which $xS = (0)$; analogously, the *right annihilator* $\mathfrak{R}(S)$ of the set S is the set of all elements $x \in R$ for which $Sx = (0)$. Clearly, $\mathfrak{L}(S)$, $\mathfrak{R}(S)$, if they do not coincide with the entire algebra, are closed left and right ideals respectively; moreover, it is obvious that

$$S \subset \mathfrak{L}(\mathfrak{R}(S)) \quad \text{and} \quad S \subset \mathfrak{R}(\mathfrak{L}(S)) \tag{1a}$$

$$\mathfrak{L}(S_1) \subset \mathfrak{L}(S_2), \quad \mathfrak{R}(S_1) \subset \mathfrak{R}(S_2) \quad \text{for} \quad S_1 \supset S_2 \tag{1b}$$

and if S_1, S_2 are linear subspaces in R, then

$$\mathfrak{L}(S_1+S_2) = \mathfrak{L}(S_1) \cap \mathfrak{L}(S_2), \quad \mathfrak{R}(S_1+S_2) = \mathfrak{R}(S_1) \cap \mathfrak{R}(S_2). \tag{1c}$$

In some cases in this section, we shall find it convenient to consider also the entire algebra R as an ideal. We shall call the algebra R an *improper* (*right, left, and two-sided*) *ideal.* Then we can say that $\mathfrak{L}(\mathfrak{R}(S))$ and $\mathfrak{R}(\mathfrak{L}(S))$ are closed (proper or improper) left and right ideals, containing S.

The algebra R is called an *annihilator algebra* if

$$\mathfrak{L}(R) = \mathfrak{R}(R) = (0) \tag{2}$$

and

$$\mathfrak{L}(I_r) \neq (0), \tag{3a}$$

$$\mathfrak{R}(I_l) \neq (0) \tag{3b}$$

for any closed left ideal I_l and arbitrary closed right ideal I_r.

The algebra R is called a *dual algebra* if for every closed (proper or improper) left ideal I_l and for every closed (proper or improper) right ideal I_r,

$$\mathfrak{L}(\mathfrak{R}(I_l)) = I_l, \tag{4a}$$

$$\mathfrak{R}(\mathfrak{L}(I_r)) = I_r. \tag{4b}$$

It follows directly from these definitions that *every dual algebra is also an annihilator algebra.* In fact, setting $I_l = (0)$, $I_r = (0)$ in (4a), (4b), and taking into consideration the fact that $\mathfrak{R}((0)) = R$, $\mathfrak{L}((0)) = R$, we obtain (2). Further, $\mathfrak{R}(I_l) = (0)$ cannot hold for the ideal I_l because then we would have $\mathfrak{L}(\mathfrak{R}(I_l)) = \mathfrak{L}((0)) = R \neq I_l$, which contradicts (4a); consequently, $\mathfrak{R}(I_l) \neq 0$ and thus (3a) is proved. (3b) is proved analogously.

I. *For an arbitrary subset S of a dual algebra, $\mathfrak{R}(\mathfrak{L}(S))$ and $\mathfrak{L}(\mathfrak{R}(S))$ are minimal closed (proper or improper) right and left ideals, respectively, containing S.*

In fact, if, for instance, I_r is the minimal closed (proper or improper) right ideal, containing S, then $I_r = \mathfrak{R}(\mathfrak{L}(I_r)) \supset \mathfrak{R}(\mathfrak{L}(S))$; since $\mathfrak{R}(\mathfrak{L}(S))$ also is a closed (proper or improper) right ideal, containing S, we have $I_r = \mathfrak{R}(\mathfrak{L}(S))$.

II. *If I_r, J_r are closed right ideals in the dual algebra R, then*

$$\mathfrak{L}(I_r \cap J_r) = \overline{\mathfrak{L}(I_r)+\mathfrak{L}(J_r)}; \tag{5a}$$

analogously, for closed left ideals,

$$\mathfrak{R}(I_l \cap J_l) = \overline{\mathfrak{R}(I_l)+\mathfrak{R}(J_l)}. \tag{5b}$$

Proof. Since $\mathfrak{L}(I_r)+\mathfrak{L}(J_r)$ is a (proper or improper) left ideal, then, in virtue of I, (4a) and (4b),

$$\overline{\mathfrak{L}(I_r)+\mathfrak{L}(J_r)} = \mathfrak{L}\{\mathfrak{R}[\mathfrak{L}(I_r)+\mathfrak{L}(J_r)]\} = \mathfrak{L}\{\mathfrak{R}[\mathfrak{L}(I_r)] \cap \mathfrak{R}[\mathfrak{L}(J_r)]\} = \mathfrak{L}(I_r \cap J_r).$$

(5b) is proved analogously.

III. *If R is a dual algebra, then*

$$x \in \overline{xR} \quad \textit{and} \quad x \in \overline{Rx}$$

for every $x \in R$.

Proof. We shall prove that $x \in \overline{xR}$; the relation $x \in \overline{Rx}$ is proved analogously. Since $\overline{xR} = \mathfrak{R}(\mathfrak{L}(\overline{xR}))$, it suffices to prove that $\mathfrak{L}(\overline{xR})x = (0)$. But if $y \in \mathfrak{L}(\overline{xR})$, then $yxR = 0$. Consequently, $yx \in \mathfrak{L}(R) = (0)$, $yx = 0$; this means that $\mathfrak{L}(\overline{xR})x = (0)$.

2. Ideals in an annihilator algebra. Everywhere in this subsection, R will denote an annihilator algebra with continuous quasi-inverse (see subsection 6, § 8).

I. *If $I_l = \{yp-y: y \in R\}$, then $\mathfrak{R}(I_l) = \{x \in R;\ px = x\}$.*

Proof. If $x \in \mathfrak{R}(I_l)$, then $y(px-x) = (yp-y)x = 0$ for all $y \in R$, so that in virtue of (2), subsection 1, $px-x = 0$. Conversely, if $px-x = 0$, then

$$(yp-y)x = y(px-x) = 0 \quad \text{and} \quad x \in \mathfrak{R}(I_l).$$

Clearly, the analogous proposition holds for the ideal $I_r = \{py-y: y \in R\}$.

II. *If the element $-p$ does not have a left quasi-inverse, then there exists an element $x \neq 0$ such that $px = x$.*

Proof. Suppose $-p$ does not have a left quasi-inverse; then the set $I_l = \{yp-y: y \in R\}$ is a regular left ideal in R (see VII, subsection 4, § 7). Consequently, its closure $\bar{I}_l$ also is a regular left ideal in R (see IV, subsection 3, I, subsection 6, § 8 and VI, subsection 4, § 7), and therefore $\mathfrak{R}(\bar{I}_l) \neq (0)$. Suppose x is a nonzero element in $\mathfrak{R}(\bar{I}_l)$; then also $x \in \mathfrak{R}(I_l)$ and in virtue of I, $px = x$.

A left ideal I_l is said to be *minimal* if it does not contain left ideals, distinct from (0) and I_l; minimal right and two-sided ideals are defined analogously. Further, a closed left ideal I_l is called a *minimal closed ideal* if I_l does not contain closed left ideals distinct from (0) and I_l; minimal closed right and two-sided ideals are defined similarly. A closed left ideal is called a *maximal closed ideal* if it is not contained in any other closed left ideal. In an analogous manner, we define minimal closed right and two-sided ideals. An element $p \neq 0$ of the algebra R is said to be *idempotent* if $p^2 = p$.

THEOREM 1. *If M_r is a maximal closed right ideal in R and $\mathfrak{L}(M_r)$ intersects the radical of the algebra R only in the zero element, then $\mathfrak{L}(M_r)$ contains an idempotent p. Moreover, $\mathfrak{L}(M_r) = Rp$ and*

$$M_r = \{x-px: x \in R\}. \tag{1}$$

Proof. Since $\mathfrak{L}(M_r) \neq (0)$, there exists an element $a \neq 0$, $a \in \mathfrak{L}(M_r)$. But then

$M_r \subset \mathfrak{R}((a)) \neq R$ (see (2), subsection 1) and since M_r is maximal,

$$\mathfrak{R}((a)) = M_r. \tag{2}$$

In virtue of the condition of the theorem, the element a does not belong to the radical; consequently, for some λ and some $y \in R$ the element $-p = \lambda a + ya$ does not have a left quasi-inverse (see the definition of the radical of an algebra without identity in subsection 5, § 7) and hence $p \neq 0$; clearly, $p \in \mathfrak{L}(M_r)$. In virtue of II, there exists an element $x \neq 0$ such that $px = x$ and therefore $(p^2 - p)x = 0$. We shall prove that $p^2 - p = 0$; by the same token, the first assertion of the theorem will be proved. We shall assume the contrary, that $p^2 - p \neq 0$. Since $p^2 - p \in \mathfrak{L}(M_r)$, then in virtue of (2) (with a replaced by $p^2 - p$), we have $x \in M_r$, and hence $x = px = 0$, in contradiction with what we stated above. Consequently, $p^2 = p$.

Since $p \in \mathfrak{L}(M_r)$ and $p \neq 0$, then in virtue of (2) (with a replaced by p), we have $M_r = \mathfrak{R}((p)) = \{x - px \colon x \in R\}$; in fact, $p(x - px) = px - p^2x = 0$ and if $py = 0$, then $y = y - py$. It follows, in virtue of I and the fact that p is idempotent, that $\mathfrak{L}(M_r) = Rp$.

COROLLARY 1. *Under the same conditions as in Theorem* 1, M_r *is a maximal right ideal and* $\mathfrak{L}(M_r)$ *is a minimal left ideal. Moreover,* pR *is a minimal right ideal, and* $\mathfrak{L}(pR)$ *is a maximal left ideal.*

Proof. In virtue of Theorem 1, M_r is regular; moreover, by assumption, M_r is a maximal closed right ideal. Consequently, M_r is a maximal right ideal (see IV, subsection 3, I, subsection 6, § 8 and V and VI, subsection 4, § 7). Further, in virtue of (1),

$$R = pR \dotplus M_r, \tag{3}$$

and this equality is a decomposition of the algebra R into the direct sum of the right ideals pR and M_r. Since M_r is maximal, we conclude from (3) that pR is minimal. (Everywhere in this section, $\mathfrak{M} \dotplus \mathfrak{N}$ denotes the *direct sum* of the linearly independent subspaces $\mathfrak{M}$, $\mathfrak{N}$ (see subsection 3, § 1), i.e. the set of all vectors $x + y$, $x \in \mathfrak{M}$, $y \in \mathfrak{N}$; thus, the notation $\mathfrak{M} \dotplus \mathfrak{N}$ already connotes that $\mathfrak{M}$, $\mathfrak{N}$ are linearly independent.)

We set $I_l = \mathfrak{L}(pR)$. If $y \in I_l$, then $ypR = 0$; consequently, $yp = 0$ and $y = y - yp$. From this we conclude that $I_l = \{y - yp \colon y \in R\}$ and, consequently, in virtue of I, $\mathfrak{R}(I_l) = pR$.

We shall prove that I_l is a maximal closed left ideal. In fact, suppose J_l is a closed left ideal containing I_l; then $(0) \neq \mathfrak{R}(J_l) \subset pR$. Since pR is minimal, it follows from this that $\mathfrak{R}(J_l) = pR$, i.e. $J_l pR = (0)$. But then in virtue of (2), subsection 1, $J_l p = (0)$, i.e. $J_l = \{y - yp \colon y \in R\} = I_l$. Consequently, I_l is a maximal closed left ideal; since I_l is regular, it is also a maximal left ideal. But then it follows from the relation $R = I_l \dotplus Rp$ that $\mathfrak{L}(M_r) = Rp$ is a minimal left ideal.

THEOREM 2. *If I_l is a minimal closed left ideal or a minimal left ideal in R which is not contained entirely in the radical, then I_l contains an idempotent p and*

$$I_l = Rp, \quad \mathfrak{R}(Rp) = \{x - px: x \in R\}.$$

Proof. By assumption, I_l contains an element x_0 which does not belong to the radical; consequently, for some λ and some $y \in R$, the element $-p = \lambda x_0 + yx_0$ (which also belongs to I_l) does not have a left quasi-inverse and hence $p \neq 0$. In virtue of II, it follows from this that there exists an element $a \neq 0$ such that $pa = a$. On the other hand, the set $J_l = \{x \in I_l: xa = 0\}$ is a left ideal (which is closed if I_l is closed), contained in I_l, and not coinciding with I_l, because $p \notin J_l$, $p \in I_l$. In virtue of the minimality of the ideal I_l, this is possible only when $J_l = (0)$, so that $xa \neq 0$ for $x \in I_l$, $x \neq 0$. But since $p^2 - p \in I_l$ and $(p^2 - p)a = 0$, we have $p^2 - p = 0$ and so p is idempotent.

We note that Rp is the set of all elements $x \in R$ which satisfy the condition $x = xp$ and therefore Rp is a closed set. Consequently, Rp is a closed left ideal, contained in I_l, and hence $Rp = I_l$ in virtue of the minimality of I_l. But then it is obvious that $\mathfrak{R}(I_l) = \{x - px: x \in R\}$.

COROLLARY 2. *Under the same conditions, I_l is a minimal left ideal and $\mathfrak{R}(I_l)$ is a maximal right ideal. Moreover, pR is a minimal right ideal and $\mathfrak{L}(pR)$ is a maximal left ideal.*

The proof of this corollary is analogous to the proof of Corollary 1.

COROLLARY 3. *Every minimal left ideal I_l in R which is not contained in the radical is closed.*

In fact, $I_l = Rp$, and Rp is a closed set.

REMARK 1. Clearly, in all the preceding results, the roles of left and right ideals could have been interchanged.

THEOREM 3. *Every normed annihilator C^*-algebra R is dual.*

Proof. Suppose I_l is a closed left ideal in R; we shall first prove that

$$R = I_l \dotplus (\mathfrak{R}(I_l))^*. \tag{4}$$

If $x \in I_l \cap (\mathfrak{R}(I_l))^*$, then $xx^* = 0$ and, consequently, $x = 0$. Thus, $I_l \cap (\mathfrak{R}(I_l))^* = (0)$. We set $J_l = I_l \dotplus (\mathfrak{R}(I_l))^*$; if $J_l \neq R$, then J_l is a left ideal in R. This ideal is closed. In fact, suppose $x = y + z$, $y \in I_l$, $z \in (\mathfrak{R}(I_l))^*$. Then $xz^* = zz^*$, and hence $|x|\,|z^*| \geqq |z|^2$, $|x| \geqq |z|$.

Analogously, $|x| \geqq |y|$. It follows easily from these inequalities that J_l is closed. But then, in virtue of condition (3b), subsection 1, there exists a nonzero element $a \in R$ such that $J_l a = (0)$ and, consequently, also $I_l a = (0)$, $(\mathfrak{R}(I_l))^* a = (0)$.

It follows from the first equality that $a \in \mathfrak{R}(I_l)$, $a^* \in (\mathfrak{R}(I_l))^*$, and then the second equality yields $a^*a = 0$, which is impossible. Therefore $J_l = R$ and (4) is proved.

One proves analogously that

$$R = I_r \dot{+} (\mathfrak{L}(I_r))^* \tag{5}$$

for an arbitrary closed right ideal I_r. Setting $I_r = \mathfrak{R}(I_l)$ in (5), we obtain

$$R = \mathfrak{R}(I_l) \dot{+} (\mathfrak{L}(\mathfrak{R}(I_l)))^*,$$

or, applying the involution operation to both members of this relation,

$$R = (\mathfrak{R}(I_l))^* \dot{+} \mathfrak{L}(\mathfrak{R}(I_l)). \tag{6}$$

Since $I_l \subset \mathfrak{L}(\mathfrak{R}(I_l))$, comparison of relations (6) and (4) yields that $I_l = \mathfrak{L}(\mathfrak{R}(I_l))$. Applying the involution operation, we conclude that also $I_r = \mathfrak{R}(\mathfrak{L}(I_r))$ for an arbitrary closed right ideal I_r, and the theorem is proved.

REMARK 2. We note that for Theorem 3 to be valid in reality only one of the conditions (3), subsection 1, suffices.

In fact, it is easily seen that each of these conditions is obtained from the other by applying the involution operation, and condition (2), subsection 1, is satisfied in every completely regular algebra.

3. Semisimple annihilator algebras. In this subsection, R will denote a semisimple annihilator algebra with continuous quasi-inverse. It follows from the results of subsection 2 that in the algebra R every minimal left ideal is closed and contains an idempotent and that the annihilator of a maximal closed right ideal is a minimal left ideal, and the annihilator of a minimal left ideal is a maximal regular right ideal.

THEOREM 4. *The sum of all minimal left (or right) ideals of an algebra R is dense in R.* (By the sum of the left ideals $I_l^{(\alpha)}$ is meant the set of all possible finite sums of elements $x_\alpha \in I_l^{(\alpha)}$. The sum of right and two-sided ideals is defined analogously.)

Proof. Suppose S is the sum of all the minimal left ideals in R and that $\bar{S}$ is its closure. If $\bar{S} \neq R$, then $\bar{S}$ is a closed left ideal in R and therefore there exists an element $x \neq 0$ such that $\bar{S}x = (0)$. Then x belongs to all the right annihilators of all the minimal left ideals, i.e. to the intersection of all maximal regular right ideals. But according to III′, subsection 5, § 7, this intersection is the radical of the algebra R and hence $= (0)$ in virtue of the semisimplicity of R. Consequently, $x = 0$, and we have arrived at a contradiction. Hence $\bar{S} = R$.

I. *If the (left, right or two-sided) ideal I in R satisfies the condition $I^2 = (0)$, then $I = (0)$.*

Proof. Suppose, for instance, that I is a left ideal in R and that $I^2 = (0)$. It follows that $(\alpha x + yx)^2 = 0$ for all numbers α and all $x \in I$, $y \in R$, because also $\alpha x + yx \in I$. But then the element $z = \alpha x + yx$ has a left quasi-inverse, equal to $-z$, and therefore x belongs to the radical of the algebra R (see subsection 5, § 7).

In virtue of the semisimplicity of the algebra R, we conclude from this that $x = 0$ and, consequently, $I = (0)$. The assertion for right ideals is proved analogously.

II. *If I_r is a minimal right ideal in R then the closed two-sided ideal $[I_r]$ generated by the ideal I_r is a minimal closed two-sided ideal in R.*

Proof. Suppose J is a closed two-sided ideal contained in $[I_r]$; then $I_r \cap J$ is a right ideal, contained in the minimal right ideal I_r, and consequently either $I_r \cap J = I_r$, i.e. $I_r \subset J$, or $I_r \cap J = (0)$.

In the first case, $[I_r] \subset J$ and therefore $[I_r] = J$. In the second case, $I_r J \subset I_r \cap J = (0)$ and, consequently, $I_r \subset \mathfrak{L}(J)$. But, as is easily seen, $\mathfrak{L}(J)$ is a closed two-sided ideal and hence also $[I_r] \subset \mathfrak{L}(J)$. But then $J \subset \mathfrak{L}(J)$; consequently $J^2 = (0)$ and, in virtue of I, $J = (0)$.

Thus, in the second case $J = (0)$, and the minimality of the ideal $[I_r]$ is proved.

Two idempotents p_1, p_2 are said to be *orthogonal* if $p_1 p_2 = p_2 p_1 = 0$. A system $\{p\}$ of idempotents is said to be *orthogonal* if every two distinct idempotents of this system are orthogonal. An idempotent p is said to be *irreducible* if it is not the sum of two mutually orthogonal idempotents.

III. *If pR is a minimal right ideal (Rp is a minimal left ideal), then p is an irreducible idempotent.*

Proof. Suppose p is reducible, $p = p_1 + p_2$, where $p_1 p_2 = p_2 p_1 = 0$. Then $p_1 R = pp_1 R \subset pR$, $p_2 R = pp_2 R \subset pR$ and

$$pR = p_1 R \dotplus p_2 R. \tag{1}$$

In fact, $px = p_1 x + p_2 x$, and if $p_1 x + p_2 y = 0$, then, multiplying both members of the last equality on the left by p_1 and p_2, we obtain that $p_1 x = p_2 y = 0$. But it follows from (1) that pR is not a minimal right ideal.

IV. *If pR is a minimal right ideal and $a \in R$, then either $apR = (0)$ or $apR = p'R$ is a minimal right ideal.*

Proof. A left regular representation $x \to ax$ maps pR onto apR; the kernel of the mapping is a right ideal, contained in pR. Consequently, either this kernel is (0) and the mapping is one-to-one, or it coincides with pR and then $apR = (0)$.

In the first case, apR is a minimal right ideal in R and hence $apR = p'R$ for some irreducible p'.

V. *Every right ideal I_r in R contains a minimal right ideal and, consequently, it contains an irreducible idempotent p.*

Proof. Suppose I_r does not contain any minimal right ideals and let pR be a minimal right ideal. Then $pR \cap I_r$ is a right ideal, contained in the minimal ideal pR and not coinciding with pR because pR is not contained in I_r. Consequently, $pR \cap I_r = (0)$.

In virtue of IV, for arbitrary $a \in R$, apR is either (0) or a minimal right ideal; therefore also $apR \cap I_r = (0)$ for all $a \in R$. Hence $ap \cap I_r = app \cap I_r \subset apR \cap I_r = (0)$ for all $a \in R$, i.e. $Rp \cap I_r = (0)$. Since $I_r Rp \subset Rp \cap I_r$ then also $I_r Rp = (0)$.

But in virtue of Corollary 2 and Remark 1, subsection 2, if pR runs through all minimal right ideals, then Rp runs through all minimal left ideals so that $I_r Rp = (0)$ for all minimal left ideals Rp. According to Theorem 4, it follows from this that $I_r R = (0)$ and hence $I_r = (0)$.

VI. *If p is an irreducible idempotent, then pR and Rp are minimal right and left ideals, respectively.*

Proof. Suppose p is an irreducible idempotent. We shall assume that the right ideal pR is not minimal. In virtue of V, pR contains a minimal right ideal $p'R$, which does not coincide with pR, because pR is not minimal; in particular, $p' = p'^2 = pa$ for some $a \in R$. Then the element $p'' = p'p = pap$ belongs to $p'R$ and commutes with p, because $pp'' = p^2ap = pap = p''$ and $p''p = pap^2 = pap = p''$; moreover, $p''^2 = p'pp'' = p'p'' = p'p = p''$. Also, $p'' \neq 0$, because $p''p' = p'pp' = (pa)p(pa) = (pa)^2 = p'^2 = p' \neq 0$; consequently, p'' is an idempotent contained in $p'R$.

Further, also $p-p'' \neq 0$ (because $p''R \subset p'R \neq pR$) and $p''(p-p'') = p''-p'' = 0$. But then, setting $p = p''+(p-p'')$, we obtain, contrary to the irreducibility of the idempotent p, a representation of p in the form of the sum of two mutually orthogonal idempotents. Consequently, pR is minimal.

VII. *The left and right annihilators of a closed two-sided ideal I coincide.*

Proof. First of all, we note that in virtue of I, $I \cap \mathfrak{R}(I) = (0)$ because $I \cap \mathfrak{R}(I)$ is a right ideal, satisfying the condition $J^2 = (0)$. Consequently, also $\mathfrak{R}(I)I = (0)$, i.e. $\mathfrak{R}(I) \subset \mathfrak{L}(I)$; analogously, $\mathfrak{L}(I) \subset \mathfrak{R}(I)$ and therefore $\mathfrak{L}(I) = \mathfrak{R}(I)$.

VIII. *For an arbitrary closed two-sided ideal I, the ideal $I+\mathfrak{L}(I) = I+\mathfrak{R}(I)$ is dense in R.*

Proof. If $I+\mathfrak{L}(I)$ is not dense in R, then its closure is an ideal in R and, consequently, there exists an element $x \neq 0$ such that $x(I+\mathfrak{L}(I)) = (0)$, from which it follows that $(\lambda x+yx)I = (0)$, $(\lambda x+yx)\mathfrak{L}(I) = (0)$. But from the first equality it follows that $\lambda x+yx \in \mathfrak{L}(I)$, and consequently from the second that $(\lambda x+yx)^2 = 0$ for all $y \in R$ and all λ; but this is impossible in a semisimple algebra for $x \neq 0$ (see the proof of Proposition I).

IX. *Every minimal closed two-sided ideal I in R is an annihilator algebra; if, moreover, R is a dual algebra then I is also a dual algebra.*

Proof. Suppose $x \in I$ and $Ix = (0)$; then $x = 0$, because $I \cap \mathfrak{R}(I) = (0)$ (see the proof of Proposition VII).

Analogously, if $xI = (0)$ and $x \in I$, then $x = 0$. Consequently, condition (2), subsection 1, is satisfied.

Now suppose J_l, J_r are closed left and right ideals in I. We must prove that $\mathfrak{R}(J_l) \cap I \neq (0)$ and that $\mathfrak{L}(J_r) \cap I \neq (0)$. We note that J_l is also a closed left ideal in R. In fact, $(I+\mathfrak{L}(I))J_l = IJ_l \subset J_l$, and since $I+\mathfrak{L}(I)$ is dense in R, then also $RJ_l \subset J_l$. We set $I_l = J_l+\mathfrak{L}(I) = J_l+\mathfrak{R}(I)$; then either I_l is dense in R or $\mathfrak{R}(I_l) \neq (0)$.

Consider the first case. Suppose $a \in I$; then $aI_l = aJ_l + a\mathfrak{R}(I) = aJ_l \subset J_l$ and therefore also $aR = a\bar{I}_l \subset J_l$. This means that $IR \subset J_l$ and *a fortiori* that $I^2 \subset J_l$. But I^2 is a two-sided ideal, distinct from (0) (in virtue of the semisimplicity of the algebra R; see I), contained in the minimal closed two-sided ideal I. Consequently, $\bar{I}^2 = I$; therefore also $I \subset J_l$ and then $J_l = I$, contrary to our assumption. Thus, the first case is impossible and $\mathfrak{R}(I_l) \neq (0)$. In virtue of V, $\mathfrak{R}(I_l)$ contains an irreducible idempotent p. We shall prove that $p \in I$. Suppose $p \notin I$ and let $[pR]$ be a two-sided ideal generated by the minimal right ideal pR. Then $[pR] \cap I$ is a closed two-sided ideal contained in the minimal closed two-sided ideals $[pR]$ (see II) and I, and hence either coincides with both of them or equals zero. But the first case is impossible because $[pR]$ contains the element $p^2 = p \notin I$; consequently, $[pR] \cap I = (0)$. But then also $pI = (0)$ (because $pI \subset [pR] \cap I$), i.e. $p \in \mathfrak{L}(I) \subset I_l$. Since, on the other hand, $p \in \mathfrak{R}(I_l)$, we arrive at the impossible equality $p^2 = 0$. Thus $\mathfrak{R}(I_l) \cap I \neq (0)$ and *a fortiori* $\mathfrak{R}(J_l) \cap I \neq (0)$. One proves analogously that $\mathfrak{L}(J_r) \cap I \neq (0)$.

We now assume that R is a semisimple dual algebra; we shall prove that I is also dual.

We shall show that if $x \in I$ and $[\mathfrak{L}(J_r) \cap I]x = (0)$, then $x \in J_r$. In virtue of (1a), subsection 1, by the same token it will be proved that condition (4b), subsection 1, is satisfied in I; it is proved analogously that also condition (4a), subsection 1, is satisfied in I.

In virtue of the duality of the algebra R, it will suffice to show that $\mathfrak{L}(J_r)x = (0)$. By assumption,

$$x \in \mathfrak{R}[\mathfrak{L}(J_r) \cap I] = \overline{\mathfrak{R}[\mathfrak{L}(J_r)] + \mathfrak{R}(I)} = \overline{J_r + \mathfrak{R}(I)} = \overline{J_r + \mathfrak{L}(I)} \tag{2}$$

(see (5b), subsection 1, and VII). But $(J_r + \mathfrak{L}(I))I = J_r I \subset J_r$ because J_r is a right ideal in I; consequently, also $(\overline{J_r + \mathfrak{L}(I)})I \subset J_r$. It follows, in virtue of (2), that $xI \subset J_r$, and therefore $\mathfrak{L}(J_r)xI \subset \mathfrak{L}(J_r)J_r = (0)$. Moreover, $\mathfrak{L}(J_r)x\mathfrak{R}(I) = (0)$, because $x \in I$; consequently, also $\mathfrak{L}(J_r)x(I + \mathfrak{R}(I)) = (0)$. Since $I + \mathfrak{R}(I)$ is dense in R (see VIII), then also $\mathfrak{L}(J_r)xR = (0)$, and, consequently, $\mathfrak{L}(J_r)x \subset \mathfrak{L}(R) = (0)$.

THEOREM 5. *Every semisimple annihilator algebra R with continuous quasi-inverse is the completion of the direct sum of all its minimal closed two-sided ideals which are simple annihilator algebras and in the case of a dual algebra R, those which are simple dual algebras.* [A sum of two-sided ideals I^α (see Theorem 4 at the beginning of this subsection) is said to be *direct* if the I^α are linearly independent subspaces in R and $I^{\alpha_1}I^{\alpha_2} = (0)$ for $\alpha_1 \neq \alpha_2$.] [In this section, we shall call the algebra R *simple* if it contains no *closed* two-sided ideals.]

Proof. In virtue of V, every minimal closed two-sided ideal I in R contains a minimal right ideal I_r and, consequently, it coincides with $[I_r]$ (see II). Conversely, in virtue of II, for an arbitrary minimal right ideal I_r the ideal $[I_r]$ is a minimal closed two-sided ideal. According to IX, the $[I_r]$ are annihilator algebras and they are dual algebras provided R is dual. Moreover, the $[I_r]$ are simple algebras because otherwise every closed two-sided ideal in $[I_r]$ would be a closed two-sided ideal in R (see the proof of Proposition IX), contained in the minimal $[I_r]$, which is impossible.

According to Theorem 4, the sum of all minimal right ideals I_r, and *a fortiori* of all minimal closed two-sided ideals $[I_r]$, is dense in R. It remains to prove that this sum is direct. But if $[I_r] \neq [I'_r]$ then

$$I_r I'_r \subset [I_r] \cap [I'_r] = (0), \tag{3}$$

because $[I_r] \cap [I'_r]$ is a closed two-sided ideal, contained in the minimal ideals $[I_r]$, $[I'_r]$ and not coinciding with them. Suppose

$$0 = x_1 + x_2 + \cdots + x_n, \tag{4}$$

where $x_k \in [I_r^{(k)}]$, $k = 1, 2, \cdots, n$, and $[I_r^{(1)}], \cdots, [I_r^{(n)}]$ are pairwise distinct. Multiplying both members of (4) on the left by $I_r^{(k)}$ and taking (3) into consideration, we obtain that $[I_r^{(k)}]x_k = (0)$, and consequently, $(x_k R)^2 \subset [I_r^{(k)}] \cdot (x_k R) = (0)$. It follows that $x_k R = (0)$ in virtue of the semisimplicity of the algebra R, and hence $x_k = 0$.

COROLLARY. *Suppose in addition to the conditions of Theorem 5, R is a symmetric algebra in which involution $x \to x^*$ is continuous and satisfies the condition*

$$x^*x = 0 \quad \textit{if, and only if, } x = 0. \tag{5}$$

Then all minimal closed two-sided ideals I and R are symmetric.

Proof. The mapping $x \to x^*$ maps the minimal closed two-sided ideal I into the minimal closed two-sided ideal I^*; if I^* were $\neq I$, then we would have $I^*I = (0)$ and, consequently, $x^*x = 0$ for all $x \in I$, which contradicts condition (5).

The following two theorems form a partial converse of Theorem 5.

THEOREM 6. *Suppose $\{R_\alpha\}$ is a family of dual algebras and let R be the algebra of all complexes $x = \{x_\alpha\}$, $x_\alpha \in R_\alpha$, having only a finite number of nonzero elements x_α, where the operations in R are defined in the usual way by the formulas*:

$$\lambda x = \{\lambda x_\alpha\}, \quad x + y = \{x_\alpha + y_\alpha\}, \quad xy = \{x_\alpha y_\alpha\}$$

with $x = \{x_\alpha\}$, $y = \{y_\alpha\}$.

Suppose an arbitrary topology is defined in R in which R becomes a topological algebra and whose restriction to each R_α is weaker than or coincides with the initial topology in R_α. Then R is a dual algebra.

Proof. We agree not to distinguish between the element x_{α_0} and the complex $x = \{x_\alpha\}$ in which $x_\alpha = 0$ for $\alpha \neq \alpha_0$. Then we can state that R consists of all possible finite sums of elements $x_\alpha \in R_\alpha$, where $x_\alpha x_\beta = x_\beta x_\alpha = 0$ for $\alpha \neq \beta$.

Now suppose I_r is a (proper or improper) closed right ideal in R, and that $I_r^{(\alpha)}$ is the set of all α-th components of elements $\{x_\alpha\} \in I_r$. If $x_\alpha \in I_r^{(\alpha)}$ and $y_\alpha \in R_\alpha$, then $y_\alpha x_\alpha \in I_r$, so that $R_\alpha I_r^{(\alpha)} \subset I_r$. But, in virtue of III, subsection 1, $R_\alpha I_r^{(\alpha)}$ is dense in $I_r^{(\alpha)}$ in the topology of R_α and, consequently, also $I_r^{(\alpha)} \subset I_r$. Therefore, $I_r^{(\alpha)} = I_r \cap R_\alpha$ and $I_r^{(\alpha)}$ is a (proper or improper) closed ideal in R_α.

Clearly, $\mathfrak{L}(I_r)$ consists of those, and only those, $y = \{y_\alpha\}$ for which $y_\alpha I_r^{(\alpha)} = (0)$, i.e. for which y_α is contained in the left annihilator $\mathfrak{L}_\alpha(I_r^{(\alpha)})$ of the ideal $I_r^{(\alpha)}$ in the algebra R_α. We find, analogously, that $\mathfrak{R}(\mathfrak{L}(I_r))$ consists of those, and only those, $z = \{z_\alpha\}$, for which $z_\alpha \in \mathfrak{R}_\alpha(\mathfrak{L}_\alpha(I_r^{(\alpha)}))$. But since the algebras R_α are dual and the ideals $I_r^{(\alpha)}$ are closed, we have $\mathfrak{R}_\alpha(\mathfrak{L}_\alpha(I_r^{(\alpha)})) = I_r^{(\alpha)}$ and, consequently, $\mathfrak{R}(\mathfrak{L}(I_r))$ consists of all $z = \{z_\alpha\}$ where $z_\alpha \in I_r^{(\alpha)}$, i.e. $\mathfrak{R}(\mathfrak{L}(I_r)) = I_r$.

Analogously, $\mathfrak{L}(\mathfrak{R}(I_l)) = I_l$ for an arbitrary closed (proper or improper) left ideal I_l in R.

THEOREM 7. *Suppose R is a topological algebra satisfying the following conditions:*

1) $x \in \overline{xR}$ *and* $x \in \overline{Rx}$ *for an arbitrary* $x \in R$;

2) *R contains a two-sided left ideal I, dense in R, which is a dual algebra in some topology which is stronger than the topology in R or coincides with the topology in R restricted to I.*

Then R is a dual algebra.

Proof. Suppose I_r, I_l are closed (proper or improper) right and left ideals, respectively, in R; in virtue of condition 2), $I_r \cap I$ is then a closed (proper or improper) right ideal in I. Moreover, $I_r \cap I$ is dense in I_r in the sense of the topology in R; in fact, $I_r I \subset I_r \cap I$ and, in the sense of the topology in R, $I_r I$ is dense in $I_r R$ (because by assumption I is dense in R), and $I_r R$ is dense in I_r in virtue of condition 1). Thus,

$$\overline{I_r \cap I} = I_r. \tag{6}$$

It follows from this that the left annihilator of the ideal $I_r \cap I$ in I is exactly $\mathfrak{L}(I_r) \cap I$, where $\mathfrak{L}$, as also $\mathfrak{R}$ below, denote the left and right annihilators, respectively, in the entire algebra R.

Analogously, the right annihilator in I of the ideal $\mathfrak{L}(I_r) \cap I$ is exactly $\mathfrak{R}(\mathfrak{L}(I_r)) \cap I$. In virtue of the duality of the algebra I and the fact that the (proper or improper) ideal $I_r \cap I$ is closed in I, we conclude from this that $\mathfrak{R}(\mathfrak{L}(I_r)) \cap I = I_r \cap I$. But in virtue of (6), the closures in R of the left and right members of this relation are $\mathfrak{R}(\mathfrak{L}(I_r))$ and I_r. Therefore $\mathfrak{R}(\mathfrak{L}(I_r)) = I_r$ and analogously $\mathfrak{L}(\mathfrak{R}(I_l)) = I_l$.

4. Simple annihilator algebras. In this subsection, R will denote a simple annihilator algebra with continuous quasi-inverse.

I. *If p is an irreducible idempotent, then the algebra pRp is isomorphic to the field of complex numbers.*

Proof. We shall prove that pRp is a field; as a subalgebra of the algebra R with continuous quasi-inverse, it will be a field with continuous inverse and, consequently, it will be isomorphic to the field of complex numbers (see subsections 5–6, § 8).

Clearly, p is the identity of the algebra pRp.

Suppose $x \in pRp$ and $x \neq 0$; we must prove that x has an inverse in pRp. To this end, we note that Rx is a left ideal in R which is distinct from (0) (in virtue of condition (2), subsection 1) and is contained in the minimal ideal Rp; consequently, $Rx = Rp$. From this we conclude that there exists an element $y \in R$ for which $yx = p^2 = p$. But in this case

$$(pyp)x = (pyp)(pxp) = pypxp = pyx = pp = p,$$

i.e. pyp is a left inverse of the element x in the algebra pRp. One proves analogously that x has a right inverse in pRp.

II. *R contains a maximal orthogonal system $\{p_\alpha\}$ of irreducible idempotents.* (See subsection 3, above.) *The sum of the corresponding minimal left ideals Rp_α, and also of the minimal right ideals $p_\alpha R$, is dense in R.*

Proof. In virtue of V, subsection 3, R contains irreducible idempotents; applying Zorn's lemma, we conclude that there exists a maximal orthogonal system $\{p_\alpha\}$ of irreducible idempotents in R.

Suppose I_l is the sum of all left ideals Rp_α; if $\bar{I}_l \neq R$, then $\bar{I}_l$ is a closed left ideal and hence $\mathfrak{R}(\bar{I}_l)$ is a right ideal, $\neq (0)$. But then, according to V, subsection 3, $\mathfrak{R}(\bar{I}_l)$ contains an irreducible idempotent which is orthogonal to all the p_α; but the latter is impossible in virtue of the fact that the system $\{p_\alpha\}$ is maximal. Consequently, the sum of all Rp_α is dense in R and similarly the sum of all $p_\alpha R$ is dense in R.

THEOREM 8. *Every simple annihilator algebra R with continuous quasi-inverse can be mapped isomorphically into the algebra of continuous linear operators in a locally convex linear space in such a way that:*

1) *the image of the algebra R under this isomorphism contains all finite-dimensional operators* [recall that a linear operator is said to be *finite-dimensional* if its range is finite-dimensional (see subsection 6, § 4)];

2) *there exists a dense subalgebra R_1 in R whose image under this isomorphism consists of finite-dimensional operators.*

If, moreover, R is a Banach algebra, then this isomorphism can be realized in the form of a continuous isomorphism in the algebra of all operators which are limits, in the sense of the operator norm, of finite-dimensional operators in some Banach space.

Proof. Suppose $\{p_\alpha\}$ is a maximal orthogonal system of irreducible idempotents in R and let p_1 be a fixed idempotent in $\{p_\alpha\}$. The set Rp_1R is a two-sided ideal, distinct from (0), in R; since the algebra R is simple, we conclude from this that

$$\overline{Rp_1R} = R, \tag{1}$$

and consequently $p_\alpha Rp_1Rp_\alpha \neq 0$. But $p_\alpha Rp_1Rp_\alpha \subset p_\alpha Rp_\alpha = \{\lambda p_\alpha\}$; therefore there exist $x, y \in R$ (depending in general on α) such that

$$p_\alpha x p_1 y p_\alpha = p_\alpha. \tag{2}$$

We set

$$p_{\alpha 1} = p_\alpha x p_1, \; p_{1\alpha} = p_1 y p_\alpha, \; p_{\alpha\beta} = p_{\alpha 1} p_{1\beta}. \tag{3}$$

Then

$$p_{\alpha\alpha} = p_\alpha, \quad p_{\alpha_1\beta_1} p_{\alpha_2\beta_2} = \begin{cases} 0 & \text{for } \beta_1 \neq \alpha_2, \\ p_{\alpha_1\beta_2} & \text{for } \beta_1 = \alpha_2. \end{cases} \tag{4}$$

In fact, in virtue of formula (3) and the fact that the element p_1 is idempotent, the equality $p_{\alpha\alpha} = p_\alpha$ is simply equality (2). Further, $p_{1\alpha}p_{\alpha 1} \in p_1 R p_1 = \{\lambda p_1\}$ and, consequently, $p_{1\alpha}p_{\alpha 1} = \lambda p_1$ for some λ. Multiplying both members of this equality on the left by $p_{\alpha 1}$ and on the right by $p_{1\alpha}$, we obtain that $p_\alpha^2 = \lambda p_\alpha$, i.e. $p_\alpha = \lambda p_\alpha$, so that $\lambda = 1$ and $p_{1\alpha}p_{\alpha 1} = p_1$. The lower relation in the second equation in (4) easily follows from this. The upper relation follows directly from the orthogonality of the system $\{p_\alpha\}$.

The set $p_\alpha R p_\beta$ consists of multiples of the element $p_{\alpha\beta}$. In fact, $p_\alpha x p_{\beta\alpha} \in p_\alpha R p_\alpha = \{\lambda p_\alpha\}$ and, consequently, $p_\alpha x p_{\beta\alpha} = \lambda p_\alpha$ for some λ. Multiplying both members of this equality on the right by $p_{\alpha\beta}$ and using the second relation in (4), we obtain that $p_\alpha x p_\beta = \lambda p_{\alpha\beta}$. From this and relation (4) it follows that the sum of all $p_\alpha R p_\beta$ is a subalgebra of the algebra R. We denote this subalgebra by R_1. It is dense in R. In fact, the sum of all $p_\alpha R p_\beta$ contains the linear hull of the set $(\sum_\alpha p_\alpha R)(\sum_\beta R p_\beta)$, dense in the linear hull of the set $(\sum_\alpha p_\alpha R)R$, which in its turn is dense in the linear hull of the set R^2, because $\sum_\alpha p_\alpha R$, $\sum_\beta R p_\beta$ are dense in R and the product xy is continuous with respect to each of the factors. On the other hand, the linear hull of the set R^2 is a two-sided ideal in R and therefore dense in R. Consequently, R_1 is dense in R.

It also follows from (1) that $\overline{xRp_1R} \neq (0)$ for arbitrary $x \neq 0$; it follows that $xRp_1R \neq (0)$ and

$$xRp_1 \neq (0) \quad \text{for arbitrary } x \neq 0. \tag{5}$$

We assign to each element $x \in R$ the operator $A_x y = xy$ of a left regular representation in the subspace Rp_1. In virtue of (5), the representation $x \to A_x$ is an isomorphism.

If $x = p_{\alpha\beta}$, then in virtue of formula (4) the operator A_x maps the one-dimensional subspace $p_\beta R p_1$ into $p_\alpha R p_1$ and all the subspaces $p_{\beta'} R p_1$, with $\beta' \neq \beta$, into zero. Since the sum of all $p_\beta R p_1$ is dense in Rp_1, we conclude from this that the range of the operator A_x, for $x = p_{\alpha\beta}$, is one-dimensional. But then the operator A_x is finite-dimensional for an arbitrary $x \in R_1$, i.e. the image of the algebra R_1 under the isomorphism $x \to A_x$ consists of finite-dimensional operators.

We shall now prove that all finite-dimensional operators are found among the operators A_x, $x \in R$. Clearly, it suffices to prove that all one-dimensional operators are found among the operators A_x, $x \in R$.

Suppose A is a one-dimensional operator in Rp_1 and let b be an element in Rp_1 which is such that $Ab \neq 0$. We denote by $\mathfrak{N}$ the set of all vectors in Rp_1 on which A vanishes. Then $Rp_1 = \{\lambda b\} + \mathfrak{N}$. Moreover, $\mathfrak{L}(\mathfrak{N}) \neq (0)$. In fact, if $\mathfrak{L}(\mathfrak{N}) = (0)$, then also $\mathfrak{L}(\mathfrak{N}R) = \mathfrak{L}(\mathfrak{N}) = (0)$; since the closed linear hull $[\mathfrak{N}R]$ of the set $\mathfrak{N}R$ is a (proper or improper) right ideal in R, we conclude from this that $[\mathfrak{N}R] = R$. Since $\mathfrak{N} = \mathfrak{N}p_1$, we then have $[\mathfrak{N}p_1 Rp_1] = Rp_1$; in virtue of the relation $p_1 Rp_1 = \{\lambda p_1\}$, this leads to $\mathfrak{N} = Rp_1$, which is impossible.

Thus $\mathfrak{L}(\mathfrak{N}) \neq (0)$; suppose a is a nonzero element in $\mathfrak{L}(\mathfrak{N})$. Then $ab \neq 0$. In fact, if $ab = 0$ then $aRp_1 = a[\{\lambda b\} + \mathfrak{N}] = (0)$, which contradicts (5). But in this case Rab is a left ideal, distinct from (0), contained in the minimal left ideal Rp_1, and hence $Rab = Rp_1$. In particular, there exists an $x \in R$ such that $xab = Ab$. This means that the operators A_{xa} and A coincide on the element b; since, moreover, $A_{xa}\mathfrak{N} = xa\mathfrak{N} = (0) = A\mathfrak{N}$ (because $a \in \mathfrak{L}(\mathfrak{N})$), we have $A_{xa} = A$, and the first part of the theorem is proved.

We shall now assume that R is a Banach algebra. Then Rp_1, as a closed subspace in R, is a Banach space and all the operators A_x are bounded operators in a Banach space. In virtue of the relation $|A_x| \leqq |x|$ (see (2), subsection 7, § 9), the isomorphism $x \rightarrow A_x$ is continuous and every operator A_x is the limit, in the sense of the operator norm, of finite-dimensional operators A_{x_n}, $x_n \in R_1$.

REMARK. *Suppose that in addition to the conditions of Theorem* 8, *R is a symmetric algebra in which* $x^*x + y^*y = 0$ *only for* $x = y = 0$. *Then the elements* $p_{\alpha\beta}$ *can be chosen so that* $p^*_{\alpha\beta} = p_{\beta\alpha}$.

Proof. We shall first prove that all the p_α can be assumed to be Hermitian.

Suppose p is an irreducible idempotent; then $pRp = \{\lambda p\}$ and, consequently, $pp^*p = \lambda p$ for some λ. It follows that

$$pp^*pp^* = \lambda pp^*. \tag{6}$$

Since $p \neq 0$, then $pp^* \neq 0$ and $pp^*pp^* = (pp^*)^*pp^* \neq 0$; consequently, also $\lambda \neq 0$. Moreover, λ is a real number because the left and right members in (6) are Hermitian. We set $p' = \lambda^{-1}pp^*$; then $p' \neq 0$, $p'^* = p'$ and in virtue of (6), $p'^2 = p'$; consequently, p' is a Hermitian idempotent. This idempotent is irreducible. In fact, $p'R = \lambda^{-1}pp^*R$ is a right ideal, contained in the minimal right ideal pR and distinct from (0) (because $p = \lambda^{-1}pp^*p \in p'R$); consequently, $p'R = pR$. But then, in virtue of V, subsection 3, it follows from this that every right and every left ideal contains an irreducible Hermitian idempotent. From this, as in II, we conclude that there exists a maximal orthogonal system $\{p_\alpha\}$ of irreducible Hermitian idempotents in R.

Let the $p_{1\alpha}$ be constructed for this system $\{p_\alpha\}$ as in the proof of Theorem 8. Then $p_{1\alpha}p^*_{1\alpha} \in p_1 Rp_1$; consequently,

$$p_{1\alpha} p^*_{1\alpha} = \lambda_{p_1} \tag{7}$$

for some λ. Since $p_{1\alpha}p_{1\alpha}^*$ and p_1 are Hermitian, λ is real and not equal to 0 because $p_{1\alpha} \neq 0$; hence $p_{1\alpha}p_{1\alpha}^* \neq 0$. Moreover, λ can not be negative. In fact, if $\lambda < 0$, then (7) can be written in the form $p_{1\alpha}p_{1\alpha}^*+(\sqrt{|\lambda|}p_1)(\sqrt{|\lambda|}p_1)^* = 0$, which is impossible by assumption. Thus, $\lambda > 0$, and replacing $p_{1\alpha}$ by $\sqrt{\lambda}p_{1\alpha}$, we can assume that $p_{1\alpha}p_{1\alpha}^* = p_1$. Then $p_{1\alpha}^*p_{1\alpha}$ is an idempotent in $p_\alpha R p_\alpha$ and hence coincides with p_α. We set $p_{\alpha 1} = p_{1\alpha}^*$ and $p_{\alpha\beta} = p_{\alpha 1}p_{1\beta}$; it is not difficult to verify that $(p_{\alpha\beta})^* = p_{\beta\alpha}$ and that relations (4) are satisfied.

5. H*-algebras. R is called an *H^*-algebra* (or a *Hilbert algebra*) if

1) R is a symmetric Banach algebra;
2) R is a Hilbert space;
3) the norm in the algebra R coincides with the norm in the Hilbert space R;
4) $(xy, z) = (y, x^*z)$ for all $x, y, z \in R$;
5) $x^*x \neq 0$ for $x \neq 0$.

Since $|x^*| = |x|$, $(x^*, x^*) = (x, x)$; from this, using the expression for the inner product in terms of the norm, we conclude that $(x^*, y^*) = (y, x)$. Combining this relation with property 4), we obtain $(xy, z) = (x, zy^*)$.

An example of an H^*-algebra is the set $X(\mathfrak{A})$ of all matrices $x = ||x_{\alpha\beta}||$, $\alpha, \beta \in \mathfrak{A}$ (where $\mathfrak{A}$ is a fixed set of indices), satisfying the condition $\sum |x_{\alpha\beta}|^2 < \infty$, in which the operations are defined as the corresponding operations on matrices, and the inner product by the formula

$$(x, y) = \omega \sum_{\alpha, \beta} x_{\alpha\beta}\bar{y}_{\alpha\beta},$$

where ω is a fixed number $\geqq 1$.

Further on, we shall show that every H^*-algebra is isomorphic to the direct and orthogonal sum of the algebras $X(\mathfrak{A})$. Everywhere in this subsection, R will denote an H^*-algebra.

To each element $x \in R$ there corresponds an operator A_x in the Hilbert space R, namely the operator of left regular representation $A_x y = xy$; then condition 4) means that $(A_x)^* = A_{x^*}$.

If $A_x = 0$, then $xy = 0$ for all $y \in R$. Setting $y = x^*$, we obtain that $xx^* = 0$. In virtue of condition 5), we conclude from this that $x^* = 0$ and, consequently, $x = 0$. Therefore, *the left regular representation $x \to A_x$ is a symmetric isomorphism* of the algebra R into the algebra of bounded operators in R. Since every symmetric algebra of bounded operators in a Hilbert space is semisimple (see II, subsection 1, § 24), we conclude from this that R is a semisimple algebra. Moreover, if $x^*x+y^*y = 0$ for $x, y \in R$, then $x = 0$, $y = 0$ because the same fact holds for operators in Hilbert space.

I. *If I_r, I_l are (proper or improper) closed right and left ideals in R, then $\mathfrak{L}(I_r)$ and $\mathfrak{R}(I_l)$ are the orthogonal complements of the sets I_r^*, I_l^* in the Hilbert space R.*

Proof. We note first of all that $I_r R$ and RI_l are dense in I_r and I_l respectively. In fact, if, for example, $I_r R$ were not dense in I_r, then there would exist a $y_0 \in I_r$, $y_0 \neq 0$, such that $(I_r R, y_0) = 0$. This would imply that $0 = (RI_r^*, y_0^*) = (R, y_0^* I_r)$, $y_0^* I_r = (0)$ and, in particular, $y_0^* y_0 = 0$, which contradicts condition 5). Now the equality $\mathfrak{L}(I_r) = R \ominus I_r^*$ follows from the following chain of equivalent relations:

$$x \in \mathfrak{L}(I_r),\ xI_r = (0),\ (xI_r, R) = 0,\ (x, RI_r^*) = 0,$$
$$(x, I_r^*) = 0,\ x \in R \ominus I_r^*.$$

It can be proved analogously that $\mathfrak{R}(I_l) = R \ominus I_l^*$.

II. *Every H^*-algebra is dual.*

Proof. If I_l, I_r are (proper or improper) closed left and right ideals in R, then in virtue of I,

$$\mathfrak{R}(\mathfrak{L}(I_r)) = \mathfrak{R}(R \ominus I_r^*) = R \ominus (R \ominus I_r) = I_r,$$

and analogously $\mathfrak{L}(\mathfrak{R}(I_l)) = I_l$.

THEOREM 9. *Every H^*-algebra is the direct, and at the same time the orthogonal, sum of its minimal closed two-sided ideals which are simple H^*-algebras.*

Proof. In virtue of Theorem 5, subsection 3, and the corollary to this theorem, the minimal closed two-sided ideals I in R are simple H^*-algebras and the completion of their direct sum is R. Therefore, it will suffice to show that the distinct minimal closed two-sided ideals I, I' are mutually orthogonal subspaces in the Hilbert space R. To this end, we note that $II' = (0)$ and, consequently, in virtue of I,

$$I' \subset \mathfrak{R}(I) = R \ominus I^* = R \ominus I.$$

THEOREM 10. *Every simple H^*-algebra is completely isomorphic to some algebra $X(\mathfrak{A})$.*

Proof. Suppose $p_{\alpha\beta}$ are the elements constructed in the proof of Theorem 8; according to the remark at the end of subsection 4, we can assume that $p_{\alpha\beta}^* = p_{\beta\alpha}$. From the proof of Theorem 8 it also follows that the sum of the one-dimensional subspaces $p_\alpha R p_\beta = \{\lambda p_{\alpha\beta}\}$ is dense in R. These one-dimensional subspaces are mutually orthogonal because

$$(p_{\alpha\beta}, p_{\alpha'\beta'}) = (p_{\alpha\beta}, p_{\alpha'} p_{\alpha'\beta'} p_{\beta'}) = (p_{\alpha'} p_{\alpha\beta} p_{\beta'}, p_{\alpha'\beta'}) = (p_{\alpha'} p_\alpha p_{\alpha\beta} p_\beta p_{\beta'}, p_{\alpha'\beta'}) = 0 \quad (1)$$

for $\alpha' \neq \alpha$ or $\beta' \neq \beta$. Consequently, the direct sum of these one-dimensional Hilbert subspaces is all of R, i.e. the elements $p_{\alpha\beta}$ form a complete orthogonal system in R.

We note further that

$$(p_{\alpha\beta}, p_{\alpha\beta}) = (p_{\alpha 1} p_{1\beta}, p_{\alpha 1} p_{1\beta}) = (p_{1\beta} p_{\beta 1}, p_{1\alpha} p_{\alpha 1}) = (p_1, p_1). \quad (2)$$

To each element $x \in R$ we assign a matrix $||x_{\alpha\beta}||$ which is such that $x =$

$\sum_{\alpha,\beta} x_{\alpha\beta}p_{\alpha\beta}$. From the properties of the elements $p_{\alpha\beta}$ and relations (1) and (2), it follows that for $x = \sum_{\alpha,\beta} x_{\alpha\beta}p_{\alpha\beta}$ and $y = \sum_{\alpha,\beta} y_{\alpha\beta}p_{\alpha\beta}$,

$$(x, y) = \omega \sum_{\alpha,\beta} x_{\alpha\beta}\bar{y}_{\alpha\beta},$$

where $\omega = (p_1, p_1)$, and that the correspondence $x \to ||x_{\alpha\beta}||$ is a symmetric isomorphism of the algebra R onto the algebra $X(\mathfrak{A})$.

Theorems 9 and 10 imply the following theorem.

THEOREM 11. *Every H^*-algebra R is completely isomorphic to the direct and orthogonal sum of the algebras $X(\mathfrak{A})$.*

THEOREM 12. *For a simple H^*-algebra R, the following assertions are equivalent:*

a) *R is finite-dimensional;*

b) *R is an algebra with identity;*

c) *the center of the algebra R is distinct from* (0).

Proof. If R is finite-dimensional, then the maximal system $\{p_\alpha\}$ of irreducible idempotents is finite and the element $p = \sum_\alpha p_\alpha$ is an idempotent, satisfying the conditions

$$x = \sum_\alpha xp_\alpha = xp; \quad x = \sum_\alpha p_\alpha x = px,$$

i.e. p is the identity in R. Consequently, a) implies b). Further, if R contains the identity p, then the center of the algebra R contains p and therefore it is $\neq (0)$; consequently, b) implies c).

Now suppose the center of the algebra R is distinct from (0) and let $x \neq 0$ be an element of the center of the algebra R. Since $xp_\alpha = (xp_\alpha)p_\alpha = p_\alpha xp_\alpha$, then in virtue of I, subsection 4, $xp_\alpha = c_\alpha p_\alpha$, where c_α is a number. It follows that $x = \sum_\alpha xp_\alpha = \sum_\alpha c_\alpha p_\alpha$. But $c_\beta p_{\beta\delta} = c_\beta p_\beta p_{\beta\delta} = xp_\beta p_{\beta\delta} = xp_{\beta\delta} = p_{\beta\delta}x = p_{\beta\delta}p_\delta x = p_{\beta\delta}c_\delta p_\delta = c_\delta p_{\beta\delta}$, and therefore $c_\beta = c_\delta$. Thus, all the coefficients c_α in the sum $\sum_\alpha c_\alpha p_\alpha$ are equal, which is possible only when $\{p_\alpha\}$ is a finite system and R is finite-dimensional. Consequently, c) implies a), and the theorem is proved.

At the same time we see that every element of the center of the algebra R has the form $x = \sum cp_\alpha = c\sum p_\alpha = cp$, where p is the identity of the algebra.

6. Completely regular dual algebras.

THEOREM 13. *The set R of all compact operators in the fixed Hilbert space $\mathfrak{H}$ is a simple dual algebra.*

Proof. Suppose $\{\varphi_\alpha\}$ is a fixed complete orthonormal system in $\mathfrak{H}$. We denote by I the set of all operators A on $\mathfrak{H}$ whose matrix $||a_{\alpha\beta}||$ in the system $\{\varphi_\alpha\}$ satisfies the condition

$$\sum_{\alpha,\beta} |a_{\alpha\beta}|^2 < \infty.$$

Such operators A are compact (see V, subsection 6, § 4; it is easily seen that this is also true in the non-separable space l^2), and consequently I, with the inner product

$$(A, B) = \sum_{\alpha, \beta} a_{\alpha\beta} \bar{b}_{\alpha\beta}, \quad A, B \in I,$$

where $||a_{\alpha\beta}||$, $||b_{\alpha\beta}||$ are the matrices of the operators A, B, forms an H^*-, and hence a dual, algebra. We shall now show that Theorem 7, subsection 3, is applicable to R and I.

We first note that I is a two-sided ideal in $\mathfrak{B}(\mathfrak{H})$ and hence also in R. In fact, if $A \in I$, $C \in \mathfrak{B}(\mathfrak{H})$, and $C = ||c_{\alpha\beta}||$ is the matrix of the operator C in the system $\{\varphi_\alpha\}$ then the matrix of the operator CA in this same system is given by the formula $||\sum_\gamma c_{\alpha\gamma} a_{\gamma\beta}||$ and in view of the fact that the operator C is bounded,

$$\sum_\beta \sum_\alpha |\sum_\gamma c_{\alpha\gamma} a_{\gamma\beta}|^2 \leqq |C|^2 \sum_\beta \sum_\alpha |a_{\alpha\beta}|^2 < \infty.$$

Consequently, $CA \in I$, and hence $AC = (C^*A^*)^* \in I$ also. But in virtue of V, subsection 1, § 22, R is a simple algebra and, consequently, I is dense in R in the sense of the operator norm. Further, the norm of the operator $A \in I$ is less than or equal to its norm in the Hilbert space I; in fact, if $x = \sum_\alpha x_\alpha \varphi_\alpha$ is an arbitrary element in $\mathfrak{H}$, then in virtue of the Cauchy-Bunyakovsky inequality,

$$|Ax|^2 = \sum_\alpha |\sum_\beta a_{\alpha\beta} x_\beta|^2 \leqq \sum_\alpha \sum_\beta |a_{\alpha\beta}|^2 |x_\beta|^2 \leqq (A, A) \sum_\beta |x_\beta|^2 = (A, A)|x|^2,$$

so that $|A| \leqq \sqrt{(A, A)}$. From this we conclude that the restriction to I of the topology of the algebra R is weaker than or coincides with the topology of the H^*-algebra I.

Finally, suppose $A \in R$ and that $A = UH$ is the canonical decomposition of the operator A, and let $P(\lambda)$ be the spectral function of the operator H. We set $P_\varepsilon = 1 - P(\varepsilon)$, $\varepsilon > 0$. Then P_ε is finite-dimensional and, consequently, $P_\varepsilon \in R$; $HP_\varepsilon \to H$ in the sense of the operator norm and hence $AP_\varepsilon = UHP_\varepsilon \to UH = A$ as $\varepsilon \to 0$ (see VI, subsection 4, § 17). This means that $A \in \overline{AR}$. Replacing A by A^*, we obtain that $A^* \in \overline{A^*R}$, i.e, that $A \in \overline{RA}$.

Thus the algebras R and I satisfy all the conditions of Theorem 7, subsection 3; consequently R is a dual algebra.

The following theorem shows that the algebra of all compact operators in a Hilbert space is the universal model of all simple dual C^*-algebras.

THEOREM 14. *Every simple dual C^*-algebra R is completely isomorphic to the algebra of all compact operators in some Hilbert space.*

Proof. Suppose p_α, $p_{\alpha\beta}$ are the same as in the proof of Theorem 8, subsection 4. According to the remark at the end of subsection 4, we can assume that $p_\alpha^* = p_\alpha$ and that $p_{\alpha\beta}^* = p_{\beta\alpha}$. Further, according to Theorem 8, subsection 4, the algebra R is isomorphic to the algebra of operators of left regular representation $A_a x = ax$

in the space Rp_1. But if $x, y \in Rp_1$, then $y^*x \in p_1 Rp_1 = \{\lambda p_1\}$ so that $y^*x = \lambda p_1$ for some λ. We define the inner product in Rp_1 by means of the formula $(x, y) = \lambda$ so that $y^*x = (x, y)p_1$. It is easily verified that all the properties of the inner product will be satisfied; in particular, it follows from the relations $x^*x = (x, x)p_1 = (x, x)p_1^2$ that $(x, x) \geqq 0$, because the elements of the form x^*x form a cone in the completely regular algebra R (see subsection 2, § 24). Moreover, $(x, x) = 0$ implies that $x^*x = 0$ and hence that $x = 0$.

We note that in virtue of the relations $p_1^2 = p_1$ and $p_1^* = p_1$, $|p_1| = |p_1|^2$ and hence $|p_1| = 1$. From this it follows that for $x \in Rp_1$

$$|x|^2 = |x^*x| = (x, x)|p_1| = (x, x),$$

so that the Hilbert space norm in Rp_1 coincides with the initial norm. Further, for $a \in R$, $x, y \in Rp_1$

$$(ax, y)p_1 = y^*ax = (a^*y)^*x = (x, a^*y)p_1;$$

it follows that $(ax, y) = (x, a^*y)$, i.e. $(A_a x, y) = (x, A_{a^*}y)$. This means that $A_{a^*} = (A_a)^*$, i.e. that the correspondence $a \to A_a$ is a symmetric isomorphism of the C^*-algebra R into the ring of bounded operators in Rp_1. On the basis of Theorem 3, subsection 1, § 24, we conclude from this that this isomorphism preserves norm, i.e. $|A_a| = |a|$.

But then, in virtue of Theorem 8, subsection 4, every operator A_a is the limit, in the sense of the operator norm, of finite-dimensional operators and therefore it is compact. On the other hand, in virtue of the same theorem, all finite-dimensional operators and hence all their limits also (because the set of all operators A_a, $a \in R$, is a complete algebra), i.e. all the compact operators, occur among the operators A_a.

We now go over to arbitrary completely regular dual algebras. To this end, we consider the following general method of constructing completely regular dual algebras.

Suppose T is a discrete space; adjoining the infinitely distant point to it we can assume it to be a compact space. We shall assume that every $t \in T$ is assigned a completely regular algebra R_t; we denote by $C_\infty(T, R_t)$ the set of all vector-valued functions $x = x(t)$, where $x(t) \in R_t$, satisfying the condition $|x(t)| \to 0$ as $t \to \infty$. This set forms a C^*-algebra if for $x = \{x(t)\}$, $y = \{y(t)\}$ we set

$$\lambda x = \{\lambda x(t)\}, \quad x+y = \{x(t)+y(t)\}, \quad xy = \{x(t)y(t)\},$$
$$x^* = \{x(t)^*\}, \; |x| = \sup_{t \in T} |x(t)|.$$

Applying Theorems 5, 6 and 7, subsection 3, we can prove that

I. *If all the algebras R_t are dual, then $C_\infty(T, R_t)$ is a dual algebra.*

II. *Every dual C^*-algebra is an algebra of the form $C_\infty(T, R_t)$, where all the R_t are minimal closed two-sided ideals in R and, consequently, simple dual C^*-algebras.*

§ 26. Algebras of vector-valued functions

1. Definition of an algebra of vector-valued functions. Suppose T is a topological space and let every point $t \in T$ be assigned a Banach algebra R_t. We denote by $\mathfrak{C}(T, R_t)$ the set of all vector-valued functions $x = \{x(t)\}$ having the following properties:

1) $x(t) \in R_t$ for $t \in T$;

2) $|x(t)|$ is a bounded continuous function on T.

An *algebra of vector-valued functions, generated by the space T and the algebras R_t*, is any subset $\mathfrak{R}$ in $\mathfrak{C}(T, R_t)$ which forms a Banach algebra with respect to the operations

$$\alpha x = \{\alpha x(t)\}, \quad x+y = \{x(t)+y(t)\}, \quad xy = \{x(t)y(t)\} \tag{1}$$

and the norm

$$|x| = \sup_{t \in T} |x(t)|, \tag{2}$$

where $x = \{x(t)\}$, $y = \{y(t)\}$.

Any algebra of vector-valued functions, generated by the space T and the algebras R_t, will be denoted by $\mathfrak{R}(T, R_t)$.

If, for example, T is a discrete space, consisting of a finite number, namely, of n points $t_1, \cdots, t_n$, then, setting $x_k = x(t_k)(\in R_{t_k})$, we can identify the function $x = \{x(t)\}$ with the system $x = \{x_1, x_2, \cdots, x_n\}$. Clearly, condition 2) becomes superfluous and $\mathfrak{C}(T, R_t)$ itself becomes a Banach algebra, and formulas (1) and (2) for the operations in the algebra and for the norm assume the form

$$\alpha x = \{\alpha x_1, \alpha x_2, \cdots, \alpha x_n\}, \quad x+y = \{x_1+y_1, y_2+y_2, \cdots, x_n+y_n\},$$
$$xy = \{x_1 y_1, x_2 y_2, \cdots, x_n y_n\}, \quad |x| = \max\{|x_1|, \cdots, |x_n|\}.$$

In this case, the algebra $\mathfrak{C}(T, R_t)$ is called the *direct sum of the algebras* $R_{t_1}, R_{t_2}, \cdots, R_{t_n}$ and is denoted by $R_{t_1}+R_{t_2}+ \cdots +R_{t_n}$.

If, in particular, all the algebras R_t coincide with a fixed Banach algebra R, then this direct sum assumes the form $R+R+ \cdots +R$.

We obtain a generalization of this case if T is an arbitrary topological space and all the algebras R_t coincide, as before, with a fixed Banach algebra R; in this case, the simplest example of the algebra $\mathfrak{R}(T, R)$ will be the set of all continuous vector-valued functions $x = x(t)$ with values in R which are bounded in norm. This set will be denoted by $C(T, R)$. If, in particular, R is the field of complex numbers, then $C(T, R)$ coincides with the algebra $C(T)$ of all bounded continuous numerical functions on T; but if R is the field of real numbers then $C(T, R)$ coincides with the algebra $C^r(T)$ of all bounded continuous real-valued functions on T.

Suppose $\mathscr{F}$ is a family of numerical functions defined on T. The algebra $\mathfrak{R}(T, R_t)$ is said to be *closed with respect to the multiplication by functions in* $\mathscr{F}$ if $\{x(t)\} \in \mathfrak{R}(T, R_t)$ and $\{\alpha(t)\} \in \mathscr{F}$ imply that also $\{\alpha(t)x(t)\} \in \mathfrak{R}(T, R_t)$. In particular, it is meaningful to say that the algebra $\mathfrak{R}(T, R_t)$ is closed with respect to multiplication by functions in $C(T)$ or by functions in $C^r(T)$.

Let S be an arbitrary set in $\mathfrak{R}(T, R_t)$. The set of values $x(t_0)$ assumed by all the functions $x = \{x(t)\} \in S$ at a fixed point $t_0 \in T$ is called the *projection of the set S on R_{t_0}* and is denoted by S_{t_0}; clearly, $S_{t_0} \subset R_{t_0}$, where in the general case S_{t_0} may not coincide with R_{t_0}. In particular, the projection $\mathfrak{R}_{t_0}(T, R_t)$ of the algebra $\mathfrak{R}(T, R_t)$ on R_{t_0} may not coincide with R_{t_0}.

2. Ideals in an algebra of vector-valued functions.

THEOREM 1. *Suppose T is a compact Hausdorff space at each of whose points a Banach algebra R_t is defined, and let $\mathfrak{R}(T, R_t)$ be an algebra satisfying the conditions:*

1) $\mathfrak{R}_t(T, R_t) = R_t$ *for all* $t \in T$;

2) *for arbitrary continuous real-valued numerical function $\alpha(t)$ and arbitrary element $x = \{x(t)\}$ of the algebra $\mathfrak{R}(T, R_t)$ the function $\{\alpha(t)x(t)\}$ belongs to the closed ideal in $\mathfrak{R}(T, R_t)$ generated by the element x.*

Then every closed (right, left, two-sided) ideal I in $\mathfrak{R}(T, R_t)$ is the set of all vector-valued functions $x = \{x(t)\} \in \mathfrak{R}(T, R_t)$, satisfying the condition that $x(t) \in I_t$ for all $t \in T$, where the I_t are closed (right, left respectively two-sided), proper or improper ideals in R_t.

Proof. Suppose I'_t is the projection of the ideal I on R_t; we set $I_t = \bar{I}'_t$. Clearly, I_t is a (proper or improper) closed ideal in R_t. Let $y = \{y(t)\}$ be an arbitrary element in $\mathfrak{R}(T, R_t)$ which satisfies the condition

$$y(t) \in I_t \text{ for all } t \in T. \tag{1}$$

We shall prove that $y \in I$. Since $I_t = \bar{I}'_t$, for every $\varepsilon > 0$ there exists a function $x_\tau = \{x_\tau(t)\}$ in I such that $|x_\tau(\tau) - y(\tau)| < \varepsilon$. By continuity, we also have $|x_\tau(t) - y(t)| < \varepsilon$ in some neighborhood $U(\tau)$ of the point τ. Among the neighborhoods $U(\tau)$ there are a finite number $U(\tau_1), \cdots, U(\tau_n)$ which form an open covering of the space T. Suppose $\alpha_1(t), \cdots, \alpha_n(t)$ is the corresponding partition of unity in terms of continuous nonnegative numerical functions (see II, subsection 2, § 15); in virtue of condition 2),

$$\{\sum_{k=1}^{n} \alpha_k(t) x_{\tau_k}(t)\} \in I. \tag{2}$$

On the other hand,

$$|\sum_{j=1}^{n} \alpha_j(t) x_{\tau_j}(t) - y(t)| < \varepsilon \text{ for all } t \in T. \tag{3}$$

In fact, suppose t is an arbitrary point in the space T and that t lies in the neighborhoods $U(\tau_1), \cdots, U(\tau_k)$ but not in the neighborhoods $U(\tau_{k+1}), \cdots, U(\tau_n)$. Then $\alpha_{k+1}(t) = \cdots = \alpha_n(t) = 0$ and $|x_{\tau_j}(t) - y(t)| < \varepsilon$, $j = 1, \cdots, k$. It follows that

$$|\sum_{j=1}^{n} \alpha_j(t)x_{\tau_j}(t) - y(t)| = |\sum_{j=1}^{k} \alpha_j(t)[x_{\tau_j}(t) - y(t)]|$$
$$\leqq \sum_{j=1}^{k} \alpha_j(t)|x_{\tau_j}(t) - y(t)| < \varepsilon \sum_{j=1}^{k} \alpha_j(t) = \varepsilon.$$

Since I is closed, it follows from (2) and (3) that $y \in I$. On the other hand, every $y \in I$ satisfies condition (1) because $y(t) \in I'_t \subset I_t$. This completes the proof of the theorem.

COROLLARY. *Suppose T and R_t are the same as in Theorem* 1. *Then*:

1) *Every maximal closed (right, left, two-sided) ideal in $\mathfrak{R}(T, R_t)$ consists of all the elements $x = \{x(t)\}$ of the algebra $\mathfrak{R}(T, R_t)$ whose values at a fixed point t_0 belong to some fixed maximal closed (right, left respectively two-sided) ideal of the algebra R_{t_0}.*

2) *If each of the algebras R_t is simple then every closed two-sided ideal in $\mathfrak{R}(T, R_t)$ consists of all elements $x = \{x(t)\}$ of the algebra $\mathfrak{R}(T, R_t)$ which vanish on some closed subset in T.*

If additional assumptions are made with respect to the algebras R_t then condition 2) of Theorem 1 can be weakened; namely, the following proposition is valid.

I. *Suppose T is a compact Hausdorff space and that the following conditions are satisfied*:

1) $\mathfrak{R}_t(T, R_t) = R_t$ *for all* $t \in T$;

2) *for every* $t \in T$ *and every* $y \in R_t$, $y \in \overline{yR_t}$;

3) $\mathfrak{R}(T, R_t)$ *is closed with respect to multiplication by functions in* $C^r(T)$.

Then for an arbitrary $x = \{x(t)\} \in \mathfrak{R}(T, R_t)$ and arbitrary continuous real-valued numerical function $\alpha(t)$ the element $\{\alpha(t)x(t)\}$ belongs to the closed ideal I_x in $\mathfrak{R}(T, R_t)$ generated by the element $x = \{x(t)\}$.

Proof. Suppose $x = \{x(t)\} \in \mathfrak{R}(T, R_t)$ and that $\varepsilon > 0$. In virtue of conditions 1), 2), for every $\tau \in T$ there exists an element $y_\tau = \{y_\tau(t)\} \in \mathfrak{R}(T, R_t)$ such that $|x(\tau)y_\tau(\tau) - x(\tau)| < \varepsilon$. By continuity, also $|x(t)y_\tau(t) - x(t)| < \varepsilon$ in some neighborhood $U(\tau)$. We choose a finite covering $U(\tau_1). \cdots, U(\tau_n)$ of the space T and the corresponding partition of unity in terms of continuous nonnegative functions $\beta_1(t), \cdots, \beta_n(t)$ and set $z(t) = \alpha(t) \sum_{j=1}^{n} \beta_j(t)y_{\tau_j}(t)$. By condition 3), $z = \{z(t)\} \in \mathfrak{R}(T, R_t)$. Repeating the reasoning followed in the proof of Theorem 1, we conclude that

$$|\sum_{j=1}^{n} \beta_j(t)y_{\tau_j}(t)x(t) - x(t)| < \varepsilon \text{ for all } t \in T,$$

and hence

$$\sup_{t \in T} |z(t)x(t) - \alpha(t)x(t)| < \varepsilon \sup_{t \in T} |\alpha(t)|. \tag{4}$$

Since $\{z(t)x(t)\} \in I_x$, it follows from (4) that $\{\alpha(t)x(t)\} \in I_x$, which concludes the proof of the theorem.

We note that condition 2) of Proposition I is obviously satisfied if each of the algebras R_t contains the identity.

Combining Proposition I with Theorem 1, we arrive at the following result.

II. *Suppose T is a compact Hausdorff space and that R is a Banach algebra such that $z \in \overline{zR}$ for every $z \in R$. Then every closed ideal I in $C(T, R)$ is the set of all continuous vector-valued functions $x = \{x(t)\}$ satisfying the condition*

$$x(t) \in I_t \text{ for all } t \in T,$$

where the I_t are (proper or improper) closed ideals in R.

REMARK. In Proposition II, the compact space T can be replaced by a locally compact space and the algebra $C(T, R)$ by the algebra $C_\infty(T, R)$ of all continuous vector-valued functions $x = x(t)$ with values in R, these values being equal to zero at infinity. In fact, suppose T' is a compact extension of the space T obtained by adjoining the infinitely distant point. Then $C_\infty(T, R)$ can be considered as an algebra $\mathfrak{R}(T', R_t)$ in which $R_t = R$ for $t \in T$ and $R_\infty = (0)$.

3. Conditions for a vector-valued function to belong to an algebra. A vector-valued function $y = \{y(t)\}$ is said to *locally belong to the algebra* $\mathfrak{R}(T, R_t)$ *at the point* $\tau \in T$ if there exist a vector-valued function $\{x_\tau(t)\} \in \mathfrak{R}(T, R_t)$ and a neighborhood $U(\tau)$ such that

$$y(t) = x_\tau(t) \text{ for all } t \in U(\tau).$$

A vector-valued function $y = \{y(t)\}$ is said to *locally belong to the algebra* $\mathfrak{R}(T, R_t)$ if it locally belongs to this algebra at all points $\tau \in T$.

Repeating the reasoning followed in the proof of Theorem 1, subsection 2, § 15, we arrive at the following theorem.

THEOREM 2. *Suppose T is a compact Hausdorff space and that $\mathfrak{R}(T, R_t)$ is closed with respect to multiplication by all continuous real-valued functions $\alpha(t)$ with values in the closed interval* $[0, 1]$. *Then every vector-valued function $y = \{y(t)\}$ which locally belongs to the algebra $\mathfrak{R}(T, R_t)$ lies in this algebra.*

A vector-valued function $y = \{y(t)\}$, $y(t) \in R_t$, is said to be *continuous with respect to the algebra* $\mathfrak{R}(T, R_t)$ if for an arbitrary function $x = \{x(t)\}$ in $\mathfrak{R}(T, R_t)$ the numerical function $|y(t) - x(t)|$ is continuous. If, for example, all the algebras R_t coincide with a fixed algebra R and $\mathfrak{R}(T, R)$ is a subalgebra in $C(T, R)$, then every vector-valued function $y = \{y(t)\}$, $y(t) \in R$, which is continuous on T, is also continuous with respect to $\mathfrak{R}(T, R)$.

THEOREM 3. *Suppose T is a compact Hausdorff space and that the algebra $\mathfrak{R}(T, R_t)$ satisfies the conditions*:

1) $\mathfrak{R}_t(T, R_t) = R_t$ *for all* $t \in T$;

2) *$\mathfrak{R}(T, R_t)$ is closed with respect to multiplication by all continuous numerical functions on T with values in* [0, 1].

Then every vector-valued function $y = \{y(t)\}$ which is continuous with respect to $\mathfrak{R}(T, R_t)$ is contained in $\mathfrak{R}(T, R_t)$.

Proof. Suppose $y = \{y(t)\}$ is continuous with respect to $\mathfrak{R}(T, R_t)$. In virtue of condition 1), for every $\tau \in T$ there exists a vector-valued function $x_\tau = \{x_\tau(t)\} \in \mathfrak{R}(T, R_t)$ such that $x_\tau(\tau) = y(\tau)$. Then, by continuity, $|y(t)-x_\tau(t)| < \varepsilon$ in some neighborhood $U(\tau)$. We choose a finite covering $U(\tau_1), \cdots, U(\tau_n)$ if the space T and the corresponding partition of unity in terms of continuous nonnegative functions $\alpha_1(t), \cdots, \alpha_n(t)$. Repeating the reasoning followed in the proof of Theorem 1, subsection 2, we conclude that

$$|y(t)-\sum_{k=1}^{n} \alpha_k(t)x_{\tau_k}(t)| < \varepsilon \text{ for all } t \in T. \tag{1}$$

Since, by condition 2), the vector-valued function $\{\sum_{k=1}^{n} \alpha_k(t)x_{\tau_k}(t)\}$ belongs to the algebra $\mathfrak{R}(T, R_t)$, and this algebra is complete, it follows from (1) that $\{y(t)\}$ also belongs to the algebra $\mathfrak{R}(T, R_t)$.

If, in particular, $R_t = R$ and $\mathfrak{R}(T, R_t) \subset C(T, R)$, then we arrive at the following result.

COROLLARY. *Suppose T is a compact Hausdorff space and that $\mathfrak{R}$ is a closed subalgebra of the algebra $C(T, R)$ which satisfies the following conditions:*

1) *$\mathfrak{R}_t = R$ for all $t \in T$;*

2) *$\mathfrak{R}$ is closed with respect to multiplication by all continuous numerical functions on T with values in* [0, 1].

Then $\mathfrak{R} = C(T, R)$.

4. The case of completely regular algebras. We shall now consider the case when each of the algebras R_t is symmetric. The algebra $\mathfrak{R}(T, R_t)$ is then said to be symmetric if $\{x(t)\} \in \mathfrak{R}(T, R_t)$ implies that also $\{[x(t)]^*\} \in \mathfrak{R}(T, R_t)$. If $\mathfrak{R}(T, R_t)$ is a symmetric algebra then we define involution in this algebra by setting $x^* = \{[x(t)]^*\}$ for $x = \{x(t)\}$; clearly, all the involution axioms will be satisfied and the following proposition is valid.

I. *If each of the algebras R_t is completely regular, then the symmetric algebra $\mathfrak{R}(T, R_t)$ is also completely regular.*

We recall now the following result. Suppose R is a C^*-algebra with identity and that x is a Hermitian element in the algebra R. Then for an arbitrary function $f(\lambda)$, continuous on the spectrum of the element x, there exists an element $f(x) \in R$; this element $f(x)$ is the limit in norm of the sequence $\langle p_n(x)\rangle$, where $\langle p_n(\lambda)\rangle$ is an arbitrary sequence of polynomials which converges uniformly to $f(\lambda)$ on the spectrum of the element x. If, in particular, $f(0) = 0$, then $f(x)$ belongs to a closed (right, left or two-sided) ideal containing x. In fact, it suffices to apply Corollary 6, subsection 2, § 16, to the Banach algebra with identity, generated by the element x, and to note that we can also assume $p_n(0) = 0$ when $f(0) = 0$.

In Propositions II, III, below, R is a C^*-algebra with identity.

II. *If I is a closed symmetric two-sided ideal in R and $x^\wedge$ is the image of the Hermitian element x of the algebra R under the natural homomorphism $R \to R/I$, then for an arbitrary function $f(\lambda)$, continuous on the spectrum of the element x,*

$$f(x^\wedge) = [f(x)]^\wedge. \tag{1}$$

In fact, for an arbitrary polynomial $p(\lambda)$, $p(x^\wedge) = [p(x)]^\wedge$; passing to the limit, taking into consideration the fact that the spectrum of $x^\wedge$ is contained in the spectrum of x and using the fact that the homomorphism $R \to R/I$ is continuous, we obtain (1).

III. *Suppose $\{I_\alpha\}$ is a set of closed symmetric two-sided ideals in the algebra R, whose intersection $= (0)$, and that x_α is the image of the Hermitian element x of the algebra R under the natural homomorphism $R \to R/I_\alpha$. If the spectrum of each of the elements x_α is nonnegative, then the spectrum of the element x is also nonnegative.*

Proof. We set

$$f(\lambda) = \begin{cases} 0 & \text{for } \lambda \geqq 0, \\ -\lambda & \text{for } \lambda < 0, \end{cases}$$

$$\varphi(\lambda) = \begin{cases} \lambda & \text{for } \lambda > 0, \\ 0 & \text{for } \lambda \leqq 0. \end{cases}$$

Then the spectrum of each of the elements $f(x)$, $\varphi(x)$ is nonnegative and $x = \varphi(x) - f(x)$. By assumption, $f(x)_\alpha = f(x_\alpha) = 0$; consequently, $f(x)$ belongs to all the I_α, and hence $f(x) = 0$. But then $x = \varphi(x) - f(x) = \varphi(x)$ so that the spectrum of the element x is nonnegative.

IV. *Suppose R_t is a C^*-algebra with identity and that $\mathfrak{R}(T, R_t)$ is a symmetric algebra with identity. Then for an arbitrary function $f(\lambda)$, which is continuous on the spectrum of the Hermitian element $x = \{x(t)\} \in \mathfrak{R}(T, R_t)$, we have $f(x) = \{f(x(t))\}$, and if the spectrum of $x(t)$ is nonnegative for every $t \in T$ the spectrum of the element x is also nonnegative.*

Proof. Suppose I_t is the set of all elements of the algebra $\mathfrak{R}(T, R_t)$ which vanish at the point t. Then I_t is a closed symmetric two-sided ideal in $\mathfrak{R}(T, R_t)$ and the intersection of all the I_t equals (0); consequently, it suffices to apply Propositions II and III to the present case.

V. *Suppose the algebra $\mathfrak{R}(T, R_t)$ is the same as in Proposition IV and that $x = \{x(t)\}$ is a Hermitian element of this algebra. Then for every $\tau \in T$ and every $\varepsilon > 0$ there exists a neighborhood $U(\tau)$ such that for $t \in U(\tau)$ the spectrum of the element $x(t)$ lies in the ε-neighborhood of the spectrum of the element $x(\tau)$.*

Proof. We denote the spectrum of the element $x(t)$ by S_t and the ε-neighborhood of the set S_τ by U. Suppose $f(\lambda)$ is a continuous function which equals zero on S_τ and unity in the exterior of U. Then $f(x(\tau)) = 0$ and by continuity $|f(x(t))| < 1$ in some neighborhood $U(\tau)$. But then for $t \in U(\tau)$ the entire spectrum of the element $f(x(t))$ lies in the interval $(-1, 1)$, and hence S_t lies in U. In fact, if some point λ_0 of S_t did lie in the exterior of U, then the spectrum of $f(x(t))$ would contain a point $f(\lambda_0) = 1$, which is impossible.

REMARK 1. In the proof, we used the continuity of the function in the algebra $\mathfrak{R}(T, R_t)$ [namely, the function $f(x(t))$] only at those points t_0, where $x(t_0) = 0$. Consequently, the assertion of Proposition V remains valid under the weaker assumption that the function $x(t)$ be continuous only at those points t_0, where $x(t_0) = 0$.

THEOREM 4. *Suppose the symmetric algebra $\mathfrak{R}(T, R_t)$ with identity satisfies the conditions*:

1) *T is a compact Hausdorff space*;
2) *each R_t is complete and completely regular*;
3) *for arbitrary distinct $t_1, t_2 \in T$ there exists a vector-valued function $x = \{x(t)\}$ in the algebra $\mathfrak{R}(T, R_t)$ such that $x(t_1) = 0$, $x(t_2) = e$.* (We agree to denote the identity in every algebra R_t by the same symbol e; this will not lead to confusion at any time.)

Then $\mathfrak{R}(T, R_t)$ contains all the vector-valued functions $\{\alpha(t)e\}$, where $\alpha(t)$ is a continuous numerical function on T.

Proof. Suppose $t_1, t_2 \in T$ and $t_1 \neq t_2$; by assumption, there exists a function $x = \{x(t)\}$ in $\mathfrak{R}(T, R_t)$ such that $x(t_1) = 0$, $x(t_2) = e$. Setting $y = x^*x$, we obtain a Hermitian vector-valued function $y = \{y(t)\}$ for which also $y(t_1) = 0$, $y(t_2) = e$. It follows that $|y(t)| < \varepsilon$ in some neighborhood $U(t_1)$. Suppose $f(\lambda)$ is a continuous real-valued function which equals zero in the neighborhood $|\lambda| < \varepsilon$ of the point $\lambda = 0$ and to unity for $\lambda = 1$. Then $z = f(y)$ is a Hermitian vector-valued function in the algebra $\mathfrak{R}(T, R_t)$, which equals zero in the neighborhood $U(t_1)$ and to e at the point t_2.

We now consider a closed set $F \subset T$ and a point $t_2 \notin F$. In virtue of what was just stated, for an arbitrary point $\tau \in F$ there exists in the algebra $\mathfrak{R}(T, R_t)$ a vector-valued function $z_\tau = \{z_\tau(t)\}$, equal to unity at t_2 and to zero in some neighborhood $U(\tau)$. There are a finite number $U(\tau_1), \cdots, U(\tau_n)$ of these neighborhoods which form a covering of the set F. Setting $z = (z_{\tau_1} z_{\tau_2} \cdots z_{\tau_n})^*(z_{\tau_1} z_{\tau_2} \cdots z_{\tau_n})$, we obtain a Hermitian vector-valued function $z = \{z(t)\}$ in the algebra $\mathfrak{R}(T, R_t)$ which equals zero on F and unity at t_2.

Now suppose F_1 and F_2 are disjoint closed sets. In virtue of what we proved above, for each point $\tau \in F_2$ there exists in the algebra $\mathfrak{R}(T, R_t)$ a Hermitian vector-valued function $z_\tau = \{z_\tau(t)\}$ which equals zero on F_1 and unity at the point τ.

Setting $h_\tau = e - z_\tau$, we obtain a Hermitian vector-valued function $h_\tau = \{h_\tau(t)\}$ in the algebra $\mathfrak{R}(T, R_t)$ which equals unity on F_1 and zero at the point τ. Then $|h_\tau(t)| < \varepsilon$ in some neighborhood $U(\tau)$ and hence the element $g_\tau = f(h_\tau)$ (where $f(\lambda)$ is the same function as above) is a Hermitian vector-valued function $g_\tau = \{g_\tau(t)\}$ in the algebra $\mathfrak{R}(T, R_t)$ which equals unity on F and zero in $U(\tau)$. Choosing a finite covering $U(\tau_1), \cdots, U(\tau_n)$ of the set F_2 and setting $g = (g_{\tau_1} g_{\tau_2} \cdots g_{\tau_n})^* (g_{\tau_1} g_{\tau_2} \cdots g_{\tau_n})$, we obtain a Hermitian vector-valued function $g = \{g(t)\}$ in the algebra $\mathfrak{R}(T, R_t)$ which equals unity on F_1 and zero on F_2.

Thus, *for arbitrary two disjoint closed sets F_1, F_2 from T there exists a Hermitian vector-valued function $g = \{g(t)\}$ in the algebra $\mathfrak{R}(T, R_t)$ which equals unity on F_1 and zero on F_2.*

Now suppose $\alpha(t)$ is an arbitrary continuous numerical function on T, and let $\varepsilon > 0$. To each point $\tau \in T$ there corresponds a neighborhood $U(\tau)$ in which the oscillation of the function $\alpha(t)$ is less than ε. In virtue of the normality of the space T, the neighborhood $U(\tau)$ contains a neighborhood $V(\tau)$ such that $\overline{V(\tau)} \subset U(\tau)$.

Suppose $V(\tau_1), \cdots, V(\tau_n)$ is a covering of the space T; since $\overline{V(\tau_j)}$ and $T - U(\tau_j)$ are disjoint closed sets, there exists a Hermitian vector-valued function $g_j = \{g_j(t)\}$ in the algebra $\mathfrak{R}(T, R_t)$ which equals unity on $\overline{V(\tau_j)}$ and zero on $T - U(\tau_j)$. We set $k = \sum_{j=1}^{n} g_j^* g_j$; then $k = \{k(t)\}$ is a Hermitian vector-valued function in the algebra $\mathfrak{R}(T, R_t)$, where for each $t \in T$ the spectrum of the element $k(t) - e$ is nonnegative. In virtue of IV, we conclude from this that the spectrum of $k - e$ is nonnegative and hence the element $q = k^{-\frac{1}{2}}$ exists in $\mathfrak{R}(T, R_t)$. Setting $q_j = k^{-\frac{1}{2}} g_j^* g_j k^{-\frac{1}{2}}$, we obtain the Hermitian elements q_j in the algebra $\mathfrak{R}(T, R_t)$ with nonnegative spectrum, satisfying the conditions

$$\sum_{j=1}^{n} q_j = e; \quad q_j(t) = 0 \text{ in the exterior of } U(\tau_j). \tag{2}$$

Now suppose α_j is one of the values assumed by the function $\alpha(t)$ in $U(\tau_j)$. We shall prove that

$$|\alpha(t)e - \sum_{j=1}^{n} \alpha_j q_j(t)| < \varepsilon \text{ for all } t \in T;$$

inasmuch as the algebra $\mathfrak{R}(T, R_t)$ is complete and the number $\varepsilon > 0$ is arbitrary, it will follow from this that $\alpha(t)e \in \mathfrak{R}(T, R_t)$ and the theorem will be proved.

We shall assume that the fixed point t belongs to $U(\tau_1), \cdots, U(\tau_m)$ and that it does not belong to $U(\tau_{m+1}), \cdots, U(\tau_n)$. Then

$$|\alpha(t) - \alpha_j| < \varepsilon \quad \text{for } j = 1, \cdots, m; \; q_j(t) = 0 \quad \text{for } j > m. \tag{3}$$

Further, $q_j(t)$ can be considered as a nonnegative bounded Hermitian operator in some Hilbert space $\mathfrak{H}_t$ (see Theorem 5, subsection 2, § 24). In virtue of (2) and (3), for arbitrary $\xi \in \mathfrak{H}_t$,

$$\begin{aligned}|((\alpha(t)e-\sum_{j=1}^{n}\alpha_j q_j(t))\xi,\xi)| &= |((\alpha(t)e-\sum_{j=1}^{m}\alpha_j q_j(t))\xi,\xi)| \\ &= |\sum_{j=1}^{m}[\alpha(t)-\alpha_j](q_j(t)\xi,\xi)| \leqq \sum_{j=1}^{m}|\alpha(t)-\alpha_j|(q_j(t)\xi,\xi) \\ &< \varepsilon((\sum_{j=1}^{m} q_j(t))\xi,\xi) = \varepsilon(\xi,\xi) = \varepsilon|\xi|^2,\end{aligned}$$

and consequently

$$|\alpha(t)e-\sum_{j=1}^{m}\alpha_j q_j(t)| \leqq \varepsilon.$$

REMARK 2. In the proof of Theorem 4 we used only the weakened condition of continuity of the norm of the function $|x(t)|$, where $\{x(t)\} \in \mathfrak{R}(T, R_t)$, consisting of the following: if $x(t_0) = 0$ at some point $t_0 \in T$, then for arbitrary $\varepsilon > 0$ there exists a neighborhood $U(t_0)$ in which $|x(t)| < \varepsilon$. Therefore the assertion of the theorem also remains valid in the case when all the functions of the algebra $\mathfrak{R}(T, R_t)$ satisfy only this weakened continuity condition (of course, with the retention of all the other conditions of this theorem).

REMARK 3. The conditions of Theorem 4 may be modified in the following way: *Suppose*

1) *T is a compact Hausdorff space*;

2) *for all points $t \in T$, except a fixed point t_0, R_t is a C^*-algebra, and $R_{t_0} = (0)$*;

3) *for any two points $t_1, t_2 (t_1 \neq t_2, t_2 \neq t_0)$ there exists a vector-valued function $x = \{x(t)\}$ in $\mathfrak{R}(T, R_t)$ such that $x(t_1) = 0$, $x(t_2) = e$*;

4) *all the vector-valued functions of the algebra $\mathfrak{R}(T, R_t)$ satisfy the weakened continuity condition and, moreover, they are such that if $x(\tau) = e$ at some point $\tau \in T$, then for every $\varepsilon > 0$ there exists a neighborhood $U(\tau)$ in which $|x(t)-e| < \varepsilon$.*

Then the algebra $\mathfrak{R}(T, R_t)$ contains all the vector-valued functions $\{\alpha(t)e\}$, where $\alpha(t)$ is a continuous numerical function on T which vanishes at the point t_0.

In fact, if we adjoin the identity $e = \{e\}$ to $\mathfrak{R}(T, R_t)$ we obtain the algebra $\mathfrak{R}'(T, R'_t)$, where $R'_t = R_t$ for $t \neq t_0$, and R'_{t_0} is the algebra of scalars. $\mathfrak{R}'(T, R'_t)$ will satisfy all the conditions of Theorem 4, where the weakened continuity condition on the norm of any vector-valued function $x = \{x(t)\}$ in $\mathfrak{R}'(T, R'_t)$ follows from condition 4). Therefore $\mathfrak{R}'(T, R'_t)$ contains all the functions $\{\alpha(t)e\}$ where $\alpha(t)$ is an arbitrary continuous numerical function on T. Since $\mathfrak{R}'(T, R'_t)$ is obtained from $\mathfrak{R}(T, R_t)$ by the adjunction of the identity, we conclude from this that $\mathfrak{R}(T, R_t)$ contains all the functions $\{\alpha(t)e\}$ where $\alpha(t)$ is a continuous numerical function on T which vanishes at the point t_0. Now suppose T' is the locally compact space, obtained from T by discarding the point t_0; the algebra $\mathfrak{R}(T, R_t)$ can then be considered as an algebra $\mathfrak{R}(T', R_t)$ all vector-valued functions of which vanish at infinity. Thus, we can also assume that T is a locally compact space and that t_0 is its point at infinity.

THEOREM 5. *Suppose the symmetric algebra* $\mathfrak{R}(T, R_t)$ *satisfies the conditions*:

1) *T is a compact Hausdorff space*;

2) *every R_t is a symmetric Banach algebra consisting of compact operators in some Hilbert space* $\mathfrak{H}_t$;

3) *for any two distinct points $t_1, t_2 \in T$ and arbitrary elements $x_1 \in R_{t_1}$, $x_2 \in R_{t_2}$ there exists a function $x = \{x(t)\}$ in the algebra $\mathfrak{R}(T, R_t)$ which assumes the values x_1, x_2 at the points t_1, t_2 respectively.*

Then the algebra $\mathfrak{R}(T, R_t)$ is closed under multiplication by functions of $C^r(T)$.

Proof. Suppose $x_0 = \{x_0(t)\} \in \mathfrak{R}(T, R_t)$ and $\alpha(t) \in C^r(T)$; we are required to prove that $\{\alpha(t)x_0(t)\} \in \mathfrak{R}(T, R_t)$. We set $y = x_0^* x_0$ and consider $y(t_0)$ for the fixed point $t_0 \in T$. By assumption, $y(t_0)$ is a compact and moreover a positive definite operator in some Hilbert space $\mathfrak{H}_{t_0}$. Consequently, the spectrum of $y(t_0)$ consists of the point 0 and of a finite or denumerable number of positive characteristic values which do not have a nonzero limit point.

We first consider the case of an infinite set of distinct characteristic values; they form a decreasing sequence $\lambda_1, \lambda_2, \lambda_3, \cdots$, which converges to zero. Then, for given $\varepsilon > 0$, there exists a λ_r such that $\lambda_r < \varepsilon$. We choose two numbers β and γ such that $\lambda_r > \beta > \gamma > \lambda_{r+1}$ and denote by $f(\lambda)$ a continuous numerical function which equals unity for $\lambda \geqq \beta$ and zero for $\lambda \leqq \gamma$ and which is linear in the interval $[\gamma, \beta]$. We set $p = f(y)$; then $p = \{p(t)\} \in \mathfrak{R}(T, R_t)$ and $p(t_0)$ is a projection operator in $\mathfrak{H}_{t_0}$. Moreover, the spectrum of the Hermitian operator $y(t_0)-p(t_0)y(t_0)$ lies in the interval $[0, \varepsilon]$ and therefore $|y(t_0) - p(t_0)y(t_0)| < \varepsilon$. By continuity, we also have that $|y(t) - p(t)y(t)| < \varepsilon$ in some neighborhood $U_1(t_0)$.

Further, in virtue of V, there exists a neighborhood $U_2(t_0)$ for all points t of which the spectrum of the operator $y(t)$ lies in the exterior of $[\gamma, \beta]$ and hence $p(t)$ is a projection operator. Suppose $U(t_0)$ is a neighborhood of the point t_0 whose closure is contained in $U_1(t_0) \cap U_2(t_0)$. Then $p(t)$ is a projection operator and

$$|y(t)-p(t)y(t)| < \varepsilon \quad \text{for all } t \in \overline{U(t_0)}. \tag{4}$$

We shall prove that this situation also holds in case the operator $y(t_0)$ has only a finite number of positive characteristic values. Suppose λ_0 is the smallest of these positive characteristic values; we choose two numbers β, γ such that $0 < \gamma < \beta < \lambda_0$, and define a continuous function $f(\lambda)$ as above. Again setting $p = f(y)$, we obtain that $y(t_0)-p(t_0)y(t_0) = 0$. Further, repeating the preceding line of reasoning, we again arrive at (4).

We shall now show that the multiplication of a vector-valued function $\{p(t)\}$ by an arbitrary continuous real-valued function $\alpha(t)$ is possible in $\overline{U(t_0)}$; specifically, for any such function $\alpha(t)$ these exists a vector-valued function $z = \{z(t)\}$ in the algebra $\mathfrak{R}(T, R_t)$ such that $z(t) = \alpha(t)p(t)$ for all $t \in \overline{U(t_0)}$. To this end, we denote the set of all vector-valued functions $x = \{x(t)\}$ in $\mathfrak{R}(T, R_t)$ which satisfy the condition

$$p(t)x(t) = x(t)p(t) = x(t) \quad \text{for all } t \in \overline{U(t_0)} \tag{5}$$

by $\mathfrak{R}_1$. Clearly, $\mathfrak{R}_1$ is a Banach subalgebra of the algebra $\mathfrak{R}(T, R_t)$. Further, we denote by $\mathfrak{R}_1^{\wedge}$ the set of all vector-valued functions $x = \{x(t)\}$ in $\mathfrak{R}_1$ considered only on $\overline{U(t_0)}$. The mapping $x \to x^{\wedge}$ which assigns to each function $x = \{x(t)\}$ in $\mathfrak{R}_1$ its restriction $x^{\wedge} = \{x^{\wedge}(t)\}$ to $\overline{U(t_0)}$ is a symmetric homomorphism of the C^*-algebra $\mathfrak{R}_1$ onto the completely regular algebra $\mathfrak{R}_1^{\wedge}$; consequently, $\mathfrak{R}_1^{\wedge}$ is complete (see Corollary, subsection 3, § 24). On the other hand, $\mathfrak{R}_1^{\wedge}$ is an algebra of the form $\mathfrak{R}(\overline{U(t_0)}, R_t')$, where R_t' is the set of all elements x of the algebra R_t which satisfy the condition $xp(t) = p(t)x = x(t)$. From this and condition 3), we conclude that $\mathfrak{R}_1^{\wedge}$ satisfies all the conditions of Theorem 4, where $p(t)$ is the identity in R_t' and p is the identity in $\mathfrak{R}_1^{\wedge}$. On the basis of Theorem 4, the algebra $\mathfrak{R}_1^{\wedge}$ contains the function $\{\alpha(t)p(t)\}$; this means that $\mathfrak{R}(T, R_t)$ has a vector-valued function $z = \{z(t)\}$, which coincides with $\alpha(t)p(t)$ in $\overline{U(t_0)}$.

We now modify the function $z = \{z(t)\}$ so that it vanishes in the exterior of $\overline{U(t_0)}$. Suppose $\tau \notin U(t_0)$ and $t_1 \in U(t_0)$; by construction, the spectrum of the operator $p(\tau)$ consists of a finite number of positive characteristic values with finite multiplicities. Suppose q_0 is a projection operator onto the direct sum of the characteristic subspaces corresponding to these characteristic values; since q_0 is finite-dimensional, it is compact and, furthermore, $p(\tau)q_0 = q_0 p(\tau) = p(\tau)$, $q_0 \in R_\tau$. In virtue of condition 3), there exists a Hermitian element $q = \{q(t)\}$ in $\mathfrak{R}(T, R_t)$ such that $q(t_1) = 0$, $q(\tau) = q_0$. Suppose $\varphi(\lambda)$ is a continuous real-valued function which vanishes in a neighborhood of the point $\lambda = 0$ and is equal to unity at the point $\lambda = 1$. Setting $k = \varphi(q)$, we obtain a vector-valued function $k = \{k(t)\}$ in $\mathfrak{R}(T, R_t)$, which vanishes in some neighborhood $U(t_1)$ and equals q_0 at the point τ. Suppose $V(t_0)$ is a neighborhood of the point t_0 such that $\overline{V(t_0)} \subset U(t_0)$. Carrying out the preceding reasoning for each point $t_1 \in \overline{V(t_0)}$, and then selecting a finite covering of the set $\overline{V(t_0)}$ by the corresponding neighborhoods $U(t_1)$ and multiplying together the corresponding vector-valued functions $k = \{k(t)\}$, we obtain a vector-valued function $\tilde{k} = \{\tilde{k}(t)\}$ which vanishes on $\overline{V(t_0)}$ and equals q_0 at the point τ. We set $\tilde{p} = p - \tilde{k}p$; then $\tilde{p} = \{\tilde{p}(t)\}$ is a vector-valued function in $\mathfrak{R}(T, R_t)$, which equals $p(t)$ for $t \in \overline{V(t_0)}$ and zero at the point τ. Further, setting $v = \tilde{p}^*\tilde{p}$, we obtain a Hermitian vector-valued function $v = \{v(t)\}$, possessing the same property. Then considering $\varphi(v)$, where $\varphi(\lambda)$ is the same function as above, and applying the preceding reasoning to v and $T - U(t_0)$ instead of to k and $\overline{V(t_0)}$, we obtain a Hermitian vector-valued function $h = \{h(t)\}$ in the algebra $\mathfrak{R}(T, R_t)$, which equals $p(t)$ in $\overline{V(t_0)}$ and vanishes in $T - U(t_0)$.

Now suppose $\beta(t)$ is an arbitrary continuous real-valued function on T which vanishes in the exterior of $V(t_0)$. By what was proved above, the algebra $\mathfrak{R}(T, R_t)$ contains a vector-valued function equal to $\beta(t)p(t)$ in $\overline{U(t_0)}$; multiplying this vector-valued function by $h = \{h(t)\}$, we obtain a vector-valued function in the algebra

$\mathfrak{R}(T, R_t)$, which equals $\{\beta(t)p(t)\}$ everywhere in T. In other words, $\{\beta(t)p(t)\} \in \mathfrak{R}(T, R_t)$ for an arbitrary continuous real-valued function $\beta(t)$ on T which vanishes in the exterior of $V(t_0)$.

We shall now construct the vector-valued function $p = \{p(t)\}$ and the corresponding neighborhood $V(t_0)$ for each point $t_0 \in T$. From these neighborhoods we choose a finite covering $V(t_1), \cdots, V(t_n)$ and denote the corresponding vector-valued functions p by $p_1, \cdots, p_n$. Suppose $\beta_1(t), \cdots, \beta_n(t)$ is the partition of unity in terms of continuous real-valued functions corresponding to the covering $V(t_1), \cdots, V(t_n)$. Then $\mathfrak{R}(T, R_t)$ contains the vector-valued functions

$$g_k = \{\alpha(t)\beta_k(t)x_0(t)p_k(t)\}, \quad k = 1, 2, \cdots, n.$$

We set $g = g_1 + \cdots + g_n$. Then $g \in \mathfrak{R}(T, R_t)$ and

$$g(t)-\alpha(t)x_0(t) = \sum_{k=1}^{n} [g_k(t)-\alpha(t)\beta_k(t)x_0(t)] = \sum_{k=1}^{n} \alpha(t)\beta_k(t)x_0(t)[p_k(t)-e];$$

it follows that

$$[g(t)-\alpha(t)x_0(t)]^*[g(t)-\alpha(t)x_0(t)] = |\alpha(t)|^2 \sum_{j,k=1}^{n} \beta_k(t)\beta_j(t)[p_k(t)-e]y(t)[p_j(t)-e] \quad (6)$$

(where we recall that $y = x_0^*x_0$). But in virtue of (4),

$$|[y(t)]^{\frac{1}{2}}[p_j(t)-e]|^2 = |y(t)-y(t)p_j(t)| < \varepsilon \text{ in } V(t_j)$$

and hence

$$\beta_j(t)|[y(t)]^{\frac{1}{2}}[p_j(t)-e]| \leqq \beta_j(t)\sqrt{\varepsilon} \quad \text{in all of } T.$$

Therefore, we conclude from (6) that

$$|g(t)-\alpha(t)x_0(t)|^2 \leqq \sup_{t\in T} |\alpha(t)| \cdot \sum_{k,j=1}^{n} \varepsilon\beta_k(t)\beta_j(t) = \sup_{t\in T} |\alpha(t)| \cdot \varepsilon.$$

In virtue of the completeness of the algebra $\mathfrak{R}(T, R_t)$, this means that $\{\alpha(t)x_0(t)\} \in \mathfrak{R}(T, R_t)$, and the theorem is proved.

REMARK 4. In Theorem 5, the space T can be considered to be locally compact. In fact, this case reduces to the case considered in Theorem 5 upon adjunction of the infinitely distant point ∞ to the space T and the adjunction of the algebra $R_\infty = (0)$ to the algebras R_t.

Now, the combination of Theorems 4 and 5 with Theorem 3 yields the following result, which is the non-commutative generalization of the Stone theorem (see Theorem 1, subsection 10, § 2).

THEOREM 6. *Suppose T is a locally compact Hausdorff space and that R is a C^*-algebra which either contains the identity or consists of compact operators in some*

Hilbert space. Suppose, further, that $\mathfrak{R}_\infty(T, R)$ is the algebra of all vector-valued functions $x = \{x(t)\}$ with values in R, which are continuous on T and tend to zero at infinity. Then, if $\mathfrak{R}_1$ is a closed symmetric subalgebra in $\mathfrak{R}_\infty(T, R)$, whose vector-valued functions assume arbitrary values in R at any two distinct points $t_1, t_2 \in T$, then $\mathfrak{R}_1$ coincides with $\mathfrak{R}_\infty(T, R)$.

GLIMM [1] obtained the following generalization of Stone's theorem to non-commutative algebras: *If R is a C^*-algebra with identity, R_1 a closed subalgebra in R, Q the weak closure of the set of all normalized indecomposable positive functionals on R and if for any $f_1, f_2 \in Q$, $f_1 \neq f_2$ there exists an element $x_0 \in R_1$ such that $f_1(x_0) \neq f_2(x_0)$, then $R_1 = R$.* [Concerning other generalizations of Stone's theorem, see BOLUS [1], JEWETT [1] and FARRELL [1].]

5. Continuous analogue of the Schur lemma. If each R_t is a symmetric Banach algebra of operators in some Hilbert space $\mathfrak{H}_t$ and $\mathfrak{R}(T, R_t)$ is a symmetric algebra, then for arbitrary fixed $t_0 \in T$ the correspondence $x \to x(t_0)$ is a representation of the algebra $\mathfrak{R}(T, R_t)$ in the space $\mathfrak{H}_{t_0}$. The results of the preceding section make it possible to carry Schur's lemma over to such representations, where we must make a number of assumptions concerning these representations, which assumptions are satisfied, after all, for the representations of a large number of important concrete algebras (in this connection, see, for example, GELFAND and NAIMARK [7], page 202, and NAIMARK [7, 9]).

We shall first prove the following proposition.

I. *Suppose $\mathfrak{C}(\mathfrak{H})$ is the algebra of all compact operators in the Hilbert space $\mathfrak{H}$. Then every closed symmetric irreducible subalgebra R of the algebra $\mathfrak{C}(\mathfrak{H})$ coincides with $\mathfrak{C}(\mathfrak{H})$.*

Proof. We denote the set of all projection operators $P \in R$ by R^P. Obviously, all operators in R^P are finite-dimensional. If H is a Hermitian operator and $H \in R$, then H has the form $H = \sum_{k=1}^{\infty} \lambda_k P_k$. Taking the continuous function $f(\lambda)$ which equals unity at the point λ_{k_0} and vanishes at all the remaining points λ_k, we see that $P_{k_0} = f(H) \in R$. Consequently, R is the minimal closed algebra containing R^P, and therefore we also have that R^P is an irreducible set.

Suppose now that $\{e_\alpha\}$ is a fixed complete orthonormal system in $\mathfrak{H}$; we denote the set of all operators whose matrix $||a_{\alpha\beta}||$ in this orthonormal system satisfies the condition $\sum_{\alpha,\beta} |a_{\alpha\beta}|^2 < \infty$ by $\mathfrak{C}'(\mathfrak{H})$, and we set $R' = R \cap \mathfrak{C}'(\mathfrak{H})$. Then R' is an H^*-algebra which contains R^P and therefore it is irreducible. But every H^*-algebra is the orthogonal sum of its minimal closed two-sided ideals; since R' is irreducible, there can be only one ideal in this sum and hence R' is completely isomorphic to some matrix algebra $X(\mathfrak{A})$ (see subsection 5, § 25). The isomorphism $X(\mathfrak{A}) \to R'$ can be considered as an irreducible representation of the algebra $X(\mathfrak{A})$. But then it follows from the line of reasoning pursued in the proof of Theorem 2, subsection 2, § 22 that R', and hence also R^P, contains all the projection operators onto finite-dimensional subspaces, and therefore $R = \mathfrak{C}(\mathfrak{H})$.

THEOREM 7. *Suppose T is a locally compact Hausdorff space, $\mathfrak{C}$ is the set of all compact operators in a fixed Hilbert space $\mathfrak{H}$, and that $\mathfrak{R}_\infty(T, \mathfrak{C})$ is the set of all vector-valued functions with values in $\mathfrak{C}$ which are continuous on T and tend to zero at infinity. Suppose, further, that $\mathfrak{R}_1$ is a closed symmetric subalgebra in $\mathfrak{R}_\infty(T, \mathfrak{C})$ which has the following properties:*

1) *for every $t_0 \in T$ the correspondence $x \to x(t_0)$ is an irreducible representation of the algebra $\mathfrak{R}_1$;*

2) *for arbitrary $t_1, t_2 \in T$, $t_1 \neq t_2$ the representations $x \to x(t_1)$, $x \to x(t_2)$* are *not equivalent.*

Then $\mathfrak{R}_1 = \mathfrak{R}_\infty(T, \mathfrak{C})$.

Proof. In virtue of condition 1), $\mathfrak{R}_{1t}$ is an irreducible subalgebra in $\mathfrak{C}$; as the image of a C^*-algebra $\mathfrak{R}_1$ under a symmetric homomorphism, $\mathfrak{R}_{1t}$ is closed and hence coincides with $\mathfrak{C}$ (see I and the corollary in subsection 3, § 24).

We shall prove that for arbitrary $x_1, x_2 \in \mathfrak{C}$ and $t_1, t_2 \in T$, $t_1 \neq t_2$, there exists a vector-valued function $x = \{x(t)\} \in \mathfrak{R}_1$ such that $x(t_1) = x_1$, $x(t_2) = x_2$. The assertion of the theorem will then follow from Theorem 6, subsection 3. We denote the set of all vector-valued functions $x = \{x(t)\} \in \mathfrak{R}_1$ which satisfy the condition $x(t_1) = 0$ by I_1 and we set $\mathfrak{C}_1 = I_{1t_2}$. Then $\mathfrak{C}_1 = \mathfrak{C}$. In fact, otherwise $\mathfrak{C}_1$ would be a closed two-sided ideal in $\mathfrak{C}$ and hence $\mathfrak{C}_1 = (0)$ (see V, subsection 1, § 22). This means that $x(t_1) = 0$, $x \in \mathfrak{R}_1$ imply that $x(t_2) = 0$ also. But then the correspondence $x(t_1) \to x(t_2)$ for $x \in \mathfrak{R}_1$ is single-valued and hence is a nonzero representation of the algebra $\mathfrak{C}$. On the basis of Theorem 2, subsection 2, § 22, this representation is equivalent to the identity representation, i.e. to the representation $x(t_1) \to x(t_1)$. This means, contrary to condition 2), that the representations $x \to x(t_1)$ and $x \to x(t_2)$ of the algebra $\mathfrak{R}_1$ are equivalent. This proves the equality $\mathfrak{C}_1 = \mathfrak{C}$. This equality means that for arbitrary $x_2 \in \mathfrak{C}$ there exists a function $x_2(t) \in \mathfrak{R}_1$ such that $x_2(t_1) = 0$, $x_2(t_2) = x_2$. Analogously, there exists a function $x_1(t) \in \mathfrak{R}_1$ such that $x_1(t_1) = x_1$, $x_1(t_2) = 0$. Then the function $x(t) = x_1(t) + x_2(t)$ will satisfy the requirements set down, and the theorem is proved.

Now suppose $\mathfrak{H}$ is a separable Hilbert space, T is a locally compact Hausdorff space, and that μ is a measure on T. For the sake of simplicity in the exposition, we shall assume everywhere in the remainder of this subsection that the infinitely distant point in T has a countable neighborhood basis $U_1, U_2, \cdots$; clearly, it follows from this that the union of the denumerable number of compact sets $Q_1 = T - U_1$, $Q_2 = T - U_2, \cdots$ is the entire space T:

$$\bigcup_{n=1}^{\infty} Q_n = T. \tag{1}$$

This condition may be replaced by a weaker condition, requiring only that T be the union $T = T_0 \cup \bigcup_\alpha T_\alpha$ of a locally zero set T_0 and (of perhaps a nondenumerable number) of summable sets T_α ($T_0 \cap T_\alpha = \phi$) which are such that every summable set intersects no more than a denumerable number of the sets T_α (see subsection 16, § 6). The following results remain valid under these assumptions if the expressions "set of measure zero" and "almost everywhere" are replaced by the expressions "locally zero set" and "locally almost everywhere" (see subsection 9, § 6). We leave the task of carrying over to this case the proofs given below to the reader as a useful exercise.

A vector-valued function $\xi = \{\xi(t)\}$, $t \in T$, with values in $\mathfrak{H}$ will be called *μ-measurable* (or simply measurable) if for an arbitrary $\eta \in \mathfrak{H}$ the numerical function $(\xi(t), \eta)$ is μ-measurable. If the vector-valued functions $\xi = \{\xi(t)\}$ and $\eta = \{\eta(t)\}$ are μ-measurable, then the numerical functions $(\xi(t), \eta(t))$, $|\xi(t)|$ are measurable. The assertion follows directly from the formulas

$$(\xi(t), \eta(t)) = \sum_k |(\xi(t), e_k)|^2, \quad |\xi(t)| = \sqrt{(\xi(t), \xi(t))}, \tag{1'}$$

where $\{e_k\}$ is a complete orthonormal system in $\mathfrak{H}$ (see IV, V and VII, subsection 10, § 6). We denote by $\mathscr{H}$ the set of all μ-measurable vector-valued functions $\xi = \{\xi(t)\}$, $t \in T$, with values in $\mathfrak{H}$, satisfying the condition $\int_T |\xi(t)|^2 d\mu < \infty$. (Two vector-valued functions $\xi(t)$, $\eta(t) \in \mathscr{H}$ are not assumed to be distinct if $\int_T |\xi(t) - \eta(t)|^2 d\mu = 0$.) If we set

$$\alpha\xi = \{\alpha\xi(t)\}, \; \xi + \eta = \{\xi(t) + \eta(t)\}, \; (\xi, \eta) = \int (\xi(t), \eta(t)) d\mu$$

for $\xi = \{\xi(t)\}, \eta = \{\eta(t)\} \in \mathscr{H}$, then $\mathscr{H}$ becomes a Hilbert space; its completeness follows from the completeness of the space L^2 (see subsection 11, § 6). In fact, if we take a complete orthonormal system $\{e_k\}$ in $\mathfrak{H}$ and put $\xi_k(t) = (\xi(t), e_k)$, then $\xi(t) \to \xi_k(t)$ will be an isometric mapping of the space $\mathscr{H}$ onto the direct sum of a finite or denumerable number of copies of the space $L^2(T)$ (see (1) and subsection 6, § 5). This Hilbert space $\mathscr{H}$ is called the *direct integral* of the space $\mathfrak{H}$ with respect to the measure μ and is denoted by $\int_T \mathfrak{H} d\mu$. (Another, more general, definition of the direct integral is given further on in subsection 1, § 41.)

If T is discrete and the measure of every point equals unity then, obviously, $\int_T \mathfrak{H} d\mu$ coincides with the usual direct sum $\sum_T \oplus \mathfrak{H}$ of $|T|$ copies of the space $\mathfrak{H}$, where $|T|$ is the cardinality of T (see subsection 6, § 5).

We shall now consider the operator-valued function $A = \{A(t)\}$, where $A(t)$ is an operator in $\mathfrak{H}$ with domain $\mathfrak{H}$; such an operator-valued function $A = \{A(t)\}$ is said to be μ-measurable if:

1) it is defined for all $t \in T$ with the possible exclusion of a set of μ-measure zero;

2) for arbitrary μ-measurable vector-valued function $\xi = \{\xi(t)\}$ the vector-valued function $\{A(t)\xi(t)\}$ is μ-measurable.

II. *If $\{A(t)\}$ is a μ-measurable operator-valued function then $|A(t)|$ is a measurable numerical function.*

Proof. Suppose $\langle \xi_n \rangle$ is a denumerable dense sequence in $\mathfrak{H}$. Then $|A(t)| = \sup_n |A(t)\xi_n|/|\xi_n|$ and $|A(t)|$ is measurable, being the supremum of the sequence of μ-measurable functions $|A(t)\xi_n|/|\xi_n|$ (see VI, subsection 10, § 6).

III. *If* $\{A(t)\}$ *is a* μ*-measurable operator-valued function then the formula* $A\{\xi(t)\} = \{A(t)\xi(t)\}$ *defines a bounded operator* A *in* $\mathscr{H}$ *if, and only if,* $|A(t)|$ *is an essentially bounded function. In this case,* $|A| = |||A(t)|||_\infty$. (The notation $||x(t)||_\infty$ is explained in subsection 13, § 6.)

Proof. We set $c = |||A(t)|||_\infty$; if $c < \infty$, then

$$|A\xi|^2 = \int |A(t)\xi(t)|^2 d\mu \leqq c^2 \int |\xi(t)|^2 d\mu = c^2|\xi|^2.$$

Consequently, A is a bounded operator and

$$|A| \leqq c. \tag{2}$$

Conversely, suppose A is bounded and let $\langle \xi_n \rangle$ be a denumerable dense sequence in $\mathfrak{H}$. We denote by S_n the set of all points t for which

$$|A(t)\xi_n| > |A|\,|\xi_n| \tag{3}$$

and by S the set of all those points t for which

$$|A(t)| > |A|. \tag{4}$$

Then

$$S = \bigcup_{n=1}^{\infty} S_n. \tag{5}$$

In fact, if $t \notin S$, i.e. inequality (4) does not hold, then inequality (3) is not satisfied for any n and hence $t \notin \bigcup_{n=1}^{\infty} S_n$. Conversely, if $t \notin \bigcup_{n=1}^{\infty} S_n$, then $|A(t)\xi_n| \leqq |A|\,|\xi_n|$ for all $n = 1, 2, 3, \cdots$, and hence $|A(t)| \leqq |A|$, because $\langle \xi_n \rangle$ is dense in $\mathfrak{H}$; this means that $t \notin S$, and (5) is proved.

We shall prove that each S_n is a set of μ-measure zero; it will then follow from (5) that S is also a set of μ-measure zero and hence

$$c \leqq |A|. \tag{6}$$

Suppose $\alpha(t)$ is the characteristic function of the set S_n and let $\beta(t)$ be an arbitrary μ-measurable numerical function satisfying the condition $0 < \int_T |\beta(t)|^2 d\mu < \infty$; then $\xi_n(t) = \beta(t)\alpha(t)\xi_n \in \mathscr{H}$ and, if $\mu(S_n) > 0$,

$$|A\{\xi_n(t)\}|^2 = \int_{S_n} |A(t)\xi_n|^2 |\beta(t)|^2 d\mu > |A|^2 \int_{S_n} |\xi_n|^2 |\beta(t)|^2 d\mu. \tag{7}$$

But inequality (7) contradicts the definition of the norm of the operator A. Consequently, $\mu(S_n) = 0$ and (6) is proved. Combining inequalities (2) and (6), we obtain that $|A| = c = |||A(t)|||_\infty$.

IV. *Every bounded operator* A *in* $\mathscr{H} = \int_T \mathfrak{H} d\mu$ *which commutes with all the operators* $L_\beta = \{\beta(t)1\}$, $\beta(t) \in L^\infty(T)$ *has the form* $A = \{A(t)\}$, *where* $\{A(t)\}$ *is a* μ*-measurable essentially bounded operator-valued function.*

Proof. Suppose $\{e_n\}$ is a complete orthonormal system in $\mathfrak{H}$ and that $\alpha_m(t)$ is the characteristic function of the set Q_m in (1). We set

$$\xi_{nm} = \{\alpha_m(t)e_n\}, \quad A^*\xi_{nm} = \eta_{nm} = \{\eta_{nm}(t)\}.$$

Then $L_{\alpha_m}\xi_{nm} = \xi_{nm}$ and therefore $A^*\xi_{nm} = A^*L_{\alpha_m}\xi_{nm} = L_{\alpha_m}A^*\xi_{nm}$, i.e.

$$\eta_{nm}(t) = \alpha_m(t)\eta_{nm}(t).$$

Further, for an arbitrary vector-valued function $\zeta = \{\zeta(t)\} \in \mathscr{H}$ we set $A\zeta = \zeta' = \{\zeta'(t)\}$. Then

$$(\zeta, \eta_{nm}) = (\zeta, A^*\xi_{nm}) = (A\zeta, \xi_{nm}) = (\zeta', \xi_{nm}),$$

i.e.

$$\int_T (\zeta(t), \eta_{nm}(t))d\mu = \int_T (\zeta(t), \eta_{nm}(t))\alpha_m(t)d\mu = \int_T (\zeta'(t), e_n)\alpha_m(t)d\mu. \tag{8}$$

We define the operator $B(t)$ in $\mathfrak{H}$ by setting $B(t)e_n = \eta_{nm}(t)$ for $t \in Q_m$ and we then extend it linearly to finite linear combinations of vectors e_n; then (8) can be re-written in the form

$$\int_T (\zeta(t), B(t)e_n)\alpha_m(t)d\mu = \int_T (\zeta'(t), e_n)\alpha_m(t)d\mu. \tag{9}$$

By assumption, $AL_\beta = L_\beta A$, and therefore $AL_\beta\zeta = L_\beta A\zeta = L_\beta\zeta'$; consequently, we may replace $\zeta(t)$ and $\zeta'(t)$ by $\beta(t)\zeta(t)$ and $\beta(t)\zeta'(t)$ in formula (9). But then we obtain that

$$\int_T (\zeta(t), B(t)e_n)\beta(t)\alpha_m(t)d\mu = \int_T (\zeta'(t), e_n)\beta(t)\alpha_m(t)d\mu$$

for all $\beta(t) \in L^\infty(T)$; from this we conclude that

$$(\zeta(t), B(t)e_n) = (\zeta'(t), e_n) \tag{10}$$

on Q_m with the exclusion of a set S_{nm} of μ-measure zero. But then (10) will also be satisfied on Q_m for all $n = 1, 2, 3, \cdots$ in the exterior of the set $S_m = \bigcup_{n=1}^\infty S_{nm}$ which is also of μ-measure zero; therefore, (10) holds for $n = 1, 2, 3, \cdots$ for all $t \in \bigcup_{m=1}^\infty (Q_m - S_m) = T - S$ where $S \subset \bigcup_{m=1}^\infty S_m$; consequently, S is also of μ-measure zero. Hence $\zeta(t) \in \mathfrak{D}_{B^*(t)}$ and $B^*(t)\zeta(t) = \zeta'(t)$ for every $t \in T - S$.

Setting $A(t) = B^*(t)$, we obtain that $A\{\zeta(t)\} = \{A(t)\zeta(t)\}$ for all $\{\zeta(t)\} \in \mathscr{H}$. [At first glance it may appear that we could have set $\zeta'(t) = A(t)\zeta(t)$ right away. The situation is this, however, that in the general case, $\zeta'(t_0)$ may depend on *all* the values of the vector-valued function $\{\zeta(t)\}$ and not only on its value $\zeta(t_0)$ at a given point t_0 (see subsection 15, § 5, where the analogous situation holds for the discrete space T). The formula $B^*(t)\zeta(t) = \zeta'(t)$, proved above, shows that in the case of an operator A which commutes with all the operators L_β, the value $\zeta'(t_0)$ will indeed depend only on $\zeta(t_0)$ and that for $t \notin S$ the vector $B(t)e_n$ does not depend on the choice of the set Q_m containing t.] In virtue of III, $\{A(t)\}$ is an essentially bounded operator-valued function, and this completes the proof of Proposition IV.

V. *Every operator* L_β, *where* $\beta(t) \in L^\infty(T)$, *is a contact point in the weak operator topology of the set of operators* L_α, $\alpha(t) \in C_\infty(T)$ (see Example 2, subsection 3, § 8).

Proof. The space $L^\infty(T)$ is the conjugate to $L^1(T)$ (subsection 16, § 6), where $C_\infty(T)$ is dense in $L^\infty(T)$ in the weak topology; in fact, if $f(t) \in L^1(T)$, then the equality $\int x(t)\alpha(t)d\mu = 0$ for all $\alpha(t) \in C_\infty(T)$ is possible only for $x = 0$ (see II, subsection 11, § 3). Therefore, for every $\beta \in L^\infty(T)$, $x_j(t) \in L^1(T)$, $j = 1, \cdots, n$ and $\varepsilon > 0$, there exists a function $\alpha(t) \in C_\infty(T)$ such that $|\int [\beta(t)-\alpha(t)]x_j(t)d\mu| < \varepsilon$. Setting here $x_j(t) = (y_j(t), z_j(t))$, where $y_j, z_j \in \mathscr{H}$, we obtain that

$$|((L_\beta - L_\alpha)y_j, z_j)| = \left|\int (\beta(t)-\alpha(t))(y_j(t), z_j(t))d\mu\right| < \varepsilon,\ j = 1, \cdots, n,$$

and this means that L_β is a contact point in the weak topology of the set of operators L_α. (See subsection 4, § 29.)

VI. *If the bounded operator A in* $\mathscr{H}$ *commutes with all the operators* L_α, $\alpha \in C_\infty(T)$, *then it also commutes with all the operators* L_β, $\beta \in L^\infty(T)$.

The assertion follows directly from V, subsection 1, § 8 and V.

We shall now assume that for every $t \in T$ there is given a representation $x \to A_x(t)$ of the symmetric Banach algebra R into the space $\mathfrak{H}$. If for each $x \in R$, the operator-valued function $A_x = \{A_x(t)\}$ is μ-measurable and essentially bounded, then A_x is an operator in $\mathscr{H}$ and the correspondence $x \to A_x$ is a representation of the algebra R in the space $\mathscr{H}$. It is called the *direct integral of the representations* $x \to A_x(t)$ *with respect to the measure* μ and is denoted by $x \to \int_T A_x(t)d\mu$.

THEOREM 8 (CONTINUOUS ANALOGUE OF THE SCHUR LEMMA). *Suppose the representation* $x \to A_x$ *of the symmetric Banach algebra* R *is the direct integral* $x \to \int_T A_x(t)d\mu$ *of the representations* $x \to A_x(t)$ *of this algebra in the separable Hilbert space* $\mathfrak{H}$. *Suppose, moreover, that the following conditions are satisfied*:

1) *for every* $x \in R$, $A_x(t)$ *is an operator-valued function on* T *which is continuous relative to the operator norm and which equals zero at infinity*;

2) *for all* $x \in R$ *and* $t \in T$, *the operator* $A_x(t)$ *is compact*;

3) *every representation* $x \to A_x(t)$ *is irreducible*;

4) *for arbitrary* $t_1, t_2 \in T$, $t_1 \neq t_2$, *the representations* $x \to A_x(t_1)$, $x \to A_x(t_2)$ *are not equivalent.*

Then every bounded operator B *in* $\mathscr{H} = \int_T \mathfrak{H} d\mu$ *which commutes with all the operators* A_x, $x \in R$ *has the form* $B = \{\beta(t)1\}$, *where* $\beta(t) \in L^\infty(T)$.

Proof. Suppose $\mathfrak{R}_1$ is the image of the algebra R under the mapping $x \to A_x$ and $\overline{\mathfrak{R}}_1$ is the closure of $\mathfrak{R}_1$ in $\mathfrak{R}_\infty(T, \mathfrak{C})$; then $\overline{\mathfrak{R}}_1$ is a symmetric Banach subalgebra in

$\mathfrak{R}_\infty(T, \mathfrak{C})$ which satisfies the conditions 1), 2) of Theorem 7. Therefore $\overline{\mathfrak{R}}_1 = \mathfrak{R}_\infty(T, \mathfrak{C})$. Consequently, the operator B commutes with all the operators in $\mathfrak{R}_\infty(T, \mathfrak{C})$.

Let $\{\varphi_n\}$ be a complete orthonormal system in $\mathfrak{H}$ and suppose P_n is a projection operator in $\mathfrak{H}$ onto the subspace enveloping $\varphi_1, \varphi_2, \cdots, \varphi_n$. Then $P_n\xi \to \xi$ as $n \to \infty$ for all $\xi \in \mathfrak{H}$ relative to the norm in $\mathfrak{H}$. We define the operator $\mathfrak{P}_n$ in $\mathscr{H}$ by setting $\mathfrak{P}_n\xi = \{P_n\xi(t)\}$ for $\xi = \{\xi(t)\} \in \mathscr{H}$. It is easily seen that $\mathfrak{P}_n$ is a projection operator in $\mathscr{H}$ and that $\mathfrak{P}_n\xi \to \xi$ as $n \to \infty$ for all $\xi \in \mathscr{H}$ relative to the norm in $\mathscr{H}$. Since $L_\beta \mathfrak{P}_n = \{\beta(t)P_n)\} \in \mathfrak{R}_\infty(T, \mathfrak{C})$, the operator B commutes with all the operators $L_\beta \mathfrak{P}_n$, i.e. $BL_\beta \mathfrak{P}_n = L_\beta \mathfrak{P}_n B$. Passing to the limit as $n \to \infty$, we obtain that B commutes with all the operators

$$L_\beta = \{\beta(t)1\}, \ \beta(t) \in C_\infty(T).$$

On the basis of IV and VI, we conclude from this that $B = \{B(t)\}$, where $B(t)$ is an essentially bounded measurable operator-valued function with values in $\mathfrak{C}$. The commutativity of the operator B with the operators $A = \{A(t)\} \in \mathfrak{R}_\infty(T, \mathfrak{C})$ means that

$$A(t)B(t) = B(t)A(t) \tag{11}$$

for almost all t.

Suppose $A^{(pq)}$ is an operator in $\mathfrak{H}$ whose matrix $||a_{jk}^{(pq)}||$ in a fixed complete orthonormal system has the form

$$a_{jk}^{(pq)} = \begin{cases} 0 \text{ for } j \neq p \ \text{ or } \ k \neq q, \\ 1 \text{ for } j = p \text{ and } k = q. \end{cases} \tag{12}$$

Clearly, $A^{(pq)} \in \mathfrak{C}$ and every operator in $\mathfrak{C}$ is the limit in the operator norm of finite linear combinations of operators $A^{(pq)}$. We apply (11) to the operator-valued function (which $\in \mathfrak{R}_\infty(T, \mathfrak{C})$)

$$A_x(t) = \alpha_n(t)A^{(pq)} \text{ for all } t \in T,$$

where $\alpha_n(t)$ is a fixed continuous numerical function on T, which satisfies the following conditions (see (1)): $\alpha_n(t) > 0$ for all $t \in Q_n$, $\alpha_n(t) \to 0$ as $t \to \infty$. We obtain that in the exterior of some set N_{pqn} of μ-measure zero, $\alpha_n(t)A^{(pq)}B(t) = \alpha_n(t)B(t)A^{(pq)}$, and consequently

$$A^{(pq)}B(t) = B(t)A^{(pq)} \text{ for } t \in Q_n - N_{pqn}. \tag{13}$$

Set $N_n = \bigcup_{p,q=1}^\infty N_{pqn}$; then N_n is of μ-measure zero, $\mu N_n = 0$, and for $t \in Q_n - N_n$, relation (13) will be satisfied for all $p, q = 1, 2, \cdots$, and hence also $AB(t) = B(t)A$ for all $A \in \mathfrak{C}$. But the latter is possible only when $B(t) = \beta(t)1$ for $t \in Q_n - N_n$. It follows from this, in virtue of (1), that $B(t) = \beta(t)1$ for $t \in T - \bigcup_{n=1}^\infty N_n$, where $\mu(\bigcup_{n=1}^\infty N_n) = 0$, and hence $B = \{\beta(t)1\}$.

COROLLARY 1. *Suppose the representation $x \to A_x$ is the same as in Theorem* 8. *Then every subspace $\mathfrak{M}$ in $\mathscr{H} = \int_T \mathfrak{H} d\mu$, which is invariant with respect to all the operators A_x, is the set of all vectors $\xi = \{\xi(t)\} \in \mathscr{H}$ which satisfy the condition $\xi(t) = 0$ for almost all $t \in \Delta$, where Δ is a fixed μ-measurable set in T.*

Proof. Suppose P is a projection operator onto the invariant subspace $\mathfrak{M}$. Then P commutes with all the A_x and hence it has the form $P = \{\beta(t)1\}$, where $\beta(t) \in L^\infty(T)$. Since $P^2 = P$, we have $\beta^2(t) = \beta(t)$ for almost all t, and therefore

$$\beta(t) = 0 \text{ or } 1 \quad \text{for almost all } t \in T. \tag{14}$$

We set $\Delta = \{t: \beta(t) = 0\}$; since $\beta(t)$ is μ-measurable, Δ is μ-measurable. The space $\mathfrak{M}$ is the set of all vectors $\xi = \{\xi(t)\}$ for which $P\xi = \xi$, i.e. $\beta(t)\xi(t) = \xi(t)$. In virtue of (14), this reduces to the condition: $\xi(t) = 0$ for $t \in \Delta$.

COROLLARY 2. *Suppose the representation $x \to A_x$ is the same as in Theorem* 8 *and that $\mathfrak{M}$ is a subspace in $\mathscr{H} = \int_T \mathfrak{H} d\mu$, which is invariant with respect to all the operators A_x. If $\mathfrak{M}$ contains the vector $\xi = \{\xi(t)\}$, satisfying the condition*

$$\xi(t) \neq 0 \quad \textit{for almost all } t \in T, \tag{15}$$

then $\mathfrak{M} = \mathscr{H}$.

In fact, it follows from (15) that in this case Δ is a set of μ-measure zero.

6. **Structure space of a completely regular algebra.** Suppose Π is the structure space of the C^*-algebra R (see subsection 3, § 15) and that $P \in \Pi$. We denote the quotient algebra R/P by R_P; by the same token, we assign the algebra R_P to the point P for each $P \in \Pi$. On the basis of Theorem 6, subsection 3, § 24, each R_P is also a C^*-algebra. We denote, further, the image of the element x of the algebra R under the natural homomorphism $R \to R_P$ by $x(P)$. We then obtain the vector-valued function $x = x(P)$ with values in R_P. Clearly, the correspondence $x \to \{x(P)\}$ thus established possesses the following properties:

$$\alpha x \to \{\alpha x(P)\},\ x+y \to \{x(P)+y(P)\},\ xy \to \{x(P)y(P)\},\ x^* = \{[x(P)]^*\}. \tag{1}$$

Moreover,

$$|x| = \sup_{P \in \Pi} |x(P)|. \tag{2}$$

In fact, let us momentarily set $|x|_1 = \sup_{P\in\Pi} |x(P)|$ and denote by R_1 the algebra of vector-valued functions $\{x(P)\}$ with norm $|x|_1$. Then the correspondence $x \to x(P)$ is a symmetric isomorphism of the C^*-algebra R into the completely regular algebra R_1 and hence it preserves norm (see Theorem 3, subsection 1, § 24); consequently $|x|_1 = |x|$.

It can also be shown (see KAPLANSKY [13], page 234), that $\sup |x(P)|$ is attained at some point $P_0 \in \Pi$.

I. *Suppose x is a Hermitian element of a C^*-algebra R with structure space Π. Then for an arbitrary closed set S of real numbers, the set Q of all those points $P \in \Pi$ for which the spectrum of $x(P)$ lies in S is closed in Π.*

Proof. Suppose $P_0 \notin Q$; then the spectrum of the element $x(P_0)$ contains a point $\alpha \notin S$. Suppose $f(\lambda)$ is a continuous real-valued function which vanishes on S and equals unity at the point α. Then $y = f(x)$ is an element of the algebra R which equals zero on Q and is distinct from zero at the point P_0. But this means that $P_0 \notin hk(Q) = \overline{Q}$, and hence $\overline{Q} = Q$.

THEOREM 9. *The structure space Π of a C^*-algebra R with identity is Hausdorff if, and only if, the function $|x(P)|$ is continuous on Π for every element $x \in R$.*

Proof. Suppose all the functions $|x(P)|$, $x \in R$, are continuous on Π; we shall prove that Π is Hausdorff. If $P_1 \neq P_2$, then there exists an element x in R which belongs to P_1 but not to P_2. This means that $x(P_1) = 0$, $x(P_2) \neq 0$. We set $\varepsilon = \frac{1}{3}|x(P_2)|$ and denote the sets of all points P on which $|x(P)| < \varepsilon$ and $|x(P)| > 2\varepsilon$ by U_1 and U_2, respectively. Since the function $|x(P)|$ is continuous, U_1 and U_2 are disjoint open sets, containing P_1 and P_2, respectively; consequently Π is Hausdorff.

We shall now prove the converse assertion. Suppose Π is Hausdorff. We shall first prove the continuity of the function $|x(P)|$ at those points P_0 for which $|x(P_0)| = 0$. We denote by I the closure of the set of all functions $y(P)$, $y \in R$, which vanish in some neighborhood of the point P_0; then I is a closed two-sided ideal in R, coinciding with P_0. In fact, I is the intersection of the primitive ideals containing it and hence if $I \neq P_0$, then I is also contained in some other primitive ideal P_1. Since Π is Hausdorff, there exists a neighborhood $U(P_0)$ whose closure does not contain P_1, i.e. $P_1 \notin hk(U(P_0))$. This means that R contains an element $y = y(P)$ which is equal to zero on $U(P_0)$ and is distinct from zero at P_1. But then $y \in I \subset P_1$, and hence $y(P_1) = 0$. The contradiction thus obtained shows that $I = P_0$. In other words, every element $x \in R$ which is equal to zero at the point P_0 is the limit of a sequence of elements $x_n \in R$ which vanish in some neighborhood $U_n(P_0)$. But then the inequality $|x-x_n| < \varepsilon$, for $n > N$, yields: $|x(P)| < \varepsilon$ in U_N, i.e. the function $|x(P)|$ is continuous at the point P_0. Therefore Proposition V, subsection 4 (see Remark 1, subsection 4) is applicable to the algebra R.

Now suppose that $|x(P_0)| = r > 0$. We can assume that x is Hermitian; otherwise it can be replaced by the element $y = x^*x$ for which $|y(P_0)| = |x(P_0)|^2 = r^2 > 0$. But then on the basis of Proposition V, subsection 4, $|x(P)| < r+\varepsilon$ in some neighborhood $U(P_0)$. On the other hand, in virtue of I, the set of points P at which $|x(P)| \leqq r-\varepsilon$ is closed and therefore the set of those points P at which $|x(P)| > r-\varepsilon$ is open.

REMARK. It can be shown (see KAPLANSKY [13], page 234) that the structure space is compact provided the condition of Theorem 9 is satisfied. Theorem 9 also remains valid for a C^*-algebra R without identity; in this case, the structure space of the algebra R is locally compact. In connection with Theorem 9 there arises the interesting problem of finding simple conditions under which the structure space of a completely regular algebra is Hausdorff. In this connection, there are at present only isolated results of a particular character (see KAPLANSKY [13] and DIXMIER [17, 20]).

The results of subsections 1, 2, § 23 and I, subsection 3, § 23 are due to GELFAND and NAIMARK [1, 6], RAIKOV [7] and NAIMARK [2]; Theorem 3, subsection 1, § 24 is due to GELFAND and NAIMARK [1, 6]; another proof of Theorem 3, subsection 1, § 24, based on the application of Theorem 1, subsection 1, § 24, was later given by KAPLANSKY [9]. Proposition II, subsection 3, § 23 and Theorem 2, subsection 4, § 3 are due to RAIKOV [7]; Theorem 1, subsection 3, § 23 to NAIMARK [2]; Corollaries 1–4 to I, subsection 3, § 23, Propositions IV, V, subsection 3, § 23 (excluding Theorem 1) to SONIS [1, 2] (for the case of a normal element, Proposition V, subsection 3, § 23, was obtained earlier by NAIMARK [2] and independently by SEGAL [5]). Theorems 1, 2, 6, § 24, and the corollaries to them, and also the results of subsections 1–4 and 6, § 26 are due to KAPLANSKY [9, 13] (concerning the further development of this theory, see FELL [2]); the results of subsection 5, § 26 to NAIMARK [7]; the results of subsection 5, § 25 to AMBROSE [2]; the remaining results of § 25 to KAPLANSKY [3] and BONSALL and GOLDIE [2]. Theorem 6, § 24 was also proved, independently, by SEGAL [6].

In a recent work, DIXMIER and DOUADY [1] investigated the topological structure of the algebras $\mathfrak{R}(T, R_t)$.

Theorem 4, § 24 is an answer to the question posed by GELFAND and NAIMARK [1]; a short proof of it, based on combining the results of KELLEY and VAUGHT [1] and Kaplansky's remarks on them, is to be found in Schatz's review of Fukamiya's paper in Mathematical Reviews, Volume 14, 1953, p. 884 (see FUKAMIYA [1]).

Theorem 5, § 25, with the additional condition of complete symmetry on the algebra, was first proved by GELFAND and NAIMARK [1]. Another proof of this theorem was subsequently given by LUMER [3]. He also proved that *if in a symmetric Banach algebra R with identity,*

$$\frac{|x^*x|}{|x^*||x|} = 1+o(r)$$

as $r = |e-x| \to 0$, *then R is completely symmetric and completely regular.* Other possibilities of weakening or modifying the axioms, defining acompletely regular algebra, were investigated by ARENS [2], VIDAV [3] and YOOD [1].

Numerous other results of the theory of completely regular algebras are collected in the monographs of DIXMIER [20] and RICKART [6].

Recently JOHNSON [1] gave an example of a commutative semisimple annihilator Banach algebra which is not dual.

CHAPTER VI

GROUP ALGEBRAS

§ 27. Topological groups

1. Definition of a group. A set $\mathfrak{G}$ is called a *group* if there is defined a (generally non-commutative) product $g_1 g_2$ of any two elements g_1, g_2 in $\mathfrak{G}$, satisfying the following conditions:

α) $g_1 g_2 \in \mathfrak{G}$ for any two $g_1, g_2 \in \mathfrak{G}$;

β) $(g_1 g_2)g_3 = g_1(g_2 g_3)$ for any $g_1, g_2, g_3 \in \mathfrak{G}$;

γ) there exists a unique element e in $\mathfrak{G}$ such that $eg = ge = g$ for all $g \in \mathfrak{G}$; e is called the *identity element of the group* $\mathfrak{G}$;

δ) for every element $g \in \mathfrak{G}$ there exists a unique element in $\mathfrak{G}$, which is denoted by g^{-1}, such that $gg^{-1} = g^{-1}g = e$; the element g^{-1} is called the *inverse* of g.

Clearly, g is the inverse of g^{-1} so that $(g^{-1})^{-1} = g$.

A group $\mathfrak{G}$ is said to be *commutative* if $g_1 g_2 = g_2 g_1$ for all $g_1, g_2 \in \mathfrak{G}$.

A mapping $g \to g'$ of the group $\mathfrak{G}$ into the group $\mathfrak{G}'$ is called a *homomorphism* if

$$(g_1 g_2)' = g_1' g_2' \quad \text{for all } g_1, g_2 \in \mathfrak{G}.$$

If this mapping is one-to-one, then it is called an *isomorphism*. Two groups $\mathfrak{G}$, $\mathfrak{G}'$ are said to be *isomorphic* if there exists an isomorphism of $\mathfrak{G}$ onto $\mathfrak{G}'$.

The set of all elements $g \in \mathfrak{G}$ which map into the identity element under the homomorphism $g \to g'$ is called the *kernel* of this homomorphism. Clearly, a homomorphism is an isomorphism if, and only if, its kernel consists of the identity element only.

EXAMPLES. 1. The set R^1 of all real numbers is a group if multiplication in R^1 is defined as the addition of real numbers; this group is called the *additive group of the real numbers*. The identity element of this group is the number zero and the inverse element of the number x is the number $-x$.

The additive group C^1 of the complex numbers is defined analogously. Clearly, both groups are commutative.

2. The set R_0^1 of all nonzero real numbers forms a group if multiplication in R_0^1 is defined as the usual multiplication of numbers. This group is called the *multiplicative group of the real numbers*. The identity of this group is the number 1 and the inverse of the number x is the number $1/x$.

The multiplicative group C_0^1 of the complex numbers is defined analogously. It is obvious that both groups are commutative.

3. The set A_n of all complex matrices of order n with determinant equal to unity is a group if multiplication is defined in A_n as the multiplication of matrices. This group is called the *complex unimodular group of order n*. The identity of this group is the identity matrix and the inverse ele-

ment to the matrix a is the inverse matrix a^{-1}. The real unimodular group is defined analogously. These groups are non-commutative when $n \geqq 2$.

4. Suppose $\mathfrak{A}$ is an arbitrary set. A *transformation* S of the set $\mathfrak{A}$ is any 1-1 mapping of $\mathfrak{A}$ onto itself. The product $S_1 S_2$ of two transformations S_1, S_2 is the transformation which results from the successive application of first the transformation S_2 and then the transformation S_1. It can easily be verified that, with this definition of product, the set of all transformations of a given set $\mathfrak{A}$ is a group. Its identity element is the identity transformation which assigns to each element $\alpha \in \mathfrak{A}$ the same element α.

5. We now consider the special case when $\mathfrak{A} = (-\infty, \infty)$ and S consists of only those transformations which are given by the formula $x' = \alpha x + \beta$, where $\alpha > 0$, and β is an arbitrary real number. The product $S_1 S_2$ of two such transformations

$$S_1: x' = \alpha_1 x + \beta_1,$$

$$S_2: x' = \alpha_2 x + \beta_2$$

is a transformation

$$x' = \alpha_1(\alpha_2 x + \beta_2) + \beta_1 = \alpha_1 \alpha_2 x + (\alpha_1 \beta_2 + \beta_1)$$

of the same form and the set of all transformations $x' = \alpha x + \beta$, $\alpha > 0$, is a group. It is called the *group of linear transformations of the real line.* [Specifically, it is called the group of linear transformations of the real line, leaving direction invariant.]

6. Suppose $\mathfrak{A}$ is a circle. We consider all possible transformations of the circle $\mathfrak{A}$ obtained by rotating it about the center through an arbitrary angle. The set of all such transformations forms a group which is called the *group of rotations of a circle.* Clearly, this group is commutative. Its elements can be given by the angles of rotation of the circle, where angles which differ by a multiple of 2π define the same element of this group.

2. Subgroups. A set $\mathfrak{G}_1 \subset \mathfrak{G}$ is called a *subgroup* of the group $\mathfrak{G}$ if $g, h \in \mathfrak{G}_1$ implies that $gh^{-1} \in \mathfrak{G}_1$. In particular, if $h = g$, then we obtain that $e \in \mathfrak{G}_1$, and therefore $g, h \in \mathfrak{G}_1$ implies that $h^{-1} \in \mathfrak{G}_1$ and $gh \in \mathfrak{G}_1$. Consequently, the set $\mathfrak{G}_1$ is also a group if the group operation in $\mathfrak{G}_1$ is taken to be the same as that in $\mathfrak{G}$.

Thus, R^1 and R_0^1 are subgroups of C^1 and C_0^1, respectively (see Examples 1, 2, subsection 1).

Suppose H is a subgroup of the group $\mathfrak{G}$; every set Hg_0 (i.e. the set of all elements hg_0, $h \in H$) is called a *right coset of the group* $\mathfrak{G}$ *modulo the subgroup* H; left cosets are defined analogously. Two elements $g_1, g_2 \in \mathfrak{G}$ belong to the same right coset modulo H if, and only if, $g_2 = hg_1$, where $h \in H$, i.e. if, and only if, $g_2 g_1^{-1} \in H$. Moreover, every element $g \in \mathfrak{G}$ belongs to some right coset, namely to the coset Hg. From this it is clear that the entire group $\mathfrak{G}$ decomposes into right cosets modulo H. We denote the right coset containing g by $g^\wedge$; any element g in a given coset will be called a *representative* of this coset. Upon multiplication on the right by some element g_0, every right coset goes over into another right coset; consequently, this multiplication realizes a transformation in the set of these cosets.

We agree to denote this transformation by $\bar{g}_0$ and to write $g^{\wedge\prime} = g^{\wedge}\bar{g}_0$ if $g^{\wedge\prime}$ is the coset obtained from $g^{\wedge}$ upon multiplication on the right by g_0.

Analogously, multiplication on the left by the element $g_0 \in \mathfrak{G}$ realizes a transformation in the set of left cosets.

3. Definition and simplest properties of a topological group. $\mathfrak{G}$ is called a *topological group* if:

α) $\mathfrak{G}$ is a group;

β) $\mathfrak{G}$ is a Hausdorff topological space;

γ) the functions $f(g) = g^{-1}$ and $f(g, h) = gh$ are continuous (the second with respect to both variables g, h).

Thus, the groups R^1, C^1, R_0^1, C_0^1, A_n in the examples of subsection 1 are topological groups if we define a topology in them in the usual way (see the examples in subsections 1, 2, § 2).

We note that every group $\mathfrak{G}$ can be considered to be a topological group if neighborhoods of the element $g \in \mathfrak{G}$ are considered to be all possible sets which contain g. Such a topology in $\mathfrak{G}$ is called the *discrete topology* and in this case the group $\mathfrak{G}$ itself is called a *discrete group*.

If $\mathfrak{G}$ is a topological group, then for every fixed $g_0 \in \mathfrak{G}$, the mapping $g \to gg_0$, called a *right translation*, is a homeomorphism of the group $\mathfrak{G}$ onto itself. Therefore all the neighborhoods of an arbitrary element $g_0 \in \mathfrak{G}$ have the form Ug_0, where the U's are neighborhoods of the identity e. [Ug_0 denotes the set of all elements gg_0, $g \in U$. Analogously, S^{-1} denotes the set of all elements g^{-1}, $g \in S$, and ST denotes the set of all elements $g_1 g_2$, $g_1 \in S$, $g_2 \in T$.] An analogous assertion holds with regard to *left translations* $g \to g_0 g$. Furthermore, the mapping $g \to g^{-1}$ is also a homeomorphism of the group $\mathfrak{G}$ onto itself; therefore if U is a neighborhood of the identity, then U^{-1} is also a neighborhood of the identity.

I. *Every neighborhood U of the identity contains a symmetric neighborhood V, i.e. a neighborhood which satisfies the condition $V^{-1} = V$.*

In fact, $V = U \cap U^{-1}$ is such a neighborhood.

II. *Every neighborhood U of the identity contains a neighborhood V such that $VV \subset U$.*

Proof. Since the product $g_1 g_2$ is continuous for $g_1 = e$, $g_2 = e$, there exist neighborhoods V_1, V_2 of the identity for which $V_1 V_2 \subset U$; then the neighborhood $V = V_1 \cap V_2$ satisfies the condition $VV \subset U$.

Analogously, for an arbitrary natural number n there exists a neighborhood V such that $V^n \subset U$; here, V^n denotes $VV \cdots V$ (n times).

III. *If Q is a compact set in the group $\mathfrak{G}$ and U is an open set containing Q, then there exists a neighborhood W of the identity such that $WQ \subset U$.*

Proof. To every $g \in Q$ there corresponds a neighborhood V of the identity such that $Vg \subset U$ and then a neighborhood W of the identity such that $W^2 \subset V$;

since Q is compact, there exists a finite number of elements $g_1, g_2, \cdots, g_n \in Q$ and neighborhoods $W_1, W_2, \cdots, W_n$ of the identity such that $W_1 g_1, W_2 g_2, \cdots, W_n g_n$ form a covering of the set Q. The neighborhood $W = W_1 \cap W_2 \cap \cdots \cap W_n$ has the required property. In fact, if $g \in Q$, then $g \in W_k g_k$ for some $k = 1, 2, \cdots, n$, and therefore $Wg \subset WW_k g_k \subset W_k^2 g_k \subset V_k g_k \subset U$.

IV. *If Q_1, Q_2 are compact sets in $\mathfrak{G}$, then $Q_1 Q_2$ is compact.*

In fact, $Q_1 Q_2$ is the image of the compact set $Q_1 \times Q_2 \subset \mathfrak{G} \times \mathfrak{G}$ under the continuous mapping $\langle g_1, g_2 \rangle \rightarrow g_1 g_2$ (see IV, subsection 6, § 2).

A topological group is said to be *locally compact* (*compact*) if it is a locally compact (compact) topological space. Thus, the groups in the examples of subsection 1 are locally compact whereas the group in Example 6 (the group of rotations of a circle) is compact.

We denote the set of all continuous numerical functions on the group $\mathfrak{G}$, which are equal to zero in the exterior of a compact set (where each function has its own compact set), by $L = L(\mathfrak{G})$.

V. *If $\mathfrak{G}$ is locally compact, then every function $x(g) \in L(\mathfrak{G})$ is uniformly continuous, i.e. for every $\varepsilon > 0$ there exists a neighborhood V of the identity such that $|x(g_1) - x(g_2)| < \varepsilon$ for $g_1 g_2^{-1} \in V$.*

Proof. Suppose $x(g) = 0$ in the exterior of a compact set Q and that U is a symmetric neighborhood of the identity with compact closure. The set W of points g for which $|x(gg_1) - x(g_1)| < \varepsilon$ for all $g_1 \in \overline{U}Q$ is open (see V, subsection 12, § 2) and contains the identity. If $g \in U$, then $x(gg_1)$ and $x(g_1)$ vanish for $g_1 \notin \overline{U}Q$ and therefore $|x(gg_1) - x(g_1)| < \varepsilon$ for $g \in V = W \cap U$.

A set $\mathfrak{G}_1 \subset \mathfrak{G}$ is called a *topological subgroup* of the topological group $\mathfrak{G}$ if:

a) $\mathfrak{G}_1$ is a subgroup of the group $\mathfrak{G}$;

b) $\mathfrak{G}_1$ is a topological subspace of the topological space $\mathfrak{G}$.

A topological subgroup is said to be *closed* if it is a closed subset of a topological space $\mathfrak{G}$. Thus, R^1 and R_0^1 are closed subgroups of C^1 and C_0^1 respectively.

4. Invariant integrals and invariant measures on a locally compact group. Suppose $\mathfrak{G}$ is a locally compact group. If $x(g) \in L = L(\mathfrak{G})$, then the functions $x^h(g) = x(gh)$ and $x_h(g) = x(h^{-1}g)$ also belong to L for arbitrary $h \in \mathfrak{G}$. In fact, this follows directly from the fact that the mappings $g \rightarrow hg$ and $g \rightarrow gh$ are homeomorphisms.

The integral I (see § 6) on $L(\mathfrak{G})$ is called a *right invariant integral* on the group $\mathfrak{G}$ if

$$I(x^h) = I(x) \tag{1}$$

for all $h \in \mathfrak{G}$ and all $x \in L(\mathfrak{G})$; a left invariant integral on $\mathfrak{G}$ is defined analogously. The measures, defined by right and left invariant integrals, are called *right* and *left invariant measures*, respectively. If we denote these measures by μ_r and μ_l

respectively, we can rewrite condition (1) in the form

$$\int x(gh)d\mu_r(g) = \int x(g)d\mu_r(g) \tag{1'}$$

for the measure μ_r, and in the form

$$\int x(hg)d\mu_l(g) = \int x(g)d\mu_l(g) \tag{2}$$

for the measure μ_l. Clearly, these conditions mean that

$$\mu_r(\Delta h) = \mu_r(\Delta), \quad \mu_l(h\Delta) = \mu_l(\Delta) \tag{3}$$

for an arbitrary measurable set Δ and any $h \in \mathfrak{G}$.

5. Existence of an invariant integral on a locally compact group. We shall now assume that the group $\mathfrak{G}$ is locally compact. We shall prove that in this case there exists a right invariant integral on the group $\mathfrak{G}$ and that, to within a constant factor, it is unique.

We denote by L^+ the set of all nonnegative continuous functions on $\mathfrak{G}$ which vanish in the exterior of a compact set (where each function has its own compact set). Suppose $f, \varphi \in L^+$ and that $\varphi \not\equiv 0$; we shall consider all possible finite systems of elements $g_i \in \mathfrak{G}$ and numbers $c_i \geqq 0$ such that

$$f(g) \leqq \sum_i c_i \varphi(gg_i) \quad \text{for all } g \in \mathfrak{G}; \tag{1}$$

we denote the greatest lower bound of the sums $\sum_i c_i$ over all such systems by $(f: \varphi)$.

Such systems exist. In fact, since $\varphi \not\equiv 0$, $\varphi \geqq 0$, there exists a neighborhood U with compact closure on which the greatest lower bound m of the values of the function φ is positive. Suppose Q is a compact set in the exterior of which $f = 0$. Then Q can be covered by a finite number of sets Ug_i^{-1}, and if $f(g) \leqq M$, then

$$f(g) \leqq \frac{M}{m} \sum_i \varphi(gg_i).$$

I. *The number* $(f: \varphi)$ *satisfies the conditions*:

1) $(f^h: \varphi) = (f: \varphi)$;

2) $(cf: \varphi) = c(f: \varphi)$ *for an arbitrary* $c \geqq 0$;

3) $(f_1 + f_2: \varphi) \leqq (f_1: \varphi) + (f_2: \varphi)$;

and furthermore

4) $(f: \psi) \leqq (f: \varphi)(\varphi: \psi)$;

5) $(f_1: \varphi) \leqq (f_2: \varphi)$ *for* $f_1 \leqq f_2$.

Proof. Relations 1)−3) and 5) are obvious. We shall prove relation 4). If $f(g) \leqq \sum_i c_i\varphi(gg_i)$ and $\varphi(g) \leqq \sum_j d_j\psi(gh_j)$, then $f(g) \leqq \sum_{i,j} c_i d_j \psi(gg_i h_j)$.

II. *If* $f \not\equiv 0$, *then* $(f:\varphi) > 0$.

In fact, it follows from (1) that $\sup f \leqq \sup \varphi \sum_i c_i$, and therefore $(f:\varphi) \geqq \sup f/\sup \varphi$. We now choose a fixed function $f_0 \in L^+$, which does not vanish identically, and set

$$I_\varphi(f) = \frac{(f:\varphi)}{(f_0:\varphi)}.$$

It follows from property 4) that

$$\frac{1}{(f_0:f)} \leqq \frac{(f:\varphi)}{(f_0:\varphi)} \leqq (f:f_0), \tag{2}$$

i.e.

$$\frac{1}{(f_0:f)} \leqq I_\varphi(f) \leqq (f:f_0).$$

Furthermore, we deduce from properties 1)−3) and 5) in I that

α) $I_\varphi(f^h) = I_\varphi(f)$;

β) $I_\varphi(cf) = cI_\varphi(f)$ for $c \geqq 0$; (3)

γ) $I_\varphi(f_1+f_2) \leqq I_\varphi(f_1)+I_\varphi(f_2)$;

δ) $I_\varphi(f_1) \leqq I_\varphi(f_2)$ for $f_1 \leqq f_2$.

We denote the segment $[1/(f_0:f), (f:f_0)]$ by Δ_f and the topological product $\Pi_{f\in L^+, f\neq 0}\, \Delta_f$ of all the Δ_f by Δ. According to Tikhonov's theorem (see II, subsection 12, § 2), Δ is a compact space. By (2), to every function $\varphi \in L^+$, $\varphi \neq 0$, there corresponds a point $I_\varphi = \{I_\varphi(f)\}$ of the space Δ. For every neighborhood $V = V(e)$ of the element $e \in \mathfrak{G}$, we set

$$\Phi_V = \{\varphi \in L^+ : \varphi \neq 0, \varphi(x) = 0 \text{ for all } x \notin V\}$$

and

$$M_V = \{I_\varphi : \varphi \in \Phi_V\}.$$

The sets M_V form a centralized system in Δ. In fact, let $M_{V_1}, \cdots, M_{V_n}$ be a finite number of such sets. $V = \bigcap_{j=1}^n V_j$ is a neighborhood of the element e,

$$\Phi_V = \bigcap_{j=1}^n \Phi_{V_j},$$

and hence $M_V = \bigcap_{j=1}^n M_{V_j}$. Therefore, the sets M_V have a common contact point $I = \{I(f)\}$ (see III, subsection 6, § 2). By the very definition of the topology in the product of spaces, this means that for all nonzero $f_1, \cdots, f_n \in L^+$ and $\varepsilon > 0$ and every neighborhood $V = V(e)$ there exists a function $\varphi \in \Phi_V$, $\varphi \neq 0$, such that

$$|I(f_j) - I_\varphi(f_j)| < \varepsilon \text{ for } j = 1, \cdots, n. \tag{4}$$

Further, we set, by definition, $I(0) = 0$ and prove that $I(f)$ possesses the following properties:

a) $I(f) > 0$ for $f \in L^+, f \neq 0$;

b) $I(f^g) = I(f)$;

c) $I(cf) = cI(f)$ for $c \geqq 0$;

d) $I(f_1+f_2) = I(f_1)+I(f_2)$;

e) $I(f_1) \leqq I(f_2)$, $f_1 \leqq f_2$ for all $f, f_1, f_2 \in L^+$.

Property a) follows from the fact that for $f \neq 0$ the point $I \in \Delta$ and hence $I(f) \in \Delta_f$, but $0 \notin \Delta_f$. In the proof of properties b)–d), we shall assume that $f \neq 0, f_1 \neq 0, f_2 \neq 0$, $c > 0$, for otherwise they are trivial.

Setting $n = 2, f_1 = f, f_2 = f^g$ in (4) and taking property α) of (3) into consideration, we obtain

$$|I(f)-I_\varphi(f)| < \varepsilon, \quad |I(f^g)-I_\varphi(f)| = |I(f^g)-I_\varphi(f^g)| < \varepsilon,$$

from which it follows that $|I(f)-I(f^g)| < 2\varepsilon$ for arbitrary $\varepsilon > 0$; consequently, $I(f^g) = I(f)$.

Analogously, setting $n = 2, f_1 = f, f_2 = cf$ in (4) and using property β) of (3), we can prove property c); setting $n = 2$ and using property b) of (3), we can prove property e); finally, setting $n = 3, f_3 = f_1+f_2$ and using γ) of (3), we can prove that

$$I(f_1+f_2) \leqq I(f_1)+I(f_2). \tag{5}$$

Therefore, to prove property c) it remains to show that the reverse inequality is also valid. We first prove the following lemma.

Lemma. *Let f, h_1, $h_2 \in L^+$ and $h_1+h_2 \leqq 1$; then*

$$I(fh_1)+I(fh_2) \leqq I(f). \tag{6}$$

Proof. The assertion is trivial if one of the functions f, h_1, h_2 equals 0 (see property β) in (3)). Therefore, we can assume that none of the functions f, h_1, h_2 equals zero. Let $\varepsilon > 0$. Since h_1 and h are uniformly continuous on $\mathfrak{G}$ (see V, subsection 3), there exists a neighborhood $V = V(e)$ such that

$$|h_j(g)-h_j(g')| < \varepsilon \quad \text{for } g \in Vg',\ j = 1, 2. \tag{7}$$

Let $\varphi \in \Phi_V$ and

$$f(g) \leqq \sum_i c_i \varphi(gg_i^{-1}) \tag{8}$$

(we find it convenient to write g_i^{-1} instead of g_i here). If, for given g, $gg_i^{-1} \notin V$, then $\varphi = 0$; therefore, (8) can be rewritten in the form

$$f(g) \leqq \sum_i{}' c_i \varphi(gg_i^{-1}), \tag{9}$$

where the prime means that for every given g the sum contains only those terms for which $gg_i^{-1} \in V$. Multiplying both members of (9) by $h_i(g)$, we obtain, in virtue of (7),

$$f(g)h_j(g) \leqq \sum_i{}' c_i \varphi(gg_i^{-1})h_j(g) \leqq \sum_i{}' c_i \varphi(gg_i^{-1})[h_j(g_i)+\varepsilon] \tag{10}$$

because in the sum (9) $gg_i^{-1} \in V$, i.e. $g \in Vg_i$. It follows from (10) that

$$f(g)h_j(g) \leqq \sum_i c_i[h_j(g_i)+\varepsilon]\varphi(gg_i^{-1}). \tag{11}$$

But inequality (11) is an inequality of the form (1) for fh_j, where the roles of c_i and g_i are played by $c_i[h_j(g_i)+\varepsilon]$ and g_i^{-1} respectively. Therefore,

$$(fh_j\colon \varphi) \leqq \sum_i c_i[h_j(g_i)+\varepsilon], \quad j = 1, 2.$$

Summing with respect to j and taking into account that $h_1+h_2 \leqq 1$, we conclude that

$$(fh_1\colon \varphi)+(fh_2\colon \varphi) \leqq \sum_i c_i(1+\varepsilon) = (1+\varepsilon)\sum c_i, \tag{12}$$

and then taking the greatest lower bound in (12) over all c_i and g_i in (8), that

$$(fh_1\colon \varphi)+(fh_2\colon \varphi) \leqq (f\colon \varphi)(1+\varepsilon). \tag{13}$$

Dividing both members of (13) by $(f_0\colon \varphi)$, we obtain

$$I_\varphi(fh_1)+I_\varphi(fh_2) \leqq I_\varphi(f)(1+\varepsilon). \tag{14}$$

Now setting $n = 3, f_1 = fh_1, f_2 = fh_2, f_3 = f$ in (4) and using (14), we arrive at the inequalities

$$I(fh_1)+I(fh_2) < I_\varphi(fh_1)+I_\varphi(fh_2)+2\varepsilon \leqq$$
$$\leqq I_\varphi(f)(1+\varepsilon)+2\varepsilon < (I(f)+\varepsilon)(1+\varepsilon)+2\varepsilon,$$

from which, since $\varepsilon > 0$ is arbitrary, it follows that

$$I(fh_1)+I(fh_2) \leqq I(f).$$

We turn now to the proof of property d).

Let $f_1, f_2 \in L^+, f_1 \neq 0, f_2 \neq 0$, where $f_1 = 0$ and $f_2 = 0$ outside a compact set Q in $\mathfrak{G}$. According to the Urysohn lemma (see II, subsection 9, § 2), there exists a function $f' \in L^+$ which is equal to 1 on Q.

For a given $\varepsilon > 0$, we set

$$f = f_1+f_2+\varepsilon f', \quad h_1 = f_1/f, \quad h_2 = f_2/f \tag{15}$$

(assuming $h_1 = h_2 = 0$ at those points where $f = 0$). Then $h_1, h_2 \in L^+$, $h_1 \neq 0$,

$h_2 \neq 0$, $h_1+h_2 \leqq 1$, so that, according to the lemma (see (6)),

$$I(fh_1)+I(fh_2) \leqq I(f) = I(f_1+f_2+\varepsilon f') \leqq I(f_1+f_2)+\varepsilon I(f'),$$

i.e., in virtue of (15),

$$I(f_1)+I(f_2) \leqq I(f_1+f_2)+\varepsilon I(f').$$

Since $\varepsilon > 0$ is arbitrary, it follows that

$$I(f_1)+I(f_2) \leqq I(f_1+f_2),$$

which in combination with inequality (5) proves property d).

Now if $f \in L^r$, then $f = f_1 - f_2$, where $f_1, f_2 \in L^+$ (see subsection 1, § 6), and we set, by definition,

$$I(f) = I(f_1)-I(f_2). \tag{16}$$

If also $f = f_1'-f_2'$, where $f_1', f_2' \in L^+$, then $f_1+f_2' = f_1'+f_2$, and hence $I(f_1)+I(f_2') = I(f_1')+I(f_2)$, from which it follows that $I(f_1)-I(f_2) = I(f_1')-I(f_2')$; consequently, formula (16) defines $I(f)$ uniquely. Finally, for $f = f_1+if_2$, $f_1, f_2 \in L^r$, we set

$$I(f) = I(f_1)+iI(f_2). \tag{17}$$

It follows from properties a), c) and d) that $I(f)$ is an integral on L which is not identically equal to zero, and from property b) that $I(f)$ is right invariant. This also proves the *existence* of a right invariant integral.

We shall now prove its uniqueness to within a multiplier.

Let I' be any right invariant integral on $\mathfrak{G}$ which is not identically equal to zero, and let μ_r and ν_r be measures defined by the integrals I and I' respectively.

Let $f_0, f \in L^+$, $f_0 \neq 0$, $f \neq 0$.

We set

$$\varphi(g') = \frac{f(g'^{-1})}{\int f_0(g'^{-1}g)d\mu_r(g)}; \tag{18}$$

it follows that

$$f(g') = \varphi(g'^{-1})\int f_0(g'g)d\mu_r(g). \tag{19}$$

Making use of the uniform continuity of the function f_0 and property a) of the integral I, it is not difficult to show that the denominator in (18) is continuous and > 0; therefore, $\varphi \in L^+$.

We set

$$\psi(g) = \int f_0(g')\varphi(gg'^{-1})d\nu_r(g'); \tag{20}$$

then $\psi \in L^+$. [The continuity of ψ follows from the uniform continuity of the function φ; moreover, if $f_0 = 0$ outside of Q_0 and $\varphi = 0$ outside of Q, where Q_0 and Q are compact, then $\psi = 0$ outside the compact set QQ_0 (see IV, subsection 3).] Integrating both members of (20) with respect to $d\mu_r(g)$, applying Theorem 5, subsection 18, § 6, and using the right invariance of the measure μ_r, we obtain

$$I(\psi) = \int \psi(g)d\mu_r(g) = \int f_0(g') \left[\int \varphi(gg'^{-1})d\mu_r(g) \right] dv_r(g')$$
$$= \int f_0(g') \left[\int \varphi(g)d\mu_r(g) \right] dv_r(g') = I(\varphi)I'(f_0). \quad (21)$$

On the other hand, in virtue of the right invariance of the measure v_r, we can substitute $g'g$ for g' in (20); then (20) assumes the form

$$\psi(g) = \int f_0(g'g)\varphi(g'^{-1})dv_r(g')$$

and, in virtue of (19),

$$I(\psi) = \int \varphi(g'^{-1}) \left[\int f_0(g'g)d\mu_r(g) \right] dv_r(g) = \int f(g)dv_r(g) = I'(f). \quad (22)$$

Comparing (21) and (22), we conclude that

$$I'(f) = I'(f_0)I(\varphi). \quad (23)$$

In particular, (23) must hold for $I' = I$, so that

$$I(f) = I(f_0)I(\varphi).$$

Therefore,

$$\frac{I'(f)}{I(f)} = \frac{I'(f_0)}{I(f_0)},$$

i.e.

$$I'(f) = cI(f), \quad (24)$$

where $c = I'(f_0)/I(f_0) > 0$. Clearly (see (16) and (17)), relation (24) remains valid for all $f \in L$. This proves the following theorem.

THEOREM. *On every locally compact group $\mathfrak{G}$ there exists a right invariant integral which is unique to within a constant multiplier.*

[H. CARTAN [1] gave another proof of this theorem (see the first edition of this book), without using Zorn's lemma (on which the proof of Tikhonov's theorem is based); concerning other proofs also see BREDON [1] and KELLEY and MORSE [1].]

The existence and uniqueness to within a multiplier of a left invariant integral is proved analogously.

III. *A right invariant measure of every open set U is positive and that of every compact set Q is finite.*

Proof. Suppose V is an open set, whose closure is compact and contained in U, and that f is a function in L^+ which vanishes in the exterior of $\bar{V}$ and satisfies the conditions $f \not\equiv 0, f \leqq 1$. Then $\mu_r(U) \geqq I(f) > 0$.

Further, suppose $f \in L^+$ and that $f = 1$ on Q; then

$$\mu_r(Q) \leqq I(f) < \infty.$$

If

$$I_r(x) = \int x(g)d\mu_r(g)$$

s a right invariant integral on the locally compact group $\mathfrak{G}$, then

$$I_r'(x) = \int x(hg)d\mu_r(g)$$

for every fixed $h \in \mathfrak{G}$ is also a right invariant integral and therefore

$$\int x(hg)d\mu_r(g) = \lambda(h)\int x(g)d\mu_r(g), \tag{25}$$

where $\lambda(h)$ is some numerical function. It is easily seen that

$$\lambda(h_1 h_2) = \lambda(h_1)\lambda(h_2), \quad \lambda(e) = 1.$$

Moreover, it follows from the properties of the function $x \in L$ (see V, subsection 3) and of the functional I_r that $\lambda(h)$ is continuous.

Setting

$$I_l(x) = \int x(g)\lambda(g)d\mu_r(g),$$

we obtain the left invariant integral I_l. In fact, applying (25) to $x(g)\lambda(g)$ instead of to $x(g)$, we conclude that

$$\lambda(h)\int x(hg)\lambda(g)d\mu_r(g) = \int x(hg)\lambda(hg)d\mu_r(g) = \lambda(h)\int x(g)\lambda(g)d\mu_r(g),$$

from which it follows that

$$\int x(hg)\lambda(g)d\mu_r(g) = \int x(g)\lambda(g)d\mu_r(g).$$

This shows that for a suitable normalization of the measures,

$$\frac{d\mu_l(g)}{d\mu_r(g)} = \lambda(g). \tag{26}$$

Moreover,

$$\int x(g^{-1})d\mu_r(g) = \int x(g)\lambda(g)d\mu_r(g). \tag{27}$$

In fact, setting $x^*(g) = x(g^{-1})$, we have that

$$\int x(g_0 g^{-1})d\mu_r(g) = \int x^*(gg_0^{-1})d\mu_r(g) = \int x^*(g)d\mu_r(g) = \int x(g^{-1})d\mu_r(g).$$

Consequently, the left member in (27) is a left invariant integral, and therefore

$$\int x(g^{-1})d\mu_r(g) = c\int x(g)\lambda(g)d\mu_r(g).$$

Applying this equality to the functions x which are nonzero only in neighborhoods where $|\lambda(g)-1| < \varepsilon$ and satisfy the conditions: $x(g^{-1}) = x(g)$, $\int x(g)d\mu_r(g) = 1$, we conclude that $c = 1$.

IV. *The group $\mathfrak{G}$ is compact if, and only if, its invariant measures are finite.*

Proof. If $\mathfrak{G}$ is compact, then $1 \in L(\mathfrak{G})$ and therefore $\mu_l(\mathfrak{G}) = I_l(1) < \infty$ and $\mu_r(\mathfrak{G}) = I_r(1) < \infty$. Conversely, if $\mathfrak{G}$ is not compact, then there exists a neighborhood V of the identity with compact closure such that $\mathfrak{G}$ cannot be covered by a finite number of translations of the neighborhood V. Consequently, there exists an infinite sequence of elements $g_n \in \mathfrak{G}$ such that $g_n \notin \bigcup_{k=1}^{n-1} g_k V$. Suppose U is a symmetric neighborhood of the identity such that $UU \subset V$. Then the sets $g_n U$ do not intersect and have the same positive left invariant measure; therefore $\mu_l(\mathfrak{G}) = \infty$; that $\mu_r(\mathfrak{G}) = \infty$ is proved analogously.

A group $\mathfrak{G}$ is called *unimodular* if $\lambda(g) \equiv 1$. In the case of a unimodular group, a right invariant integral is also left invariant and the measure defined by it is said to be *two-sided invariant* or simply *invariant*. Thus, for an invariant measure μ, equalities (1′), (2) in subsection 4 and (27) assume the form

$$\int x(hg)d\mu(g) = \int x(gh)d\mu(g) = \int x(g^{-1})d\mu(g).$$

Clearly, every commutative group is unimodular. Furthermore,

V. *A compact group is unimodular.*

In fact, setting $x(g) \equiv 1$ in (25), we have

$$\mu_r(\mathfrak{G}) = \lambda(h)\mu_r(\mathfrak{G}), \ \lambda(h) \equiv 1.$$

Suppose now that U is a symmetric neighborhood of the identity which is such that $\overline{U}$ is compact. We set $\mathfrak{G}_0 = \bigcup_{n=1}^{\infty} \overline{U}^n$, so that $\mathfrak{G}_0$ is the union of a denumerable number of compact sets $\overline{U}^n$. Clearly, $\mathfrak{G}_0$ is a subgroup of the group $\mathfrak{G}$. Moreover, $\mathfrak{G}_0$ is both open and closed. In fact, if $g_0 \in \mathfrak{G}_0$, then $g_0 \in \overline{U}^n$ for some n, $g_0 U$ is a neighborhood of the element g_0, and $g_0 U \subset \overline{U}^{n+1} \subset \mathfrak{G}_0$; consequently, $\mathfrak{G}_0$ is open. On the other hand, $\overline{\mathfrak{G}}_0 \subset U\mathfrak{G}_0 = \mathfrak{G}_0$, and therefore $\mathfrak{G}_0$ is closed.

But then the entire group $\mathfrak{G}$ is the union (of perhaps a nondenumerable number) of disjoint right cosets $\{\mathfrak{G}_0 g_\alpha\}$ (see subsection 20), each of which is both a closed and an open set and also the union of a denumerable number of compact sets. If V is an arbitrary open set, then $V = \bigcup_\alpha V_\alpha$, where the $V_\alpha = V \cap \mathfrak{G}_0 g_\alpha$ are open sets. But, in virtue of III, the measure of a nonvoid open set is positive; moreover, $\mu(V) = \sum_\alpha \mu(V_\alpha)$ (see I, subsection 5, § 6); consequently, if V is summable, then V cannot intersect more than a denumerable number of the sets $\mathfrak{G}_0 g_\alpha$. On the other hand, for every summable set $A \subset \mathfrak{G}$ there exists an open summable set $V \supset A$ (see IX, subsection 8, § 6); therefore, every summable set intersects no more than a denumerable number of the sets $\mathfrak{G}_0 g_\alpha$.

Since $\mathfrak{G}_0 g_\alpha = \bigcup_{n=1}^{\infty} (\overline{U}^{n+1} g_\alpha - \overline{U}^n g_\alpha) \cup \overline{U} g_\alpha$ and each of the sets $\overline{U} g_\alpha$, $\overline{U}^{n+1} g_\alpha - \overline{U}^n g_\alpha$ is summable, each of the $\mathfrak{G}_0 g_\alpha$ is the union of at most a denumerable number of disjoint sets of which one is a zero set and the remaining are compact (see X, subsection 8, § 6). Consequently,

VI. *Every locally compact group satisfies the conditions of Remark* 1, *subsection* 16, § 6.

EXAMPLES. 1. In the group R^1, an invariant integral has the form $\int_{-\infty}^{\infty} x(t)dt$; in fact, $\int_{-\infty}^{\infty} x(t+t_0)dt = \int_{-\infty}^{\infty} x(t)dt$. Analogously, an invariant integral in the group C^1 has the form $\int_{-\infty}^{\infty} \int_{-\infty}^{\infty} x(\xi, \eta) d\xi d\eta$, $\zeta = \xi + i\eta$; in the groups R_0^1 and C_0^1 we have

$$\int_{-\infty}^{\infty} x(t) \frac{dt}{t} \quad \text{and} \quad \int_{-\infty}^{\infty} \int_{-\infty}^{\infty} x(\xi, \eta) \frac{d\xi\, d\eta}{|\xi + i\eta|^2},$$

respectively.

2. Suppose K is the group of real triangular matrices of the second order:

$$K = \left\{ \begin{Vmatrix} \lambda^{-1} & \mu \\ 0 & \lambda \end{Vmatrix} \right\}.$$

Then the left and right invariant integrals on K are defined uniquely, to within a constant multiplier, by the formulas

$$I_l(x) = \int_{-\infty}^{\infty} \int_{-\infty}^{\infty} x(\lambda, \mu) d\lambda\, d\mu, \quad I_r(x) = \int_{-\infty}^{\infty} \int_{-\infty}^{\infty} x(\lambda, \mu) |\lambda|^{-2} d\lambda\, d\mu.$$

§ 28. Definition and fundamental properties of a group algebra

1. Definition of a group algebra. We shall first consider the case where the group $\mathfrak{G}$ consists of a finite number of elements; such a group is said to be *finite*. We denote the set of all formal linear combinations of elements of the group $\mathfrak{G}$ by $R(\mathfrak{G})$. Each such linear combination is defined by the coefficients of the elements of the group $\mathfrak{G}$; we denote by $x(g)$ the coefficient of $g \in \mathfrak{G}$; we can then say that $R(\mathfrak{G})$

consists of all possible sums $x = \sum_g x(g)g$ where the $x(g)$ are all possible functions on $\mathfrak{G}$ and the summation runs over the entire group $\mathfrak{G}$. Two such linear combinations $x = \sum_g x(g)g$, $y = \sum_g y(g)g$ will be assumed to be equal if, and only if, $x(g) \equiv y(g)$; in particular, we shall write $x = 0$ if, and only if, $x(g) \equiv 0$.

We define the operations of scalar multiplication, addition, and multiplication in $R(\mathfrak{G})$ by setting

$$\lambda x = \sum_g [\lambda x(g)]g, \quad x+y = \sum_g [x(g)+y(g)]g,$$
$$xy = \sum_{g_1}\sum_{g_2} x(g_1)y(g_2)g_1 g_2 \tag{1}$$

with $x = \sum_g x(g)g$, $y = \sum_g y(g)g$. Then $R(\mathfrak{G})$ becomes an algebra with identity equal to the identity element of the group $\mathfrak{G}$, i.e. the combination $x = \sum x(g)g$, in which $x(e) = 1$ and $x(g) = 0$ for $g \neq e$.

The algebra $R(\mathfrak{G})$ is called the *group algebra* of the finite group $\mathfrak{G}$.

Instead of considering the linear combinations $x = \sum_g x(g)g$, we can consider the functions $x = x(g)$ themselves and treat $R(\mathfrak{G})$ as the set of all these functions; in this case, the operations of scalar multiplication and addition are defined as usual:

$$\lambda x = \lambda x(g), \quad x+y = x(g)+y(g).$$

In order to clarify the definition of multiplication in $R(\mathfrak{G})$, we set $g_1 g_2 = g$ in (1) and interchange the order of summation; we then obtain

$$xy = \sum_g (\sum_{g_1} x(g_1)y(g_1^{-1}g))g.$$

Consequently, the rule of multiplication in $R(\mathfrak{G})$ is defined by the formula

$$xy(g) = \sum_{g_1} x(g_1)y(g_1^{-1}g). \tag{1'}$$

The multiplication operation, defined by (1′), is usually called *convolution*. (In those cases where it will be necessary to emphasize that we are dealing with convolution and not with the usual multiplication of functions, we shall write $x \cdot y$ instead of xy.)

We can now carry over the definition of a group algebra to an arbitrary locally compact group. Suppose $\mathfrak{G}$ is a locally compact group and that μ is a left invariant measure in $\mathfrak{G}$. Suppose $L^1 = L^1(\mathfrak{G})$ is the L^1-space defined by this measure (see subsection 7, § 6), i.e. the set of all μ-measurable functions $a = a(g)$ for which $\int |a(g)| d\mu(g) < \infty$. Then L^1 is a Banach space with operations $\lambda a = \lambda a(g)$, $a+b = a(g)+b(g)$. But L^1 can also be made into a normed algebra by defining the multiplication operation in L^1, in analogy with (1′), by the formula

$$(a \cdot b)(g) = \int a(g_1)b(g_1^{-1}g)d\mu(g_1).$$

In order to justify this definition, we must establish that $(a \cdot b)(g)$ also belongs to L^1 and that $||a \cdot b||_1 \leqq ||a||_1 \, ||b||_1$. Moreover, we must establish that

$$\begin{cases} \alpha(a \cdot b) = (\alpha a) \cdot b, \; a \cdot (\alpha b) = \alpha(a \cdot b), \; (a \cdot b) \cdot c = a \cdot (b \cdot c), \\ (a+b) \cdot c = a \cdot c + b \cdot c, \; a \cdot (b+c) = a \cdot b + a \cdot c \end{cases} \tag{2}$$

for all $a, b, c \in L^1$ and all complex numbers α (see subsection 1, § 7). But it follows from the Fubini theorem (see subsection 18, § 6) and the left invariance of measure that

$$\begin{aligned} ||a \cdot b||_1 &= \int \left| \int a(g_1) b(g_1^{-1} g) d\mu(g_1) \right| d\mu(g) \\ &\leqq \int \left[\int |a(g_1)| |b(g_1^{-1} g)| d\mu(g_1) \right] d\mu(g) \\ &= \int \left[\int |b(g_1^{-1} g)| d\mu(g) \right] |a(g_1)| d\mu(g_1) \\ &= \int \left[\int |b(g)| d\mu(g) \right] |a(g_1)| d\mu(g_1) = ||b||_1 \int |a(g_1)| d\mu(g_1) = ||b||_1 \, ||a||_1, \end{aligned}$$

and relations (2) are easily verified directly. [In this case, the existence of the iterated integral implies the existence of the multiple integral. In fact, suppose $a \geqq 0$, $b \geqq 0$, $a, b \in L^1(\mathfrak{G})$; then there exist functions a_1, b_1 which are the limits of nondecreasing sequences of upper semi-continuous functions and a_2, b_2 which are the limits of nonincreasing sequences of lower semi-continuous functions such that $a_1 \leqq a \leqq a_2$, $b_1 \leqq b \leqq b_2$ and $I_l(a_1) = I_l(a) = I_l(a_2)$, $I_l(b_1) = I_l(b) = I_l(b_2)$. Suppose J_l is the integral on $\mathfrak{G} \times \mathfrak{G}$ corresponding to the measure $\mu_l \times \mu_l$. Then

$$\begin{gathered} a_1(g_1) b_1(g_1^{-1} g_2), \quad a_2(g_1) b_2(g_1^{-1} g_2) \in L^1(\mathfrak{G}_1 \times \mathfrak{G}_2), \\ a_1(g_1) b_1(g_1^{-1} g_2) \leqq a(g_1) b(g_1^{-1} g_2) \leqq a_2(g_1) b_2(g_1^{-1} g_2) \end{gathered} \tag{3}$$

and

$$J_l(a_1(g_1) b_1(g_1^{-1} g_2)) = J_l(a_2(g_1) b_2(g_1^{-1} g_2)) = I_l(a) I_l(b).$$

In fact, the function $a_1(g_1) b_1(g_1^{-1} g_2)$ is the limit of a nondecreasing sequence of lower semi-continuous functions, and the function $a_2(g_1) b_2(g_1^{-1} g_2)$ is the limit of a nonincreasing sequence of upper semi-continuous functions on $\mathfrak{G} \times \mathfrak{G}$; therefore for the function $a_2(g_1) b_2(g_1^{-1} g_2)$ the existence of the iterated integral implies the existence of the multiple integral (see I, subsection 18, § 6 and a) in the proof of Proposition X, subsection 18, § 6). Consequently, $J_l(a_2(g_1) b_2(g_1^{-1} g_2)) = I_l(a_2) I_l(b_2) = I_l(a) I_l(b) < \infty$. Since $a_1(g_1) b_1(g_1^{-1} g_2)$ is J_l-measurable and $a_1(g_1) b_1(g_1^{-1} g_2) \leqq a_2(g_1) b_2(g_1^{-1} g_2)$, then also $a_1(g_1) b_1(g_1^{-1} g_2) \in L^1(J_l)$ and hence $J_l(a_1(g_1) b_1(g_1^{-1} g_2)) = I_l(a_1) I_l(b_1)$. But then it follows from (3) that $a(g_1) b(g_1^{-1} g_2)$ is J_l-measurable and $J_l(a(g_1) b(g_1^{-1} g_2)) = I_l(a) I_l(b)$ exists so that the assertion is proved for

$a \geqq 0, b \geqq 0, a, b \in L^1(\mathfrak{G})$. It easily follows that the assertion is valid for arbitrary $a, b \in L^1(\mathfrak{G})$.]

We now set

$$a^*(g) = \lambda^{-1}(g)\overline{a(g^{-1})},$$

where $a(g) \in L^1$. In virtue of (26) and (27), subsection 5, § 27,

$$\|a^*\|_1 = \int |a^*(g)| d\mu(g) = \int |a(g^{-1})| \lambda^{-1}(g) d\mu(g)$$

$$= \int |a(g^{-1})| d\mu_r(g) = \int |a(g)| \lambda(g) d\mu_r(g) = \int |a(g)| d\mu(g) = \|a\|_1 < \infty,$$

so that $a^* \in L^1$ and $\|a^*\|_1 = \|a\|_1$. It is also easily verified that all the other involution axioms are satisfied, including the relation $(xy^*) = y^*x^*$; consequently, with this definition of involution, $L^1(\mathfrak{G})$ becomes a symmetric Banach algebra.

The algebra $L^1(\mathfrak{G})$ contains the identity if, and only if, the group $\mathfrak{G}$ is discrete.

Proof. If $\mathfrak{G}$ is discrete, then the individual elements of the group $\mathfrak{G}$ are sets having the same positive invariant measure which can be assumed equal to unity. The identity of the algebra $L^1(\mathfrak{G})$ will then be the function

$$e(g) = \begin{cases} 1 & \text{for } g = e, \\ 0 & \text{for } g \neq e. \end{cases}$$

Conversely, suppose $L^1(\mathfrak{G})$ contains the identity $e(g)$; we shall first prove that then the measures of nonvoid open sets have a positive greatest lower bound. In fact, if this were not the case there would exist a neighborhood U of the identity of the group $\mathfrak{G}$ such that $\int_U |e(g)| d\mu_l(g) < \varepsilon$ (see XI, subsection 10, § 6). But then, choosing a symmetric neighborhood V of the identity of the group, satisfying the condition $V^2 \subset U$, we would have

$$\xi_V(g) = (e \cdot \xi_V)(g) = \int e(g')\xi_V(g'^{-1}g) d\mu_l(g')$$

$$= \int_{gV} e(g') d\mu_l(g') \leqq \int_U |e(g')| d\mu_l(g') < \varepsilon$$

for the characteristic function $\xi_V(g)$ of the set V, where $g \in V$; but this contradicts the equality $\xi_V(g) = 1$.

Thus the measure of every nonvoid open set remains greater than some $a > 0$. But then an open set whose closure is compact can consist of only a finite number of elements because otherwise its measure would be larger than na for arbitrary natural number n. Consequently every single point is an open set and hence the group $\mathfrak{G}$ is discrete.

Thus, discarding the case of a discrete group, we can assume that $L^1(\mathfrak{G})$ does not contain the identity. Adjoining the identity to $L^1(\mathfrak{G})$ we obtain a symmetric Banach algebra with identity, which we denote by $R(\mathfrak{G})$ and call the *group algebra of the group* $\mathfrak{G}$.

But if the group $\mathfrak{G}$ is discrete, then its group algebra will be assumed to be $L^1(\mathfrak{G})$ so that in this case $R(\mathfrak{G}) = L^1(\mathfrak{G})$.

2. Some properties of the group algebra.

I. *The elements*

$$x^{g_0}(g) = x(gg_0), \quad x_{g_0}(g) = x(g_0^{-1}g),$$

with $x \in L^1(\mathfrak{G})$, *of the algebra* $L^1(\mathfrak{G})$ *are continuous functions of* g_0 *in the sense of the norm in* $L^1(\mathfrak{G})$.

Proof. In virtue of V, subsection 3, § 27, the assertion holds for all functions $x \in L(\mathfrak{G})$; on the other hand, $L(\mathfrak{G})$ is dense in $L^1(\mathfrak{G})$ (see subsection 7, § 6). Therefore, first choosing $x' \in L(\mathfrak{G})$ so that $||x-x'||_1 < \varepsilon/3$ and then a neighborhood U of the identity so that $||x'_g - x'_h||_1 < \varepsilon/3$ for $g \in Uh$, we have

$$||x_g - x_h||_1 \leqq ||x_g - x'_g||_1 + ||x'_g - x'_h||_1 + ||x'_h - x_h||_1 < \frac{\varepsilon}{3} + \frac{\varepsilon}{3} + \frac{\varepsilon}{3} = \varepsilon.$$

II. *The algebra* $L^1(\mathfrak{G})$ *contains a set which approximates the identity.*

Proof. To each neighborhood U of the identity we assign a nonnegative function $z_U \in L^1(\mathfrak{G})$ which vanishes in the exterior of U and is such that $\int z_U(g)d\mu(g) = 1$. The functions z_U form a set which approximates the identity. In fact, for $x \in L^1(\mathfrak{G})$,

$$(z_U \cdot x)(g) = \int z_U(g_1)x(g_1^{-1}g)d\mu(g_1),$$

i.e.

$$z_U \cdot x = \int z_U(g_1)x_{g_1}\, d\mu(g_1).$$

It follows that

$$||z_U \cdot x - x||_1 \leqq \int z_U(g_1)||x_{g_1} - x||_1\, d\mu(g_1) = \int_U z_U(g_1)||x_{g_1} - x||_1\, d\mu(g_1).$$

We choose U so that $||x_{g_1} - x||_1 < \varepsilon$ for $g_1 \in U$ (see I); then

$$||z_U \cdot x - x||_1 \leqq \varepsilon \int_U z_U(g_1)d\mu(g_1) = \varepsilon.$$

It is proved analogously that $||x \cdot z_U - x||_1 \to 0$; it is only that the proof is somewhat complicated by the necessity of introducing the function $\lambda(g)$ (see subsection 5, § 27) and we leave the detailed execution of the proof to the reader.

REMARKS. 1) *The assertions of Propositions I and II remain valid if in them the space* $L^1(\mathfrak{G})$ *is replaced by the space* $L^2(\mathfrak{G})$ (*taking in this connection* U *with compact* *closure $\bar{U}$). In particular, for an arbitrary function $x(g) \in L^2(\mathfrak{G})$,*

$$||x - x \cdot z_U||_2 < \varepsilon$$

for suitable choice of neighborhood U of the identity element.

2) *If $x(g) \in L(\mathfrak{G})$, then for arbitrary $\varepsilon > 0$, there exists a neighborhood U of the identity such that*

$$|(z_U \cdot x)(g) - x(g)| < \varepsilon \text{ for all } g \in \mathfrak{G}.$$

In fact, choosing U such that $\bar{U}$ is compact and $|x(g_1^{-1}g) - x(g)| < \varepsilon$ for $g_1 \in U$ (see V, subsection 3, § 27), we have that

$$|(z_U \cdot x)(g) - x(g)| = |\int_U z_U(g_1)[x(g_1^{-1}g) - x(g)]d\mu(g_1)|$$
$$< \varepsilon \int_U z_U(g_1)d\mu(g_1) = \varepsilon.$$

III. *If $x_1, x_2 \in L^2(\mathfrak{G})$, then $(x_1 \cdot x_2)(g)$ is a continuous function of g.*

Proof. We put $y(g) = \overline{x_2(g^{-1})}$. In virtue of the Remark 1), above, the element $y_{g_0}(g) = y(g_0^{-1}g) = \overline{x_2(g^{-1}g_0)}$ is a continuous function of g_0 in the sense of the norm in $L^2(\mathfrak{G})$ and therefore the inner product

$$(x_1, y_{g_0}) = \int x_1(g)\overline{y_{g_0}(g)}d\mu_l(g) = \int x_1(g)x_2(g^{-1}g_0)d\mu_l(g) = (x_1 \cdot x_2)(g_0)$$

is a continuous function of g_0.

IV. *A closed subspace I in $L^1(\mathfrak{G})$ is a left (right) ideal in $L^1(\mathfrak{G})$ if, and only if, it is invariant under left (right) translations.*

Proof. Suppose z_U is the same as in Proposition II. If I is a left ideal in $L^1(\mathfrak{G})$ and $x \in I$, then also $z_{Ug_0}x \equiv (z_U)_{g_0}x \in I$. But

$$(z_U)_{g_0} \cdot x(g) \equiv z_{Ug_0} \cdot x(g) = \int z_U(g_0^{-1}g')x(g'^{-1}g)d\mu(g')$$
$$= \int z_U(g')x(g'^{-1}g_0^{-1}g)d\mu(g') = (z_U \cdot x)_{g_0}(g).$$

It follows, in virtue of II, that

$$(z_U)_g \cdot x \equiv z_{Ug} \cdot x = (z_U \cdot x)_g \to x_g;$$

consequently, $x_g \in I$, i.e. I is invariant under left translations.

Conversely, suppose I is invariant under any left translation and let $x \in I$. We first assume that $z' \in L(\mathfrak{G})$ and vanishes in the exterior of the compact set Q. Then the function $z'(g_1)x_{g_1}$ is continuous in norm in L^1 (in virtue of I) and vanishes in the exterior of Q. Consequently, repeating the usual arguments, it can be shown that

$$z' \cdot x = \int_{\mathfrak{G}} z'(g_1)x_{g_1}\, d\mu_l(g_1) = \int_Q z'(g_1)x_{g_1}\, d\mu_l(g_1)$$

is the limit in norm of finite sums of the form

$$\sum_k z'(g_k)x_{g_k}\, \mu_l(\Delta_k) \in I,$$

and therefore $z' \cdot x \in I$ inasmuch as I is closed. But since $L(\mathfrak{G})$ is dense in $L^1(\mathfrak{G})$ (see subsection 7, § 6), for every function $z \in L^1(\mathfrak{G})$ and every $\varepsilon > 0$ there exists a function $z' \in L(\mathfrak{G})$ such that $||z'-z||_1 < \varepsilon$. Then $||z' \cdot x - z \cdot x||_1 < \varepsilon||x||$ and, consequently, $z \cdot x \in I$. The assertion for a right ideal is proved analogously.

V. *The group algebra of a locally compact topological group is a reduced algebra.*

Proof. For an arbitrary function $a \in L^1(\mathfrak{G})$, we define the operator A_a in the Hilbert space $L^2(\mathfrak{G})$ by setting

$$A_a \xi = a \cdot \xi \quad \text{for } \xi \in L^2(\mathfrak{G}). \tag{1}$$

We shall prove that A_a is bounded. To this end, we set $T_{g_1}\xi = \xi(g_1^{-1}g)$; the operator T_{g_1} is unitary because it is a one-to-one mapping of $L^2(\mathfrak{G})$ onto itself and it is isometric in virtue of the fact that

$$||T_{g_1}\xi||_2^2 = \int |\xi(g_1^{-1}g)|^2 d\mu(g) = \int |\xi(g)|^2 d\mu(g) = ||\xi||_2^2.$$

Formula (1) can be rewritten in the form

$$A_a\xi = \int a(g_1)T_{g_1}\xi d\mu(g_1),$$

from which it follows that

$$||A_a\xi||_2 \leqq \int |a(g_1)|\, ||T_{g_1}\xi||_2\, d\mu(g_1) = \int |a(g_1)| d\mu(g_1)\, ||\xi||_2 = ||a||_1\, ||\xi||_2.$$

Consequently, A_a is bounded and $|A_a| \leqq ||a||_1$. It is easily verified that the correspondence $a \to A_a$ is a representation of the algebra $L^1(\mathfrak{G})$.

Setting

$$\lambda e + a \to \lambda 1 + A_a,$$

we obtain a representation of the entire algebra $R(\mathfrak{G})$. Therefore, for an arbitrary function $\xi \in L^2(\mathfrak{G})$,

$$f(\lambda e + a) = \lambda(\xi, \xi) + (A_a\xi, \xi)$$

is a positive functional on the algebra $R(\mathfrak{G})$.

Now suppose $f(a^*a) = 0$ for an arbitrary positive functional f; then, in particular, $||A_a\xi||_2 = 0$ for all $\xi \in L^2(\mathfrak{G})$. This means that whatever the function $\xi(g) \in L^2(\mathfrak{G})$

we have

$$\int a(g_1)\xi(g_1^{-1}g)d\mu(g_1) = 0$$

for almost all $g \in \mathfrak{G}$. From this, taking the characteristic function of an arbitrary summable set for ξ, we conclude that $a(g) = 0$ almost everywhere on $\mathfrak{G}$. Hence, if $a \neq 0$, then among the functionals $f(x) = (A_x\xi, \xi)$ there exists one for which $f(a^*a) \neq 0$. But if $a = 0$, $\lambda \neq 0$, then such a functional is $f(\lambda e + a) = \lambda$. Thus, $R(\mathfrak{G})$ is a reduced algebra.

We have at the same time proved the following proposition.

VI. *If* $x_1 \in L^1(\mathfrak{G})$, $x_2 \in L^2(\mathfrak{G})$, *then* $x_1 \cdot x_2 \in L^2(\mathfrak{G})$.

VII. *The group algebra of a locally compact group is semisimple.*

In fact, the radical is always contained in the reducing ideal (see VII, subsection 2, § 18); and the latter is (0) in virtue of V.

In the papers of WENDEL [1, 2], GREENLEAF [1], STRICHARTZ [1], EDWARDS [5], and other authors, homomorphisms and isomorphisms of the group algebras $L^1(\mathfrak{G})$ (and $L^p(\mathfrak{G})$ if $\mathfrak{G}$ is compact) are investigated. STRICHARTZ [1] proved that every isometric isomorphism U of the algebra $L^p(\mathfrak{G}_1)$ onto the algebra $L^p(\mathfrak{G}_2)$ ($\mathfrak{G}_1$, $\mathfrak{G}_2$ are compact for $p \neq 1$) for $p \neq 2$ has the form $(UF)(g) = J\chi(g)F(Ag)$, where $A: \mathfrak{G}_2 \to \mathfrak{G}_1$ is an isomorphism of topological groups, $\chi(g)$ is a homomorphism of the topological group $\mathfrak{G}_2$ onto the group of rotations of the circle and J is a constant $= 1$ for $p \neq 1$.

§ 29. Unitary representations of a locally compact group and their relationship with the representations of the group algebra

1. Unitary representations of a group. A *unitary representation* of a group $\mathfrak{G}$ is any homomorphism of the group $\mathfrak{G}$ into the group of unitary operators in a Hilbert space $\mathfrak{H}$. In other words, a unitary representation is a mapping $g \to U_g$ of the group $\mathfrak{G}$ into the group of unitary operators U, for which

$$U_{g_1}U_{g_2} = U_{g_1g_2}, \ U_e = 1.$$

Suppose $\mathfrak{G}$ is a locally compact group. A unitary representation of the group $\mathfrak{G}$ is said to be *weakly measurable* if the function

$$\varphi(g) = (U_g\xi, \eta)$$

is measurable for arbitrary vectors $\xi, \eta \in \mathfrak{H}$ and *weakly continuous* if the function $\varphi(g) = (U_g\xi, \eta)$ is continuous for arbitrary $\xi, \eta \in \mathfrak{H}$. A unitary representation is said to be *continuous* at $g = g_0$ if

$$|U_g\xi - U_{g_0}\xi| \to 0 \quad \text{as } g \to g_0$$

for all vectors $\xi \in \mathfrak{H}$.

If a representation is continuous for $g = e$, then it is continuous for all other g. In fact, this follows from the equality

$$|U_g\xi - U_{g_0}\xi| = |U_{g_0}(U_{g_0^{-1}g}\xi - \xi)| = |(U_{g_0^{-1}g}\xi - \xi)|.$$

2. Relationship between representations of a group and of the group algebra. Suppose $x \to A_x$ is a representation of the group algebra $R(\mathfrak{G})$ in the space $\mathfrak{H}$. We denote by $\mathfrak{N}$ the set of all vectors ξ such that $A_a\xi = 0$ for all $a \in L^1(\mathfrak{G})$. $\mathfrak{N}$ is an invariant subspace. In fact, suppose $x = \lambda e + a_1$, $a_1 \in L^1(\mathfrak{G})$. Then

$$ax = a(\lambda e + a_1) = \lambda a + aa_1 \in L^1(\mathfrak{G}).$$

Consequently,

$$A_a A_x \xi = A_{ax}\xi = 0$$

for $\xi \in \mathfrak{N}$. This means $\xi \in \mathfrak{N}$ implies that $A_x\xi \in \mathfrak{N}$ for all $x \in R(\mathfrak{G})$, i.e. that $\mathfrak{N}$ is an invariant subspace. In this invariant subspace $\mathfrak{N}$, a representation has the trivial form

$$A_{\lambda e + a} = \lambda 1. \tag{1}$$

The representation, defined by formula (1), is called the *degenerate* representation. The corresponding representation of the algebra $L^1(\mathfrak{G})$ is also called the *degenerate* representation; it maps all elements of the algebra $L^1(\mathfrak{G})$ into zero. If $\mathfrak{N} \neq (0)$, then we can consider our representation only in $\mathfrak{H} \ominus \mathfrak{N}$. In the sequel, we shall assume that this splitting off of the subspace $\mathfrak{N}$ has already been done, i.e. that $\mathfrak{N} = (0)$. In this case, we shall say that the given representation $x \to A_x$ *does not contain the degenerate representation.*

THEOREM 1. *To each representation $x \to A_x$ of the group algebra $R(\mathfrak{G})$, not containing the degenerate representation, there corresponds a continuous unitary representation $g \to U_g$ of the group $\mathfrak{G}$. Conversely, to each weakly continuous unitary representation $g \to U_g$ of the group $\mathfrak{G}$ there corresponds a representation $x \to A_x$ of its group algebra $R(\mathfrak{G})$, not containing the degenerate representation. These representations are connected by the formula*

$$A_{\lambda e + a} = \lambda 1 + \int a(g) U_g \, d\mu(g).$$

(Concerning the integration of operator-valued functions, see subsection 19, § 6.)

Proof. It will suffice to prove the theorem for a cyclic representation, because every representation is the direct sum of cyclic representations. Thus, suppose $x \to A_x$ is a cyclic representation of the algebra $R(\mathfrak{G})$; we denote a cyclic vector of this representation by ξ_0. Suppose $a = a(g) \in L^1(\mathfrak{G})$; we set $a_{g_0}(g) = a(g_0^{-1}g)$. Clearly, a_{g_0} is also an element of $L^1(\mathfrak{G})$.

We shall prove that

$$b_{g_0}^* \cdot a_{g_0} = b^* \cdot a \tag{2}$$

for any two elements a, b in $L^1(\mathfrak{G})$. In fact, in virtue of relation (32), subsection 5, § 27, and the left invariance of measure,

$$b^* \cdot a = \int \overline{b(g_1^{-1})}\lambda(g_1^{-1})a(g_1^{-1}g)d\mu(g_1) = \int \overline{b(g_1)}a(g_1 g)d\mu(g_1)$$

$$= \int \overline{b(g_0^{-1}g_1)}a(g_0^{-1}g_1 g)d\mu(g_1) = \int \overline{b_{g_0}(g_1)}a_{g_0}(g_1 g)d\mu(g_1) = b_{g_0}^* \cdot a_{g_0}.$$

We denote by $\mathfrak{H}'$ the set of all elements ξ of the form $\xi = A_a\xi_0$, $a \in L^1(\mathfrak{G})$. $\mathfrak{H}'$ is dense in $\mathfrak{H}$. In fact, otherwise there would exist a vector $\xi_1 \neq 0$, orthogonal to $\mathfrak{H}'$, i.e. $(\xi_1, A_a\xi_0) = 0$ for all $a \in L^1(\mathfrak{G})$. In this equality, we set $a = b^* \cdot x$, where $x \in R(\mathfrak{G})$, $b \in L^1(\mathfrak{G})$. We then obtain that

$$(A_b\xi_1, A_x\xi_0) = (\xi_1, A_{b^* \cdot x}\xi_0) = 0.$$

Thus, the vector $A_b\xi_1$ is orthogonal to all the vectors $A_x\xi_0$, $x \in R(\mathfrak{G})$. Since ξ_0 is a cyclic vector, the vectors $A_x\xi_0$ form a set dense in $\mathfrak{H}$. Consequently, $A_b\xi_1 = 0$ for all $b \in L^1(\mathfrak{G})$. This means that $\xi_1 \in \mathfrak{N}$. But the latter contradicts our basic assumption that $\mathfrak{N} = (0)$.

Thus $\mathfrak{H}'$ is dense in $\mathfrak{H}$. We now introduce the operator U_{g_0} in $\mathfrak{H}'$, by setting

$$U_{g_0}\xi = A_{a_{g_0}}\xi_0$$

for $\xi = A_a\xi_0$, $a \in L^1(\mathfrak{G})$. This definition does not depend on the choice of a. In fact, if $\xi = A_a\xi_0 = A_b\xi_0$, $a, b \in L^1(\mathfrak{G})$, then $A_{a-b}\xi_0 = 0$ and therefore $A_c\xi_0 = 0$ for all c from the closed left ideal I_l in $L^1(\mathfrak{G})$, which is generated by $a-b$. But then also $a_{g_0} - b_{g_0} \in I_l$ (see IV, subsection 2, § 28) and therefore $A_{a_{g_0}}\xi_0 = A_{b_{g_0}}\xi_0 = U_{g_0}\xi$. The operator U_{g_0} maps $\mathfrak{H}'$ onto itself. Moreover, the operator U_{g_0} is isometric. In fact, it follows from equality (2) that

$$(U_{g_0}\xi, U_{g_0}\xi) = (A_{a_{g_0}}\xi_0, A_{a_{g_0}}\xi_0) = (A_{a^*_{g_0} \cdot a_{g_0}}\xi_0, \xi_0)$$

$$= (A_{a^* \cdot a}\xi_0, \xi_0) = (A_a\xi_0, A_a\xi_0) = (\xi, \xi).$$

Therefore the operator U_{g_0} is uniquely extendible to a unitary operator in the space $\mathfrak{H}$.

Clearly, $a_{g_1 g_2} = (a_{g_2})_{g_1}$. From this it follows that $U_{g_1 g_2} = U_{g_1}U_{g_2}$. Therefore the mapping $g \to U_g$ is a unitary representation of the group $\mathfrak{G}$ in the space $\mathfrak{H}$.

This representation is continuous. In fact, in virtue of I, subsection 2, § 28,

$$||a_{g_0} - a||_1 = \int |a(g_0^{-1}g) - a(g)|d\mu(g) \to 0 \quad \text{as } g_0 \to e.$$

On the other hand, every representation of a symmetric Banach algebra is continuous (see Theorem 1, subsection 3, § 17); consequently, we also have that

$$|A_{a_{g_0}} - A_a| \to 0 \quad \text{as } g_0 \to e.$$

It follows that

$$|A_{a_{g_0}}\xi_0 - A_a\xi_0| \to 0 \quad \text{as } g_0 \to e,$$

i.e.

$$|U_{g_0}\xi - \xi| \to 0 \quad \text{as } g_0 \to e$$

for all elements ξ in $\mathfrak{H}'$. Since $\mathfrak{H}'$ is dense in $\mathfrak{H}$ and $|U_{g_0}| = 1$, this relation holds in the entire space $\mathfrak{H}$.

This proves that the representation $g \to U_g$ is continuous for the element $g = e$, and, consequently, also for any other element.

We shall now consider how the initial representation of the group algebra is obtained from this representation of the group $\mathfrak{G}$.

Suppose $a_0(g) \in L^1(\mathfrak{G})$. We set

$$B_{a_0} = \int a_0(g) U_g \, d\mu(g).$$

We shall prove that $A_{a_0} = B_{a_0}$. We set $f(x) = (A_x\xi_0, \xi_0)$. Then, for $\xi = A_a\xi_0$, $\eta = A_b\xi_0$, $a, b \in L^1(\mathfrak{G})$,

$$(B_{a_0}\xi, \eta) = \int a_0(g_1)(U_{g_1}\xi, \eta)d\mu(g_1) = \int a_0(g_1)(U_{g_1}A_a\xi_0, A_b\xi_0)d\mu(g_1)$$

$$= \int a_0(g_1)(A_{b^*}A_{a_{g_1}}\xi_0, \xi_0)d\mu(g_1) = \int a_0(g_1) f(b^* \cdot a_{g_1})d\mu(g_1).$$

Since f is a continuous functional in $L^1(\mathfrak{G})$, the last expression is equal to $f(b^* \cdot \int a_0(g_1)a_{g_1}d\mu(g_1))$. On the other hand,

$$\int a_0(g_1)a_{g_1}d\mu(g_1) = \int a_0(g_1)a(g_1^{-1}g)d\mu(g_1) = a_0 \cdot a.$$

Therefore

$$(B_{a_0}\xi, \eta) = f(b^* \cdot (a_0 \cdot a)) = (A_{b^*a_0a}\xi_0, \xi_0) = (A_{a_0}A_a\xi_0, A_b\xi_0) = (A_{a_0}\xi, \eta).$$

Since the vectors ξ, η under consideration form a dense set in $\mathfrak{H}$, it follows from this that $A_{a_0} = B_{a_0}$.

Thus $A_{a_0} = \int a_0(g)U_g d\mu(g)$ and, consequently,

$$A_{\lambda e + a} = \lambda 1 + \int a(g) U_g \, d\mu(g). \tag{3}$$

We have thus proved that to every representation of the group algebra $R(\mathfrak{G})$, not containing the degenerate representation, there corresponds a representation of the group $\mathfrak{G}$, from which this representation of the algebra $R(\mathfrak{G})$ is obtained by formula (3).

Conversely, suppose U_g is a weakly continuous unitary representation of the group $\mathfrak{G}$. We set

$$(A_a\xi, \eta) = \int a(g)(U_g\xi, \eta)d\mu(g) \tag{4}$$

for $a \in L^1(\mathfrak{G})$. This equality defines a bounded operator A_a in $\mathfrak{H}$ because

$$\left|\int a(g)(U_g\xi, \eta)d\mu(g)\right| \leqq \int |a(g)|d\mu(g)\, |\xi|\, |\eta|.$$

It is easily seen that the correspondence $a \to A_a$ is a representation of the algebra $L^1(\mathfrak{G})$. Setting

$$A_{\lambda e+a} = \lambda 1 + A_a, \quad a \in L^1(\mathfrak{G}),$$

we obtain a representation of the algebra $R(\mathfrak{G})$. Here, $\mathfrak{N} = (0)$, i.e. the representation $x \to A_x$ does not contain the degenerate representation. In fact, if $\xi \neq 0$, then $(\xi, \xi) \neq 0$. We choose a neighborhood V of the identity of the group $\mathfrak{G}$ so that $\bar{V}$ is compact, $\mu(V) \leqq 1$ and $|(U_g\xi, \xi)-(\xi, \xi)| < \frac{1}{2}(\xi, \xi)$ for $g \in V$. Then if $a(g)$ is the characteristic function of the set V, we have

$$\begin{aligned} |(A_a\xi, \xi)-(\xi, \xi)| &= \left|\int_V (U_g\xi, \xi)d\mu(g)-(\xi, \xi)\right| \\ &< \int_V |(U_g\xi, \xi)-(\xi, \xi)|d\mu(g)+[1-\mu(V)](\xi, \xi) \\ &\leqq \tfrac{1}{2}\mu(V)(\xi, \xi)+[1-\mu(V)](\xi, \xi) < (\xi, \xi), \end{aligned}$$

and therefore $A_a\xi \neq 0$. This completes the proof of the theorem.

In virtue of this theorem, to the representation of the algebra $R(\mathfrak{G})$ just constructed there corresponds a continuous unitary representation of the group $\mathfrak{G}$ – we denote it by U'_g – such that

$$A_a = \int a(g)U'_g d\mu(g), \quad a \in L^1(\mathfrak{G}).$$

But then

$$(A_a\xi, \eta) = \int a(g)(U'_g\xi, \eta)d\mu(g).$$

Comparison with equality (4) yields

$$\int a(g)(U_g\xi, \eta)d\mu(g) = \int a(g)(U'_g\xi, \eta)d\mu(g).$$

This equality holds for an arbitrary summable function $a(g)$; consequently, for almost all g,

$$(U_g\xi, \eta) = (U'_g\xi, \eta). \tag{5}$$

But both members of (5) are continuous functions on $\mathfrak{G}$ and therefore (5) holds for all $g \in \mathfrak{G}$. So we have proved the following corollary.

COROLLARY. *Every weakly continuous unitary representation of a group $\mathfrak{G}$ is continuous.*

REMARK 1. If the representation space $\mathfrak{H}$ is separable, then the second assertion of Theorem 1 holds for weakly measurable representations $g \to U_g$ and $U_g = U'_g$ for almost all $g \in \mathfrak{G}$.

In fact, we have only to show that $\mathfrak{N} = (0)$.

Let $\{\xi_n\}$ be a complete orthonormal system in $\mathfrak{H}$. If $\xi \in \mathfrak{N}$, then $\int a(g)(U_g\xi, \xi_m)d\mu(g) = 0$ for all $a(g) \in L^1(\mathfrak{G})$ and therefore $(U_g\xi, \xi_m) = 0$ for all $g \in \mathfrak{G} - S_m$, where $\mu(S_m) = 0$. Then $(U_g\xi, \xi_m) = 0$ for $m = 1, 2, 3, \cdots$ if $g \in \mathfrak{G} - S$, where $S = \bigcup_{m=1}^{\infty} S_m$ is also of μ-measure zero.

Since $\{\xi_m\}$ is complete, it follows that $U_g\xi = 0$ for $g \in \mathfrak{G} - S$ and therefore $\xi = 0$. So $\mathfrak{N} = (0)$. Further, in virtue of (5), $(U_g\xi_n, \xi_m) = (U'_g\xi_n, \xi_m)$ for $g \in \mathfrak{G} - \gamma_{nm}$, where $\mu(\gamma_{nm}) = 0$. Therefore $(U_g\xi_n, \xi_m) = (U'_g\xi_n, \xi_m)$ for all $n, m = 1, 2, \cdots$ for $g \in \mathfrak{G} - \bigcup_{n,m} \gamma_{nm}$, i.e. almost everywhere on $\mathfrak{G}$.

Consequently, $U_g = U'_g$ almost everywhere on the group $\mathfrak{G}$.

We have thus proved the following corollary.

COROLLARY 2. *In a separable Hilbert space, every weakly measurable unitary representation of the group $\mathfrak{G}$ coincides with its continuous unitary representation almost everywhere.*

REMARK 2. In the formulation of Theorem 1, the algebra $R(\mathfrak{G})$ can obviously be replaced by the algebra $L^1(\mathfrak{G})$; in this connection, it is natural to call the representation $a \to A_a$ of the algebra $L^1(\mathfrak{G})$ degenerate if $A_a = 0$ for all $a \in L^1(\mathfrak{G})$.

3. Completeness theorem. For a unitary representation of the group $\mathfrak{G}$ one can, as for representations of an algebra, introduce the concepts of an invariant subspace, irreducibility, and equivalence.

The validity of the following proposition is easily verified.

A subspace $\mathfrak{M}$ is invariant with respect to the unitary representation $g \to U_g$ of the group $\mathfrak{G}$ if, and only if, it is invariant with respect to the corresponding representation $x \to A_x$ of the group algebra $R(\mathfrak{G})$.

It follows from this that *a unitary representation of the group $\mathfrak{G}$ is irreducible if, and only if, the corresponding representation of the group algebra $R(\mathfrak{G})$ is irreducible.*

In subsections 3 and 4, the expression "unitary representation" will mean "continuous unitary representation".

Further, it is also easily verified that

Two unitary representations of a group $\mathfrak{G}$ are equivalent if, and only if, the corresponding representations of the group algebra are equivalent.

It follows from these propositions that the results of subsections 1, 2, 6, § 17 and of § 21 carry over to unitary representations of groups. In this connection, the direct sum of unitary representations of a group is defined in a way analogous to how this is done for the direct sum of representations of an algebra.

A set of representations of a group $\mathfrak{G}$ is said to be *complete* if for any element $g_0 \neq e$ of the group $\mathfrak{G}$ there exists a representation $g \to U_g$ in this set such that $U_{g_0} \neq 1$.

THEOREM 2. *Every locally compact group has a complete set of irreducible unitary representations.*

In fact, there exists an element $a \in L^1(\mathfrak{G})$ such that $a_{g_0} - a \neq 0$. (It suffices, for instance, to take the characteristic function of a neighborhood V of the identity for the function a, where V satisfies the conditions: 1) $\bar{V}$ is compact; 2) $g_0 V \cap V = \phi$.) According to Theorem 3, subsection 4, § 19, there exists an irreducible representation $x \to A_x$ of the group algebra $R(\mathfrak{G})$ for which $A_{a_{g_0} - a} \neq 0$, i.e. $A_{a_{g_0}} \neq A_a$. Therefore there exists a vector ξ_0 satisfying the condition $A_{a_{g_0}}\xi_0 \neq A_a\xi_0$. We set $A_a\xi_0 = \xi$, $A_{a_{g_0}}\xi_0 = U_{g_0}\xi$; then the last inequality becomes $U_{g_0}\xi \neq \xi$. Consequently, an irreducible unitary representation $g \to U_g$ constructed with respect to the representation $x \to A_x$ of the algebra $R(\mathfrak{G})$ satisfies the requirements set down.

4. Examples. a) UNITARY REPRESENTATIONS OF THE GROUP OF LINEAR TRANSFORMATIONS OF THE REAL LINE. Suppose $\mathfrak{G}_r$ is the group of linear transformations $y = \alpha x + \beta$ of the real line, where α is an arbitrary positive number, β is an arbitrary real number (see Example 5, subsection 1, § 27). This group contains the following two commutative subgroups: 1) the group $\mathfrak{T}$ of translations $x \to x+\beta$ whose elements will be denoted by t_β; 2) the group $\mathfrak{S}$ of similarity transformations $x \to \alpha x$ whose elements will be denoted by s_α.

Every other element of the group $\mathfrak{G}_r$ is expressible in the form of the product of an element in $\mathfrak{T}$ by an element in $\mathfrak{S}$. Therefore a representation of the group $\mathfrak{G}_r$ will be determined if we know the operators T_β and S_α which correspond to the elements t_β and s_α of these subgroups.

The group $\mathfrak{G}_r$ has a series of one-dimensional representations which are obtained if we set $T_\beta = 1$, $S_\alpha = \chi(\alpha)1$, where $\chi(\alpha) = \alpha^{i\rho}$ and ρ is an arbitrary real number. Distinct values of ρ are assigned non-equivalent representations of this series.

We shall now construct a representation of the group $\mathfrak{G}_r$ which is not one-dimensional. To this end, we denote by H^+ the set of all functions $\varphi(\lambda)$ with summable square in the interval $0 \leqq \lambda < \infty$ with the usual operations and inner product. To each element g: $x \to \alpha x + \beta$ of the group $\mathfrak{G}_r$ we assign the operator

$$U_g^+\varphi(\lambda) = e^{i\lambda\beta}\varphi(\lambda\alpha)\cdot\alpha^{\frac{1}{2}}$$

in the space H^+. In particular,

$$T_\beta^+\varphi(\lambda) = e^{i\lambda\beta}\varphi(\lambda), \quad S_\alpha^+\varphi(\lambda) = \varphi(\lambda\alpha)\cdot\alpha^{\frac{1}{2}}.$$

It is easily seen that U_g^+ is a unitary operator in H^+ and that $g \to U_g^+$ is a representation of the group $\mathfrak{G}_r$. This representation is irreducible. In fact, suppose A is an operator in H^+ which commutes with all the operators U_g^+; we are required to prove that it is a multiple of the identity. A commutes with all the operators T_β^+, i.e. with the operators of multiplication by $e^{i\lambda\beta}$. Consequently, it commutes with all the linear combinations of such operators and their limits in the sense of the weak topology in $\mathfrak{B}(\mathfrak{H})$ (see Example 2, subsection 3, § 8), i.e. with an arbitrary operator of multiplication by an essentially bounded function. [In fact, the space $L^\infty(-\infty, \infty)$ is the conjugate to $L^1(-\infty, \infty)$ (subsection 16, § 6), where linear combinations of the functions $e^{i\lambda\beta}$ form a dense set in $L^\infty(-\infty, \infty)$ in the sense of the weak topology; this follows from the fact that for $x(\lambda) \in L^1(-\infty, \infty)$ the equality $\int_{-\infty}^{\infty} x(\lambda)e^{i\lambda\beta}d\lambda = 0$ for all real β is impossible, except when $x = 0$ (see II, subsection 11, § 3, Corollary 3). Consequently, for an arbitrary function $f(\lambda) \in L^\infty(-\infty, \infty)$ and arbitrary $\varepsilon > 0$, $x(\lambda) \in L^1(-\infty, \infty)$ there exists a linear combination $f_1(\lambda) = \sum_{\nu=1}^n c_\nu e^{i\lambda\beta_\nu}$ such that $|\int_{-\infty}^{\infty}[f(\lambda)-f_1(\lambda)]x(\lambda)d\lambda| < \varepsilon$. If we set $x(\lambda) = y(\lambda)\overline{z(\lambda)}$, where $y, z \in L^2(-\infty, \infty)$, in this inequality, we obtain $|\int_{-\infty}^{\infty}[f(\lambda)-f_1(\lambda)]y(\lambda)\overline{z(\lambda)}d\lambda| < \varepsilon$. This means that the operator of multiplication by $f(\lambda)$ is the limit in the weak operator topology of operators of multiplication by $f_1(\lambda) = \sum_{\nu=1}^n c_\nu e^{i\beta\lambda_\nu}$.] But then A itself is an operator of multiplication by an essentially bounded measurable function (see IV, subsection 5, § 26): $A\varphi(\lambda) = \omega(\lambda)\varphi(\lambda)$, where $\omega(\lambda)$ is an essentially bounded measurable function. It follows that

$$AS_\alpha^+\varphi(\lambda) = \omega(\lambda)\varphi(\lambda\alpha)\alpha^{\frac{1}{2}},$$

$$S_\alpha^+A\varphi(\lambda) = \omega(\lambda\alpha)\varphi(\lambda\alpha)\alpha^{\frac{1}{2}}.$$

Therefore it follows from the commutativity of the operators S_α^+ and A that $\omega(\lambda\alpha) = \omega(\lambda)$ for almost all λ. This is possible only when $\omega(\lambda) =$ constant; consequently, A is an operator of multiplication by a constant, i.e. it is a multiple of the identity. This completes the proof of the irreducibility of the representation $g \to U_g^+$.

The set H^- of all measurable functions $\varphi(\lambda)$ with summable square on the interval $-\infty < \lambda \leqq 0$ can be considered analogously. Setting

$$U_g^-\varphi(\lambda) = e^{i\lambda\beta}\varphi(\lambda\alpha)\alpha^{\frac{1}{2}},$$

we also obtain an irreducible representation $g \to U_g^-$ of the group $\mathfrak{G}_r$.

It can be shown that every irreducible unitary representation of the group $\mathfrak{G}_r$ is equivalent to one of the representations in the one-dimensional series or to one of the representations U_g^+, U_g^-. Moreover, it turns out that every unitary representation of the group $\mathfrak{G}_r$ decomposes into the direct integral of representations, equivalent to these irreducible representations (for the proof of these propositions, see Gelfand and Naimark [3]).

If one goes over from the functions $\varphi(\lambda)$ to their Fourier transforms

$$f(x) = \frac{1}{\sqrt{2\pi}}\int_{-\infty}^{\infty}\varphi(\lambda)e^{i\lambda x}d\lambda,$$

the spaces H^+, H^- map onto the spaces $\mathfrak{H}^+$, $\mathfrak{H}^-$ of functions $f(x)$ with summable square on the axis $-\infty < x < +\infty$, which are limit values of functions, analytic in the upper and lower half-planes respectively. Moreover, in each of these spaces, the operators U_g^+ and U_g^- will be defined by the same formula

$$U_g^\pm f(x) = \alpha^{-\frac{1}{2}}f(\alpha^{-1}(x+\beta)).$$

b) Unitary Representations of the Proper Lorentz Group. We denote the group of all second order matrices $g = \binom{\alpha\beta}{\gamma\delta}$ with complex elements whose determinant equals unity by $\mathfrak{G}_2$.

To each such matrix g we assign the fourth order matrix

$$a = \begin{Vmatrix} \Re(\alpha\bar{\delta}+\bar{\gamma}\beta) & -\Im(\alpha\bar{\delta}-\bar{\gamma}\beta) & \Re(\alpha\bar{\gamma}-\beta\bar{\delta}) & \Re(\alpha\bar{\gamma}+\beta\bar{\delta}) \\ \Im(\alpha\bar{\delta}+\bar{\gamma}\beta) & \Re(\alpha\bar{\delta}-\bar{\gamma}\beta) & \Im(\alpha\bar{\gamma}-\beta\bar{\delta}) & \Im(\alpha\bar{\gamma}+\beta\bar{\delta}) \\ \Re(\alpha\bar{\beta}-\gamma\bar{\delta}) & -\Im(\alpha\bar{\beta}-\gamma\bar{\delta}) & \dfrac{\alpha\bar{\alpha}-\beta\bar{\beta}-\gamma\bar{\gamma}+\delta\bar{\delta}}{2} & \dfrac{\alpha\bar{\alpha}+\beta\bar{\beta}-\gamma\bar{\gamma}-\delta\bar{\delta}}{2} \\ \Re(\alpha\bar{\beta}+\gamma\bar{\delta}) & -\Im(\alpha\bar{\beta}+\gamma\bar{\delta}) & \dfrac{\alpha\bar{\alpha}-\beta\bar{\beta}+\gamma\bar{\gamma}-\delta\bar{\delta}}{2} & \dfrac{\alpha\bar{\alpha}+\beta\bar{\beta}+\gamma\bar{\gamma}+\delta\bar{\delta}}{2} \end{Vmatrix}$$

with real elements. It is easily verified that the mapping $g \to a$ is a homomorphism of the group $\mathfrak{G}_2$ onto the group of all fourth order matrices a with real entries, which leave the quadratic form $\xi_1^2+\xi_2^2+\xi_3^2-\xi_4^2$ invariant and satisfy the conditions: $\det a = 1$, $a_{44} \geqq 1$ (see, for instance, NAIMARK [12], § 9). This group is called the *proper Lorentz group*. The kernel of the homomorphism $g \to a$ consists of the matrices

$$\begin{pmatrix} 1 & 0 \\ 0 & 1 \end{pmatrix}, \quad \begin{pmatrix} -1 & 0 \\ 0 & -1 \end{pmatrix};$$

therefore the matrices $\left(\begin{smallmatrix} \alpha & \beta \\ \gamma & \delta \end{smallmatrix}\right)$, $\left(\begin{smallmatrix} -\alpha & -\beta \\ -\gamma & -\delta \end{smallmatrix}\right)$, and only these, map onto the same matrix a.

From this it follows that every unitary representation $a \to U_a$ of the proper Lorentz group can be assigned a unitary representation $g \to U_g$ of the group $\mathfrak{G}_2$. Namely, it suffices to set $U_g = U_a$, where a is the image of the element g. Conversely, to every unitary representation $g \to U_g$ of the group $\mathfrak{G}_2$ one can assign a single-valued or double-valued unitary representation of the Lorentz group. (Here, "double-valued" means that to each element $g \in \mathfrak{G}_2$ there correspond two operators $U_g^{(k)}$, $k = 1, 2$, with $U_{g_1}^{(k)}U_{g_2}^{(l)} = U_{g_1g_2}^{(j)}$ for arbitrary $g_1, g_2 \in \mathfrak{G}_2$, $k, l = 1, 2$, and certain $j = 1, 2$, which depend on k and l.) Therefore one can consider the representations of the group $\mathfrak{G}_2$ instead of the representations of the Lorentz group. These representations can be described in the following way.

We denote by $\mathfrak{H}$ the set of all measurable functions $f(z) = f(x+iy)$ with summable square on the entire complex plane. We introduce the operations of addition, scalar multiplication, and formation of the inner product in $\mathfrak{H}$ in the usual way. Then $\mathfrak{H}$ becomes a Hilbert space.

For $f(z) \in \mathfrak{H}$, we set

$$U_g f(z) = |\beta z+\delta|^{n+i\rho-2}(\beta z+\delta)^{-n} f\left(\frac{\alpha z+\gamma}{\beta z+\delta}\right), \tag{1}$$

where n is an arbitrary integer, ρ is an arbitrary real number, and U_g is a unitary operator in $\mathfrak{H}$. In fact,

$$|U_g f|^2 = \int |U_g f(z)|^2 dxdy = \int |\beta z+\delta|^{-4} \left| f\left(\frac{\alpha z+\gamma}{\beta z+\delta}\right)\right|^2 dxdy;$$

$$|f|^2 = \int |f(z_1)|^2 dx_1\, dy_1 = \int |\beta z+\delta|^{-4} \left| f\left(\frac{\alpha z+\gamma}{\beta z+\delta}\right)\right|^2 dxdy;$$

the last equality is obtained by making the substitution $z_1 = (\alpha z+\gamma)/(\beta z+\delta)$. Thus $|U_g f|^2 = |f|^2$; moreover, the operator U_g is a one-to-one mapping of $\mathfrak{H}$ onto itself; consequently U_g is a unitary operator. It is easily seen that $g \to U_g$ is a unitary representation of the group $\mathfrak{G}_2$. Thus, every pair of numbers (n, ρ) determines a representation of the group $\mathfrak{G}_2$. The set of all these representations is called the *principal series of representations of the group* $\mathfrak{G}_2$.

All representations of the principal series are irreducible.

To prove this, we go over from the functions $f(z)$ to their Fourier transforms with respect to x and y:

$$\varphi(w) = \frac{1}{2\pi}\int f(z)e^{i\Re(\bar{z}w)}dxdy.$$

Further, we denote the elements $\begin{pmatrix}1 & \beta\\ 0 & 1\end{pmatrix}$ and $\begin{pmatrix}1/\delta & 0\\ 0 & \delta\end{pmatrix}$ of the group $\mathfrak{G}_2$ by s_β and t_δ, respectively. We shall consider the operator U_g as an operator in the space of functions $\varphi(w)$; then it follows from the general formula (1) that

$$U_{s_\beta}\varphi(w) = e^{-i\Re(\bar{\beta}w)}\varphi(w);$$

$$U_{t_\delta}\varphi(w) = |\delta|^{n+i\rho+2}\delta^{-n}\varphi(w\bar{\delta}^2).$$

Every bounded operator A which commutes with all the operators U_g commutes, in particular, with all the operators U_{s_β} and U_{t_δ}. From this it follows, as in Example a), that A is an operator of multiplication by a number, i.e. it is a multiple of the identity. This completes the proof of the fact that the representation $g \to U_g$ is irreducible.

The group $\mathfrak{G}_2$ has still other irreducible representations which are not equivalent to the representations of the principal series. They may be described as follows.

Suppose ρ is a real number in the interval $0 < \rho \leqq 2$ and let $\mathfrak{H}'_\rho$ be the set of all measurable functions $f(z) = f(x+iy)$ such that

$$|f(z)| \leqq \frac{C}{(1+|z|^2)^{1+\rho/2}}$$

for some constant C depending on f.

We introduce addition and scalar multiplication in $\mathfrak{H}'_\rho$ in the usual way; moreover, we define the inner product in $\mathfrak{H}'_\rho$ by the formula

$$(f_1, f_2)_\rho = \int |z_1 - z_2|^{-2+\rho} f_1(z_1)\overline{f_2(z_2)}dx_1\,dy_1\,dx_2\,dy_2\,.$$

Then all the inner product axioms will be satisfied for $0 < \rho < 2$. In fact, to verify this we need only the positive definiteness, i.e. the inequality

$$(f, f)_\rho \geqq 0.$$

To prove this, it suffices to go over from the function $f(z)$ to its Fourier transform

$$\varphi(w) = \frac{1}{2\pi}\int f(z)e^{i\Re(\bar{z}w)}dxdy;$$

then

$$(f, f)_\rho = 2^{-1+\rho}\frac{\Gamma\left(\frac{\rho}{2}\right)}{\Gamma\left(1-\frac{\rho}{2}\right)}\int |w|^{-\rho}|\varphi(w)|^2 dudv \quad (w = u+iv).$$

(For a detailed derivation of this formula, see NAIMARK [12], § 12.)

$\mathfrak{H}'_\rho$ thus becomes a pre-Hilbert space. Its completion will be denoted by $\mathfrak{H}_\rho$.

We now introduce the operator

$$U'_g f(z) = |\beta z+\delta|^{-2-\rho} f\left(\frac{\alpha z+\gamma}{\beta z+\delta}\right) \tag{2}$$

in $\mathfrak{H}'_\rho$. It is easily verified that U'_g maps $\mathfrak{H}'_\rho$ isometrically onto itself. It can therefore be uniquely extended to a unitary operator in all of $\mathfrak{H}_\rho$. We denote this operator again by U'_g. The correspondence $g \to U'_g$ is a unitary representation of the group $\mathfrak{G}_2$ in the space $\mathfrak{H}_\rho$. To each value ρ in the interval $0 < \rho < 2$ there corresponds a representation of this sort. The set of all these representations is called the *complementary series* of representations of the group $\mathfrak{G}_2$.

All representations of the complementary series are irreducible.

The proof is analogous to the proof of the irreducibility of the representations in the principal series.

The representations in the complementary series reduce in a natural way to the representations of the principal series when $n = 0$, $\rho = 0$. Formula (2) for U'_g is obtained formally from formula (1) for U_g when $n = 0$ if in the latter we replace ρ by ρi. For $\rho = 2$, we obtain

$$(f_1, f_2)_2 = \int f_1(z)dxdy \int \overline{f_2(z)}dxdy.$$

Thus the element f must be considered to be equal to zero if $\int f(z)dxdy = 0$; consequently, $\mathfrak{H}_2$ is one-dimensional. Its elements are uniquely defined by the numbers

$$\xi = \int f(z)dxdy.$$

Moreover,

$$U'_g \xi = \int |\beta z+\delta|^{-4} f\left(\frac{\alpha z+\gamma}{\beta z+\delta}\right) dxdy = \int f(z)dxdy = \xi,$$

i.e. U'_g is the identity operator.

Thus, for $\rho = 2$, the space $\mathfrak{H}_\rho$ is one-dimensional and the representation consists in this that each element of the group $\mathfrak{G}_2$ is assigned the identity operator. This representation is called the *identity representation.*

It can be shown that every irreducible unitary representation of the group $\mathfrak{G}_2$, which is distinct from the identity representation, is equivalent to one of the representations of the principal or the complementary series and, moreover, that every unitary representation of the group $\mathfrak{G}_2$ is the direct integral of such irreducible representations (see GELFAND and NAIMARK [4]).

For a generalization of the first result to non-unitary representations, see NAIMARK [12], § 15. Another realization of unitary and non-unitary irreducible representations of the group $\mathfrak{G}_2$ was later given in the book of GELFAND, VILENKIN and GRAEV [1] (also see ZHELOBENKO [1–3]).

We note further that the representations of the principal series, corresponding to the pairs (n, ρ) and $(-n, -\rho)$, are equivalent. In all the remaining cases, distinct representations of the principal and complementary series are pairwise nonequivalent.

We denote the set of all unitary matrices of the group $\mathfrak{G}_2$ by $\mathfrak{B}$; clearly, $\mathfrak{B}$ is a subgroup of the group $\mathfrak{G}_2$. Representations of the principal series, corresponding to $n = 0$, possess the following property.

In the space $\mathfrak{H}$ of representations $g \to U_g$ there exists a vector $f_0 = f_0(z)$ which is invariant with respect to all the operators U_v, $v \in \mathfrak{B}$.

To prove this, we first note that every matrix $v \in \mathfrak{V}$ has the form

$$v = \begin{pmatrix} \alpha & \beta \\ -\bar{\beta} & \bar{\alpha} \end{pmatrix}, \quad |\alpha|^2 + |\beta|^2 = 1.$$

In virtue of formula (1) for the operator U_g, the condition for invariance of the vector $f_0(z)$ yields

$$|\beta z + \bar{\alpha}|^{-2+i\rho} f_0 \left(\frac{\alpha z - \bar{\beta}}{\beta z + \bar{\alpha}} \right) = f_0(z). \tag{2'}$$

Setting $z = 0$, we obtain

$$|\alpha|^{-2+i\rho} f_0 \left(-\frac{\bar{\beta}}{\bar{\alpha}} \right) = c,$$

where $c = f_0(0)$. Now replacing $-\bar{\beta}/\bar{\alpha}$ by z, we have

$$|\alpha|^2 = \frac{|\alpha|^2}{|\alpha|^2 + |\beta|^2} = \frac{1}{1 + \left| \frac{\beta}{\alpha} \right|^2} = \frac{1}{1 + |z|^2}.$$

Consequently,

$$f_0(z) = c(1 + |z|^2)^{-1 + i\rho/2}. \tag{2''}$$

We see that for a given representation in $\mathfrak{H}$ *there exists an invariant vector* $f_0(z)$ *which is unique to within a constant multiplier.*

[The preceding line of reasoning is of a heuristic nature and serves to find the vector $f_0(z)$ since *a priori* $f(z)$ can be discontinuous and one cannot talk about its value $f(0)$. However, it is not difficult to verify directly that the right member of (2″) satisfies condition (2). On the other hand, setting $f_0(z) = \omega(z)(1+|z|^2)^{-1+i\rho/2}$ and substituting into (2′), we find that $\omega((\alpha z - \bar{\beta})/(\beta z - \bar{\alpha})) = \omega(z)$. Consequently, $\omega(z) = $ constant. From this it follows that *every* function $f_0(z)$ in $\mathfrak{H}$, satisfying condition (2′), is given by formula (2″).]

Taking $c = 1/\sqrt{\pi}$, we obtain the normalized vector

$$f_0(z) = \frac{1}{\sqrt{\pi}} (1 + |z|^2)^{-1 + i\rho/2}.$$

We now set

$$\varphi(g) = (U_g f_0, f_0).$$

The function $\varphi(g)$ is called the *spherical function corresponding to the given representation* $g \to U_g$. We shall calculate this function. We note first of all that every matrix $g \in \mathfrak{G}_2$ can be represented in the form $g = vh$, where $v \in \mathfrak{V}$ and h is a positive definite Hermitian matrix. Further, the matrix h can be reduced to the diagonal form, i.e. it can be represented in the form $h = v_2^{-1} \varepsilon v_2$, where ε is the diagonal matrix

$$\varepsilon = \begin{pmatrix} \lambda^{-1} & 0 \\ 0 & \lambda \end{pmatrix}, \quad \lambda > 0$$

and $v_2 \in \mathfrak{V}$.

The matrix ε is determined by the matrix g to within a permutation of the diagonal elements. Consequently, it can be uniquely defined by the condition $\lambda \geqq 1$. In the sequel we shall assume that this condition is satisfied. Thus, we finally have that

$$g = v_1 \varepsilon v_2, \; v_1, v_2 \in \mathfrak{B}.$$

It follows, in virtue of the invariance of the vector f_0, that

$$\varphi(g) = (U_{v_1} U_\varepsilon U_{v_2} f_0, f_0) = (U_\varepsilon U_{v_2} f_0, U^*_{v_1} f_0) = (U_\varepsilon f_0, f_0) = \varphi(\varepsilon);$$

consequently, it will suffice to calculate $\varphi(\varepsilon)$.

Since

$$U_\varepsilon f_0(z) = \frac{1}{\sqrt{\pi}}(1+|z|^2\lambda^{-4})^{-1+i\rho/2}\lambda^{-2+i\rho},$$

we have

$$\begin{aligned}\varphi(\varepsilon) &= \frac{1}{\pi}\int (1+|z|^2\lambda^{-4})^{-1+i\rho/2}\lambda^{-2+i\rho}(1+|z|^2)^{-1-i\rho/2}dxdy \\ &= \frac{1}{\pi}\lambda^{-2+i\rho}\int_0^{2\pi}\int_0^\infty (1+r^2\lambda^{-4})^{-1+i\rho/2}(1+r^2)^{-1-i\rho/2}rdrd\theta \\ &= \lambda^{-2+i\rho}\int_0^\infty (1+\xi\lambda^{-4})^{-1+i\rho/2}(1+\xi)^{-1-i\rho/2}d\xi.\end{aligned}$$

If, in the last integral, we make the substitution

$$x = \frac{1-\lambda^4}{1+\xi},$$

we obtain

$$\begin{aligned}\varphi(\varepsilon) &= \lambda^{-i\rho+2}(1-\lambda^4)^{-1}\int_0^{1-\lambda^4}(1-x)^{-1+i\rho/2}dx \\ &= -\frac{2\lambda^{-i\rho+2}(1-\lambda^4)^{-1}}{i\rho}(\lambda^{2i\rho}-1) = \frac{2}{i\rho}\frac{\lambda^{i\rho}-\lambda^{-i\rho}}{\lambda^2-\lambda^{-2}}.\end{aligned}$$

In the sequel we shall find it convenient to take $t = \ln \lambda$ as the parameter. Then the formula for $\varphi(\varepsilon)$ takes on the form

$$\varphi(\varepsilon) = \frac{2 \sin \rho t}{\rho \sinh 2t}. \tag{3}$$

It can be shown analogously that in the space $\mathfrak{H}_\rho$ of representations $g \to U'_g$ of the complementary series there exists an invariant vector with respect to U'_v, $v \in \mathfrak{B}$. The corresponding spherical function has the form

$$\varphi(\varepsilon) = \frac{2 \sinh \rho t}{\rho \sinh 2t}.$$

However, we shall not need this in the sequel. (For the generalization of the results discussed here to representations of various classes of groups and, in particular, for the general theory of spherical functions see the survey article by Mackey [9] and also the paper by Godement [9].)

c) Example of a Group Algebra which is not Completely Symmetric. We denote the group algebra of the group $\mathfrak{G}_2$ considered in Example b) by $R(\mathfrak{G}_2)$. We shall show that $R(\mathfrak{G}_2)$ is not a completely symmetric algebra.

In the proof, we denote the set of all functions $a(g)$ which are summable on $\mathfrak{G}_2$ by $L^1(\mathfrak{G}_2)$ and by $\mathfrak{A}'$ the set of all functions $a(g) \in L^1_*(\mathfrak{G}_2)$ which satisfy the condition

$$a(vg) = a(gv) = a(g) \quad \text{for all } v \in \mathfrak{V}.$$

[Here we retain the same notation as in Example b).]

$\mathfrak{A}'$ is a subalgebra of the algebra $R(\mathfrak{G}_2)$. In fact, it suffices to verify that for $a_1, a_2 \in \mathfrak{A}'$, the function

$$a(g) = \int a_1(g_1)a_2(g_1^{-1}g)d\mu(g_1)$$

also belongs to $\mathfrak{A}'$, where $d\mu(g)$ denotes the differentra of the two-sided invariant measure in $\mathfrak{G}_2$. But this follows from the equalities

$$a(gv) = \int a_1(g_1)a_2(g_1^{-1}gv)d\mu(g_1) = \int a_1(g_1)a_2(g_1^{-1}g)d\mu(g_1) = a(g),$$

$$a(vg) = \int a_1(g_1)a_2(g_1^{-1}vg)d\mu(g_1) = \int a_1(g_1)a_2(g_1^{-1}g)d\mu(g_1) = a(g).$$

Therefore the set of elements of the form $\lambda e+a$, $a \in \mathfrak{A}'$ is an, obviously closed, subalgebra with identity of the algebra $R(\mathfrak{G}_2)$; we denote this subalgebra by $\mathfrak{A}$. It is commutative. In fact, it suffices to prove that, for $a_1, a_2 \in \mathfrak{A}'$,

$$\int a_1(g_1)a_2(g_1^{-1}g)d\mu(g_1) = \int a_2(g_1)a_1(g_1^{-1}g)d\mu(g_1). \tag{4}$$

We denote the matrix which is the Hermitian adjoint of the matrix g by g^*. Then $a(g^*) = a(g)$ for any function $a(g) \in \mathfrak{A}'$. In fact, setting $g = v_1 \varepsilon v_2$, we have

$$a(g^*) = a(v_2^{-1}\varepsilon v_1^{-1}) = a(\varepsilon) = a(v_1 \varepsilon v_2) = a(g).$$

Moreover, for any function $a(g) \in L^1(\mathfrak{G}_2)$,

$$\int a(g)d\mu(g) = \int a(g^*)d\mu(g).$$

This follows from the fact that if the matrix g is defined by the parameters β, γ, δ, then

$$d\mu(g) = \frac{d\mu(\beta)d\mu(\gamma)d\mu(\delta)}{|\delta|^2}.$$

It follows from these remarks that for $a_1, a_2 \in \mathfrak{A}'$,

$$\int a_1(g_1)a_2(g_1^{-1}g)d\mu(g_1) = \int a_1(g_1^*)a_2(g^*g_1^{*-1})d\mu(g_1) = \int a_1(g_1)a_2(g^*g_1^{-1})d\mu(g_1).$$

Since the last integral is a function in $\mathfrak{A}'$, it is equal to

$$\int a_1(g_1)a_2(gg_1^{-1})d\mu(g_1) = \int a_1(g_1^{-1})a_2(gg_1)d\mu(g_1) = \int a_1(g_1^{-1}g)a_2(g_1)d\mu(g_1).$$

This proves equality (4) and hence the commutativity of the algebra $\mathfrak{A}$.

We shall prove that $\mathfrak{A}$ is not a completely symmetric algebra. According to subsection 1, § 14, it suffices to prove that $\mathfrak{A}$ contains non-symmetric maximal ideals. To this end, we shall find all the maximal ideals of the algebra $\mathfrak{A}$.

We first express the integral with respect to g in terms of integrals with respect to v_1, ε, v_2. We denote the set of all diagonal unitary matrices of the second order by Γ and by $\bar{\mathfrak{V}}$ the set of all right cosets of the group $\mathfrak{V}$ modulo Γ, and these right cosets themselves by $\bar{v}$. Multiplication by v_0 is equivalent to a transformation in the space $\bar{\mathfrak{V}}$; we denote by $\bar{v}v_0$ the coset into which $\bar{v}$ maps under this transformation.

In the representation $g = v_1 \varepsilon v_2$, the matrices v_1, v_2 are not defined uniquely by the matrix g. Namely, $g = v_1 \tau^{-1} \varepsilon \tau v_2$, where $\tau \in \Gamma$, is a representation of the same form. Normalizing v_2 in some definite way, for instance, so that $\alpha \geqq 0$, we choose a specific representative v_2 in almost every coset $\bar{v}_2$. Then the matrices v_1, ε will be determined uniquely by the matrix g, where v_1 runs through the entire group $\mathfrak{V}$ and v_2 can be identified with the coset $\bar{v}_2$.

Suppose $d\mu_0(\bar{v}_2)$ is any differentiable measure in $\bar{\mathfrak{V}}$. [A measure μ_0 on $\bar{\mathfrak{V}}$ is said to be *differentiable* if $d\mu_0(\bar{v}g_0)$ is differentiable with respect to $d\mu_0(\bar{v})$ for all $g_0 \in \mathfrak{G}_2$.] Suppose, moreover, that $d\mu(v_1)$, $d\mu(\varepsilon)$ are invariant measures in the group $\mathfrak{V}$ and in the multiplicative group of positive numbers, respectively. We choose $d\mu(v_1)$ so that $\mu(\mathfrak{V}) = 1$. Then we may, obviously, set

$$\int f(g)d\mu(g) = \int f(v_1 \varepsilon v_2)\omega(v_1, \varepsilon, \bar{v}_2)d\mu(v_1)d\mu(\varepsilon)d\mu_0(\bar{v}_2), \tag{5}$$

where $\omega(v_1, \varepsilon, \bar{v}_2)$ is the Jacobian arising from passing from g to $v_1, \varepsilon, \bar{v}_2$. Since $d\mu(g)$ is left invariant,

$$\int f(vg)d\mu(g) = \int f(g)d\mu(g),$$

i.e.

$$\int f(vv_1 \varepsilon v_2)\omega(v_1, \varepsilon, \bar{v}_2)d\mu(v_1)d\mu(\varepsilon)d\mu_0(\bar{v}_2)$$
$$= \int f(v_1 \varepsilon v_2)\omega(v_1, \varepsilon, \bar{v}_2)d\mu(v_1)d\mu(\varepsilon)d\mu_0(\bar{v}_2). \tag{6}$$

But in virtue of the left invariance of $d\mu(v_1)$, the first integral equals

$$\int f(v_1 \varepsilon v_2)\omega(v^{-1}v_1, \varepsilon, \bar{v}_2)d\mu(v_1)d\mu(\varepsilon)d\mu_0(\bar{v}_2);$$

therefore it follows from equality (6) that $\omega(v^{-1}v_1, \varepsilon, \bar{v}_2) = \omega(v_1, \varepsilon, \bar{v}_2)$ almost everywhere with respect to v_1 because f is an arbitrary function. Thus, ω does not depend on v_1, and we set $\omega = \omega(\varepsilon, \bar{v}_2)$. Further, in virtue of the right invariance of $d\mu(g)$,

$$\int f(gv)d\mu(g) = \int f(g)d\mu(g),$$

i.e.

$$\int f(v_1 \varepsilon v_2')\omega(\varepsilon, \bar{v}_2)d\mu(v_1)d\mu(\varepsilon)d\mu_0(\bar{v}_2) = \int f(v_1 \varepsilon v_2)\omega(\varepsilon, \bar{v}_2)d\mu(v_1)d\mu(\varepsilon)d\mu_0(\bar{v}_2), \quad (7)$$

where v_2' is a normalized representative of the coset $\bar{v}_2 v$. In the second integral, we make the change of variables $\bar{v}_2 \to v_2'$, i.e. $\bar{v}_2 \to \bar{v}_2 v$, and denote the Jacobian of this change of variables by $\lambda(\bar{v}_2, v)$. Then this integral can be rewritten in the form

$$\int f(v_1 \varepsilon v_2')\omega(\varepsilon, \bar{v}_2 v)\lambda(\bar{v}_2, v)d\mu(v_1)d\mu(\varepsilon)d\mu_0(\bar{v}_2);$$

therefore it follows from (7) that

$$\omega(\varepsilon, \bar{v}_2) = \omega(\varepsilon, \bar{v}_2 v)\lambda(\bar{v}_2, v).$$

We assign the fixed value $\bar{v}_2^0$ to $\bar{v}_2$ and then set $\bar{v}_2^0 v = \bar{v}$. The last equality can then be written in the form

$$\omega(\varepsilon, \bar{v}) = \omega(\varepsilon)\lambda^{-1}(v),$$

where

$$\omega(\varepsilon) = \omega(\varepsilon, \bar{v}_2^0), \quad \lambda(v) = \lambda(\bar{v}_2^0, v).$$

We now set $d\mu(\bar{v}_2) = \lambda^{-1}(v)d\mu_0(\bar{v}_2)$; then equality (5) can be written in the form

$$\int f(g)d\mu(g) = \int f(v_1 \varepsilon v_2)\omega(\varepsilon)d\mu(v_1)d\mu(\varepsilon)d\mu(\bar{v}_2).$$

In particular, for $a \in \mathfrak{A}'$,

$$||a||_1 = \int |a(g)|d\mu(g) = \int |a(\varepsilon)|\omega(\varepsilon)d\mu(\varepsilon), \quad (8)$$

because the constant $\int d\mu(\bar{v}_2)$ can be included as a factor in ω. Further, we shall find the rule for multiplication in $\mathfrak{A}'$. To this end, we consider one of the representations $g \to U_g$ of the principal series of the group $\mathfrak{G}_2$ with $n = 0$. Suppose f_0 is a normalized vector which is invariant with respect to all the operators U_v, $v \in \mathfrak{V}$. We set

$$f = \int U_v U_\varepsilon f_0 \, d\mu(v).$$

The vector f is also invariant with respect to all the operators U_v because

$$U_{v_0} f = U_{v_0} \int U_v U_\varepsilon f_0 \, d\mu(v) = \int U_{v_0 v} U_\varepsilon f_0 \, d\mu(v)$$

$$= \int U_v U_\varepsilon f_0 \, d\mu(v) = f.$$

Therefore f differs from f_0 by a multiplier: $f = cf_0$. In order to find the multiplier c, we note that

$$c = (f, f_0) = \int (U_v U_\varepsilon f_0, f_0) d\mu(v) = \int (U_\varepsilon f_0, U_{v^{-1}} f_0) d\mu(v)$$

$$= (U_\varepsilon f_0, f_0) \int d\mu(v) = (U_\varepsilon f_0, f_0) = \varphi(\varepsilon),$$

i.e. c coincides with the spherical function corresponding to the given representation.

Thus, the formula

$$\int U_v U_\varepsilon f_0 \, d\mu(v) = \varphi(\varepsilon) f_0$$

is valid. It follows that for $a \in \mathfrak{A}'$,

$$A_a f_0 = \int a(g) U_g f_0 \, d\mu(g) = \int a(\varepsilon)\omega(\varepsilon) U_v U_\varepsilon f_0 \, d\mu(v) d\mu(\varepsilon)$$

$$= \int a(\varepsilon)\omega(\varepsilon)\varphi(\varepsilon) d\mu(\varepsilon) f_0,$$

i.e. f_0 is a characteristic vector of the operator A_a of the corresponding representation of the algebra $\mathfrak{A}$. It follows that

$$A_{a_1 a_2} f_0 = A_{a_1} A_{a_2} f_0 = \int a_1(\varepsilon)\omega(\varepsilon)\varphi(\varepsilon) d\mu(\varepsilon) \int a_2(\varepsilon)\omega(\varepsilon)\varphi(\varepsilon) d\mu(\varepsilon) f_0;$$

consequently the correspondence

$$\lambda e + a \to \lambda + \int a(\varepsilon)\omega(\varepsilon)\varphi(\varepsilon) d\mu(\varepsilon) \tag{9}$$

is a homomorphism of the algebra $\mathfrak{A}$ into the field of complex numbers.

The matrix ε can be given in terms of the parameter $t = \ln \lambda$. [Recall that $\varepsilon = \begin{pmatrix} \lambda^{-1} & 0 \\ 0 & \lambda \end{pmatrix}$, $\lambda \geqq 1$.] Since $\lambda \geqq 1$, we have $t \geqq 0$. We set

$$f(t) = \frac{a(\varepsilon)\omega(\varepsilon)}{\sinh 2t}.$$

By the same token, we map $\mathfrak{A}'$ onto the set R_0' of all functions $f(t)$ which, in virtue of (8), satisfy the condition

$$||f|| = \int |f(t)| \sinh 2t \, dt < \infty,$$

where $||f|| = ||a||$. We shall now carry over to R_0' all the operations in $\mathfrak{A}'$ and see how they are expressed in R_0'. Clearly, addition and scalar multiplication are defined as usual. To find the multiplication rule in R_0' we note that, in virtue of formula (3), correspondence (9) for a spherical function can now be written in the form

$$f(t) \to \int_0^\infty f(t) \frac{2 \sin \rho t}{\rho} dt. \tag{10}$$

Hence this correspondence must be a homomorphism of the algebra R_0' into the field of complex numbers. From this it follows that the rule for multiplication in R_0' is defined by the formula

$$f(u) = \int_0^\infty \int_{|t-u|}^{t+u} f_1(s) f_2(t) ds dt.$$

In fact,

$$\int_0^\infty f(u) \frac{2 \sin \rho u}{\rho} du = \int_0^\infty \left(\int_0^\infty \int_{|t-u|}^{t+u} f_1(s) f_2(t) ds dt \right) \frac{2 \sin \rho u}{\rho} du$$

$$= \int_0^\infty \int_0^\infty f_1(s) f_2(t) \left(\int_{|s-t|}^{s+t} \frac{2 \sin \rho u}{\rho} du \right) ds dt \tag{11}$$

$$= \int_0^\infty \int_0^\infty f_1(s) f_2(t) \frac{4 \sin \rho s \cdot \sin \rho t}{\rho^2} ds dt = \int_0^\infty f_1(s) \frac{2 \sin \rho s}{\rho} ds \cdot \int_0^\infty f_2(t) \frac{2 \sin \rho t}{\rho} dt.$$

Since correspondence (10) is a homomorphism of the algebra R_0', the last expression coincides with $\int_0^\infty (f_1 \cdot f_2)(u)(2/\rho) \sin \rho u \, du$. Therefore it follows from equality (11) that $f = f_1 \cdot f_2$, because the functions $\sin \rho u$ constitute a complete system.

Finally, we note that involution in R_0' is defined by means of the formula $f^*(t) = \overline{f(t)}$. This follows from the fact that $a^*(g) = \overline{a(g^{-1})} = \overline{a(u_2^{-1} \varepsilon^{-1} u_1^{-1})} = \overline{a(\varepsilon)}$. It follows from the formulas obtained for $f_1 \cdot f_2$ and f^* that the algebra $\mathfrak{A}$ is completely isomorphic to the algebra R_0, considered in Example b), subsection 3, § 20. Therefore, the maximal ideals of the algebras R_0 and $\mathfrak{A}$ coincide. But, as we have already seen, the algebra R_0 contains non-symmetric maximal ideals. Consequently, $\mathfrak{A}$ is not a completely symmetric algebra.

We shall now prove that $R(\mathfrak{G}_2)$ is not a completely symmetric algebra. To this end, we note that if the element $e+a$, $a \in \mathfrak{A}'$, has an inverse in the algebra $R(\mathfrak{G}_2)$, then this inverse belongs to the algebra $\mathfrak{A}$, i.e. it has the form $e+b$, $b \in \mathfrak{A}'$. In fact, suppose $e+b$, $b \in L^1(\mathfrak{G}_2)$, is the inverse of the element $e+a$, $a \in \mathfrak{A}'$. Then

$$(e+a)(e+b) = e,$$

from which it follows that $b = -a-a \cdot b$, i.e. that

$$b(g) = -a(g) - \int a(g_1) b(g_1^{-1} g) d\mu(g_1).$$

But then for $v \in \mathfrak{B}$,

$$b(vg) = -a(vg) - \int a(g_1) b(g_1^{-1} vg) d\mu(g_1) = -a(g) - \int a(vg_1) b(g_1^{-1} g) d\mu(g_1)$$

$$= -a(g) - \int a(g_1) b(g_1^{-1} g) d\mu(g_1) = b(g),$$

because $a \in \mathfrak{A}'$ by assumption. Using the equality $(e+b)(e+a) = e$, we obtain analogously that $b(gv) = b(g)$. Consequently, $b \in \mathfrak{A}'$, $e+b \in \mathfrak{A}$. Since $\mathfrak{A}$ is not a completely symmetric algebra, there exists an element a in $\mathfrak{A}$ such that $e+a^*a$ does not have an inverse in the algebra $\mathfrak{A}$. According to the remark just made, above, it also does not have an inverse in the algebra $R(\mathfrak{G}_2)$. Consequently, $R(\mathfrak{G}_2)$ is not completely symmetric.

§ 30. Positive definite functions

1. Positive definite functions and their relationship with unitary representations. A function $\varphi(g)$, defined on the group $\mathfrak{G}$, is said to be *positive definite* if

$$\sum_{k,l=1}^{n} \varphi(g_l^{-1}g_k)\lambda_k\bar{\lambda}_l \geqq 0 \tag{1}$$

for all finite systems $g_1, \cdots, g_n$ of elements of the group $\mathfrak{G}$ and complex numbers $\lambda_1, \cdots, \lambda_n$.

Every positive definite function satisfies the conditions

$$\varphi(e) \geqq 0, \tag{2}$$

$$\varphi(g^{-1}) = \overline{\varphi(g)}, \tag{3}$$

$$|\varphi(g)| \leqq \varphi(e). \tag{4}$$

In fact, setting $n = 1$, $g_1 = e$, $\lambda_1 = 1$ in (1), we obtain relation (2). Further, setting $n = 2$, $g_1 = g$, $g_2 = e$, $\lambda_1 = 1$, $\lambda_2 = \lambda$ in (1), we obtain

$$\varphi(e)+\varphi(g)\bar{\lambda}+\varphi(g^{-1})\lambda+\varphi(e)|\lambda|^2 \geqq 0; \tag{5}$$

from this we conclude that $\varphi(g)\bar{\lambda}+\varphi(g^{-1})\lambda$ is a real number for arbitrary complex λ. Taking $\lambda = 1$ and $\lambda = i$, we obtain that $\varphi(g)+\varphi(g^{-1})$ and $i[\varphi(g^{-1})-\varphi(g)]$ are real numbers. This is possible only when $\varphi(g^{-1}) = \overline{\varphi(g)}$. We shall now prove inequality (4). In virtue of (2), only the following cases are possible: a) $\varphi(e) = 0$; then we obtain from (5) with $\lambda = -\varphi(g)$ that $2|\varphi(g)|^2 = 0$, $\varphi(g) = 0$; b) $\varphi(e) > 0$; setting $\lambda = -\varphi(g)/\varphi(e)$ in (5) we obtain $\varphi(e)-|\varphi(g)|^2/\varphi(e) \geqq 0$ and, consequently, $|\varphi(g)|^2 \leqq [\varphi(e)]^2$; hence inequality (4) is proved for both cases.

[We note further the following inequality pointed out by M. Krein [9]:

$$|\varphi(g)-\varphi(h)|^2 \leqq 2\varphi(e)[\varphi(e)-\Re\varphi(gh^{-1})].$$

To prove this, it suffices to put $n = 3$, $g_1 = g$, $g_2 = h$, $g_3 = e$, $\lambda_1 = \lambda_3 = 1$, $\lambda_2 = -1$ in (1).]

If $g \to U_g$ is a given unitary representation of the group $\mathfrak{G}$ then a positive definite function can be constructed by setting

$$\varphi(g) = (U_g\xi_0, \xi_0),$$

where $\xi_0 \neq 0$ is a fixed vector of the representation space. In fact,

$$\sum_{k,l=1}^{n} \varphi(g_l^{-1}g_k)\lambda_k\bar{\lambda}_l = \sum_{k,l=1}^{n} (U_{g_l}^{-1}U_{g_k}\xi_0, \xi_0)\lambda_k\bar{\lambda}_l = (\sum_{k=1}^{n} \lambda_k U_{g_k}\xi_0, \sum_{k=1}^{n} \lambda_k U_{g_k}\xi_0) \geqq 0.$$

Moreover, if the representation $g \to U_g$ is continuous, then the function $\varphi(g)$ is continuous.

Conversely, suppose $\varphi(g) \not\equiv 0$ is an arbitrary positive definite function. To it there corresponds a unitary representation $g \to U_g$ of the group $\mathfrak{G}$ which is such that $\varphi(g) = (U_g \xi_0, \xi_0)$.

In fact, suppose $\mathfrak{L}$ is the set of all functions $x(g)$ on the group $\mathfrak{G}$ which assume only a finite number of nonzero values. We define the operations of addition and scalar multiplication in $\mathfrak{L}$ in the usual way; then $\mathfrak{L}$ becomes a linear space. Further, we define the bilinear form

$$(x, y) = \sum_{g, h \in \mathfrak{G}} \varphi(h^{-1}g)x(g)\overline{y(h)}$$

in $\mathfrak{L}$. In virtue of the fact that the function φ is positive definite, relations (1) and (3) hold; consequently,

$$\overline{(y, x)} = (x, y) \text{ and } (x, x) \geqq 0;$$

however, it may occur that $(x, x) = 0$ also for $x \not\equiv 0$. Suppose $\mathfrak{N}$ is the set of all $x \in \mathfrak{L}$ for which $(x, x) = 0$; then $\mathfrak{N}$ is a subspace in $\mathfrak{L}$ and (x, y) defines the inner product (ξ, η) in the quotient space $\mathfrak{H}' = \mathfrak{L}/\mathfrak{N}$ according to the formula

$$(\xi, \eta) = (x, y), \text{ where } x \in \xi, y \in \eta$$

(see I, subsection 1, § 5). Consequently, the completion $\mathfrak{H}$ of the space $\mathfrak{H}'$ with respect to the norm $|\xi| = \sqrt{(\xi, \xi)}$ is a Hilbert space.

To each element $g_0 \in \mathfrak{G}$ we assign the operator T_{g_0} according to the rule $T_{g_0}x(g) = x(g_0^{-1}g)$. Then

$$(T_{g_0}x, T_{g_0}y) = \sum_{g, h} \varphi(h^{-1}g)x(g_0^{-1}g)\overline{y(g_0^{-1}h)} = \sum_{g, h} \varphi(h^{-1}g_0^{-1}g_0 g)x(g)\overline{y(h)} = (x, y).$$

Consequently $\mathfrak{N}$ is invariant with respect to T_{g_0}, and hence T_{g_0} can also be considered as an operator in $\mathfrak{H}'$, with $(T_{g_0}\xi, T_{g_0}\eta) = (\xi, \eta)$. But then T_{g_0} is isometric in $\mathfrak{H}'$ and hence it is uniquely extendible to an isometric operator in $\mathfrak{H}$; we denote this isometric operator in $\mathfrak{H}$ by U_{g_0}. Clearly, the correspondence $g \to U_g$ is a representation of the group $\mathfrak{G}$; since $U_{g^{-1}}$ is the operator inverse to U_g, U_g is unitary; consequently, $g \to U_g$ is a unitary representation of the group $\mathfrak{G}$.

Suppose $x_0(g)$ is a function on $\mathfrak{G}$ defined by the formula

$$x_0(g) = \begin{cases} 1 & \text{for } g = e, \\ 0 & \text{for } g \neq e, \end{cases}$$

and that ξ_0 is the coset modulo $\mathfrak{N}$ which contains x_0. Then

$$(U_{g_0}\xi_0, \xi_0) = (T_{g_0}x_0, x_0) = \sum_{g, h} \varphi(h^{-1}g)x_0(g_0^{-1}g)\overline{x_0(h)} = \varphi(g_0).$$

Consequently, the above-constructed representation $g \to U_g$ satisfies all the requirements set down. We note that ξ_0 is a cyclic vector of the representation $g \to U_g$.

If the function φ is continuous, then the representation $g \to U_g$ constructed with respect to it is also continuous.

In fact, if φ is continuous, then for fixed $\xi, \eta \in \mathfrak{H}'$ and $x \in \xi$, $y \in \eta$ the expression

$$(U_{g_0}\xi, \eta) = (T_{g_0}x, y) = \sum_{g,h} \varphi(h^{-1}g)x(g_0^{-1}g)\overline{y(h)} = \sum_{g,h} \varphi(h^{-1}g_0 g)x(g)\overline{y(h)}$$

is a continuous function of g_0; since $\mathfrak{H}'$ is dense in $\mathfrak{H}$ and $|U_{g_0}| = 1$, this implies the continuity of the function $(U_g\xi, \eta)$ for all $\xi, \eta \in \mathfrak{H}$.

We have thus proved the following theorem.

THEOREM 1. *To every continuous unitary representation $g \to U_g$ of the topological group $\mathfrak{G}$ and vector $\xi_0 \neq 0$ in the representation space there corresponds a continuous positive definite function $\varphi(g) = (U_g\xi_0, \xi_0) \not\equiv 0$. Conversely, to every continuous positive definite function $\varphi(g) \not\equiv 0$ there corresponds a continuous unitary cyclic representation $g \to U_g$ of the group $\mathfrak{G}$ with cyclic vector ξ_0 such that $\varphi(g) = (U_g\xi_0, \xi_0)$.*

2. Relationship of positive definite functions with positive functionals on a group algebra. Suppose $\mathfrak{G}$ is a locally compact group. A positive functional f_0 in the group algebra $R(\mathfrak{G})$, defined by the equality $f_0(\lambda e + a) = \lambda c$, where c is a positive constant, is said to be *degenerate*. A positive functional f in a group algebra is called *regular* if it does not dominate the degenerate functional.

LEMMA. *A cyclic representation $x \to A_x$ of a group algebra does not contain the degenerate representation if, and only if, the positive functional $f(x) = (A_x\xi_0, \xi_0)$ defining it is regular.*

Proof. Suppose the representation $x \to A_x$ contains the degenerate representation and that $\mathfrak{N}$ is an invariant subspace on which the representation $x \to A_x$ is degenerate. Setting

$$\xi_0 = \xi_1 + \xi_2, \quad \xi_1 \in \mathfrak{N}, \ \xi_2 \in \mathfrak{H} \ominus \mathfrak{N},$$

we have that

$$f(x) = (A_x\xi_0, \xi_0) = (A_x\xi_1, \xi_1) + (A_x\xi_2, \xi_2);$$

consequently, $f_0(x) = (A_x\xi_1, \xi_1)$ is a degenerate functional dominated by the functional f.

Conversely, suppose the degenerate functional f_0 is dominated by the functional f. According to Theorem 1, subsection 1, § 19,

$$f_0(x) = (A_x B\xi_0, \xi_0),$$

where B is a positive definite operator in $\mathfrak{H}$ which commutes with all the operators

A_x. Since, for $a \in L^1(\mathfrak{G})$, also $x^*a \in L^1(\mathfrak{G})$ for all $x \in R(\mathfrak{G})$ and since f_0 vanishes on $L^1(\mathfrak{G})$,

$$0 = f_0(x^*a) = (A_x^* A_a B\xi_0, \xi_0) = (A_a B\xi_0, A_x\xi_0).$$

But the vectors $A_x\xi_0$ form a dense set in $\mathfrak{H}$ and consequently

$$A_a B\xi_0 = 0 \text{ for all } a \in L^1(\mathfrak{G}).$$

Therefore, the vectors of the form $\lambda B\xi_0$ form a one-dimensional invariant subspace, in which the representation $x \to A_x$ is degenerate, in $\mathfrak{H}$.

Now suppose that $g \to U_g$ is a continuous unitary representation of the group $\mathfrak{G}$ constructed with respect to the positive definite function φ. To this unitary representation $g \to U_g$ there corresponds the representation $x \to A_x$ of the group algebra $R(\mathfrak{G})$ which does not contain the degenerate representation. But now the representation $x \to A_x$ in its turn is assigned the regular positive functional

$$\begin{aligned} f(\lambda e+a) &= \lambda(\xi_0, \xi_0)+(A_a\xi_0, \xi_0) \\ &= \lambda\varphi(e)+\int a(g)(U_g\xi_0, \xi_0)d\mu(g) \\ &= \lambda\varphi(e)+\int a(g)\varphi(g)d\mu(g). \end{aligned}$$

Conversely, to a regular positive functional f in $R(\mathfrak{G})$ there corresponds a cyclic representation $x \to A_x$ of this algebra which does not contain the degenerate representation such that

$$f(\lambda e+a) = \lambda(\xi_0, \xi_0)+(A_a\xi_0, \xi_0), \tag{1}$$

where ξ_0 is a cyclic vector of the representation $x \to A_x$. Suppose $g \to U_g$ is a continuous unitary representation of the group $\mathfrak{G}$ corresponding to this representation $x \to A_x$, such that $A_a = \int a(g)U_g d\mu(g)$. Setting this expression in (1) for A_a we obtain

$$f(\lambda e+a) = \lambda(\xi_0, \xi_0)+\int a(g)(U_g\xi_0, \xi_0)d\mu(g) = \lambda\varphi(e)+\int a(g)\varphi(g)d\mu(g),$$

where $\varphi(g) = (U_g\xi_0, \xi_0)$ is a continuous positive definite function on the group $\mathfrak{G}$.

Thus the following theorem holds.

THEOREM 2. *There exists a one-to-one correspondence between the set of all regular positive functionals $f(\lambda e+a)$ on the algebra $R(\mathfrak{G})$ and the set of all continuous positive definite functions $\varphi(g) \not\equiv 0$ on the group $\mathfrak{G}$. This correspondence is defined by the formula*

$$f(\lambda e+a) = \lambda\varphi(e)+\int a(g)\varphi(g)d\mu(g).$$

We note that the completeness theorem (see subsection 3, § 29) can also be obtained with the aid of positive definite functions (see GELFAND and RAIKOV [2]).

We denote the space L^∞ corresponding to a left invariant integral on $\mathfrak{G}$ by $L^\infty(\mathfrak{G})$ (see subsection 13, § 6). In virtue of (4), subsection 1, every continuous positive definite function on $\mathfrak{G}$ belongs to $L^\infty(\mathfrak{G})$. The space $L^\infty(\mathfrak{G})$ is the conjugate to $L^1(\mathfrak{G})$ (see subsection 16, § 6); therefore the weak topology in $L^\infty(\mathfrak{G})$ is defined by the neighborhoods $U(\varphi_0; a_1, \cdots, a_n; \varepsilon)$, $a_1, \cdots, a_n \in L^1(\mathfrak{G})$, $\varepsilon > 0$, which constitute the set of all functions $\varphi \in L^\infty(\mathfrak{G})$, satisfying the inequalities

$$\left| \int [\varphi(g) - \varphi_0(g)] a_k(g) d\mu_l(g) \right| < \varepsilon, \quad k = 1, \cdots, n.$$

A function $\varphi \in L^\infty(\mathfrak{G})$ is called *integrally positive definite* if

$$\int\int \overline{a(g_1)} a(g_1 g) \varphi(g) d\mu_l(g_1) d\mu_l(g) \geqq 0 \tag{1}$$

for any function $a \in L^1(\mathfrak{G})$.

I. *The set of all integrally pos tive definite functions is a weakly closed set in* $L^\infty(\mathfrak{G})$.

In fact, setting $b = a^* \cdot a$, we can write (1) in the form $\int b(g)\varphi(g) d\mu_l(g) \geqq 0$; on the other hand, the expression $f(\varphi) = \int b(g)\varphi(g) d\mu_l(g)$ is a linear functional in $L^\infty(\mathfrak{G})$ which is continuous in the weak topology; consequently, if inequality (1) is valid for some set of functions φ in $L^\infty(\mathfrak{G})$, then it is also valid for any weak contact point of this set.

The reasoning of subsection 1 can be carried over to integrally positive definite functions. In fact, setting

$$(a_1, a_2) = \int\int \overline{a_2(g_1)} a_1(g_1 g) \varphi(g) \ \mu_l(g_1) d\mu_l(g), \tag{2}$$

we obtain a bilinear form in $L^1(\mathfrak{G})$ which satisfies the condition $(a, a) \geqq 0$. Consequently, $(a_1 + \lambda a_2, a_1 + \lambda a_2) \geqq 0$ for arbitrary $a_1, a_2 \in L^1(\mathfrak{G})$ and any complex λ; from this we easily conclude that the form (a_1, a_2) is Hermitian. Therefore it defines an inner product (ξ, η) in the quotient space $\mathfrak{H}' = L^1(\mathfrak{G})/\mathfrak{N}$, where $\mathfrak{N}$ is the subspace consisting of all functions $a \in L^1(\mathfrak{G})$ satisfying the condition $(a, a) = 0$. The completion $\mathfrak{H}'$ with respect to the norm $|\xi| = \sqrt{(\xi, \xi)}$ is a Hilbert space which we shall denote by $\mathfrak{H}$.

We set $T_{g_0} a(g) = a(g_0^{-1} g)$; from formula (2) and the left invariance of measure, we conclude that $(T_{g_0} a_1, T_{g_0} a_2) = (a_1, a_2)$. From this, as in subsection 1, it follows that T_g defines a unitary operator U_g in $\mathfrak{H}$ and that the correspondence $g \to U_g$ is a unitary representation of the group $\mathfrak{G}$. This representation is continuous. In fact, if $a_1, a_2 \in L^1(\mathfrak{G})$ and $a_1 \in \xi$, $a_2 \in \eta$, then the expression

$$(U_{g_0}\xi, \eta) = (T_{g_0}a_1, a_2) = \int \overline{a_2(g_1)}a_1(g_0^{-1}g_1 g)\varphi(g)d\mu_l(g_1)d\mu_l(g)$$

is a continuous function of g_0 because, in virtue of I, subsection 2, § 28, the vector-valued function

$$b_{g_0} = \{b_{g_0}(g)\} = \left\{\int \overline{a_2(g_1)}T_{g_1^{-1}g_0}a_1(g)d\mu_l(g_1)\right\} = \left\{\int \overline{a_2(g_1)}a_1(g_0^{-1}g_1 g)d\mu_l(g_1)\right\}$$

[here the notation $b_{g_0}(g)$ does not mean $b(g_0^{-1}g)$] with values in $L^1(\mathfrak{G})$ is continuous in the sense of the norm in $L^1(\mathfrak{G})$ and

$$(U_{g_0}\xi, \eta) = \int b_{g_0}(g)\varphi(g)d\mu_l(g) = f(b_{g_0}),$$

where $f(b) = \int b(g)\varphi(g)d\mu_l(g)$ is a continuous linear functional in $L^1(\mathfrak{G})$. [In fact, we choose a neighborhood $U(g_0)$ so that $\|T_{g_0'}a_1 - T_{g_0}a_1\| < \varepsilon$ for $g_0' \in U(g_0)$ (see I, subsection 2, § 28). Then, for $g_0' \in U(g_0)$,

$$\|b_{g_0'} - b_{g_0}\|_1 = \int \left| \int \overline{a_2(g_1)}[a_1(g_0'^{-1}g_1 g) - a_1(g_0^{-1}g_1 g)]d\mu_l(g_1) \right| d\mu_l(g)$$

$$\leqq \int |a_2(g_1)| \left[\int |a_1(g_0'^{-1}g_1 g) - a_1(g_0^{-1}g_1 g)| d\mu_l(g)\right] d\mu_l(g_1)$$

$$= \int |a_2(g_1)| \left[\int |a_1(g_0'^{-1}g) - a_1(g_0^{-1}g)| d\mu_l(g)\right] d\mu_l(g_1)$$

$$= \int |a_2(g_1)| \, \|T_{g_0'}a_1 - T_{g_0}a_1\|_1 \, d\mu_l(g_1) < \varepsilon \int |a_2(g_1)| \, d\mu_l(g_1).$$

This proves the continuity in norm of b_{g_0}.]

Since $\mathfrak{H}'$ is dense in $\mathfrak{H}$ and $|U_g| = 1$, it follows from this that the function $(U_g\xi, \eta)$ is continuous for all $\xi, \eta \in \mathfrak{H}$. Thus,

II. *Every integrally positive definite function φ on the group $\mathfrak{G}$ defines a continuous unitary representation $g \to U_g$ of the group $\mathfrak{G}$.*

To this representation there corresponds a representation of the group algebra $R(\mathfrak{G})$ of the group $\mathfrak{G}$ which does not contain the degenerate representation and, consequently, a regular positive functional f in $R(\mathfrak{G})$ such that

$$f(a_2^* a_1) = (a_1, a_2) = \iint \overline{a_2(g_1)}a_1(g_1 g)\varphi(g)d\mu_l(g_1)d\mu_l(g). \tag{3}$$

But, according to Theorem 2,

$$f(a_2^* a_1) = \iint \overline{a_2(g_1)}a_1(g_1 g)\varphi_1(g)d\mu_l(g_1)d\mu_l(g),$$

where φ_1 is a continuous positive definite function. Comparison of this formula with (3) shows that $\varphi(g) = \varphi_1(g)$ locally almost everywhere on $\mathfrak{G}$; consequently,

III. *Every integrally positive definite function on $\mathfrak{G}$ is equal to a continuous positive definite function locally almost everywhere on $\mathfrak{G}$.*

Conversely,

IV. *Every continuous positive definite function is also an integrally positive definite function.*

The assertion follows directly from Theorem 2, according to which $f(a) = \int a(g)\varphi(g)d\mu_l(g)$ is a positive functional in $L^1(\mathfrak{G})$, and therefore

$$\int\int \overline{a_2(g_1)}a(g_1 g)\varphi(g)d\mu_l(g_1)d\mu_l(g) = f(a^*a) \geqq 0.$$

3. Regular sets. Suppose φ_1, φ_2 are continuous positive definite functions on the locally compact group $\mathfrak{G}$ and that f_1, f_2 are the corresponding positive functionals. The function φ_1 will be said to be *dominated* by the function φ_2, and in this case we shall write $\varphi_1 \prec \varphi_2$, if $\varphi_2 - \varphi_1$ is a positive definite function, i.e. if $f_2 - f_1$ is a positive functional. In particular, a positive definite function is called *elementary* if every positive definite function dominated by the function φ is a multiple of φ. Thus, φ is elementary if, and only if, the corresponding positive functional is indecomposable and, consequently, if, and only if, the unitary representation of the group corresponding to it is irreducible.

I. *Suppose $\varphi_2(g) = (U_g\xi_0, \xi_0)$, where $g \to U_g$ is a cyclic representation of the group $\mathfrak{G}$ with the cyclic vector ξ_0. The relation $\varphi_1 \prec \varphi_2$ holds if, and only if, $\varphi_1(g) = (BU_g\xi_0, \xi_0)$, where B is a positive definite operator with norm less than or equal to unity which commutes with all the operators U_g.*

Proof. Suppose f_1, f_2 are regular positive functionals on $R(\mathfrak{G})$ corresponding to the positive definite functions φ_1, φ_2 so that

$$f_1(a) = \int a(g)\varphi_1(g)d\mu_l(g), \quad f_2(a) = \int a(g)\varphi_2(g)d\mu_l(g) \tag{1}$$

for all $a \in L^1(\mathfrak{G})$. The relation $\varphi_1 \prec \varphi_2$ is equivalent to saying that $f_2 - f_1$ is a positive functional which in turn is equivalent to the equality

$$f_1(a) = (A_a B\xi_0, \xi_0) \tag{2}$$

for all $a \in L^1(\mathfrak{G})$, where B is a positive definite operator with norm less than or equal to unity which commutes with all the operators A_a, or, what amounts to the same thing, with all the operators U_g (see Theorem 1, subsection 1, § 19). But in virtue of (1), formula (2) means that

$$\int a(g)\varphi_1(g)d\mu_l(g) = (A_a B\xi_0, \xi_0) = \int a(g)(BU_g\xi_0, \xi_0)d\mu_l(g),$$

and consequently, in view of the fact that the function $a \in L^1(\mathfrak{G})$ is arbitrary, that $\varphi_1(g) = (BU_g\xi_0, \xi_0)$. (See Theorem 1, subsection 2, § 29.)

We now set $H = \sqrt{B}$. Since ξ_0 is a cyclic vector, there exists a linear combination

$$\xi = \sum_{k=1}^{n} \lambda_k U_{g_k}\xi_0$$

such that $|H\xi_0-\xi| < \varepsilon$ for arbitrary $\varepsilon > 0$. Then

$$|(BU_g\xi_0, \xi_0)-(U_g\xi, \xi)| = |(U_gH\xi_0, H\xi_0)-(U_g\xi, \xi)|$$
$$\leqq |(U_gH\xi_0, H\xi_0)-(U_g\xi, H\xi_0)|+|(U_g\xi, H\xi_0)-(U_g\xi, \xi)|$$
$$\leqq |U_g(H\xi_0-\xi)|\cdot|H\xi_0|+|U_g\xi|\cdot|H\xi_0-\xi| < \varepsilon(|H\xi_0|+|\xi|) < \varepsilon(2|H\xi_0|+\varepsilon).$$

On the other hand,

$$(U_g\xi, \xi) = \sum_{p,q} \lambda_p\bar{\lambda}_q\varphi_2(g_q^{-1}gg_p);$$

consequently, the inequality obtained above can be written in the form

$$|\varphi_1(g)-\sum_{p,q}\lambda_p\bar{\lambda}_q\varphi_2(g_q^{-1}gg_p)| < \varepsilon(2||H\xi_0||+\varepsilon).$$

Thus,

II. *If $\varphi_1 \prec \varphi_2$, then φ_1 is the uniform limit on the group $\mathfrak{G}$ of the functions*

$$\varphi(g) = \sum_{p,q}\lambda_p\bar{\lambda}_q\varphi_2(g_q^{-1}gg_p).$$

We denote by P the set of all continuous positive definite functions φ on $\mathfrak{G}$ and by P_0 the set of all functions in P which satisfy the condition that $\varphi(e) \leqq 1$; P_0 is a weakly closed convex subset of the unit sphere in $L^\infty(\mathfrak{G})$ and hence P_0 is compact in the weak topology in $L^\infty(\mathfrak{G})$ (see III, subsection 7, § 3).

In fact, suppose φ_0 is a weak contact point of the set P_0; in virtue of IV and I, subsection 2, φ_0 is an integrally positive definite function and we may assume that $\varphi_0 \in P$ (see III, subsection 2). We must prove that $\varphi_0(e) \leqq 1$. Let us assume the contrary; suppose $\varphi_0(e) = 1+\varepsilon$, where $\varepsilon > 0$. Since $|\varphi_0(g)-\varphi_0(e)| < \varepsilon/2$ in some neighborhood U of the identity e, where it can be assumed that $\bar{U}$ is compact, then, setting

$$a(g) = \begin{cases} \dfrac{1}{\mu_l(\bar{U})} \text{ for } g \in \bar{U}, \\ 0 \text{ for } g \notin \bar{U}, \end{cases}$$

we obtain that for $\varphi \in P_0$

$$\left|\int a(g)[\varphi_0(g)-\varphi(g)]d\mu_l(g)\right| = \frac{1}{\mu_l(\bar{U})}\left|\int_{\bar{U}}[\varphi_0(g)-\varphi(g)]d\mu_l(g)\right|$$
$$= \frac{1}{\mu_l(\bar{U})}\left|\varphi_0(e)\mu_l(\bar{U})+\int_{\bar{U}}[\varphi_0(g)-\varphi_0(e)]d\mu_l(g)-\int_{\bar{U}}\varphi(g)d\mu_l(g)\right|$$
$$\geqq \frac{1}{\mu_l(\bar{U})}\left[\varphi_0(e)\mu_l(\bar{U}) - \int_{\bar{U}}|\varphi_0(g)-\varphi_0(e)|d\mu_l(g)-\int_{\bar{U}}\varphi(e)d\mu_l(g)\right]$$
$$\geqq \varphi_0(e)-\frac{\varepsilon}{2}-\varphi(e) \geqq (1+\varepsilon)-\frac{\varepsilon}{2}-1 = \frac{\varepsilon}{2}.$$

This means, contrary to our assumption, that the neighborhood $U(\varphi_0; a; \varepsilon/2)$ does not contain any function of the set P_0. Consequently, $\varphi_0(e) \leqq 1$.

The set $\mathfrak{A} \subset P_0$ is called a *regular subset* of the set P_0 if

1° $\mathfrak{A}$ is convex and weakly closed;

2° $\varphi_1 \in P$, $\varphi_2 \in \mathfrak{A}$, $\varphi_1 \prec \varphi_2$ imply that $\varphi_1 \in \mathfrak{A}$;

3° $\varphi \in \mathfrak{A}$ implies that $\varphi(g)/\varphi(e) \in \mathfrak{A}$.

It follows from 2° that the function $\varphi \equiv 0 \in \mathfrak{A}$; 0 is an extreme point of this set because if $\varphi_1, \varphi_2 \in P$ and $t\varphi_1 + (1-t)\varphi_2 = 0$, $0 < t < 1$, then $t\varphi_1(e) + (1-t)\varphi_2(e) = 0$ and, consequently, $\varphi_1(e) = \varphi_2(e) = 0$, $\varphi_1 = \varphi_2 = 0$.

Clearly, the entire set P_0 is also a regular subset of itself.

A positive definite function φ is said to be *normalized* if $\varphi(e) = 1$.

III. *If $\mathfrak{A}$ is a regular subset of P_0 then the set of all its extreme points, which are distinct from* 0, *coincides with the set of all elementary normalized positive definite functions belonging to it.*

Proof. Suppose φ is an extreme point of $\mathfrak{A}$ and let $\varphi \neq 0$. Since $0 \in \mathfrak{A}$, $\{1/\varphi(e)\}\varphi \in \mathfrak{A}$, then also the entire segment $[0, \{1/\varphi(e)\}\varphi] \in \mathfrak{A}$; the function φ is extreme and, hence, it must coincide with the endpoint $\{1/\varphi(e)\}\varphi$ of this segment. It follows that $\varphi(e) = 1$, i.e. that φ is normalized.

We shall prove that φ is elementary. Suppose $\varphi_1 \prec \varphi$ and $\varphi_1 \neq \varphi$, $\varphi_1 \neq 0$; then $\varphi - \varphi_1 \prec \varphi$, $\varphi - \varphi_1 \neq 0$; consequently, in virtue of 2°,

$$\varphi_1 \in \mathfrak{A}, \quad \varphi - \varphi_1 \in \mathfrak{A}.$$

In virtue of 3°, the functions

$$\psi_1 = \frac{1}{\varphi_1(e)}\varphi_1, \quad \psi_2 = \frac{1}{\varphi(e) - \varphi_1(e)}[\varphi - \varphi_1] = \frac{\varphi - \varphi_1}{1 - \varphi_1(e)}$$

belong to the set $\mathfrak{A}$. The equality

$$\varphi = \varphi_1(e)\psi_1 + [1 - \varphi_1(e)]\psi_2$$

shows that φ belongs to the segment $[\psi_1, \psi_2]$. Since, by assumption, φ is extreme, either $\varphi = \psi_1$ or $\varphi = \psi_2$. In each of these cases, $\varphi_1 = \varphi_1(e)\varphi$. This means that the function φ is elementary.

Conversely, suppose $\varphi \in \mathfrak{A}$ is an elementary normalized function. We shall prove that φ is extreme in $\mathfrak{A}$. Suppose φ belongs to the segment $[\varphi_1, \varphi_2] \subset \mathfrak{A}$, i.e. that

$$\varphi = \lambda\varphi_1 + (1-\lambda)\varphi_2, \quad \varphi_1, \varphi_2 \in \mathfrak{A}, \ 0 \leqq \lambda \leqq 1.$$

Then $\lambda\varphi_1 \prec \varphi$; hence if $\lambda \neq 0$, then, since φ is elementary, we must have $\varphi_1 = \mu\varphi$. Analogously, if $\lambda < 1$, then $\varphi_2 = \nu\varphi$. Consequently, for $0 < \lambda < 1$,

$$\varphi = \lambda\mu\varphi + (1-\lambda)\nu\varphi,$$

from which it follows that

$$1 = \lambda\mu + (1-\lambda)\nu.$$

Since $0 \leqq \mu \leqq 1$, $0 \leqq \nu \leqq 1$ (because $\varphi_1, \varphi_2 \in P_0$) this is possible only if $\mu = \nu = 1$, i.e. $\varphi_1 = \varphi_2 = \varphi$, which completes the proof of Proposition III. Now applying Theorem 1, subsection 9, § 3, to the regular subset $\mathfrak{A}$ of the set P_0, we obtain

THEOREM 3 (GELFAND and RAIKOV [2]). *Suppose $\mathfrak{A}$ is a regular subset of the set P_0. Then every function $\varphi \in \mathfrak{A}$ is a weak contact point in $L^\infty(\mathfrak{G})$ of functions of the form*

$$\sum \lambda_k \varphi_k(g), \quad \lambda_k \geqq 0, \ \sum \lambda_k \leqq \varphi(e),$$

where the φ_k are elementary normalized functions in $\mathfrak{A}$.

In particular, this theorem holds for the set P_0 and hence also for the set P.

4. Trigonometric polynomials on a group. A *trigonometric polynomial* on a group $\mathfrak{G}$ is any function of the form

$$\lambda_1 \varphi_1 + \cdots + \lambda_n \varphi_n,$$

where the φ_k are elementary normalized functions in P. The numbers λ_k are called the *coefficients* of the trigonometric polynomial. If, in particular, $\mathfrak{G}$ is the additive group of all real numbers x, then, as we shall see below (see Example a), subsection 1, § 31), a trigonometric polynomial on this group coincides with the usual trigonometric polynomial $\Sigma_k \lambda_k e^{i\alpha_k x}$.

I. *If f is a trigonometric polynomial and $x \in L^1(\mathfrak{G})$, then the function*

$$f_1(g) = \int f(g_1) x(g_1^{-1} g) d\mu_l(g_1)$$

is also a trigonometric polynomial.

It will be sufficient to prove this proposition for the case when f is an elementary positive definite function. Then

$$f(g) = (U_g \xi_0, \xi_0),$$

where $g \to U_g$ is an irreducible unitary representation of the group $\mathfrak{G}$; hence

$$f_1(g) = \int (U_{g_1} \xi_0, \xi_0) x(g_1^{-1} g) d\mu_l(g_1) = \int (U_{gg_1} \xi_0, \xi_0) x(g_1^{-1}) d\mu_l(g_1) = (U_g \xi, \xi_0),$$

where

$$\xi = \int x(g_1^{-1}) U_{g_1} \xi_0 \, d\mu_l(g_1).$$

Therefore the function f_1 can be represented in the form of a linear combination of the functions

$$\varphi_1(g) = (U_g(\xi+\xi_0), \xi+\xi_0), \qquad \varphi_3(g) = (U_g(\xi+i\xi_0), \xi+i\xi_0),$$
$$\varphi_2(g) = (U_g(\xi-\xi_0), \xi-\xi_0), \qquad \varphi_4(g) = (U_g(\xi-i\xi_0), \xi-i\xi_0),$$

each of which is an elementary positive definite function. This means that f_1 is a trigonometric polynomial. In an exactly analogous manner, it can be proved that

II. *Every right or left translation of a trigonometric polynomial is a trigonometric polynomial.*

5. Spectrum. Suppose $\mathfrak{S}$ is the set of bounded continuous functions on $\mathfrak{G}$. We shall say that the function f is *approximated by functions in* $\mathfrak{S}$ *uniformly on every compact set* if for an *arbitrary* compact set $Q \subset \mathfrak{G}$ and $\varepsilon > 0$ there exists a function $\varphi \in \mathfrak{S}$ such that

$$|f(g) - \varphi(g)| < \varepsilon \text{ on } Q.$$

Suppose $\varphi \in P_0$. We denote by $\mathfrak{A}_\varphi$ the set of all functions of the form

$$f(g) = \sum_{p,q} \alpha_p \bar{\alpha}_q \varphi(g_q^{-1} g g_p)$$

and finite sums of such functions, belonging to P_0. Further, we denote by $\tilde{\mathfrak{A}}_\varphi$ the set of all functions in P_0 which can be uniformly approximated on every compact set by functions in $\mathfrak{A}_\varphi$.

Lemma 1. $\tilde{\mathfrak{A}}_\varphi$ *is weakly closed.*

[This lemma follows from a theorem of Raikov [8].]

Proof. Suppose θ belongs to the weak closure of the set $\tilde{\mathfrak{A}}_\varphi$; we shall prove that $\theta \in \tilde{\mathfrak{A}}_\varphi$. Suppose $y \in L^1(\mathfrak{G})$. For brevity, we set (see subsection 4, § 27)

$$y_{g'}(g) = y(g'^{-1}g)$$

and

$$\psi_y(g) = \iint \psi(g_2^{-1}gg_1)y(g_1)\overline{y(g_2)}d\mu_l(g_1)d\mu_l(g_2).$$

If g' runs through the compact set $Q \subset \mathfrak{G}$, then $y_{g'}$ runs through a precompact set $S \subset L^1(\mathfrak{G})$. In fact, it suffices to show that for arbitrary $\varepsilon > 0$ the set S contains a finite ε-net (see I, subsection 14, § 2). Since $y_{g'}$ is a continuous function of g' relative to the norm in $L^1(\mathfrak{G})$, then to every point $g'_0 \in Q$ there corresponds a neighborhood $U(g'_0)$, such that $||y_g - y_{g'_0}||_1 < \varepsilon$ for $g \in U(g'_0)$. In virtue of the compactness of the set Q, a finite covering $U(g'_1), \cdots, U(g'_n)$ of the set Q can be selected from these neighborhoods. Clearly, the points $y_{g'_1}, \cdots, y_{g'_n}$ form an ε-net in the set S.

Thus, S is precompact and, consequently, the set y^*S of all functions $z_g = y^*y_g$, $g \in Q$, is also precompact. But, according to Proposition III, subsection 4, § 19, the weak contact point θ of the set $\tilde{\mathfrak{A}}_\varphi$ is also a contact point of it relative to the uniform-on-compacta topology. Therefore, there exists a function $\psi \in \tilde{\mathfrak{A}}_\varphi$ such that

$$\left| \int [\psi(g') - \theta(g')] z_g(g') d\mu_l(g') \right| < \varepsilon$$

for all $g \in Q$ which, as is easily seen, means that

$$|\psi_y(g) - \theta_y(g)| < \varepsilon$$

for all $g \in Q$.

In other words, *the function θ_y is uniformly approximated on every compact set by functions ψ_y, $\psi \in \tilde{\mathfrak{A}}_\varphi$.* Clearly, for $\theta_y(e) \leqq 1$, it can also be assumed that $\psi_y(e) \leqq 1$.

Suppose $g \to U_g$ is a unitary representation with cyclic vector ξ_0 such that $\psi(g) = (U_g\xi_0, \xi_0)$; then $\psi_y(g) = (U_g A_y \xi_0, A_y \xi_0)$. But, since the vector ξ_0 is cyclic, $A_y\xi_0$ is the limit in norm of vectors of the form

$$\sum_k \alpha_k U_{g_k} \xi_0;$$

therefore $\psi_y(g)$ is the limit, uniform on the entire group $\mathfrak{G}$, of functions

$$\sum_{j,k} \alpha_k \bar{\alpha}_j (U_g U_{g_k} \xi_0, U_{g_j} \xi_0) = \sum_{j,k} \psi(g_j^{-1} g g_k) \alpha_k \bar{\alpha}_j \in \tilde{\mathfrak{A}}_\varphi.$$

Consequently, $\psi_y(g) \in \tilde{\mathfrak{A}}_\varphi$ and therefore also $\theta_y \in \tilde{\mathfrak{A}}_\varphi$. On the other hand, for given $\varepsilon > 0$ and compact $Q \subset \mathfrak{G}$, there exists a neighborhood U of the identity in $\mathfrak{G}$ such that

$$|\theta(g_2^{-1}gg_1) - \theta(g)| < \varepsilon$$

for $g_1, g_2 \in U$, $g \in Q$ [because the set of all points $\langle g_1, g_2 \rangle \in \mathfrak{G} \times \mathfrak{G}$ for which $|\theta(g_2^{-1}gg_1) - \theta(g)| < \varepsilon$ for all $g \in Q$ is open and contains the identity $\langle e, e \rangle \in \mathfrak{G} \times \mathfrak{G}$; see V, subsection 12, § 2]. Moreover, it can be assumed that $\bar{U}$ is compact. Choosing $y \in L^1(\mathfrak{G})$ so that $y(g) \geqq 0$, $y(g) = 0$ in the exterior of $\bar{U}$ and

$$\int y(g) d\mu_l(g) = 1,$$

we have

$$|\theta_y(g)-\theta(g)| \leqq \int_{\bar{U}}\int_{\bar{U}}|\theta(g_2^{-1}gg_1)-\theta(g)|y(g_1)y(g_2)d\mu_l(g_1)d\mu_l(g_2)$$
$$< \varepsilon\int_{\bar{U}}\int_{\bar{U}}y(g_1)y(g_2)d\mu_l(g_1)d\mu_l(g_2) = \varepsilon$$

for $g \in Q$. Hence, $\theta \in \tilde{\mathfrak{A}}_\varphi$.

Lemma 2. *$\tilde{\mathfrak{A}}_\varphi$ is a regular subset of P_0.*

Proof. We must prove that $\tilde{\mathfrak{A}}_\varphi$ satisfies conditions 1°, 2°, 3° in the definition of a regular subset (see subsection 3, above). Clearly the set $\mathfrak{A}_\varphi$ is convex; consequently, $\tilde{\mathfrak{A}}_\varphi$ is also convex. Further, it is obvious that $\psi \in \mathfrak{A}_\varphi$, $\psi \neq 0$ imply that $\psi/\psi(e) \in \mathfrak{A}_\varphi$ and hence this same property also holds for $\psi \in \tilde{\mathfrak{A}}_\varphi$, i.e. condition 3° is satisfied.

Suppose $\psi \in \tilde{\mathfrak{A}}_\varphi$, $\theta \in P$, $\theta \prec \psi$; we shall prove that $\theta \in \tilde{\mathfrak{A}}_\varphi$. Suppose Q is a compact set in $\mathfrak{G}$ and that $\varepsilon > 0$. According to Proposition II, subsection 3, there exists a function of the form

$$\sum_{p,q}\beta_p\bar{\beta}_q\psi(g_q^{-1}gg_p)$$

such that

$$|\sum_{p,q}\beta_p\bar{\beta}_q\psi(g_q^{-1}gg_p)-\theta(g)| < \varepsilon$$

for all $g \in \mathfrak{G}$. On the other hand, $\psi \in \tilde{\mathfrak{A}}_\varphi$; consequently, there exists a function $\varphi' \in \mathfrak{A}_\varphi$ such that

$$|\psi(g)-\varphi'(g)| < \frac{\varepsilon}{\sum_{p,q}|\beta_p\bar{\beta}_q|}$$

for all $g \in \bigcup_{p,q} g_q^{-1}Qg_p$, because the latter set is compact in $\mathfrak{G}$. It follows that

$$|\sum_{p,q}\beta_p\bar{\beta}_q\psi(g_q^{-1}gg_p)-\sum_{p,q}\beta_p\bar{\beta}_q\varphi'(g_q^{-1}gg_p)| < \varepsilon$$

for $g \in Q$; consequently,

$$|\theta(g)-\sum_{p,q}\beta_p\bar{\beta}_q\varphi'(g_q^{-1}gg_p)| < 2\varepsilon. \tag{1}$$

It remains to show that the function

$$\varphi''(g) = \sum_{p,q}\beta_p\bar{\beta}_q\varphi'(g^{-1}gg_p)$$

can be chosen to be an element of $\mathfrak{A}_\varphi$, i.e. so that $\varphi''(e) \leqq 1$. To this end, it suffices to take a set Q containing e. Inequality (1) will then be valid for $g = e$ also, i.e.

$$|\theta(e)-\varphi''(e)| < 2\varepsilon.$$

Since $\theta \prec \psi$, $\theta(e) \leqq 1$; consequently, $\varphi''(e) < 1+2\varepsilon$. Therefore it is sufficient to replace the function $\varphi''(g)$ by the function $\varphi''(g)/(1+2\varepsilon)$.

Thus, condition 2° is also satisfied. Finally, that $\tilde{\mathfrak{A}}_\varphi$ is weakly closed was proved in Lemma 1.

The *spectrum* of the function $\varphi \in P$ is the set of all normalized elementary positive definite functions in $\tilde{\mathfrak{A}}_{\varphi/\varphi(e)}$.

THEOREM 4 (GODEMENT [3]). *Every function $\varphi \in P$ can be uniformly approximated on every compact set by means of trigonometric polynomials with positive coefficients in elements of the spectrum of the function φ.*

Proof. Without loss of generality, we can assume that $\varphi(e) = 1$. Applying Theorem 3, subsection 3, to $\tilde{\mathfrak{A}}_\varphi$ we obtain that φ is a weak contact point of the function $\psi(g) = \Sigma_p \lambda_p \varphi_p(g)$, where the functions φ_p belong to the spectrum of the function φ, and $\lambda_p \geqq 0$, $\Sigma_p \lambda_p \leqq 1$. It remains to replace the weak topology by the uniform-on-compacta topology. To this end, we note that for $y \in L^1(\mathfrak{G})$ the function $\varphi_y(g)$ is approximated uniformly on compact sets by the functions $\psi_y(g) \in \tilde{\mathfrak{A}}_\varphi$ (see the proof of Lemma 1). But since $\psi(g) = \Sigma_p \lambda_p \varphi_p(g)$, we have that $\psi_y(g) = \Sigma_p \lambda_p \varphi_{p,y}(g)$, where $\varphi_{p,y} \in \tilde{\mathfrak{A}}_\varphi$. Suppose $g \to U_g$ is a unitary representation of the group $\mathfrak{G}$ which is such that $\varphi_p(g) = (U_g \xi_0, \xi_0)$; since φ_p is an elementary positive definite function, the representation $g \to U_g$ is irreducible. Therefore the function $\varphi_{p,y}(g) = (U_g A_y \xi_0, A_y \xi_0)$ is also elementary positive definite. From this we conclude that $\varphi_{p,y}$ belongs to the spectrum of φ and that $\psi_y(g)$ is a trigonometric polynomial with positive coefficients λ_p. Consequently, it only remains to note that the function φ can be uniformly approximated on every compact set of $\mathfrak{G}$ by the functions $\varphi_y(g)$ (see the end of the proof of Lemma 1).

THEOREM 5 (GODEMENT [3]). *Every continuous function on $\mathfrak{G}$ can be uniformly approximated on every compact set by means of trigonometric polynomials.*

Proof. Suppose $f \in L(\mathfrak{G})$; then $f \cdot f^*$ is a positive definite function. Consequently, $f \cdot f^*$ can be approximated by trigonometric polynomials according to Theorem 4. For $f_1, f_2 \in L(\mathfrak{G})$, the function $f_1 \cdot f_2^*$ is a linear combination of the functions $(f_1 \pm f_2) \cdot (f_1 \pm f_2)^*$, $(f_1 \pm if_2) \cdot (f_1 \pm if_2)^*$; consequently, Theorem 4 can also be generalized to functions $f_1 \cdot f_2^*$, where $f_1, f_2 \in L(\mathfrak{G})$. But the function f can be uniformly approximated on $\mathfrak{G}$ by the functions $f_1 \cdot f$ (see Remark 2 to II, subsection 2, § 28); hence, Theorem 5 is valid for all functions $f \in L(\mathfrak{G})$. From this it follows that this theorem holds for all continuous functions because every continuous function f can be uniformly approximated on compact sets by functions $f \in L(\mathfrak{G})$.

§ 31. Harmonic analysis on commutative locally compact groups

1. Maximal ideals of the group algebra of a commutative group; characters. Suppose now that $\mathfrak{G}$ is a commutative locally compact group. It is easily seen that its group algebra $R(\mathfrak{G})$ is also commutative. We shall find all maximal ideals of this algebra. One such maximal ideal is $M_0 = L^1(\mathfrak{G})$ from which $R(\mathfrak{G})$ is obtained by adjunction of the identity. We shall find all maximal ideals $M \neq M_0$.

If $M \neq M_0$, then there exists a function $b \in L^1(\mathfrak{G})$ such that $b(M) \neq 0$; by normalization, if necessary, we may assume that $b(M) = 1$. We set

$$b_{g_0}(g) = b(g_0^{-1} g)$$

and

$$\chi(g_0) = b_{g_0}(M). \tag{1}$$

Then (see I, subsection 2, § 11)

$$|\chi(g_0)| \leqq \|b_{g_0}\|_1 = \|b\|_1; \quad \chi(e) = 1. \tag{2}$$

Moreover,

$$b_{g_1} \cdot b_{g_2} = b \cdot b_{g_1 g_2}. \tag{3}$$

In fact, since the group $\mathfrak{G}$ is commutative,

$$(b_{g_1} \cdot b_{g_2})(g) = \int b_{g_1}(h)b_{g_2}(h^{-1}g)d\mu(h)$$

$$= \int b(g_1^{-1}h)b(g_2^{-1}h^{-1}g)d\mu(h) = \int b(h)b(g_2^{-1}h^{-1}g_1^{-1}g)d\mu(h)$$

$$= \int b(h)b(g_2^{-1}g_1^{-1}h^{-1}g)d\mu(h) = \int b(h)b_{g_1g_2}(h^{-1}g)d\mu(h) = (b \cdot b_{g_1g_2})(g).$$

It follows from (3) that $b_{g_1}(M)b_{g_2}(M) = b(M)b_{g_1g_2}(M) = b_{g_1g_2}(M)$, i.e. according to (1), that

$$\chi(g_1)\chi(g_2) = \chi(g_1g_2). \tag{4}$$

But then, applying (2) to g_0^n, $n = \pm 1, \pm 2, \pm 3, \cdots$, instead of to g_0, we obtain that $|\chi(g_0)|^n \leqq ||b||_1$ for all $n = \pm 1, \pm 2, \cdots$; now this is possible only when

$$|\chi(g)| = 1. \tag{5}$$

Finally, it follows from the relation

$$|\chi(g)-\chi(g_0)| \leqq ||b_g - b_{g_0}||_1 \to 0 \quad \text{as } g \to g_0$$

that $\chi(g)$ is a uniformly continuous function on $\mathfrak{G}$.

Every continuous function χ on $\mathfrak{G}$ which satisfies conditions (4) and (5) is called a *character* of the group $\mathfrak{G}$. Clearly, a character can be considered to be a unitary, namely a one-dimensional, representation of the group $\mathfrak{G}$.

We have thus constructed the character $\chi = \chi_M$ of the group $\mathfrak{G}$ with respect to the given maximal ideal $M \neq M_0$.

I. 1) *The character χ_M does not depend on the choice of the function $b \in L^1(\mathfrak{G})$ for which $b(M) = 1$;*

2) *if $\{z_U\} = \{z_U(g)\}$ is a set which approximates the identity and $z_{U,g_0} = z_U(g_0^{-1}g)$, then $z_{U,g}(M)$ tends to $\chi_M(g)$ for every fixed $M \neq M_0$.*

[See subsection 2, § 28; $\{z_U\}$ is considered as a set directed downward by the relation $z_U \succ z_V$ for $U \subset V$.]

Proof. Let $a, b \in L^1(\mathfrak{G})$ and $a(M) = b(M) = 1$. Applying the homomorphism $a \to a(M)$ to the easily verified relation $a \cdot b_g - a_g \cdot b$, we obtain $a(M)b_g(M) = a_g(M)b(M)$, i.e. $b_g(M) = a_g(M)$; this proves assertion 1). From it follows, in particular, that $z_{U,g}(M) = \chi(g)z_U(M)$. On the other hand,

$$|z_U(M)-1| = |z_U(M)a(M)-a(M)| \leqq ||z_U \cdot a-a||_1 \to 0.$$

We shall now show how to reconstruct the ideal M, or, what amounts to the same thing, the function $x(M)$, given $\chi_M(g)$. For fixed $M \neq M_0$, $a(M)$ is a bounded linear functional in $L^1(\mathfrak{G})$ and hence

$$a(M) = \int a(g)\alpha_M(g)d\mu(g), \tag{6}$$

where $\alpha_M(g) \in L^\infty(\mathfrak{G})$ (see IX, subsection 5, § 27). Hence

$$\begin{aligned} a(M) &= \lim_U \int (a \cdot z_U)(g)\alpha_M(g)d\mu(g) \\ &= \lim_U \int\int a(g')z_U(g'^{-1}g)\alpha_M(g)d\mu(g) \\ &= \lim_U \int a(g')\chi_M(g')d\mu(g') \cdot \int z_U(g)\alpha_M(g)d\mu(g) \\ &= \lim_U z_U(M) \int a(g')\chi_M(g')d\mu(g') \end{aligned}$$

and, in virtue of assertion 2) of Proposition I,

$$a(M) = \int a(g')\chi(g')d\mu(g')$$

for $\chi = \chi_M$.

Hence, for $x = \lambda e + a$,

$$x(M) = (\lambda e + a)(M) = \lambda + \int a(g)\chi(g)d\mu(g). \tag{7}$$

Conversely, a direct verification shows that for an arbitrary character $\chi(g)$ of the group $\mathfrak{G}$, formula (7) defines a homomorphism $x \to x(M)$ of the algebra $R(\mathfrak{G})$ into the field of complex numbers, i.e. a maximal ideal of this algebra.

We have thus proved the following theorem.

THEOREM 1. *There exists a one-to-one correspondence between the set of all maximal ideals $M \neq L^1(\mathfrak{G})$ of the group algebra $R(\mathfrak{G})$ of the commutative group $\mathfrak{G}$ and the set of all characters χ of this group. This correspondence is defined by the formula*

$$x(M) = (\lambda e + a)(M) = \lambda + \int a(g)\chi(g)d\mu(g).$$

From formula (1) it follows, in virtue of Proposition I, that

$$\chi(g) = \frac{a_g(M)}{a(M)}$$

for an arbitrary function $a \in L^1(\mathfrak{G})$ which satisfies the condition $a(M) \neq 0$.

COROLLARY 1. *The group algebra of a commutative locally compact group is completely symmetric.*

Proof. Since $\chi(g^{-1}) = \overline{\chi(g)}$, it follows from (7) that $x^*(M) = \overline{x(M)}$ for an arbitrary maximal ideal M.

We saw above (see Example c), subsection 4, § 29) that this assertion is in general not true for a non-commutative group algebra.

COROLLARY 2. *Every elementary normalized positive definite function on a commutative group is a character of this group.*

Proof. To the elementary normalized positive definite function φ there corresponds the elementary positive functional

$$f(\lambda e+a) = \lambda+\int a(g)\varphi(g)d\mu(g) \tag{8}$$

in $R(\mathfrak{G})$. Since $R(\mathfrak{G})$ is commutative, this functional is determined by some maximal ideal M in $R(\mathfrak{G})$ by the formula $f(\lambda e+a) = \lambda+a(M)$. (See the Remark to Theorem 3, subsection 2, § 20.) It follows, in virtue of (7), that $f(\lambda e+a) = \lambda+\int a(g)\chi(g)d\mu(g)$, and comparison with (8) yields that $\varphi(g) = \chi(g)$.

COROLLARY 3. *For any two elements $g_1 \neq g_2$ of the group $\mathfrak{G}$ there exists a character χ_0 such that $\chi_0(g_1) \neq \chi_0(g_2)$.*

Proof. The assertion is a particular case of the general completeness theorem (Theorem 2, subsection 3, § 29). It can also be proved directly as follows. We set $g_0 = g_2^{-1}g_1$; then $g_0 \neq e$; by the Urysohn lemma, there exists a function $x \in L(\mathfrak{G})$ such that $x(g_0) = 1$, $x(e) = 0$ and, hence, setting $x_{g_0}(g) = x(g_0^{-1}g)$, we have that $x_{g_0} \neq x$. Since $R(\mathfrak{G})$ is a semisimple algebra (see VII, subsection 2, § 28), there exists a maximal ideal M_0 in the algebra $R(\mathfrak{G})$ such that $x_{g_0}(M_0) \neq x(M_0)$. If we set $x^\wedge(\chi) = \int x(g)\chi(g)d\mu(g)$, then this means that $\hat{x}_{g_0}(\chi_0) \neq x^\wedge(\chi_0)$, i.e. $\chi_0(g_0)x^\wedge(\chi_0) \neq x^\wedge(\chi_0)$, where χ_0 is the character of the group $\mathfrak{G}$ that corresponds to the ideal M_0. It follows that $\chi_0(g_0) \neq 1$, i.e. that $\chi_0(g_2^{-1}g_1) \neq 1$, and hence $\chi_0(g_1) \neq \chi_0(g_2)$.

COROLLARY 4. *Every continuous function f on the commutative locally compact group $\mathfrak{G}$ can be uniformly approximated on every compact set by linear combinations of the characters of this group.*

The assertion follows directly from Corollary 2 and Theorem 5, subsection 5, § 30. It also follows from Stone's theorem (see subsection 10, § 2) because, in virtue of Corollary 3, the linear combinations of characters form a symmetric algebra of the functions which separate points on $\mathfrak{G}$.

EXAMPLES. a) Suppose $\mathfrak{G}$ is the additive group of all real numbers t, $-\infty < t < \infty$. Then the characters are the continuous functions $\chi(g) = \chi(t)$ which satisfy the conditions

$$\chi(t_1+t_2) = \chi(t_1)\chi(t_2), \quad |\chi(t)| = 1.$$

But, as is easily seen, a continuous function $\chi(t)$ satisfying these conditions must have the form

$$\chi(t) = e^{it\alpha},$$

where α may be an arbitrary fixed real number. In this case, the algebra $L^1(\mathfrak{G})$ reduces to the algebra $L^1(-\infty, \infty)$ of all summable measurable functions $a(t)$, with

$$||a||_1 = \int_{-\infty}^{\infty} |a(t)|dt, \; (a \cdot b)(t) = \int_{-\infty}^{\infty} a(t_1)b(t-t_1)dt_1$$

and

$$a^*(t) = \overline{a(-t)}.$$

Theorem 1 asserts that in this case every maximal ideal M of the algebra $R(\mathfrak{G})$, which is distinct from $M_0 = L^1(-\infty, \infty)$, is defined by the formula

$$x(M) = (\lambda e + a)(M) = \lambda + \int_{-\infty}^{\infty} a(t)e^{it\alpha}dt.$$

In particular,

$$a(M) = \int_{-\infty}^{\infty} a(t)e^{it\alpha}dt,$$

so that the transition from $a(t)$ to $a(M)$ is, in this case, the Fourier transform. Therefore, in general, the transition from $a(g)$ to $a(M)$ can be considered as a generalization of the Fourier transform to the case of an arbitrary commutative locally compact group.

Applying Corollary 4 to the additive group of all real numbers, we arrive at the following theorem.

Every continuous function $f(t)$, $-\infty < t < \infty$, can be uniformly approximated on every finite interval by linear combinations of the functions $e^{i\alpha t}$.

b) Suppose $\mathfrak{G}$ is the additive (discrete) group of all integers $n = 0, \pm 1, \pm 2, \cdots$; the characters are the functions $\chi(g) = \chi(n)$ which satisfy the conditions

$$\chi(n_1+n_2) = \chi(n_1)\chi(n_2), \quad |\chi(n)| = 1.$$

Setting $\chi(1) = e^{i\alpha}$, it is easily seen that then $\chi(n) = e^{in\alpha}$, where α is determined to within a term which is a multiple of 2π.

The group algebra $R(\mathfrak{G})$ in this case is the set of all sequences $\langle a_n \rangle$ such that

$$\sum_{n=-\infty}^{\infty} |a_n| < \infty$$

where

$$(a \cdot b)_n = \sum_{p=-\infty}^{\infty} a_p b_{n-p}, \; (a^*)_n = \bar{a}_{-n}, \; ||a||_1 = \sum_{n=-\infty}^{\infty} |a_n|.$$

Theorem 1 asserts that all maximal ideals of this algebra are defined by the formula

$$a(M) = \sum_{n=-\infty}^{\infty} a_n e^{in\alpha}$$

(which, after all, can also be proved directly).

c) Suppose $\mathfrak{G}$ is the group of rotations of a circle: then all possible characters on $\mathfrak{G}$ are given by the formula

$$\chi(t) = e^{int}, \quad n = 0, \pm 1, \pm 2, \cdots; \; 0 \leqq t \leqq 2\pi.$$

Applying Corollary 4 to the group of rotations of a circle, we obtain that:

Every continuous function $f(t)$, $0 \leqq t \leqq 2\pi$, satisfying the condition $f(0) = f(2\pi)$, can be uniformly approximated in $[0, 2\pi]$ by linear combinations of the functions e^{int}, $n = 0, \pm 1, \pm 2, \cdots$.

2. Group of characters. If χ_1, χ_2 are characters of the group $\mathfrak{G}$, then the formula $\chi(g) = \chi_1(g)\chi_2(g)$ determines a character χ of the group $\mathfrak{G}$. This character is called the *product* of the characters χ_1, χ_2 and is denoted by $\chi_1\chi_2$. Hence, the set $\overline{\mathfrak{G}}$ of all characters of the group $\mathfrak{G}$ constitutes a commutative group; this group is called the *character group* of the group $\mathfrak{G}$.

Suppose $\mathfrak{M}$ is the space of maximal ideals of the algebra $R(\mathfrak{G})$. In subsection 1, we showed that there exists a one-to-one correspondence between the characters $\chi \in \overline{\mathfrak{G}}$ and the maximal ideals $M \neq M_0$ of the algebra $R(\mathfrak{G})$; this enables us to carry over to $\overline{\mathfrak{G}}$ the topology of the locally compact space $\mathfrak{M} - M_0$, for which the neighborhoods in $\overline{\mathfrak{G}}$ are to be considered to be the images of the neighborhoods in $\mathfrak{M} - M_0$ under the mapping $M \rightarrow \chi$. Then $\overline{\mathfrak{G}}$ becomes a locally compact space. We note that by the very definition of a neighborhood in $\mathfrak{M}$, a neighborhood of the character $\chi_0 \in \overline{\mathfrak{G}}$ is the set of all characters χ which satisfy the inequalities

$$|x_k(M) - x_k(M_0)| = \left| \int a_k(g)[\chi(g) - \chi_0(g)]d\mu(g) \right| < \varepsilon \tag{1}$$

($x_k = \lambda_k e + a_k$) for fixed $a_1, \cdots, a_n \in L^1(\mathfrak{G})$, and these neighborhoods form a neighborhood basis in $\overline{\mathfrak{G}}$.

I. *The character $\chi(g)$ is a continuous function in both the variables $g \in \mathfrak{G}$ and $\chi \in \overline{\mathfrak{G}}$ simultaneously.*

Proof. Suppose χ_1 is an arbitrary character of the group $\mathfrak{G}$ and that M_1 is the maximal ideal corresponding to it. Since $M_1 \neq M_0$, there exists an $a(g) \in L^1(\mathfrak{G})$ such that $a(M_1) \neq 0$. The inequality

$$|a_{g_1}(M) - a_{g_2}(M_1)| \leqq |a_{g_1}(M) - a_{g_2}(M)| + |a_{g_2}(M) - a_{g_2}(M_1)|$$
$$\leqq \|a_{g_1} - a_{g_2}\|_1 + |a_{g_2}(M) - a_{g_2}(M_1)|$$

shows that $a_g(M)$ is a continuous function in both the variables $g \in \mathfrak{G}$ and $M \in \mathfrak{M} - M_0$ simultaneously. Consequently, $\chi(g) = a_g(M)/a(M)$ is also a continuous function in both the variables g and M simultaneously, or, what amounts to the same thing, with respect to both variables g, χ at any point (g_1, χ_1).

II. *The topology in $\overline{\mathfrak{G}}$ coincides with the uniform-on-compacta topology, i.e. the topology of uniform convergence on compact sets.*

Proof. The theorem means that the sets $U(\chi_0; Q; \varepsilon)$ of characters χ which satisfy the condition $|\chi(g) - \chi_0(g)| < \varepsilon$ for $g \in Q$, where Q is compact, 1) are open and 2) form a neighborhood basis in $\overline{\mathfrak{G}}$.

Assertion 1) follows directly from V, subsection 12, § 2. To prove assertion 2), we consider the neighborhood $U(\chi_0; a_1, \cdots, a_n; \varepsilon)$ determined by inequalities (1). We choose a compact set Q so that

$$\int_{\mathfrak{G}-Q} |a_k(g)| d\mu(g) < \frac{\varepsilon}{4}, \quad k = 1, 2, \cdots, n,$$

and we then take

$$\delta < \frac{\varepsilon}{2\|a_k\|_1}, \quad k = 1, 2, \cdots, n.$$

Then

$$\left| \int a_k(g)[\chi(g) - \chi_0(g)] d\mu(g) \right| \leqq \int_Q |a_k(g)| |\chi(g) - \chi_0(g)| d\mu(g)$$
$$+ \int_{\mathfrak{G}-Q} |a_k(g)| \, |\chi(g) - \chi_0(g)| d\mu(g) < \delta \|a_k\|_1 + 2\frac{\varepsilon}{4} < \varepsilon$$

for $\chi \in U(\chi_0; Q; \delta)$; consequently, $U(\chi_0; Q; \delta) \subset U(\chi_0; a_1, \cdots, a_n; \varepsilon)$ and assertion 2) is also proved.

III. *With the topology in $\overline{\mathfrak{G}}$ defined above, $\overline{\mathfrak{G}}$ is a topological group.*

Proof. If $\chi \in U(\chi_0; Q; \varepsilon/2)$, $\chi' \in U(\chi_0'; Q; \varepsilon/2)$, then, for $g \in Q$,

$$|\chi(g)\chi'(g) - \chi_0(g)\chi_0'(g)| \leqq |\chi(g) - \chi_0(g)| \, |\chi'(g)|$$
$$+ |\chi'(g) - \chi_0'(g)| \, |\chi_0(g)| = |\chi(g) - \chi_0(g)| + |\chi'(g) - \chi_0'(g)| < \varepsilon,$$

i.e. $\chi\chi' \in U(\chi_0\chi_0'; Q; \varepsilon)$. This means that the product $\chi\chi'$ is continuous. The continuity of transition to the inverse character $\chi^{-1}(g) = \overline{\chi(g)}$ is obvious.

In the sequel, we shall understand the character group $\overline{\mathfrak{G}}$ to be the topological group defined in this way.

3. Positive definite functions on a commutative group. Suppose φ is a continuous positive definite function on the commutative locally compact group $\mathfrak{G}$. According to Theorem 2, subsection 2, § 30, to this function there corresponds a regular positive functional

$$f(\lambda e + a) = \lambda\varphi(e) + \int a(g)\varphi(g) d\mu(g) \tag{1}$$

in the algebra $R(\mathfrak{G})$. On the other hand, this positive functional is uniquely representable in the form

$$f(x) = \int x(M) d\sigma(M), \tag{2}$$

where σ is the measure on $\mathfrak{M}$ defined by the functional f (see Corollary 1, subsection 4, § 20). Moreover, $\sigma(M_0) = 0$ (where, as before, M_0 denotes the ideal $L^1(\mathfrak{G})$ in $R(\mathfrak{G})$). In fact, if $\sigma(M_0) > 0$ then, writing formula (2) in the form

$$f(x) = \int_{\mathfrak{M}-M_0} x(M)d\sigma(M) + x(M_0)\sigma(M_0),$$

we obtain that the degenerate functional $f_0(x) = x(M_0) = \lambda$ for $x = \lambda e + a$, $a \in L^1(\mathfrak{G})$, is dominated by the functional f; this, however, contradicts the regularity of the functional f. Thus, formula (2) can be written in the form

$$f(x) = \int_{\mathfrak{M}'} x(M)d\sigma(M), \tag{2'}$$

where $\mathfrak{M}' = \mathfrak{M} - M_0$. In virtue of the correspondence $M \leftrightarrow \chi$, established in subsection 1, σ can also be considered as a measure on $\overline{\mathfrak{G}}$. But, for $x = \lambda e + a$,

$$x(M) = \lambda + \int a(g)\chi(g)d\mu(g).$$

Substituting this expression into formula (2′) and comparing with (1), we obtain that

$$\lambda\varphi(e) + \int a(g)\varphi(g)d\mu(g) = \lambda\sigma(\overline{\mathfrak{G}}) + \int \left[\int a(g)\chi(g)d\mu(g)\right] d\sigma(\chi).$$

It follows that

$$\sigma(\overline{\mathfrak{G}}) = \varphi(e)$$

and

$$\int a(g)\varphi(g)d\mu(g) = \int \left[\int a(g)\chi(g)d\mu(g)\right] d\sigma(\chi). \tag{3}$$

Interchanging the order of integration in the right member of equality (3) and taking into consideration the fact that the function $a(g) \in L^1(\mathfrak{G})$ is arbitrary and the continuity of the functions $\chi(g)$ and $\int \chi(g)d\sigma(\chi)$, we conclude that

$$\varphi(g) = \int \chi(g)d\sigma(\chi).$$

It is easily seen that, conversely, every function φ, defined by this formula, is positive definite. This proves the following theorem.

THEOREM 2 (RAIKOV [3]). *Every continuous positive definite function φ on the commutative locally compact group $\mathfrak{G}$ is uniquely representable in the form*

$$\varphi(g) = \int \chi(g)d\sigma(\chi), \tag{4}$$

where σ is a measure on $\overline{\mathfrak{G}}$ which satisfies the condition $\sigma(\overline{\mathfrak{G}}) = \varphi(e)$. Conversely, *for an arbitrary measure σ on $\overline{\mathfrak{G}}$, satisfying the condition $\sigma(\overline{\mathfrak{G}}) < \infty$, formula* (4) *defines a continuous positive definite function $\varphi(g)$ on the group $\mathfrak{G}$.*

Examples. a) Suppose $\mathfrak{G}$ is the additive group of the real numbers; in this case, the continuous positive definite functions on $\mathfrak{G}$ will be the continuous functions $\varphi(t)$, $-\infty < t < \infty$, satisfying the condition

$$\sum_{p,q=1}^{n} \varphi(t_p - t_q)\lambda_p \bar{\lambda}_q \geqq 0 \tag{5}$$

for arbitrary real $t_1, \cdots, t_n$ and any complex $\lambda_1, \cdots, \lambda_n$ $(n = 1, 2, 3, \cdots)$. But in the present case, $\overline{\mathfrak{G}}$ can be considered as the additive group of the real numbers with the correspondence between χ and α being established by the formula $\chi(t) = e^{it\alpha}$. Applying Theorem 2, we arrive at the following theorem due to Bochner [1].

Every continuous function $\varphi(t)$, satisfying condition (5), *can be represented in the form*

$$\varphi(t) = \int_{-\infty}^{\infty} e^{it\alpha} d\sigma(\alpha),$$

where σ is a measure on the real line which satisfies the condition that $\sigma(-\infty, \infty) = \varphi(0)$.

b) Suppose $\mathfrak{G}$ is the additive group of all integers; the positive definite functions on $\mathfrak{G}$ will be the sequences $\langle\varphi(n)\rangle = \langle\varphi_n\rangle$, satisfying the condition

$$\sum_{p,q=-n}^{n} \varphi_{p-q}\lambda_p \bar{\lambda}_q \geqq 0 \tag{6}$$

for all complex $\lambda_1, \cdots, \lambda_n$ $(n = 1, 2, 3, \cdots)$. But in this case, $\overline{\mathfrak{G}}$ can be identified with the points on the unit circle or with the points α in the interval $[0, 2\pi)$, where the correspondence between χ and α is given by the formula

$$\chi(n) = e^{in\alpha}.$$

Applying Theorem 2, we obtain a representation of the form

$$\varphi_n = \int_0^{2\pi} e^{in\alpha} d\sigma(\alpha) \tag{7}$$

for φ_n.

The problem of determining the measure σ, with respect to a prescribed sequence $\langle\varphi_n\rangle$, for which (7) is satisfied, is called the *trigonometric moment problem*. Thus, the application of Theorem 2 leads to the following theorem due to Herglotz [1].

The trigonometric moment problem for a given sequence $\langle\varphi_n\rangle$ is solvable if, and only if, this sequence satisfies condition (6).

4. Inversion formula and Plancherel's theorem for commutative groups. We denote the set of all continuous positive definite functions on a given commutative locally compact group $\mathfrak{G}$ by $P = P(\mathfrak{G})$, and by $[L^1 \cap P]$ the linear hull of the functions in $L^1 \cap P$, where $L^1 = L^1(\mathfrak{G})$.

Lemma. *$[L^1 \cap P]$ is dense in $L^1 = L^1(\mathfrak{G})$ and in $L^2 = L^2(\mathfrak{G})$.*

Proof. We set $L = L(\mathfrak{G})$; the functions of the form $y^* \cdot x$, $x, y \in L$, form a set which is dense in L in the sense of the norm in L^1 (see II, subsection 2, § 28). But then they also constitute a dense set in L^1 because L is dense in L^1. On the other hand, $y^* \cdot x$ is a linear combination of the four positive definite functions $(x \pm y)^* \cdot (x \pm y)$, $(x \pm iy)^* \cdot (x \pm iy)$ and hence belongs to $[L^1 \cap P]$; consequently, $[L^1 \cap P]$ is dense in L^1. It is proved analogously that $[L^1 \cap P]$ is dense in L^2.

We now set

$$x^\wedge(\chi) = \int x(g)\chi(g)d\mu(g) \tag{1}$$

for an arbitrary function $x \in L^1$ and any character $\chi \in \overline{\mathfrak{G}}$; this integral exists and is a continuous function $x^\wedge$ on the group $\overline{\mathfrak{G}}$. The function $x^\wedge$ is called the *Fourier transform* of the function $x \in L^1$. In virtue of the results of subsection 1, $x^\wedge(\chi)$ is simply $x(M)$, where M is the maximal ideal of the algebra $R(\mathfrak{G})$ which corresponds to the character χ. From this it follows that

$$(x_1+x_2)^\wedge(\chi) = x_1^\wedge(\chi)+x_2^\wedge(\chi), \quad (\lambda x)^\wedge(\chi) = \lambda x^\wedge(\chi),$$
$$(x^{*\wedge})(\chi) = \overline{x^\wedge(\chi)}, \quad (x_1 \cdot x_2)^\wedge(\chi) = x_1^\wedge(\chi)x_2^\wedge(\chi),$$

and, in particular, that

$$(x^*x)^\wedge(\chi) = |x^\wedge(\chi)|^2.$$

The validity of these formulas can also be verified directly.

If $\mathfrak{G}$ is the additive group of the real numbers t, then $\chi(g) = \chi(t) = e^{it\alpha}$, and formula (1) goes over into the usual Fourier transform

$$x^\wedge(\alpha) = \int_{-\infty}^{\infty} x(t)e^{it\alpha}dt.$$

We denote an invariant measure on the group $\overline{\mathfrak{G}}$ by $\mu(\chi)$, and by $L^1(\overline{\mathfrak{G}})$ the space of functions summable on $\overline{\mathfrak{G}}$ with respect to the measure $\mu(\chi)$. The following theorem (which generalizes the classical theorem on the Fourier integral) shows how one reconstructs the function x knowing its Fourier transform $x^\wedge$.

THEOREM 3. *If $x \in [L^1 \cap P]$, then $x^\wedge \in L^1(\overline{\mathfrak{G}})$ and*

$$x(g) = \int x^\wedge(\chi)\overline{\chi(g)}d\mu(\chi),$$

where $\mu(\chi)$ is an invariant measure on the group $\overline{\mathfrak{G}}$ which has been suitably normalized.

Proof. We denote the set of all functions $x \in L^1$ which are uniformly continuous on the group $\mathfrak{G}$ by I. It is easily seen that I is an ideal in $L^1(\mathfrak{G})$. In I, we define the linear functional

$$f(x) = x(e) \quad \text{for } x \in I.$$

This functional is positive on I because for $x = y^* \cdot y$, $y \in I$,

$$x(g) = \int \overline{y(g_1)} y(g_1 g) d\mu(g_1),$$

and therefore

$$x(e) = \int |y(g_1)|^2 d\mu(g_1) \geqq 0.$$

Theorem 5, subsection 4, § 20 can be applied to this functional f because $L^1(\mathfrak{G})$ is a completely symmetric commutative algebra (see Corollary 1, subsection 1), and I is dense in $L^1(\mathfrak{G})$. On the basis of this theorem, there exists a measure μ, defined on $\mathfrak{M}' = \mathfrak{M} - M_0$, and hence also on $\overline{\mathfrak{G}}$, such that $p^{\wedge} \in L^1(\overline{\mathfrak{G}})$ and

$$f(px) = \int p^{\wedge}(\chi) x^{\wedge}(\chi) d\mu(\chi) \tag{2}$$

for an arbitrary function $p \in I$, which is positive with respect to f. But any function in $L^1 \cap P$ will be such a function because, for $p \in L^1 \cap P$,

$$f(px^*x) = \int\int p(g_1') \overline{x(g_1)} x(g_1 g_1') d\mu(g_1) d\mu(g_1')$$

$$= \int\int p(g_1^{-1} g_2) \overline{x(g_1)} x(g_2) d\mu(g_1) d\mu(g_2) \geqq 0$$

(see IV, subsection 2, § 30). Moreover, condition (3) in subsection 4, § 20 is also satisfied because $L^1(\mathfrak{G})$ contains a set which approximates the identity (see II, subsection 2, § 28) and $f(px) = \int p(g^{-1}) x(g) d\mu(g)$ is a continuous functional of x in $L^1(\mathfrak{G})$ (see V, subsection 2, § 10). Therefore, (2) holds for an arbitrary function $p \in L^1 \cap P$, and hence also for any function $p \in [L^1 \cap P]$. Replacing x by x^* in (2), substituting in (2) for $f(px^*)$ and $x^{*\wedge}(\chi) = \overline{x^{\wedge}(\chi)}$ their expressions from (1) and (2), and interchanging the order of integration in the right member of (2), we obtain that

$$\int p(g) \overline{x(g)} d\mu(g) = \int \overline{x(g)} \left[\int p^{\wedge}(\chi) \overline{\chi(g)} d\mu(\chi) \right] d\mu(g).$$

It follows, in virtue of the fact that the function $x \in L^1$ is arbitrary, that

$$p(g) = \int p^{\wedge}(\chi) \overline{\chi(g)} d\mu(\chi);$$

in particular,

$$p(e) = \int p^{\wedge}(\chi) d\mu(\chi).$$

It remains to prove that $d\mu(\chi)$ is an invariant measure on $\overline{\mathfrak{G}}$. To this end, we note that if $p \in L^1 \cap P$, then for an arbitrary $\chi_0 \in \overline{\mathfrak{G}}$ we also have that $p\chi_0 \in L^1 \cap P$ and the Fourier transform of this function is

$$\int p(g)\chi_0(g)\chi(g)d\mu(g) = \int p(g)(\chi\chi_0)(g)d\mu(g) = p\hat{}(\chi\chi_0).$$

It follows that

$$\int p\hat{}(\chi)d\mu(\chi) = p(e) = p(e)\chi_0(e) = \int p\hat{}(\chi\chi_0)d\mu(\chi). \tag{3}$$

Since the functions $p \in [L^1 \cap P]$ form a dense set in L^1, their Fourier transfo\u0131ms $p\hat{}$ form a dense set in $C_0(\overline{\mathfrak{G}})$ (see Corollary 3, subsection 3, § 14); consequently, it follows from (3) that $d\mu(\chi)$ is an invariant measure in $\overline{\mathfrak{G}}$, and the theorem is proved.

COROLLARY 1. *If $p \in L^1 \cap P$, then $p\hat{}(\chi) \geqq 0$.*

Proof. If $p \in L^1 \cap P$, then, according to Theorem 3, $p\hat{} \in L^1(\overline{\mathfrak{G}})$ and $p(g) = \int \overline{\chi(g)}p\hat{}(\chi)d\mu(\chi)$; therefore for an arbitrary function $x \in L^1(\mathfrak{G})$,

$$\int x(g)p(g)d\mu(g) = \int \left(\int x(g)\overline{\chi(g)}d\mu(g)\right) p\hat{}(\chi)d\mu(\chi)$$

$$= \int x\hat{}(\bar{\chi})p\hat{}(\chi)d\mu(\chi) = \int x\hat{}(\chi)p\hat{}(\bar{\chi})d\mu(\chi).$$

Replacing x by x^*x, we conclude that

$$\int |x\hat{}(\chi)|^2 p\hat{}(\bar{\chi})d\mu(\chi) \geqq 0 \tag{4}$$

because $f(x) = \int x(g)p(g)d\mu(g)$ is a positive functional in $L^1(\mathfrak{G})$ (see Theorem 2, subsection 2, § 30). But the functions $x\hat{}$, $x \in L^1(\mathfrak{G})$ form a dense set in $C_0(\overline{\mathfrak{G}})$; therefore it follows from (4) that $\int \varphi(\chi)p\hat{}(\bar{\chi})d\mu(\chi) \geqq 0$ for an arbitrary non-negative function $\varphi \in C_0(\overline{\mathfrak{G}})$. But this is possible only when $p\hat{}(\chi) \geqq 0$.

COROLLARY 2. *If $\varphi \in L^1(\overline{\mathfrak{G}})$, $\varphi(\chi) \geqq 0$ and the function $x(g) = \int \overline{\chi(g)}\varphi(\chi)d\mu(\chi)$ belongs to $L^1(\mathfrak{G})$, then*

$$\varphi(\chi) = \int x(g)\chi(g)d\mu(g) = x\hat{}(\chi)$$

almost everywhere on $\overline{\mathfrak{G}}$.

Proof. In virtue of Theorem 2, subsection 3, $x \in P$, and hence $x \in L^1 \cap P$. On the basis of Theorem 3, we conclude that

$$x(g) = \int \overline{\chi(g)}x\hat{}(\chi)d\mu(\chi),$$

so that

$$\int \overline{\chi(g)}\varphi(\chi)d\mu(\chi) = \int \overline{\chi(g)}x^{\wedge}(\chi)d\mu(\chi).$$

It follows that for an arbitrary function $y \in L^1(\mathfrak{G})$ we have

$$\int y^{\wedge}(\bar{\chi})\varphi(\chi)d\mu(\chi) = \int \left(\int y(g)\overline{\chi(g)}d\mu(g)\right)\varphi(\chi)d\mu(\chi)$$
$$= \int y(g)\left(\int \overline{\chi(g)}\varphi(\chi)d\mu(\chi)\right)d\mu(g) = \int y(g)\left(\int \overline{\chi(g)}x^{\wedge}(\chi)d\mu(\chi)\right)d\mu(g)$$
$$= \int y^{\wedge}(\bar{\chi})x^{\wedge}(\chi)d\mu(\chi).$$

From this equality we conclude that $\varphi(\chi) = x^{\wedge}(\chi)$ almost everywhere on $\overline{\mathfrak{G}}$, because the functions $y^{\wedge}(\bar{\chi})$ form a dense set in $C_0(\overline{\mathfrak{G}})$.

THEOREM 4. *The Fourier transform $x \to x^{\wedge}$ is an isometric mapping of a set dense in $L^2(\mathfrak{G})$ onto a set dense in $L^2(\overline{\mathfrak{G}})$, and hence it is uniquely extendible to an isometric mapping of $L^2(\mathfrak{G})$ onto $L^2(\overline{\mathfrak{G}})$.*

Proof. Suppose I, f, P are the same as above. Setting $x = q^*$, where $q \in [L^1 \cap P]$, in (2), we have that

$$\int p(g)\overline{q(g)}d\mu(g) = f(pq^*) = \int p^{\wedge}(\chi)\overline{q^{\wedge}(\chi)}d\mu(\chi);$$

consequently, the Fourier transform $p \to p^{\wedge}$ is an isometric mapping of the subset $[L^1 \cap P]$ of the space $L^2(\mathfrak{G})$ into the space $L^2(\overline{\mathfrak{G}})$. Since $[L^1 \cap P]$ is dense in $L^2(\mathfrak{G})$, this mapping is uniquely extendible to an isometric operator T which maps the space $L^2(\mathfrak{G})$ into the space $L^2(\overline{\mathfrak{G}})$.

It remains to show that T maps $L^2(\mathfrak{G})$ *onto* $L^2(\overline{\mathfrak{G}})$, for which it suffices to establish that the image of the space $L^2(\mathfrak{G})$ under the mapping T is dense in $L^2(\overline{\mathfrak{G}})$. To this end, we shall find the operator T^*. Suppose

$$\varphi \in L^1(\overline{\mathfrak{G}}) \cap L^2(\overline{\mathfrak{G}}) \text{ and } x \in [L^1 \cap P];$$

setting $\varphi' = T^*\varphi$, we have that $(Tx, \varphi) = (x, \varphi')$, i.e.

$$\iint x(g)\chi(g)\overline{\varphi(\chi)}d\mu(g)d\mu(\chi) = \int x(g)\overline{\varphi'(g)}d\mu(g).$$

Since $[L^1 \cap P]$ is dense in $L^2(\mathfrak{G})$, it follows that

$$\varphi'(g) = (T^*\varphi)(g) = \int \varphi(\chi)\overline{\chi(g)}d\mu(\chi). \tag{5}$$

We shall now assume that $\varphi, \psi \in L^1(\overline{\mathfrak{G}}) \cap L^2(\overline{\mathfrak{G}})$ and that $\varphi \geqq 0, \psi \geqq 0$. Then we also have that $\varphi \cdot \psi \in L^1(\overline{\mathfrak{G}}) \cap L^2(\overline{\mathfrak{G}})$ (see VI, subsection 2, § 28) and it easily follows from formula (5) that

$$T^*(\varphi \cdot \psi) = \varphi'\psi' \in L^1(\mathfrak{G}) \cap L^2(\mathfrak{G}),$$

because $\varphi' = T^*\varphi \in L^2(\mathfrak{G})$, $\psi' = T^*\psi \in L^2(\mathfrak{G})$. (Here and below, in this section, $\varphi \cdot \psi$ denotes the *convolution* and $\varphi'\psi'$ the *product* of the functions φ, ψ and φ', ψ' respectively.) Moreover, $\varphi \cdot \psi \in L^1(\overline{\mathfrak{G}})$ and $\varphi \cdot \psi \geqq 0$; on the basis of Corollary 2, we conclude from this that $\varphi \cdot \psi = T(\varphi'\psi') \in \overline{TL^2(\mathfrak{G})}$.

Suppose now that φ, ψ are arbitrary functions in $L^1(\overline{\mathfrak{G}}) \cap L^2(\overline{\mathfrak{G}})$. Each of them can be represented in the form of a linear combination of the four nonnegative functions $\varphi_j, \psi_j \in L^1(\overline{\mathfrak{G}}) \cap L^2(\overline{\mathfrak{G}})$, and hence $\varphi \cdot \psi$ can be represented in the form of a linear combination of the sixteen nonnegative functions $\varphi_j \cdot \psi_k$ which, in virtue of the preceding discussion belong to $\overline{TL^2(\mathfrak{G})}$. Consequently, $\varphi \cdot \psi \in \overline{TL^2(\mathfrak{G})}$ also. But the functions $\varphi \cdot \psi$ form a dense set in $L^2(\overline{\mathfrak{G}})$ (see Remark 1 in subsection 2, § 28); consequently, $TL^2(\mathfrak{G})$ is dense in $L^2(\overline{\mathfrak{G}})$ and the theorem is proved completely.

We note that the equality $\overline{TL^2(\mathfrak{G})} = L^2(\overline{\mathfrak{G}})$ is also a consequence of the continuous analogue of the Schur lemma (see subsection 5, § 26).

Theorem 4 generalizes the Plancherel theorem on the classical Fourier integral and hence is called the Plancherel theorem for a commutative group $\mathfrak{G}$.

COROLLARY 3. *The sets S_1 and S_2 of all functions x in $L^1(\mathfrak{G})$ and $L^2(\mathfrak{G})$, respectively, for which $x^\wedge$ vanishes in the exterior of a compact set (one for each function $x^\wedge$) are dense in $L^1(\mathfrak{G})$ and $L^2(\mathfrak{G})$ respectively.*

Proof. S_2 contains the inverse image under the isometric mapping T of the set $L(\overline{\mathfrak{G}})$ which is dense in $L^2(\overline{\mathfrak{G}})$; consequently, S_2 is dense in $L^2(\mathfrak{G})$.

If $x \in L^1(\mathfrak{G})$, then $x = yz$, where $y, z \in L^2(\mathfrak{G})$. [One can, for instance, set $y = |x|^{\frac{1}{2}}$, $z = 0$ for $x = 0$ and $z = |x|^{-\frac{1}{2}}x$ for $x \neq 0$.] Consequently, there exist $y_1, z_1 \in S_2$ such that $\|y-y_1\|_2 < \varepsilon$, $\|z-z_1\|_2 < \varepsilon$; then $y_1, z_1 \in L^2(\mathfrak{G})$; hence $y_1 z_1 \in L^1(\mathfrak{G})$ and furthermore $(y_1 z_1)^\wedge = y_1^\wedge \cdot z_1^\wedge = 0$ in the exterior of the compact set $Q_1 \cdot Q_2$ if $y_1^\wedge = 0$ in the exterior of the compact set Q_1, and $z_1^\wedge = 0$ in the exterior of the compact set Q_2. Therefore $y_1 z_1 \in S_1$ and on the basis of the Cauchy-Bunyakovsky inequality,

$$\|yz-y_1z_1\|_1 \leqq \|y\|_2\|z-z_1\|_2+\|z_1\|_2 \cdot \|y-y_1\|_2 < \varepsilon(\|y\|_2+\|z\|_2+\varepsilon),$$

so that S_1 is dense in $L^1(\mathfrak{G})$.

We denote the set of all finite linear combinations of functions $x \cdot y$, where $x, y \in L(\mathfrak{G})$ by $\mathfrak{B} = \mathfrak{B}(\mathfrak{G})$. Clearly, $\mathfrak{B}(\mathfrak{G})$ is a subalgebra in $L^1(\mathfrak{G})$. In virtue of II and Remark 2, subsection 2, § 28, $\mathfrak{B}(\mathfrak{G})$ is dense in $L^1(\mathfrak{G})$ and in $L(\mathfrak{G})$ relative to the norms $\|x\|_1$ and $\|x\|_\infty$, respectively.

COROLLARY 4. *The image $T\mathfrak{B}$ of the algebra $\mathfrak{B}$ under the Fourier transform T is dense in $L^1(\overline{\mathfrak{G}})$.*

Proof. According to Theorem 4, the functions $x^\wedge = Tx$, $x \in L(\mathfrak{G})$, form a dense set in $L^2(\overline{\mathfrak{G}})$; consequently, the functions $x^\wedge y^\wedge = T(x \cdot y)$, belonging to $T\mathfrak{B}$, form a dense set in $L^1(\overline{\mathfrak{G}})$ (see the end of the proof of Corollary 3).

5. Separation property of the set $[\mathbf{L}^1 \cap \mathbf{P}]$. We set $\mathfrak{A} = \mathfrak{A}(\mathfrak{G}) = [L_1 \cap P]$; then $\mathfrak{A}$ is a symmetric subalgebra of the algebra L^1. In fact, if $x, y \in \mathfrak{A}$, then $x \cdot y$ is a linear combination of the functions $(x^* \pm y)^* \cdot (x^* \pm y)$, $(x^* \pm iy)^* \cdot (x^* \pm iy)$ belonging to $L^1 \cap P$, and hence $x \cdot y \in \mathfrak{A}$.

LEMMA. *For arbitrary compact set $F \subset \mathfrak{G}$ and open set $U \supset F$ there exists in the algebra $\mathfrak{A}$ a function x which equals unity on F and vanishes in the exterior of U.*

Proof. We choose a symmetric neighborhood V of the identity such that $\mu(V) < \infty$ and (see III, subsection 3, § 27) $V^2F \subset U$. Let y, z be the characteristic functions of the sets V and VF respectively; we set $x = \{1/\mu(V)\}y^* \cdot z$. Then $x \in \mathfrak{A}$ and

$$x(g) = \frac{1}{\mu(V)}\int y(g_1)z(g_1 g)d\mu(g_1) = \frac{1}{\mu(V)}\int_V z(g_1 g)d\mu(g_1)$$

$$= \frac{1}{\mu(V)}\int_{Vg} z(g_1)d\mu(g_1); \quad (1)$$

for $g \in F$, we have $Vg \subset VF$, and hence

$$x(g) = \frac{1}{\mu(V)}\int_{Vg} d\mu(g_1) = \frac{1}{\mu(V)}\mu(Vg) = 1.$$

Further, for $g \notin U$ we also have that $g \notin V^2F$, and therefore Vg does not intersect VF; from this and (1) we conclude that $x(g) = 0$ for $g \notin U$. Consequently, the function x satisfies the requirements set down.

From the lemma just proved, it follows, in particular, that:

For any two distinct points $g_1, g_2 \in \mathfrak{G}$ there exists a function $x \in \mathfrak{A}$, equal to unity at g_1 and to zero at g_2.

In fact, it suffices to set $F = g_1$, $U = \mathfrak{G} - g_2$ in the preceding lemma.

6. Duality theorem. Suppose $\mathfrak{G}$ is a locally compact commutative group and that $\overline{\mathfrak{G}}$ is the group of its characters. For fixed $g \in \mathfrak{G}$, the function $f_g(\chi) = \chi(g)$ satisfies the conditions

$$|f_g(\chi)| = 1, \quad f_g(\chi_1\chi_2) = (\chi_1\chi_2)(g) = \chi_1(g)\chi_2(g) = f_g(\chi_1)f_g(\chi_2)$$

and consequently it is a character of the group $\overline{\mathfrak{G}}$. Moreover,

$$f_{g_1 g_2}(\chi) = \chi(g_1 g_2) = \chi(g_1)\chi(g_2) = f_{g_1}(\chi)f_{g_2}(\chi);$$

consequently, the correspondence $g \to f_g$ is a homomorphism of the group $\mathfrak{G}$ into the group $\overline{\overline{\mathfrak{G}}}$, i.e. into the character group of $\overline{\mathfrak{G}}$.

THEOREM 5 (PONTRYAGIN'S [2, 6] DUALITY THEOREM). *The correspondence $g \to f_g$ is an isomorphism and a homeomorphism of the group $\mathfrak{G}$ onto the group $\overline{\overline{\mathfrak{G}}}$ of characters of the group $\overline{\mathfrak{G}}$.*

Proof. 1) According to Corollary 3, subsection 1, for $g_1 \neq g_2$ there exists a character $\chi_0 \in \overline{\mathfrak{G}}$ such that $\chi_0(g_1) \neq \chi_0(g_2)$. This means that $f_{g_1}(\chi_0) \neq f_{g_2}(\chi_0)$, and therefore the correspondence $g \to f_g$ is one-to-one. Consequently, this correspondence is an isomorphism of the group $\mathfrak{G}$ into the group $\overline{\overline{\mathfrak{G}}}$, and we can therefore assume that $\mathfrak{G} \subset \overline{\overline{\mathfrak{G}}}$ and set $\chi(g) = f_g(\chi) = g(\chi)$. We must prove that then $\mathfrak{G} = \overline{\overline{\mathfrak{G}}}$ and that the topologies in $\mathfrak{G}$ and $\overline{\overline{\mathfrak{G}}}$ coincide.

2) We again denote the algebra of all linear combinations of the $x \cdot y$, $x, y \in L(\mathfrak{G})$ by $\mathfrak{B}$ (see end of subsection 4). Clearly, $\mathfrak{B} \subset \mathfrak{A}(\mathfrak{G})$ (see subsection 5); consequently, on the basis of Theorem 3, every function $x \in \mathfrak{B}$ has the form

$$x(g) = \int x^\wedge(\chi)\overline{\chi(g)}d\mu(\chi), \tag{1}$$

where $x^\wedge \in L^1(\overline{\mathfrak{G}})$, $x^\wedge \in \mathfrak{B}^\wedge = T\mathfrak{B}$. On the other hand, on the basis of the Urysohn lemma, the functions $x \in L(\mathfrak{G})$ separate points in $\mathfrak{G}$ and, moreover, for an arbitrary point $g_0 \in \mathfrak{G}$ there exists a function $x \in L(\mathfrak{G})$ such that $x(g_0) \neq 0$. Since $\mathfrak{B}$ is dense in $L(\mathfrak{G})$ in the sense of the norm $||x||_\infty$, the functions in $\mathfrak{B}$ possess the same property; furthermore, since $\mathfrak{B} \subset L(\mathfrak{G})$, all the functions in $\mathfrak{B}$ vanish at the infinitely distant point. From this we conclude that the weak topology in $\mathfrak{G}$ determined by the algebra $\mathfrak{B}$ coincides with the initial topology in $\mathfrak{G}$ (see II, subsection 11, § 2).

In virtue of (1), this means that the sets $U(g_0; x_1^\wedge, \cdots, x_n^\wedge; \varepsilon)$ in $\mathfrak{G}$, defined by inequalities of the form

$$\left|\int x_j^\wedge(\chi)[\overline{\chi(g)}-\overline{\chi(g_0)}]d\mu(\chi)\right| < \varepsilon \quad (j = 1, 2, \cdots, n, x_j^\wedge \in \mathfrak{B}^\wedge, \varepsilon > 0), \tag{2}$$

form a neighborhood basis in $\mathfrak{G}$. But inequality (2) can be written in the form

$$\left|\int x_j^\wedge(\chi)[\overline{g(\chi)}-\overline{g_0(\chi)}]d\mu(\chi)\right| < \varepsilon;$$

consequently, according to the definition of the topology in the character group (see subsection 2), the sets $U(g_0; x_1^\wedge, \cdots, x_n^\wedge; \varepsilon)$ are neighborhoods in the group $\mathfrak{G}$, considered as a subspace of the topological space $\overline{\overline{\mathfrak{G}}}$, and they form a neighborhood basis there because $\mathfrak{B}^\wedge$ is dense in $L^1(\overline{\mathfrak{G}})$ in virtue of Corollary 4, subsection 4.

Thus, $\mathfrak{G}$ is a locally compact subspace in $\bar{\bar{\mathfrak{G}}}$.

3) We now prove that $\mathfrak{G}$ is dense in $\bar{\bar{\mathfrak{G}}}$. If $\mathfrak{G}$ is not dense in $\bar{\bar{\mathfrak{G}}}$, then $\bar{\bar{\mathfrak{G}}}$ contains a neighborhood with compact closure which does not intersect $\mathfrak{G}$, and on the basis of the lemma in subsection 5 there exists a function $x(\bar{\bar{g}}) = (y \cdot z)(\bar{\bar{g}})$, y, $z \in L^1(\bar{\bar{\mathfrak{G}}}) \cap L^2(\bar{\bar{\mathfrak{G}}})$, $y \geqq 0$, $z \geqq 0$, which is not identically zero and equals zero on $\mathfrak{G}$. Repeating the reasoning followed in the proof of the Plancherel theorem (with $\bar{\mathfrak{G}}$ and $\bar{\bar{\mathfrak{G}}}$ instead of $\mathfrak{G}$ and $\bar{\mathfrak{G}}$; see subsection 4), we conclude that

$$x(\bar{\bar{g}}) = \int x^{\wedge}(\chi)\bar{\bar{g}}(\chi)d\mu(\chi), \tag{3}$$

where $x^{\wedge} \in L^1(\bar{\mathfrak{G}}) \cap L^2(\bar{\mathfrak{G}})$, and therefore

$$x(g) = \int x^{\wedge}(\chi)\chi(g)d\mu(\chi).$$

But, by assumption, $x(g) = 0$ on $\mathfrak{G}$ and, consequently,

$$\overline{x(g)} = \int \overline{x^{\wedge}(\chi)}\,\overline{\chi(g)}d\mu(\chi) = 0 \quad \text{for all } g \in \mathfrak{G},$$

i.e. $T^*\overline{x^{\wedge}} = 0$. It follows that $x^{\wedge} = 0$ and, in virtue of (3), $x(\bar{\bar{g}}) = 0$, which contradicts the definition of the function $x(\bar{\bar{g}})$. Thus, $\mathfrak{G}$ is dense in $\bar{\bar{\mathfrak{G}}}$.

4) Finally, we shall prove that $\bar{\bar{\mathfrak{G}}} = \mathfrak{G}$. By what was proved in 2), $\mathfrak{G}$ is a locally compact subspace in $\bar{\bar{\mathfrak{G}}}$; therefore $\bar{\bar{\mathfrak{G}}}$ contains a neighborhood $\bar{\bar{U}}$ of the identity such that $U = \bar{\bar{U}} \cap \mathfrak{G}$ has a compact closure $\tilde{U}$ in $\mathfrak{G}$.

Since the embedding of $\mathfrak{G}$ in $\bar{\bar{\mathfrak{G}}}$ is continuous, $\tilde{U}$ is compact and hence also closed in $\bar{\bar{\mathfrak{G}}}$. On the other hand, U is dense in $\bar{\bar{U}}$ and hence $\bar{\bar{U}} \subset \tilde{U}$. Now let $\bar{\bar{g}}_0$ be an arbitrary element in $\bar{\bar{\mathfrak{G}}}$. By what was proved in 3), its neighborhood $\bar{\bar{g}}_0 \bar{\bar{U}}^{-1}$ contains an element $g \in \mathfrak{G}$. But then $g_0 \in g\bar{\bar{U}} \subset g_0\tilde{U} \subset \mathfrak{G}$, and the theorem is proved.

The duality theorem was first established by PONTRYAGIN [2] on the basis of a profound investigation of the structure of the groups under consideration. An analytic proof, based on the general theory of Fourier transforms on locallly compact commutative groups, was first given by RAIKOV [4].

7. Unitary representations of commutative groups.

THEOREM 6. *Every unitary representation $g \to T_g$ of a locally compact commutative group $\mathfrak{G}$ is given by the formula*

$$T_g = \int \overline{\chi(g)}dP(\chi),$$

where $P(\Delta)$ is a spectral measure on $\bar{\mathfrak{G}}$.

Proof. To the unitary representation $g \to T_g$ of the group $\mathfrak{G}$ into the space $\mathfrak{H}$ there corresponds the representation $x \to T_x$ of its group algebra $R(\mathfrak{G})$ and, hence, also of the completion $\widetilde{R(\mathfrak{G})}$ of this algebra relative to the minimal regular norm. But every such representation can be written in the form (see II, subsection 4, § 17)

$$T_x = \int x(M) dP(M), \tag{1}$$

where $P(\Delta)$ is a spectral measure in the space $\mathfrak{M}$ of maximal ideals of the algebra $\widetilde{R(\mathfrak{G})}$ and hence also of $R(\mathfrak{G})$ (because $R(\mathfrak{G})$ is a completely symmetric algebra; see Corollary 1, subsection 1, § 31 and the corollary in subsection 2, § 20). Moreover, $P(\{M_0\}) = 0$ (where, as above, $M_0 = L^1(\mathfrak{G})$) because otherwise the representation $x \to T_x$ would contain the degenerate representation. In virtue of the correspondence $M \leftrightarrow \chi$ between the maximal ideals $\neq M_0$ and the characters, $P(\Delta)$ can also be considered as a spectral measure on $\overline{\mathfrak{G}}$ and then formula (1) can be written in the form

$$\int x(g) T_g \, d\mu(g) = \int x^\wedge(\chi) dP(\chi) = \int \left[\int x(g) \overline{\chi(g)} d\mu(g) \right] dP(\chi)$$

$$= \int x(g) \left[\int \overline{\chi(g)} dP(\chi) \right] d\mu(g),$$

from which it follows that $T_g = \int \overline{\chi(g)} dP(\chi)$.

8. Theorems of Tauberian type.

THEOREM 7. *The group algebra of a locally compact commutative group is regular.*

Proof. Since $\mathfrak{M} - M_0$ is homeomorphic to the group $\overline{\mathfrak{G}}$, it will suffice to establish that for every compact set $F \subset \overline{\mathfrak{G}}$ and every point $\chi_0 \notin F$ there exists a function $x \in L^1(\mathfrak{G})$ such that $x^\wedge = 1$ on F and $x^\wedge = 0$ at the point χ_0. We shall prove a stronger assertion, namely that if F is a compact set and U is an open set in $\overline{\mathfrak{G}}$ and if $F \subset U$, then there exists a function $x \in L^1(\mathfrak{G})$ such that $x^\wedge = 1$ on F and $x^\wedge = 0$ in the exterior of U.

In fact, according to the lemma in subsection 5, applied to the group $\overline{\mathfrak{G}}$, there exists a function $x^\wedge \in \mathfrak{A}(\overline{\mathfrak{G}})$ which possesses these properties. But then, according to Theorems 3 and 5, $x^\wedge$ is the Fourier transform of a function $x \in L^1(\mathfrak{G})$, so that this function x satisfies the requirements set down.

THEOREM 8. *If $\mathfrak{G}$ is a locally compact commutative group, then every closed ideal in $L^1(\mathfrak{G})$ is contained in a maximal regular ideal.*

The assertion of the theorem follows directly from Theorem 7, Corollary 3, subsection 4, the corollary in subsection 5, § 15, and the fact that the algebra $L^1(\mathfrak{G})$ is semisimple [see VII, subsection 2, § 28 and (3), subsection 5, § 7].

COROLLARY 1. *If the function* $x \in L^1(\mathfrak{G})$ *is such that its Fourier transform* $x^\wedge$ *vanishes nowhere on* $\overline{\mathfrak{G}}$, *then the translations of the function* x *generate all of* $L^1(\mathfrak{G})$.

Proof. By assumption, x does not belong to any maximal regular ideal of the algebra $L^1(\mathfrak{G})$. On the other hand, in virtue of IV, subsection 2, § 28, the closed subspace $\mathfrak{L}$, generated by translations of the function x, is an ideal in $L^1(\mathfrak{G})$ or it is all of $L^1(\mathfrak{G})$. But if $\mathfrak{L}$ is an ideal in $L^1(\mathfrak{G})$, then, according to Theorem 8, this ideal and hence also the function x are contained in some maximal regular ideal of the algebra $L^1(\mathfrak{G})$, which contradicts the assumption. Consequently, $\mathfrak{L} = L^1(\mathfrak{G})$.

COROLLARY 2 (GENERALIZED WIENER TAUBERIAN THEOREM). *Let* $\mathfrak{G}$ *be a locally compact commutative group which is not compact, and suppose* x *is a function in* $L^1(\mathfrak{G})$ *such that* $x^\wedge$ *vanishes nowhere on* $\overline{\mathfrak{G}}$. *Then: if* $y \in L^\infty$ *is such that* $x \cdot y$ *vanishes at infinity, then* $z \cdot y$ *vanishes at infinity for all* $z \in L^1(\mathfrak{G})$.

Proof. The set I of all functions $z \in L^1(\mathfrak{G})$ for which $z \cdot y$ vanishes at infinity is, obviously, a linear subspace in $L^1(\mathfrak{G})$. We set $z^{g_0} = z(gg_0)$; then I is invariant under the translations $z \to z^{g_0}$ because if $z \cdot y$ vanishes at infinity, then $z^{g_0} \cdot y = (z \cdot y)^{g_0}$ also possesses this property. Moreover, I is closed. In fact, if $z \in \bar{I}$, then for given $\varepsilon > 0$ there exists a function $z' \in I$ for which

$$||z-z'||_1 < \frac{\varepsilon}{2(||y||_\infty)}.$$

Since $y \cdot z'$ vanishes at infinity, there exists a compact set $Q \subset \mathfrak{G}$ such that

$$|(y \cdot z')(g)| < \frac{\varepsilon}{2} \text{ in the exterior of } Q.$$

But

$$|(y \cdot z)(g)-(y \cdot z')(g)| \leqq ||y||_\infty \cdot ||z-z'||_1 < \frac{\varepsilon}{2};$$

therefore $|(y \cdot z)(g)|$ is also $< \varepsilon$ in the exterior of Q. Consequently, $y \cdot z$ also vanishes at infinity. This means that $z \in I$; hence I is closed.

But then either I is an ideal in $L^1(\mathfrak{G})$ or $I = L^1(\mathfrak{G})$. In virtue of Theorem 8, the first possibility is excluded because I contains the element x which is not contained in any maximal regular ideal. Consequently, $I = L^1(\mathfrak{G})$ which concludes the proof.

COROLLARY 3 (WIENER'S [1] TAUBERIAN THEOREM). *If* $x \in L^1(-\infty, \infty)$ *and* $x^\wedge(\alpha)$ *does not vanish for any* α, *and* y *is a function in* $L^\infty(-\infty, \infty)$ *such that* $(x \cdot y)(t) \to 0$ *as* $t \to +\infty$, *then* $(z \cdot y)(t) \to 0$ *as* $t \to +\infty$ *for every function* $z \in L^1(-\infty, \infty)$.

This corollary cannot be considered as a particular case of Corollary 2 because here we are dealing with a one-sided limit. However, the proof remains the same as for Corollary 2.

Theorem 8 means that the hull $h(I)$ of the closed ideal I cannot be the void set. In connection with this theorem, the more subtle problem, which has not yet been completely solved, arises of the classification of closed ideals in the algebra $L^1(\mathfrak{G})$, in particular, the problem of determining the conditions under which a closed ideal in the algebra $L^1(\mathfrak{G})$ is equal to the kernel of its hull (see subsection 3, § 15).

[The latter problem is closely related to the problem of the possibility of *spectral synthesis* on a group, i.e. to the problem of the conditions under which a space of functions on a group which is invariant under translations on a group is generated by the "elementary" functions (analogues of the functions $x^r \exp(-ixz)$ for the additive group R^1) contained in it (see, e.g., RUDIN [12] and ELLIOTT [1, 2]). For the structure of closed ideals in the algebra $L^1(\mathfrak{G})$ and in certain of its subalgebras, see, e.g., WARNER [1].]

The fact that this is not always true was shown in an example by L. SCHWARTZ [1] (see subsection 2, § 16). To answer the question formulated above, we shall first prove the following lemma.

LEMMA 1. *Suppose a regular semisimple Banach algebra R*

1) *contains a set which approximates the identity*;

2) *is such that the set R' of all functions $x(M)$ of the algebra R which equal zero, each in the exterior of some compact set, is dense in R.*

Then the algebra R satisfies condition (D) (see subsection 4, § 15) *at infinity*. (Of course, this lemma is meaningful only for an algebra R without identity and, consequently, for the case when $\mathfrak{M}$ is not compact.)

Proof. We must prove that for every $x \in R$ and every $\varepsilon > 0$ there exists a function $y = y(M) \in R'$, for which $|xy-x| < \varepsilon$. In virtue of condition 1), there exists a $u \in R$ for which $|xu-x| < \varepsilon/2$; further, in virtue of condition 2), there exists a function $y = y(M) \in R'$ such that $|y-u| < \varepsilon/(2|x|)$. Then $|xy-xu| < \varepsilon/2$ and hence $|xy-x| < \varepsilon$.

COROLLARY 4. *Suppose R is the same as in the preceding lemma. If the closed ideal I in R has a compact hull then I contains every element $x \in R$ which is such that $h(I) \subset \operatorname{Int} h((x))$.*

Proof. According to the preceding lemma, there exists a $y \in R'$ for which $|xy-x| < \varepsilon$. But, in virtue of Proposition III, subsection 4, § 15, we conclude from the condition $h(I) \subset \operatorname{Int} h((x))$ that $x(M)$, and hence also $x(M)y(M)$, locally belongs to the ideal I at every finite point (see subsection 5, § 15); moreover, $x(M)y(M)$ vanishes in some neighborhood of the infinitely distant point and hence it also locally belongs to the ideal I at the infinitely distant point; consequently (see Theorem 3″, subsection 4, § 15), $xy \in I$. Since I is closed and ε is an arbitrary positive number, we also have that $x \in I$.

To verify that condition (D) is satisfied at finite points, we shall prove the following lemma.

LEMMA 2. *Suppose $\mathfrak{G}$ is a locally compact commutative group. For every compact set $Q \subset \mathfrak{G}$ and every $\varepsilon > 0$, there exists a function $z \in L^1(\mathfrak{G})$ such that:*

1) $z^\wedge \equiv 1$ *in some neighborhood of the identity of the group* $\overline{\mathfrak{G}}$;
2) $||z||_1 < 2$;
3) $||z - z_g||_1 < \varepsilon$ *for all* $g \in Q$.

Proof. Suppose U is a symmetric neighborhood of the identity in $\overline{\mathfrak{G}}$ whose closure is compact and let V be another such neighborhood whose closure is contained in U and which satisfies the condition $\mu(U)/\mu(V) < 4$. Let $u^\wedge$, $v^\wedge$ be the characteristic functions of the sets U, V, respectively, and let u, v be their inverse images in $L^2(\mathfrak{G})$ under the Fourier transform. Then the function $z(g) = u(g)v(g)/\mu(V)$ belongs to $L^1(\mathfrak{G})$ and

$$||z||_1 \leqq \frac{1}{\mu(V)} ||u||_2 ||v||_2 = \frac{1}{\mu(V)} ||u^\wedge||_2 ||v^\wedge||_2 = \left[\frac{\mu(U)}{\mu(V)}\right]^{\frac{1}{2}} < 2.$$

Moreover, if W is a neighborhood of the identity in $\overline{\mathfrak{G}}$ which is such that $VW \subset U$, then, for $\chi \in W$,

$$z^\wedge(\chi) = \frac{1}{\mu(V)} (u^\wedge \cdot v^\wedge)(\chi) = \frac{1}{\mu(V)} \int u^\wedge(\chi') v^\wedge(\chi'^{-1}\chi) d\mu(\chi') = 1.$$

We denote the set of all $\chi \in \overline{\mathfrak{G}}$ satisfying the condition $|1-\chi(g)| < \varepsilon/4$ for all $g \in Q$ by U'. Clearly, U' is an open set (see V, subsection 12, § 2) which contains the identity of the group $\overline{\mathfrak{G}}$, and hence it is possible to choose $U \subset U'$. But then for all $g \in Q$,

$$||u - u_g||_2^2 = \int |u^\wedge(\chi)[1-\chi(g)]|^2 d\mu(\chi) < \mu(U) \left(\frac{\varepsilon}{4}\right)^2$$

and analogously,

$$||v - v_g||_2^2 < \mu(V) \left(\frac{\varepsilon}{4}\right)^2.$$

It follows that

$$\begin{aligned} ||z - z_q||_1 &= \frac{1}{\mu(V)} ||u(v - v_q) + v_g(u - u_g)||_1 \\ &\leqq \frac{1}{\mu(V)} (||u||_2 ||v - v_g||_2 + ||v_g||_2 ||u - u_g||_2) \\ &< \frac{1}{\mu(V)} \frac{\varepsilon}{4} 2[\mu(U)\mu(V)]^{\frac{1}{2}} < \varepsilon \end{aligned}$$

for all $g \in Q$, so that the function $z(g)$ constructed above satisfies all the requirements set down.

COROLLARY 5. *Suppose U and* $z = z_U$ *are the same as in Lemma 2. If* $x \in L^1(\mathfrak{G})$ *and* $x^\wedge(e^\wedge) = 0$, *then* $x \cdot z_U \to 0$ *as U decreases indefinitely.*

Proof. For given $\delta > 0$, we choose a symmetric set Q and a number $\varepsilon > 0$ in Lemma 2 so that

$$\int_{\mathfrak{G}-Q} |x(g)| dg < \frac{\delta}{8}, \ \varepsilon < \frac{\delta}{2||x||_1}.$$

Since

$$x^\wedge(e^\wedge) = \int x(g) d\mu(g) = 0,$$

we have

$$(x \cdot z)(g) = \int x(g') z(g'^{-1}g) d\mu(g') = \int x(g')[z(g'^{-1}g) - z(g)] d\mu(g').$$

Consequently,

$$||x \cdot z||_1 \leqq \int |x(g')| \, ||z_{g'^{-1}} - z||_1 \, d\mu(g')$$

$$= \int_Q |x(g')| \, ||z_{g'^{-1}} - z||_1 \, d\mu(g') + \int_{\mathfrak{G}-Q} |x(g')| \, ||z_{g'^{-1}} - z||_1 \, d\mu(g')$$

$$< ||x||_1 \varepsilon + \frac{\delta}{8} 4 < \delta.$$

COROLLARY 6. *There exists a uniformly bounded net of functions* $u \in L^1(\mathfrak{G})$ *which are such that* $u^\wedge \equiv 0$ *in some neighborhood of the point* $e^\wedge$ *and* $x \cdot u \to x$ *for all* $x \in L^1(\mathfrak{G})$, *satisfying the condition* $x^\wedge(e^\wedge) = 0$.

Proof. Suppose v runs through a set in $L^1(\mathfrak{G})$ which approximates the identity; we set $u = v - z \cdot v$, where z is the same as in Lemma 2. Then $||u||_1 \leqq 3$ and $u^\wedge = v^\wedge - v^\wedge z^\wedge = 0$ in a neighborhood where $z^\wedge \equiv 1$. Moreover (see Corollary 5),

$$||x - x \cdot u||_1 \leqq ||x - x \cdot v||_1 + ||x \cdot z||_1 ||v||_1 \to 0.$$

Corollary 6 means that condition (D) is satisfied at the point $e^\wedge$; from this, by means of a translation, we conclude that it is satisfied at all other points of the group $\mathfrak{G}$. According to Lemma 1, condition (D) is also satisfied at infinity. Thus, Shilov's theorem (see Theorem 5, subsection 4, § 15) is applicable to $L^1(\mathfrak{G})$, and we arrive at the following result.

THEOREM 9. *Suppose I is a closed ideal in* $L^1(\mathfrak{G})$ *and that x is a function in* $L^1(\mathfrak{G})$, *belonging to* $kh(I)$. *If the intersection of the boundary of the set* $h(x)$ *with* $h(I)$ *contains no nonvoid perfect set, then* $x \in I$.

COROLLARY 7. *If I is a closed ideal in $L^1(\mathfrak{G})$ whose hull is discrete (i.e. consists of isolated points only) then $I = kh(I)$.*

9. The case of a compact group.

THEOREM 10. *A locally compact commutative group $\mathfrak{G}$ is compact if, and only if, $\overline{\mathfrak{G}}$ is discrete.*

Proof. If $\mathfrak{G}$ is discrete, then $L^1(\mathfrak{G})$ contains the identity (see subsection 1, § 28); but then the group $\overline{\mathfrak{G}}$ is homeomorphic to the space of all maximal ideals of a Banach algebra with identity and hence is compact. Conversely, if $\overline{\mathfrak{G}}$ is compact, then the function $x_0^\wedge \equiv 1$ belongs to $L^1(\overline{\mathfrak{G}}) \cap P(\overline{\mathfrak{G}})$ and the function $x_0 \in L^1(\mathfrak{G})$ corresponding to it is the identity in $L^1(\mathfrak{G})$. But we saw in subsection 1, § 28 that this is possible only if $\mathfrak{G}$ is discrete. The assertion of the theorem is now obtained if we interchange the roles of $\overline{\mathfrak{G}}$ and $\mathfrak{G}$, using the duality theorem.

THEOREM 11. *If a commutative group $\mathfrak{G}$ is compact and the invariant measure on it is normalized so that $\mu(\mathfrak{G}) = 1$, then in the inversion theorem the invariant measure on $\overline{\mathfrak{G}}$ must be normalized so that the measure at each point equals unity.*

Proof. Suppose the measure μ on $\overline{\mathfrak{G}}$ is normalized so that the measure at each point equals c. Then the function

$$u^\wedge(\chi) = \begin{cases} \dfrac{1}{c} & \text{for } \chi = e^\wedge, \\ 0 & \text{for } \chi \neq e^\wedge \end{cases}$$

is the identity in $L^1(\overline{\mathfrak{G}})$. Since to the convolution in $L^1(\overline{\mathfrak{G}})$ there corresponds multiplication by the function $x(g)$, the corresponding function is $u(g) \equiv 1$; it follows that

$$u^\wedge(\chi) = \int \chi(g) d\mu(g) = \begin{cases} \dfrac{1}{c} & \text{for } \chi = e^\wedge, \\ 0 & \text{for } \chi \neq e^\wedge. \end{cases} \tag{1}$$

In particular, for $\chi = e^\wedge$,

$$\frac{1}{c} = \int d\mu(g) = \mu(\mathfrak{G}).$$

Consequently, $\mu(\mathfrak{G}) = 1$ if, and only if, $c = 1$.

It follows from (1), with $c = 1$, that

$$\int \chi_1(g)\overline{\chi_2(g)} d\mu(g) = \int (\chi_1 \chi_2^{-1})(g) d\mu(g) = \begin{cases} 1 \text{ for } \chi_1 = \chi_2, \\ 0 \text{ for } \chi_1 \neq \chi_2. \end{cases}$$

This proves the following corollary.

COROLLARY 1. *The characters of a compact commutative group form an orthonormal system in* $L^2(\mathfrak{G})$.

THEOREM 12. *The characters* χ_α *of a compact commutative group form a complete orthonormal system in* $L^2(\mathfrak{G})$ *and the expansion (in a generalized Fourier series)*

$$x(g) = \sum_\alpha (x, \chi_\alpha)\chi_\alpha(g) \tag{2}$$

of the function $x \in L^2(\mathfrak{G})$ *is the inverse Fourier transform.*

Proof. If $x \in L^2(\mathfrak{G})$, then $x^\wedge \in L^2(\overline{\mathfrak{G}})$, and hence at most a denumerable number of values $x^\wedge(\chi_\alpha)$, say, $x^\wedge(\chi_1), x^\wedge(\chi_2), \cdots$, are distinct from zero. Setting

$$c_n = x^\wedge(\bar{\chi}_n) = \int x(g)\overline{\chi_n(g)}d\mu(g)$$

and applying the Plancherel formula, we have

$$\int |x(g)|^2 d\mu(g) = \int |x^\wedge(\chi)|^2 d\mu(\chi) = \sum_n |c_n|^2.$$

This means (see IX, subsection 4, § 5) that the characters χ constitute a complete system in $L^2(\mathfrak{G})$. Further, the continuous function $x_n(g) = \sum_1^n c_k\chi_k(g)$ is the inverse Fourier transform of the function

$$x_n^\wedge(\chi) = \begin{cases} c_k & \text{for } \chi = \chi_k,\ k \leqq n, \\ 0 & \text{otherwise.} \end{cases}$$

Since $x_n^\wedge(\chi) \to x^\wedge(\chi)$ in the sense of the norm in $L^2(\overline{\mathfrak{G}})$, $x_n \to x$ in the sense of the norm in $L^2(\mathfrak{G})$, and (2) is the inverse Fourier transform.

If $\mathfrak{G}$ is the group of rotations of a circle, then the characters are $\chi_n = e^{int}$, $n = 0, \pm 1, \pm 2, \cdots$, and we obtain the theorem on the completeness of the system of functions $\chi_n = e^{int}$, $n = 0, \pm 1, \pm 2, \cdots$, in $L^2(0, 2\pi)$ as a particular case.

COROLLARY 2. *Every continuous function on a compact commutative group can be uniformly approximated by finite linear combinations of characters.*

The assertion follows directly from Corollary 4, subsection 1.

10. Spherical functions. One of the generalizations of the concept of character to non-commutative groups are the spherical functions. Suppose $\mathfrak{G}$ is a locally compact group and that K is a compact subgroup of $\mathfrak{G}$.

A *spherical function on* $\mathfrak{G}$ corresponding to the subgroup K is any elementary positive definite function φ on $\mathfrak{G}$ satisfying the condition

$$\varphi(k_1 g k_2) = \varphi(g) \quad \text{for all } k_1, k_2 \in K. \tag{1}$$

If $g \to U_g$ is an irreducible unitary representation of the group $\mathfrak{G}$ for which $\varphi(g) = (U_g\xi_0, \xi_0)$, then, as is easily seen, relation (1) is equivalent to the condition $U_k\xi_0 = \xi_0$ for all $k \in K$. Thus, *an elementary positive definite function is a spherical function if, and only if, the space of the corresponding irreducible representation* $g \to U_g$ *contains a vector* $\xi_0 \neq 0$, *satisfying the condition* $U_k\xi_0 = \xi_0$ *for all* $k \in K$.

A spherical function φ can also be considered as a function on the manifold X of right cosets Kg of the group $\mathfrak{G}$ modulo the subgroup K. Namely, we set

$$\varphi(x_1, x_2) = \varphi(g_1 g_2^{-1}) \text{ for } x_1, x_2 \in X, x_1 = Kg_1, x_2 = Kg_2. \tag{2}$$

It follows from (1) that this definition does not depend on the choice of the elements $g_1 \in x_1$, $g_2 \in x_2$.

To the elements $g_0 \in \mathfrak{G}$ there correspond the transformations $Kg \to Kgg_0$ in the space X; we shall write these transformations in the form $x \to xg_0$. It follows from (2) that $\varphi(x_1 g, x_2 g) = \varphi(x_1, x_2)$ for all $g \in \mathfrak{G}$; moreover, $\varphi(x_1, x_2)$ is a positive definite kernel. [$\varphi(x_1, x_2)$ is a positive definite kernel provided $\sum_{j,k=1}^{n} \varphi(x_j, x_k)\xi_j\bar{\xi}_k \geq 0$ for arbitrary $x_1, x_2, \cdots, x_n \in X$, arbitrary complex numbers $\xi_1, \xi_2, \cdots, \xi_n$ and any natural number n.] If we fix the point $x_0 \in X$, then the function $\psi(x) = \varphi(x, x_0)$ will be invariant with respect to all transformations g which leave the point x_0 fixed, i.e. which will be constant on "spheres" with "center" at the point x_0.

We now place an additional condition on the group $\mathfrak{G}$ and on K, namely, we require that there exist an automorphism $g \to g'$ in the group $\mathfrak{G}$ (i.e. an isomorphism of the group $\mathfrak{G}$ onto the group $\mathfrak{G}$) such that: 1) $g'' = g$, 2) $k' = k$ for all $k \in K$. Everywhere in the remainder of this subsection we shall assume that this condition is satisfied. (This condition is satisfied, for instance, if X is a symmetric Riemannian space and $\mathfrak{G}$ is the group of motions in X.)

In this case, the spherical functions are described with the aid of the maximal ideals of the following commutative algebra. Suppose R_0' is the set of all functions $x \in L^1(\mathfrak{G})$, satisfying the condition

$$x(k_1 g k_2) = x(g) \quad \text{for all } k_1, k_2 \in K. \tag{3}$$

It is easily seen that under the assumption that the automorphism indicated above exists, R_0' is a commutative subalgebra of the algebra $L^1(\mathfrak{G})$. (The proof of this fact, as well as proof of other propositions formulated here, is a generalization of the corresponding arguments in Example c), subsection 4, § 29.) Suppose R_0 is the algebra obtained from R_0' upon adjunction of the identity. There exists a one-to-one correspondence $M \to \varphi_M$ between the set of all symmetric maximal ideals, distinct from R_0', of the algebra R_0 and the set of all spherical functions; this correspondence is given by the formula $x(M) = \int x(g)\varphi_M(g)d\mu(g)$ for all $x \in R_0'$. Combining this result with Theorem 3, subsection 2, § 20, we conclude that *every positive definite function* φ, *which satisfies condition* (1), *is representable in the form*

$$\varphi(g) = \int_{\mathfrak{M}} \varphi_M(g)d\sigma(M), \tag{4}$$

where $\mathfrak{M}$ *is the space of symmetric maximal ideals of the algebra* R_0 *and* σ *is a measure on* $\mathfrak{M}$.

We denote the set of two-sided cosets $s = KgK$ of the group $\mathfrak{G}$ modulo the subgroup K, i.e. the sets s of all elements $k_1 g k_2$ with fixed g, by S; in virtue of what we stated above, these cosets s can be considered as "spheres" in X. Condition (1) means that the function φ must be constant on every coset, and hence it can be considered as a function $\varphi(s)$ of the coset s.

Analogously, every function $f(g) \in R_0'$ may be considered as a function $f(s)$ on S.

Passing from the integral on $\mathfrak{G}$ to the integral on $K \times K \times S$, we conclude that the product $f = f_1 \cdot f_2$ in the algebra R_0 must be defined by the formula

$$f(s) = \int_S f_1(s_1)f_2(s_2)a(s_1, s_2, s)ds_1\, ds_2, \tag{5}$$

where $a(s_1, s_2, s)$ is some function of s_1, s_2, s. From this we conclude that *for spherical functions* φ_M, *the following multiplication rule holds*:

$$\varphi_M(s_1)\varphi_M(s_2) = \int_S a(s_1, s_2, s)\varphi_M(s)ds. \tag{6}$$

We now assume that $\mathfrak{G}$ is a semisimple Lie group; in this case, the spherical functions satisfy differential equations which are obtained in the following way. Suppose $\{e_1, e_2, \cdots, e_n\}$ is a basis of the Lie algebra Γ of the group $\mathfrak{G}$, and that $E_1, E_2, \cdots, E_n$ are the corresponding differential operators in X (i.e. Lie operators consisting of infinitesimal translations). We denote by Q the algebra of all formal polynomials in $e_1, \cdots, e_n$ with relations corresponding to the commutation relations in Γ.

We consider the elements $P(e_1, e_2, \cdots, e_n)$ of the center of the algebra Q; then the $P(E_1, E_2, \cdots, E_n)$ are differential operators in X which commute with all the transformations $x \to xg$. (It can be shown that the center of the algebra Q consists of all polynomials $P = a1 + \Sigma\, a^i e_i + \Sigma\, a^{ik} e_i e_k + \cdots$ with symmetric coefficients $a^{ik}, a^{ikl}, \cdots$ for which the forms $\Sigma\, a^i \xi_i$, $\Sigma\, a^{ik}\xi_i\xi_k, \cdots$ are invariant with respect to transformations of the adjoined group.) Since the functions $\varphi_M(x)$ define the irreducible representations of the group $\mathfrak{G}$, it follows from this that $P(E_1, E_2, \cdots, E_n)\varphi_M(x) = \lambda\varphi_M(x)$, where λ is a number.

Among the elements of the center of the algebra Q, there is a finite number $P_1, \cdots, P_m$ such that every other element of the center is a polynomial in $P_1, \cdots, P_m$. Setting $\Delta_i = P_i(E_1, \cdots, E_n)$, $i = 1, 2, \cdots, m$, we obtain that the functions $\varphi_M(x)$ satisfy the following equations:

$$\Delta_i \varphi_M(x) = \lambda_i \varphi_M(x), \quad i = 1, 2, \cdots, m. \tag{7}$$

If, for example, $\mathfrak{G}$ is the complex unimodular group of the second order, and K is a unitary subgroup of $\mathfrak{G}$, then the spherical functions coincide with the spherical functions computed in Example b), subsection 4, § 29; X can in this case be interpreted as the Lobachevsky plane and $\mathfrak{G}$ as the group of motions in the Lobachevsky plane. In this case, all the elements of the center are expressible in terms of one of them, and the corresponding operator Δ is the Laplace operator on the Lobachevsky plane.

11. The generalized translation operation. Suppose T is a locally compact Hausdorff space, that μ is a measure on T, and that $L^1(T)$ is the L^1 space constructed with respect to this measure. We consider the family of bounded linear operators A^s in $L^1(T)$, possessing the following properties:

1) for every $t \in T$ there exists an operator A^{t*} in $L^1(T)$ such that

$$\int A^{t*}x(t) \cdot \overline{y(t)}d\mu = \int x(t)\overline{A^t y(t)}d\mu \text{ for all } x, y \in L(T);$$

2) the numerical functions $|A^t|$, $|A^{t*}|$ are μ-measurable and bounded on every compact set;

3) $A^s x(t)$, $A^{s*}x(t)$ belong to $L(T \times T)$ for an arbitrary function $x \in L(T)$ (see subsections 1, 18, § 6 for the notation $L(T)$ and $L(T \times T)$);

4) $A^s x(t)$, as a function of s, belongs to $L^1(T)$ for $x(t) \in L^1(T)$ for almost every $t \in T$;

5) there exists a sequence of real-valued functions $x_n(t) \in L^1(T)$ such that for arbitrary function $f(x) \in L^1(T)$,

$$\lim_{n\to\infty} \int A^s f(t)x_n(t)d\mu(t) = f(s), \quad \lim_{n\to\infty} \int A^s f(t)x_n(s)d\mu(s) = f(t),$$

$$\lim_{n\to\infty} \int A^{s*} f(t)x_n(s)d\mu(s) = f(t), \quad \lim_{n\to\infty} \int A^{s*} f(t)x_n(t)d\mu(t) \in L^1(T)$$

exist; we set

$$f^*(s) = \lim_{n\to\infty} \int \overline{A^{s*}f(t)x_n(t)d\mu(t)}; \tag{1}$$

6) $A_s^\tau A_t^s x(t) = A_t^s A_t^\tau x(t)$ (if x is a function of several arguments $t, t_1, t_2, \cdots$, then the lower index in A_t^s denotes that this operator is applied to x as a function of the variable t);

7) $A_t^s A_t^\tau x(t) = A_t^\tau A_t^s x(t)$;

8) $A_t^{s*} A_t^\tau x(t) = A_t^\tau A_t^{s*} x(t)$,

where relations 6)–8) hold for every $\tau, s \in T$ for almost all $t \in T$.

Every family of operators A^t satisfying conditions 1)–6) is called a *generalized translation operation*; if, furthermore, conditions 7) and 8) are satisfied, then the translation operation is called *commutative* and *normal* respectively.

A simple example of a generalized translation operation is the case where T is a locally compact group and μ is a left invariant measure on T; then the operator of left translation $A^s x(t) = x(st)$ will satisfy conditions 1)–6); in this connection, as is easily seen, $A^{s*}x(t) = x(s^{-1}t)$. If, moreover, the group T is commutative, then conditions 7)–8) will also be satisfied.

A less trivial example will be introduced further on, at the end of this subsection.

In the sequel, we shall consider only generalized translation operations satisfying conditions 7)–8); moreover, for simplicity of exposition, we shall further assume that $|A_t^{s*}| \leqq C$, where C is some constant. Then $L^1(T)$ can be transformed into a complete normed symmetric algebra by setting

$$|x| = C \int |x(t)| d\mu(t), \ (x \cdot y)(t) = \int A_t^{s*} x(t) y(s) d\mu(s) \tag{2}$$

and defining involution by formula (1); then the associativity of multiplication follows from condition 6). Adjoining the identity, we obtain a complete normed symmetric algebra R with identity; it can be shown that it follows from conditions 7) and 8) that R is a commutative completely symmetric algebra. Repeating the line of reasoning followed in subsection 1, it can be shown that the maximal ideals, distinct from $L^1(T)$, in the algebra R, are given by the formula

$$(\lambda e + x)(M) = \lambda + \int x(t) \overline{\varphi(t, M)} d\mu(t), \tag{3}$$

where $\varphi(t, M)$ is a continuous function on $T \times \mathfrak{M}$, satisfying the condition

$$A_t^s \varphi(t, M) = \varphi(s, M) \varphi(t, M); \tag{4}$$

conversely, every such function defines, by formula (3), a maximal ideal in the algebra R.

A continuous function $p(t)$ is called *positive definite with respect to the operation* A^s if

$$\sum_{j,k=1}^{n} A_{t_j}^{t_k *} p(t_j) \xi_j \bar{\xi}_k \geqq 0$$

for arbitrary points $t_1, \cdots, t_n \in T$ and arbitrary complex numbers $\xi_1, \cdots, \xi_n$. In this case, the formula $F(\lambda e + x) = \lambda + \int p(t) x(t) d\mu(t)$ defines a positive functional on R. Applying Corollary 1, subsection 4, § 20 (cf. the analogous reasoning in subsection 3) to this functional, we conclude that:

THEOREM 13. *A function $p(t)$ is continuous and positive definite with respect to A^s if, and only if,* *it is representable in the form*

$$p(t) = \int_{\mathfrak{M}} \overline{\varphi(t, M)} d\sigma(M),$$

where σ is some measure on $\mathfrak{M}$.

We now set $M_0 = L^1(T)$ and $\mathfrak{M}' = \mathfrak{M} - M_0$; repeating essentially the line of reasoning followed in subsection 4, we arrive at the following result.

THEOREM 14 (PLANCHEREL'S THEOREM FOR THE GENERALIZED TRANSLATION OPERATION). *There exists a unique measure ν on $\mathfrak{M}'$ such that the formulas*

$$F(M) = \int_T f(t)\overline{\varphi(t, M)} d\mu(t), f(t) = \int_{\mathfrak{M}'} F(M)\varphi(t, M) d\nu(M) \tag{5}$$

realize mutually inverse isometric mappings of $L^2_\mu(T)$ onto $L^2_\nu(\mathfrak{M}')$ and of $L^2_\nu(\mathfrak{M}')$ onto $L^2_\mu(T)$, where the integrals in (5) *converge in norm in $L^2_\nu(\mathfrak{M}')$ and $L^2_\mu(T)$ respectively.*

We now set

$$\begin{aligned} B^{N*}_M F(M) &= \int \overline{\varphi(t, M)}\varphi(t, N) f(t) d\mu(t) \\ \text{for } f(t) &= \int_{\mathfrak{M}'} F(M)\varphi(t, M) d\nu(M),\ N \in \mathfrak{M}'; \end{aligned} \tag{6}$$

if B^{N*}_M and the operator B^N_M adjoint to it satisfy conditions 1)–8) (with $\mathfrak{M}'$ and ν instead of T and μ) and if also $|B^{N*}_M| \leqq C_1$, then in virtue of the preceding discussion we can construct a normed algebra, which we shall denote by R_1, with respect to the operation B^N_M. For an arbitrary fixed $t \in T$, the formula $(\lambda e + F(M))(t) = \lambda + \int F(M)\varphi(t, M) d\nu(M)$ defines a maximal ideal in R_1 and, consequently, it can be assumed that the space T_1 of maximal ideals of the algebra R_1 contains T. Suppose ν' is the measure on $T' = T_1 - L^1_\nu(\mathfrak{M}')$, which is the analogue to the measure ν on $\mathfrak{M}'$. It can be shown (see LEVITAN [4]) that if the algebra R_1 is regular and $\nu'(T - T) = 0$, then $T' = T$, i.e. the duality law holds, generalizing to this case the duality theorem for locally compact commutative groups.

A non-trivial example of the generalized translation operation is obtained if we denote by $A^y_x f(x)$ the solution of the equation

$$\frac{\partial^2 u}{\partial x^2} - \frac{\partial^2 u}{\partial y^2} - [\rho(x) - \rho(u)]u = 0, 0 \leqq x < \infty, \tag{7}$$

satisfying the initial conditions $u(x, 0) = f(x)$, $u_y(x, 0) = 0$.

It can be shown (see, for example. POVZNER [7]) that if $\rho(x)$ is real-valued, continuous on the half-axis $0 \leqq x < \infty$, and satisfies the condition $\rho(x) = O(1/x^{3+\varepsilon})$, $\varepsilon > 0$, as $x \to +\infty$, then all the conditions 1)–8) will be satisfied and $|A^y_x| \leqq C$, where C is some constant. Consequently, all the preceding results are applicable to the operator A^y_x; the corresponding function $\varphi(x, M)$ is a solution of the equation $d^2u/dx^2 - \rho(x)u + \lambda u = 0$ for some real λ, satisfying the initial conditions $\varphi(0) = 1$, $\varphi'(0) = 0$, and the mutually inverse formulas (5) in the generalized Plancherel theorem go over into the formulas

$$F(\lambda) = \int_0^\infty f(x)\varphi(x, \lambda) dx, f(x) = \int_{-\infty}^\infty F(\lambda)\varphi(x, \lambda) d\nu(\lambda), \tag{8}$$

which realize mutually inverse isometric mappings of $L^2(0, \infty)$ onto $L^2_v(-\infty, \infty)$ and of $L^2_v(-\infty, \infty)$ onto $L^2(0, \infty)$, respectively.

We note that formulas (8) are in reality true for wider classes of functions $\rho(x)$, namely, they are valid for arbitrary real-valued measurable functions $\rho(x)$, summable on every finite interval $(0, c)$, $c > 0$. But at the present time, this general result is not obtainable by the methods of algebra theory; this is explained by the fact that here instead of symmetric normed algebras it would be necessary to consider symmetric topological algebras the theory of which, however, is not yet very well developed.

§ 32. Representations of compact groups

In the case of a compact group $\mathfrak{G}$ the space $L^2(\mathfrak{G})$ is an H^*-algebra (see below, subsection 1); the application of the general theory of H^*-algebras (see subsection 5, § 25) to $L^2(\mathfrak{G})$ leads to a number of results concerning unitary representations of compact groups.

1. The algebra $\mathbf{L^2(\mathfrak{G})}$. Suppose $\mathfrak{G}$ is a compact group; then the invariant measure $\mu(\mathfrak{G})$ of the entire group is finite; we shall assume it to be normalized so that $\mu(\mathfrak{G}) = 1$. Then

$$||x||_1 \leqq ||x||_2 \leqq ||x||_\infty$$

and

$$||x \cdot y||_\infty \leqq ||x||_2||y||_2 .$$

From these inequalities it follows, in particular, that $L^2(\mathfrak{G})$ is a Banach algebra because

$$||x \cdot y||_2 \leqq ||x \cdot y||_\infty \leqq ||x||_2 \, ||y||_2 .$$

It is easily verified directly that $L^2(\mathfrak{G})$ satisfies all the remaining H^*-algebra axioms so that the general theory of such algebras is applicable to $L^2(\mathfrak{G})$. However it turns out that, furthermore, all the minimal two-sided ideals of the algebra $L^2(\mathfrak{G})$ are finite-dimensional. This result is obtained from the following lemmas.

Lemma 1. *A continuous function z is in the center of the algebra $L^2(\mathfrak{G})$ if, and only if, $z(g_1 g_2) \equiv z(g_2 g_1)$.*

Proof. The function z belongs to the center of $L^2(\mathfrak{G})$ if, and only if,

$$\int z(g_1)x(g_1^{-1}g)d\mu(g_1) = \int x(g_1)z(g_1^{-1}g)d\mu(g_1) \tag{1}$$

for all $x(g) \in L^2(\mathfrak{G})$. Replacing g_1 by gg_1 in the first integral, and g_1 by g_1^{-1} in the second, we obtain that

$$\int [z(gg_1) - z(g_1 g)]x(g_1^{-1})d\mu(g_1) = 0. \tag{2}$$

Inasmuch as the function $x \in L^2(\mathfrak{G})$ is arbitrary and the function z is continuous, equality (2) (and hence also (1)) holds if, and only if, $z(gg_1)-z(g_1g) \equiv 0$.

LEMMA 2. *Every closed two-sided ideal $I \subset L^2(\mathfrak{G})$, distinct from* (0), *contains a nonzero element of the center of the algebra $L^2(\mathfrak{G})$, which is a continuous function on the group $\mathfrak{G}$.*

Proof. (The proof given here is due to SEGAL [4].) We choose a nonzero element $y \in I$ and set $x = y \cdot y^*$. Then the function x is continuous on $\mathfrak{G}$. We set

$$z(g) = \int x(g_1 g g_1^{-1}) d\mu(g_1).$$

Then z is continuous, $z(e) = x(e) = \int |y(g)|^2 d\mu(g) > 0$, so that $z(g) \not\equiv 0$. Moreover, it follows from the invariance of measure that $z(g'g) = z(gg')$, so that z belongs to the center of the algebra $L^2(\mathfrak{G})$.

On the other hand, $x(g) \in I$, and hence also $x_{g_1}^{g_1}(g) = x(g_1 g g_1^{-1}) \in I$ (see IV, subsection 2, § 28). Therefore, also

$$z = \int x_{g_1}^{g_1} d\mu(g_1) \in I.$$

Now applying Theorems 5 and 12, subsections 3 and 5, § 25, to a minimal two-sided ideal $I \subset L^2(\mathfrak{G})$ and taking Lemma 2 into consideration, we conclude that:

LEMMA 3. *Every minimal two-sided ideal $I \subset L^2(\mathfrak{G})$ is finite-dimensional.*

But then application of the results of subsection 5, § 25 leads to the following theorem.

THEOREM 1. *If $\mathfrak{G}$ is a compact group, then $L^2(\mathfrak{G})$ is the direct, and at the same time the orthogonal, sum of its minimal two-sided ideals I_α, each of which is finite-dimensional and, consequently, completely isomorphic to the algebra of all matrices of a fixed finite order* (*a matrix for each ideal*).

2. Representations of a compact group.

THEOREM 2. *If $g \to T_g$ is a continuous representation of the compact group $\mathfrak{G}$ in the Hilbert space $\mathfrak{H}$, then there exists another inner product in $\mathfrak{H}$ defining a norm equivalent to the initial norm relative to which all the operators T_g are unitary.*

Proof. We set

$$(x, y)_1 = \int (T_g x, T_g y) d\mu(g)$$

for $x, y \in \mathfrak{H}$. It is easily seen that $(x, y)_1$ satisfies all the inner product axioms; in particular, if

$$(x, x)_1 = \int (T_g x, T_g x) d\mu(g) = 0,$$

then $(T_g x, T_g x) \equiv 0$ because $(T_g x, T_g x)$ is a continuous function of g and is $\geqq 0$. Setting $g = e$, we obtain that

$$(x, x) = (T_e x, T_e x) = 0, x = 0.$$

Further, it follows from the invariance of the measure μ that

$$(T_{g_0} x, T_{g_0} y)_1 = \int (T_g T_{g_0} x, T_g T_{g_0} y) d\mu(g)$$

$$= \int (T_{gg_0} x, T_{gg_0} y) d\mu(g) = \int (T_g x, T_g y) d\mu(g) = (x, y)_1;$$

consequently, the operator T_{g_0} is unitary relative to the inner product $(x, y)_1$. We set

$$C = \sup_{g \in \mathfrak{G}} |T_g|,$$

where $|T_g|$ is the norm of the operator T_g relative to the inner product (x, y). [$C < \infty$. In fact, $\sup_{g \in \mathfrak{G}} |T_g x| < \infty$ for every $x \in \mathfrak{H}$ because $|T_g x|$ is a continuous function on the compact group $\mathfrak{G}$. By the well-known Banach-Steinhaus theorem (see, e.g., LYUSTERNIK and SOBOLEV [1]), it follows from this that $C < \infty$.] Then

$$|x|_1^2 = (x, x)_1 = \int |T_g x|^2 d\mu(g) \leqq C^2 |x|^2 \int d\mu(g) = C^2 |x|^2.$$

On the other hand, integrating both members of the inequality

$$|x|^2 = |T_{g^{-1}} T_g x|^2 \leqq C^2 |T_g x|^2$$

with respect to g, we obtain

$$|x|^2 \leqq C^2 |x|_1^2.$$

Consequently, the norms $|x|$ and $|x|_1$ are equivalent.

In the sequel, we shall consider only unitary representations of the group $\mathfrak{G}$. According to the general theory (see subsection 2, § 29), the unitary representations of the group $\mathfrak{G}$ are completely described by the representations of its group algebra, or, what amounts to the same thing, by the representations

$$x \to A_x, \; A_x = \int x(g) U_g d\mu(g) \tag{1}$$

of the algebra $L^1(\mathfrak{G})$. (Recall that we are considering only symmetric representations of symmetric algebras.) But in case the group $\mathfrak{G}$ is compact, $L^2(\mathfrak{G}) \subset L^1(\mathfrak{G})$; consequently, every unitary representation of the group $\mathfrak{G}$ also generates a representation of the algebra $L^2(\mathfrak{G})$. Conversely, every representation $x \to A_x$ of the algebra $L^2(\mathfrak{G})$, not containing a degenerate representation, is obtained by formula (1) from some unitary representation of the group $\mathfrak{G}$; in order to convince ourselves of this, it suffices to repeat the line of reasoning followed in subsection 2, § 29, this time for the algebra $L^2(\mathfrak{G})$. Therefore, it suffices to study representations of the algebra $L^2(\mathfrak{G})$.

THEOREM 3. *Every representation $x \to A_x$ of an H^*-algebra R which does not contain the degenerate representation decomposes uniquely into the direct sum of representations which are one-to-one representations of certain minimal two-sided ideals I_α of this algebra.*

Proof. We denote by H the space of the representation under consideration, and by H_α the closure of the linear hull of all vectors $A_x\xi$, $\xi \in H$, $x \in I_\alpha$. Certain of the H_α equal (0); discarding them, we may obviously consider only the $H_\alpha \neq (0)$. All these subspaces H_α are mutually orthogonal.

In fact, suppose I_α, I_β are two distinct, and hence orthogonal, minimal two-sided ideals of the algebra R. If $x \in I_\alpha$, $y \in I_\beta$, then $x^*y = 0$ and therefore for $\xi, \eta \in H$,

$$(A_x\xi, A_y\eta) = (\xi, A_{x^*y}\eta) = 0.$$

Since $x \to A_x$ does not contain the degenerate representation, the invariant subspace $\mathfrak{N} \subset H$ of all vectors on which all the operators A_x equal zero is $= (0)$. Therefore, H is the direct sum of all the H_α and, consequently, the initial representation of the algebra R is the direct sum of representations of the ideals I_α in the spaces H_α. These representations are isomorphisms because the I_α are simple algebras.

We note that if x_α is the projection on I_α of the element $x \in R$, then $A_x\xi = A_{x_\alpha}\xi$ for $\xi \in H_\alpha$; in fact, for $\xi = A_y\eta$, $y \in I_\alpha$, $\eta \in H$,

$$A_{x-x_\alpha}\xi = A_{x-x_\alpha}A_y\eta = A_{(x-x_\alpha)y}\eta = 0,$$

because $(x-x_\alpha)y = 0$. Since the linear hull of the vectors $A_y\eta$ is dense in H_α, $A_{x-x_\alpha}\xi = 0$ for all $\xi \in H_\alpha$; consequently, $A_x\xi = A_{x_\alpha}\xi$ for $\xi \in H_\alpha$.

In the theorem just proved, the representations of the ideals I_α in H_α are in general reducible; therefore, we shall occupy ourselves with the further decomposition of the one-to-one representations of the simple algebra I_α.

Suppose I is one of the ideals I_α. In anticipation of later applications to compact groups we shall make the additional assumption that I is finite-dimensional. Then I contains an orthogonal basis of elements p_{jk}, $j, k = 1, 2, \cdots, n$, satisfying the conditions

$$p_{jk}^* = p_{kj}, (p_{jk}, p_{jk}) = (p_{11}, p_{11}), p_{jk}p_{j_1k_1} = \begin{cases} p_{jk_1} \text{ for } k = j_1, \\ 0 \quad \text{for } k \neq j_1, \end{cases} \tag{2}$$

and the left regular representation

$$T_a x \to ax, \quad x \in Ip_{11}$$

is a symmetric isomorphism of the algebra I onto the algebra $\mathfrak{B}(\mathfrak{H}_n)$ of all linear operators in the finite-dimensional Hilbert space $\mathfrak{H}_n = Ip_{11}$ (see subsection 5, § 25).

Consequently, the representation $x \to A_x$ of the algebra I can also be considered as a representation of the algebra $\mathfrak{B}(\mathfrak{H}_n)$.

But the algebra $\mathfrak{B}(\mathfrak{H}_n)$ is a particular case of the algebra of all compact operators. On the basis of Theorem 2, subsection 2, § 22, every representation of this algebra is the direct sum of irreducible representations each of which is equivalent to the identity or zero representation.

We thus arrive at the following theorem.

THEOREM 4. *Every irreducible unitary representation of a compact group is finite-dimensional; every unitary representation of a compact group is the direct orthogonal sum of its irreducible (and hence, finite-dimensional) representations.*

We now consider one of the irreducible representations $g \to U_g$ of the compact group $\mathfrak{G}$. By what we stated above, the representation $x \to A_x$ of the algebra $L^2(\mathfrak{G})$, generated by the unitary representation $g \to U_g$, reduces to the representation of some minimal two-sided ideal I of this algebra and hence to a representation of the algebra $\mathfrak{B}(\mathfrak{H}_n)$, where $\mathfrak{H}_n = Ip_{11}$; the latter representation in its turn is equivalent to the identity representation of the algebra $\mathfrak{B}(\mathfrak{H}_n)$. Consequently (to within a unitary equivalence), we can assume that our representation of the algebra $\mathfrak{B}(\mathfrak{H}_n)$ is simply the identity representation of this algebra.

This means that for $a, x \in I$, $\xi = xp_{11}$, we have

$$A_a \xi = a\xi,$$

and consequently

$$U_{g_0} A_a \xi = a_{g_0} \xi = a\xi_{g_0}, \tag{3}$$

where $a_{g_0}(g) = a(g_0^{-1}g)$, $\xi_{g_0}(g) = \xi(g_0^{-1}g)$.

In (3), we set $a = p$, where $p = p_{11}+p_{22}+ \cdots +p_{nn}$ is the identity in the algebra I. We obtain that

$$U_{g_0} \xi = \xi_{g_0}, \text{ i.e. } U_{g_0} \xi(g) = \xi(g_0^{-1}g). \tag{4}$$

Consequently,

I. *Every irreducible representation of a compact group $\mathfrak{G}$ can be realized as a representation in some minimal left ideal of the algebra $L^2(\mathfrak{G})$, in such a way that the operators of the representation will be the left translation operators.*

The functions $p_{11}(g), p_{21}(g), \cdots, p_{n1}(g)$ form an orthogonal basis in Ip_{11} and, consequently, the functions

$$\frac{1}{\omega} p_{11}(g), \frac{1}{\omega} p_{21}(g), \cdots, \frac{1}{\omega} p_{n1}(g), \text{ where } \omega = \sqrt{(p_{11}, p_{11})}, \tag{5}$$

form an orthonormal basis in Ip_{11}.

Suppose $||c_{jk}(g)||$ is the matrix of the operator U_g in the basis (5) so that, by (4),

$$\frac{1}{\omega} p_{j1}(g_0^{-1}g) = \sum_{k=1}^{n} c_{kj}(g_0) \frac{1}{\omega} p_{k1}(g).$$

From this and (2), we conclude that

$$c_{kj}(g_0) = \frac{1}{\omega^2} \int p_{j1}(g_0^{-1}g)\overline{p_{k1}(g)}d\mu(g)$$

$$= \frac{1}{\omega^2} \int \overline{p_{k1}(g)p_{1j}(g^{-1}g_0)d\mu(g)} = \frac{1}{\omega^2} \overline{p_{kj}(g_0)}. \tag{6}$$

We shall now find ω. To this end, we note that

$$\sum_{j=1}^{n} |c_{kj}(g)|^2 = 1$$

since the matrix $||c_{jk}(g)||$ is unitary; i.e. in virtue of (6),

$$\frac{1}{\omega^4} \sum_{j=1}^{n} |p_{kj}(g)|^2 = 1.$$

Integrating both members of this equality and taking (2) and the definition of the number ω into consideration, we obtain $\omega^{-2}n = 1$, from which it follows that $\omega^2 = n$. Thus,

II. *The matrix elements $c_{kj}(g)$ of an irreducible unitary representation $g \to U_g$ of a compact group $\mathfrak{G}$ with respect to the orthonormal basis $n^{-\frac{1}{2}}p_{11}, n^{-\frac{1}{2}}p_{21}, \cdots, n^{-\frac{1}{2}}p_{n1}$* are given by the formula

$$c_{kj}(g) = \frac{1}{n} \overline{p_{kj}(g)}, j, k = 1, 2, \cdots, n. \tag{7}$$

The function

$$\chi(g) = c_{11}(g)+c_{22}(g)+ \cdots +c_{nn}(g) = \frac{1}{n}[\overline{p_{11}(g)}+\overline{p_{22}(g)}+ \cdots +\overline{p_{nn}(g)}],$$

i.e. the trace of the matrix $||c_{jk}(g)||$, is called the *character* of the given irreducible representation $g \to U_g$. It is clear from this definition that unitarily equivalent representations have the same character and that the character χ belongs to the center of the minimal two-sided ideal I, defined by the given irreducible representation. In this connection,

$$\chi(e) = c_{11}(e)+c_{22}(e)+ \cdots +c_{nn}(e) = 1+1+ \cdots +1 = n,$$

so that $\chi(g) \not\equiv 0$.

On the other hand, two distinct minimal two-sided ideals I_α, I_β are orthogonal; consequently, the characters χ_α, χ_β of the corresponding irreducible representations are also orthogonal, and hence $\chi_\alpha(g) \not\equiv \chi_\beta(g)$. From this we conclude that the corresponding irreducible representations are not equivalent.

We have thus proved the following theorem.

THEOREM 5. *If $\mathfrak{G}$ is a compact group then every minimal two-sided ideal I_α in $L^2(\mathfrak{G})$ determines an irreducible unitary representation $g \to U_g^{(\alpha)}$ of the group $\mathfrak{G}$ and every irreducible unitary representation of the group $\mathfrak{G}$ is equivalent to one of the representations $g \to U_g^{(\alpha)}$. The representations $g \to U_g^{(\alpha)}$, corresponding to distinct ideals I_α, are not equivalent and the characters corresponding to them are orthogonal. The matrix elements $c_{jk}^{(\alpha)}(g)$ of these representations form a complete orthogonal system in $L^2(\mathfrak{G})$.*

The last assertion of the theorem follows from formulas (7), because the p_{kj} form an orthogonal basis in the corresponding ideal I_α, and $L^2(\mathfrak{G})$ is the orthogonal sum of the ideals I_α. It also follows from this that for two non-equivalent irreducible representations $g \to ||c_{jk}(g)||$, $g \to ||c'_{jk}(g)||$,

$$\int c_{jk}(g)\overline{c'_{j'k'}(g)}d\mu(g) = 0 \tag{8}$$

and that

$$\int c_{jk}(g)\overline{c_{j'k'}(g)}d\mu(g) = \begin{cases} \dfrac{1}{n} \text{ for } j = j',\ k = k', \\ 0 \text{ for } j \neq j' \text{ or } k \neq k', \end{cases} \tag{9}$$

where n is the dimension of the representation.

Relations (8) and (9) are called the *orthogonality relations*.

From these relations and the last assertion of Theorem 5, it follows that

$$\int |x(g)|^2 d\mu(g) = \sum_\alpha \sum_{j,k=1}^{n_\alpha} n_\alpha \left| \int x(g)\overline{c_{jk}^{(\alpha)}(g)}d\mu(g) \right|^2 \tag{10}$$

for an arbitrary function $x \in L^2(\mathfrak{G})$. But the numbers $x_{kj}^{(\alpha)} = \int x(g)\overline{c_{jk}^{(\alpha)}(g)}d\mu(g)$ are the matrix elements of the operator

$$T_x^{(\alpha)} = \int x(g) U_g^{(\alpha)*} d\mu(g) \tag{11}$$

with respect to the basis $n_\alpha^{-\frac{1}{2}}p_{11}, n_\alpha^{-\frac{1}{2}}p_{21}, \cdots, n_\alpha^{-\frac{1}{2}}p_{n_\alpha 1}$; consequently, the sum

$$\sum_{j,k=1}^{n_\alpha} |x_{kj}^{(\alpha)}|^2 = \sum_{j,k=1}^{n_\alpha} \left| \int x(g)\overline{c_{jk}^{(\alpha)}(g)}d\mu(g) \right|^2$$

is the trace $S(T_x^{(\alpha)}T_x^{(\alpha)*})$ of the operator $T_x^{(\alpha)}T_x^{(\alpha)*}$, and formula (10) assumes the form

$$\int |x(g)|^2 d\mu(g) = \sum_{\alpha} n_\alpha S(T_x^{(\alpha)}T_x^{(\alpha)*}). \tag{12}$$

Formula (12) is called the *Plancherel formula for a compact group.* It can be considered as the analogue of the Plancherel formula for a commutative group if we assume that the role of the Fourier transform is played by the transition from the function x to the operator $T_x^{(\alpha)}$ according to formula (11).

3. Tensor product of representations. Suppose $g \to U_g$, $g \to U'_g$ are representations of the group $\mathfrak{G}$ in the finite-dimensional spaces R_n, R'_m of dimensions n, m respectively. We can assume that the spaces R_n, R'_m are realized as the spaces of all ordered n- respectively m-tuples $x = \langle x_1, x_2, \cdots, x_n \rangle$, $y = \langle y_1, y_2, \cdots, y_m \rangle$ of complex numbers $x_1, x_2, \cdots, x_n$ respectively $y_1, y_2, \cdots, y_m$ with the usual definitions of the operations in them, and the operators of our representations as linear transformations

$$x'_j = \sum_{k=1}^{n} u_{jk}(g)x_k;\ y'_\mu = \sum_{\nu=1}^{m} u'_{\mu\nu}(g)y_\nu;\ j = 1, 2, \cdots, n;\ \mu = 1, 2, \cdots, m \tag{1}$$

in these spaces.

We denote by $R_{n,m}$ the vector space of dimension nm, consisting of all systems $x = \{x_{j\mu}\}$, $j = 1, 2, \cdots, n$; $\mu = 1, 2, \cdots, m$, of complex numbers $x_{j\mu}$ in which the operations are defined in the usual manner:

$$\lambda x = \{\lambda x_{j\mu}\},\ x+y = \{x_{j\mu}+y_{j\mu}\}$$

with $x = \{x_{j\mu}\}$, $y = \{y_{j\mu}\}$. The *tensor product* $U_g \times U'_g$ of the operators U_g, U'_g (i.e. of the transformations (1)) is the linear transformation

$$x'_{j\mu} = \sum_{k=1}^{n} \sum_{\nu=1}^{m} u_{jk}(g)u'_{\mu\nu}(g)x_{k\nu}. \tag{2}$$

The correspondence $g \to U_g \times U'_g$ *is a representation of the group* $\mathfrak{G}$. In fact, to the product $g_1 g_2$ there corresponds the successive application of first the transformations (1), corresponding to g_2, and then the transformations (1), corresponding to g_1, and hence, also the successive application of first the transformation (2), corresponding to g_2, and then the transformation (2), corresponding to g_1.

The representation $g \to U_g \times U'_g$ is called the *tensor* (or the *Kronecker*) product of the representations $g \to U_g$, $g \to U'_g$.

Suppose now that the representations $g \to U_g$, $g \to U'_g$ are unitary; without loss of generality, we can assume that the inner product in R_n, R'_m is given by the formulas

$$(x, x') = \sum_{k=1}^{n} x_k \overline{x'_k},\ (y, y') = \sum_{\mu=1}^{m} y_\mu \overline{y'_\mu},$$

respectively. We define the inner product in $R_{n,m}$ by the formula

$$(x, x') = \sum_{k=1}^{n} \sum_{\mu=1}^{m} x_{k\mu} \overline{x'_{k\mu}};$$

a simple computation shows that then the tensor representation $g \to U_g \times U'_g$ is a unitary representation in the space $R_{n,m}$. Even if the representations $g \to U_g$, $g \to U'_g$ are irreducible, the representation $g \to U_g \times U'_g$ is, in general, reducible. One of the important problems, having applications in theoretical physics, is to find, for various concrete groups, formulas which realize the decomposition of the tensor product of two irreducible representations into irreducible representations.

4. Duality theorem for a compact group. Suppose now that $\mathfrak{G}$ is a compact group. If $g \to U(g)$ and $g \to V(g)$ are two finite-dimensional unitary representations of the group $\mathfrak{G}$, then their tensor product $g \to U(g) \times V(g)$ is also a unitary representation of this group. From this it follows that the linear hull R of the matrix elements of all finite-dimensional unitary representations of the group $\mathfrak{G}$ with the usual definitions of the operations forms a commutative algebra of functions on $\mathfrak{G}$; this algebra is called the *representation algebra* of the group $\mathfrak{G}$.

If $g \to ||u_{jk}(g)||$ is a unitary representation of the group $\mathfrak{G}$, given in the matrix form, then the correspondence $g \to ||\overline{u_{jk}(g)}||$, where $\overline{u_{jk}(g)}$ is the complex conjugate of the function $u_{jk}(g)$, is also a unitary representation of the group $\mathfrak{G}$; consequently, the representation algebra R also contains the complex conjugate of every function in it.

We now consider all irreducible representations $g \to U_g^{(\alpha)}$ or, in the matrix form, $g \to ||c_{jk}^{(\alpha)}(g)||$, constructed in subsection 2. Every finite-dimensional unitary representation $g \to U_g$ is equivalent to the direct sum of a finite number of representations $g \to ||c_{jk}^{(\alpha)}(g)||$; from this we conclude that the matrix elements $c_{jk}^{(\alpha)}(g)$ form a basis in R. In particular, the tensor product $g \to U_g^{(\alpha)} \times V_g^{(\beta)}$ of any two of these irreducible representations is equivalent to the direct sum of certain of the representations $g \to U_g^{(\alpha)}$. This can be written symbolically in the form

$$U_g^{(\alpha)} \times U_g^{(\beta)} = S^{-1}(U_g^{(\alpha_1)} \dot{+} U_g^{(\alpha_2)} \dot{+} \cdots \dot{+} U_g^{(\alpha_p)})S, \tag{1}$$

where S is a unitary numerical matrix, and $\alpha_1, \alpha_2, \cdots, \alpha_p$ are certain values of the index α; S, p and $\alpha_1, \cdots, \alpha_p$ depend, in general, on α and β. It follows from (1) that

$$\chi_\alpha(g)\chi_\beta(g) = \chi_{\alpha_1}(g) + \chi_{\alpha_2}(g) + \cdots + \chi_{\alpha_p}(g), \tag{2}$$

where $\chi_\beta(g)$ is a character of the representation $g \to U_g^{(\beta)}$; integrating (2), we obtain

$$\int \chi_\alpha(g)\chi_\beta(g)d\mu(g) = \int \chi_{\alpha_1}(g)d\mu(g) + \int \chi_{\alpha_2}(g)d\mu(g) + \cdots + \int \chi_{\alpha_p}(g)d\mu(g). \tag{3}$$

But, by what we proved in subsection 2, the left member of (3) is either unity or zero depending on whether the equalities $c_{jk}^{(\beta)}(g) = \overline{c_{jk}^{(\alpha)}(g)}$ holds or not; further, the integral $\int \chi_\alpha(g)d\mu(g)$ is unity or zero depending on whether the representation $g \to U_g^{(\alpha)}$ is the identity representation or not. The representation $g \to U_g$ is called the *identity representation* if U_g is the identity operator in a one-dimensional space for arbitrary $g \in \mathfrak{G}$; clearly, the character of the identity representation is identically equal to unity.

It follows from this discussion that the right member of formula (1) contains the identity representation if, and only if, $c_{jk}^{(\beta)}(g) = \overline{c_{jk}^{(\alpha)}(g)}$ and, in this case, the identity representation is contained in formula (1) exactly once.

A symmetric commutative algebra R is called a *square block-algebra* if it contains a basis which can be decomposed into disjoint sets U_α, with n_α^2 elements, arranged in the form of square matrices $U_\alpha = ||u_{jk}^{(\alpha)}||$, $j, k = 1, 2, \cdots, n_\alpha$, in such a way that the following conditions are satisfied:

1) the set consisting of only the identity element e is one of the sets U_α;

2) to each set U_α there corresponds a numerical matrix S_α of order n_α which is such that $S_\alpha^{-1}||u_{jk}^{(\alpha)*}||S_\alpha$ is also one of the sets U_α;

3) for any two sets U_α, U_β there exists a numerical unitary matrix $S_{\alpha\beta}$ such that

$$U_\alpha \times U_\beta = S_{\alpha\beta}^{-1}(U_{\alpha_1}+U_{\alpha_1}+\cdots+U_{\alpha_p})S_{\alpha\beta} \tag{4}$$

for certain $\alpha_1, \cdots, \alpha_p$ [$U_\alpha \times U_\beta$ denotes the matrix $||u_{jk}^{(\alpha)}u_{\mu\nu}^{(\beta)}||$; thus, formula (4) gives the rule for multiplying the basis elements and hence also for multiplying arbitrary elements of the algebra R];

4) the set $\{e\}$ either is not contained in the right member of (4) at all, or it is contained exactly once; the last case holds if, and only if, $||u_{jk}^{(\beta)}|| = S_\alpha^{-1}||u_{jk}^{(\alpha)*}||S_\alpha$, where S_α is some numerical unitary matrix;

5) for an arbitrary set $U_\alpha = ||u_{jk}^{(\alpha)}||$ the condition

$$||u_{jk}^{(\alpha)}||\ ||u_{kj}^{(\alpha)*}|| = ||\delta_{jk}e|| \tag{5}$$

is satisfied, where

$$\delta_{jk} = \begin{cases} 1 \text{ for } j = k, \\ 0 \text{ for } j \neq k. \end{cases}$$

The sets U_α are called the *blocks* of the given square block-algebra.

The preceding discussion shows that *the representation algebra of a locally compact group is a square block-algebra.*

Suppose now that R is an arbitrary square block-algebra. A functional f on R is called *elementary* if it realizes a symmetric homomorphism of the algebra R into the field of complex numbers.

In virtue of condition (5), every elementary functional f generates a mapping $c_{jk}^{(\alpha)} = f(u_{jk}^{(\alpha)})$ of the blocks U_α into unitary matrices, $f(U_\alpha) = ||c_{jk}^{(\alpha)}||$, satisfying the same relations (4) as the blocks themselves. It is easily seen that, conversely, every mapping $U_\alpha \to f(U_\alpha)$ of blocks into unitary matrices of the same orders n_α, satisfying the same relations (4) as the blocks U_α themselves, is generated by some elementary functional. We can therefore define the product f_1f_2 of two elementary functionals by setting $f = f_1f_2$, if $f(U_\alpha) = f_1(U_\alpha)f_2(U_\alpha)$ for all blocks U_α. It is easily verified that with this definition of multiplication the set $\mathfrak{G}(R)$ of all elementary functionals on R forms a group; moreover, the identity of the group $\mathfrak{G}(R)$ will be the functional f which assigns to each block U_α the identity matrix of order n_α. A topology can be introduced in the group $\mathfrak{G}(R)$ by taking the neighborhood basis to be all possible sets of functionals $f \in \mathfrak{G}(R)$ which satisfy the inequalities

$$|f(U_{\alpha_k}) - f_0(U_{\alpha_k})| < \varepsilon,$$
$$k = 1, 2, \cdots, n,$$

for all possible fixed f_0, $\varepsilon > 0$ and $\alpha_1, \alpha_2, \cdots, \alpha_n$.

It is easily verified that then $\mathfrak{G}(R)$ becomes a topological group. Moreover, it follows from a line of reasoning analogous to that used in the proof of Theorem 2, subsection 3, § 11, that with this definition of the topology, $\mathfrak{G}(R)$ is a compact space; consequently, $\mathfrak{G}(R)$ is a compact topological group. This group is called the *representation group* of the block-algebra R.

It turns out that the correspondence between compact groups and square block-algebras established in this way is dual in the following sense:

THEOREM 6 (M. KREIN'S [8, 9] DUALITY THEOREM). *If $\mathfrak{G}$ is a compact group and R is its representation algebra, then the representation group $\mathfrak{G}(R)$ of the algebra R is topologically isomorphic to the initial group and the isomorphism between $\mathfrak{G}$ and $\mathfrak{G}(R)$ is given by the formula*

$$f(U_\alpha) = ||c_{jk}^{(\alpha)}(g)||,$$

where $g \to ||c_{jk}^{(\alpha)}(g)||$ *is an irreducible representation of the group* $\mathfrak{G}$, *corresponding to the block* U_α *of the algebra* R.

If R *is a square block-algebra and* $\mathfrak{G} = \mathfrak{G}(R)$ *is its representation group, then a symmetric isomorphism can be established between the algebra* R *and the representation algebra* $R(\mathfrak{G})$ *of the group* $\mathfrak{G}$ *under which the blocks* U_α *of the algebra* R *go over into a complete system* $g \to ||c_{jk}^{(\alpha)}(g)||$ *of irreducible representations of the group* $\mathfrak{G}$.

(For the proof of this theorem, the reader is referred to M. KREIN [9]; we note that the first part of Theorem 6 was obtained earlier, independently, by TANNAKA [1].)

Theorem 6 can be considered as the analogue of Pontryagin's duality theorem (see subsection 6, § 31) for a locally compact commutative group.

A further generalization of the duality theorem to arbitrary (not necessarily commutative) groups was subsequently obtained by STINESPRING [2]. The general theory of duality has been worked out by KATS [1, 3]. In this theory, there is constructed a class Ω of objects (the so-called *algebra groups*) which possess the following properties: 1) Ω contains all locally compact groups, 2) together with each object ω, the class Ω also contains the dual object ω^*, 3) ω^{**} is isomorphic to ω. Concerning duality theory, also see KELLEY [4], TAKESAKI [1] and TATSUUMA [1].

The results of §§ 28, 29 and subsections 1–2, § 30, are due basically to Gelfand, Raikov and Naimark (see GELFAND and RAIKOV [2], GELFAND and NAIMARK [2–4, 6], NAIMARK [2]); some of these results were later obtained independently by GODEMENT [3] and some by SEGAL [4]. The results of subsections 3–5, § 30 are due to GELFAND and RAIKOV [2], RAIKOV [8], and GODEMENT [3]; of subsections 1–3 and 6, § 31 to GELFAND and RAIKOV [1], RAIKOV [2, 3, 6]; of subsection 10, § 31 to GELFAND [8, 9]; of subsection 4, § 31 to M. KREIN [7]; of subsection 4, § 32 to TANNAKA [1] and M. KREIN [8, 9]. The results of subsection 4, § 31 were also obtained independently and somewhat earlier by WEIL [1] who, however, made use of the refined Pontryagin-van Kampen theorems on the structure of locally compact commutative groups in his proofs (see also POVZNER [1, 4]).

The Plancherel theorem was generalized in the papers by GELFAND and NAIMARK [4, 5, 7] first for the case of the group of all complex unimodular matrices of the second order, and later to matrices of the n-th order. Later, this result of Gelfand and Naimark was generalized to other non-commutative groups also (see GELFAND and GRAEV [1], HARISH-CHANDRA [2, 5], SEGAL [9]).

The theory of positive definite kernels and algebras generated by them, generalizing the theory of positive definite functions, is due to M. KREIN [9]. The generalization of a number of results of Gelfand and Raikov to groups which are the direct product of a compact group with a locally compact commutative group is given in LYUBARSKY [1].

The theory of generalized translation operations (subsection 11, § 31) is due to LEVITAN [1–10] (the notion of generalized translation operations was introduced by DELSARTE [1, 2]) and its application to Sturm-Liouville differential operators (subsection 11, § 31) is due to Levitan and Povzner (see LEVITAN [1–8], LEVITAN and POVZNER [1], and POVZNER [2, 3, 7]).

Theorem 6, subsection 7, § 31 is due to NAIMARK [1]; it was obtained independently and somewhat later in many other works (see, for instance, AMBROSE [1]).

Theorems of Tauberian type were obtained by GODEMENT [2] by applying the results of Gelfand and Shilov (see § 15).

The theory of representations of compact groups was worked out by PETER and WEYL [1].

In conclusion, we note that much additional information related to the material of Chapters II-VI can be found in the monographs of GELFAND, RAIKOV and SHILOV [1], DIXMIER [20], HEWITT and ROSS [1], and RICKART [6].

CHAPTER VII

ALGEBRAS OF OPERATORS IN HILBERT SPACE

§ 33. Various topologies in the algebra $\mathfrak{B}(\mathfrak{H})$

Let $\mathfrak{B}(\mathfrak{H})$ denote, as usual, the algebra of all bounded linear operators in a fixed Hilbert space $\mathfrak{H}$. A topology can be defined in $\mathfrak{B}(\mathfrak{H})$ in various ways, with respect to which it is transformed into a topological algebra.

We shall now consider the most important of these topologies.

1. Weak topology. Let us recall (see Example 2, subsection 3, § 8) that a *weak neighborhood* $U(A_0;f_1, \cdots, f_n, \varphi_1, \cdots, \varphi_n; \varepsilon)$ of the operator A_0 is the set of all operators $A \in \mathfrak{B}(\mathfrak{H})$ which satisfy the inequalities

$$|((A-A_0)f_k, \varphi_k)| < \varepsilon, k = 1, 2, \cdots, n$$

with fixed $\varepsilon > 0$ and fixed $f_1, \cdots, f_n, \varphi_1, \cdots, \varphi_n \in \mathfrak{H}$. The topology in which all possible neighborhoods $U(A_0;f_1, \cdots, f_n, \varphi_1, \cdots, \varphi_n; \varepsilon)$ form a neighborhood basis is called the *weak topology* in $\mathfrak{B}(\mathfrak{H})$.

It is easy to verify that $\mathfrak{B}(\mathfrak{H})$ is a topological algebra in this topology.

Since $|((A-A_0)f_k, \varphi_k)| = |((A^*-A_0^*)\varphi_k, f_k)|$, the transition from A to A^* is continuous in the weak topology.

The closure of a set $S \subset \mathfrak{B}(\mathfrak{H})$ in the weak topology will be denoted by $\bar{S}^1$.

2. Strong topology. Suppose $\varphi_1, \cdots, \varphi_s$ are elements in the space $\mathfrak{H}$ and that ε is a positive number. A *strong neighborhood* $V(A_0; \varphi_1, \cdots, \varphi_s; \varepsilon)$ of the operator A_0 is the set of all operators A which satisfy the inequalities

$$|(A-A_0)\varphi_k| < \varepsilon, k = 1, \cdots, s.$$

The topology in the algebra $\mathfrak{B}(\mathfrak{H})$, in which all possible neighborhoods $V(A_0; \varphi_1, \cdots, \varphi_s; \varepsilon)$ form a neighborhood basis, is called the *strong topology*.

The closure of a set $S \subset \mathfrak{B}(\mathfrak{H})$ in the strong topology will be denoted by $\bar{S}^2$.

The convergence of the sequence $\langle A_n \rangle$ to the operator A in this topology means that $|A_nf - Af| \to 0$ for an arbitrary vector $f \subset \mathfrak{H}$. This type of convergence is called *strong convergence* and in this case the operator A is called the *strong limit* of the sequence $\langle A_n \rangle$. $\mathfrak{B}(\mathfrak{H})$ is a Hausdorff topological space in this topology also, and the expressions αA, $A+B$, AB are continuous (the last for each of the factors with the other factor fixed); consequently $\mathfrak{B}(\mathfrak{H})$ is a topological algebra in this topology also. However, in this topology the transition from A to A^* is no longer always continuous.

In order to verify this, it suffices to show that there exists a sequence of operators $\langle A_n \rangle$ which converges strongly to zero, but that at the same time $\langle A_n^* \rangle$ does not converge strongly to zero. Suppose, for example, $\mathfrak{H}$ is separable. In order to construct such a sequence, we choose an orthonormal basis $\{\varphi_1, \varphi_2, \varphi_3, \cdots\}$ in $\mathfrak{H}$. We define a bounded linear operator U by the equalities

$$U\varphi_n = \varphi_{n-1}, \quad n = 2, 3, 4, \cdots; \quad U\varphi_1 = 0,$$

and we set $A_n = U^n$. Then if f is an arbitrary vector in $\mathfrak{H}$, we can write

$$f = \sum_{k=1}^{\infty} \alpha_k \varphi_k, \quad \alpha_k = (f, \varphi_k), \quad \sum_{k=1}^{\infty} |\alpha_k|^2 < \infty;$$

consequently,

$$|A_n f|^2 = |\sum_{k=1}^{\infty} \alpha_k U^n \varphi_k|^2 = |\sum_{k=n+1}^{\infty} \alpha_k \varphi_{k-n}|^2 = \sum_{k=n+1}^{\infty} |\alpha_k|^2 \to 0.$$

Thus, the sequence $\langle A_n \rangle$ tends strongly to zero. On the other hand, it is clear that $U^* \varphi_n = \varphi_{n+1}$ and hence

$$|A_n^* f|^2 = |\sum_{k=1}^{\infty} \alpha_k U^{*n} \varphi_k| = |\sum_{k=1}^{\infty} \alpha_k \varphi_{k+n}|^2 = \sum_{k=1}^{\infty} |\alpha_k|^2 = |f|^2 > 0 \quad \text{if } f \neq 0. \tag{1}$$

These relations show that $\langle A_n^* f \rangle$ does not tend strongly to zero. Consequently, the transition from A to A^* is not continuous in the strong topology.

On the other hand, it is continuous in the weak topology and, therefore, the strong topology is distinct from the weak. Moreover, *the strong topology is stronger than the weak topology because every weak neighborhood contains a strong neighborhood.* In fact, if $U(A_0; \varphi_1, \cdots, \varphi_s, \psi_1, \cdots, \psi_s; \varepsilon)$ is a prescribed weak neighborhood of the operator A_0, then it contains the strong neighborhood

$$V(A_0; \varphi_1, \cdots, \varphi_s; \delta), \quad \delta = \frac{\varepsilon}{\max(|\psi_1|, \cdots, |\psi_s|)}$$

of this operator. Namely, the relation $A \in V(A_0; \varphi_1, \cdots, \varphi_s; \delta)$ means that $|(A - A_0)\varphi_k| < \delta$, $k = 1, \cdots, s$. But then

$$|((A - A_0)\varphi_k, \psi_k)| \leqq |(A - A_0)\varphi_k| \, |\psi_k| < \delta \max(|\psi_1|, \cdots, |\psi_s|) = \varepsilon,$$

i.e.

$$A \in U(A_0; \varphi_1, \cdots, \varphi_s, \psi_1, \cdots, \psi_s; \varepsilon).$$

It follows from the assertion just proved that every strong contact point is also a contact point in the weak topology, and therefore $\bar{S}^2 \subset \bar{S}^1$ (see subsections 1–3, § 2).

The converse assertion is not true, because the weak topology does not coincide with the strong.

The sequence $\langle A_n \rangle$, considered above, can also be used to prove that the product AB is not continuous in the weak topology with respect to both variables A, B. In fact, otherwise, the sequence $\langle A_n A_n^* \rangle$ would have to tend weakly to 0, whereas at the same time $A_n A_n^* = 1$.

We shall show, furthermore, that neither is the product AB continuous with respect to both variables A, B in the strong topology.

To this end, we shall choose an element f with norm equal to unity ($|f| = 1$), and a strong neighborhood of zero $V(0; f; \varepsilon)$, $0 < \varepsilon < 1$. Further, we set

$$A_{n,\delta} = \frac{1}{\delta} A_n, \quad B_{n,\delta} = \delta A_n^*, \quad \delta > 0.$$

Then

$$|A_{n,\delta} B_{n,\delta} f| = |A_n A_n^* f| = |f| = 1 > \varepsilon,$$

i.e.

$$A_{n,\delta} B_{n,\delta} \notin V(0; f; \varepsilon)$$

for arbitrary n and δ. On the other hand, for any neighborhoods $V(0; \varphi_1, \cdots, \varphi_p; \varepsilon_1)$, $V(0; \psi_1, \cdots, \psi_q; \varepsilon_2)$ whatsoever, if we take $\delta < \varepsilon_2/\max(|\psi_1|, \cdots, |\psi_q|)$ then

$$|B_{n,\delta} \psi_k| = \delta |A_n^* \psi_k| = \delta |\psi_k| < \frac{\varepsilon_2}{\max(|\psi_1|, \cdots, |\psi_q|)} |\psi_k| \leqq \varepsilon_2$$

for any n, i.e.

$$B_{n,\delta} \in V(0; \psi_1, \cdots, \psi_q; \varepsilon_2).$$

Further, choosing an n such that

$$|A_n \varphi_k| < \varepsilon_1 \delta \quad (k = 1, 2, \cdots, p),$$

we have

$$|A_{n,\delta} \varphi_k| < \varepsilon_1 \quad (k = 1, 2, \cdots, p),$$

i.e.

$$A_{n,\delta} \in U(0; \varphi_1, \cdots, \varphi_p; \varepsilon_1).$$

Thus, in any two strong neighborhoods of the zero operator there exist $A_{n,\delta}$, $B_{n,\delta}$ such that $A_{n,\delta} B_{n,\delta}$ is not in the prescribed neighborhood of 0, i.e. AB is not continuous for $A = 0$, $B = 0$.

3. Strongest topology. Suppose $f_1, f_2, \cdots$ are given elements in $\mathfrak{H}$ such that $\sum_{k=1}^{\infty} |f_k|^2 < \infty$, and that ε is a positive number. A *strongest neighborhood* $W(A_0; f_1, f_2, \cdots; \varepsilon)$ of the operator A_0 is the set of all operators A which satisfy the inequality $\sum_{k=1}^{\infty} |(A - A_0) f_k|^2 < \varepsilon^2$. The topology in the algebra $\mathfrak{B}(\mathfrak{H})$, in which a neighborhood basis is formed by all possible neighborhoods $W(A_0; f_1, f_2, \cdots; \varepsilon)$, is called the *strongest topology*. The closure of a set S in the strongest topology will be denoted by $\bar{S}^3$. In this topology, the algebra $\mathfrak{B}(\mathfrak{H})$ also turns out to be a Hausdorff space. In this topology also the expressions αA, $A + B$, AB are continuous with respect to α, A, B, where AB is continuous only with respect to each factor when the other factor is fixed. The fact that AB is not continuous with respect to both factors also in the strongest topology can be established by again considering the operators $A_{n,\delta}$ and $B_{n,\delta}$.

The strongest topology majorizes the strong topology, i.e. every strong neighborhood contains a strongest neighborhood.

In fact, if $V(A_0;f_1,\cdots,f_s;\varepsilon)$ is a prescribed strong neighborhood of the operator A_0, then

$$W(A_0;f_1,\cdots,f_s,f_{s+1},0,\cdots;\varepsilon) \subset V(A_0;f_1,\cdots,f_s;\varepsilon),$$

because it follows from the inequality $\sum_{k=1}^{\infty}|(A-A_0)f_k|^2 < \varepsilon^2$ that

$$|(A-A_0)f_k| < \varepsilon \quad (k = 1, 2, 3, \cdots).$$

Consequently,

$$\bar{S}^3 \subset \bar{S}^2.$$

It is easy to verify that the strongest topology does not coincide with the strong topology and therefore it is stronger than the latter.

4. Uniform topology. The *uniform topology* in $\mathfrak{B}(\mathfrak{H})$ is the topology which is determined by the operator norm. Consequently, all possible open balls defined by the inequalities $|A-A_0| < \varepsilon$, $\varepsilon > 0$ form a neighborhood basis in the uniform topology.

With norm equal to the operator norm, $\mathfrak{B}(\mathfrak{H})$ is a normed algebra and therefore the product AB is continuous in the uniform topology with respect to both factors. Consequently, the uniform topology is different from the preceding topologies and, as can easily be seen, it is stronger than both of them.

Thus, if we denote closure in the uniform topology by $\bar{S}^4$, we have

$$\bar{S}^4 \subset \bar{S}^3 \subset \bar{S}^2 \subset \bar{S}^1.$$

§ 34. Weakly closed subalgebras of the algebra $\mathfrak{B}(\mathfrak{H})$

1. Fundamental concepts. Suppose $\mathfrak{H}$ is a Hilbert space; the set $\mathfrak{B}(\mathfrak{H})$ of all bounded linear operators in $\mathfrak{H}$ is a symmetric Banach algebra if we define the operations in $\mathfrak{B}(\mathfrak{H})$ as the operations on operators, the norm as the operator norm, and involution as the transition to the adjoint operator. A symmetric subalgebra of the algebra $\mathfrak{B}(\mathfrak{H})$ is called a *weakly closed algebra* if it is closed in the weak topology in $\mathfrak{B}(\mathfrak{H})$. [Such algebras are also called *von Neumann algebras.*]

In the sequel we shall consider primarily weakly closed algebras.

The intersection of all weakly closed algebras, which contain a given set $S \subset \mathfrak{B}(\mathfrak{H})$, is the *minimal weakly closed algebra containing* S; it is denoted by $R(S)$.

Suppose S is an arbitrary set in $\mathfrak{B}(\mathfrak{H})$; we denote by S^* the set of all operators A^*, $A \in S$, and by S' the set of all operators in $\mathfrak{B}(\mathfrak{H})$, which commute with all the operators in $S \cup S^*$. By V, subsection 1, § 8, S' is a weakly closed algebra with identity (here and in the sequel in this chapter "identity" means the identity operator 1 in $\mathfrak{H}$); it is called *the commutant of the given set S.*

The operation of transition from S to S' has, obviously, the following properties:
1) $S' = (R_{a*}(S))' = (R(S))'$;
2) if $S_1 \subset S_2$, then $S_1' \supset S_2'$;
3) $S \subset S''$.
[Recall that $R_{a*}(S)$ denotes the minimal symmetric algebra containing the given set S (see subsection 1, § 10).]

Relation 3) shows that S'' is a weakly closed algebra which contains the set S; it therefore contains the minimal such algebra $R(S)$, i.e. $R(S) \subset S''$. Further, applying relation 3) to the set S', we have $S' \subset S'''$. On the other hand, applying the operation $'$ to both members of relation 3) and using property 2), we obtain that $S' \supset S'''$, so that $S' = S'''$. Substituting $S', S'', \cdots$ for S then yields

$$S'' = S^{IV} = \cdots, \quad S' = S''' = S^{V} = \cdots.$$

2. Principal identity. Suppose S is an arbitrary subset of the algebra $\mathfrak{B}(\mathfrak{H})$ and that $\mathfrak{N}$ is the set of all elements of the space $\mathfrak{H}$ on which the operators A and A^* vanish, whenever $A \in S$. The projection operator E_0 onto $\mathfrak{H} \ominus \mathfrak{N}$ is called the *principal identity* of the set S. Thus, by definition, the equalities $Af = A^*f = 0$ for all $A \in S$ and $E_0 f = 0$ are equivalent.

If S is an algebra, containing the identity operator, then, clearly, $\mathfrak{N} = (0)$, so that the principal identity equals 1. Further, the principal identity E_0 of the set S satisfies the equalities

$$E_0 A = AE_0 = A \tag{1}$$

for arbitrary operator $A \in S$. In fact, by the definition of the principal identity, $A(1-E_0)f = A^*(1-E_0)f = 0$ for an arbitrary vector $f \in \mathfrak{H}$, i.e. $A(1-E_0) = 0$, $A^*(1-E_0) = 0$. If we apply the involution operation to both members of the last equality, we obtain $(1-E_0)A = 0$, so that

$$A - AE_0 = A - E_0 A = 0, \quad \text{i.e.} \quad AE_0 = E_0 A = A.$$

[It follows from this that if M is an arbitrary symmetric subalgebra in $\mathfrak{B}(\mathfrak{H})$ and E_0 is its principal identity, then M is an algebra with identity in the space $E_0\mathfrak{H}$, and all operators in M vanish on $(1-E_0)\mathfrak{H}$.]

LEMMA. *The principal identity of the set S belongs to S' and S''.*

Proof. The first assertion follows from (1). Suppose further that $B \in S'$; if $E_0 f = 0$, then $Af = A^*f = 0$ for all $A \in S$. But then

$$ABf = BAf = 0, \; A^*Bf = BA^*f = 0,$$

so that $E_0 Bf = 0$. If we take $f = (1-E_0)g$, $g \in \mathfrak{H}$, we obtain that

$$E_0 B(1-E_0) = 0, \; E_0 B = E_0 BE_0.$$

Replacing B by B^* in the last equality and applying the involution operation to the resulting equality, we arrive at the equality $BE_0 = E_0 BE_0$; consequently

$$E_0 B = BE_0 = E_0 BE_0.$$

Hence, the operator E_0 commutes with any operator $B \in S'$; consequently, $E_0 \in S''$.

We shall now use this lemma to prove the following fundamental theorem.

THEOREM. *Suppose S is an arbitrary subset of the algebra $\mathfrak{B}(\mathfrak{H})$ and that E_0 is the principal identity of the set S. Then $R(S)$ is the strongest closure of the algebra $R_{a*}(S)$; $R(S)$ consists of those and only those elements $A \in S''$ which satisfy the condition*

$$E_0 A = AE_0 = A. \tag{2}$$

Proof. We denote by S_1 the set of all elements of the algebra S'' which satisfy condition (2). Suppose $A \in S_1$; we shall prove that A is a strongest contact point of the algebra $R_{a*}(S)$, i.e. that any strongest neighborhood $W(A; f_1^0, f_2^0, \cdots; \varepsilon)$ contains elements of the algebra $R_{a*}(S)$. To this end, we form the direct sum

$$\mathfrak{H}' = \mathfrak{H} \oplus \mathfrak{H} \oplus \mathfrak{H} \oplus \cdots$$

of a denumerable number of copies of the space $\mathfrak{H}$; clearly, $f_0' = \langle f_1^0, f_2^0, \cdots \rangle \in \mathfrak{H}'$. For an arbitrary bounded linear operator B, we have $\sum_{k=1}^{\infty} |Bf_k^0|^2 \leqq |B|^2 \sum_{k=1}^{\infty} |f_k^0|^2$ so that the vector

$$f_B' = \langle Bf_1^0, Bf_2^0, \cdots \rangle$$

is also an $\in \mathfrak{H}'$. We shall call f_B' the *image* of the operator B in the space $\mathfrak{H}'$.

Suppose $\mathfrak{C}'$ is the set of all images of operators of the algebra $R_{a*}(S)$ and that $\overline{\mathfrak{C}}'$ is the closure of the set $\mathfrak{C}'$ relative to the norm in $\mathfrak{H}'$. Clearly, $\mathfrak{C}'$ is linear, so that $\overline{\mathfrak{C}}'$ is linear and closed. Suppose E' is a projection operator in $\mathfrak{H}'$ on $\overline{\mathfrak{C}}'$; then the correspondence $E' \sim ||E_{t,s}||$, $t, s = 1, 2, 3, \cdots$ (see subsection 15, § 5) holds. If C is an arbitrary element of the algebra $R_{a*}(S)$, then $C' \sim ||\delta_{ts} C||$, where $t, s = 1, 2, 3, \cdots$ and

$$\delta_{ts} = \begin{cases} 1 & \text{for } t = s, \\ 0 & \text{for } t \neq s, \end{cases}$$

is an operator in $\mathfrak{H}'$ and $C' f_B' = \langle CBf_1^0, CB_2^0, \cdots \rangle = f_{CB}'$. Therefore, if $B \in R_{a*}(S)$, then $f_{CB}' \in \mathfrak{C}'$, so that the operator C' maps $\mathfrak{C}'$ into $\mathfrak{C}'$, and hence $\overline{\mathfrak{C}}'$ into $\overline{\mathfrak{C}}'$. Since $E'f' \in \overline{\mathfrak{C}}'$ for an arbitrary $f' \in \mathfrak{H}'$, it follows from this that $C'E'f' \in \overline{\mathfrak{C}}'$, i.e.

$$C'E' = E'C'E'. \tag{3}$$

Replacing C by C^* in (3) (where C^* is also an $\in R_{a*}(S)$) and making use of the relation $C^{*\prime} = C'^*$ (see subsection 15, § 5), we obtain that

$$C'^*E' = E'C'^*E', E'C' = (E'C'^*E')^* = E'C'E' = C'E'.$$

The last relation is equivalent to the relations $CE_{ts} = E_{ts}C$, $t, s = 1, 2, 3, \cdots$; since C is an arbitrary element of the algebra $R_{a*}(S)$, we have $E_{ts} \in (R_{a*}(S))' = S'$.

We now return to the given element A and show that

$$f'_A \in \overline{\mathfrak{C}}'.$$

Since, by definition, $f'_B \in \mathfrak{C}'$, $B \in R_{a*}(S)$ implies that $E'f'_B = f'_B$, i.e.

$$Bf_t^0 = \sum_{s=1}^{\infty} E_{ts} Bf_s^0 = \sum_{s=1}^{\infty} BE_{ts} f_s^0, \quad B(f_t^0 - \sum_{s=1}^{\infty} E_{ts} f_s^0) = 0$$

for all operators $B \in R_{a*}(S)$. But then, by the definition of the principal identity E_0, we also have

$$E_0(f_t^0 - \sum_{s=1}^{\infty} E_{ts} f_s^0) = 0,$$

and consequently, in virtue of condition (2),

$$A(f_t^0 - \sum_{s=1}^{\infty} E_{ts} f_s^0) = AE_0(f_t^0 - \sum_{s=1}^{\infty} E_{ts} f_s^0) = 0.$$

(The set S and the algebra $R_{a*}(S)$ obviously have the same principal identity.) Since $A \in S''$, the last equality can be written in the form

$$Af_t^0 - \sum_{s=1}^{\infty} E_{ts} Af_s^0 = 0, \quad \text{i.e. } E'f'_A = f'_A,$$

so that $f'_A \in \overline{\mathfrak{C}}'$. But then, for arbitrary $\varepsilon > 0$, there exists an element in $\mathfrak{C}'$, whose distance from f'_A is less than ε. As an element of the space $\mathfrak{C}'$ it is representable in the form f'_B, $B \in R_{a*}(S)$; therefore, recalling the definition of distance in the space $\mathfrak{H}'$, we obtain

$$\sum_{k=1}^{\infty} |Af_k^0 - Bf_k^0|^2 = |f'_A - f'_B|^2 < \varepsilon^2,$$

i.e.

$$B \in W(A; f_1^0, f_2^0, \cdots; \varepsilon).$$

Thus, the set S_1 of all elements $A \in S''$, which satisfy condition (2), is contained in the strongest closure of the algebra $R_{a*}(S)$, and hence in its weak closure $R(S)$ also. On the other hand, it is clear that $S \subset S_1$, and therefore also $R_{a*}(S) \subset S_1$. Since the algebra S_1 is weakly closed, it follows that also $R(S) \subset S_1$. (The algebra S_1 is weakly closed in virtue of the weak continuity of the products AE_0, E_0A for variable A.) Thus, $R(S) = S_1 = \overline{R_{a*}(S)}^3$, and the theorem is proved.

COROLLARY 1. *If R is a symmetric subalgebra of the algebra $\mathfrak{B}(\mathfrak{H})$, then its weak, strong, and strongest closures coincide*:

$$\bar{R}^1 = \bar{R}^2 = \bar{R}^3. \tag{4}$$

In fact, according to the preceding theorem, $\bar{R}^1 = \bar{R}^3$; on the other hand, $\bar{R}^1 \supset \bar{R}^2 \supset \bar{R}^3$.

COROLLARY 2. *If M is a weakly closed algebra, containing the identity, then $M'' = M$.*

In fact, if we set $S = M$, we have $E_0 = 1$; consequently, $S_1 = M''$, $M = R(M) = S_1 = M''$.

COROLLARY 3. *The set of all commutants coincides with the set of all weakly closed algebras with identity.*

In fact, every algebra S' contains the identity; on the other hand, every weakly closed algebra M which contains the identity is representable in the form $M = (M')'$.

COROLLARY 4. *For an arbitrary set $S \subset \mathfrak{B}(\mathfrak{H})$, we have $S'' = R(S, 1)$.*

In fact, since the algebra $R(S, 1)$ contains the identity, by 1), subsection 1 and Corollary 2, we have

$$R(S, 1) = (R(S, 1))'' = ((S, 1)')' = (S')' = S''.$$

COROLLARY 5. *Suppose H is a bounded Hermitian operator and that $P(\lambda)$ is its spectral function. A necessary and sufficient condition for the operator H to belong to a given weakly closed algebra M is that all the operators $P(\lambda)$, $\lambda < 0$, and $1-P(\lambda)$, $\lambda \geqq 0$, belong to the algebra M.*

(Here and in the sequel we shall assume the spectral functions to be continuous from the right.)

Proof. The sufficiency follows from the relations

$$H = \int_\alpha^\beta \lambda dP(\lambda) = \lim_{\lambda_k - \lambda_{k-1} \to 0} \sum_{k=1}^n \lambda_k [P(\lambda_k) - P(\lambda_{k-1})], \quad P(\lambda_k) - P(\lambda_{k-1}) \in M,$$

for $\lambda_k \neq 0$, where (α, β) is an interval which contains the entire spectrum of the operator H, $\alpha = \lambda_0 < \lambda_1 < \cdots < \lambda_n = \beta$ (0 is a division point λ_j if $\alpha\beta \leqq 0$), and the limit exists in the uniform topology in $\mathfrak{B}(\mathfrak{H})$. Conversely, if $H \in M$, then according to the properties of the spectral function, $P(\lambda) \in R(H)''$. Suppose E_0 is the principal identity of the algebra $R(H)$; then $H(1-E_0) = 0$, and consequently

$$0 = |H(1-E_0)f|^2 = \int_\alpha^\beta \lambda^2 d|P(\lambda)(1-E_0)f|^2,$$

so that also $P(\Delta)(1-E_0) = 0$ whenever the interval $\Delta = (\alpha', \beta')$ does not contain the point 0. Applying the involution operation to both members of this equality, we obtain that also $(1-E_0)P(\Delta) = 0$ so that according to the preceding theorem, $P(\Delta) \in R(H) \subset M$. If we take $\alpha' = \lambda > 0$, $\beta' > \beta$, it follows that $1-P(\lambda) \in M$ and by continuity from the right this relation is valid for $\lambda \geqq 0$; for $\alpha' < \alpha$, $\beta' = \lambda < 0$, we obtain that $P(\lambda) \in M$.

COROLLARY 6. *If M^P is the set of all projection operators in the weakly closed algebra M, then $R(M^P) = M$.*

Proof. Since $M^P \subset M$, we have $R(M^P) \subset M$, so that $[R(M^P)]^P \subset M^P$. On the other hand, obviously $[R(M^P)]^P \supset M^P$, so that $[R(M^P)]^P = M^P$. Thus, the algebras $R(M^P)$ and M contain the same projection operators, and hence, according to Corollary 5, they contain the same Hermitian operators. But every element $A \in M$ is representable in the form $A = H_1 + iH_2$, where $H_1 = (A+A^*)/2$, $H_2 = (A-A^*)/2i$ are Hermitian operators in the algebra M. Consequently, the algebras M and $R(M^P)$ coincide.

COROLLARY 7. *The set M^U of all unitary operators in the weakly closed algebra M is nonvoid if and only if M contains the identity, and in this case $R(M^U) = M$.*

Proof. If $U \in M^U$, then $1 = U^*U \in M$, so that M contains the identity. Conversely, if M contains the identity, then it follows from $P \in M^P$ that $2P-1 \in M^U$. In fact, $2P-1 \in M$, and, in virtue of the relations $(2P-1)^*(2P-1) = (2P-1)(2P-1)^* = 4P^2-4P+1 = 1$, $2P-1$ is a unitary operator. It follows that

$$P = \tfrac{1}{2}[(2P-1)+1] \in R(M^U), \quad M^P \subset R(M^U), \quad M = R(M^P) \subset R(M^U).$$

On the other hand, $M^U \subset M$ implies that $R(M^U) \subset M$ so that $R(M^U) = M$.

COROLLARY 8. *The principal identity E_0 of the weakly closed algebra M belongs to $M \cap M'$.*

In fact, $E_0 \in M''$ and $A = E_0$ satisfies condition (2) so that $E_0 \in M$ according to the preceding theorem; furthermore, $E_0 \in M'$ according to the preceding lemma.

3. Center. Let us recall (see subsection 3, § 7) that the *center* Z_M of the algebra M is the set of all elements M that commute with any element in M. Clearly, $Z_M = M \cap M'$, so that Z_M is an algebra and furthermore it is commutative. If the algebra M contains the identity, then Z_M is nonvoid, namely, it contains the algebra of all products $\alpha 1$, where the α's are scalars. The latter algebra will be called the *algebra of scalars* and it will be denoted by $(\alpha 1)$.

If M is a weakly closed algebra, then $Z_M = M \cap M'$ is also a weakly closed algebra. The weakly closed algebra M is called a *factor* if Z_M consists only of $(\alpha 1)$, i.e. if $M \cap M' = (\alpha 1)$. The algebra $\mathfrak{B}(\mathfrak{H})$ is an example of a factor. In fact, since $(\alpha 1)' = \mathfrak{B}(\mathfrak{H})$, then in virtue of Corollary 2, we have $\mathfrak{B}(\mathfrak{H})' = (\alpha 1)'' = (\alpha 1)$, $\mathfrak{B}(\mathfrak{H}) \cap \mathfrak{B}(\mathfrak{H})' = (\alpha 1)$. Another example of a factor is the algebra of scalars $(\alpha 1)$. In the sequel, we shall consider less trivial examples of factors. If M is a factor, then M' is also a factor because the equality $M \cap M' = (\alpha 1)$ can be written in the form $M'' \cap M' = (\alpha 1)$.

Another limiting case is a commutative algebra. If M is a commutative algebra then $Z_M = M$. Conversely, if $Z_M = M$, then the algebra M is commutative because it then coincides with the commutative algebra Z_M.

4. Factorization. The *union* $R(M_1, M_2, \cdots)$ of the weakly closed algebras $M_1, M_2, \cdots$ is the minimal weakly closed algebra generated by the set-theoretic union $M_1 \cup M_2 \cup \cdots$. (The set of algebras $M_1, M_2, \cdots$ is not assumed to be denumerable.) It is easily seen that

$$(R(M_1, M_2, \cdots))' = M_1' \cap M_2' \cap \cdots, \tag{1}$$

and if the algebra $R(M_1, M_2, \cdots)$ contains the identity then

$$R(M_1, M_2, \cdots) = (M_1' \cap M_2' \cap \cdots)'. \tag{2}$$

In fact, according to relation 1), subsection 1, $(R(M_1, M_2, \cdots))' = (M_1 \cup M_2 \cup \cdots)' = M_1' \cap M_2' \cap \cdots$. But relation (2) is obtained, in virtue of Corollary 2, from (1) by application of the $'$ operation to both members.

The set $\{M_1, M_2, \cdots\}$ of weakly closed algebras, all distinct from (0), is called a *factorization* if the elements of distinct algebras of this set commute and if their union $R(M_1, M_2, \cdots)$ coincides with the algebra $\mathfrak{B}(\mathfrak{H})$. The individual algebras $M_1, M_2, \cdots$ of the factorization are called its *elements*. It is easily seen that the elements of a factorization are factors. In fact, if $\{M_1, M_2, \cdots\}$ is a factorization, then $M_i \subset M_k'$, $i \neq k$ and, consequently, by (1),

$$M_i \cap M_i' \subset M_i' \bigcap_{k \neq i} M_k' = \bigcap_i M_i' = (R(M_1, M_2, \cdots))' = (\alpha 1).$$

Therefore, if E_0 is the principal identity of the algebra M_i then, according to Corollary 8, $E_0 = \alpha 1$, so that either $\alpha = 1$ or $\alpha = 0$. But in the latter case, $E_0 = 0$ and for an arbitrary operator $A \in M_i$ we have $A = E_0 A = 0$, $M_i = (0)$, which contradicts our assumption. Therefore $\alpha = 1$, and the algebra M_i contains the identity and $M_i \cap M_i' = (\alpha 1)$. Conversely, *if M is a factor, then $\{M, M'\}$ is a factorization*, because

$$R(M, M') = \{R(M, M')\}'' = (M \cup M')'' = (M' \cap M'')' = (M' \cap M)' = (\alpha 1)' = \mathfrak{B}(\mathfrak{H}).$$

A factorization of the form $\{M, M'\}$ is called a *coupled* factorization.

Murray and von Neumann [1] showed that not every factorization $\{M_1, M_2\}$ is coupled, and they found some sufficient conditions under which it can be asserted that $\{M_1, M_2\}$ is a coupled factorization.

§ 35. Relative equivalence

1. Operators and subspaces adjoined to an algebra. Suppose M is a weakly closed algebra with identity, $\mathfrak{S}$ is a subset of the space $\mathfrak{H}$, and that R is an operator (which is not necessarily bounded) in $\mathfrak{H}$. We shall say that the set $\mathfrak{S}$ and the operator R are *adjoined to the algebra M* (in their original paper, Murray and von Neumann [1] use the term "belong"), and we write $\mathfrak{S} \eta M$, $R \eta M$, if the set $\mathfrak{S}$ is invariant with respect to every unitary operator in M' and the operator R commutes with every such operator, respectively.

If $R \in \mathfrak{B}(\mathfrak{H})$, i.e. the operator R is bounded, then the relation $R \eta M$ is equivalent to the relation $R \in M$, because the relation $R \eta M$ then means that $R \in (M'^U)' = (M')' = M$. But if R is a closed linear operator with domain dense in $\mathfrak{H}$, and $R = WH$ is its canonical decomposition (see II, subsection 1, § 21), then in virtue of the uniqueness of this decomposition, the equality $U^*RU = U^*WUU^*HU$

is the canonical decomposition of the operator U^*RU and, consequently, the equality $U^*RU = R$ means that $U^*WU = W$, $U^*HU = H$, i.e. $W \in M$, $H\eta M$.

Further, if the set $\mathfrak{S}$ is linear and closed, then the relation $\mathfrak{S}\eta M$ is equivalent to the relation $P_{\mathfrak{S}} \in M$, where $P_{\mathfrak{S}}$ is a projection operator onto $\mathfrak{S}$. In fact, the relation $\mathfrak{S}\eta M$ means that $U\mathfrak{S} \subset \mathfrak{S}$, whatever the unitary operator $U \in M'$. Replacing U by U^* everywhere, we have that $U^*\mathfrak{S} \subset \mathfrak{S}$ from which it follows that $\mathfrak{S} \subset U\mathfrak{S}$, so that $\mathfrak{S} = U\mathfrak{S}$. But then also $U(\mathfrak{H} \ominus \mathfrak{S}) = \mathfrak{H} \ominus \mathfrak{S}$ and hence

$$P_{\mathfrak{S}} Ux = Ux = UP_{\mathfrak{S}}x \quad \text{for } x \in \mathfrak{S},$$

$$P_{\mathfrak{S}} Ux = 0 = UP_{\mathfrak{S}}x \quad \text{for } x \in \mathfrak{H} \ominus \mathfrak{S}.$$

Consequently, $P_{\mathfrak{S}}U = UP_{\mathfrak{S}}$ for all $U \in M'$. This means that $P_{\mathfrak{S}}\eta M$, $P_{\mathfrak{S}} \in M$.

Thus, for a closed subspace $\mathfrak{M}$, the relation $\mathfrak{M}\eta M$ means that the operator $P_{\mathfrak{M}}$ commutes with all the operators in M', i.e. that $\mathfrak{M}$ reduces all the operators in M'.

A simple example of a closed subspace, adjoined to the algebra M, is obtained by taking a fixed element $f \in \mathfrak{H}$, applying all the operators $A \in M'$ to it, and then taking the closure relative to the norm in $\mathfrak{H}$ of the set $\mathfrak{S}_f$ of all such elements Af. In fact, if U is a unitary operator belonging to M', then $UA \in M'$ and consequently $UAf \in \mathfrak{S}_f$. It follows that $U\mathfrak{S}_f \subset \mathfrak{S}_f$, $U\overline{\mathfrak{S}}_f \subset \overline{\mathfrak{S}}_f$.

We introduce the notation $\overline{\mathfrak{S}}_f = \mathfrak{M}_f^{M'}$, $P_{\overline{\mathfrak{S}}_f} = E_f^{M'}$. We define $\mathfrak{M}_f^M$ and E_f^M analogously; they are obtained if M is replaced by the factor M' and hence M' by the factor $M'' = M$.

Clearly, $f \in \mathfrak{M}_f^{M'}$ because it suffices to set $A = 1$. Thus $\mathfrak{M}_f^{M'}$ is the minimal closed subspace which reduces all the operators in M' and contains f. From this it follows that *if $\mathfrak{M}$ is a closed subspace, $\mathfrak{M}\eta M$ and $f \in \mathfrak{M}$, then also $\mathfrak{M}_f^{M'} \subset \mathfrak{M}$.* An analogous statement holds for $\mathfrak{M}_f^M$ and M'.

2. Fundamental lemma. We shall need the following lemma in the sequel.

LEMMA. *If M is a factor and $A \in M$, $A' \in M'$, then the equality $AA' = 0$ is possible if, and only if, either $A = 0$ or $A' = 0$.*

Proof. Suppose $\mathfrak{M}$ is the set of all elements $f \in \mathfrak{H}$ for which $AXf = 0$ whatever the operator $X \in M$. In particular, $Af = 0$ for $f \in \mathfrak{M}$. Clearly, $\mathfrak{M}$ is a closed subspace in $\mathfrak{H}$. Let P be a projection operator on $\mathfrak{M}$; we shall prove that P commutes with all the elements in M and in M', i.e. that $\mathfrak{M}$ reduces all operators in M and in M'.

Since the algebras M and M' contain A^* together with any A, it suffices to show that the subspace $\mathfrak{M}$ is invariant with respect to all the operators in M and in M'. Suppose $f \in \mathfrak{M}$; for an arbitrary operator $B \in M$ we also have $XB \in M$ and, consequently, $AXBf = 0$; this means that $Bf \in \mathfrak{M}$, so that the subspace $\mathfrak{M}$ is invariant with respect to all the operators in M. But if $f \in \mathfrak{M}$, $B \in M'$, then, since the operator B commutes with the operators A and X, $AXBf = BAXf = 0$. From this we conclude that $\mathfrak{M}$ is also invariant with respect to all the operators in M'. But then $P \in M \cap M' = (\alpha 1)$, so that either $P = 0$ or $P = 1$, i.e. either $\mathfrak{M} = (0)$ or $\mathfrak{M} = \mathfrak{H}$.

Since $AA' = 0$,

$$AXA'f = AA'Xf = 0$$

for arbitrary $f \in \mathfrak{H}$, $X \in M$, and consequently $A'f \in \mathfrak{M}$. It follows, in the first case, that $A'f = 0$ whatever the element $f \in \mathfrak{H}$, i.e. $A' = 0$. Further, for arbitrary $g \in \mathfrak{H}, f \in \mathfrak{M}$,

$$(A^*g, f) = (g, Af) = 0,$$

so that $A^*g \perp \mathfrak{M}$. It follows that, in the second case, $A^*g = 0$, $A^* = 0$, and hence that $A = 0$ also.

3. Definition of relative equivalence. Everywhere in the remainder of this section we shall assume that M is a factor.

The closed subspaces $\mathfrak{M}$ and $\mathfrak{N}$ are said to be *equivalent with respect to M* if there exists a partially isometric operator U in the algebra M with initial domain $\mathfrak{M}$ and terminal domain $\mathfrak{N}$ (see subsection 14, § 5). In this case, we shall write $\mathfrak{M} \sim \mathfrak{N}(\cdots M)$, and also $P_{\mathfrak{M}} \sim P_{\mathfrak{N}}(\cdots M)$. If $\mathfrak{M} \sim \mathfrak{N}(\cdots M)$, then, obviously, $\mathfrak{M}$ and $\mathfrak{N}$ have the same dimension. The converse is not true *a priori* because it is necessary not simply to map $\mathfrak{M}$ isometrically onto $\mathfrak{N}$ but rather to carry out this mapping with the aid of an operator of the algebra M. $\mathfrak{M} \sim \mathfrak{N}(\cdots M)$ implies that $\mathfrak{M}\eta M$ and $\mathfrak{N}\eta M$ because $P_{\mathfrak{M}} = U^*U$, $P_{\mathfrak{N}} = UU^*$, so that the concept of equivalence is meaningful only for subspaces adjoined to the algebra M. Further, it is obvious that if $\mathfrak{M}, \mathfrak{N}, \mathfrak{P}\eta M$, then $\mathfrak{M} \sim \mathfrak{M}$; $\mathfrak{M} \sim \mathfrak{N}$ whenever $\mathfrak{N} \sim \mathfrak{M}$; $\mathfrak{M} \sim \mathfrak{N}$, $\mathfrak{N} \sim \mathfrak{P}$ imply that $\mathfrak{M} \sim \mathfrak{P}$. Finally, the relations $\mathfrak{M}_\alpha \sim \mathfrak{N}_\alpha$, $\alpha \in \mathfrak{A}$, $\mathfrak{M}_\alpha \perp \mathfrak{M}_\beta$, $\mathfrak{N}_\alpha \perp \mathfrak{N}_\beta$ with $\alpha \neq \beta$ imply that $\sum_{\alpha\in\mathfrak{A}} \oplus \mathfrak{M}_\alpha \sim \sum_{\alpha\in\mathfrak{A}} \oplus \mathfrak{N}_\alpha$ because if U_α is a partially isometric operator in the algebra M with initial domain $\mathfrak{M}_\alpha$ and terminal domain $\mathfrak{N}_\alpha$, then $U = \sum_{\alpha\in\mathfrak{A}} U_\alpha$ is a partially isometric operator in the algebra M with initial domain $\sum_{\alpha\in\mathfrak{A}} \oplus \mathfrak{M}_\alpha$ and terminal domain $\sum_{\alpha\in\mathfrak{A}} \oplus \mathfrak{N}_\alpha$.

Examples of closed subspaces which are equivalent with respect to M are the closures of the ranges $\mathfrak{R}(X)$, $\mathfrak{R}(X^*)$ of the operators X and X^*, where X is an arbitrary closed linear operator with domain dense in $\mathfrak{H}$, adjoined to the algebra M. (As a matter of convenience, we shall write $\mathfrak{D}(X)$ and $\mathfrak{R}(X)$ instead of $\mathfrak{D}_X$ and $\mathfrak{R}_X$ in this chapter.) In fact, if $X = WH$ is the canonical decomposition of the operator X then W is a partially isometric operator in the algebra M with initial domain $\overline{\mathfrak{R}(X^*)}$ and terminal domain $\overline{\mathfrak{R}(X)}$ (see subsection 1 and II, subsection 1, § 21).

4. Comparison of closed subspaces. We shall now make use of the notion of equivalence to order the set of all closed subspaces $\mathfrak{M}\eta M$.

Suppose $\mathfrak{M}$ and $\mathfrak{N}$ are two closed subspaces adjoined to the factor M; we shall say that the subspace $\mathfrak{M}$ *is not greater than* the subspace $\mathfrak{N}$ with respect to M, and we shall write $\mathfrak{M} \lesssim \mathfrak{N}(\cdots M)$ (or $P_{\mathfrak{M}} \lesssim P_{\mathfrak{N}}(\cdots M)$), if $\mathfrak{M}$ is equivalent to a subset of the subspace $\mathfrak{N}$. We shall say that the subspace $\mathfrak{M}$ *is less than* the subspace $\mathfrak{N}$, and we shall write $\mathfrak{M} < \mathfrak{N}(\cdots M)$, $P_{\mathfrak{M}} < P_{\mathfrak{N}}(\cdots M)$, if $\mathfrak{M} \lesssim \mathfrak{N}$ and $\mathfrak{M}$ is not equivalent to $\mathfrak{N}$.

In the sequel the term "subspace" will mean *closed* subspace.

I. *If $\mathfrak{M}$ and $\mathfrak{N}$ are completely arbitrary subspaces adjoined to the factor M, then we always have that either $\mathfrak{M} \lesssim \mathfrak{N}(\cdots M)$ or $\mathfrak{N} \lesssim \mathfrak{M}(\cdots M)$.*

Proof. We can obviously assume that $\mathfrak{M} \neq (0)$ and $\mathfrak{N} \neq (0)$. We shall prove that in this case there exist subspaces $\mathfrak{M}_1$ and $\mathfrak{N}_1$ in $\mathfrak{M}$ and $\mathfrak{N}$ respectively, where $\mathfrak{M}_1$ and $\mathfrak{N}_1$ are distinct from (0), are also adjoined to the algebra M, and are equivalent with respect to M. To this end, we choose an element $f \neq 0$ in $\mathfrak{M}$. Then $E_f^M \in M'$, $E_f^M \neq 0$; since, moreover, $P_{\mathfrak{N}} \in M$, $P_{\mathfrak{N}} \neq 0$, then according to the lemma in subsection 2 we also have that $P_{\mathfrak{N}}E_f^M \neq 0$. But $P_{\mathfrak{N}}E_f^M$ is a projection operator onto $\mathfrak{N} \cap \mathfrak{M}_f^M$ and consequently $\mathfrak{N} \cap \mathfrak{M}_f^M \neq (0)$. We choose an element g in the space $\mathfrak{N} \cap \mathfrak{M}_f^M$, whose norm equals unity. Since $g \in \mathfrak{M}_f^M$ there exists an operator A in the algebra M such that $|g-Af| < 1$ and, consequently, such that *a fortiori* $|P_{\mathfrak{N}}g-P_{\mathfrak{N}}Af| < 1$. But $P_{\mathfrak{N}}g = g$, $P_{\mathfrak{M}}f = f$ and therefore the last inequality may be written in the form $|g-P_{\mathfrak{N}}AP_{\mathfrak{M}}f| < 1$. This inequality shows that the operator $X = P_{\mathfrak{N}}AP_{\mathfrak{M}} \neq 0$ because otherwise we should have $|g| < 1$ whereas, by assumption, $|g| = 1$. We set $\mathfrak{M}_1 = \overline{\mathfrak{R}(X^*)}$, $\mathfrak{N}_1 = \overline{\mathfrak{R}(X)}$; then $\mathfrak{M}_1 \neq (0)$, $\mathfrak{N}_1 \neq (0)$ and $\mathfrak{M}_1 = \overline{\mathfrak{R}(P_{\mathfrak{M}}A^*P_{\mathfrak{N}})} \subset \mathfrak{M}$, $\mathfrak{N}_1 = \overline{\mathfrak{R}(P_{\mathfrak{N}}AP_{\mathfrak{M}})} \subset \mathfrak{N}$. Since $X = P_{\mathfrak{N}}AP_{\mathfrak{M}} \in M$, the subspaces $\mathfrak{M}_1$ and $\mathfrak{N}_1$ are equivalent with respect to M (see subsection 3).

In virtue of the relations

$$P_{\mathfrak{M}\ominus\mathfrak{M}_1} = P_{\mathfrak{M}}-P_{\mathfrak{M}_1} \in M, \quad P_{\mathfrak{N}\ominus\mathfrak{N}_1} = P_{\mathfrak{N}}-P_{\mathfrak{N}_1} \in M,$$

the subspaces $\mathfrak{M} \ominus \mathfrak{M}_1$, $\mathfrak{N} \ominus \mathfrak{N}_1$ are also adjoined to the algebra M. If, furthermore, they are both distinct from (0), then, applying the preceding result to them, we obtain that there exist subspaces $\mathfrak{M}_2$ and $\mathfrak{N}_2$ in $\mathfrak{M} \ominus \mathfrak{M}_1$ and $\mathfrak{N} \ominus \mathfrak{N}_1$, respectively, which are distinct from (0) and are equivalent with respect to M. Moreover, it is clear that $\mathfrak{M}_1 \perp \mathfrak{M}_2$ and $\mathfrak{N}_1 \perp \mathfrak{N}_2$. Repeating this reasoning and applying the Zorn lemma, we can construct subspaces $\mathfrak{M}_\alpha$, $\mathfrak{N}_\alpha$, distinct from (0), such that $\mathfrak{M}_\alpha \sim \mathfrak{N}_\alpha(\cdots M)$ and such that either $\sum_\alpha \oplus \mathfrak{M}_\alpha = \mathfrak{M}$ or $\sum_\alpha \oplus \mathfrak{N}_\alpha = \mathfrak{N}$. Since $\sum_\alpha \oplus \mathfrak{M}_\alpha \sim \sum_\alpha \oplus \mathfrak{N}_\alpha(\cdots M)$ we have that $\mathfrak{M} \lesssim \mathfrak{N}$ in the first case and that $\mathfrak{N} \lesssim \mathfrak{M}$ in the second.

II. *If $\mathfrak{M} \lesssim \mathfrak{N}$ and $\mathfrak{N} \lesssim \mathfrak{M}$ then $\mathfrak{M} \sim \mathfrak{N}$.*

The proof of this proposition is analogous to the proof of the Bernstein theorem on the comparison of powers. Suppose

$$\mathfrak{M} \sim \mathfrak{N}' \subset \mathfrak{N}, \tag{1}$$

$$\mathfrak{N} \sim \mathfrak{M}' \subset \mathfrak{M}, \tag{2}$$

and let $U \in M$ be a partially isometric operator with initial domain $\mathfrak{N}$ and terminal domain $\mathfrak{M}'$. It maps the subspace $\mathfrak{N}' \subset \mathfrak{N}$ isometrically onto some subspace $\mathfrak{M}'' \subset \mathfrak{M}'$. Therefore $UP_{\mathfrak{N}'}$ is a partially isometric operator in the algebra M with initial domain $\mathfrak{N}'$ and terminal domain $\mathfrak{M}''$ so that $\mathfrak{N}' \sim \mathfrak{M}''$. From this and (1) it follows that $\mathfrak{M} \sim \mathfrak{M}''$. Suppose V is a partially isometric operator in the algebra M with initial domain $\mathfrak{M}$ and terminal domain $\mathfrak{M}''$; we set $\mathfrak{M}^{(2\nu)} = V^\nu\mathfrak{M}$, $\mathfrak{M}^{(2\nu+1)} = V^\nu\mathfrak{M}'$ (so that $\mathfrak{M}^{(0)} = \mathfrak{M}$, $\mathfrak{M}^{(1)} = \mathfrak{M}'$, $\mathfrak{M}^{(2)} = \mathfrak{M}''$). It is easily seen that $\mathfrak{M}^{(0)} \supset \mathfrak{M}^{(1)} \supset \mathfrak{M}^{(2)} \supset \cdots$ and that $\mathfrak{M}^{(\nu)} \eta M$, $\nu = 0, 1, 2, 3, \cdots$; moreover, the operator V maps the space $\mathfrak{M}^{(p)}$ isometrically onto the space $\mathfrak{M}^{(p+2)}$, the space $\mathfrak{M}^{(p+1)}$ onto the space $\mathfrak{M}^{(p+3)}$ and, consequently, the space $\mathfrak{M}^{(p)} \ominus \mathfrak{M}^{(p+1)}$ onto the space $\mathfrak{M}^{(p+2)} \ominus \mathfrak{M}^{(p+3)}$. From this it follows, as above, that $\mathfrak{M}^{(p)} \ominus \mathfrak{M}^{(p+1)} \sim \mathfrak{M}^{(p+2)} \ominus \mathfrak{M}^{(p+3)}(\cdots M)$. Therefore the subspaces in odd positions in the equality

$$\mathfrak{M}^{(0)} = [\mathfrak{M}^{(0)} \ominus \mathfrak{M}^{(1)}] \oplus [\mathfrak{M}^{(1)} \ominus \mathfrak{M}^{(2)}] \oplus [\mathfrak{M}^{(2)} \ominus \mathfrak{M}^{(3)}] \oplus \cdots \oplus \mathfrak{M}^{(0)} \cap \mathfrak{M}^{(1)} \cap \mathfrak{M}^{(2)} \cap \cdots \tag{3}$$

are equivalent to the subspaces in even positions in the equality

$$\mathfrak{M}^{(1)} = [\mathfrak{M}^{(1)} \ominus \mathfrak{M}^{(2)}] \oplus [\mathfrak{M}^{(2)} \ominus \mathfrak{M}^{(3)}] \oplus [\mathfrak{M}^{(3)} \ominus \mathfrak{M}^{(4)}] \oplus \cdots \oplus \mathfrak{M}^{(1)} \cap \mathfrak{M}^{(2)} \cap \mathfrak{M}^{(3)} \cap \cdots, \tag{4}$$

and subspaces in even positions in equality (3) are equal, and hence also equivalent, to the subspaces in odd positions in equality (4). Furthermore, $\mathfrak{M}^{(0)} \cap \mathfrak{M}^{(1)} \cap \mathfrak{M}^{(2)} \cap \cdots = \mathfrak{M}^{(1)} \cap \mathfrak{M}^{(2)} \cap \mathfrak{M}^{(3)} \cap \cdots$. Consequently, $\mathfrak{M}^{(0)} \sim \mathfrak{M}^{(1)}$, i.e. $\mathfrak{M} \sim \mathfrak{M}'$. On the other hand, in virtue of (2), $\mathfrak{M}' \sim \mathfrak{N}$. Therefore $\mathfrak{M} \sim \mathfrak{N}$.

III. *For every pair of subspaces* $\mathfrak{M}$ *and* $\mathfrak{N}$, *adjoined to the factor* M, *one, and only one, of the following three relations holds*:

$$\mathfrak{M} \sim \mathfrak{N}(\cdots M), \mathfrak{M} < \mathfrak{N}(\cdots M), \mathfrak{M} > \mathfrak{N}(\cdots M).$$

In fact, in virtue of Proposition I, either $\mathfrak{M} \sim \mathfrak{N}' \subset \mathfrak{N}$ or $\mathfrak{N} \sim \mathfrak{M}' \subset \mathfrak{M}$. If both these relations hold, then, in virtue of Proposition II, $\mathfrak{M} \sim \mathfrak{N}$. But if only the first or only the second of these relations holds, then $\mathfrak{M} < \mathfrak{N}$ or $\mathfrak{N} < \mathfrak{M}$ respectively.

IV. *If* $\mathfrak{M} \sim \mathfrak{N}$ *then the relations* $\mathfrak{M} < \mathfrak{P}$, $\mathfrak{M} \sim \mathfrak{P}$, $\mathfrak{M} > \mathfrak{P}$ *are equivalent respectively to the relations* $\mathfrak{N} < \mathfrak{P}$, $\mathfrak{N} \sim \mathfrak{P}$, $\mathfrak{N} > \mathfrak{P}$.

In fact, relation $\mathfrak{M} \lesssim \mathfrak{P}$ means that $\mathfrak{M} \sim \mathfrak{P}' \subset \mathfrak{P}$. But then also $\mathfrak{N} \sim \mathfrak{P}' \subset \mathfrak{P}$, i.e. $\mathfrak{N} \lesssim \mathfrak{P}$. If, in this connection, $\mathfrak{N} \sim \mathfrak{P}$, then also $\mathfrak{M} \sim \mathfrak{P}$; therefore, the relation $\mathfrak{M} < \mathfrak{P}$ implies the relation $\mathfrak{N} < \mathfrak{P}$. Further, $\mathfrak{P} \leqq \mathfrak{M}$ means that $\mathfrak{P} \sim \mathfrak{M}' \subset \mathfrak{M}$; in virtue of the relation $\mathfrak{N} \sim \mathfrak{M}$ it follows from this that there exists a subspace

$\mathfrak{N}'$ in $\mathfrak{N}$ which is equivalent to the subspace $\mathfrak{M}'$. But then $\mathfrak{P} \sim \mathfrak{N}' \subset \mathfrak{N}$, i.e. $\mathfrak{P} \lesssim \mathfrak{N}$. Again, as above, we conclude from this that the relation $\mathfrak{P} < \mathfrak{N}$ is equivalent to the relation $\mathfrak{P} < \mathfrak{M}$.

V. *If* $\mathfrak{M} < \mathfrak{N}$, $\mathfrak{N} < \mathfrak{P}$ *then* $\mathfrak{M} < \mathfrak{P}$.

It follows from the relations $\mathfrak{M} < \mathfrak{N}$, $\mathfrak{N} \sim \mathfrak{P}' \subset \mathfrak{P}$, in virtue of Proposition IV, that $\mathfrak{M} < \mathfrak{P}' \subset \mathfrak{P}$, i.e. that $\mathfrak{M} \sim \mathfrak{P}'' \subset \mathfrak{P}' \subset \mathfrak{P}$. Thus, $\mathfrak{M} \lesssim \mathfrak{P}$; if we should have $\mathfrak{M} \sim \mathfrak{P}$ then, in virtue of Proposition IV, we would have $\mathfrak{P} < \mathfrak{N}$, which contradicts the assumption. Therefore $\mathfrak{M} < \mathfrak{P}$.

We note further the following simple propositions.

VI. *a*) *If* $\mathfrak{M}$, $\mathfrak{N} \eta M$ *and* $\mathfrak{M} \subset \mathfrak{N}$, *then* $\mathfrak{M} \lesssim \mathfrak{N}$;

b) $(0) \lesssim \mathfrak{M}$ *where* $(0) \sim \mathfrak{M}$ *if, and only if,* $\mathfrak{M} = (0)$;

c) $\mathfrak{M} \lesssim \mathfrak{H}$.

VII. *If* $\mathfrak{M}$, $\mathfrak{N} \eta M$, *then* $\mathfrak{M} \dotplus \mathfrak{N} \eta M$ *and*

$$(\mathfrak{M} \dotplus \mathfrak{N}) \ominus \mathfrak{N} \lesssim \mathfrak{M}. \tag{5}$$

(In distinction to the notation in § 25, everywhere in this chapter we shall denote the minimal *closed* subspace containing $\mathfrak{M}$ and $\mathfrak{N}$ by $\mathfrak{M} \dotplus \mathfrak{N}$; this will obviously be the set of all elements $f+g$, $f \in \mathfrak{M}$, $g \in \mathfrak{N}$ and their strong limits.)

Proposition VI is clear; to prove Proposition VII we note that the space $(\mathfrak{M} \dotplus \mathfrak{N}) \ominus \mathfrak{N}$ is the closure of the set of all elements of the form

$$P_{\mathfrak{H} \ominus \mathfrak{N}}(f+g) = P_{\mathfrak{H} \ominus \mathfrak{N}} f = P_{\mathfrak{H} \ominus \mathfrak{N}} P_{\mathfrak{M}} h, \quad f \in \mathfrak{M}, \ g \in \mathfrak{N}, \ h \in \mathfrak{H}.$$

Thus,

$$(\mathfrak{M} \dotplus \mathfrak{N}) \ominus \mathfrak{N} = \overline{\mathfrak{R}(P_{\mathfrak{H} \ominus \mathfrak{N}} P_{\mathfrak{M}})} \sim \overline{\mathfrak{R}((P_{\mathfrak{H} \ominus \mathfrak{N}} P_{\mathfrak{M}})^*)} = \overline{\mathfrak{R}(P_{\mathfrak{M}} P_{\mathfrak{H} \ominus \mathfrak{N}})} \subset \mathfrak{M},$$

from which relation (5) also follows. Thus, $(\mathfrak{M} \dotplus \mathfrak{N}) \ominus \mathfrak{N} \eta M$ and, consequently, also $\mathfrak{M} \dotplus \mathfrak{N} \eta M$.

5. Finite and infinite subspaces. A subspace $\mathfrak{M}$, adjoined to the factor M, is said to be *infinite* if it is equivalent to a proper subspace of itself, and *finite* otherwise. A projection operator $P \in M$ is said to be *finite* if $P\mathfrak{H}$ is *finite*, and *infinite* if $P\mathfrak{H}$ is infinite.

Clearly, the subspace (0) is finite. Further,

I. *If* $\mathfrak{M} \lesssim \mathfrak{N}$ *and* $\mathfrak{N}$ *is finite then* $\mathfrak{M}$ *is also finite.*

Proof. Suppose $\mathfrak{M} \sim \mathfrak{N}' \subset \mathfrak{N}$ and let $U \in M$ be a partially isometric operator with initial domain $\mathfrak{M}$ and terminal domain $\mathfrak{N}'$. If the subspace $\mathfrak{M}$ is infinite, then $\mathfrak{M} \sim \mathfrak{M}' \subsetneqq \mathfrak{M}$; therefore, setting $U\mathfrak{M}' = \mathfrak{N}''$, we have $\mathfrak{N}'' \sim \mathfrak{M}' \sim \mathfrak{M} \sim \mathfrak{N}'$, $\mathfrak{N}'' \subsetneqq \mathfrak{N}'$. It follows that

$$\mathfrak{N} = \mathfrak{N}' \oplus [\mathfrak{N} \ominus \mathfrak{N}'] \sim \mathfrak{N}'' \oplus [\mathfrak{N} \ominus \mathfrak{N}'] \subsetneqq \mathfrak{N};$$

consequently, $\mathfrak{N}$ is infinite, in contradiction with our assumption.

Thus, either all the subspaces $\mathfrak{M}\eta M$ (in particular, the entire space $\mathfrak{H}$) are finite, or there exist infinite subspaces $\mathfrak{M}\eta M$, and then the entire space $\mathfrak{H}$ is infinite. But every finite subspace is smaller than any infinite subspace.

In the further study of subspaces ηM, we shall use a method analogous to laying off equal segments.

II. *If $\mathfrak{M}$ and $\mathfrak{N}$ are subspaces adjoined to the factor M, then the subspace $\mathfrak{M}$ can be represented in the form*

$$\mathfrak{M} = \Big(\sum_{\alpha\in\mathfrak{A}} \oplus \mathfrak{N}_\alpha\Big) \oplus \mathfrak{P}, \tag{1}$$

where $\mathfrak{A}$ is some index set, $\mathfrak{N}_\alpha \sim \mathfrak{N}$ for all $\alpha \in \mathfrak{A}$ and $\mathfrak{P} < \mathfrak{N}$.

Proof. If $\mathfrak{M} < \mathfrak{N}$, then we obtain (1) by setting $\mathfrak{A} = \phi$ and $\mathfrak{P} = \mathfrak{M}$; and if $\mathfrak{M} \sim \mathfrak{N}$, then we obtain (1) by setting $\mathfrak{A} = \{1\}$, $\mathfrak{N}_1 = \mathfrak{M}$, $\mathfrak{P} = (0)$. Therefore, in the sequel we shall assume that $\mathfrak{N} < \mathfrak{M}$. Then $\mathfrak{N} \sim \mathfrak{N}_1 \subsetneqq \mathfrak{M}$; if $\mathfrak{N} > \mathfrak{M} \ominus \mathfrak{N}_1$, then the representation $\mathfrak{M} = \mathfrak{N}_1 \oplus [\mathfrak{M} \ominus \mathfrak{N}_1]$ is already the one sought ($\mathfrak{A} = \{1\}$, $\mathfrak{P} = \mathfrak{M} \ominus \mathfrak{N}_1$). Otherwise, $\mathfrak{N} \sim \mathfrak{N}_2 \subset \mathfrak{M} \ominus \mathfrak{N}_1$. Applying Zorn's lemma, we conclude that there exists a maximal system $\{\mathfrak{N}_\alpha, \alpha \in \mathfrak{A}\}$ of mutually orthogonal subspaces $\mathfrak{N}_\alpha$ in $\mathfrak{M}$ which are equivalent to the subspace $\mathfrak{N}$. Setting $\mathfrak{P} = \mathfrak{M} \ominus \sum_{\alpha\in\mathfrak{A}} \oplus \mathfrak{N}_\alpha$ and taking into consideration the fact that the system $\{\mathfrak{N}_\alpha, \alpha \in \mathfrak{A}\}$ is maximal, we have $\mathfrak{P} < \mathfrak{N}$ and $\mathfrak{M} = (\sum_{\alpha\in\mathfrak{A}} \oplus \mathfrak{N}_\alpha) \oplus \mathfrak{P}$.

III. *If the set $\mathfrak{A}$ is infinite, then we can set $\mathfrak{P} = (0)$ in equality* (1).

In fact, suppose $\mathfrak{A}' = \mathfrak{A} - \{1\}$; since the sets $\mathfrak{A}$ and $\mathfrak{A}'$ are of the same power,

$$\mathfrak{M} = \Big(\sum_{\alpha\in\mathfrak{A}} \oplus \mathfrak{N}_\alpha\Big) \oplus \mathfrak{P} \sim \Big(\sum_{\alpha\in\mathfrak{A}'} \oplus \mathfrak{N}_\alpha\Big) \oplus \mathfrak{P} \lesssim \Big(\sum_{\alpha\in\mathfrak{A}'} \oplus \mathfrak{N}_\alpha\Big) \oplus \mathfrak{N}_1 \lesssim \mathfrak{M}.$$

If follows that $\mathfrak{M} \sim \sum_{\alpha\in\mathfrak{A}} \oplus \mathfrak{N}_\alpha$; suppose U is a partially isometric operator in the algebra M with initial domain $\mathfrak{M}$ and terminal domain $\sum_{\alpha\in\mathfrak{A}} \oplus \mathfrak{N}_\alpha$. Setting $\mathfrak{N}'_\alpha = U^*\mathfrak{N}_\alpha$, we have $\mathfrak{N}'_\alpha \perp \mathfrak{N}'_\beta$ for $\alpha \neq \beta$; $\mathfrak{N}'_\alpha \sim \mathfrak{N}_\alpha \sim \mathfrak{N}\ (\cdots M)$,

$$\mathfrak{M} = \sum_{\alpha\in\mathfrak{A}} \oplus \mathfrak{N}'_\alpha.$$

IV. *Every infinite subspace $\mathfrak{M}$ can be represented in the form*

$$\mathfrak{M} = \sum_{\alpha\in\mathfrak{A}} \oplus \mathfrak{N}_\alpha,$$

where the subspaces $\mathfrak{N}_\alpha$ are $\neq (0)$ and mutually equivalent, and the set $\mathfrak{A}$ is infinite.

Proof. By assumption, $\mathfrak{M} \sim \mathfrak{M}' \subsetneqq \mathfrak{M}$; suppose U is a partially isometric operator in the algebra M with initial domain $\mathfrak{M}$ and terminal domain $\mathfrak{M}'$. We set

$$\mathfrak{M}^{(n)} = U^n\mathfrak{M}; \quad \mathfrak{N}'_n = \mathfrak{M}^{(n)} \ominus \mathfrak{M}^{(n+1)}, \quad n = 0, 1, 2, \cdots.$$

Then

$$\mathfrak{M}^{(n+1)} \subsetneqq \mathfrak{M}^{(n)}, \quad \mathfrak{N}'_{n+1} = U\mathfrak{N}'_n;$$

consequently, $\mathfrak{N}_n' \neq (0)$, all the subspaces $\mathfrak{N}_n'$ are mutually equivalent, and

$$\mathfrak{M} = (\sum_{n=1}^{\infty} \oplus \mathfrak{N}_n') \oplus \mathfrak{M}^{(0)} \cap \mathfrak{M}^{(1)} \cap \mathfrak{M}^{(2)} \cap \cdots .$$

Applying Proposition II to the subspaces $\mathfrak{M}^{(0)} \cap \mathfrak{M}^{(1)} \cap \mathfrak{M}^{(2)} \cap \cdots$ and $\mathfrak{N}_0'$ (instead of to $\mathfrak{M}$ and $\mathfrak{N}$), we obtain that

$$\mathfrak{M}^{(0)} \cap \mathfrak{M}^{(1)} \cap \mathfrak{M}^{(2)} \cap \cdots = (\sum_{\alpha \in \mathfrak{A}'} \oplus \mathfrak{N}_\alpha') \oplus \mathfrak{P}', \quad \mathfrak{P}' < \mathfrak{N}_0',$$

and

$$\mathfrak{M} = (\sum_{\alpha \in \mathfrak{A}} \oplus \mathfrak{N}_\alpha') \oplus \mathfrak{P}', \quad \mathfrak{A} = \{0, 1, 2, 3, \cdots\} \bigcup \mathfrak{A}'.$$

Since the set $\mathfrak{A}$ is infinite, we have, according to Proposition III, that

$$\mathfrak{M} = \sum_{\alpha \in \mathfrak{A}} \oplus \mathfrak{N}_\alpha, \quad \text{where } \mathfrak{N}_\alpha \sim \mathfrak{N}_\beta \text{ for all } \alpha, \beta \in \mathfrak{A}.$$

V. *Every infinite subspace $\mathfrak{M}$ contains a subspace $\mathfrak{N}$ such that $\mathfrak{M} \sim \mathfrak{N} \sim \mathfrak{M} \ominus \mathfrak{N}$.*

Proof. Decomposing the set $\mathfrak{A}$ in Proposition IV into two infinite subsets $\mathfrak{A}_1$ and $\mathfrak{A}_2$ of the same power and setting $\mathfrak{N} = \sum_{\alpha \in \mathfrak{A}_1} \oplus \mathfrak{N}_\alpha$, we have

$$\sum_{\alpha \in \mathfrak{A}} \oplus \mathfrak{N}_\alpha \sim \sum_{\alpha \in \mathfrak{A}_1} \oplus \mathfrak{N}_\alpha \sim \sum_{\alpha \in \mathfrak{A}_2} \oplus \mathfrak{N}_\alpha,$$

i.e.

$$\mathfrak{M} \sim \mathfrak{N} \sim \mathfrak{M} \ominus \mathfrak{N}.$$

VI. *If $\mathfrak{H}$ is separable, then all infinite subspaces of $\mathfrak{H}$ are mutually equivalent.*

Proof. Suppose the subspaces $\mathfrak{M}$ and $\mathfrak{M}'$ are infinite; representing $\mathfrak{M}$ according to Proposition IV, we have $\mathfrak{M} = \sum_{\alpha \in \mathfrak{A}} \oplus \mathfrak{N}_\alpha$ where the set $\mathfrak{A}$ is denumerable. Further, applying Proposition II to the subspaces $\mathfrak{M}'$ and $\mathfrak{N}_\alpha$, we have

$$\mathfrak{M}' = (\sum_{\alpha \in \mathfrak{A}'} \oplus \mathfrak{N}_\alpha) \oplus \mathfrak{P}',$$

where the set $\mathfrak{A}'$ is finite (perhaps void) or denumerable and $\mathfrak{P}' < \mathfrak{N}_\alpha$. If the set $\mathfrak{A}'$ is finite, then obviously $\mathfrak{M}' \lesssim \mathfrak{M}$; but if $\mathfrak{A}'$ is denumerable, then we can assume that $\mathfrak{P}' = (0)$, so that $\mathfrak{M}' \sim \mathfrak{M}$. Thus, in either of these cases, $\mathfrak{M}' \lesssim \mathfrak{M}$. Interchanging the roles of $\mathfrak{M}$ and $\mathfrak{M}'$, we conclude that $\mathfrak{M}' \sim \mathfrak{M}$.

VII. *The sum of a finite number of finite subspaces is finite.*

The proof of this proposition will be split up into the following four steps:

1⁰ *If the subspaces $\mathfrak{M}$, $\mathfrak{N}$, $\mathfrak{P}\eta M$ are closed, and if $\mathfrak{M} \perp \mathfrak{N}$, $\mathfrak{P} \subset \mathfrak{M} \oplus \mathfrak{N}$, then $\mathfrak{P}$ can be represented in the form $\mathfrak{P} = \mathfrak{M}' \oplus \mathfrak{N}' \oplus \mathfrak{P}'$, where $\mathfrak{M}'$, $\mathfrak{N}'$, $\mathfrak{P}'$ are mutually orthogonal subspaces which are adjoined to the algebra M; $\mathfrak{M}' \subset \mathfrak{M}$, $\mathfrak{N}' \subset \mathfrak{N}$ and $\mathfrak{P}'$ is the range of the operator $1+A$, where A is a closed linear operator adjoined to the algebra M with domain orthogonal to $\mathfrak{M}'$, and range orthogonal to $\mathfrak{N}'$, which is such that the equality $Af = 0$ is possible if, and only if, $f = 0$, and where $\mathfrak{P}' \sim \overline{\mathfrak{D}(A)} \sim \overline{\mathfrak{R}(A)}$.*

We set $\mathfrak{M}' = \mathfrak{M} \cap \mathfrak{P}$, $\mathfrak{N}' = \mathfrak{N} \cap \mathfrak{P}$, $\mathfrak{P}' = \mathfrak{P} \ominus [\mathfrak{M}' \oplus \mathfrak{N}']$. If $f \in \mathfrak{P}'$, then $f \in \mathfrak{P} \subset \mathfrak{M} \oplus \mathfrak{N}$, so that the element f can be represented in the form $f = g+h$, where $g \in \mathfrak{M}$, $h \in \mathfrak{N}$. Since $f \perp \mathfrak{M}'$, $h \in \mathfrak{N} \perp \mathfrak{M}'$, then also $g = f-h \perp \mathfrak{M}'$; analogously, $h \perp \mathfrak{N}'$. We set $h = Ag$; this equality defines the operator A uniquely. In fact, it follows from the equality $g = 0$ that $h = f \in \mathfrak{P}'$ and hence $h \in \mathfrak{P} \cap \mathfrak{N} = \mathfrak{N}'$; since, on the other hand, $h \perp \mathfrak{N}'$, we have $h = 0$. Analogously, it follows from the equality $h = 0$ that $g = 0$ also.

It easily follows from the fact that the set $\mathfrak{P}'$ is linear, closed, and ηM, that the operator A has the same properties. We set $\mathfrak{M}'' = \overline{\mathfrak{D}(A)}$, $\mathfrak{N}'' = \overline{\mathfrak{R}(A)}$, $A_1 = (1+A)P_{\mathfrak{M}''}$. Then the operator $A_1 \eta M$ is also defined on a set dense in $\mathfrak{H}$ and consequently

$$\mathfrak{P}' = \mathfrak{R}(1+A) = \mathfrak{R}(A_1) \sim \overline{\mathfrak{R}(A_1^*)} = \mathfrak{H} \ominus \{f : A_1 f = 0\}$$
$$= \mathfrak{H} \ominus \{f : P_{\mathfrak{M}''} f = 0\} = \mathfrak{H} \ominus (\mathfrak{H} \ominus \mathfrak{M}'') = \mathfrak{M}''.$$

(Here, the expression $\{f: A_1 f = 0\}$ denotes the null-space of the operator A_1, i.e. the set of all those elements f for which $A_1 f = 0$.) Analogously, $\mathfrak{P}' \sim \mathfrak{N}''$.

2^0 *If the subspaces $\mathfrak{M}$, $\mathfrak{N}$, $\mathfrak{P}$ are the same as in Proposition 1^0, then either $\mathfrak{P} \lesssim \mathfrak{M}$ or $(\mathfrak{M} \oplus \mathfrak{N}) \ominus \mathfrak{P} \lesssim \mathfrak{N}$.*

Applying Proposition 1^0 to $\mathfrak{M}$, $\mathfrak{N}$, $\mathfrak{P}$, we obtain the subspaces $\mathfrak{M}_1'$, $\mathfrak{M}_1''$, $\mathfrak{N}_1'$, $\mathfrak{N}_1''$, $\mathfrak{P}_1'$ and the operator A_1; then applying Proposition 1^0 to $\mathfrak{M}, \mathfrak{N}, (\mathfrak{M} \oplus \mathfrak{N}) \ominus \mathfrak{P}$, we obtain the subspaces $\mathfrak{M}_2'$, $\mathfrak{M}_2''$, $\mathfrak{N}_2'$, $\mathfrak{N}_2''$, $\mathfrak{P}_2'$ and the operator A_2.

It is easily seen that $\mathfrak{M}_1' \perp \mathfrak{M}_2'$, $\mathfrak{N}_1' \perp \mathfrak{N}_2'$. We set

$$\mathfrak{M}_0 = \mathfrak{M} \ominus (\mathfrak{M}_1' \oplus \mathfrak{M}_2'), \quad \mathfrak{N}_0 = \mathfrak{N} \ominus (\mathfrak{N}_1' \oplus \mathfrak{N}_2').$$

If $f \in \mathfrak{D}(A_1)$, then $f+A_1 f \in \mathfrak{P}_1' \subset \mathfrak{P}$, $f+A_1 f \perp \mathfrak{M}_2' \subset (\mathfrak{M} \oplus \mathfrak{N}) \ominus \mathfrak{P}$. On the other hand, $A_1 f$, as an element of the subspace $\mathfrak{N}$, is also orthogonal to $\mathfrak{M}_2' \subset \mathfrak{M}$. It follows that $f \perp \mathfrak{M}_2'$ and, consequently, $\mathfrak{M}_1'' \perp \mathfrak{M}_2'$ and $\mathfrak{M}_1'' \subset \mathfrak{M}_0$. Analogously, $\mathfrak{M}_2'' \subset \mathfrak{M}_0$ and $\mathfrak{N}_1'', \mathfrak{N}_2'' \subset \mathfrak{N}_0$. We now consider the subspaces $\mathfrak{N}_1'$ and $\mathfrak{M}_2'$; one of the relations $\mathfrak{N}_1' \lesssim \mathfrak{M}_2'$, $\mathfrak{M}_2' \lesssim \mathfrak{N}_1'$ holds for these subspaces. In the first case, there exists a subspace $\mathfrak{M}_2 \subset \mathfrak{M}_2'$ such that $\mathfrak{N}_1' \sim \mathfrak{M}_2 \subset \mathfrak{M}_2'$; consequently,

$$\mathfrak{P} = \mathfrak{M}_1' \oplus \mathfrak{N}_1' \oplus \mathfrak{P}_1' \sim \mathfrak{M}_1' \oplus \mathfrak{M}_2 \oplus \mathfrak{M}_1'' \subset \mathfrak{M}_1' \oplus \mathfrak{M}_2' \oplus \mathfrak{M}_0 = \mathfrak{M}, \quad \mathfrak{P} \lesssim \mathfrak{M};$$

in the second case, there exists a subspace $\mathfrak{N}_1 \subset \mathfrak{N}_1'$ such that $\mathfrak{M}_2' \sim \mathfrak{N}_1 \subset \mathfrak{N}_1'$; consequently,

$$(\mathfrak{M} \oplus \mathfrak{N}) \ominus \mathfrak{P} = \mathfrak{M}_2' \oplus \mathfrak{N}_2' \oplus \mathfrak{P}_2' \sim \mathfrak{N}_1 \oplus \mathfrak{N}_2' \oplus \mathfrak{N}_2'' \subset \mathfrak{N}_1' \oplus \mathfrak{N}_2' \oplus \mathfrak{N}_0 = \mathfrak{N},$$
$$(\mathfrak{M} \oplus \mathfrak{N}) \ominus \mathfrak{P} \lesssim \mathfrak{N}.$$

3^0 *The sum of two mutually orthogonal finite subspaces is finite.*

Suppose $\mathfrak{M} \perp \mathfrak{N}$ and that $\mathfrak{M}$, $\mathfrak{N}$ are finite. If $\mathfrak{M} \oplus \mathfrak{N}$ is infinite, then, according to Proposition V, there exists a subspace $\mathfrak{P}$ in $\mathfrak{M} \oplus \mathfrak{N}$ such that $\mathfrak{M} \oplus \mathfrak{N} \sim \mathfrak{P} \sim (\mathfrak{M} \oplus \mathfrak{N}) \ominus \mathfrak{P}$. According to Proposition 2^0, either $\mathfrak{P} \lesssim \mathfrak{M}$ or

$(\mathfrak{M} \oplus \mathfrak{N}) \ominus \mathfrak{P} \lesssim \mathfrak{N}$; consequently, either $\mathfrak{M}$ or $\mathfrak{N}$ is not smaller than the infinite subspace $\mathfrak{M} \oplus \mathfrak{N}$, which is impossible because $\mathfrak{M}$ and $\mathfrak{N}$ are finite by assumption.

4^0 *The sum* $\mathfrak{M} \dotplus \mathfrak{N}$ *of two finite subspaces* $\mathfrak{M}$ *and* $\mathfrak{N}$ *is finite.*

Set $\mathfrak{P} = (\mathfrak{M} \dotplus \mathfrak{N}) \ominus \mathfrak{N}$; according to VII, subsection 4, $\mathfrak{P} \lesssim \mathfrak{M}$ and, consequently, $\mathfrak{P}$ is finite. On the other hand, $\mathfrak{P} \perp \mathfrak{N}$ and $\mathfrak{M} \dotplus \mathfrak{N} = \mathfrak{P} \oplus \mathfrak{N}$ and therefore $\mathfrak{M} \dotplus \mathfrak{N}$ is finite in virtue of Proposition 3^0.

Now Proposition VII is easily obtained from 4^0 by induction.

§ 36. Relative dimension

1. Entire part of the ratio of two subspaces. We shall now show that with each finite subspace there can be associated a nonnegative number possessing the properties of dimension. In this connection, it turns out that in certain cases dimension can assume arbitrary (even irrational) values.

We recall first of all (Proposition II, subsection 5, § 35) that the representation

$$\mathfrak{M} = (\sum_{\alpha \in \mathfrak{A}} \oplus \mathfrak{N}_\alpha) \oplus \mathfrak{P}$$

is valid for given subspaces $\mathfrak{M}, \mathfrak{N} \eta M$ $(\mathfrak{N} \neq (0))$, where $\mathfrak{N}_\alpha \sim \mathfrak{N}$ and $\mathfrak{P} < \mathfrak{N}$ (if $\mathfrak{M} < \mathfrak{N}$, then the set $\mathfrak{A}$ is void). We now assume that the subspace $\mathfrak{N}$ is finite and consider what the set $\mathfrak{A}$ must be so that the subspace $\mathfrak{M}$ is also finite.

LEMMA. *If the subspace* $\mathfrak{N}$ *is finite and* $\neq (0)$, *then the set* $\mathfrak{A}$ *is finite in the representation*

$$\mathfrak{M} = (\sum_{\alpha \in \mathfrak{A}} \oplus \mathfrak{N}_\alpha) \oplus \mathfrak{P}, \quad \mathfrak{N}_\alpha \sim \mathfrak{N}, \ \mathfrak{P} < \mathfrak{N} \tag{1}$$

if, and only if, the subspace $\mathfrak{M}$ *is finite. If* $\mathfrak{M}$ *is finite, the number of elements in the set* $\mathfrak{A}$ *does not depend on the choice of representation of the form* (1).

Proof. Necessity is a direct consequence of Proposition VII, subsection 5, § 35. Conversely, suppose the subspace $\mathfrak{M}$ is finite. If the set $\mathfrak{A}$ were infinite, then, setting $\mathfrak{M}' = \mathfrak{N}_1 \oplus \mathfrak{N}_2 \oplus \mathfrak{N}_3 + \cdots$, we would have

$$\mathfrak{M}' = \mathfrak{N}_1 \oplus \mathfrak{N}_2 \oplus \mathfrak{N}_3 \oplus \cdots \subset \mathfrak{M},$$
$$\mathfrak{M}' = \mathfrak{N}_1 \oplus \mathfrak{N}_2 \oplus \mathfrak{N}_3 \oplus \cdots \sim \mathfrak{N}_2 \oplus \mathfrak{N}_3 \oplus \cdots \subsetneqq \mathfrak{M}',$$

i.e. $\mathfrak{M}'$ would be an infinite subspace which is $\lesssim$ a finite subspace $\mathfrak{M}$, which is impossible in virtue of Proposition I, subsection 5, § 35.

Suppose the number of these subspaces is k and let

$$\mathfrak{M} = \mathfrak{N}_1' \oplus \mathfrak{N}_2' \oplus \cdots \oplus \mathfrak{N}_l' \oplus \mathfrak{P}' \tag{2}$$

be a second such representation; if, for instance, $k < l$, then in virtue of the assumption $\mathfrak{P} < \mathfrak{N} \sim \mathfrak{N}_{k+1}'$, there exists a subspace $\mathfrak{K}$ such that $\mathfrak{P} \sim \mathfrak{K} \subset \mathfrak{N}_{k+1}'$; it follows that

$$\mathfrak{M} = \mathfrak{N}_1 \oplus \cdots \oplus \mathfrak{N}_k \oplus \mathfrak{P} \sim \mathfrak{N}_1' \oplus \cdots \oplus \mathfrak{N}_k' \oplus \mathfrak{K} \subsetneqq \mathfrak{M},$$

which contradicts the fact that the subspace $\mathfrak{M}$ is finite.

Representation (1) will in this case be written symbolically in the form

$$\mathfrak{M} = k\mathfrak{N} \dotplus \mathfrak{P}; \tag{3}$$

the number k will be called the *entire part of the ratio of* $\mathfrak{M}$ *to* $\mathfrak{N}$ and it will be denoted by $[\mathfrak{M}/\mathfrak{N}]$.

We note that equality (3) is also meaningful for $\mathfrak{M} \lesssim \mathfrak{N}$; if $\mathfrak{M} \sim \mathfrak{N}$, then $k = 1$, $\mathfrak{P} = (0)$; but if $\mathfrak{M} < \mathfrak{N}$, then $k = 0$, $\mathfrak{P} = \mathfrak{M}$.

We also note the following, almost obvious, inequalities:

$$\left[\frac{\mathfrak{N}}{\mathfrak{M}}\right] \cdot \left[\frac{\mathfrak{M}}{\mathfrak{P}}\right] \leqq \left[\frac{\mathfrak{N}}{\mathfrak{P}}\right] \leqq \left(\left[\frac{\mathfrak{N}}{\mathfrak{M}}\right] + 1\right)\left(\left[\frac{\mathfrak{M}}{\mathfrak{P}}\right] + 1\right) \tag{4}$$

and, for $\mathfrak{N} \perp \mathfrak{M}$,

$$\left[\frac{\mathfrak{M}}{\mathfrak{P}}\right] + \left[\frac{\mathfrak{N}}{\mathfrak{P}}\right] \leqq \left[\frac{\mathfrak{M} \oplus \mathfrak{N}}{\mathfrak{P}}\right] \leqq \left[\frac{\mathfrak{M}}{\mathfrak{P}}\right] + \left[\frac{\mathfrak{N}}{\mathfrak{P}}\right] + 1; \tag{5}$$

we leave their proof to the reader.

2. The case when a minimal subspace exists. We shall assume that in the space $\mathfrak{H}$ there exists a *minimal subspace adjoined to the factor* M, i.e. a subspace $\mathfrak{M}_0 \neq (0)$ and ηM, such that $\mathfrak{N} < \mathfrak{M}_0$, $\mathfrak{N}\eta M$ imply $\mathfrak{N} = (0)$. If in relation (3) we replace the subspace $\mathfrak{N}$ by such a minimal subspace $\mathfrak{M}_0$, then in virtue of the condition $\mathfrak{P} < \mathfrak{M}_0$ we shall have $\mathfrak{P} = (0)$, i.e. $\mathfrak{M} = k\mathfrak{M}_0$. The number k will be called the *ratio of* $\mathfrak{M}$ *to* $\mathfrak{M}_0$ and it will be denoted by $\mathfrak{M}/\mathfrak{M}_0$. If $\mathfrak{M}$ and $\mathfrak{N} \neq (0)$ are arbitrary finite subspaces, then their ratio will be defined by the equality

$$\frac{\mathfrak{M}}{\mathfrak{N}} = \frac{\dfrac{\mathfrak{M}}{\mathfrak{M}_0}}{\dfrac{\mathfrak{N}}{\mathfrak{N}_0}}.$$

It is then easy to verify the validity of the following assertions:

1° $\dfrac{\mathfrak{M}}{\mathfrak{N}} = \dfrac{\mathfrak{M}'}{\mathfrak{N}}$ if and only if $\mathfrak{M} \sim \mathfrak{M}'$;

2° $\dfrac{\mathfrak{M}}{\mathfrak{N}} < \dfrac{\mathfrak{M}'}{\mathfrak{N}}$ if and only if $\mathfrak{M} < \mathfrak{M}'$;

3° $\dfrac{\mathfrak{M}}{\mathfrak{N}} = 0$ if and only if $\mathfrak{M} = (0)$;

4° $\dfrac{\mathfrak{M}}{\mathfrak{N}} = 1$ if and only if $\mathfrak{M} \sim \mathfrak{N}$;

5° $\dfrac{\mathfrak{M}}{\mathfrak{N}} \cdot \dfrac{\mathfrak{N}}{\mathfrak{P}} = \dfrac{\mathfrak{M}}{\mathfrak{P}}$;

6° $\dfrac{\mathfrak{M} \oplus \mathfrak{P}}{\mathfrak{N}} = \dfrac{\mathfrak{M}}{\mathfrak{N}} + \dfrac{\mathfrak{P}}{\mathfrak{N}}$ for $\mathfrak{M} \perp \mathfrak{P}$.

3. The case when a minimal subspace does not exist. We now assume that the space $\mathfrak{H}$ contains no minimal subspace but that finite subspaces exist in $\mathfrak{H}$. We shall then construct a sequence of finite subspaces $\mathfrak{M}_1, \mathfrak{M}_2, \mathfrak{M}_3, \cdots$ such that $[\mathfrak{M}_k/\mathfrak{M}_{k+1}] \geqq 2$. To this end, we start with an arbitrary finite subspace $\mathfrak{M}_1$. Since $\mathfrak{M}_1$ is not minimal, there exists a subspace $\mathfrak{N} < \mathfrak{M}_1$, i.e. $\mathfrak{N} \sim \mathfrak{P} \subsetneqq \mathfrak{M}_1$, which is different from (0). In virtue of Proposition III, subsection 4, § 35, either $\mathfrak{N} \lesssim \mathfrak{M}_1 \ominus \mathfrak{P}$ or $\mathfrak{M}_1 \ominus \mathfrak{P} \lesssim \mathfrak{N}$. In the first case, we set $\mathfrak{M}_2 = \mathfrak{N}$ and in the second case, we set $\mathfrak{M}_2 = \mathfrak{M}_1 \ominus \mathfrak{P}$; then $[\mathfrak{M}_1/\mathfrak{M}_2] \geqq 2$. Applying the same method to the subspace $\mathfrak{M}_2$, we obtain $\mathfrak{M}_3$, and so on; in this way we construct a sequence $\mathfrak{M}_1, \mathfrak{M}_2, \mathfrak{M}_3, \cdots$.

Suppose $\mathfrak{M}$, $\mathfrak{N}$ are arbitrary finite subspaces, distinct from (0). We shall prove that

$$\lim_{n\to\infty} \frac{\left[\dfrac{\mathfrak{M}}{\mathfrak{M}_n}\right]}{\left[\dfrac{\mathfrak{N}}{\mathfrak{M}_n}\right]}$$

exists. The inequalities (see (4), subsection 1)

$$\left[\frac{\mathfrak{N}}{\mathfrak{M}_{n+1}}\right] \geqq \left[\frac{\mathfrak{N}}{\mathfrak{M}_n}\right]\left[\frac{\mathfrak{M}_n}{\mathfrak{M}_{n+1}}\right] \geqq 2\left[\frac{\mathfrak{N}}{\mathfrak{M}_n}\right]$$

imply that either $[\mathfrak{N}/\mathfrak{M}_n] = 0$, $n = 1, 2, 3, \cdots$, or that $\lim_{n\to\infty}[\mathfrak{N}/\mathfrak{M}_n] = \infty$. In the first case, $\mathfrak{N} < \mathfrak{M}_n$, $n = 1, 2, 3, \cdots$; but then from the inequalities

$$\left[\frac{\mathfrak{M}_n}{\mathfrak{M}_{n+m}}\right] \geqq \left[\frac{\mathfrak{M}_n}{\mathfrak{M}_{n+1}}\right]\left[\frac{\mathfrak{M}_{n+1}}{\mathfrak{M}_{n+2}}\right] \cdots \left[\frac{\mathfrak{M}_{n+m-1}}{\mathfrak{M}_{n+m}}\right] \geqq 2^m$$

we conclude that $\mathfrak{M}_1 \gtrsim 2^{m-1}\mathfrak{M}_m \gtrsim 2^{m-1}\mathfrak{N}$, and hence that $[\mathfrak{M}_1/\mathfrak{N}] \geqq 2^{m-1}$ for an arbitrary natural number m, which is impossible. Thus, $\lim_{n\to\infty}[\mathfrak{N}/\mathfrak{M}_n] = \infty$ and analogously, $\lim_{n\to\infty}[\mathfrak{M}/\mathfrak{M}_n] = \infty$. But then it follows from the inequalities (see (4), subsection 1)

$$\frac{\left[\frac{\mathfrak{M}}{\mathfrak{M}_{n+m}}\right]}{\left[\frac{\mathfrak{N}}{\mathfrak{M}_{n+m}}\right]} \leqq \frac{\left[\frac{\mathfrak{M}}{\mathfrak{M}_n}\right]+1}{\left[\frac{\mathfrak{N}}{\mathfrak{M}_n}\right]} \cdot \frac{\left[\frac{\mathfrak{M}_n}{\mathfrak{M}_{n+m}}\right]+1}{\left[\frac{\mathfrak{M}_n}{\mathfrak{M}_{n+m}}\right]} \leqq \frac{\left[\frac{\mathfrak{M}}{\mathfrak{M}_n}\right]+1}{\left[\frac{\mathfrak{N}}{\mathfrak{M}_n}\right]} \cdot \frac{2^m+1}{2^m}$$

that

$$\overline{\lim_{n\to\infty}} \frac{\left[\frac{\mathfrak{M}}{\mathfrak{M}_n}\right]}{\left[\frac{\mathfrak{N}}{\mathfrak{M}_n}\right]} = \overline{\lim_{m\to\infty}} \frac{\left[\frac{\mathfrak{M}}{\mathfrak{M}_{n+m}}\right]}{\left[\frac{\mathfrak{N}}{\mathfrak{M}_{n+m}}\right]} \leqq \frac{\left[\frac{\mathfrak{M}}{\mathfrak{M}_n}\right]+1}{\left[\frac{\mathfrak{N}}{\mathfrak{M}_n}\right]}. \tag{1}$$

Since $\lim_{n\to\infty}[\mathfrak{N}/\mathfrak{M}_n] = \infty$, it follows from inequality (1) that

$$\underline{\lim}_{n\to\infty} \frac{\left[\frac{\mathfrak{M}}{\mathfrak{M}_n}\right]}{\left[\frac{\mathfrak{N}}{\mathfrak{M}_n}\right]} \geqq \underline{\lim}_{n\to\infty} \frac{\left[\frac{\mathfrak{M}}{\mathfrak{M}_n}\right]+1}{\left[\frac{\mathfrak{N}}{\mathfrak{M}_n}\right]} \geqq \overline{\lim_{n\to\infty}} \frac{\left[\frac{\mathfrak{M}}{\mathfrak{M}_n}\right]}{\left[\frac{\mathfrak{N}}{\mathfrak{M}_n}\right]}.$$

Therefore

$$\underline{\lim}_{n\to\infty} \frac{\left[\frac{\mathfrak{M}}{\mathfrak{M}_n}\right]}{\left[\frac{\mathfrak{N}}{\mathfrak{M}_n}\right]} = \overline{\lim_{n\to\infty}} \frac{\left[\frac{\mathfrak{M}}{\mathfrak{M}_n}\right]}{\left[\frac{\mathfrak{M}}{\mathfrak{M}_n}\right]} \tag{2}$$

and $\lim_{n\to\infty}[\mathfrak{M}/\mathfrak{M}_n]/[\mathfrak{N}/\mathfrak{M}_n]$ exists. This limit is finite (in virtue of (1)) and $\geqq 0$; since finiteness must not be disturbed upon interchanging $\mathfrak{M}$ and $\mathfrak{N}$, equality to zero is ruled out. Thus, $0 < \lim_{n\to\infty}[\mathfrak{M}/\mathfrak{M}_n]/[\mathfrak{N}/\mathfrak{M}_n] < \infty$.

We shall call this limit the *ratio of* $\mathfrak{M}$ *to* $\mathfrak{N}$ and denote it by $\mathfrak{M}/\mathfrak{N}$. When $\mathfrak{M} = (0)$, we set $\mathfrak{M}/\mathfrak{N} = 0$.

We shall show further on that the ratio $\mathfrak{M}/\mathfrak{N}$ does not depend on the choice of the sequence $\mathfrak{M}_1, \mathfrak{M}_2, \mathfrak{M}_3, \cdots$; but for the time being we shall consider all such ratios, specifying this sequence in any manner whatever. As in the case of the existence of a minimal subspace, the ratio $\mathfrak{M}/\mathfrak{N}$ possesses properties $1°-6°$, subsection 2; they follow easily from the analogous properties for the expression $[\mathfrak{M}/\mathfrak{N}]$ and inequalities (4), (5), subsection 1.

4. Existence and properties of relative dimension. The results of subsections 2 and 3 can be formulated in the form of the following fundamental theorem.

THEOREM 1. *There exists a function* $D(\mathfrak{M})$, *defined for all closed subspaces adjoined to the factor* M *and satisfying the following conditions*:

1*) $D(\mathfrak{M}) = 0$ *if* $\mathfrak{M} = (0)$, $D(\mathfrak{M}) > 0$ *if* $\mathfrak{M} \neq (0)$;

2*) $D(\mathfrak{M}) = \infty$ *if and only if* $\mathfrak{M}$ *is infinite*;

3*) *if* $\mathfrak{M} \sim \mathfrak{N}$, *then* $D(\mathfrak{M}) = D(\mathfrak{N})$;

4*) *if* $\mathfrak{M} \perp \mathfrak{N}$, *then* $D(\mathfrak{M} \oplus \mathfrak{N}) = D(\mathfrak{M}) + D(\mathfrak{N})$;

5*) *if* $\mathfrak{M} < \mathfrak{N}$ *and* $\mathfrak{M}$ *is finite, then* $D(\mathfrak{M}) < D(\mathfrak{N})$ [we note that in the case of a separable space $\mathfrak{H}$, the relation $\mathfrak{M} < \mathfrak{N}$ already implies that $\mathfrak{M}$ is finite (see VI, subsection 5, § 35)].

Conditions 1*, 3*, 4* *define the function* $D(\mathfrak{M})$ *uniquely to within a constant multiplier, provided it assumes at least one finite positive value.*

Proof. If the space $\mathfrak{H}$ contains no finite subspaces distinct from (0), then the function $D(\mathfrak{M})$ assumes only two values, namely, $D(\mathfrak{M}) = 0$ for $\mathfrak{M} = (0)$ and $D(\mathfrak{M}) = \infty$ for $\mathfrak{M} \neq (0)$. Obviously, conditions 1*–5* will then be satisfied. Discarding this trivial case, we shall assume that $\mathfrak{H}$ contains finite subspaces $\neq (0)$. Suppose $\mathfrak{N}$ is one of them. Then we set $D(\mathfrak{M}) = \mathfrak{M}/\mathfrak{N}$ for finite $\mathfrak{M}$ and we set $D(\mathfrak{M}) = \infty$ for infinite $\mathfrak{M}$. The function $D(\mathfrak{M})$ thus obtained satisfies all the conditions 1*–5*.

Now suppose $D(\mathfrak{M})$ is an arbitrary function satisfying conditions 1*, 3*, 4*. Then $D(\mathfrak{M}) = \infty$ if $\mathfrak{M}$ is infinite. In fact, if $\mathfrak{M}$ is infinite, then there exists a subspace $\mathfrak{M}_1 \subsetneqq \mathfrak{M}$ which is equivalent to the space $\mathfrak{M}$; from this and from conditions 4*, 3* and 1* we conclude that $D(\mathfrak{M}) = D(\mathfrak{M}_1) + D(\mathfrak{M} \ominus \mathfrak{M}_1) = D(\mathfrak{M}) + D(\mathfrak{M} \ominus \mathfrak{M}_1)$, where $D(\mathfrak{M} \ominus \mathfrak{M}_1) > 0$. But this is possible if, and only if, $D(\mathfrak{M}) = \infty$. Now suppose $\mathfrak{M}$ is finite and let $\mathfrak{N}$ be a subspace for which $D(\mathfrak{N})$ is finite and > 0. $\mathfrak{N}$ is finite in virtue of what we just proved; in virtue of condition 1*, $\mathfrak{N} \neq (0)$. Applying properties 1*, 3*, 4* to decomposition (3), subsection 1, we obtain that

$$D(\mathfrak{M}) = kD(\mathfrak{N}) + D(\mathfrak{P}) \leqq (k+1)D(\mathfrak{N});$$

consequently, $D(\mathfrak{M})$ is finite. Therefore conditions 1*, 3*, 4* imply 2*.

If the space $\mathfrak{H}$ contains a minimal subspace, $\mathfrak{M}_0$, then for an arbitrary finite $\mathfrak{M}$ we have

$$\mathfrak{M} = \mathfrak{M}_1 \oplus \cdots \oplus \mathfrak{M}_k, \text{ where } \mathfrak{M}_j \sim \mathfrak{M}_0 \; (j = 1, 2, \cdots, k) \text{ and } k = \frac{\mathfrak{M}}{\mathfrak{M}_0}.$$

From this, in virtue of 4* and 3*, we have

$$D(\mathfrak{M}) = D(\mathfrak{M}_1) + \cdots + D(\mathfrak{M}_k) = kD(\mathfrak{M}_0) = \frac{\mathfrak{M}}{\mathfrak{M}_0} D(\mathfrak{M}_0);$$

consequently (taking 1* into consideration), $D(\mathfrak{M})/D(\mathfrak{M}_0) = \mathfrak{M}/\mathfrak{M}_0$ and

$$\frac{D(\mathfrak{M})}{D(\mathfrak{N})} = \frac{\dfrac{\mathfrak{M}}{\mathfrak{M}_0}}{\dfrac{\mathfrak{N}}{\mathfrak{M}_0}} = \frac{\mathfrak{M}}{\mathfrak{N}}. \tag{1}$$

We now assume that the space $\mathfrak{H}$ contains no minimal element; we choose a sequence $\mathfrak{M}_1, \mathfrak{M}_2, \mathfrak{M}_3, \cdots$ such that $[\mathfrak{M}_n/\mathfrak{M}_{n+1}] \geqq 2$ $(n = 1, 2, 3, \cdots)$. Then for arbitrary finite subspaces $\mathfrak{M}$ and $\mathfrak{N}$, both distinct from (0), we shall have

$$\mathfrak{M} = \left[\frac{\mathfrak{M}}{\mathfrak{M}_n}\right] \mathfrak{M}_n \dotplus \mathfrak{P}_n, \quad \mathfrak{N} = \left[\frac{\mathfrak{N}}{\mathfrak{M}_n}\right] \mathfrak{M}_n \dotplus \mathfrak{K}_n; \quad \mathfrak{P}_n < \mathfrak{M}_n, \quad \mathfrak{K}_n < \mathfrak{M}_n;$$

consequently, in virtue of condition 4*,

$$D(\mathfrak{M}) = \left[\frac{\mathfrak{M}}{\mathfrak{M}_n}\right] D(\mathfrak{M}_n) + D(\mathfrak{P}_n), \quad D(\mathfrak{N}) = \left[\frac{\mathfrak{N}}{\mathfrak{M}_n}\right] D(\mathfrak{M}_n) + D(\mathfrak{K}_n). \tag{2}$$

On the other hand, the relation $\mathfrak{P}_n < \mathfrak{M}_n$ implies that $\mathfrak{P}_n \sim \mathfrak{M}_n' \subsetneqq \mathfrak{M}_n$; therefore, in virtue of 4*, 1* and 3*,

$$D(\mathfrak{M}_n) = D(\mathfrak{M}_n') + D(\mathfrak{M}_n \ominus \mathfrak{M}_n') > D(\mathfrak{M}_n') = D(\mathfrak{P}_n),$$

so that, in virtue of (2),

$$D(\mathfrak{M}) < \left(\left[\frac{\mathfrak{M}}{\mathfrak{M}_n}\right] + 1\right) D(\mathfrak{M}_n), \quad D(\mathfrak{N}) > \left[\frac{\mathfrak{N}}{\mathfrak{M}_n}\right] D(\mathfrak{M}_n),$$

and

$$\frac{D(\mathfrak{M})}{D(\mathfrak{N})} < \frac{\left[\frac{\mathfrak{M}}{\mathfrak{M}_n}\right] + 1}{\left[\frac{\mathfrak{N}}{\mathfrak{M}_n}\right]}.$$

Passing to the limit as $n \to \infty$, we obtain that

$$\frac{D(\mathfrak{M})}{D(\mathfrak{N})} \leqq \frac{\mathfrak{M}}{\mathfrak{N}}.$$

Since, analogously, $D(\mathfrak{N})/D(\mathfrak{M}) \leqq \mathfrak{N}/\mathfrak{M}$, we consequently have (on the basis of 5° and 4°, subsection 2; see also the end of subsection 3) that $D(\mathfrak{M})/D(\mathfrak{N}) = \mathfrak{M}/\mathfrak{N}$.

Thus, equality (1) also holds in the case when a minimal subspace in $\mathfrak{H}$ does not exist. From this we conclude, first, that the ratio $\mathfrak{M}/\mathfrak{N}$ does not depend on the choice of the minimal subspace or of the sequence $\mathfrak{M}_1, \mathfrak{M}_2, \mathfrak{M}_3, \cdots$. Second, if $D_1(\mathfrak{M})$ and $D_2(\mathfrak{M})$ are two distinct functions, satisfying conditions 1*, 3*, 4*, then

$$\frac{D_1(\mathfrak{M})}{D_1(\mathfrak{N})} = \frac{D_2(\mathfrak{M})}{D_2(\mathfrak{N})} = \frac{\mathfrak{M}}{\mathfrak{N}}$$

for finite $\mathfrak{M}$ and $\mathfrak{N}$, from which we have that

$$\frac{D_1(\mathfrak{M})}{D_2(\mathfrak{M})} = \frac{D_1(\mathfrak{N})}{D_2(\mathfrak{N})} = \text{constant}.$$

Thus we have proved that conditions 1*, 3*, 4* define the function $D(\mathfrak{M})$ to within a constant multiplier.

The function $D(\mathfrak{M})$, defined by conditions 1*, 3*, 4*, is called the *relative dimension.* If it is necessary in any particular instance to emphasize that the dimension is taken with respect to the factor M, then we shall write $D_M(\mathfrak{M})$. The choice of the constant multiplier will be called the *normalization* of the function $D(\mathfrak{M})$.

Relative dimension can also be considered as a function defined on projection operators $P \subset M$, by setting $D(P) = D(\mathfrak{M})$ for $P = P_{\mathfrak{M}}$. Conditions 1*–5* then assume the following form:

1**) $D(P) = 0$ for $P = 0$, $D(P) > 0$ for $P \neq 0$;

2**) $D(P) = \infty$ if and only if P is infinite;

3**) if $P_1 \sim P_2$, then $D(P_1) = D(P_2)$;

4**) if $P_1 P_2 = 0$, then $D(P_1+P_2) = D(P_1)+D(P_2)$;

5**) if $P_1 < P_2$ and P_1 is finite, then $D(P_1) < D(P_2)$.

I. *If $\mathfrak{M}_1, \mathfrak{M}_2, \cdots$ is an infinite sequence of mutually orthogonal subspaces, adjoined to the algebra M, then*

$$D(\mathfrak{M}_1 \oplus \mathfrak{M}_2 \oplus \cdots) = D(\mathfrak{M}_1)+D(\mathfrak{M}_2)+ \cdots. \tag{3}$$

Proof. We first consider the case when $\mathfrak{M}_1 \oplus \mathfrak{M}_2 \oplus \cdots$ is a finite subspace. We can obviously assume that all the $D(\mathfrak{M}_k)$ are > 0. Since

$$\mathfrak{M}_1 \oplus \mathfrak{M}_2 \oplus \cdots \oplus \mathfrak{M}_n < \mathfrak{M}_1 \oplus \mathfrak{M}_2 \oplus \cdots,$$

we have

$$D(\mathfrak{M}_1)+D(\mathfrak{M}_2)+ \cdots +D(\mathfrak{M}_n) < D(\mathfrak{M}_1 \oplus \mathfrak{M}_2 \oplus \cdots);$$

consequently,

$$D(\mathfrak{M}_1)+D(\mathfrak{M}_2)+ \cdots \leqq D(\mathfrak{M}_1 \oplus \mathfrak{M}_2 \oplus \cdots) \tag{4}$$

and the series in the left member of inequality (4) converges. It remains to prove that the equality sign holds in (4).

Let us assume the contrary: suppose

$$\varepsilon = D(\mathfrak{M}_1 \oplus \mathfrak{M}_2 \oplus \cdots)-[D(\mathfrak{M}_1)+D(\mathfrak{M}_2)+ \cdots] > 0. \tag{5}$$

Since the series on the left in inequality (4) converges, $D(\mathfrak{M}_n) \to 0$ and, consequently, there exists an index ν such that $0 < D(\mathfrak{M}_\nu) < \varepsilon$. We set $\mathfrak{M}_\nu = \mathfrak{N}$. Further, we can write

$$\begin{aligned}\varepsilon &= D(\mathfrak{M}_1 \oplus \cdots \oplus \mathfrak{M}_n)+D(\mathfrak{M}_{n+1} \oplus \mathfrak{M}_{n+2} \oplus \cdots)\\ &\quad -[D(\mathfrak{M}_1)+ \cdots +D(\mathfrak{M}_n)]-[D(\mathfrak{M}_{n+1})+D(\mathfrak{M}_{n+2})+ \cdots]\\ &= D(\mathfrak{M}_{n+1} \oplus \mathfrak{M}_{n+2} \oplus \cdots)-[D(\mathfrak{M}_{n+1})+D(\mathfrak{M}_{n+2})+ \cdots];\end{aligned}$$

consequently

$$D(\mathfrak{N}) < D(\mathfrak{M}_{n+1} \oplus \mathfrak{M}_{n+2} \oplus \cdots) - [D(\mathfrak{M}_{n+1}) + D(\mathfrak{M}_{n+2}) + \cdots] \\ < D(\mathfrak{M}_{n+1} \oplus \mathfrak{M}_{n+2} \oplus \cdots). \tag{6}$$

On the other hand, we can choose an index n so large that

$$D(\mathfrak{M}_{n+1}) + D(\mathfrak{M}_{n+2}) + \cdots < D(\mathfrak{N}). \tag{7}$$

Then, in particular, $D(\mathfrak{M}_{n+1}) < D(\mathfrak{N})$ and, consequently, $\mathfrak{M}_{n+1} < \mathfrak{N}$; therefore, there exists a subspace $\mathfrak{N}_{n+1} \subset \mathfrak{N}$ such that $\mathfrak{M}_{n+1} \sim \mathfrak{N}_{n+1} \subsetneqq \mathfrak{N}$ and inequality (7) can be written in the form

$$D(\mathfrak{N}_{n+1}) + D(\mathfrak{M}_{n+2}) + D(\mathfrak{M}_{n+3}) + \cdots < D(\mathfrak{N}_{n+1}) + D(\mathfrak{N} \ominus \mathfrak{N}_{n+1}),$$

from which it follows that

$$D(\mathfrak{M}_{n+2}) + D(\mathfrak{M}_{n+3}) + \cdots < D(\mathfrak{N} \ominus \mathfrak{N}_{n+1}).$$

Repeating the same process, we construct subspaces $\mathfrak{N}_{n+2}, \mathfrak{N}_{n+3}, \cdots$ such that $\mathfrak{N}_k \sim \mathfrak{M}_k$, $k = n+1, n+2, \cdots$, and $\sum_{k=n+1}^{\infty} \oplus \mathfrak{N}_k \subset \mathfrak{N}$. But then

$$\sum_{k=n+1}^{\infty} \oplus \mathfrak{M}_k \sim \sum_{k=n+1}^{\infty} \oplus \mathfrak{N}_k \subset \mathfrak{N},$$

and hence

$$D(\mathfrak{M}_{n+1} \oplus \mathfrak{M}_{n+2} \oplus \cdots) \leqq D(\mathfrak{N}), \tag{8}$$

which contradicts inequality (6). Thus, inequality (5) is impossible; consequently, equality (3) holds.

It follows from the same discussion that the series $\sum_{k=1}^{\infty} D(\mathfrak{M}_k)$ diverges if the subspace $\sum_{k=1}^{\infty} \oplus \mathfrak{M}_k$ is infinite. In fact, if this series did converge, then from inequality (7), we would arrive, as above, at inequality (8), from which it follows that the subspace $\sum_{k=n+1}^{\infty} \oplus \mathfrak{M}_k$, and hence also the subspace $\sum_{k=1}^{\infty} \oplus \mathfrak{M}_k$, is finite.

II. *If* $\mathfrak{M}_1 \subset \mathfrak{M}_2 \subset \mathfrak{M}_3 \subset \cdots$ *and* $\mathfrak{M}_k \eta M$, $k = 1, 2, 3, \cdots$, *then*

$$D(\sum_{k=1}^{\infty} \cdot \mathfrak{M}_k) = \lim_{n \to \infty} D(\mathfrak{M}_n).$$

[Here and below $\sum_{k=1}^{\infty} \cdot \mathfrak{M}_k$ denotes the closed linear hull of the set $\bigcup_{n=1}^{\infty} \mathfrak{M}_k$.]

Proof. Setting $\mathfrak{N}_1 = \mathfrak{M}_1$, $\mathfrak{N}_k = \mathfrak{M}_k - \mathfrak{M}_{k-1}$, $k = 2, 3, \cdots$, we have:

$$\sum_{k=1}^{\infty} \cdot \mathfrak{M}_k = \mathfrak{N}_1 \oplus \mathfrak{N}_2 \oplus \cdots; \quad \mathfrak{M}_k = \mathfrak{N}_1 \oplus \cdots \oplus \mathfrak{N}_k.$$

It follows that

$$D(\sum_{k=1}^{\infty} \cdot \mathfrak{M}_k) = D(\mathfrak{N}_1) + D(\mathfrak{N}_2) + \cdots = \lim_{n \to \infty} \sum_{k=1}^{n} D(\mathfrak{N}_k) \\ = \lim_{n \to \infty} D(\mathfrak{N}_1 \oplus \cdots \oplus \mathfrak{N}_n) = \lim_{n \to \infty} D(\mathfrak{M}_n).$$

III. *If* $\mathfrak{M}_1 \supset \mathfrak{M}_2 \supset \mathfrak{M}_3 \supset \cdots$, $\mathfrak{M}_1$ *is finite and all the* $\mathfrak{M}_n \eta M$, *then*

$$D(\mathfrak{M}_1 \cap \mathfrak{M}_2 \cap \mathfrak{M}_3 \cap \cdots) = \lim_{n\to\infty} D(\mathfrak{M}_n).$$

This proposition reduces to the preceding one, because

$$\mathfrak{M}_1 \ominus \mathfrak{M}_2 \subset \mathfrak{M}_1 \ominus \mathfrak{M}_3 \subset \cdots.$$

IV. *If the subspaces* $\mathfrak{N}_k$, $k = 1, 2, 3, \cdots$ *are finite and the series* $D(\mathfrak{N}_1)+D(\mathfrak{N}_2)+\cdots$ *converges, then the subspace* $\mathfrak{N}_1 \dotplus \mathfrak{N}_2 \dotplus \mathfrak{N}_3 \dotplus \cdots$ *is finite and*

$$D(\mathfrak{N}_1 \dotplus \mathfrak{N}_2 \dotplus \mathfrak{N}_3 \dotplus \cdots) \leqq D(\mathfrak{N}_1)+D(\mathfrak{N}_2)+D(\mathfrak{N}_3)+\cdots.$$

Proof. First of all, according to Proposition VI, subsection 4, § 35, $D(\mathfrak{N}_1 \dotplus \mathfrak{N}_2)-D(\mathfrak{N}_2) \leqq D(\mathfrak{N}_1)$, and consequently,

$$D(\mathfrak{N}_1 \dotplus \mathfrak{N}_2) \leqq D(\mathfrak{N}_1)+D(\mathfrak{N}_2).$$

By induction, we obtain from this that

$$D(\mathfrak{N}_1 \dotplus \cdots \dotplus \mathfrak{N}_n) \leqq D(\mathfrak{N}_1)+\cdots+D(\mathfrak{N}_n)$$

for an arbitrary natural number n. We now set $\mathfrak{M}_n = \mathfrak{N}_1 \dotplus \mathfrak{N}_2 \dotplus \cdots \dotplus \mathfrak{N}_n$. According to Proposition II,

$$D(\mathfrak{N}_1 \dotplus \mathfrak{N}_2 \dotplus \mathfrak{N}_3 \dotplus \cdots)$$
$$= \lim_{n\to\infty} D(\mathfrak{M}_n) \leqq \lim_{n\to\infty} [D(\mathfrak{N}_1)+D(\mathfrak{N}_2)+\cdots+D(\mathfrak{N}_n)] = D(\mathfrak{N}_1)+D(\mathfrak{N}_2)+\cdots.$$

5. The range of the relative dimension; classification of factors. We denote the range of the relative dimension $D(\mathfrak{M})$ by Δ. We shall see what the set Δ must be. In any case, the set Δ contains the number 0 and consists of nonnegative numbers. If the space $\mathfrak{H}$ contains no finite subspaces, distinct from (0), then $\Delta = \{0, \infty\}$. Setting this trivial case aside for the time being, we note that the set Δ also possesses the following two properties:

1^0 if $\alpha, \beta \in \Delta$, $\alpha > \beta$, then $\alpha-\beta \in \Delta$;

2^0 if $\alpha_k, \alpha \in \Delta$ and the (finite or denumerable) sum

$$\sum_k \alpha_k \text{ is } \leqq \alpha, \text{ then } \sum_k \alpha_k \in \Delta.$$

In fact, if $D(\mathfrak{N}) = \alpha$, $D(\mathfrak{M}) = \beta$, then the inequality $D(\mathfrak{N}) > \mathfrak{D}(\mathfrak{M})$ implies that $\mathfrak{M} < \mathfrak{N}$, i.e. that $\mathfrak{M} \sim \mathfrak{N}' \subsetneqq \mathfrak{N}$. But then

$$\alpha = D(\mathfrak{N}) = D(\mathfrak{N}')+D(\mathfrak{N} \ominus \mathfrak{N}') = \beta+D(\mathfrak{N} \ominus \mathfrak{N}'),$$

so that

$$\alpha-\beta = D(\mathfrak{N} \ominus \mathfrak{N}') \in \Delta.$$

If, furthermore, $\alpha_k, \alpha \in \Delta$, then we take an arbitrary $\mathfrak{M}_1 \eta M$ such that $D(\mathfrak{M}_1) = \alpha_1$, and an $\mathfrak{M}\eta M$ such that $\alpha = D(\mathfrak{M})$. Repeating, further, the line of reasoning followed at the end of the proof of Proposition I, subsection 4, we construct subspaces $\mathfrak{M}_2, \mathfrak{M}_3, \cdots$ such that $\mathfrak{M}_j \perp \mathfrak{M}_k$ for $j \neq k$, $D(\mathfrak{M}_k) = \alpha_k$, and $\sum_k \oplus \mathfrak{M}_k \subset \mathfrak{M}$. According to I, subsection 4, it follows from this that

$$\sum_k \alpha_k = \sum_k D(\mathfrak{M}_k) = D(\sum_k \oplus \mathfrak{M}_k) \in \Delta.$$

We shall first assume that the set Δ contains a smallest positive element ε. If $\alpha \in \Delta$ and $\alpha \neq \infty$, then $n\varepsilon \leqq \alpha < (n+1)\varepsilon$ for some n. Setting $\alpha_1 = \alpha_2 = \cdots = \alpha_n = \varepsilon$ in property 2^0, we obtain that $n\varepsilon \in \Delta$ and consequently, in virtue of property 1^0, $\alpha - n\varepsilon \in \Delta$. But $\alpha - n\varepsilon < \varepsilon$, so that in virtue of the fact that the positive number $\varepsilon \in \Delta$ is minimal, we must have $\alpha - n\varepsilon = 0$, $\alpha = n\varepsilon$. If $m\varepsilon \in \Delta$, then, according to 2^0, all the $k\varepsilon$, $k = 1, 2, \cdots, m-1$, also belong to Δ. Therefore, the set Δ consists of all numbers of the form $k\varepsilon$, $k = 1, 2, \cdots, m$, where m is a natural number or ∞.

We shall now assume that the set Δ does not contain a smallest positive element. Let $\varepsilon = \inf x$, $x \in \Delta$, $x > 0$. We shall prove that $\varepsilon = 0$. In fact, if $\varepsilon > 0$, then Δ contains a positive element α which is less than 2ε. Clearly, $\alpha > \varepsilon$, because in case $\alpha = \varepsilon$, the number ε would be the smallest positive element in the set Δ. But it follows from the inequality $\alpha > \varepsilon$ that Δ contains a positive element β less than α. We thus have that $\varepsilon < \beta < \alpha < 2\varepsilon$; consequently, $\alpha - \beta < \varepsilon$, $\alpha - \beta > 0$, $\alpha - \beta \in \Delta$ (in virtue of property 1^0), which contradicts the definition of the number ε.

Thus $\varepsilon = 0$. But then for an arbitrary $\delta > 0$ there exists an element $\alpha \in \Delta$ such that $0 < \alpha < \delta$. Suppose β is an arbitrary positive element of the set Δ and that $0 < \gamma_1 < \gamma_2 < \beta$; we shall prove that there is at least one element of the set Δ between γ_1 and γ_2. To this end, we set $\delta = \gamma_2 - \gamma_1$; then for some $n = 0, 1, 2, 3, \cdots$ we shall have $n\alpha \leqq \gamma_1 < (n+1)\alpha < \gamma_2$. As above, it follows from this that $(n+1)\alpha \in \Delta$, so that $(n+1)\alpha$ is the element sought. We shall further prove that any number γ between 0 and β belongs to the set Δ. In fact, choosing a sequence $\beta_1 < \beta_2 < \beta_3 < \cdots$ such that $\beta_n \to \gamma$, $\beta_n < \beta$, and then numbers $\gamma_k \in \Delta$ such that $\beta_{k-1} < \gamma_k < \beta_k$, we have that $\gamma_k \to \gamma$, i.e.

$$\gamma = \gamma_1 + (\gamma_2 - \gamma_1) + (\gamma_3 - \gamma_2) + \cdots < \beta.$$

According to property 2^0 it follows from this that $\gamma \in \Delta$.

Thus only the following cases are possible:

a) The space $\mathfrak{H}$ is finite: then $\bar{\alpha} = D(\mathfrak{H})$ is the largest element of the set Δ and Δ is the interval $[0, \bar{\alpha}]$. Normalizing the function $D(\mathfrak{M})$ so that $D(\mathfrak{H}) = 1$, we obtain that $\Delta = [0, 1]$.

b) The space $\mathfrak{H}$ is infinite: suppose $\mathfrak{N}$ is an arbitrary finite closed subspace distinct from (0). In virtue of IV, subsection 5, § 35, we then have that $\mathfrak{H} = \sum_{\alpha \in \mathfrak{A}} \oplus \mathfrak{N}_\alpha$, where $\mathfrak{N}_\alpha \sim \mathfrak{N}$ and $\mathfrak{A}$ is infinite. From this we conclude that

Δ contains arbitrarily large numbers $D(\mathfrak{N}_{\alpha_1} \oplus \cdots \oplus \mathfrak{N}_{\alpha_n}) = nD(\mathfrak{N})$; consequently, $\Delta = [0, \infty]$.

We have thus proved the following theorem.

THEOREM 2. *The range of the relative dimension is one of the following sets:*

(I_n) *the set of all numbers* $k\varepsilon$, $k = 1, 2, \cdots, n$;

(I_∞) *the set of all* $k\varepsilon$, $k = 1, 2, 3, \cdots, \infty$;

(II_1) *the interval* $[0, \bar{\alpha}]$;

(II_∞) *the interval* $[0, \infty]$;

(III) *the numbers* $0, \infty$.

We shall say that M is a factor of type (I_n), (I_∞), (II_1), (II_∞), (III) if the case of this designation holds for the set Δ, respectively. The types (I_n), (I_∞) are called *discrete*, (II_1), (II_∞) are called *continuous*, (I_n), (II_1) *finite*, and (I_∞), (II_∞) *infinite*. The type (III) is called *totally infinite*. None of these types is void, i.e. there exist factors of any of these types: examples of factors of the various types will be given below (see § 38).

We note further that in the cases (I_n), (I_∞), (II_1) one can normalize the function $D(\mathfrak{M})$ so that $\varepsilon = 1$, $\bar{\alpha} = 1$. Such a normalization of the function $D(\mathfrak{M})$ will be called the *standard* normalization.

6. Invariance of factor type under symmetric isomorphisms.

THEOREM 3. *The types* (I_n), (I_∞), (II_1), (II_∞), (III) *and the function* $D(\mathfrak{M})$, *to within a multiplier, and also the concepts of equivalence, finiteness, infiniteness, and minimality of projection operators are invariant under an arbitrary symmetric isomorphism of algebras.*

Proof. The relation $E \sim F(\cdots M)$ means that there exists an operator $U \in M$ such that

$$UU^*U = U, \ U^*U = E, \ UU^* = F; \tag{1}$$

further, the fact that the projection operator E is infinite means that there exists an operator $F \in M$ such that

$$F^2 = F^* = F, \ EF = F, \ F \neq E, \ F \sim E, \tag{2}$$

and the minimality of the operator E means that there is no operator $F \in M$ which satisfies the conditions

$$F^2 = F^* = F, \ EF = F, \ F \neq 0, \ F \neq E. \tag{3}$$

Since relations (1), (2), (3) are invariant with respect to a symmetric isomorphism of algebras, the same is true of the concepts of equivalence, finiteness, infiniteness, and minimality. This implies the invariance of types. Since, further, $D_M(E)$ is characterized in terms of the relations $E = 0$, $E \sim F$, $G = E+F$,

$EF = 0$, the function $D_M(E)$ is also invariant to within a multiplier. This completes the proof of the theorem.

The question arises whether the concepts listed above form a complete system of invariants of a factor with respect to symmetric isomorphisms. The answer to this question is in the affirmative in the case of a factor of type (I_n) or (I_∞) and negative in the case of a factor of type (II_1) or (II_∞) (see subsections 3 and 6, § 38). The problem of finding a complete system of invariants for factors of type (II_1) or (II_∞) has not yet been solved.

§ 37. Relative trace

1. Definition of trace. The *trace* of an operator in a finite-dimensional space is the sum of all its characteristic values, where each characteristic value is counted as many times as indicated by its multiplicity.

The results of the preceding section enable us to generalize this concept to arbitrary factors of a finite type. Suppose M is a factor of a finite type, that A is a Hermitian operator in M, and finally that $P(\lambda)$ is the spectral function of the operator A. Then $P(\lambda) \in M$ and, consequently, $D_M(P(\lambda))$ is meaningful and turns out to be a nondecreasing function of λ. If $|A| < C$, then the function $P(\lambda)$ and, consequently, also $D_M(P(\lambda))$, is a constant in the exterior of the interval $[-C, C]$. Therefore

$$\int_{-\infty}^{+\infty} \lambda dD_M(P(\lambda)) = \int_{-C}^{C} \lambda dD_M(P(\lambda))$$

exists. This number will be called the *relative trace* of the operator A and it will be denoted by $T_M(A)$. Here we shall take the function $D_M(E)$ in its standard normalization.

If, for instance, M is a factor of a discrete type, then the function $P(\lambda)$, being for each value of λ the sum of minimal mutually orthogonal projection operators, can increase only in jumps, and its points of growth λ_n are characteristic values of the operator A. Therefore $T_M(A) = \sum_n k_n \lambda_n$, where k_n is the corresponding jump of the function $D_M(P(\lambda))$. If, in particular, M is the algebra of all operators in a finite-dimensional (namely, N-dimensional) space, then k_n is $1/N$ times the multiplicity of the characteristic value λ_n and $T_M(A)$ coincides with the usual trace, multiplied by $1/N$.

In order to deduce the basic properties of the trace we first note the following lemma.

LEMMA. *If* $D(\mathfrak{M}) > D(\mathfrak{N})$, *then* $\mathfrak{M} \cap (\mathfrak{H} \ominus \mathfrak{N}) \neq (0)$.

Proof. By Proposition VII, subsection 4, § 35,

$$D([\mathfrak{N} + (\mathfrak{H} \ominus \mathfrak{M})] \cap \mathfrak{M}) = D([\mathfrak{N} + (\mathfrak{H} \ominus \mathfrak{M})] \ominus (\mathfrak{H} \ominus \mathfrak{M})) \leqq D(\mathfrak{N}) < D(\mathfrak{M}),$$

so that the space $[\mathfrak{N} + (\mathfrak{H} \ominus \mathfrak{M})] \cap \mathfrak{M}$ is a proper subset of the space $\mathfrak{M}$. It follows that

$$(0) \neq \mathfrak{M} \ominus \{[\mathfrak{N} \dotplus (\mathfrak{H} \ominus \mathfrak{M})] \cap \mathfrak{M}\} = \mathfrak{M} \cap (\mathfrak{H} \ominus [\mathfrak{N} \dotplus (\mathfrak{H} \ominus \mathfrak{M})])$$
$$= \mathfrak{M} \cap [(\mathfrak{H} \ominus \mathfrak{N}) \cap \mathfrak{M}] = (\mathfrak{H} \ominus \mathfrak{N}) \cap \mathfrak{M}.$$

2. Properties of the trace. Everywhere in the remainder of this section we shall assume that M is a factor of a finite type and that the function $D(\mathfrak{M})$ is normalized so that $D(\mathfrak{H}) = 1$. We introduce the notation

$$\varepsilon = \varepsilon(\alpha) = \inf \lambda, \ 0 \leqq \alpha \leqq 1,$$

under the condition that

$$D_M(P(\lambda)) \geqq \alpha,$$

where $P(\lambda)$ is the spectral function of some Hermitian operator A in M. The number $\varepsilon(\alpha)$ will be called the *characteristic value of the operator A with index α.*

If, for example, $\mathfrak{H}$ is an n-dimensional space and M is the set of all linear operators in $\mathfrak{H}$, then $P(\lambda)$ and hence also $\varepsilon(\alpha)$ vary in jumps. Here, the possible values of the function $D_M(P(\lambda))$ are the numbers $0, 1/n, 2/n, \cdots, 1$, and $\varepsilon(k/n)$ is the k-th (considering multiplicity) characteristic value (in the usual sense) of the operator A. Thus, in this case our definition of a characteristic value coincides with the usual one, and the index is equal to the usual index multiplied by $1/n$.

In the continuous case (II_1), however, the index of a characteristic value can be any number in the interval $[0, 1]$.

In the theory of finite Hermitian matrices, it is proved that the characteristic values of the operator A are related in a definite way to the maximum of the form (Af, f) under the condition that $|f| = 1$ (this is the so-called minimax principle; see, for instance, Courant and Hilbert [1], Chapter I). We shall show that this property generalizes to all factors of finite type.

I. *If A is a Hermitian operator in M, then*

$$\varepsilon(\alpha) = \inf_{D_M(\mathfrak{M}) \geqq \alpha} \{ \sup_{|f|=1, f \in \mathfrak{M}} (Af, f) \}. \tag{1}$$

Proof. We denote the right member of equality (1) by λ_0; by the definition of the number $\varepsilon = \varepsilon(\alpha)$, the inequality $D_M(P(\lambda)) \geqq \alpha$ holds for $\lambda > \varepsilon$. Consequently, in virtue of the continuity from the right of the function $P(\lambda)$, we also have that $D_M(P(\varepsilon)) \geqq \alpha$. We set $P(\varepsilon)\mathfrak{H} = \mathfrak{M}_0$; then for $f \in \mathfrak{M}_0$, $|f| = 1$,

$$(Af, f) = \int_{-\infty}^{\varepsilon} \lambda d(P(\lambda)f, f) \leqq \varepsilon |f|^2 = \varepsilon;$$

it follows, by the definition of the number λ_0, that $\lambda_0 \leqq \varepsilon$. We now assume that $\lambda_0 < \varepsilon$. Then there exists a subspace $\mathfrak{M} \eta M$ such that

$$D_M(\mathfrak{M}) \geqq \alpha \quad \text{and} \quad \sup_{|f|=1, f \in \mathfrak{M}} (Af, f) < \varepsilon.$$

Hence, choosing the number ε_1 so that $\sup_{|f|=1, f\in\mathfrak{M}}(Af,f) < \varepsilon_1 < \varepsilon$, setting $\mathfrak{M}_1 = P(\varepsilon_1)\mathfrak{H}$ and taking the definition of $\varepsilon = \varepsilon(\alpha)$ into consideration, we have:

$$D_M(\mathfrak{M}_1) = D_M(P(\varepsilon_1)) < \alpha \leqq D_M(\mathfrak{M}).$$

According to the lemma in subsection 1, $\mathfrak{M}$ contains an element f_0, with norm equal to unity, which is orthogonal to $\mathfrak{M}_1$. For this element,

$$(Af_0, f_0) = \int_{\varepsilon_1}^{+\infty} \lambda d(E(\lambda)f_0, f_0) \geqq \varepsilon_1 |f_0|^2 = \varepsilon_1 .$$

But this contradicts the choice of ε_1. This completes the proof of Proposition I.

The trace $T_M(A)$ is expressed in terms of the function $\varepsilon(\alpha)$ by the formula

$$T_M(A) = \int_0^1 \varepsilon(\alpha) d\alpha. \tag{2}$$

In fact, suppose $[a, b]$ is an arbitrary interval, containing the interval $[-|A|, |A|]$ in its interior. Consider any decomposition of the interval $(a, b]$ into subintervals $(\lambda_{k-1}, \lambda_k]$, $k = 1, 2, \cdots, n$, where $a = \lambda_0 < \lambda_1 < \cdots < \lambda_{n-1} < \lambda_n = b$, and set $\alpha_k = D_M(P(\lambda_k))$. We obtain a decomposition of the interval $(0, 1]$ into subintervals $(\alpha_{k-1}, \alpha_k]$, $k = 1, 2, \cdots, n$; in this connection, if $P(\lambda)$ is constant on the subinterval $(\lambda_{k-1}, \lambda_k]$, then the corresponding "interval" $(\alpha_{k-1}, \alpha_k]$ is the void set.

Suppose $(\lambda_{k-1}, \lambda_k]$ is an interval on which $P(\lambda) \not\equiv$ constant; then $D_M(P(\lambda_k)) \geqq D_M(P(\lambda))$ for $\lambda_{k-1} < \lambda \leqq \lambda_k$, where the $>$ sign holds for some of these numbers λ. From this, taking into consideration the definition of the function $\varepsilon(\alpha)$, we conclude that $\lambda_{k-1} < \varepsilon(\alpha_k) \leqq \lambda_k$, and hence the left member of the equality

$$\sum_{k=1}^{n} \varepsilon(\alpha_k)[D_M(P(\lambda_k)) - D_M(P(\lambda_{k-1}))] = \sum_{k=1}^{n} \varepsilon(\alpha_k)(\alpha_k - \alpha_{k-1}) \tag{3}$$

is an integral sum for $\int_{-\infty}^{\infty} \lambda d D_M(P(\lambda))$. On the other hand, if all the $\lambda_k - \lambda_{k-1}$ are $< \delta$, then

$$\left| \int_0^1 \varepsilon(\alpha) d\alpha - \sum_{k=1}^{n} \varepsilon(\alpha_k)(\alpha_k - \alpha_{k-1}) \right| \leqq \sum_{k=1}^{n} \int_{\alpha_{k-1}}^{\alpha_k} |\varepsilon(\alpha) - \varepsilon(\alpha_k)| d\alpha < \delta \sum_{k=1}^{n} \int_{\alpha_{k-1}}^{\alpha_k} d\alpha = \delta;$$

consequently, passing to the limit in (3) as $\max_k(\lambda_k - \lambda_{k-1}) \to 0$ we obtain relation (2).

Formula (2) is a generalization of the definition of the trace of an operator in finite-dimensional space as the sum of all its characteristic values.

II. *The trace $T_M(A)$ is a continuous function of the operator A in the sense of the uniform topology.* [It can also be proved (see MURRAY and VON NEUMANN [1], II, Theorem IV) that the trace $T_M(A)$ is a continuous function of the operator A in the weak topology.]

Proof. We note first of all that if $(A_1f,f) \leqq (A_2f,f)$ for all $f \in \mathfrak{H}$, then $T_M(A_1) \leqq T_M(A_2)$. In fact, in virtue of equality (1), we have in this case that $\varepsilon_1(\alpha) \leqq \varepsilon_2(\alpha)$, where $\varepsilon_1(\alpha)$, $\varepsilon_2(\alpha)$ are the characteristic values of the operators A_1, A_2 with index α, so that our assertion follows from equality (2). It can be proved analogously that $T_M(A+\lambda 1) = T_M(A)+\lambda$, where λ is an arbitrary real number. Now if $|A-B| < \varepsilon$, then $(Af,f) \leqq ((B+\varepsilon 1)f,f)$ and, consequently, $T_M(A) \leqq T_M(B+\varepsilon 1) = T_M(B)+\varepsilon$, $T_M(A)-T_M(B) < \varepsilon$. Interchanging A and B, we obtain that also $T_M(B)-T_M(A) < \varepsilon$, so that $|T_M(A)-T_M(B)| < \varepsilon$.

THEOREM 1. *Suppose M is a factor of a finite type; then there exists one, and only one, function $T(A)$, defined for all Hermitian operators A in M and satisfying the following conditions:*

1° $T(1) = 1$;

2° $T(\alpha A) = \alpha T(A)$ *for all real* α;

3° $T(A+B) = T(A)+T(B)$, *if A and B commute*;

4° $T(A) \geqq 0$, *if A is a positive definite operator*;

5° $T(U^{-1}AU) = T(A)$ *if U is a unitary operator in M.*

This function is the relative trace $T_M(A)$.

Proof. First of all, the trace $T_M(A)$ satisfies the conditions 1°–5°. In fact, conditions 1°, 4° follow from the definition, 2° with $\alpha \geqq 0$ from equalities (1) and (2), and then for $\alpha < 0$ from the relations

$$T_M(-A) = \int_{-\infty}^{\infty} \lambda dD_M(1-P(-\lambda-0)) = -\int_{-\infty}^{\infty} \lambda dD_M(P(-\lambda-0))$$
$$= -\int_{-\infty}^{\infty} \lambda dD_M(P(\lambda)) = -T_M(A),$$

where $P(\lambda)$ is the spectral function of the operator A and, consequently, as is easily seen, $1-P(-\lambda-0)$ is the spectral function of the operator $-A$. Finally, 5° follows from the equalities

$$T_M(U^{-1}AU) = \int_{-\infty}^{\infty} \lambda dD_M(U^{-1}P(\lambda)U) = \int_{-\infty}^{\infty} \lambda dD_M(P(\lambda)) = T_M(A).$$

It remains to show that the function $T_M(A)$ satisfies condition 3°. If the operators A, B commute and $E(\lambda)$, $F(\lambda)$ are their spectral functions, then $E(\lambda)$ and $F(\mu)$ also commute. Furthermore, the equalities $A = \int_{-\infty}^{+\infty} \lambda dE(\lambda)$, $B = \int_{-\infty}^{+\infty} \mu dF(\mu)$ imply that the operators A and B are the limits in the uniform topology of sums of the form

$$A_1 = \sum_k a_k E(\lambda_k), \quad B_1 = \sum_k b_k F(\mu_k),$$

where the a_k, b_k are real numbers. In virtue of Proposition II, it therefore suffices to prove that 3° holds for the operators A_1 and B_1. In other words, it suffices to prove that

$$T_M\left(\sum_{k=1}^{p} c_k E_k\right) = \sum_{k=1}^{p} c_k T_M(E_k), \tag{4}$$

if all the projection operators E_k are pairwise commutative and the c_k are real numbers.

We consider all the terms F_j of the product

$$[E_1+(1-E_1)][E_2+(1-E_2)]\cdots[E_p+(1-E_p)];$$

they are mutually orthogonal, their sum equals 1, and each operator E_k is the sum of such operators F_j; therefore, it suffices to prove equality (4) for the operators F_j, or, equivalently, it can be assumed that the operators E_k are mutually orthogonal and that their sum equals 1. But then equality (4) follows directly from the definition of the trace because the spectral function $G(\lambda)$ of the operator $\sum_{k=1}^{p} c_k E_k$ is defined by the equality $G(\lambda) = \sum_{\lambda_k \leq \lambda} E_k$.

Thus, the trace $T_M(A)$ satisfies conditions $1°-5°$. Conversely, suppose that an arbitrary function $T(A)$, defined on all Hermitian operators $A \in M$, satisfies conditions $1°-5°$. We set $D(P) = T(P)$ for the projection operator $P \in M$. It follows from condition 3° that the function $D(P)$ satisfies condition 4* of Theorem 1, subsection 4, § 36. Furthermore, $D(P) \geqq 0$. It follows from equality 2° that $D(P) = 0$ for $P = 0$. Further, if $E \sim F(\ldots M)$, then M contains a unitary operator U such that $F = U^{-1}EU$ [because $D_M(1-E) = 1-D_M(E) = D_M(1-F)$]. But then, according to condition 5°, $D(F) = D(E)$, so that condition 3* of Theorem 1, subsection 4, § 36, is also satisfied. It remains to verify that the equality $D(P) = 0$ implies that also $P = 0$. But if $P \neq 0$, then, writing out the decomposition $1 = P_1+P_2+\cdots+P_k+P_0$, $P_k \sim P$, $P_0 \lesssim P$, $P_iP_k = 0$ for $i \neq k$, we would obtain that $T(1) = D(1) = 0$, which contradicts condition 1°. According to Theorem 1, subsection 4, § 36, $D(P) = aD_M(P)$. Setting $P = 1$, we obtain that $a = 1$; consequently, $D(P) = D_M(P)$. Further, suppose A, B are commutative Hermitian operators in M which are such that $|A-B| < \varepsilon$; then the operator $\varepsilon 1-(A-B)$ is nonnegative and, consequently,

$$T(\varepsilon 1-(A-B)) \geqq 0.$$

It follows that

$$\varepsilon-[T(A)-T(B)] \geqq 0,$$

and analogously that $\varepsilon-[T(B)-T(A)] \geqq 0$; consequently $|T(A)-T(B)| < \varepsilon$; thus, the function $T(A)$ is continuous in the uniform topology on every set of mutually commutative Hermitian operators. Now if A is an arbitrary Hermitian operator in the algebra M and $P(\lambda)$ is its spectral function, then

$$A = \int_{-\infty}^{\infty} \lambda dP(\lambda) = \lim_{\lambda_k-\lambda_{k-1}\to 0} \sum_k \lambda_k[P(\lambda_k)-P(\lambda_{k-1})],$$

so that

$$T(A) = \lim_{\lambda_k - \lambda_{k-1} \to 0} \sum \lambda_k [T(P(\lambda_k)) - T(P(\lambda_{k-1}))]$$

$$= \lim_{\lambda_k - \lambda_{k-1} \to 0} \sum \lambda_k [D_M(P(\lambda_k)) - D_M(P(\lambda_{k-1}))] = \int_{-\infty}^{\infty} \lambda d D_M(P(\lambda)) = T_M(A).$$

THEOREM 2. *If there exists a function* $T(A)$, *defined for all Hermitian operators in the factor* M, *satisfying conditions* $1° - 4°$ *of the preceding theorem and the additional conditions*:

5′ $T(AB) = T(BA)$ *if* $A, B \in M$, AB *and* BA *are Hermitian operators*;

6′ $P = 0$, *if* $T(P) = 0$ *and* P *is a projection operator*;

then M *is a factor of a finite type and* $T(A)$ *is the relative trace in* M.

Proof. We set $D(P) = T(P)$ for the projection operator $P \in M$. From condition 5′ with $A = U^{-1}H$, $B = U$, where U is unitary and H is a Hermitian operator in M, we conclude that

$$T(U^{-1}HU) = T(H), \tag{5}$$

i.e. that the function $T(A)$ satisfies condition 5^0 of Theorem 1. But then the function $D(P)$ satisfies conditions 1*, 3*, 4* of Theorem 1, subsection 4, § 36, so that according to this theorem $D(P)$ is the relative dimension. Since $D(1) = T(1) = 1$ by assumption, M is a factor of finite type. In virtue of Theorem 1, it follows from this and (5) that $T(A)$ is the relative trace in M.

It can be proved (see MURRAY and VON NEUMANN [1], II, Theorem III) that the trace $T(A)$ in the algebra M is given by the formula

$$T(A) = \sum_{k=1}^{m} (Ag_k, g_k),$$

where the g_k are fixed elements in $\mathfrak{H}$. [The number m of these elements g_k is determined by means of the condition $(1/m)C \leqq 1$, where C is the constant in Theorem 4, subsection 7, § 38, below; in particular, for $C \leqq 1$ there exists an element $g \in \mathfrak{H}$ such that $T(A) = (Ag, g)$.] From this it follows, in particular, that $T(A+B) = T(A) + T(B)$ also in the case when the operators A and B do not commute.

[A simple proof of this last result was subsequently given by KADISON [8] (see also KADISON [9], where this proof is generalized to weakly closed algebras which need not be factors). Another proof was given by DIXMIER (see, for instance, DIXMIER [12]).]

Now if A is an arbitrary operator in M, then we can set $A = H_1 + iH_2$, $T(A) = T(H_1) + iT(H_2)$, where H_1, H_2 are Hermitian operators in M. The function $T(A)$ obtained in this way is defined for all operators $A \in M$ and satisfies the following conditions:

1° $T(\lambda A + \mu B) = \lambda T(A) + \mu T(B)$ for arbitrary complex numbers λ, μ;

2° $T(A^*A) \geqq 0$;

3° $T(AB) = T(BA)$.

3. Trace in factors of types (I_∞) and (II_∞). The trace $T(A)$ can also be defined in factors M of the type (I_∞) or (II_∞), but then the function $T(A)$ will not be defined for all operators $A \in M$ (see MURRAY and VON NEUMANN [1]). To this end, we introduce the concept of rank of an operator.

The *rank* of the operator $A \in M$ is the number $r(A) = D_M(\overline{\mathfrak{R}(A)})$. Since $\overline{\mathfrak{R}(A)} \sim \overline{\mathfrak{R}(A^*)}$ (see subsection 3, § 35), $r(A) = r(A^*)$.

Suppose A is an operator of finite rank; then there exists a finite subspace $\mathfrak{M}$, for example $\mathfrak{R}(A) \dotplus \mathfrak{R}(A^*)$, such that $\mathfrak{R}(A)$ and $\mathfrak{R}(A^*)$ are contained in $\mathfrak{M}$. We then set

$$T_M(A) = D(\mathfrak{M}) T_{M_{(\mathfrak{M})}}(A_{(\mathfrak{M})}), \tag{1}$$

where $A_{(\mathfrak{M})}$ is the restriction of the operator A to $\mathfrak{M}$, and $M_{(\mathfrak{M})}$ is the set of restrictions to $\mathfrak{M}$ of all the operators in M, which are reduced by the space $\mathfrak{M}$ (for more details concerning this see below in subsection 1, § 38). It can be proved that the right member of (1) does not depend on the choice of the subspace $\mathfrak{M}$.

By the same token, the trace $T_M(A)$ is defined for all operators A of finite rank. We define a new norm $||A||$ for such operators by setting

$$||A|| = \sqrt{T_M(A^*A)}$$

(if A is of finite rank, then A^*A is also of finite rank). By passing to the limit with respect to this norm, we can consider the function $T_M(A)$ on a more extensive class of operators $A \in M$; such operators are called *normable*. A projection operator is normable if and only if it is finite.

§ 38. Structure and examples of some types of factors

1. The mapping $M \to M_{(\mathfrak{M})}$. Let $\mathfrak{M}$ be a subspace distinct from (0); we shall consider those operators A in the algebra M which are reduced by the subspace $\mathfrak{M}$; their restriction to $\mathfrak{M}$ is denoted by $A_{(\mathfrak{M})}$, and the set of all operators $A_{(\mathfrak{M})}$ will be denoted by $M_{(\mathfrak{M})}$. We note the following properties of the transition from M to $M_{(\mathfrak{M})}$.

I. *If $\mathfrak{M} \eta M$, $E = P_{\mathfrak{M}}$, then $(M')_{(\mathfrak{M})}$ is the set of those (bounded) linear operators in the space $\mathfrak{M}$ which commute with all the operators*

$$C = EB \text{ in } \mathfrak{M}, \ B \in M. \tag{1}$$

Proof. Clearly, every operator in $(M')_{(\mathfrak{M})}$ commutes with any operator C of the form (1); therefore we assume that some operator A in the space $\mathfrak{M}$ commutes with any operator C of the form (1) and prove that $A \in (M')_{(\mathfrak{M})}$. We set $\alpha = |A|$, $H = \sqrt{\alpha^2 1 - A^*A}$; then the operator H also commutes with all the operators C of the form (1). Hence, if $B_1, \cdots, B_n$ are arbitrary operators in the algebra M and $f_1, \cdots, f_n$ are arbitrary elements in $\mathfrak{M}$, then

$$|\sum_{j=1}^n B_j H f_j|^2 + |\sum_{j=1}^n B_j A f_j|^2 = (\sum_{j=1}^n B_j H f_j, \sum_{k=1}^n B_k H f_k)$$
$$+ (\sum_{j=1}^n B_j A f_j, \sum_{k=1}^n B_k A f_k) = \sum_{j,k=1}^n ((HEB_k^* B_j H + A^* EB_k^* B_j A) f_j, f_k)$$
$$= \sum_{j,k=1}^n (EB_k^* B_j (H^2 + A^*A) f_j, f_k) = \alpha^2 \sum_{j,k=1}^n (B_k^* B_j f_j, f_k) = \alpha^2 |\sum_{j=1}^n B_j f_j|^2.$$

($EHf_k = Hf_k$ inasmuch as $Hf_k \in \mathfrak{M}$.) It follows that

$$|\sum_{j=1}^{n} B_j A f_j| \leqq \alpha |\sum_{j=1}^{n} B_j f_j|. \tag{2}$$

We set

$$A_0 \sum_{j=1}^{n} B_j f_j = \sum_{j=1}^{n} B_j A f_j; \tag{3}$$

if $\sum_{j=1}^{n} B_j f_j = 0$, then, in virtue of the inequality (2), we also have that $\sum_{j=1}^{n} B_j A f_j = 0$; therefore the operator A_0 is uniquely defined by equality (3) on the set $\mathfrak{S}$ of all elements of the form $\sum_{j=1}^{n} B_j f_j$. In virtue of inequality (2), it is bounded; it can therefore be extended by continuity to the closure $\mathfrak{M}_0$ of the set $\mathfrak{S}$. Further, $A_0 \eta M'$, because if $B \in M$, then

$$A_0 B \sum_{j=1}^{n} B_j f_j = A_0 \sum_{j=1}^{n} B B_j f_j = \sum_{j=1}^{n} B B_j A f_j = B \sum_{j=1}^{n} B_j A f_j = B A_0 \sum_{j=1}^{n} B_j f_j$$

and by continuity the analogous equality holds in the entire space $\mathfrak{M}_0$. It follows, in particular, that $\mathfrak{M}_0 \eta M'$, so that $A_1 = A_0 P_{\mathfrak{M}_0} \in M'$. On the other hand, $\mathfrak{M}_0 \supset \mathfrak{M}$; therefore $A_{1(\mathfrak{M})} = A$, $A \in (M')_{(\mathfrak{M})}$.

Clearly, the set of operators C of the form (1) coincides with $M_{(\mathfrak{M})}$, and hence the set of all bounded linear operators in the space $\mathfrak{M}$ which commute with all the operators C is $(M_{(\mathfrak{M})})'$. Thus,

II. *If $\mathfrak{M} \eta M$ then $(M_{(\mathfrak{M})})' = (M')_{(\mathfrak{M})}$.*

In particular, it follows from this that $M_{(\mathfrak{M})}$ and $(M')_{(\mathfrak{M})}$ are weakly closed algebras.

III. *If $\mathfrak{M}$ and E are the same as in Proposition I, then the correspondence $A \to A_{(\mathfrak{M})}$ is a symmetric isomorphism between*:

1° *the algebra of all operators A in M which are such that $EA = AE = A$, and the algebra $M_{(\mathfrak{M})}$*;

2° *the algebras M' and $(M')_{(\mathfrak{M})}$.*

Proof. If $A \in M$ and $EA = AE = A$, then $A = 0$ on $\mathfrak{H} \ominus \mathfrak{M}$; therefore the operators A and $A_{(\mathfrak{M})}$ are connected by the relation $A = A_{(\mathfrak{M})} E$. This implies 1°.

If, further, $A, B \in M'$, then the equality $A_{(\mathfrak{M})} = B_{(\mathfrak{M})}$ means that $(A - B)E = 0$; but then, according to the fundamental lemma, subsection 2, § 35, $A - B = 0$, $A = B$, i.e. the correspondence between the algebras M' and $(M')_{(\mathfrak{M})}$ is one-to-one. 2° follows easily from this.

From 2° we deduce that $(M')_{(\mathfrak{M})}$ and hence also $M_{(\mathfrak{M})} = (M'_{(\mathfrak{M})})'$ (see II) are factors.

IV. *If $\mathfrak{M}$ and E are the same as in Proposition I and F runs through all projection operators in M which are such that $FE = EF = F$, and F' runs through all the projection operators in M', then*:

1° *$F_{(\mathfrak{M})}$ runs through all the projection operators in $M_{(\mathfrak{M})}$, where the relation*

$F_{(\mathfrak{M})} \sim G_{(\mathfrak{M})}(\ldots M_{(\mathfrak{M})})$ *is equivalent to the relation* $F \sim G(\ldots M)$;

2^0 $F'_{(\mathfrak{M})}$ *runs through all the projection operators in* $(M_{(\mathfrak{M})})'$, *where the relation* $F'_{(\mathfrak{M})} \sim G'_{(\mathfrak{M})}(\ldots (M_{(\mathfrak{M})})')$ *is equivalent to the relation* $F' \sim G'(\ldots M')$.

Proof. Assertion 2° follows directly from the fact that the algebras M' and $(M')_{(\mathfrak{M})}$ are symmetrically isomorphic. Further, the first part of assertion 1° follows from assertion 1° of Proposition III; in order to prove the second part, we note that the relations

$$F \sim G(\ldots M) \tag{4}$$

and

$$F_{(\mathfrak{M})} \sim G_{(\mathfrak{M})}(\ldots M_{(\mathfrak{M})}) \tag{5}$$

respectively mean there exist partially isometric operators $U \in M$ and $U_{(\mathfrak{M})} \in M_{(\mathfrak{M})}$ such that

$$F = U^*U, \quad G = UU^*, \quad F_{(\mathfrak{M})} = U^*_{(\mathfrak{M})} U_{(\mathfrak{M})}, \quad G_{(\mathfrak{M})} = U_{(\mathfrak{M})} U^*_{(\mathfrak{M})}; \tag{6}$$

here the operator U is uniquely defined by the operator $U_{(\mathfrak{M})}$ if $UE = EU = U$ and then, in virtue of assertion 1°, Proposition III, we also have that $F = U^*U$, $G = UU^*$. Thus, relation (4) follows from relation (5). In order to prove the converse assertion, it suffices to show that the operator U in relations (6) satisfies the condition $UE = EU = U$. The latter, however, follows from the fact that $F\mathfrak{H} \subset \mathfrak{M}$, $G\mathfrak{H} \subset \mathfrak{M}$.

It follows from Proposition IV that the functions

$$D_M^{(\mathfrak{M})}(F_{(\mathfrak{M})}) = D_M(F) \quad \text{with } FE = F$$

and

$$D_{M'}^{(\mathfrak{M})}(F'_{(\mathfrak{M})}) = D_{M'}(F')$$

are relative dimensions in $M_{(\mathfrak{M})}$ and $M'_{(\mathfrak{M})}$ respectively. It follows that

V. *If* $\mathfrak{M}$ *is finite then* $M_{(\mathfrak{M})}$ *is a factor of finite type.*

2. Matrix description of factors of types (I) and (II). Two algebras M_1 and M_2 in the spaces $\mathfrak{H}_1$ and $\mathfrak{H}_2$ are said to be *spatially isomorphic* if there exists an isometric mapping of $\mathfrak{H}_1$ onto $\mathfrak{H}_2$ under which M_1 maps into M_2. Clearly, every spatial isomorphism is also a complete isomorphism; the converse, as we shall see below (beginning of subsection 3), is in general not true. We shall show how certain types of factors can be described to within a spatial isomorphism.

Suppose M is a factor of type (I) or (II) and let $\mathfrak{M}$ be a finite subspace ηM. We set $n = D_M(\mathfrak{H})/D_M(\mathfrak{M})$. In the case of a factor M of infinite type, $n = \infty$; but in the case of a factor M of finite type, we shall assume that the space $\mathfrak{M}$ is such that n is an integer. Then the space $\mathfrak{H}$ can be represented in the form

$$\mathfrak{H} = \sum_\alpha \oplus \mathfrak{M}_\alpha, \tag{1}$$

where the $\mathfrak{M}_\alpha$ are mutually orthogonal subspaces equivalent to the subspace $\mathfrak{M}$, the number of which is n, where the number of these subspaces is infinite in case $n = \infty$. (If $\mathfrak{H}$ is separable, then the number of subspaces $\mathfrak{M}_\alpha$ is at most denumerable; but in general this number can also be nondenumerable.) By the definition of equivalence, there exist partially isometric operators $U_\alpha \in M$ with initial domain $\mathfrak{M}_\alpha$ and terminal domain $\mathfrak{M}$.

Suppose

$$H = \mathfrak{M} \oplus \mathfrak{M} \oplus \mathfrak{M} \oplus \cdots$$

is the direct sum of the same number of spaces, each coinciding with $\mathfrak{M}$; the operator U, defined by the relations

$$U\mathfrak{M}_\alpha = U_\alpha \mathfrak{M}_\alpha,$$

maps $\mathfrak{H}$ isometrically onto H; our goal is to clarify how the algebras M and M' in the space H are described.

We begin with the algebra M'. According to Proposition III, subsection 1, the correspondence $A \to A_{(\mathfrak{M}_\alpha)}$ is a symmetric isomorphism between the algebras M' and $(M')_{(\mathfrak{M}_\alpha)}$. Since $\mathfrak{M}_\alpha \eta M$, the operators $A \in M'$ are reduced by all of these subspaces. Therefore, corresponding to the decomposition (1), every operator $A \in M'$ can be written in the diagonal matrix form

$$A \sim ||\delta_{\alpha\beta} A_\alpha||$$

where A_α is the restriction of the operator A to the space $\mathfrak{M}_\alpha$.

Suppose A_0 is the restriction of the operator $A \in M'$ to the space $\mathfrak{M}$.

Under the mapping U each of the operators A_α goes over into the same operator A_0 in the space $\mathfrak{M}$. In fact, if $f \in \mathfrak{M}_\alpha$, then $A_\alpha f = Af$. Under the mapping U, the element f maps into $U_\alpha f \in \mathfrak{M}$, and the operator A_α into the operator $U_\alpha A_\alpha U_\alpha^*$. Furthermore,

$$U_\alpha^* U_\alpha f = P_{\mathfrak{M}_\alpha} f = f, \quad A_\alpha f = Af;$$

consequently,

$$U_\alpha A_\alpha U_\alpha^* U_\alpha f = U_\alpha A_\alpha f = U_\alpha Af = AU_\alpha f = A_0 U_\alpha f,$$

because the operators U_α and A commute in view of the relations $U_\alpha \in M$, $A \in M'$.

Thus, under the mapping U, each of the operators $A \in M'$ is described by the diagonal matrix

$$A \sim ||\delta_{\alpha\beta} A_0|| \tag{2}$$

with the same diagonal elements A_0, where A_0 is the restriction o the operator A to $\mathfrak{M}$.

In virtue of the relation $M = M''$, the algebra M in the space $\mathfrak{H}$ must consist of all matrices $B \sim ||B_{\alpha\beta}||$ (bounded, if they are infinite), where the $B_{\alpha\beta}$ are operators in $\mathfrak{M}$, which commute with all the matrices of the form (2). The condition that these matrices commute reduces to the equalities

$$A_0 B_{\alpha\beta} = B_{\alpha\beta} A_0,$$

and hence to the condition $B_{\alpha\beta} \in ((M')_{(\mathfrak{M})})'$. According to Proposition II, subsection 1, the latter algebra coincides with $(M_{(\mathfrak{M})})'' = M_{(\mathfrak{M})}$.

We have thus proved the following theorem.

THEOREM 1. *Suppose M is a factor of type* (I) *or* (II) *and that $\mathfrak{M}\eta M$ is a finite subspace such that*

$$n = \frac{D_M(\mathfrak{H})}{D_M(\mathfrak{M})}$$

is an integer in case the factor M is of finite type. (Hence $\mathfrak{M}$ is an arbitrary finite subspace in the case of a factor M of infinite type.) *Then the algebra M is spatially isomorphic to the algebra of all matrices $A \sim ||A_{\alpha\beta}||$ of order n (which are bounded for $n = \infty$), where $A_{\alpha\beta} \in M_{(\mathfrak{M})}$. Furthermore, the algebra M' will consist of all diagonal matrices $A \sim ||\delta_{\alpha\beta} A_0||$, $A_0 \in M'_{(\mathfrak{M})}$.*

3. Description of factors of type (I). Suppose $\mathfrak{H}_0$ is an arbitrary finite-dimensional or infinite-dimensional Hilbert space. The algebra of all bounded operators in $\mathfrak{H}_0$ will be denoted by $\mathfrak{B}_0$. We form the direct sum

$$\mathfrak{H} = \sum \oplus \mathfrak{H}_0$$

where the number of copies of $\mathfrak{H}_0$ is arbitrary. To each bounded operator $A \in \mathfrak{B}_0$ in $\mathfrak{H}_0$ there is assigned a diagonal operator $A^{(1)}$ in $\mathfrak{H}$ which is defined by the matrix $||\delta_{pq} A||$, $\delta_{pq} = \{\begin{smallmatrix} 0, & p \neq q, \\ 1, & p = q. \end{smallmatrix}$ In other words, the operator $A^{(1)}$ is defined in $\mathfrak{H}$ by the equalities

$$A^{(1)}\langle f_p \rangle = \langle A f_p \rangle.$$

Clearly, $A^{(1)}$ is a bounded operator in $\mathfrak{H}$ with the same norm as A. The set of all operators $A^{(1)}$, $A \in \mathfrak{B}_0$, will be denoted by $\mathfrak{B}_0^{(1)}$. Since the correspondence $A \sim A^{(1)}$ is a complete isomorphism of the algebras $\mathfrak{B}_0$ and $\mathfrak{B}_0^{(1)}$, $\mathfrak{B}_0^{(1)}$ is a factor. [We note that the isomorphism $A \sim A^{(1)}$ will not be spatial. In fact, the spatial isomorphism would then map $\mathfrak{B}_0'$ isomorphically onto $\mathfrak{B}_0^{(1)\prime}$ which is impossible, because $\mathfrak{B}_0' = (\alpha 1)$, but $\mathfrak{B}_0^{(1)\prime} \neq (\alpha 1)$.] Every factor which is spatially isomorphic to the algebra $\mathfrak{B}_0^{(1)}$ will be called a *direct factor*. (Murray and von Neumann give another, equivalent definition of a direct factor based on the theory of the direct product of spaces developed by them. For lack of space, we shall not devote any time to this theory and refer the reader to the papers by MURRAY and VON NEUMANN [1], I, and VON NEUMANN [7], and also to the book by DIXMIER [14], where this theory is discussed.)

It follows directly from this definition that

I. *Every direct factor is completely isomorphic to the algebra $\mathfrak{B}_0$.*

All subspaces of the space $\mathfrak{H}_0$ are adjoined to $\mathfrak{B}_0$; they are equivalent with respect to $\mathfrak{B}_0$ if, and only if, they have the same dimension, and the relative dimension coincides with the usual dimension. In particular, a subspace is finite relative to $\mathfrak{B}_0$ if and only if it is finite-dimensional in the ordinary sense. It follows that in the standard normalization, $D_{\mathfrak{B}_0}(\mathfrak{M})$ assumes the values $1, 2, 3, \cdots, n$ if $\mathfrak{H}_0$ is finite-dimensional, and the values $1, 2, 3, \cdots, \infty$ if $\mathfrak{H}_0$ is infinite-dimensional. (Here, obviously, any one-dimensional subspace will be minimal.) Since the function $D(\mathfrak{M})$ is invariant with respect to the complete isomorphism of algebras, we have, on the basis of I, that

II. *Every direct factor is of discrete type.*

We shall now show that the converse assertion is also true, i.e. that *every factor of a discrete type is a direct factor.* To this end, we shall first prove the following proposition.

III. *The space* $\mathfrak{M}_f^M$ $(f \neq 0)$ *is minimal if, and only if,* $\mathfrak{M}_f^M \cap \mathfrak{M}_f^{M'} = \{\alpha f\}$, *so that the spaces* $\mathfrak{M}_f^M$ *and* $\mathfrak{M}_f^{M'}$ *are minimal simultaneously.*

Proof. Suppose $\mathfrak{M}_f^{M'}$ is minimal; we set $E = E_f^{M'}$, $E' = E_f^M$. Then $E \in M$, $E' \in M'$; consequently, the operators E and E' commute. Moreover, $Ef = f$; it follows that

$$\mathfrak{M}_f^{M'} \cap \mathfrak{M}_f^M = E\mathfrak{M}_f^M = \overline{\{EAf: A \in M\}} = \overline{\{EAEf: A \in M\}}.$$

Suppose $\overline{M}$ is the set of all those operators in the algebra M which satisfy the condition $EB = BE = B$. Clearly, $\overline{M}$ is an algebra and consists of all operators EAE, $A \in M$. Therefore, $\mathfrak{M}_f^{M'} \cap \mathfrak{M}_f^M = \overline{\{Bf : B \in \overline{M}\}}$. If P is a projection operator in the algebra $\overline{M}$, then $EP = PE = P$, i.e. $P \leqq E$. Since the operator E is minimal, it follows from this that either $P = 0$ or that $P = E$. Thus (see Corollary 6, subsection 2, § 34) $\overline{M}^P = \{0, E\}$, $\overline{M} = R(\overline{M}^P) = R(0, E) = (\alpha E)$ and

$$\mathfrak{M}_f^{M'} \cap \mathfrak{M}_f^M = \{\alpha Ef\} = \{\alpha f\}. \tag{1}$$

Conversely, suppose relation (1) holds and that the space $\mathfrak{M}_f^M$ is not minimal; then there exists a subspace $\mathfrak{N}$, distinct from (0) and adjoined to the algebra M', such that $\mathfrak{N} \subsetneqq \mathfrak{M}_f^M$. We set $\mathfrak{N}_1 = \mathfrak{M}_f^M \ominus \mathfrak{N}$; then $\mathfrak{N}_1 \neq (0)$, $\mathfrak{N}_1 \eta M'$. According to the fundamental lemma, subsection 2, § 35, it follows from this that $P_{\mathfrak{N}} E \neq 0$, $P_{\mathfrak{N}_1} E \neq 0$, i.e. that

$$\mathfrak{N} \cap \mathfrak{M}_f^{M'} \neq (0), \quad \mathfrak{N}_1 \cap \mathfrak{M}_f^{M'} \neq (0);$$

on the other hand, $\mathfrak{N} \cap \mathfrak{M}_f^{M'} \subset \mathfrak{M}_f^M \cap \mathfrak{M}_f^{M'} = \{\alpha f\}$, $\mathfrak{N}_1 \cap \mathfrak{M}_f^{M'} \subset \mathfrak{M}_f^M \cap \mathfrak{M}_f^{M'} = \{\alpha f\}$ so that $f \in \mathfrak{N} \cap \mathfrak{M}_f^{M'}$, $f \in \mathfrak{N}_1 \cap \mathfrak{M}_f^{M'}$. But the latter is impossible because these subspaces are mutually orthogonal.

Now III implies that

IV. *The factors* M *and* M' *either both are or both are not of type* (I).

Suppose now that M is a factor of type (I) and, consequently, that M' is also a factor of type (I). Let $\mathfrak{M}$ be the minimal subspace adjoined to M'. We apply Theorem 1, subsection 2, to the factor M' and subspace $\mathfrak{M}$. Then $M'' = M$ assumes the role of the algebra M'; consequently the algebra M consists of all diagonal matrices of the form $A \sim ||\delta_{pq}A_0||$, where $A_0 \in M_{(\mathfrak{M})} = (M'_{(\mathfrak{M})})'$.

Suppose F runs through all projection operators in M' which satisfy the condition $F \leqq P_{\mathfrak{M}}$; according to Proposition IV, subsection 1, $F_{\mathfrak{M}}$ then runs through all projection operators in $(M')_{(\mathfrak{M})}$. But in view of the minimality of the space $\mathfrak{M}$ with respect to M', the operator F assumes only two values: $F = 0$ and $F = P_{\mathfrak{M}}$; consequently the algebra $(M')_{(\mathfrak{M})}$ contains no projection operators except 0 and 1. According to Corollary 6, subsection 2, § 34 we conclude from this that $(M')_{(\mathfrak{M})} = R(0, 1) = (\alpha 1)$. Therefore $M_{(\mathfrak{M})} = ((M')_{(\mathfrak{M})})' = (\alpha 1)' = \mathfrak{B}_0$, where $\mathfrak{B}_0$ denotes the algebra of all bounded linear operators in the space $\mathfrak{M}$.

We have thus proved

THEOREM 2. *A factor is of discrete type if, and only if, it is direct, and hence if, and only if, it is completely isomorphic to the algebra* $\mathfrak{B}_0$ *of all bounded linear operators in some finite-dimensional or infinite-dimensional Hilbert space* $\mathfrak{H}_0$. *Moreover, if* $\mathfrak{H}_0$ *is finite-dimensional, then* M *is of the finite type* (I_n). *From this to follows that all factors of a given finite type* (I_n), $n = 1, 2, \cdots$, *and also all factors of the type* (I_∞) *in a separable space are completely isomorphic to each ather so that in the case* (I_n) *and in the separable case* (I_∞) *the relative dimension forms a complete system of invariants with respect to complete isomorphism of algebras.*

4. Structure of factors of type (II_∞). Theorem 1, subsection 2, enables us to describe the factors of type (II_∞) with the aid of factors of type (II_1). Namely, suppose M is a factor of type (II_∞) and that $\mathfrak{M}$ is a finite subspace adjoined to the algebra M. According to Proposition V, subsection 1, the algebra $M_{(\mathfrak{M})}$ is a factor of type (II_1). Therefore, applying Theorem 1, subsection 2, to the factor M and the subspace $\mathfrak{M}$, we obtain the following theorem.

THEOREM 3. *Every factor* M *of type* (II_∞) *is spatially isomorphic to the algebra of all bounded infinite matrices* $A \sim ||A_{\alpha\beta}||$, *where the* $A_{\alpha\beta}$ *are operators in a fixed factor of type* (II_1).

This theorem reduces the study of factors of type (II_∞) to the study of factors of type (II_1).

5. Example of a factor of type (II_1). The problem of determining the structure of factors of type (II_1) has not yet been solved. Only isolated examples are available and it is known that non-isomorphic factors of type (II_1) exist.

Here we shall give the simplest example of a factor of type (II_1). Suppose $\mathfrak{G}$ is a denumerable discrete group satisfying condition (A):

(A) the class $\mathfrak{C}_a$, $a \neq e$, of all elements $c^{-1}ac$, $c \in \mathfrak{G}$, is infinite.
[An example of such a group is the group of all transformations $x' = ax+b$, $a > 0$, with rational coefficients a, b.]

We form the Hilbert space $\mathfrak{H}$ whose elements are all possible systems $f = \{x_a, a \in \mathfrak{G}\}$ of complex numbers x_a which satisfy the condition

$$\sum_{a\in\mathfrak{G}} |x_a|^2 < \infty,$$

where the inner product of two vectors $f = \{x_a, a \in \mathfrak{G}\}$ and $g = \{y_a, a \in \mathfrak{G}\}$ is defined by the formula

$$(f, g) = \sum_{a\in\mathfrak{G}} x_a \bar{y}_a.$$

We introduce the operators U_{a_0}, V_{a_0}, $a_0 \in \mathfrak{G}$, in the space $\mathfrak{H}$, by setting

$$U_{a_0}\{x_a, a \in \mathfrak{G}\} = \{x_{aa_0}, \quad a \in \mathfrak{G}\},$$
$$V_{a_0}\{x_a, a \in \mathfrak{G}\} = \{x_{a_0^{-1}a}, \quad a \in \mathfrak{G}\}.$$

Clearly, U_{a_0}, V_{a_0} are unitary operators in $\mathfrak{H}$.

Suppose M_1 is the algebra of all bounded linear operators A which commute with all the operators V_{a_0}, and let M_2 be the algebra of all bounded linear operators B which commute with all the operators U_{a_0}. Every operator U_{a_0} commutes with each of the operators V_{b_0}; therefore

$$U_{a_0} \in M_1, \quad V_{a_0} \in M_2 \quad \text{for all } a_0 \in \mathfrak{G}. \tag{1}$$

Every bounded operator A in $\mathfrak{H}$ can be represented in the form of a bounded numerical matrix $A \sim ||\eta_{a,b}||_{a,b\in\mathfrak{G}}$. If $A \in M_1$, then, as can easily be seen, the condition that A commutes with the operators V_{a_0} yields $\eta_{a_0a, a_0b} = \eta_{a,b}$. Setting $a_0 = a^{-1}$ in this last equality, we obtain

$$\eta_{a,b} = \eta_{e,a^{-1}b} = \eta_{a^{-1}b},$$

where we have introduced the notation $\eta_c = \eta_{e,c}$.

Thus, the matrix of the operator $A \in M_1$ must have the form

$$A \sim ||\eta_{a^{-1}b}||_{a,b\in\mathfrak{G}}.$$

Analogously, the matrix of the operator $B \in M_2$ must have the form

$$B \sim ||\zeta_{ab^{-1}}||_{a,b\in\mathfrak{G}}.$$

We shall prove that $M_1' = M_2$ and that $M_2' = M_1$. To this end, we note that every matrix $\eta = ||\eta_{a^{-1}b}||$ commutes with every matrix $\zeta = ||\zeta_{ab^{-1}}||$. In fact,

$$\eta\zeta = ||\sum_c \eta_{a^{-1}c}\zeta_{cb^{-1}}||, \quad \zeta\eta = ||\sum_c \zeta_{ac^{-1}}\eta_{c^{-1}b}||.$$

Setting $c = ac'^{-1}b$ in the first sum, we obtain

$$\sum_{c} \eta_{a^{-1}c}\zeta_{cb^{-1}} = \sum_{c'} \eta_{a^{-1}ac'^{-1}b}\zeta_{ac'^{-1}bb^{-1}} = \sum \eta_{c'^{-1}b}\zeta_{ac'^{-1}};$$

consequently, $\eta\zeta = \zeta\eta$. It follows that

$$M_2 \subset M_1', \quad M_1 \subset M_2'. \tag{2}$$

On the other hand, it follows from relation (1) that

$$M_1 = \{R(V_{a_0}: a_0 \in \mathfrak{G})\}' \supset M_2', \tag{3}$$

$$M_2 = \{R(U_{a_0}: a_0 \in \mathfrak{G})\}' \supset M_1'. \tag{4}$$

Comparing relations (3), (4) and (2), we see that

$$M_1 = M_2' = R(U_{a_0}: a \in \mathfrak{G}), \quad M_2 = M_1' = R(V_{a_0}: a_0 \in \mathfrak{G}).$$

We shall now find the intersection $M_1 \cap M_2$. If the operator A belongs to this intersection, then its matrix $||\eta_{a,b}||$ must satisfy both of the following conditions:

$$\eta_{a_0a,\, a_0b} = \eta_{aa_0,\, ba_0} = \eta_{a,\, b}\,.$$

It follows from the first condition, as we saw, that $\eta_{a,b} = \eta_{a^{-1}b}$. Application of the second condition then yields

$$\eta_{a_0^{-1}a^{-1}ba_0} = \eta_{a^{-1}b},$$

i.e. that the function η_a, $a \in \mathfrak{G}$, is constant on every class $\mathfrak{C}_a$. On the other hand, we must have

$$\sum_{c \in \mathfrak{G}} |\eta_c|^2 = \sum_{c \in \mathfrak{G}} |\eta_{e,c}|^2 < \infty \tag{5}$$

in view of the fact that the matrix $||\eta_{a,b}||$ is bounded. Since, according to condition (A), the class $\mathfrak{C}_a$ is infinite for $a \neq e$, condition (5) holds only in the case when the constant value of the function η_c on $\mathfrak{C}_a$ equals zero. Thus, $\eta_c = 0$ for $c \neq e$; consequently, $\eta_{a,b} = \delta_{a,b}\eta_e$, where

$$\delta_{a,b} = \begin{cases} 1 & \text{for } a = b, \\ 0 & \text{for } a \neq b. \end{cases}$$

This means that the algebra $M_1 \cap M_2 = M_1 \cap M_1' = M_2' \cap M_2$ consists of the scalars $\alpha 1$, i.e. that M_1, M_2 are factors.

We shall prove that they are factors of type (II_1). Consider, for instance, the factor M_1. To each operator $A \sim ||\eta_{a^{-1}b}||$ in M_1 we assign the number $T(A) = \eta_e$ and prove that the function $T(A)$ obtained in this way satisfies all the conditions of Theorem 2, subsection 2, § 37. First, it is obvious that $T(1) = 1$, $T(\alpha A) = \alpha T(A)$, $T(A+B) = T(A)+T(B)$, so that conditions 1^0, 2^0, 3^0 of Theorem 1, subsection 2, § 37, are satisfied. Further, if A is a positive definite Hermitian operator in M_1, then $A = B^*B$, where $B \in M_1$. Suppose $A \sim ||\eta_{a^{-1}b}||$, $B \sim ||\zeta_{a^{-1}b}||$; then

$$\eta_{a^{-1}b} = \sum_c \bar{\zeta}_{c^{-1}a} \zeta_{c^{-1}b};$$

consequently, for $a = b = e$,

$$T(A) = \eta_e = \sum_c |\zeta_{c^{-1}}|^2 \geqq 0,$$

where the equality sign is possible only when $\zeta_c = 0$, i.e. when $A = 0$. This also verifies conditions 4° and 6′. It can be proved analogously that $T(AB) = T(BA)$ for arbitrary $A, B \in M_1$.

Thus, the function $T(A)$ satisfies all the conditions of Theorem 2, subsection 2, § 37. Consequently, according to this theorem, M_1 is factor of finite type. On the other hand, the case (I_n) is impossible here because M_1 contains an infinite set of linearly independent elements U_{a_0}. Hence M_1 is a factor of type (II_1). It can be proved analogously that M_2 is a factor of type (II_1).

The construction of an example of a type (III) factor follows the above example, except that in this case the steps are more complicated (see von Neumann [1]). Another construction for examples of factors of types (I) and (II), based on the concept of the direct product of a finite or infinite (even nondenumerable) number of Hilbert spaces, is given in von Neumann [7].

A still more general method of constructing factors of type (II) was subsequently pointed out by Wright [1] and factors of types (II) and (III) by Dixmier [14]. Also see Sakai [1,5].

6. Approximately finite factors of type (II_1). A factor M is said to be *approximately finite* or *hyperfinite* if there exists a sequence of factors

$$M_1 \subset M_2 \subset M_3 \subset \cdots \subset M$$

of finite types (I_{n_1}), (I_{n_2}), (I_{n_3}), $\cdots$ such that $R(M_1, M_2, M_3, \cdots) = M$. Clearly, the approximate finiteness of a factor is invariant under symmetric isomorphism of algebras. It turns out (see Murray and von Neumann [1], Theorem XIII) that every factor of type (II_1) contains a subalgebra which is an approximately finite factor of type (II_1). On the other hand, it is shown in the same paper that *there also exist factors of type* (II_1) *which are not approximately finite.* This also shows that *there exist non-isomorphic factors of type* (II_1).

J. Schwartz [1, 2] constructed an example of two non-isomorphic factors of type (II) which are not approximately finite and also an example of at least three mutually non-isomorphic factors of type (III); earlier, Pukánszky [4] constructed an example of two non-isomorphic factors of type (III). Recently, Powers [1] constructed a continuum of pairwise non-isomorphic hyperfinite factors of type (III).

7. Relationship between the types of factors M and M′. We saw above (see IV, subsection 3) that M and M' both belong or both do not belong to type (I). This result admits the following generalization (see Murray and von Neumann [1], I).

Theorem 4. *The factors M and M' either both belong or both do not belong to the type* (I), (II) *or* (III) *and* $D_M(\mathfrak{M}_f^{M'}) = CD_{M'}(\mathfrak{M}_f^{M})$ *for an arbitrary* $f \neq 0$, *where C is some constant. Furthermore,* $C = 1$ *for factors of type* (I) *if we use the standard normalization for the functions* D_M *and* $D_{M'}$. *In the case of factors of type* (I) *any combination of* (I_n) *and* (I_m), $n, m = 1, 2, \cdots, \infty$ *is possible for M and M'; and in the case of factors of type* (II) *any of the combinations* (II_1) *and* (II_1), (II_1) *and* (II_∞), (II_∞) *and* (II_∞) *with arbitrary positive value of the constant C are possible for M and M'.*

8. Relationship between symmetric and spatial isomorphisms. We assign the number

$$\theta = \frac{1}{C} \frac{D_{M'}(\mathfrak{H})}{D_M(\mathfrak{H})}$$

to a factor M of type (I) or (II) where C is the constant defined by Theorem 4, subsection 7. If $D_{M'}(\mathfrak{H})$ and $D_M(\mathfrak{H})$ are not infinite simultaneously, then θ is defined and does not depend on the normalization of the functions $D_{M'}$ and D_M; but if $D_M(\mathfrak{H})$ and $D_M(\mathfrak{H})$ are both infinite then θ is understood to be the symbol $\frac{\infty}{\infty}$. It turns out (see MURRAY and VON NEUMANN [1], II, Theorem X) that *two factors M_1 and M_2 in the separable spaces $\mathfrak{H}_1$ and $\mathfrak{H}_2$ are spatially isomorphic if, and only if, the factors M_1 and M_1' are symmetrically isomorphic to the factors M_2 and M_2' respectively and that θ is the same for both factors M_1, M_2.*

For various generalizations of this result, see DIXMIER [12, 14] and GRIFFIN [1].

9. Unbounded operators, adjoined to a factor of finite type. Suppose M is a factor of finite type; we denote the set of all closed linear operators $A\eta M$ with domain $\mathfrak{D}(A)$, dense in $\mathfrak{H}$, by U_M. Then any (non-commutative) operator polynomial in U_M is defined on a set dense in $\mathfrak{H}$ and admits a closure. Operations with these closures and passage to the adjoint operator are carried out according to the usual rules of matrix algebra. Every symmetric operator $H \in U_M$ is self-adjoint and if $A_1, A_2 \in U_M$, $A_1 \subset A_2$, then $A_1 = A_2$. For the proof of all these results, see MURRAY and VON NEUMANN [1], I, Theorem XV.

§ 39. Unitary algebras and algebras with trace

1. Definition of a unitary algebra. X is called a *unitary algebra* if:

1) X is a symmetric algebra;
2) X is a pre-Hilbert space;
3) $(xy, z) = (y, x^*z)$ for arbitrary $x, y, z \in X$;
4) $(x, y) = (y^*, x^*)$ for arbitrary $x, y \in X$;
5) the operator $U_x y = xy$ is continuous on X for every $x \in X$;
6) the elements of the form xy, $x, y \in X$, form a dense set in X.

Clearly, every H^*-algebra (see subsection 5, § 25) is a unitary algebra; the converse is in general not true. The difference consists in this that a unitary algebra need not be a normed algebra with norm $|x| = \sqrt{(x, x)}$, i.e. it does not necessarily have to satisfy the inequality $|xy| \leqq |x||y|$; moreover, a unitary algebra may not be complete.

2. Definition of an algebra with trace. Suppose M is a weakly closed algebra of bounded linear operators in a Hilbert space, containing the identity; we denote the set of all positive definite Hermitian operators in M by M_+. The *trace* on M is any function $T(H)$, defined on M_+, with values in the interval $[0, +\infty]$, which satisfies the following conditions:

1) $T(UHU^{-1}) = T(H)$ for all unitary $U \in M$;
2) if $H \in M_+$ and $T(H) = 0$, then $H = 0$;
3) if $H \in M_+$ and H is the sum of a strongly convergent series of operators $H_\alpha \in M_+$, then $T(H) = \sum_\alpha T(H_\alpha)$.

An algebra M on which a trace $T(H)$ is defined is called *an algebra with trace.* Factors of types (I) and (II) are examples of algebras with trace.

3. Unitary algebras defined by the trace. Suppose M is an algebra with trace $T(H)$. An operator A in M is said to be *finite* if it is a finite linear combination of operators in M_+ with finite trace; we denote the set of all finite operators in M by M_1.

An operator A in M is said to be *normable* if $T(A^*A) < \infty$; we denote the set of all such operators in M by M_2. From the relation $(A+B)^*(A+B) \leqq 2(A^*A+B^*B)$ and the properties of the trace, we conclude that M_2 is a subspace in M; from this it easily follows that $B^*A \in M_1$ if $A, B \in M_2$. [For Hermitian operators, $A \leqq B$ means, as usual, that $(Af, f) \leqq (Bf, f)$ for all $f \in \mathfrak{H}$.] Therefore an inner product can be defined in M_2 by setting $(A, B) = T(B^*A)$. We arrive at the following theorem, the detailed proof of which is omitted (see GODEMENT [11]).

THEOREM 1. *Suppose M is an algebra with trace $T(H)$; then the set M_2 of all normable operators in M, with the natural involution $A \to A^*$ and inner product $(A, B) = T(B^*A)$, is a unitary algebra.*

4. Canonical trace in a unitary algebra. It turns out that the method of constructing a unitary algebra from its trace, described in subsection 3, enables one to obtain any unitary algebra; specifically, every unitary algebra can be embedded in another unitary algebra obtained in the manner described above, with the aid of the trace. In order to prove this, we first consider some simple properties of a unitary algebra.

Suppose X is a unitary algebra. We denote the completion of the pre-Hilbert space X by $\mathfrak{H}$; then $\mathfrak{H}$ is a Hilbert space. It follows from property 5) of a unitary algebra (see subsection 1) that the operator $U_x y = xy$, $x, y \in X$, is uniquely extendible to a bounded linear operator in $\mathfrak{H}$; this operator in $\mathfrak{H}$ will again be denoted by U_x.

Clearly, the correspondence $x \to U_x$ is a homomorphism, where property 3) of subsection 1 means that this homomorphism is symmetric.

We denote the weakly closed algebra generated by the operators U_x by $M(X)$.

Applying property 4), subsection 1, it is easily shown that the operator $V_x y = yx$ is also bounded in X and hence is uniquely extendible to a bounded linear operator in $\mathfrak{H}$; this operator in $\mathfrak{H}$ will again be denoted by V_x.

An element $f \in \mathfrak{H}$ is said to be *bounded* if there exists a bounded linear operator U_f such that $U_f x = V_x f$ for all $x \in X$. We now define the trace in $M_+(X)$ by setting

$$T(H) = \begin{cases} (f, f) \text{ if } \sqrt{H} = U_f \text{ for some bounded } f \in \mathfrak{H}, \\ +\infty \text{ otherwise.} \end{cases}$$

We leave it to the reader to prove the fact that all the trace axioms (see subsection 2) will then be satisfied (for a detailed proof, see GODEMENT [11]). Using the canonical representation $A = WH$, we obtain that the operator A in $M(X)$ is normable if, and only if, $A = U_f$ for some bounded $f \in \mathfrak{H}$, where

$$T(B^*A) = (f, \varphi)$$

for two normable $A = U_f$, $B = U_\varphi$. We then arrive at the following theorem.

THEOREM 2. *Suppose X is a unitary algebra, $\mathfrak{H}$ is the completion of the pre-Hilbert space X, and $M(X)$ is the weakly closed algebra generated by the operators U_x, $x \in X$. Then there exists in $M(X)$ a unique trace $T(A)$ possessing the following properties*:

1) *the operator $A \in M(X)$ is normable if, and only if, $A = U_f$ for some bounded $f \in \mathfrak{H}$*;

2) *for any two normable $A = U_f$, $B = U_\varphi$,*

$$T(B^*A) = (f, \varphi);$$

in particular, $(x, y) = T(U_y^* U_x)$ for arbitrary $x, y \in X$.

The trace $T(H)$, defined in this way with respect to a given unitary algebra X, is called the *canonical trace* in $M(X)$. In general, $M(X)$ may have other traces; for example, every function $T'(H) = T(AH)$, where A is a positive definite operator in the center of the algebra $M(X)$, is also a trace in $M(X)$. From this it is clear that a *necessary* condition for the trace in $M(X)$ to be unique to within a constant multiplier is that the algebra $M(X)$ be a factor. A unitary algebra X is said to be *irreducible* if $M(X)$ is a factor. In virtue of the results of § 37, irreducibility is not only a necessary, but also a sufficient, condition for the uniqueness of the trace to within a constant multiplier.

Consequently, in the case of an irreducible algebra, the canonical trace coincides, to within a multiplier, with the relative trace in $M(X)$ and Theorem 2 yields a direct method of constructing the relative trace in $M(X)$.

THEOREM 3. *Suppose R is a symmetric algebra of bounded linear operators in a Hilbert space and let $M = R''$; assume that*

a) *M is the closure of the algebra R in the strong topology*;

b) *M is a factor of type* (I) *or* (II);

c) *if S is the relative trace in M, then $S(A^*A) < +\infty$ for all $A \in R$.*

Then the algebra R, with the natural involution $A \to A^*$ and inner product $(A, B) = S(B^*A)$, is an irreducible unitary algebra. Moreover:

α) *the completion $\mathfrak{H}$ of the pre-Hilbert space R with respect to the inner product $(A, B) = S(B^*A)$ is isomorphic to the completion of the unitary algebra M_2, consisting of the normable elements of the algebra M.*

β) The algebra $M(R)$ is isomorphic to the algebra M, where the isomorphism of M onto $M(R)$ is defined by the condition that it be weakly continuous on the unit ball of the algebra M and that it assign to each $A \in R$ the operator $U_A \in M(R)$, defined on R by means of the equality $U_A B = AB$. [Recall (see beginning of this subsection) that $M(X)$ denotes the weakly closed algebra of operators generated by the operators U_x, $x \in X$.]

γ) The canonical trace T on $M(R)$ is given by the formula

$$T(U_H) = S(H) \quad \textit{for all } H \in M_+(R),$$

where $A \to U_A$ is the isomorphism, defined above, of the algebra M onto $M(R)$.

The proof consists of verifying that:

1) the correspondence $A \to U_A$, defined by the formula $U_A B = AB$, does indeed define an isomorphism of the algebra M onto $M(R)$;

2) R, with the involution $A \to A^*$ and inner product $(A, B) = S(B^*A)$, satisfies all the unitary algebra axioms;

3) the function $T(U_H) = S(H)$, $H \in M_T(R)$, satisfies all the trace axioms and the constant multiplier by which it can differ from the canonical trace is equal, in this case, to unity.

The results of §§33–38 are due to Murray and von Neumann (see MURRAY and VON NEUMANN [1], VON NEUMANN [1, 4, 8]), and of § 39 to GODEMENT [11]; a theorem, closely related to Theorem 3, § 39 (but only for factors of a finite type), is to be found in the paper by MURRAY and VON NEUMANN [1], II.

Later, the results of Murray and von Neumann were developed in numerous papers. Of these, the papers by DIXMIER [2–13] (see also the excellent book by DIXMIER [14], where a more detailed exposition of the material in this chapter is given and many new results are added), GODEMENT [8, 11, 12], KAPLANSKY [17] and SEGAL [8–13] are of the greatest interest. In Dixmier's papers, the concept of trace is generalized to arbitrary weakly closed algebras (where the values of the trace no longer are numbers but operators from the center of an algebra) and the classification [types (I), (II), (III), finite, and others] is carried over to arbitrary weakly closed algebras (which are not necessarily factors). In KAPLANSKY [17] there is given an axiomatic development of the theory (see also DIXMIER [11], SAKAI [5] and REID [1]). In Segal's papers, there is a development of the theory of non-commutative integration; moreover, Segal applies the results of §§ 33–38 to the theory of the unitary representations of a group.

Also of interest are the papers of FUGLEDE and KADISON [1–3] in which a theory of determinants for factors of finite type is constructed (see also PALLU DE LA BARRIÈRE [1]).

The results of §§ 33–39 were applied by GODEMENT [8, 11, 12] in his theory of characters of non-commutative groups which is a generalization of the results of GELFAND and NAIMARK [7]. The concept of a commutative unitary algebra, closely related to that discussed in this section, and the theory of commutative unitary algebras were first given in a paper by ROKHLIN [1].

For applications of the theory of completely regular algebras and weakly closed algebras of operators to quantum mechanics, see, e.g., BORCHERS [1, 2], DOPLICHER [1], KASTLER [1], LOUPIAS and MIRACLE-SOLE [1] and RUELLE [1].

The analogue of weakly closed algebras of operators are the Jordan algebras. A Jordan algebra is a set $\mathfrak{A}$ of bounded self-adjoint operators in a Hilbert space, satisfying the following conditions: 1) $\mathfrak{A}$ is a real linear space; 2) $\mathfrak{A}$ is closed in the weak operator topology; 3) if $a, b \in \mathfrak{A}$, then $a \circ b =$

$\frac{1}{2}(ab+ba) \in \mathfrak{A}$. Finite-dimensional Jordan algebras were studied in the paper by JORDAN, VON NEUMANN and WIGNER [1]. In STØRMER [2] there is given a classification of Jordan algebras which are an analogue of factors of type (I_n).

In a recent work of MCDUFF [1] a continuum of pairwise non-isomorphic II_1 factors is constructed (also see SAKAI [2]). Sakai constructed a continuum of pairwise non-isomorphic factors of type II_∞ (see SAKAI [2]) and of non-hyperfinite factors of type III (see SAKAI [3]).

CHAPTER VIII

DECOMPOSITION OF AN ALGEBRA OF OPERATORS INTO IRREDUCIBLE ALGEBRAS

§ 40. Formulation of the problem; canonical form of a commutative algebra of operators in Hilbert space

1. Formulation of the problem. A set S of bounded linear operators in the Hilbert space $\mathfrak{H}$ is said to be *irreducible* if there does not exist in $\mathfrak{H}$ a closed subspace, different from (0) and all of $\mathfrak{H}$, which is invariant relative to all operators $A \in S$; otherwise, S is said to be *reducible*. In particular, a symmetric algebra $R \subset \mathfrak{B}(\mathfrak{H})$ is called irreducible or reducible if it is an irreducible or reducible set respectively.

If R is a symmetric algebra of linear operators in the *finite-dimensional* Hilbert space $\mathfrak{H}$ and if R is reducible, then there exists a subspace $\mathfrak{M}$ in $\mathfrak{H}$, where $\mathfrak{M}$ is different from (0) and all of $\mathfrak{H}$, which is invariant relative to all the operators $A \in R$. But then the orthogonal complement $\mathfrak{H} \ominus \mathfrak{M}$ is also invariant relative to all the operators $A \in R$. The operators $A \in R$, restricted to $\mathfrak{M}$, form an algebra $R_{\mathfrak{M}}$ of operators in $\mathfrak{M}$, which is called the *component* of the initial algebra R in $\mathfrak{M}$. If $R_{\mathfrak{M}}$ is reducible, then we can again decompose $\mathfrak{M}$ into mutually orthogonal subspaces and the same can also be done with $\mathfrak{H} \ominus \mathfrak{M}$. Since $\mathfrak{H}$ is finite-dimensional, by repeating this process a finite number of times, we arrive at the decomposition $\mathfrak{H} = \mathfrak{M}_1 \oplus \cdots \oplus \mathfrak{M}_n$ of the space $\mathfrak{H}$ into the orthogonal sum of a finite number of subspaces which are invariant relative to R, in each of which R (specifically, the component $R_k = R_{\mathfrak{M}_k}$ of the algebra R) is irreducible. We shall say that the decomposition $\mathfrak{H} = \mathfrak{M}_1 \oplus \cdots \oplus \mathfrak{M}_k$ realizes a *decomposition of the algebra R into the irreducible algebras R_k.*

We note that corresponding to this decomposition, every element $\xi \in \mathfrak{H}$ is representable in the form

$$\xi = \langle \xi_1, \xi_2, \cdots, \xi_n \rangle, \quad \xi_k \in \mathfrak{M}_k, \tag{1}$$

and the operator $A \in R$ is given by the formula

$$A\xi = \langle A_1\xi_1, A_2\xi_2, \cdots, A_n\xi_n \rangle, \tag{2}$$

where $A_k \in R_k$, $k = 1, 2, \cdots, n$.

In the case when $\mathfrak{H}$ is an infinite-dimensional space, the preceding line of reasoning no longer applies. In order to see how to modify this line of reasoning, we consider the case of a symmetric *commutative* algebra R of bounded linear operators, which is closed in norm, in an arbitrary Hilbert space $\mathfrak{H}$. Then R is isomorphic to the algebra of all continuous functions $a(t)$ on a locally compact space T. [Here we change our notation and write t and T instead of M and $\mathfrak{M}$.] We assume further that there is a cyclic vector ξ_0 in $\mathfrak{H}$ such that the collection of all vectors $A\xi_0$, $A \in R$, is dense in $\mathfrak{H}$. In this case, $\mathfrak{H}$ is isometric to the Hilbert space $L^2_\mu(T)$ with respect to some measure μ and the operators $A \in R$ go over under the isometric mapping

$$\xi \to \xi(t) \tag{3}$$

into operators of multiplication by the continuous functions $a(t)$, which vanish at infinity:

$$A\{\xi(t)\} = \{a(t)\xi(t)\} \tag{4}$$

(see I, § 17, subsection 4).

Formulas (3) and (4) can be considered as the continuous analogues of formulas (1) and (2) respectively; here, to the vectors ξ_k there correspond the individual values $\xi(t)$, and to the operators A_k the values $a(t)$ of the function a. Since R is commutative, these values no longer are vectors but rather numbers; of course, the same would have been the case earlier if the algebra R in the finite-dimensional space were commutative.

We can therefore say that $L^2_\mu(T)$ is a continuous direct sum (or the direct integral) of one-dimensional spaces, and R is the direct integral of algebras of operators in one-dimensional (and hence irreducible) spaces. The question arises if it is possible to generalize this line of reasoning to arbitrary, not necessarily commutative algebras of operators. It is natural to expect that then we will obtain the direct integral of non-one-dimensional spaces $\mathfrak{H}_t$, and R becomes the direct integral of algebras R_t in these subspaces, where all, or at least almost all, R_t will be irreducible.

The present chapter is devoted to the precise formulation and solution of this problem.

In order not to complicate the discussion with additional difficulties, we shall restrict ourselves to the case of a separable Hilbert space; the corresponding theory for the non-separable case has not yet been completed. [Concerning this, see the articles by ADELSON-VELSKY [1, 2], TAYLOR [1] and the dissertation by PEDERSEN [1].]

We note also that throughout this entire chapter, the term "algebra" (if it is not specifically stated otherwise) will denote a symmetric algebra of bounded linear operators in a separable Hilbert space, containing the identity operator 1.

To solve the formulated problem, we will need certain supplementary information about algebras of operators, which is of interest in itself.

2. The separability lemma.

THEOREM 1. *If $\mathfrak{H}$ is separable, then every set $S \subset \mathfrak{B}(\mathfrak{H})$ is separable in the sense of strong convergence, i.e. there exists a sequence $\langle A_n \rangle \in S$ such that every operator $A \in S$ is the strong limit of some one of its subsequences.* [This result is stronger than the relation $\overline{\langle A_n \rangle}^2 \supset S$ because not every strong contact point of the sequence $\langle A_n \rangle$ is a limit of some one of its subsequences.]

Proof. Let S_n be the subset of those operators $A \in S$ which satisfy the condition $|A| \leqq n$. Obviously, $S = \bigcup_{n=1}^{\infty} S_n$ and, consequently, it suffices to prove that every set S_n is separable. Since the set S_n is bounded in the sense of the operator norm, it suffices to prove that every bounded set S is separable.

Thus, let $|A| \leqq k$ for all $A \in S$. We choose in $\mathfrak{H}$ a complete orthonormal system $\varphi_1, \varphi_2, \cdots$. Then every operator A can be represented in the form of a bounded matrix $A \sim ||a_{st}||$, $s, t = 1, 2, \cdots$ in such a way that

$$Af = \sum_{s=1}^{\infty} \Big(\sum_{t=1}^{\infty} a_{st} x_t \Big) \varphi_s$$

if

$$f = \sum_{t=1}^{\infty} x_t \varphi_t .$$

In particular, $A\varphi_t = \sum_{s=1}^{\infty} a_{st} \varphi_s$, $(A\varphi_t, \varphi_s) = a_{st}$.

Consider the collection $\mathfrak{S}$ of all operators $B \sim ||b_{st}||$ such that, for some n, $b_{st} = 0$ for $s > n$ or $t > n$ and b_{st} rational for $s \leqq n$, $t \leqq n$. This collection is, obviously, countable, so that we can write $\mathfrak{S} = \{B_1, B_2, B_3, \cdots\}$. If $C \sim ||c_{st}||$ is an arbitrary element of the algebra $\mathfrak{B}$, then, choosing $B^{(p)} \sim ||b_{st}^{(p)}||$ so that $b_{st} = 0$ for $s > p$ or $t > p$, b_{st} rational and $|c_{st} - b_{st}^{(p)}| < 1/p$, $s, t = 1, 2, \cdots, p$, we have $B^{(p)} \in \mathfrak{S}$ and

$$\begin{aligned} |(C - B^{(p)})\varphi_t|^2 &= \Big| \sum_{s=1}^{p} (c_{st} - b_{st}^{(p)}) \varphi_s + \sum_{s=p+1}^{\infty} c_{st} \varphi_s \Big|^2 \\ &= \sum_{s=1}^{p} |c_{st} - b_{st}^{(p)}|^2 + \sum_{s=p+1}^{\infty} |c_{st}|^2 < \frac{1}{p} + \sum_{s=p+1}^{\infty} |c_{st}|^2 \to 0 \end{aligned} \qquad (1)$$

as $p \to \infty$, $t = 1, 2, 3, \cdots$. The operators $B^{(p)}$ form a subsequence of $\mathfrak{S}$ so that whatever the operator $C \in \mathfrak{B}$ is, there exists a subsequence $\{B_{\nu_1}, B_{\nu_2}, B_{\nu_3}, \cdots\} \subset \mathfrak{S}$ such that $|B_{\nu_n} \varphi_t - C\varphi_t| \to 0$ as $n \to \infty$, $t = 1, 2, 3, \cdots$. We now consider operators $A \in S$ such that

$$|(A - B_n)\varphi_t| < \frac{1}{m} \quad \text{for } t = 1, 2, \cdots, m. \qquad (2)$$

If such operators A exist, then we denote one of them by $A_{m,n}$. The collection of all operators $A_{m,n}$ is a countable subset S_1 of the set S; we shall show that it possesses the property required in the theorem. Let A be an arbitrary operator in S; making use of relation (1), we choose $n_m = \nu_{p_m}$ so that

$$|(A-B_{n_m})\varphi_t| < \frac{1}{m} \quad \text{for } t = 1, 2, \cdots, m.$$

Then the operator A_{m,n_m} exists and, by definition,

$$|(A_{m,n_m}-B_{n_m})\varphi_t| < \frac{1}{m} \quad \text{for } t = 1, 2, \cdots, m,$$

so that

$$|(A-A_{m,n_m})\varphi_t| < \frac{2}{m} \quad \text{for } t = 1, 2, \cdots, m.$$

Thus, the operators $C_m = A_{m,n_m}$ in S_1 form a subsequence of the set S for which

$$|(A-C_m)\varphi_t| \to 0$$

as $m \to \infty$, $t = 1, 2, 3, \cdots$. Therefore, also $|(A-C_m)f| \to 0$ for finite linear combinations f of elements φ_t. Therefore, if g is an arbitrary element of the space $\mathfrak{H}$, then choosing f so that $|g-f| < \varepsilon/(4k)$, and after that the index N so that $|(A-C_m)f| < \varepsilon/2$ for $m > N$, we have for $m > N$ that

$$|(A-C_m)g| \leqq |(A-C_m)f|+|(A-C_m)(g-f)| < \frac{\varepsilon}{2}+(|A|+|C_m|)|g-f| < \varepsilon.$$

Thus, the sequence C_m converges strongly to the operator A; this completes the proof of the lemma.

3. Canonical form of a commutative algebra. Let R be a symmetric algebra of bounded linear operators in a separable Hilbert space $\mathfrak{H}$, containing the identity operator 1, and let $\bar{R}$ be the closure of the algebra R in the operator norm. The algebra R is called *cyclic* if there exists a vector ξ_0 in $\mathfrak{H}$ such that the set $\{A\xi_0: A \in R\}$ is dense in $\mathfrak{H}$; in this case the vector ξ_0 is called a *cyclic vector* of the algebra R. Clearly, that R is cyclic is equivalent to $\bar{R}$ being cyclic, and the cyclic vectors are the same in both algebras. It is also clear that R cyclic is equivalent to its identity representation $A \to A$ being cyclic, and ξ_0 is a cyclic vector for R if, and only if, it is a cyclic vector of this representation. Therefore, if R is not a cyclic algebra, then (see subsection 2, § 17) $\mathfrak{H}$ is the orthogonal sum

$$\mathfrak{H} = \sum_k \oplus \mathfrak{H}_k \tag{1}$$

of subspaces $\mathfrak{H}_k$, which are invariant relative to R, on each of which the restriction

R_k of the algebra R is cyclic. We shall say that R is the *orthogonal sum of the cyclic algebras* R_k and write

$$R = \sum_k \oplus R_k. \tag{2}$$

Furthermore, since the space $\mathfrak{H}$ is separable, the index k runs over a finite or denumerable set, $k = 1, 2, \cdots, n$ or $k = 1, 2, 3, \cdots$.

In the remainder of this subsection, we shall assume that R is commutative; let ξ_k^0 be a cyclic vector of the algebra R_k in $\mathfrak{H}_k$ and let A_k be the restriction of the operator $A \in \bar{R}$ to $\mathfrak{H}_k$. We denote by T the compact space of maximal ideals t of the algebra $\bar{R}$ and by μ_k the measure on T defined by the integral $f_k(A) = (A\xi_k^0, \xi_k^0)$ in $C(T)$ (see the proof of Proposition I, subsection 4, § 17).

I. *There exists an isometric mapping U_k of the space $\mathfrak{H}_k$ onto $L^2_{\mu_k}(T) = L^2(f_k)$ under which the operators $A \in \bar{R}$ map into the operators of multiplication by $A(t)$, where $A(t)$ is the value of A on the ideal $t \in T$; moreover, $\{A(t)\colon A \in \bar{R}\} = C(T)$.*

For the proof it suffices to apply Proposition I, subsection 4, § 17 to the representation $A \to A_k$ of the algebra R.

II. *If $P(\Delta)$ is the spectral measure of the algebra R, then*

$$\mu_k(\Delta) = (P(\Delta)\xi_k^0, \xi_k^0) \tag{3}$$

for every Borel set $\Delta \subset T$.

Proof. By I, we can assume that $\mathfrak{H}_k = L^2_{\mu_k}(T)$ and the assertion follows from formula (3), subsection 4, § 17 applied to the restriction of the representation $A \to A$ to the subspace $\mathfrak{H}_k$ and the vector $\xi_k^0 = \xi_k^0(t) = 1$.

Let $\tilde{\mathfrak{H}}$ be the direct sum of the spaces $L^2_{\mu_k}(T)$:

$$\tilde{\mathfrak{H}} = \sum_k \oplus L^2_{\mu_k}(T);$$

$\tilde{\mathfrak{H}}$ consists of all sequences $\tilde{\xi} = \langle \xi_k(t) \rangle$ for which $\xi_k(t)$ is a μ_k-measurable function and

$$\sum_k \int |\xi_k(t)|^2 d\mu_k < \infty; \tag{4}$$

in this connection, the operations of addition and scalar multiplication are defined componentwise and the inner product by the formula

$$(\tilde{\xi}, \tilde{\eta}) = \sum_k \int \xi_k(t)\overline{\eta_k(t)}d\mu_k \quad \text{for } \tilde{\xi} = \langle \xi_k(t) \rangle, \tilde{\eta} = \langle \eta_k(t) \rangle. \tag{5}$$

III. *If R is commutative, then there exists an isometric mapping U of the space $\mathfrak{H}$ onto $\tilde{\mathfrak{H}}$ under which the operators $A \in \bar{R}$ map into the operators $\tilde{A}$ of multiplication by $A(t)$,*

$$\tilde{A}\{\xi(t)\} = \{A(t)\xi(t)\} \text{ for } A \in \bar{R},\ \{\xi(t)\} \in \hat{\mathfrak{H}}, \tag{6}$$

and $\{A(t)\colon A \in \bar{R}\} = C(T)$.

The assertion follows directly from Proposition I and formulas (1), (2), with $U = U_k$ on $\mathfrak{H}_k$.

Again let $P(\Delta)$ be the spectral measure of the commutative algebra R and let $\xi, \eta \in \mathfrak{H}$. We shall say that the vector ξ is *dominated* by the vector η and write $\xi \prec \eta$ if the measure $\mu_\xi(\Delta) = (P(\Delta)\xi, \xi)$ is dominated by the measure $\mu_\eta(\Delta) = (P(\Delta)\eta, \eta)$; we shall call the vector $\xi_0 \in \mathfrak{H}$ *maximal* if every vector in $\mathfrak{H}$ is dominated by ξ_0.

IV. *A cyclic vector* ξ_k^0 *is maximal in* $\mathfrak{H}_k$.

Proof. By I and II, we can assume that

$$\mathfrak{H}_k = L^2_{\mu_k}(T), \quad \xi_k^0(t) \equiv 1 \quad \text{and} \quad \mu_k(\Delta) = (P(\Delta)\xi_k^0, \xi_k^0).$$

Then: if $\eta \in \mathfrak{H}_k$, then $\eta = \eta_k(t) \in L^2_{\mu_k}(T)$ and by (3), subsection 4, § 17,

$$\mu_\eta(\Delta) = (P(\Delta)\eta, \eta) = \int_\Delta |\eta(t)|^2 d\mu_k = \int |\eta(t)|^2 d\mu_{\xi_k^0}.$$

Consequently, $\mu_\eta \prec \mu_{\xi_k^0}$, i.e., $\eta \prec \xi_k^0$.

V. *There exists a maximal vector in* $\mathfrak{H}$.

Proof. We proceed from some decomposition (1)–(2); without loss of generality, we can assume that the corresponding cyclic vectors ξ_k^0 are normalized ($|\xi_k^0| = 1$); moreover, they are mutually orthogonal. Therefore, if we set

$$\xi^{(1)} = \xi_1^0 + \tfrac{1}{2}\xi_2^0 + \tfrac{1}{3}\xi_3^0 + \cdots, \tag{7}$$

then the series in the right member (if it is infinite) will converge (because $\sum_k |(1/k)\xi_k^0|^2 = \sum_k 1/k^2 < \infty$) and it defines an element $\xi^{(1)} \in \mathfrak{H}$. Now if $\eta \in \mathfrak{H}$, then, by (1), $\eta = \sum_k \eta_k$, $\eta_k \in \mathfrak{H}_k$, and, by IV, $\eta_k \prec (1/k)\xi_k$, and therefore $\eta \prec \xi^{(1)}$.

VI. *In the decomposition* (1)–(2), *the cyclic vectors* $\xi_1^0, \xi_2^0, \cdots$ *can be chosen so that* $\xi_1^0 \succ \xi_j^0$, $j = 2, 3, \cdots$, *and hence* $\mu_1 \succ \mu_j$, $j = 2, 3, \cdots$.

[It can be shown (see, e.g., STONE [1] or PLESNER [2]) that $\xi_1^0, \xi_2^0, \cdots$ can also be chosen in such a way that $\xi_1^0 \succ \xi_2^0 \succ \cdots$; however, we do not need this.]

Proof. It suffices to choose ξ_1^0 to be a maximal vector (see V).

Summarizing the preceding results, we arrive at the following theorem.

THEOREM 1. *Every commutative symmetric algebra R of bounded linear operators on a separable Hilbert space is spatially isomorphic to an algebra of operators of multiplication*

$$A\{\xi_k(t)\} = \{A(t)\xi_k(t)\}$$

in the Hilbert space $\mathfrak{H} = \sum_k \oplus L^2_{\mu_k}(T)$, *where T is the compact space of maximal ideals of the algebra* $\overline{R_a(R, 1)}$,

$$\mu_1 \succ \mu_j, \quad j = 2, 3, \cdots \tag{8}$$

of measures on T, and $A(t) \in C(T)$. *If R is closed in the operator norm and contains the identity, then* $\overline{\{A(t): A \in R\}} = C(T)$.

[Recall that $\overline{R_a(R, 1)}$ denotes the minimal algebra, closed in the operator norm, which contains R and 1.]

The realization of the algebra R, described in Theorem 1, is called the *canonical form* of R. The conditions for a spatial isomorphism of two canonical forms are considered in the monograph of Dixmier [14], Ch. II. We note further that in the canonical form the space T can be replaced by a segment of the real line (but the $A(t)$ will then, in general, not be continuous); this follows from general results of measure theory (see, e.g., Halmos [1], Ch. VIII) and can even be obtained directly (see, e.g., Akhiezer and Glazman [1], Ch. VI). We will not need these facts in the sequel.

§ 41. Direct integral of Hilbert spaces; the decomposition of an algebra of operators into the direct integral of irreducible algebras

1. Direct integral of Hilbert spaces. Let T be an arbitrary compact space, I an integral on T and μ the corresponding measure (see § 6, subsection 8). [Instead of a compact space we could consider a locally compact space T. However, no essential generalization is obtained this way since we can replace T by the space T_∞, taking $H_\infty = (0)$ (see subsection 9, § 2).]

Assume that to almost every point $t \in T$ there is set into correspondence a separable Hilbert space H_t. We introduce now the concept of *direct integral of the spaces* H_t with respect to the measure μ.

We first consider the case when all the H_t have the same dimension. Then each of them can be identified with the same fixed Hilbert space H. [We shall not consider direct integrals which are obtained for various identifications of the spaces H_t with the fixed space H to be essentially distinct.]

We consider the function $\xi = \{\xi_t\}$, $t \in T$, whose values for almost every $t \in T$ are vectors in H. This vector-valued function is said to be *measurable* if for an arbitrary vector $h \in H$, the numerical function $f(t) = (\xi_t, h)$ is measurable (with respect to the measure μ) in the usual sense.

If $\xi = \{\xi_t\}$ and $\eta = \{\eta_t\}$ are two measurable vector-valued functions, then the inner product (ξ_t, η_t) is a measurable numerical function. This follows directly from the equality

$$(\xi_t, \eta_t) = \sum_k (\xi_t, e_k)(e_k, \eta_t),$$

where $\{e_k\}$ is a fixed complete orthonormal system in H.

Let us recall (see § 26, subsection 5) that in this case the direct integral $\mathfrak{H}=\int_T H_t d\mu$ of the spaces H_t with respect to the measure μ is the Hilbert space $\mathfrak{H}$ of all measurable vector-valued functions $\xi = \{\xi_t\}$ which satisfy the condition

$$\int_T |\xi_t|^2 d\mu < \infty. \tag{1}$$

[In this connection, as usual, two such measurable functions $\xi = \{\xi_t\}$ and $\eta = \{\eta_t\}$ are not considered to be distinct if $\xi_t \neq \eta_t$ only on some set of μ-measure zero.] The operations of the addition of vectors and scalar multiplication, and the formation of the inner product are defined in $\mathfrak{H}$ by the formulas

$$\begin{gathered} \xi+\eta = \{\xi_t+\eta_t\}, \quad \alpha\xi = \{\alpha\xi_t\} \\ (\xi, \eta) = \int_T (\xi_t, \eta_t) d\mu \end{gathered} \tag{2}$$

where $\xi = \{\xi_t\}$, $\eta = \{\eta_t\}$.

The following "coordinate notation" for the direct integral is frequently useful. We choose in H some complete orthonormal system of vectors $\{e_k\}$. Let $\xi = \{\xi_t\}$ be an arbitrary vector-valued function on T with values from H. We set

$$a_k(t) = (\xi_t, e_k).$$

Thus, to each vector-valued function $\{\xi_t\}$ there corresponds a sequence of numerical functions $a_k(t)$. By virtue of VII, § 6, subsection 7, the vector-valued function $\{\xi_t\}$ belongs to the direct integral $\mathfrak{H}$ if the spaces $H_t = H$ if, and only if, the corresponding numerical functions $a_k(t)$ belong to $L^2_\mu(T)$ and satisfy the condition

$$\sum_k \int_T |a_k(t)|^2 d\mu < \infty.$$

Furthermore, the inner product of the vector-valued functions $\xi = \{\xi_t\}$ and $\eta = \{\eta_t\}$ in $\mathfrak{H}$ is written in the form

$$(\xi, \eta) = \sum_k \int_T a_k(t)\overline{b_k(t)} d\mu, \quad \text{where } a_k(t) = (\xi_t, e_k),\ b_k(t) = (\eta_t, e_k).$$

Obviously, in all the preceding definitions and discussion, we can replace T by any of its μ-measurable subsets T' and define $\int_{T'} H_t d\mu$ analogously.

Up to this point, we assumed that all the spaces H_t have the same dimension. We shall now give the definition of the direct integral of separable spaces H_t in the general case. We shall denote by $n(t)$ the dimension of the space H_t; thus, $n(t)$ is a function which assumes only the values $1, 2, \cdots$ and ∞. The family of spaces H_t will be said to be *μ-measurable* if $n(t)$ is a μ-measurable function. Since $n(t)$ takes on no more than a countable number of values, we have that the μ-measurability of the function $n(t)$ means that the entire space T decomposes into a sum of not more than a countable number of pairwise disjoint μ-measurable sets T_n, on each of which H_t has the same dimension, so that on T_n each H_t can be identified with a fixed space of dimension n. The vector-valued function $\{\xi_t\}$, defined on T, with values belonging to the corresponding Hilbert spaces, will be called μ-measurable if for every n, this function, considered only on T_n, is μ-measurable in the sense indicated above. The direct integral of the spaces H_t is defined as the collection of all μ-measurable vector-valued functions $\{\xi_t\}$ which satisfy the condition

$$\int_T |\xi_t|^2 d\mu < \infty.$$

The inner product of two such vector-valued functions $\xi = \{\xi_t\}$ and $\eta = \{\eta_t\}$ is defined by the formula

$$(\xi, \eta) = \int_T (\xi_t, \eta_t) d\mu.$$

As above, it is easily shown that in this connection the direct integral of the spaces H_t becomes a Hilbert space which we shall again denote by

$$\int_T H_t \, d\mu.$$

The coordinate notation for the direct integral, indicated earlier for the case of spaces H_t having constant dimension, is easily carried over to the general case. Each of the spaces H_t, where $n(t) = \infty$, is identified with the space l^2 of sequences $\langle x_1, x_2, \cdots \rangle$ for which $\sum_{k=1}^{\infty} |x_k|^2 < \infty$; but if $n(t)$ is equal to the finite number m, then we identify H_t with the subspace $l_m^2 \subset l^2$ of all vectors of the form $\langle x_1, x_2, \cdots, x_m, 0, 0, \cdots \rangle$. Then every element in $\mathfrak{H} = \int_T H_t \, d\mu$ is representable in the form of a sequence $\langle a_1(t), a_2(t), \cdots \rangle$ of μ-measurable functions which satisfy the following conditions:

1) $\sum_k \int_T |a_k(t)|^2 d\mu < \infty$;
2) $a_k(t) = 0 \quad$ for $k > n(t)$ if $n(t) < \infty$,

and $\mathfrak{H}$ is the collection of all such sequences.

[Dixmier [14] and Von Neumann [10] proceed from another definition of μ-measurability; however, they subsequently arrive at this same coordinate notation for the direct integral.]

It easily follows from this coordinate notation that *in $\mathfrak{H}$ there exists a sequence of vectors $\varphi_n = \{\varphi_n(t)\}$ such that for every fixed t the vectors $\varphi_n(t)$, $n = 1, 2, \cdots, n(t)$ form an orthonormal basis in H_t*. In fact, it suffices to take for the vector $\{\varphi_n(t)\}$ the sequence $\langle \varphi_{n1}(t), \varphi_{n2}(t), \cdots \rangle$, where

$$\varphi_{nk}(t) = \begin{cases} 1 & \text{for } k = n,\ k \leqq n(t), \\ 0 & \text{otherwise.} \end{cases}$$

EXAMPLES. 1. Let T consist of a finite or countable number of points, each of which has measure 1. We shall consider T as a locally compact space with the discrete topology. Then $\mathfrak{H}$ is simply the usual direct sum of Hilbert spaces.

2. Let T be the closed interval [0, 1] with the usual Lebesgue measure on it, and let each point t of this closed interval be assigned a one-dimensional space H_t. Then $\mathfrak{H}$ is the space $L^2[0, 1]$. Thus, a realization of a Hilbert space in the form of a space of functions can be considered as a representation of it in the form of the direct integral of one-dimensional spaces.

3. Again, let T be the closed interval [0, 1] with Lebesgue measure but suppose now that H_t is the space of functions $f(s) \in L^2[0, 1]$. Then $\mathfrak{H}$ is the space of square-summable functions of the two variables t and s, $0 \leqq t \leqq 1$, $0 \leqq s \leqq 1$.

2. Decomposition of a Hilbert space into a direct integral with respect to a given commutative algebra *R*. Let

$$\mathfrak{H} = \int_T H_t \, d\mu$$

be the direct integral of the spaces H_t with respect to the measure μ and let $\varphi(t) \in L^\infty = L^\infty_\mu(T)$. Every such function $\varphi(t)$ defines a bounded operator L_φ in $\mathfrak{H}$ by the formula $L_\varphi \xi = \{\varphi(t)\xi_t\}$ for all $\xi = \{\xi_t\} \in \mathfrak{H}$; in this connection,

$$|L_\varphi| = ||\varphi||_\infty . \tag{1}$$

In fact,

$$|L_\varphi \xi|^2 = \int_T |\varphi(t)|^2 |\xi_t|^2 d\mu \leqq ||\varphi||^2_\infty \int_T |\xi_t|^2 d\mu; \tag{2}$$

on the other hand, since $T_\varepsilon = \{t\colon |\varphi(t)| > ||\varphi||_\infty - \varepsilon\}$, $\varepsilon > 0$, is a set of positive measure, there exists a vector-valued function $\{\xi_t\} \in \mathfrak{H}$, $\xi \neq 0$, which equals zero outside of T_ε. For this function,

$$\begin{aligned} |L_\varphi \xi|^2 = \int_T |\varphi(t)|^2 |\xi_t|^2 d\mu &= \int_{T_\varepsilon} |\varphi(t)|^2 |\xi_t|^2 d\mu \\ &\geqq (||\varphi||_\infty - \varepsilon)^2 \int_{T_\varepsilon} |\xi_t|^2 d\mu = (||\varphi||_\infty - \varepsilon)^2 |\xi|^2. \end{aligned} \tag{3}$$

Now (1) follows from (2) and (3).

I. *Every bounded operator A in $\mathfrak{H}$ which commutes with all the operators L_φ, $\varphi \in L^\infty$, has the form $A = \{A(t)\}$, i.e. $A\{\xi_t\} = \{A(t)\xi_t\}$, where $A(t)$ is a measurable operator-valued function and $|A(t)| \in L^\infty$; moreover, $|A| = ||\,|A(t)|\,||_\infty$.* [The concept of measurable operator-valued function is defined in the general case in a way analogous to that of § 26, subsection 5.]

Proof. If dim H_t = constant on T, then the assertion coincides with Proposition IV, § 26, subsection 5. In the general case, $\mathfrak{H}$ is the direct sum $\mathfrak{H} = \sum_k \oplus \mathfrak{H}_k$ of a finite or countable number of subspaces $\mathfrak{H}_k = \int_{T_k} H_t \, d\mu$, where dim H_t = constant on T_k. In correspondence with this, A is written in the form of a matrix, $A \sim ||A_{jk}||$, where A_{jk} is a bounded operator from $\mathfrak{H}_k$ into $\mathfrak{H}_j$ and $L_\varphi = ||\delta_{jk} L_{\varphi_k}||$, where φ_k is the restriction of φ to T_k and $\delta_{jk} = 1$ for $j = k$ and 0 for $j \neq k$. The condition $AL_\varphi = L_\varphi A$ means that $A_{jk} L_{\varphi_k} = L_{\varphi_j} A_{jk}$. Taking, for $j \neq k$, $\varphi_k = 1$ and $\varphi_j = 0$, we obtain that $A_{jk} = 0$ for $j \neq k$. Further, $A_{jj} L_{\varphi_j} = L_{\varphi_j} A_{jj}$; consequently, by virtue of IV, § 26, subsection 5, $A_{jj} = \{A_j(t)\}$, $|A_j(t)| \in L_\mu^\infty(T_j)$, where $|A_{jj}| = || \, |A_j(t)| \, ||_\infty$. [The assertion of Proposition IV, subsection 5, § 26 remains valid if we replace T by any μ-measurable subset of T; this is immediately clear from the proof.] From this, setting $A(t) = A_j(t)$ on T_j, we conclude that $A = \{A(t)\}$ and $|A| = \sup_j |A_{jj}| = \sup_j || \, |A_j(t)| \, ||_\infty = || \, |A(t)| \, ||_\infty$. It is obvious that, conversely, every measurable operator-valued function $A = \{A(t)\}$ which satisfies the condition $|A(t)| \in L^\infty$ defines, according to the formula $A\{\xi_t\} = \{A(t)\xi_t\}$, a bounded operator in $\mathfrak{H}$, which commutes with all the L_φ.

It can also be rather easily established that the following assertions are valid: *The sum and product of measurable operator-valued functions are measurble functions* Moreover, *if* $A = \{A(t)\}$, $B = \{B(t)\}$ *and* $A+B = C = \{C(t)\}$, *then, for almost all* t, $C(t) = A(t)+B(t)$. Analogously, *if* $AB = C = \{C(t)\}$ *or* $A^* = C = \{C(t)\}$, *then, for almost all* t, $C(t) = A(t)B(t)$, *or* $C(t) = (A(t))^*$ *respectively.*

The operators L_φ form a commutative algebra R which together with each operator L_φ also contains the adjoint operator $L_\varphi^* = L_{\bar{\varphi}}$ (defined by the function $\bar{\varphi}$ which is the complex conjugate of φ).

II. *The algebra R is closed in the sense of the weak topology in* $\mathfrak{B}(\mathfrak{H})$.

In fact, it suffices to show that $R'' = R$ (see subsection 1, § 34). But, by virtue of I, R' consists of all operators $A = \{A(t)\}$, where $|A(t)| \in L^\infty$. From this it follows that $R \subset R'$ and hence $R' \supset R''$. Consequently, R'' consists of operators $A = \{A(t)\}$, where $|A(t)| \in L^\infty$. But the operator $A = \{A(t)\}$ in R'' must commute with all the operators in R', i.e., with all $B = \{B(t)\}$, $|B(t)| \in L_\mu^\infty$. We take, in particular, $B(t) = B^{pq}$ for all $t \in T_j$ (since dim H_t = constant on T_j, we can assume that $H_t = H_j$ for $t \in T_j$), where B^{pq} is the operator in $\mathfrak{B}(H_j)$ whose matrix $||b_{jk}^{pq}||$, with respect to some fixed orthonormal basis, is

$$b_{jk}^{pq} = \begin{cases} 0 & \text{for } j \neq p \text{ or } k \neq q, \\ 1 & \text{for } j = p \text{ and } k = q \end{cases}$$

(see the proof of Theorem 8, subsection 5, § 26); we obtain that $A(t)B^{pq} = B^{pq}A(t)$ for $t \in T_j - N_{pqj}$, where $\mu(N_{pqj}) = 0$. Setting $N_j = \bigcup_{pq} N_{pqj}$, we see that for $t \in T_j - N_j$,

$$A(t)B^{pq} = B^{pq}A(t) \quad \text{for all } p, q = 1, 2, 3, \cdots, \tag{4}$$

where $\mu(N_j) = 0$. It follows from (4) that $A(t) = \varphi_j(t)1$ for $t \in T_j - N_j$. Setting $\varphi = \varphi_j$ for $t \in T_j - N_j$, we obtain that $\varphi \in L^\infty$ and $A = L_\varphi$. Consequently, $R'' \subset R$. Since, on the other hand, $R'' \supset R$, we have that $R'' = R$.

Thus, to each direct integral of Hilbert spaces there corresponds a commutative weakly closed algebra R of bounded linear operators in $\mathfrak{H}$, which contains the identity operator. We shall call this algebra *the algebra of the given decomposition of the space $\mathfrak{H}$ into a direct integral.*

EXAMPLES. 1. $\mathfrak{H} = H_1 \oplus H_2 \oplus \cdots \oplus H_n$. The space T consists of n points. The algebra R is isomorphic to the algebra of diagonal matrices of the n-th order.

2. $\mathfrak{H}$ is the space $L^2[0, 1]$ of functions on $[0, 1]$, i.e. the direct integral of one-dimensional spaces. R is the algebra of all essentially bounded measurable functions on $[0, 1]$.

3. $\mathfrak{H}$ is the space $L^2([0, 1] \times [0, 1])$ of functions of two variables t, s, $0 \leqq t \leqq 1$, $0 \leqq s \leqq 1$, which is represented as the direct integral of spaces $L^2[0, 1]$ of functions of one variable. Here, as in Example 2, R is the algebra of all essentially bounded measurable functions on a segment.

We shall now show that the converse proposition holds, namely:

THEOREM 1. *To each commutative weakly closed algebra R of operators in a separable Hilbert space $\mathfrak{H}$, containing the identity operator, there corresponds a decomposition of the space $\mathfrak{H}$ into a direct integral and, moreover, such that R is the collection of all operators of the form L_φ, $\varphi \in L^\infty$ and T is compact with a countable neighborhood base.*

[This means that there exists an isometric mapping of the space $\mathfrak{H}$ onto a direct integral of Hilbert spaces under which R maps onto the algebra of all operators L_φ, $\varphi \in L^\infty$.]

Proof. Let $A_1, A_2, \cdots$ be a countable subset in R which is dense in the strong operator topology (see the lemma in subsection 2, § 40). Let R_1 be a symmetric subalgebra, closed in the operator norm, in $\mathfrak{B}(\mathfrak{H})$, containing the identity operator, generated by the operators $A_1, A_2, \cdots$. Then the weak closure of the algebra R_1 coincides with R. Let T be the space of maximal ideals of the algebra R_1. Since R_1 has a countable number of generators $A_1, A_1^*, A_2, A_2^*, \cdots$, T is a compact space with a countable neighborhood base (see the corollary in subsection 5, § 11). According to Theorem 1, subsection 3, § 40, there exists an isometric mapping U of the space $\mathfrak{H}$ onto the direct sum $\tilde{\mathfrak{H}} = \sum_k \oplus L^2_{\mu_k}(T)$ under which R_1 maps onto the algebra $\tilde{R}_1$ of all operators

$$\tilde{A}\{\xi_k(t)\} = \{A(t)\xi_k(t)\},\ A(t) \in C(T),\ \{\xi_k(t)\} \in \tilde{\mathfrak{H}}, \tag{5}$$

where $A(t)$ is the value of A on the ideal t and $\mu_1, \mu_2, \cdots$ are measures on T which satisfy the condition

$$\mu_1 \succ \mu_j, \quad j = 2, 3, \cdots. \tag{6}$$

Since all the measures μ_k are dominated by the measure μ_1 (which we shall now denote by μ), for every μ-measurable set Δ,

$$\mu_k(\Delta) = \int_\Delta \omega_k(t) d\mu, \tag{7}$$

where $\omega_k \in L^1_\mu$. The space $\mathfrak{H}$ consists of (finite or countable) sequences $\langle \xi_k(t) \rangle$ of μ-measurable functions which satisfy the condition

$$\sum_k \int_T |\xi_k(t)|^2 d\mu_k < \infty. \tag{8}$$

In virtue of (7), this condition can be written in the form

$$\sum_k \int_T |\xi_k(t)|^2 \omega_k(t) d\mu < \infty, \tag{9}$$

or, finally, in virtue of VII, subsection 7, § 6, in the form

$$\int_T \{\sum_k |\xi_k(t)|^2 \omega_k(t)\} d\mu < \infty. \tag{10}$$

Set $\mathfrak{N}_k = \{t: \omega_k(t) = 0\}$; the sets $\mathfrak{N}_k$ are measurable and defined to within sets of μ-measure zero.

Let $n(t)$ denote the (finite or infinite) number of all indices k for which $t \notin \mathfrak{N}_k$; $n(t)$ is the sum of at most a countable number of μ-measurable functions $1 - \chi_k(t)$, where $\chi_k(t)$ is the characteristic function of the set $\mathfrak{N}_k$. Therefore $n(t)$ is a μ-measurable function. Let $k_1(t) < k_2(t) < \cdots$ be all the numbers k for which $t \notin \mathfrak{N}_k$. Then

$$\omega_k(t) = 0 \text{ if } k \text{ is different from all } k_j(t). \tag{11}$$

Set

$$x_p(t) = \xi_{k_p(t)}(t)\sqrt{\omega_{k_p(t)}(t)} \quad \text{for} \begin{cases} p \leqq n(t) & \text{if } n(t) < \infty, \\ p = 1, 2, 3, \cdots & \text{if } n(t) = \infty \end{cases} \tag{12}$$

and

$$x_p(t) = 0 \quad \text{for } p > n(t) \text{ if } n(t) < \infty. \tag{13}$$

The function $x_p(t)$ is μ-measurable for every p. In fact, set $\mathfrak{M}_k = T - \mathfrak{N}_k$; for fixed p and k the set T_{kp} of all those points $t \in T$ for which $k_p(t) = k$ is

$$\mathfrak{M}_k \cap \bigcup (\mathfrak{M}_{k_1} \cap \cdots \cap \mathfrak{M}_{k_{p-1}}),$$

where the union is taken over all $k_1, \cdots, k_{p-1}$ for which $k_1 < \cdots < k_{p-1} < k$; therefore T_{kp} is μ-measurable. On the other hand, $x_p(t) = \xi_k(t)\sqrt{\omega_k(t)}$ on T_{kp} and hence $x_p(t)$ is μ-measurable on T_{kp}; consequently, $x_p(t)$ is μ-measurable on $T = \bigcup_{k,p} T_{kp}$. Further, from (11) and (12) we conclude that

$$\sum_p |x_p(t)|^2 = \sum_p |\xi_{k_p(t)}(t)|^2 \omega_{k_p}(t) = \sum_k |\xi_k(t)|^2 \omega_k(t)$$

and hence

$$\int_T (\sum_p |x_p(t)|^2) d\mu = \int_T \{\sum_k |\xi_k(t)|^2 \omega_k(t)\} d\mu < \infty. \tag{14}$$

Consequently, the condition

$$\sum_k |x_k(t)|^2 < \infty \tag{15}$$

may not be satisfied only for points t forming a set of μ-measure zero.

Let H_t be the set of all numerical sequences $\langle x_p \rangle$ which satisfy the conditions:

1) $x_p = 0$ for $p > n(t)$ if $n(t) < \infty$,
2) $\sum_p |x_p|^2 < \infty$.

We note that all the spaces H_t are subspaces of the Hilbert space l^2. Conditions (13) and (14) mean that the vector-valued function $x(t) = \langle x_1(t), x_2(t), \cdots \rangle$ constructed above assumes values in H_t for μ-almost every $t \in T$ and

$$\int_T (\sum |x_p(t)|^2) d\mu < \infty. \tag{16}$$

Conversely, every vector-valued function $x(t) = \langle x_1(t), x_2(t), \cdots \rangle$ with values in H_t and satisfying condition (16) can be obtained in this way from some vector-valued function $\{\xi_k(t)\} \in \tilde{\mathfrak{H}}$.

It suffices to set

$$\xi_{k_p(t)}(t) = x_p(t)[\omega_{k_p(t)}(t)]^{-\frac{1}{2}},$$

$\xi_k(t) = 0$ if k is different from all $k_p(t)$. Since $\omega_{k_p}(t) \neq 0$, this definition is legitimate; reasoning as above, we can verify that $\xi_k(t)$ is measurable for every k and $\{\xi_k(t)\} \in \tilde{\mathfrak{H}}$. From this and (14) it follows that $\{\xi_k(t)\} \to \{x_p(t)\}$ is an isometric mapping.

We have obtained a realization of the space $\tilde{\mathfrak{H}}$ as the space of all sequences of μ-measurable functions $\langle x_1(t), x_2(t), \cdots \rangle$, satisfying condition (16) and furthermore the condition $x_k(t) = 0$ for $k > n(t)$ if $n(t) < \infty$. But this is precisely the coordinate realization of the direct integral of the spaces H_t (see subsection 1); consequently,

$$\tilde{\mathfrak{H}} = \int_T H_t \, d\mu.$$

Finally, we shall prove that with such a decomposition of the space $\tilde{\mathfrak{H}}$, the algebra $\tilde{R}$ coincides with the algebra of all operators L_φ, $\varphi \in L_\mu^\infty = L_\mu^\infty(T)$.

According to Theorem 1, subsection 3, § 40, $\tilde{R}_1 = \{L_\psi: \psi \in C(T)\}$ because R_1, and hence also $\tilde{R}_1$, is closed in the operator norm. On the other hand, in virtue of V, subsection 5, § 26, every operator L_φ, $\varphi \in L_\mu^\infty(T)$, is a contact point in the weak operator topology of the set of operators L_ψ, $\psi \in C(T)$; since $\tilde{R}$ is weakly closed, $L_\varphi \in \tilde{R}$.

Thus,

$$\tilde{R} = \overline{\tilde{R}_1^1} = \overline{\{L_\psi: \psi \in C(T)\}^1} \supset \{L_\varphi: \varphi \in L_\mu^\infty(T)\}.$$

On the other hand, by II,

$$\{L_\varphi: \varphi \in L_\mu^\infty(T)\} \supset \overline{\{L_\psi: \psi \in C(T)\}^1};$$

consequently, $\tilde{R} = \{L_\varphi: \varphi \in L_\mu^\infty(T)\}$ and the theorem is proved.

3. Decomposition with respect to a maximal commutative algebra; condition for irreducibility. In the preceding subsection, we showed that every weakly closed commutative algebra R determines a decomposition of the separable space $\mathfrak{H}$ into a direct integral (here we identify $\mathfrak{H}$ with $\tilde{\mathfrak{H}}$):

$$\mathfrak{H} = \int_T H_t \, d\mu,$$

where T is a compact space with a countable neighborhood basis. With this decomposition, every operator A, belonging to R', i.e. which commutes with all the operators in R, is representable in the form of a measurable operator-valued function $A = \{A(t)\}$, where $A(t)$ is, for almost all t, a bounded operator in H_t. By the same token, every family $\mathfrak{A}$ of operators in R' is at the same time reducible to the "diagonal" form: $A = \{A(t)\}$. Let $\mathfrak{A}(t)$ be the collection of all operators $A(t)$, corresponding to $A \in \mathfrak{A}$, for fixed t. The question arises under what conditions is the family $\mathfrak{A}(t)$ irreducible for almost all t.

The following theorem gives the answer to this question.

THEOREM 2. *If R is a maximal commutative subalgebra of the algebra $\mathfrak{A}'$, then the family $\mathfrak{A}(t)$ is irreducible for almost every t; conversely, if $\mathfrak{A}$ is at most countable and $\mathfrak{A}(t)$ is irreducible for almost every t, then R is a maximal commutative subalgebra in $\mathfrak{A}'$.*

Proof. The second assertion follows easily from Proposition I, subsection 2. In fact, let $\mathfrak{A} = \{A_1, A_2, \cdots\}$ and let R not be maximal in $\mathfrak{A}'$. Then in $\mathfrak{A}'$ there exists an operator B which does not belong to the algebra R and commutes with all the operators in R. According to Proposition I, subsection 2, the operator B can be written in the form $B = \{B(t)\}$, where the operator $B(t)$ is different from a multiple of the identity on a set of positive measure; otherwise, we would have $\{B(t)\} = L_\varphi \in R$. By assumption, $B \in \mathfrak{A}'$, i.e. it commutes with all the operators $A_k = \{A_k(t)\}$ in $\mathfrak{A}$. This means that

$$A_k(t)B(t) = B(t)A_k(t) \quad \text{for } t \in T - N_k,$$

where $\mu(N_k) = 0$. Let T_1 be a set of positive measure in $T - \bigcup_k N_k$ on which the operator $B(t)$ does not reduce to multiplication by a number. Then for every $t \in T_1$, the family $\mathfrak{A}(t)$ will be reducible because all the operators $A_k(t)$ belonging to it commute with the operator $B(t)$, which is not a multiple of the identity.

We carry out the proof of the first assertion first for the case when all the H_t have the same dimension; in this case, they can be identified with a fixed space H. The proof itself for this case we shall carry out in a number of steps.

1) *In the space S of all operators A in H with norm $\leqq 1$ we can introduce a metric such that the convergence corresponding to it is equivalent to the weak convergence of operators.*

In fact, let $\{\varphi_n\}$ be a complete orthonormal system in H. For $A, B \in S$, we set

$$\rho(A, B) = \sum_{n,m=1}^{\infty} \frac{1}{2^{n+m}} |(A\varphi_n, \varphi_m) - (B\varphi_n, \varphi_m)|. \tag{1}$$

Obviously, the relation $\rho(A^{(n)}, A) \to 0$ is equivalent to the relations

$$(A^{(n)}\varphi_p, \varphi_q) \to (A\varphi_p, \varphi_q) \quad \text{for } p, q = 1, 2, \cdots. \tag{2}$$

In view of the condition $|A^{(n)}| \leqq 1$, the relations (2) are equivalent to the weak convergence of the sequence $A^{(n)}$ to A.

2) *Let Λ be the topological product of the spaces T and S. In T there exists a set N of μ-measure zero such that*

a) *$T - N$ is a Borel subset in T;*

b) *the set Q of all pairs $\langle t, B \rangle$ in Λ such that $t \in T - N$ and*

$$A(t)B = BA(t) \tag{3}$$

for all $A = \{A(t)\} \in \mathfrak{A}$ is a Borel set in Λ.

We choose a fixed orthonormal basis $\{\varphi_n\}$ in H. In this basis, the operators $A(t)$, B are defined by the matrices $A(t) \sim ||a_{pq}(t)||$, $B \sim ||b_{pq}||$, where the $a_{pq}(t)$ are measurable functions of t. Modifying each of the functions $a_{pq}(t)$ on a set N_{pq} of μ-measure zero if necessary, we can assume that these functions are Borel functions and that the $T - N_{pq}$ are Borel sets (see Appendix II); then everywhere except on the set $N = \bigcup_{p,q} N_{pq}$ of μ-measure zero, all the functions $a_{pq}(t)$ will be Borel functions, where $T - N$ is a Borel set.

Condition (3) is equivalent to the following system of equations:

$$\sum_j a_{pj}(t)b_{jq} = \sum_j b_{pj}a_{jq}(t), \quad p, q = 1, 2, \cdots. \tag{4}$$

For fixed p and q, the left and right members of (4) are Borel functions in Λ because $a_{pj}(t)$ is a Borel function which does not depend on B and $b_{pq} = (B\varphi_q, \varphi_p)$ is even a continuous function of B in S which does not depend on t. But then the set Q_{pq} of those points $\langle t, B\rangle$, $t \in T-N$, $B \in S$ for which the equality (4) is satisfied is a Borel set. Therefore $Q = \bigcap_{p,q} Q_{pq}$ is also a Borel set.

3) *There exists a countable set* $\{B_n(t)\}$ *of measurable operator-valued functions such that for almost every fixed* t *the set of all operators* $B_n(t)$ *generates the algebra* $(\mathfrak{A}(t))'$.

We choose in S a countable basis of neighborhoods $\{U_n\}$; this can be done by virtue of 1), the lemma in § 40, subsection 2 and Proposition IV, § 2, subsection 13. Let $\mathfrak{U}_n$ be the collection of all pairs $\langle t, B\rangle$, $t \in T-N$, $B \in U_n$. Obviously, $\mathfrak{U}_n$ is a Borel set in Λ. Consequently, the intersection $Q_n = Q \cap \mathfrak{U}_n$ is also a Borel set in Λ.

Let T_n be the collection of those t for which there exists an operator B such that $\langle t, B\rangle \in Q_n$. Being the projection of a Borel set, T_n is an analytic set, and, consequently, it is measurable. On the basis of the Luzin-Yankov theorem (see Appendix III), there exists a measurable operator-valued function $\{B_n(t)\}$, defined for all $t \in T_n$, such that $\langle t, B_n(t)\rangle \in Q_n$.

We extend the definition of the function $\{B_n(t)\}$ by assuming it to be equal to zero for $t \notin T_n$; obviously, it still remains measurable when this is done. We shall prove that the family $\{B_n(t)\}$, $n = 1, 2, \cdots$, thus constructed is the one sought.

By virtue of its construction itself, $B_n(t)$ commutes with any $A(t)$ for almost all t; consequently, for almost all t,

$$B_n(t) \in (\mathfrak{A}(t))'. \tag{5}$$

We shall prove that for every fixed $t \in T-N$, the operators $B_n(t)$, $n = 1, 2, \cdots$, generate the algebra $(\mathfrak{A}(t))'$. To this end, it suffices to show that for $t \in T-N$ every operator in $(\mathfrak{A}(t))'$ which is at most equal to unity in norm is a weak contact point of the operators $B_n(t)$, i.e. that every weak neighborhood in $S \cap (\mathfrak{A}(t))'$ contains at least one operator $B_n(t)$. Since the nonvoid intersections $U_n \cap (\mathfrak{A}(t))'$ form a neighborhood basis in $S \cap (\mathfrak{A}(t))'$, it suffices to prove that every such neighborhood contains at least one operator $B_n(t)$.

For some $t = t_0 \in T-N$, let

$$U_n \cap (\mathfrak{A}(t_0))' \neq \phi.$$

This means that the set of pairs of the form $\langle t_0, B\rangle$ belonging to Q_n is not void, i.e. that $t_0 \in T_n$. But then, by the construction of the function $B_n(t)$,

$$B_n(t_0) \in U_n \cap (\mathfrak{A}(t_0))',$$

which is what we were required to prove.

We can now prove the sufficiency of the condition formulated in the theorem. The irreducibility of the families $\mathfrak{A}(t)$ for almost every t means that for almost all t the algebra $(\mathfrak{A}(t))'$ consists of only the operators of multiplication by a number. To establish this fact, it suffices in turn to show that each of the operator-valued functions $\{B_n(t)\}$ constructed above is an operator of the form L_{φ_n} (i.e. an operator of multiplication by a scalar function).

But, by virtue of (5), the operator $B_n = \{B_n(t)\}$ commutes with all the operators $A = \{A(t)\}$ in $\mathfrak{A}$,i .e. it belongs to $\mathfrak{A}'$. On the other hand, it, as also every operator which can be represented in the form of an operator-valued function, commutes with all the operators in R. But then it must belong to R, for otherwise, in contradiction to the condition of the theorem, R would not be a maximal commutative subalgebra in the algebra $\mathfrak{A}'$. The theorem is thus pıoved for the case when all the H_t have the same (finite or infinite) dimension.

We shall now consider the general case. Then the entire space T decomposes into the sum of a finite or countable number of measurable sets T_k, such that for $t \in T_k$, the spaces H_t have the same dimension k; then, for $t \in T_k$, we can assume that all the spaces H_t coincide with a fixed space H of dimension k.

Set

$$\mathfrak{H}_k = \int_{T_k} H_t \, d\mu;$$

then the direct integral

$$\mathfrak{H} = \int_T H_t \, d\mu$$

can be written in the form

$$\mathfrak{H} = \sum_k \oplus \mathfrak{H}_k.$$

All the operators of the family $\mathfrak{A}$ will be reduced by each of the spaces $\mathfrak{H}_k$, i.e. they will have the form

$$\begin{pmatrix} A_1 & 0 & 0 & \cdots \\ 0 & A_2 & 0 & \cdots \\ \cdots & \cdots & \cdots & \cdots \end{pmatrix},$$

where A_k is an operator in the space $\mathfrak{H}_k$.

In an analogous manner, the operators in R can be written in the form

$$\begin{pmatrix} L_{\varphi_1} & 0 & 0 & \cdots \\ 0 & L_{\varphi_2} & 0 & \cdots \\ \cdots & \cdots & \cdots & \cdots \end{pmatrix},$$

where L_{φ_k} is the operator of multiplication by the function $\varphi_k \in L^\infty_\mu(T_k)$.

Let $\mathfrak{A}_k$ denote the collection of all A_k for $A \in \mathfrak{A}$. Obviously, the operators L_{φ_k}, considered as operators in $\mathfrak{H}_k$, form a maximal commutative subalgebra in $\mathfrak{A}'_k$. By what we have already proved above, for almost all $t \in T_k$, the set $\mathfrak{A}(t)$ of operators $A_k(t) = A(t)$, $A \in \mathfrak{A}$, is irreducible. Since $T = \bigcup_k T_k$, it follows from this that the family $\mathfrak{A}(t)$ is irreducible for almost all t. This completes the proof of the theorem.

REMARK. Taking as $\mathfrak{A}$ an arbitrary weakly closed algebra of operators and for R its center and carrying out an analogous line of reasoning, we obtain the decomposition of the algebra $\mathfrak{A}$ into the direct integral of weakly closed algebras $\tilde{\mathfrak{A}}(t)$, generated by the sets $\mathfrak{A}(t)$, and the algebras $\tilde{\mathfrak{A}}(t)$ are factors, i.e. they satisfy the condition $\tilde{\mathfrak{A}}(t) \cap (\tilde{\mathfrak{A}}(t))' = (\alpha 1)$. This result was obtained by von Neumann (see VON NEUMANN [10]). For more details, see DIXMIER [14].

4. Decomposition of a unitary representation of a locally compact group into irreducible representations. In this subsection, we shall establish the following theorem using the criterion for irreducibility obtained in subsection 3.

THEOREM 3. *Let $g \to U_g$ be a continuous unitary representation, acting in a separable Hilbert space $\mathfrak{H}$, of the locally compact group $\mathfrak{G}$ with a countable neighborhood basis. Then there exists a decomposition*

$$\mathfrak{H} = \int_T H_t \, d\mu \tag{1}$$

of the space $\mathfrak{H}$ as a direct integral and for almost every $t \in T$ there exists an irreducible continuous unitary representation $g \to U_g(t)$ of the group $\mathfrak{G}$ such that

$$U_g = \{U_g(t)\} \tag{2}$$

for all $g \in \mathfrak{G}$.

Proof. We shall first prove the assertion for a cyclic representation. Let $g \to U_g$ be a cyclic continuous unitary representation of the group $\mathfrak{G}$ and let $\xi_0 \in \mathfrak{H}$ be the corresponding cyclic vector. The representation $g \to U_g$ defines a symmetric representation $x \to A_x$ of the algebra $L^1(\mathfrak{G})$ according to the formula

$$A_x = \int x(g) U_g \, d\nu(g),$$

where ν is a left invariant measure on $\mathfrak{G}$.

Since there is a countable neighborhood basis $\{U_n\}$ with compact closures in $\mathfrak{G}$, we have that $L^1(\mathfrak{G})$ is separable; in fact, the set of all finite linear combinations of characteristic functions ξ_{U_n} with rational coefficients forms a countable set which is dense in $L^1(\mathfrak{G})$.

Let $\{x_n\}$ be an arbitrary countable set which is dense in $L^1(\mathfrak{G})$.

We can assume that this set contains together with each pair of vectors x_n and

x_m their difference $x_n - x_m$ – otherwise, we can simply adjoin all such differences to this set.

According to Theorem 2, to the family of operators A_{x_n} there corresponds a decomposition of the space $\mathfrak{H}$ in a direct integral

$$\mathfrak{H} = \int_T H_t \, d\mu,$$

which possesses the following properties:

1) Each of the operators A_{x_n} can be represented in the form

$$A_{x_n} = \{A_{x_n}(t)\};$$

2) For almost every $t \in T$, the family of operators $A_{x_n}(t)$, $n = 1, 2, 3, \cdots$ is irreducible in $\mathfrak{H}$.

Moreover,

$$|A_{x_n}| = \| \, |A_{x_n}(t)| \, \|_\infty;$$

consequently, for almost all $t \in T$,

$$|A_{x_n}(t)| \leqq |A_{x_n}|.$$

Let T_n be the set of those points t for which $|A_{x_n}(t)| > |A_{x_n}|$; deleting all the sets T_n, $n = 1, 2, \cdots$, of μ-measure zero and also those t (which also form a set of μ-measure zero) for which at least one of the functions $A_{x_n}(t)$ is not defined or for which the family $A_{x_n}(t)$, $n = 1, 2, \cdots$, is reducible, we obtain the set T' which differs from T by a set of μ-measure zero and has the following properties:

a) $A_{x_n}(t)$ is defined for every $t \in T'$ and all $n = 1, 2, \cdots$;

b) $|A_{x_n}(t)| \leqq |A_{x_n}|$ for every $t \in T'$ and all $n = 1, 2, \cdots$;

c) for every $t \in T'$, the family $A_{x_n}(t)$, $n = 1, 2, \cdots$, is irreducible in H_t.

From property b), we conclude that

$$|A_{x_n}(t)| \leqq |A_{x_n}| \leqq \|x_n\|_1, \tag{3}$$

and therefore for $t \in T'$ we can define $A_x(t)$ for all $x \in L^1(\mathfrak{G})$. In fact, for every $x \in L^1(\mathfrak{G})$ there exists a subsequence $\langle x_{n_k} \rangle$ which converges in norm to x. It follows from (3) that

$$|A_{x_n}(t) - A_{x_m}(t)| \leqq \|x_n - x_m\|_1$$

(since $x_n - x_m$ also belongs to $\langle x_n \rangle$) and therefore the sequence $A_{x_{n_k}}(t)$ converges in the operator norm to some operator in H_t, which we shall denote by $A_x(t)$. It is easy to see that $A_x(t)$ does not depend on the choice of the subsequence $x_{n_k} \to x$. Being the limit of a sequence of measurable functions $A_{x_{n_k}}(t)$, the function $A_x(t)$ is also measurable.

It follows from the equalities

$$A_{\alpha x} = \alpha A_x, \quad A_{x+y} = A_x + A_y, \quad A_{xy} = A_x A_y, \quad A_{x^*} = (A_x)^*$$

that each of the equalities

$$\begin{aligned} A_{\alpha x}(t) &= \alpha A_x(t), \quad A_{x+y}(t) = A_x(t) + A_y(t), \\ A_{xy}(t) &= A_x(t)A_y(t), \quad A_{x^*}(t) = (A_x(t))^* \end{aligned} \tag{4}$$

holds except at most on a set of points of μ-measure zero.

Let $\{\alpha_n\}$ be the set of all rational complex numbers. We denote by $T_{n,m}$ the set of all points $t \in T'$ on which at least one of the following equalities is not satisfied:

$$\begin{aligned} A_{\alpha_n x_m}(t) &= \alpha_n A_{x_m}(t), \quad A_{x_n + x_m}(t) = A_{x_n}(t) + A_{x_m}(t), \\ A_{x_n x_m}(t) &= A_{x_n}(t) A_{x_m}(t), \quad A_{x_n^*}(t) = (A_{x_n}(t))^*. \end{aligned} \tag{5}$$

By what we stated above, $T_{n,m}$ is a set of μ-measure zero and all relations (5) hold on $T'' = T' - \bigcup_{n,m} T_{n,m}$ which differs from T by a set of measure zero. But then, taking suitable subsequences of the sequences $\langle \alpha_n \rangle$, $\langle x_n \rangle$, $\langle x_m \rangle$ in relations (5) and passing to the limit, we obtain that equalities (4) hold for every $t \in T''$ and arbitrary $x, y \in L^1(\mathfrak{G})$ and arbitrary complex α. But this means that for $t \in T''$, the correspondence $x \to A_x(t)$ is a symmetric representation of the algebra $L^1(\mathfrak{G})$. This representation is irreducible because the family of operators $A_{x_n}(t)$, $n = 1, 2, 3, \cdots$ is irreducible. We shall now construct the corresponding irreducible representations $g \to U_g(t)$ of the group $\mathfrak{G}$. Let $\xi_0 = \{\xi_0(t)\}$ be a cyclic vector and let T_0 be the set of points $t \in T''$ for which $A_x(t)\xi_0(t) = 0$ for all $x \in L^1(\mathfrak{G})$. Being the intersection of the sets $T_n = \{t: t \in T'', A_{x_n}(t)\xi(t) = 0\}$, it is μ-measurable. We shall prove that $\mu(T_0) = 0$. If $\mu(T_0) > 0$, then the vectors $A_x \xi_0$, $x \in L^1(\mathfrak{G})$ are orthogonal to all vectors $\eta = \{\eta(t)\}$ in $\mathfrak{H}$ for which $\eta(t) = 0$ for $t \notin T_0$, and this contradicts the fact that the vector ξ_0 is cyclic. Setting $T''' = T'' - T_0$, we see that for $t \in T'''$ the vector $\xi(t)$ is cyclic for the operators $A_x(t)$ because the representation $x \to A_x(t)$ is irreducible. Consequently, the vectors $\xi_n(t) = A_{x_n}(x)\xi_0(t)$ form a dense set in H_t for $t \in T''$.

We introduce the operator $U_g(t)$ in H_t by setting $U_{g_0}(t)A_x(t)\xi_0(t) = A_{x_{g_0}}(x)\xi_0(t)$ for $t \in T'''$ and extending it by continuity from the vectors $A_x(t)\xi_0(t)$ to all of H_t. By virtue of Theorem 1, § 29, subsection 2, $g \to U_g(t)$ is an continuous unitary representation of the group $\mathfrak{G}$ in H_t, where $A_x(t) = \int x(g)U_g(t)d\nu(g)$ and therefore the representation $g \to U_g(t)$ is irreducible. Furthermore, $\{U_g(t)\xi_n(t)\} = \{A_{(x_n)g}(t)\xi_0(t)\}$ is a measurable vector-valued function of t; since for arbitrary $\xi = \{\xi(t)\} \in \mathfrak{H}$ the vector-valued function $\{U_g(t)\xi(t)\}$ is the limit of some sequence of vector-valued functions $\{U_g(t)\xi_{n_k}(t)\}$, i.e.

$$\int_T |U_g(t)\xi(t) - U_g(t)\xi_{n_k}(t)|^2 d\mu = \int_T |\xi(t) - \xi_{n_k}(t)|^2 d\mu \to 0,$$

the vector-valued function $\{U_g(t)\xi(t)\}$ is also measurable and it is an element of $\mathfrak{H}$, i.e. the operator-valued function $\{U_g(t)\}$ is measurable for every fixed $g \in \mathfrak{G}$. It remains to show that $U_g = \{U_g(t)\}$. To this end, it suffices to establish that for arbitrary $\xi = \{\xi(t)\}$, $\eta = \{\eta(t)\}$ in $\mathfrak{H}$,

$$(U_g \xi, \eta) = \int_T (U_g(t)\xi(t), \eta(t)) d\mu.$$

But for $\xi = \xi_n = \{A_{x_n}(t)\xi_0(t)\}$, $\eta = \xi_m = \{A_{x_m}(t)\xi_0(t)\}$ this equality is verified directly because

$$(U_g \xi_n, \xi_m) = (A_{(x_n)_g} \xi_0, A_{x_m} \xi_0) = \int_T (A_{(x_n)_g}(t)\xi_0(t), A_{x_m}(t)\xi_0(t)) d\mu$$
$$= \int_T (U_g(t)\xi_n(t), \xi_m(t)) d\mu,$$

and for arbitrary ξ, η it is obtained from this by passing to the limit on some subsequences $\xi_{n_k} \to \xi$, $\xi_{n'_k} \to \eta$. This proves the theorem for cyclic representations.

Now let $g \to U_g$ be an arbitrary continuous unitary representation in the space $\mathfrak{H}$ and let $x \to A_x$ be the corresponding representation of the algebra $L^1(\mathfrak{G})$. Then this representation is the direct sum of not more than a countable number of cyclic continuous unitary representations $g \to U_g^{(k)}$. This means that $\mathfrak{H} = \sum_k \oplus \mathfrak{H}_k$ and $U_g \xi = U_g\{\xi_k\} = \{U_g^{(k)}\xi_k\}$, where $g \to U_g^{(k)}$ is a cyclic continuous unitary representation . By what we proved above, there exist a decomposition in a direct integral

$$\mathfrak{H}_k = \int_{T_k} H_t^k d\mu_k$$

and irreducible continuous unitary representations $g \to U_g^{(k)}(t)$, defined for almost every $t \in T_k$ such that $U_g^{(k)} = \{U_g^{(k)}(t)\}$.

We set $T' = \bigcup_k T_k$ and define a topology in T' by taking a neighborhood basis to consist of all open sets of all the T_k. Clearly, T' is locally compact; further, we set $T = T'$ if T' is compact and $T = T' \cup \infty$ (see subsection 9, § 2) if T' is not compact. Then T is compact. Further, we define a measure μ on T by taking $\mu(\Delta) = \sum_k (\frac{1}{2})^k \mu(\Delta \cap T_k)$, and $\mu(\infty) = 0$ if $T = T' \cup \infty$. We now set $H_t = \tilde{H}_t^k$, $U_g(t) = U_g^{(k)}(t)$ for $t \in T_k$, where $\tilde{H}_t^k$ coincides with H_t, but the inner product in it is defined by the formula $\widetilde{(\xi, \eta)} = 2^k(\xi, \eta)$. We then obtain a decomposition $\mathfrak{H} = \int_T H_t d\mu$ and the corresponding decomposition $U_g = \{U_g(t)\}$, which satisfy all the requirements set down.

REMARKS. 1. The decomposition into irreducible representations depends on the choice of the maximal subalgebra in $(U_g)'$ and hence, in general, it is not unique. Generally speaking, even the list of irreducible representations into which a given unitary representation of a group $\mathfrak{G}$ decomposes is not unique (see, e.g., MACKEY [6] and [9]) if $\mathfrak{G}$ is not of type (I) (see subsection 5, below). On the other hand, a central decomposition (see subsection 5, below) is by its very definition unique.

2. The actual decomposition of a given representation (if this decomposition is unique) can present significant difficulties. GELFAND and GRAEV [3] developed methods of decomposition into irreducible representations which are applicable to many concrete cases. The case of the tensor product of two arbitrary irreducible representations of the group of all complex unimodular matrices of the second order was fully investigated by NAIMARK [13, 15]; certain cases of a tensor product of irreducible unitary representations of the group of all real unimodular matrices of the second order were considered by PUKÁNSZKY [5] and ROMM [1].

5. Central decompositions and factor representations. A representation $g \to U_g$ of the group $\mathfrak{G}$ in the Hilbert space $\mathfrak{H}$ is called a *factor representation* if the algebra $R(U_g, g \in \mathfrak{G})$ (see subsection 1, § 34) is a factor. [In the sequel, if it is not specifically stated otherwise, the expression "representation of the group $\mathfrak{G}$" means a continuous unitary representation of the locally compact group $\mathfrak{G}$.] It is called a representation of type (I), (II) or (III) if this factor is of type (I), (II) or (III), respectively. A group $\mathfrak{G}$ is called a group of type (I) if all its factor representations are of type (I). For example, all complex semisimple Lie groups are of type (I) (HARISH-CHANDRA [1, 4]). If, in the proof of Theorem 3, we replace the decomposition of the families $\{A_{x_n}\}$ into irreducible representations by the decomposition with respect to the center of the algebra $\{A_{x_n}\}'$ (see the remark at the end of subsection 3), then we obtain the so-called *central* decomposition $\mathfrak{H} = \int_T \mathfrak{H}(t)d\mu$, $U_g = \{U_g(t)\}$ in which almost every $g \to U_g(t)$ is a factor representation. GUICHARDET [2] showed that if in the central decomposition almost all the representations $g \to U_g(t)$ are irreducible, then almost all of them are pairwise non-equivalent. Independently, a more general result was obtained by NAIMARK [16, 17] who proved that in the central decomposition of an arbitrary unitary representation almost all the factor representations $g \to U_g(t)$ are pairwise disjoint; if, in particular, these representations are of type (I), then they are multiples of pairwise non-equivalent irreducible representations. [Two unitary representations are called *disjoint* if no part (i.e. restriction to an invariant subspace) of either of them is equivalent to a part of the other representation. Two unitary representations are called *quasi-equivalent* if no part of either of them is disjoint from the other (both definitions are due to MACKEY [6]).] The same result was obtained independently and somewhat later by ERNEST [1].

6. Representations in a space with an indefinite metric. Suppose there is introduced in a Hilbert space $\mathfrak{H}$, together with the usual inner product (ξ, η), also an indefinite inner product $[\xi, \eta] = (Jx, y)$, $J = P-Q$, where P, Q are projection operators with $P+Q = 1$; set $\kappa = \min(\dim P, \dim Q)$. The space $\mathfrak{H}$ with both inner products (ξ, η), $[\xi, \eta]$ is called a Π_κ*-space*, in particular, a *Pontryagin space* if $\kappa < \infty$. [PONTRYAGIN [5] first began the study of Π_κ-spaces with $\kappa < \infty$ in connection with some problems in the theory of differential equations which arose in the work of Sobolev. A further development of the theory of operators in Π_κ-spaces, $\kappa < \infty$, is found in the work of IOKHVIDOV and M. KREIN [1], GINZBURG and IOKHVIDOV [1], IOKHVIDOV [1], BRODSKY [1], LANGER [1], and others.] A representation $x \to A_x$ of the algebra R with involution is said to be *J-symmetric* if $A_{x^*} = (A_x)^+$, where, for $A \in \mathfrak{B}(\mathfrak{H})$, the operator A^+ is defined by the equality $[A\xi, \eta] = [\xi, A^+\eta]$; a representation $g \to U_g$ of the group $\mathfrak{G}$ is called *J-unitary* if $[U_g\xi, U_g\eta] = [\xi, \eta]$ for all $\xi, \eta \in \Pi_\kappa$. The structure of continuous J-symmetric representations of algebras with involution and J-unitary representations of groups in a Pontryagin space are investigated in the work of NAIMARK [23, 24]. Generally speaking, these representations no longer decompose into the direct integral of irreducible representations (for example, a non-unitary representation may be reducible but not completely reducible). Nonetheless, one succeeds in obtaining formulas for such representations; in these formulas, there also occur decompositions into the direct integral and irreducible representations of the group $\mathfrak{G}$, but in a more complicated way than in Theorems 2 and 3, subsections 3 and 4. In analogy with the proof of these theorems, one preliminarily investigates the structure of J-symmetric (if $A \in R$, then $A^+ \in R$) commutative subalgebras of operators in Π_κ, $\kappa < \infty$ (NAIMARK [22, 25]). In this investigation an essential role is played by the following result (NAIMARK [20]):

For every commutative J-symmetric family of bounded operators in the space Π_κ, $\kappa < \infty$, $\kappa = \dim P$, *there exists in* Π_κ *a* κ*-dimensional nonnegative subspace which is invariant with respect to all operators of the family.*

This result is the extension to families of operators of the Pontryagin-M. Krein-Iokhvidov theorem (see IOKHVIDOV and M. KREIN [1]); J-unitary representations in Π_κ, $\kappa < \infty$ of the group of unimodular matrices of the second order over various fields were studied in the work of ISMAGILOV [1].

The results of § 40 and subsections 1–3 of § 41 are due basically to J. VON NEUMANN [10]; we note that this article by von Neumann, published in 1949, was in reality completed by him in 1938. A discussion akin to that of von Neumann's was subsequently given in the article by NAIMARK and FOMIN [4]. Another discussion, independent of that of von Neumann, can be found in the articles by ADELSON-VELSKY [1, 2]; GODEMENT [7] obtained closely related results independently. A fuller discussion of these problems is given in the book by DIXMIER [14]; we followed Dixmier in our discussion in Appendix III.

Theorem 3 of subsection 4 for countable discrete groups was first obtained by A. N. Kolmogorov (presented at the 4 February 1944 meeting of the Moscow Mathematical Society) without the use of the von Neumann theory, and subsequently for general groups in a somewhat weakened formulation by MAUTNER [2] on the basis of results of von Neumann; further, MAUTNER [3] proved the assertion of Theorem 3, subsection 4, for Lie groups and in his review of this article by Mautner GODEMENT [10] gave the proof discussed here and hence showed that in reality the assertion of Theorem 3, subsection 4, is valid for an arbitrary locally compact group with a countable neighborhood basis.

Subsequently, MAUTNER [8] showed that the function $U_g(t)$ is also measurable on $\mathfrak{G} \times T$ relative to the measure $\mu \times \nu$ (see SEGAL [11] and SAKAI [1]).

A further development of the theory of direct integrals of representations is to be found in MACKEY [7] (also see the survey article by MACKEY [9]). In these papers of Mackey, the compact space T is replaced by the space (we will again denote it by T) of mutually equivalent classes of irreducible unitary representations of a locally compact group $\mathfrak{G}$ (the so-called *dual space of the group* $\mathfrak{G}$), which, as was shown by Mackey, can be made into a Borel space and, moreover, such that the mappings $t \to U_g(t)$ are Borel (see Appendix II). The theory is simplest when T is standard (this means that T is Borel-isomorphic to a Borel susbet M of a separable metric space). The latter, as was shown by GLIMM [3], holds, if, and only if, $\mathfrak{G}$ is a group of type (I). Mackey developed an analogous theory for symmetric representations of algebras with involution. These investigations are extended and generalized in the works of GLIMM [2, 3], DIXMIER [20], FELL [1–3], ERNEST [1], EFFROS [1], and other authors, and various topologies in the dual space T were also considered. In particular, DIXMIER [18] showed that for a completely regular algebra R, the following conditions are equivalent:

1) the dual space T is standard;

2) the Borel structure on T is countably separating (i.e., there exists in T a countable collection of Borel subsets which separate points on T);

3) the Borel structure on T is generated by open sets;

4) R is a GCR-algebra in the terminology of KAPLANSKY [13]. (R is called a *CCR-algebra* if for an arbitrary irreducible representation $x \to A_x$ of the algebra R, all the operators A_x are compact, and a *GCR-algebra* if R contains a composition series $\{I_\alpha\}$ of two-sided ideals in which all the quotient algebras $I_{\alpha+1}/I_\alpha$ are CCR-algebras.) We note that the theory of decomposition in a direct integral using generalized functions on a group was developed by KATS [2] and MAURIN [1].

We note also the work of VASILEV [1] in which symmetric algebras all irreducible symmetric representations of which are finite-dimensional are studied.

APPENDIX I

PARTIALLY ORDERED SETS AND ZORN'S LEMMA

A set X is said to be *partially ordered* if a binary relation $x \prec y$ is defined for certain pairs of its elements satisfying the following conditions:

1) $x \prec x$; 2) if $x \prec y$ and $y \prec x$, then $x = y$; 3) if $x \prec y$ and $y \prec z$, then $x \prec z$. Instead of $x \prec y$ we also write $y \succ x$.

A partially ordered set X is said to be *directed downward* if for all $x, y \in X$ there exists an element $z \in X$ such that $z \prec x$, $z \prec y$; a set *directed upward* is defined analogously. A mapping $\alpha \to x_\alpha$ of a directed set A into an arbitrary set X is called a *net* in X with index set A and is denoted by $\{x_\alpha\}_{\alpha \in A}$ or simply by $\{x_\alpha\}$. A partially ordered set X is said to be *linearly ordered* if for any two distinct $x, y \in X$, either $x \prec y$ or $x \succ y$. An element $x_0 \in X$ is called *maximal* in X if there does not exist an element $x \in X$, $x \neq x_0$ satisfying the condition $x \succ x_0$. [This concept differs from the concept of largest element. An element $x_0 \in X$ is called a *largest* element in X if $x \prec x_0$ for all $x \in X$. A largest element is maximal but the converse is not true in general.] An element x_0 is called an *upper bound* of the set $A \subset X$ if $x \prec x_0$ for all $x \in A$.

The following lemma, which is equivalent to Zermelo's axiom of choice (see BIRKHOFF [1], Chapter III, § 6), has many applications in functional analysis and in other branches of mathematics.

ZORN'S LEMMA. *If, in a partially ordered set X, every linearly ordered subset has an upper bound in X, then X contains a maximal element.*

Clearly, in the formulation of this lemma, upper bound can be replaced by lower bound and then X contains a minimal, rather than a maximal, element.

APPENDIX II

BOREL SPACES AND BOREL FUNCTIONS

Let T be an arbitrary set. A *Borel structure* on T is a family $\mathscr{B}$ of subsets of T, satisfying the following conditions:

1) If a finite or denumerable number of sets $B_j \in \mathscr{B}$, then also $\bigcap_j B_j \in \mathscr{B}$ and $\bigcup_j B_j \in \mathscr{B}$;

2) If $B \in \mathscr{B}$, then $T - B \in \mathscr{B}$;

3) $\phi \in \mathscr{B}$.

Each of the sets $B \in \mathscr{B}$ is called a *Borel* set. The set T together with the Borel structure $\mathscr{B}$ on it is called a *Borel space*. Every Borel set B_0 of the Borel space T is a Borel space if we consider its Borel structure to be the collection of all $B \subset \mathscr{B}$ which are contained in B_0.

If F is an arbitrary family of subsets of the set T, then there exist families of subsets which contain F and satisfy conditions 1)–3), e.g., the collection of all subsets of the set T. The intersection of all such families, containing F, is the *minimal* family which satisfies conditions 1)–3) and contains F; it is called the *Borel structure generated by the family F*.

Let T_1 and T_2 be two Borel spaces. A mapping f of the set T_1 into the set T_2 is called a *Borel mapping* if $f^{-1}(B_2)$ is a Borel set in T_1 for every Borel set B_2 in T_2. A 1-1 mapping f of the space T_1 onto the space T_2 is called a *Borel isomorphism* if f and f^{-1} are Borel mappings. Two Borel spaces T_1 and T_2 are said to be *Borel isomorphic* if there exists a Borel isomorphism of T_1 onto T_2.

A Borel structure $\mathscr{B}$ of the topological space T is said to be *generated by the topology* of this space if $\mathscr{B}$ is generated by the family of all its closed sets; it follows from conditions 2) and 3) that closed sets can be replaced by open sets. In the sequel, we shall always consider a Borel structure in the topological space T to be the Borel structure generated by its topology.

An important example is the space $\mathfrak{B}(\mathfrak{H})$ with the weak operator topology (see subsection 1, § 33).

In the remainder of Appendix II we will consider the case when T is a locally compact Hausdorff space.

If I is an integral on T and μ is the corresponding measure, then the μ-measurable sets form a family M which satisfies conditions 1)–3) and contains all open sets (see subsection 9, § 6); but $\mathscr{B}$ is the minimal family, satisfying conditions 1)–3) and containing all open sets; therefore, $\mathscr{B} \subset M$, i.e., Borel sets are measurable with respect to any integral I.

A real-valued function f is called a *Borel function* if the set $\{t: f(t) > a\}$ is a Borel set for every real a. [It is not difficult to prove that this definition is a special case of the definition of a Borel mapping, given above, when $T_2 = R^1$ with the usual topology. We leave the proof to the reader.] A complex-valued function f is called a *Borel function* if $\operatorname{Re} f(t)$ and $\operatorname{Im} f(t)$ are Borel functions. It follows from these definitions that all the Propositions II–VII of § 6, subsection 10, remain valid if one replaces "measurable functions" and "measurable sets" by "Borel functions" and "Borel sets" respectively. Moreover, real-valued functions which are upper semi-continuous or lower semi-continuous are Borel functions because in the case, for instance, of a lower semi-continuous function f, the set $\{t : f(t) > a\}$ is open.

Now let I be an integral on T, let μ be the corresponding measure and $L^1 = L^1(I)$. Applying Proposition XI, § 6, subsection 7, to the function $f \in L^1(I)$, we conclude that *for every function $f \in L^1(I)$ there exists a set $N \subset T$ of μ-measure zero and a Borel function z such that $f(t) = z(t)$ for all $t \in T - N$.*

APPENDIX III

ANALYTIC SETS

Let N be the set of all natural numbers; it is a complete separable metric space if distance is taken to be $\rho(m, n) = |m-n|$ for $m, n \in N$. Then N^N (see subsection 15, § 2) is the metric space of all sequences $m = \langle m_k \rangle$, $m_k \in N$ with distance

$$\rho(m, n) = \sum_{k=1}^{\infty} \frac{1}{2^k} \frac{|m_k - n_k|}{1 + |m_k - n_k|}, \quad m = \langle m_k \rangle,\ n = \langle n_k \rangle \in N^N.$$

In virtue of II and III, subsection 15, § 2, N^N is a complete separable metric space.

I. *Every complete separable metric space X is a continuous image of the space N^N.*

Proof. For every $\varepsilon > 0$ there exists a covering of the space X by a countable family of closed balls of diameter $< \varepsilon$ (see VI, § 2, subsection 13). We construct such a covering for $\varepsilon_1 = \frac{1}{2}$ and index the balls of this covering arbitrarily; we denote the ball with index m_1 by $f(m_1)$. Further, we cover each ball $f(m_1)$ with a denumerable number of closed balls of diameter $< \varepsilon_2 = 1/2^2$ and we index all these balls arbitrarily. We denote the intersection of the ball with index m_2 with $f(m_1)$ by $f(m_1, m_2)$. Continuing this construction, we assign to each collection $\{m_1 \cdots, m_p\}$ a closed set $f(m_1, \cdots, m_p)$ with diameter $< 1/2^p$. To each point $m = \langle m_1, m_2, \cdots \rangle \in N^N$ there will then correspond a sequence $f(m_1) \supset f(m_1, m_2) \supset \cdots$ of closed sets with diameters $\to 0$. Therefore, there exists precisely one point x, common to all these sets (see II, subsection 13, § 2), which we denote by $\varphi(m)$. We have thus constructed a mapping φ: $N^N \to X$. It is clear that every point $x \in X$ is obtained in this way. Thus, $X = \varphi(N^N)$.

If for $m, n \in N^N$ the distance $\rho(m, n) < 1/2^{p+1}$, then $m_1 = n_1, \cdots, m_p = n_p$. This means that the points $x = f(m)$ and $y = f(n)$ belong to the same set $f(m_1, m_2, \cdots, m_p)$ of diameter $< 1/2^p$ and therefore $\rho(x, y) < 1/2^p$. We have thus proved that the mapping φ is continuous.

II. *Every compact Hausdorff space T with a countable basis is metrizable and, moreover, such that with the metric introduced it is a complete separable metric space.*

Proof. Let $\{U_k\}$ be a countable basis in T. Consider all possible pairs $P = \{U_k, U_j\}$ such that $\bar{U}_k \subset U_j$ and index them arbitrarily in the form of a sequence $P_1, P_2, \cdots$ so that $P_n = \{U_{k_n}, U_{j_n}\}$, $\bar{U}_{k_n} \subset U_{j_n}$. Since T is normal, then, according to the Urysohn lemma, for each pair P_n there exists a continuous real-valued function $f_n(t)$ on T satisfying the conditions:

$$0 \leqq f_n(t) \leqq 1, f_n(t) = 0 \text{ on } \bar{U}_{k_n}, f_n(t) = 1 \text{ on } T - U_{j_n} \tag{1}$$

(see I and II, subsection 8, § 2). To each point $t \in T$ we assign the sequence $x = \langle x_1, x_2, \cdots \rangle$, where $x_n = (1/n) f_n(t)$, and set $x = \varphi(t)$. By (1), $\sum_{n=1}^{\infty} x_n^2 < \infty$, so that $x \in l^2$. Consequently, φ is a mapping of the space T into the separable metric (and even Hilbert) space l^2 (see Example 3, subsection 1, § 4). This mapping is 1-1. In fact, let $t_1 \neq t_2$. Then there exists a neighborhood U_j of the point t_1 such that $t_2 \notin U_j$. Since the space T is normal, there exists a neighborhood U_k of the point t_1 such that $\overline{U}_k \subset U_j$. But in this case (U_k, U_j) is one of the pairs P_n. Let $(U_k, U_j) = P_s$. By (1), $f_s(t_1) = 0$ because $t_1 \in U_k \subset \overline{U}_k$ and $f_2(t_2) = 1$ because $t_2 \notin U_j$. Therefore, setting $\varphi(t_1) = x^{(1)} = \{x_k{}^{(1)}\}$, $\varphi(t_2) = x^{(2)} = \{x_k{}^{(2)}\}$ we have: $x_s{}^{(1)} = 0, x_s{}^{(2)} = 1/s$, from which it follows that

$$[\rho(x^{(1)}, x^{(2)})]^2 = \sum_{k=1}^{\infty} |x_k^{(1)} - x_k^{(2)}|^2 \geqq |x_s^{(1)} - x_s^{(2)}|^2 = \frac{1}{s^2} \neq 0 \quad \text{and} \quad x^{(1)} \neq x^{(2)}.$$

We will now prove that the mapping φ is continuous. Let $t_0 \in T$ and $\varepsilon > 0$. We choose an index n_1 so that $\sum_{k=n_1+1}^{\infty} 1/k^2 < \varepsilon^2/8$, and then a neighborhood $U(t_0)$ so that $|f_k(t) - f_k(t_0)| < \varepsilon/\sqrt{2c}$ for $k = 1, 2, \cdots, n_1$, where $c^2 = \sum_{k=1}^{\infty} 1/k^2$. Then, for $t \in U(t_0)$,

$$[\rho(\varphi(t), \varphi(t_0)]^2 = \sum_{k=1}^{n_1} \frac{1}{k^2} |f_k(t) - f_k(t_0)|^2 + \sum_{k=n_1+1}^{\infty} |f_k(t) - f_k(t_0)|^2$$

$$< \frac{\varepsilon^2}{\sqrt{2c}} \sum_{k=1}^{n_1} \frac{1}{k^2} + 4 \sum_{k=n_1+1}^{\infty} \frac{1}{k^2} < \frac{\varepsilon^2}{2} + \frac{\varepsilon^2}{2} = \varepsilon^2$$

and, consequently, $\rho(\varphi(t), \varphi(t_0)) < \varepsilon$. This proves that φ is a continuous mapping. But then φ is a homeomorphism because T is compact (see V, subsection 7, § 2). Therefore $\varphi(T)$ is closed in l^2 (II, subsection 7, § 2) and hence it is a complete separable (see VI, subsection 13, § 2) metric space which is homeomorphic to the space T.

Now let X be a complete separable metric space. A set $A \subset X$ is called *analytic* (or *Suslin*) if it is a continuous image of some complete separable metric space P. The void set is analytic by definition.

It follows directly from this definition that:

1) *Every closed subset in X is an analytic set*;

2) *A continuous image* (*in a complete separable metric space*) *of an analytic set is an analytic set*;

Moreover:

3) *Every open set U in X is an analytic set.*

Proof. Set $F = X - U$ and, for $x, y \in U$,

$$\rho'(x, y) = \rho(x, y) + |[\rho(x, F)]^{-1} - [\rho(y, F)]^{-1}|, \tag{2}$$

where

$$\rho(x, F) = \inf_{z \in F} \rho(x, z).$$

Since F is closed, $\rho(x, F) > 0$ for $x \in U$. In fact, otherwise there exists a sequence

$z_n \in F$ for which $\rho(x, z_n) \to 0$. But then $x \in F$, contrary to assumption. Clearly, ρ' satisfies all the distance axioms so that U with the distance ρ' is a metric space; we denote it by U'. Since $\rho \leqq \rho'$, the identity mapping of U' onto U is continuous.

We will show that U' is complete. Let $\langle x_n \rangle$ be a fundamental sequence in U'. Since $\rho \leqq \rho'$, $\langle x_n \rangle$ is fundamental in X; consequently, it converges in X to some $x \in X$ (recall that X is complete). On the other hand, by (1) the numbers $\rho(x_n, F)$ are bounded below by a positive constant because the set $\rho'(x_n, x_m)$, $n, m = 1, 2, \cdots$ is bounded; therefore $x \in U$. Let $\varepsilon > 0$; there exists an index n_0 such that

$$\rho'(x_m, x_n) = \rho(x_m, x_n) + |[\rho(x_m, F)]^{-1} - [\rho(x_n, F)]^{-1}| < \varepsilon$$

for $n, m > n_0$. Passing to the limit in this inequality as $m \to \infty$ and for fixed $n \geqq n_0$, we obtain that

$$\rho'(x, x_n) = \rho(x, x_n) + |[\rho(x, F)]^{-1} - [\rho(x_n, F)]^{-1}| \leqq \varepsilon.$$

Thus $x_n \to x$ in U.

It remains to show that U' is separable. Since X is separable, U contains a dense sequence $\langle z_n \rangle$. It will also be dense in U'. In fact, let $x \in U'$. There exists a subsequence $\langle z_{n_k} \rangle$ for which $\rho(z_{n_k}, x) \to 0$.

But then $\rho(z_{n_k}, F) \to \rho(x, F)$ and consequently $\rho'(z_{n_k}, x) \to 0$.

[From the triangle inequality $\rho(x, z) \leqq \rho(x, y) + \rho(y, z)$ it follows, for fixed $x, y \in U$ and an arbitrary $z \in F$, that $\rho(x, F) \leqq \rho(x, y) + \rho(y, F)$ and hence $\rho(x, F) - \rho(y, F) \leqq \rho(x, y)$. Interchanging the roles of x and y, we conclude that $|\rho(x, F) - \rho(y, F)| \leqq \rho(x, y)$. In particular, $|\rho(x_n, F) - \rho(x, F)| \leqq \rho(x_n, x) \to 0$ as $n \to \infty$.]

Finally,

4) *The union and intersection of a finite or denumerable family of analytic subsets A_j of a complete separable metric space X is an analytic set.*

In fact, let $A_j = f_j(P_j)$, where P_j is a complete separable metric space with metric $\rho_j(x, y)$ – we can assume that these metric spaces are pairwise disjoint – and where f_j is a continuous mapping of P_j into X. Then $P = \bigcup_j P_j$ is also a complete separable metric space if we equip it with the distance $\rho(x, y) = 1$ for $x \in P_j, y \in P_k, j \neq k$, and $\rho(x, y) = \min\{\rho_j(x, y), 1\}$ for $x, y \in P_j$; setting $f(x) = f_j(x)$ for $x \in P_j$, we obtain that $f(\bigcup_j P_j) = \bigcup_j A_j$, f is a continuous mapping of P into X and, consequently, $\bigcup_j A_j$ is an analytic set.

Further, if the intersection of the sets A_j is void, then it is analytic by definition. Let $\bigcap_j A_j \neq \phi$ and let $P' = \prod_j P_j$ be the topological product of the spaces P_j. Then P' is a complete separable metric space (see II and III, subsection 15, § 2). Let Q be the set of all $\{x_n\} \in P'$ for which $f_m(x_m) = f_n(x_n)$ for all m, n. Then Q is closed in P' and hence it is also a complete separable metric space. Setting $p_m(\{x_n\}) = x_m$ and denoting the restriction of $f_m \cdot p_m$ to Q by f, we conclude that f is independent of m, f is continuous and $f(Q) = \bigcap_j A_j$. Consequently, $\bigcap_j A_j$ is analytic.

Now let $\mathscr{B}$ be the Borel structure in X, considered as a topological space, and let $\mathfrak{A}$ be the family of all analytic sets in X whose complements are also analytic. [We note that in general the complement of an analytic set may not be analytic.] By 4) the family $\mathfrak{A}$ satisfies conditions 1)–3) formulated in Appendix II (recall that $\phi \in \mathfrak{A}$), and by 1) and 3) $\mathfrak{A}$ contains all closed (and all open) sets in X. Since $\mathscr{B}$ is the minimal family satisfying all these conditions, $\mathscr{B} \subset \mathfrak{A}$; consequently, *every Borel set in X is an analytic set and hence a continuous image of a Borel set is an analytic set.* [It can be shown that $\mathfrak{A} = \mathscr{B}$, but we shall not need this.]

Let F be a compact topological space. Denote by $K_{\sigma\delta}$ any set in F of the form $B_1 \cap B_2 \cap B_3 \cap \cdots$, where every B_k is the union of at most a countable number of sequentially compact sets.

We note that in a complete separable metric space, compactness and sequential compactness coincide.

III. *If the analytic set A in a complete separable metric space X is precompact, then there exist a sequentially compact metric space F, a set $B \subset F$ of type $K_{\sigma\delta}$, and a continuous mapping f of the space F into X such that $A = f(B)$.*

Proof. By replacing the space X by the set $\bar{A}$, we can assume that X is sequentially compact. Let g be a continuous mapping of the space N^N onto A. Denote by $\bar{R}$ the real line with the infinitely distant point adjoined and by $\bar{R}^N$ the topological product of a countable number of copies of the space $\bar{R}$. Since $\bar{R}$ is compact, $\bar{R}^N$ is compact (see IV, subsection 1, § 2). [$\bar{R}^N$ is also a complete sequentially compact metric space in virtue of II.] The space N^N can be considered as a subspace of the sequentially compact metric space $\bar{R}^N$.

Let $Q_n \subset R$ be the union of the segments I_{kn} of length $1/n$ with midpoints at the integer points k; set $G_n = Q_n \times Q_n \times \cdots \times Q_n \times \bar{R} \times \bar{R} \times \cdots$ (Q_n n times). Then every G is the union of a countable number of closed sets $I_{nk_1} \times \cdots \times I_{nk_n} \times \bar{R} \times \bar{R} \times \cdots$ in $\bar{R}^N$ and $\bigcap_n G_n = N^N$.

Let B be the graph of g in $N^N \times X$, $\bar{B}$ its closure in the sequentially compact space $F = \bar{R}^N \times X$ and f the projection of the space F onto X. Then $A = f(B)$. Since B is closed in $N^N \times X$ (because g is continuous), $B = \bar{B} \cap (N^N \times X)$. Consequently,

$$B = [\bar{B} \cap (G_1 \times X)] \cap [\bar{B} \cap (G_2 \times X)] \cap \cdots,$$

where each $\bar{B} \cap (G_n \times X)$ is the union of a countable number of closed, and hence sequentially compact, sets in F.

IV. *Every analytic set A in a compact Hausdorff space X with a countable basis is measurable (with respect to an arbitrary integral I on X).*

Proof. By II, X can be considered to be a complete separable metric space. Since A is contained in the sequentially compact space X, A is precompact. By virtue of III, there exist a sequentially compact metric space F, a continuous mapping f of F into X and a family of sequentially compact sets $B_{n,p}$, $n, p = 1, 2, \cdots$ in F such that $A = f(B)$, where $B = B_1 \cap B_2 \cap \cdots$ and $B_n = B_{n,1} \cup B_{n,2} \cup \cdots$.

Obviously, we can assume that $B_{n,1} \subset B_{n,2} \subset \cdots$ for every n.

Let μ be the measure defined by the given integral and let $\bar{\mu}$ be the corresponding exterior measure, i.e. $\bar{\mu}(C) = \bar{I}(\xi_C)$ for an arbitrary set $C \subset X$ (see § 6, subsection 5).

We shall prove that for every $a < \bar{\mu}(A)$ there exists a sequentially compact set $C \subset B$ such that $\mu(f(C)) \geqq a$. Inasmuch as $f(C)$ is sequentially compact, Proposition IV will then be proved.

For the construction of the set C, we shall show that there exists a sequence of indices $p_1, p_2, \cdots$ such that the set $C_n = B \cap B_{1,p_1} \cap \cdots \cap B_{n,p_n}$ satisfies the condition $\bar{\mu}(f(C_n)) > a$ for every n. Suppose the B_{i,p_i} have already been defined for $i < n$. Since $C_{n-1} \subset B \subset B_n$, we have that $C_{n-1} = (C_{n-1} \cap B_{n,1}) \cup (C_{n-1} \cap B_{n,2}) \cup \cdots$ and consequently (see (3), subsection 5, § 6) $\bar{\mu}(f(C_{n-1} \cap B_{n,k})) \to \bar{\mu}(f(C_{n-1}))$ as $k \to \infty$ and therefore $\bar{\mu}(f(C_{n-1} \cap B_{n,p_n})) > a$ for some p_n.

We now set $C = B_{1,p_1} \cap B_{2,p_2} \cap \cdots$; C is nonvoid and sequentially compact since it is the intersection of a decreasing sequence of nonvoid sequentially compact sets $B_{1,p_1} \cap \cdots \cap B_{n,p_n}$; consequently, $f(C)$ is sequentially compact and it is the intersection of the sets $f(B_{1,p_1} \cap \cdots \cap B_{n,p_n})$. [In fact, setting $Q_n = B_{1,p_1} \cap \cdots \cap B_{n,p_n}$, we have that $Q = \bigcap_{n=1}^{\infty} Q_n$ and $f(Q) \subset \bigcap_{n=1}^{\infty} f(Q_n)$. Conversely, if $x \in \bigcap_{n=1}^{\infty} f(Q_n)$, then for every n there exists a $t_n \in Q_n \subset F$ such that $x = f(t_n)$. Since F is sequentially compact, there exists a subsequence $\langle t_{n_k} \rangle$ which converges to some $t_0 \in F$. Clearly, $t_0 \in \bigcap_{n=1}^{\infty} Q_n = Q$ and $f(t_0) = \lim_{k\to\infty} f(t_{n_k}) = x$ so that $x \in f(Q)$. This proves that $f(Q) = \bigcap_{n=1}^{\infty} f(Q_n)$.] From this it follows that

$$\begin{aligned} \mu(f(C)) &= \lim_{n\to\infty} \mu(f(B_{1,p_1} \cap \cdots \cap B_{n,p_n})) \\ &\geqq \lim_{n\to\infty} \bar{\mu}(f(B \cap B_{1,p_1} \cap \cdots \cap B_{n,p_n})) \\ &= \lim_{n\to\infty} \bar{\mu}(C_n) \geqq a. \end{aligned}$$

Moreover, since $B_{n,p_n} \subset B_n$, $C = \bigcap_n B_{n,p_n} \subset \bigcap_n B_n = B$.

We now return to the space N^N and order it lexicographically.

[This means that we will consider $m < n$ ($m, n \in N^N$) if $m_1 < n_1$ or $m_1 = n_1$, $m_2 < n_2, \cdots$, in general, if for some integer $k \geqq 1$, $m_j = n_j$ for $j \leqq k$ and $m_{k+1} < n_{k+1}$.]

Then:

V. *Every nonvoid closed set F in N^N has a smallest element.*

Proof. Let p_1 be the smallest of the numbers n_1 for which $\langle n_1, n_2, n_3, \cdots \rangle \in F$ for some $n_2, n_3, \cdots$; further, let p_2 be the smallest of the numbers n_2 for which $\langle p_1, n_2, n_3, \cdots \rangle \in F$ for some $n_3, n_4, \cdots$. Repeating this line of reasoning we obtain the point $p = \langle p_1, p_2, p_3, \cdots \rangle \in N^N$ which is $\leqq n$ for arbitrary $n \in F$. Moreover, it is clear from the construction that p is a contact point of the set F and hence $p \in F$.

Let T, S be topological spaces, T locally compact, I an integral on T, and μ the corresponding measure. The mapping $s = f(t)$ of the space T into S will be called μ-measurable if it is defined on a μ-measurable set $T' \subset T$ and the inverse image of every open set in S is a μ-measurable set in T.

From this it immediately follows that:

VI. *If f_1 is a continuous mapping of the space S into the space S_1 and f is a measurable mapping of the space T into S, then the mapping $f_1 \cdot f$ of the space T into S_1 is μ-measurable.*

In particular, the projection $p\langle t, s\rangle = s$ is a continuous mapping of the space $T \times S$ onto S; therefore:

VII. *The mapping f of the space T into S is measurable if the mapping $t \to \langle t, f(t)\rangle$ of the space T into $T \times S$ is measurable.*

VIII. *Let Q be the countable set of all points in N^N all coordinates of which, beginning with a certain one, equal zero. Then every open set U in N^N is the union of intervals $[m, n)$, where $m, n \in Q$.*

Proof. Let $p = \langle p_1, p_2, \cdots\rangle \in U$. Then there exists an index k such that all the points $q = \langle p_1, p_2, \cdots, p_k, q_{k+1}, q_{k+2}, \cdots\rangle \in U$. Setting $m = \langle p_1, p_2, \cdots, p_k, p_{k+1}, 0, 0, \cdots\rangle$, $n = \langle p_1, p_2, \cdots, p_k, p_{k+1}+1, 0, \cdots\rangle$, we see that $p \in [m, n) \subset U$.

IX. *Let T, X be complete separable metric spaces, T compact, $T' \subset T$ and let every point $t \in T$ be set into correspondence with the subset $X_t \subset X$, which is nonvoid for $t \in T'$, so that the collection A of all pairs $\langle t, x\rangle$, $t \in T$, $x \in X_t$, for which X_t is nonvoid is an analytic set in $T \times X$. Then there exist a μ-measurable set $T'' \subset T$ which contains T' and a μ-measurable mapping f of the set T'' into X such that $f(t) \in X_t$ for all $t \in T''$.*

Proof. By virtue of VII it suffices to construct a μ-measurable set $T'' \supset T'$ and a μ-measurable mapping φ of the set T'' into $A \subset T \times X$ such that $p_1 \varphi(t) = t$, where p_1 is projection onto the first component: $p_1\langle t, \zeta\rangle = t$. But A is a continuous image of the space $N^N : A = \psi(N^N)$; therefore $g = p_1 \cdot \psi$ is a continuous mapping of the space N^N onto some set $T'' \supset T'$; T'' is analytic and therefore in virtue of IV it is measurable. By VI, it will suffice to construct a μ-measurable mapping h of the set T'' onto N^N such that $g \cdot h$ is the identity mapping on T''. Then $\varphi = \psi \cdot h$ will satisfy the requirements set down. Since g is continuous, we have that for an arbitrary $t \in T''$, the set $g^{-1}(\{t\})$ is nonvoid and closed in N^N; let $h(t)$ be its smallest element (see V). Then $g \cdot h$ is the identity mapping on T'' and it remains to prove that h is μ-measurable. To this end, it suffices to show that $h^{-1}(U)$ is measurable for an arbitrary open set $U \subset N^N$. According to VIII, instead of doing this it suffices to establish that every set $h^{-1}\{m: m < n\}$ is measurable for every $n \in N^N$. [In fact, for arbitrary $m, n \in N^N$, one of the relations $m = n$, $m < n$, $m > n$ holds, and hence, for $m < n$, $\{p: m \leqq p < n\} = \{p: p < n\} - \{p: p < m\}$.] But this set coincides with $g\{m: m < n\}$; being a continuous image of an open set $\{m: m < n\}$ in N^N, it is analytic and therefore μ-measurable.

Theorem IX is a particular case of the Luzin-Yankov theorem (see, e.g., YANKOV [1] or ARSENIN and LYAPUNOV [1]).

LITERATURE

Adelson-Velsky, G. M.

[1] *Spectral analysis of a ring of bounded linear operators*, Dissertation, MGU, 1948 (Russian).

[2] *Spectral analysis of a ring of bounded linear operators in Hilbert space*, DAN SSSR, **67** (1949), 957–959 (Russian).

Akhiezer, N. I. and I. M. Glazman

[1] *Theory of linear operators in Hilbert space*, 2nd ed. "Nauka", Moscow, 1966 (Russian); *Theorie der linearen Operatoren im Hilbert-Raum*, Akademie-Verlag, Berlin, 1954; English edition, F. Ungar Publ. Co., New York, 1965.

Akutowicz, E. J.

[1] *Sur l'approximation spectrale*, Ann. Sci. École Norm. Sup., 3-ième série, **82** (1965) 297–325.

Allan, G. R.

[1] *On a class of locally convex algebras*, Proc. London Math. Soc., **17** (1967), 91–114.

Allen, H. S.

[1] *Commutative rings of linear transformations and infinite matrices*, Quart. J. Math., **8** (1957), 39–53.

Ambrose, W.

[1] *Spectral resolution of groups of unitary operators*, Duke Math. J. **11** (1944) 589–595.

[2] *Structure theorems for a special class of Banach algebras*, Trans. Amer. Math. Soc. **57** (1945) 364–386.

[3] *The L_2-system of a unimodular group*, Trans. Amer. Math. Soc. **65** (1949) 27–48.

Anastasio, S.

[1] *Maximal abelian subalgebras in hyperfinite factors*, Amer. J. Math. **87** (1965) 955–971.

Anzai, H.

[1] *On compact topological rings*, Proc. Imp. Acad. Tokyo **19** (1943) 613–615.

Araki, H. and E. J. Woods

[1] *Representations of the canonical commutation relations describing a nonrelativistic infinite free Bose gas*, J. Math. Phys. **4** (1963) 637–662.

Arens, R. F.

[1] *On a theorem of Gelfand and Neumark*, Proc. Nat. Acad. Sci. USA **32** (1946) 237–239.

[2] *Representation of *-algebras*, Duke Math. J. **14** (1947) 269–282.

[3] *Linear topological division algebras*, Bull. Amer. Math. Soc. **53** (1947) 623–630.

[4] *Approximation in, and representation of, certain Banach algebras*, Amer. J. Math. **71** (1949) 763–790.

[5] *A generalization of normed rings*, Pacif. J. Math. **2** (1952) 455–471.

[6] *A Banach algebra generalization of conformal mappings of the disc*, Trans. Amer. Math. Soc. **81** (1956) 501–513.

[7] *The maximal ideals of certain function algebras*, Pacif. J. Math. **8** (1958) 641–648.

[8] *The closed maximal ideals of algebras of functions holomorphic on a Riemann surface*, Ren. Cir. Mat. di Palermo **7** (1958) 245–260.

[9] *The analytic-functional calculus in commutative topological algebras*, Pacif. J. Math. **11** (1961) 405–429.

[10] *The group of invertible elements of a commutative Banach algebra*, Studia Math., Seria Spec. Z. I (1963) 21–23.

[11] *Connectivity of maximal ideal space*, Proceedings of the Internat. Congress Math., Moscow, Functional Analysis, 1966.

Arens, R. F. and A. P. Calderón

[1] *Analytic functions of several Banach algebra elements*, Ann. Math. **62** (1955) 204–216.

[2] *Analytic functions of Fourier transforms*, Segundo symposium sobre algunos problemas matemáticos que se están estudiando en Latino América, Julio 1954, 39–52, Centro de Coop. Cient. Unesco América Latina, Montevideo, 1954.

ARENS, R. F. and I. KAPLANSKY

[1] *Topological representation of algebras*, Trans. Amer. Math. Soc. **63** (1948) 457–481.

ARENS, R. F. and I. M. SINGER

[1] *Function values as boundary integrals*, Proc. Amer. Math. Soc. **5** (1954) 735–745.

[2] *Generalized analytic functions*, Trans. Amer. Math. Soc. **81** (1956) 379–393.

ARENSON, E. L.

[1] *On certain properties of algebras of continuous functions*, DAN SSSR **171** (1966) 767–769 (Russian).

ARSENIN, V. YA. and A. A. LYAPUNOV

[1] *Theory of A-sets*, UMN **5** (1950) 45–108 (Russian).

AURORA, S.

[1] *Multiplicative norms for metric rings*, Pacif. J. Math. **7** (1957) 1279–1304.

BADE, W. G.

[1] *Weak and strong limits of spectral operators*, Pacif. J. Math. **4** (1954) 393–413.

[2] *On Boolean algebras of projections and algebras of operators*, Trans. Amer. Math. Soc. **80** (1955) 345–360.

BADE, W. G. and P. C. CURTIS

[1] *Homomorphisms of commutative Banach algebras*, Amer. J. Math. **3** (1960) 589–608.

[2] *The Wedderburn decomposition of commutative Banach algebras*, Amer. J. Math. **4** (1960) 851–866.

[3] *Embedding theorems for commutative Banach algebras*, Pacif. J. Math. **18** (1966) 391–409.

BANACH, S. S.

[1] *Sur les fonctionnelles linéaires. I, II*, Studia Math. **1** (1929) 211–216, 223–239.

[2] *Théorie des opérations linéaires*, Monografie Mat., Warsaw, 1932; *Course in functional analysis*, "Radyanska shkola", Kiiv, 1948 (Ukrainian).

BARGMAN, V.

[1] *On unitary ray representations of continuous groups*, Ann. Math. **59** (1954) 1–46.

BAUER, H.

[1] *Choquetscher Rand und Integraldarstellungen*, Jahresber. DMV **69** (1967) 89–104.

BAUM, L. E.

[1] *Note on a paper of Civin and Yood*, Proc. Amer. Math. Soc. **9** (1958) 207–208.

BEAR, H. S.

[1] *Complex function algebras*, Trans. Amer. Math. Soc. **90** (1959) 383–393.

[2] *A strong maximum modulus theorem for maximal algebras*, Trans. Amer. Math. Soc. **92** (1959) 465–469.

[3] *Some boundary properties of function algebras*, Proc. Amer. Math. Soc. **11** (1960) 1–4.

BECKER, H.

[1] *Über einen Satz der Darstellungstheorie topologischer Gruppen*, Wiss. Z. Humboldt Univ., Berlin, math.-nat. Reihe **2** (1952/53) 61–66.

BEID, G. A.

[1] *A generalization on W*-algebras*, Pacif. J. Math. **15** (1965) 1019–1026.

BERBERIAN, S. K.

[1] *The regular ring of a finite AW*-algebra*, Ann. Math. **65** (1957) 224–240.

[2] *$N \times N$ matrices over an AW*-algebra*, Amer. J. Math. **80** (1958) 37–44.

BEREZANSKY, YU. M.

[1] *Certain classes of continual algebras*, DAN SSSR **72** (1950) 237–240 (Russian).

[2] *On the center of the group ring of a compact group*, DAN SSSR **72** (1950) 825–828 (Russian).

[3] *Hypercomplex systems with discrete basis*, DAN SSSR **81** (1951) 329–332 (Russian).

[4] *On the theory of almost periodic Levitan sequences*, DAN SSSR **81** (1951) 493–496 (Russian).

[5] *Hypercomplex systems with compact basis*, Ukr. matem. zh. **3** (1951) 184–203 (Russian).

[6] *On certain normed rings constructed with respect to orthogonal polynomials*, Ukr. matem. zh. **3** (1951) 412–432 (Russian).

Berezansky, Yu. M. and S. G. Krein

[1] *Continuous algebras*, DAN SSSR **72** (1950) 5–8 (Russian).

[2] *Hypercomplex systems with continuous basis*, UMN **12** (1957) 147–152 (Russian).

Berezin, F. A.

[1] *Laplace operators on semisimple Lie groups*, DAN SSSR **107** (1956) 9–12 (Russian).

[2] *Representations of complex semisimple Lie groups in a Banach space*, DAN SSSR **110** (1956) 897–900 (Russian).

[3] *The Laplace operator in semisimple Lie groups*, Uch. zap. Mosk. matem. o-va, **6** (1957) 371–463 (Russian).

[4] *Laplace operators in semisimple Lie groups and in certain symmetric spaces*, UMN **12** (1957) 152–156 (Russian).

Berezin, F. A., I. M. Gelfand, M. I. Graev and M. A. Naimark

[1] *Representations of groups*, UMN **11** (1956) 13–40 (Russian).

Berkson, E.

[1] *Some characterizations of C*-algebras*, Illinois J. Math. **101** (1966) 1–8.

Beurling, A.

[1] *Sur les intégrales de Fourier absolument convergentes et leur application à une transformation fonctionnelle*, Congrès de Math. à Helsingfors, 1938.

[2] *Un théorème sur les fonctions bornées et uniformement continues sur l'axe réel*, Acta Math. **77** (1945) 127–136.

Bingen, F., J. Tits and L. Waelbroeck

[1] *Séminaire sur les algèbres de Banach*, Centre Belge d'Algèbre et de Topologie à Université Libre de Bruxelles, 1962–63.

Birkhoff, G.

[1] *Lattice theory*, Amer. Math. Soc. Colloquium Publ., Vol. XXV, N. Y., 1948; 3rd ed., 1960; IIL, Moscow, 1952 (Russian).

Bishop, E.

[1] *Measures orthogonal to polynomials*, Proc. Nat. Acad. Sci. USA **44** (1958) 278–280.

[2] *The structure of certain measures*, Duke Math. J. **25** (1958) 283–289; Collection "Some problems in approximation theory", IIL, Moscow, 1963 (Russian).

[3] *Subalgebras of functions on a Riemann surface*, Pacif. J. Math., **8** (1958) 29–50.

[4] *Approximation by a polynomial and its derivative on certain closed sets*, Proc. Amer. Math. Soc. **9** (1958) 946–953.

[5] *Simultaneous approximation by a polynomial and its derivatives*, Proc. Amer. Math. Soc. **10** (1959) 741–743.

[6] *Some theorems concerning function algebras*, Bull. Amer. Math. Soc. **65** (1959) 77–78.

[7] *A minimal boundary for function algebras*, Pacific J. Math. **9** (1959) 629–642; Collection "Some problems in approximation theory", IIL, Moscow, 1963 (Russian).

[8] *Boundary measures of analytic differentials*, Duke Math. J. **27** (1960) 331–340; Collection "Some problems in approximation theory", IIL, Moscow, 1963 (Russian).

[9] *Mapping of partially analytic spaces*, Amer. J. Math. **83** (1961) 209–242.

[10] *Some global problems in the theory of functions of several complex variables*, Amer. J. Math. **83** (1961) 479–498.

[11] *Partially analytic spaces*, Amer. J. Math. **83** (1961) 669–692.

[12] *A generalization of the Stone-Weierstrass theorem*, Pacif. J. Math. **11** (1961) 777–783.

[13] *A general Rudin-Carleson theorem*, Proc. Amer. Math. Soc. **13** (1962) 140–143.

[14] *Analyticity in certain Banach algebras*, Trans. Amer. Math. Soc. **102** (1962) 507–544.

[15] *Holomorphic completions, analytic continuations, and the interpolation of seminorms*, Ann. Math. **78** (1963) 468–500.

[16] *Representing measures for points in a uniform algebra*, Bull. Amer. Math. Soc. **70** (1964) 121–122.

[17] *Conditions for the analyticity of certain sets*, Michigan Math. J. **11** (1964) 289–304.

[18] *Uniform Algebras*, Proc. Conf. Complex Analysis, Minneapolis, 1964, Springer, Berlin–Heidelberg–New York (1965) 272–281.

Bishop, E. and K. de Leeuw

[1] *The representation of linear functionals by measures on sets of extreme points*, Ann. Inst. Fourier **9** (1959) 305–331.

Blair, A.

[1] *Continuity of multiplication in operator algebras*, Proc. Amer. Math. Soc. **6** (1955) 209–210.

Blum, E. K.

[1] *A theory of analytic functions in Banach algebras*, Trans. Amer. Math. Soc. **78** (1955) 343–370.

Bochner, S.

[1] *Vorlesungen über Fouriersche Integrale*, Akad. Verlagsgesellschaft, Leipzig, 1932.

[2] *Completely monotone functions in partially ordered spaces*, Duke Math. J. **9** (1942) 519–526.

[3] *On a theorem of Tannaka and Krein*, Ann. Math. **43** (1942) 56–58.

Bochner, S. and R. S. Phillips

[1] *Absolutely convergent Fourier expansions for non-commutative normed rings*, Ann. Math. **43** (1942) 409–418.

Bohnenblust, H. F. and S. Karlin

[1] *Geometrical properties of the unit sphere of Banach algebras*, Ann. Math. **62** (1955) 217–229.

Bohnenblust, H. F. and A. Sobczyk

[1] *Extensions of functionals on complex linear spaces*, Bull. Amer. Math. Soc. **44** (1938) 91–93.

Bolus, M.

[1] *Sur quelques généralisations du théorème de Stone-Weierstrass*, Thèse doct. Math. Fac. sci. Univ. Nancy, 1957, 26 pages.

Bonsall, F. F.

[1] *A minimal property of the norm in some Banach algebras*, J. London Math. Soc. **29** (1954) 156–164.

[2] *On the representation of cones and semi-algebras with given generators*, Proc. London Math. Soc. **15** (1965) 39–60.

[3] *A survey of Banach algebra theory*, Bull. London Math. Soc. **2** (1970) 257–274.

Bonsall, F. F. and A. W. Goldie

[1] *Algebras which represent their linear functionals*, Proc. Cambridge Phil. Soc. **49** (1953) 1–14.

[2] *Annihilator algebras*, Proc. London Math. Soc. **4** (1954) 154–167.

Borchers, H. J.

[1] *On the structure of the algebra of field operators, I, II*, Nuovo Cimento **24** (1962) 214; Commun. Math. Phys. **1** (1965) 49–56.

[2] *On the vacuum state in quantum field theory, II*, Commun. Math. Phys. **1** (1965) 57–79.

Bourbaki, N.

[1] *Éléments de mathématiques*, XIII, Livre VI, *Intégration*, Hermann, Paris, 1952.

[2] *Elements of mathematics*, Book V, *Topological vector spaces*, IIL, Moscow, 1959 (1953, 1955) (Russian); *Éléments de mathématiques*, Livre V, *Espaces vectoriels topologiques*, Hermann, Paris, 1950.

[3] *Éléments de mathématiques*, Livre III, *Topologie générale*, Hermann, Paris, 1940–1949 (1958). [Chapters I-III and IV-VIII have appeared in Russian translation, 2 volumes.] *General topology*, Addison-Wesley, Reading, 1961.

[4] *General topology. Fundamental structures*, Fizmatgiz, Moscow, 1958 (Russian).

[5] *General topology. Numbers and groups and spaces connected with them*, Fizmatgiz, Moscow, 1959 (Russian).

Bourgin, D. G.

[1] *Approximately isometric and multiplicative transformations on continuous function rings*, Duke Math. J. **16** (1949) 385–397.

Bredon, G. E.

[1] *A new treatment of the Haar integral*, Michigan Math. J. **10** (1963) 365–373.

Brelot, M.

[1] *On the Martin boundary*, Collection of translations "Matematika" **9**, 5 (1965) 136–155 (Russian).

Broconnier, J.
[1] *L'analyse harmonique dans les groupes abéliens, I, II*, Enseign. Math. **2** (1956) 12–41 and 257–273.
Brodsky, M. L.
[1] *On the properties of an operator which maps a nonnegative part of a space with an indefinite metric into itself*, UMN **14** (1959) 147–152 (Russian).
Broise, M.
[1] *Sur les opérateurs unitaires et les fonctions de type positif*, C. R. Acad. Sci. Paris **258** (1964) 2471–2473.
Browder, A.
[1] *Cohomology of maximal ideal spaces*, Bull. Amer. Math. Soc. **76** (1961) 515–516.
[2] *On a theorem of Hoffman and Wermer*, Function Algebras, Proc. Int. Symposium on Function Algebras, Tulane Univ., 1965, 88–89, Scott, Foresman and Co., Palo Alto.
Browder, A. and J. Wermer
[1] *Some algebras of functions on an arc*, J. Math. Mech. **12** (1963) 119–130.
[2] *A method for constructing Dirichlet algebras*, Proc. Amer. Math. Soc. **15** (1964) 546–552.
Bruhat, F.
[1] *Sur certaines représentations unitaires des groupes de Lie semi-simples*, C. R. Acad. Sci. Paris **240** (1955) 2196–2198.
[2] *Sur les représentations induites des groupes de Lie*, Bull. Soc. Math. France **84** (1956) 97–205.
[3] *Distributions sur un groupe localement compact*, Bull. Soc. Math. France **89** (1961) 43–75.
Buck, R.
[1] *Generalized group algebras*, Proc. Nat. Acad. Sci. USA **36** (1950) 747–749.
Calkin, J. W.
[1] *Two-sided ideals and congruences in the ring of bounded operators in Hilbert space*, Ann. Math. **42** (1941) 839–873.
Cambes, F.
[1] *Étude des représentations tracées d'une C*-algèbre*, C. R. Acad. Sci. Paris **262** (1966) 114–117.
Cameron, R. H.
[1] *Analytic functions of absolutely convergent generalized trigonometric sums*, Duke Math. J. **3** (1937) 682–688.
Carleson, L.
[1] *On bounded analytic functions and closure problems*, Arkiv Mat. **2** (1952) 283–291.
[2] *On generators of normed rings*, Skand. Math. Kongr. Lund 1953, **12** (1954) 16–17.
[3] *Representations of continuous functions*, Math. Zeit. **66** (1957) 447–451.
[4] *Interpolation by bounded analytic functions and the Corona problem*, Ann. Math. **76** (1962) 547–559.
[5] *Mergelyan's theorem on uniform polynomial approximation*, Math. Scand. **15** (1964) 167–175.
Cartan, H.
[1] *Sur la mesure de Haar*, C. R. Acad. Sci. Paris **211** (1940) 759–762.
[2] *Idéaux et modules de fonctions analytiques de variables complexes*, Bull. Soc. Math. France **78** (1950) 29–64.
[3] *Seminaires*, Paris, 1951–1952.
Cartan, H. and R. Godement
[1] *Théorie de la dualité et analyse harmonique dans les groupes abéliens localement compacts*, Ann. Sci. École Norm. Sup. **64**, 3 (1947) 79–99.
Cartier, D. and J. Dixmier
[1] *Vecteurs analytiques dans les représentations de groupes de Lie*, Amer. J. Math. **80** (1958) 131–145.
Čech, E.
[1] *On bicompact spaces*, Ann. Math. **38** (1937) 823–844.

CHARLES, B.
[1] *Sur certains anneaux commutatifs d'opérateurs linéaires*, C. R. Acad. Sci. Paris **236** (1953) 990.
[2] *Sur l'algèbre des opérateurs linéaires*, J. Math. Pures et Appl. **33** (1954) 81–145.
CHEVALLEY, C. and O. FRINK
[1] *Bicompactness of cartesian products*, Bull. Amer. Math. Soc. **47** (1941) 612–614.
CIESIELSKY, Z. and Z. SEMADENI
[1] *Przegląd niektórych nowszych metod w teorii potencjala*, I, II, Roczniki Polsk. Tow. Matem., ser. I, Prace Matem. VIII (1967) 147–191, IX (1965) 149–168 (*Survey of some new methods in potential theory*).
CIVIN, P.
[1] *A maximum modulus property of maximal subalgebras*, Proc. Amer. Math. Soc. **10** (1959) 51–54.
CIVIN, P. and B. YOOD
[1] *Regular Banach algebras with a countable space of maximal regular ideals*, Proc. Amer. Math. Soc. **7** (1956) 1005–1010.
CODDINGTON, E. A.
[1] *Some Banach algebras*, Proc. Amer. Math. Soc. **8** (1957) 258–261.
COHEN, P. J.
[1] *On a conjecture of Littlewood and idempotent measures*, Amer. J. Math. **82** (1960) 191–212.
[2] *On homomorphisms of group algebras*, Amer. J. Math. **82** (1960) 213–226.
[3] *A note on constructive methods in Banach algebras*, Proc. Amer. Math. Soc. **12** (1961) 159–163.
CORDESSE, A.
[1] *Représentations factorielles de certains produits semidirect*, C. R. Acad. Sci. Paris **262** (1966) 1109–1112.
CORRELL, E. and M. HENRIKSEN
[1] *On rings of bounded continuous functions with values in a division ring*, Proc. Amer. Math. Soc. **7** (1956) 194–198.
COTLAR, M. and R. RICABARRA
[1] *On the existence of characters in topological groups*, Amer. J. Math. **76** (1954) 375–388.
COURANT, R. and D. HILBERT
[1] *Methods of Mathematical Physics*, Vol. I, Gostekhizdat, Moscow, 1951 (Russian); Interscience, N. Y., 1955; *Methoden der mathematischen Physik, I*, Berlin, 1937.
DANIELL, P. J.
[1] *A general form of integral*, Ann. Math. **19** (1917–1918) 279–294.
[2] *Further properties of the general integral*, Ann. Math. **21** (1919–1920) 203–220.
[3] *The integral and its generalizations*, Rice Inst. Pamphlet **8** (1921) 34–62.
VAN DANTZIG, D.
[1] *Zur topologischen Algebra. I*, Math. Ann. **107** (1932) 587–626.
DARSOW, W. F.
[1] *Positive definite functions and states*, Ann. Math. **60** (1954) 447–453.
DAVIS, C.
[1] *Generators of the ring of bounded operators*, Proc. Amer. Math. Soc. **6** (1955) 970–972.
DELSARTE, J.
[1] *Sur une extension de la formule de Taylor*, J. Math. Pures et Appl. **17** (1938) 213–230.
[2] *Une extension nouvelle de la théorie des fonctions presque-périodiques de Bohr*, Acta Math. **69** (1939) 259.
DENY, Y.
[1] *Sur l'approximation des fonctions harmoniques*, Bull. Soc. Math. France **73** (1945) 71–73.
DHOMBRES, J.
[1] *Caractérisation d'une classe de transformations semi-multiplicatives*, C. R. Acad. Sci. Paris, **264** (1967) A, 113–116.
DIEUDONNÉ, J. A.
[1] *Une généralisation des espaces compactes*, J. Math. Pures et Appl. **23** (1944) 65–76.

[2] *Recent developments in the theory of locally convex vector spaces*, Bull. Amer. Math. Soc. **59** (1953) 495–512.

[3] *Champs de vecteurs non localement triviaux*, Arch. Math. **7** (1956) 6–10.

[4] *Sur la théorie spectrale*, J. Math. Pures et Appl. **35** (1956) 175–187.

DITKIN, V. A.

[1] *Investigation of the structure of ideals in certain normed rings*, Uchen. zap. Mosk. u-ta **30** (1939) 83–130 (Russian).

DIXMIER, J.

[1] *Sur un théorème de Banach*, Duke Math. J. **15** (1948) 1057–1071.

[2] *Mesure de Haar et trace d'un opérateur*, C. R. Acad. Sci. Paris **228** (1949) 152–154.

[3] *Les anneaux d'opérateurs de classe finie*, Ann. École Norm. Sup. **66**, 3 (1949) 209–261.

[4] *Les fonctionnelles linéaires sur l'ensemble des opérateurs bornées d'un espace de Hilbert*, Ann. Math. **51** (1950) 387–408.

[5] *Applications put in this symbol dans les anneaux d'opérateurs*, C. R. Acad. Sci. Paris **230** (1950) 607–609.

[6] *Sur la réduction des anneaux d'opérateurs*, Ann. École Norm. Sup. **68**, 3 (1951) 185–202.

[7] *Sur certains espaces considérés par M. H. Stone*, Sum. Bras. Math. **11** (1951) 151–182.

[8] *Applications put in this symbol dans les anneaux d'opérateurs*, Compositio Math. **10** (1952) 1–55.

[9] *Remarques sur les applications put in this symbol*, Archiv Math. **3** (1952) 290–297.

[10] *Algèbres quasi-unitaires*, Com. Math. Helv. **26** (1952) 275–322.

[11] *Formes linéaires sur un anneau d'opérateurs*, Bull. Soc. Math. France **81** (1953) 9–39.

[12] *Sur les anneaux d'opérateurs dans les espaces hilbertiens*, C. R. Acad. Sci. Paris **238** (1954) 439–441.

[13] *Sous-anneaux abéliens maximaux dans les facteurs de type fini*, Ann. Math. **59** (1954) 279–286.

[14] *Les algèbres d'opérateurs dans l'espace hilbertien* (*Algèbres de von Neumann*), Paris, 1957.

[15] *On unitary representations of nilpotent Lie groups*, Proc. Nat. Acad. Sci. USA **43** (1957) 985–986.

[16] *Sur les représentations unitaires des groupes de Lie nilpotents, II*, Bull. Soc. Math. France **85** (1957) 325–328.

[17] *Sur les C*-algèbres*, Bull. Soc. Math. France **88** (1960) 95–112.

[18] *Sur les structures boréliennes du spectre d'une C*-algèbre*, Publ. math. Inst. hautes études sci. **6** (1960) 5–11.

[19] *Traces sur les C*-algèbres*, Ann. Inst. Fourier **13**, 1 (1963) 219–262.

[20] *Les C*-algèbres et leurs représentations*, Paris, 1964.

DIXMIER, J. and A. DOUADY

[1] *Champs continus d'espaces Hilbertiens et de C*-algèbres*, Bull. Soc. Math. France **91** (1963) 227–284.

DOMAR, Y.

[1] *Harmonic analysis based on certain commutative Banach algebras*, Acta Math. **96** (1956) 1–66.

DOMRACHEVA, G. I.

[1] *Ideals in normal subrings of the ring of continuous functions*, Uchen. map. ped. inst. im. A. I. Gertsena, Leningrad **166** (1958) 29–38 (Russian).

DONOGHUE, W. F.

[1] *The Banach algebra l^1 with an application to linear transformations*, Duke Math. J. **23** (1956) 533–537.

DOPLICHER, S.

[1] *An algebraic spectrum condition*, Comm. math. phys. **1** (1965) 1–5.

DUNCAN, J.

[1] *The continuity of the involution on Banach *-algebras*, J. London Math. Soc., **41** (1966) 701–706.

DUNFORD, N.

[1] *Resolutions of the identity for commutative B*-algebras of operators*, Acta Szeged **12** (1950) 51–56.

[2] *Spectral operators*, Pacif. J. Math. **4** (1954) 321–354.
[3] *A survey of the theory of spectral operators*, Bull. Amer. Math. Soc. **64** (1958) 217–274.
DUNFORD, N. and J. T. SCHWARTZ
[1] *Linear operators, I, II*, IIL, Moscow, 1962; "Mir", 1966 (Russian); Interscience Publishers, Inc., N. Y., 1958, 1963.
DYE, H. A.
[1] *The Radon-Nikodým theorem for finite rings of operators*, Trans. Amer. Math. Soc. **72** (1952) 243–280.
[2] *The unitary structure in finite rings of operators*, Duke Math. J. **20** (1953) 55–69.
[3] *On the geometry of projections in certain operator algebras*, Ann. Math. **61** (1955) 73–89.
DYE, H. A. and R. S. PHILLIPS
[1] *Groups of positive operators*, Canad. J. Math. **8** (1956) 462–486.
EBERLEIN, W. F.
[1] *A note on Fourier-Stieltjes transforms*, Proc. Amer. Math. Soc. **6** (1955) 310–312.
[2] *Spectral theory and harmonic analysis*, Proc. Symp. Spectral Theory and Differential Problems, Oklahoma, 1955, 209–219.
EDWARDS, R. E.
[1] *Multiplicative norms on Banach algebras*, Proc. Cambridge Phil. Soc. **47** (1951) 473–474.
[2] *On functions which are Fourier transforms*, Proc. Amer. Math. Soc. **5** (1954) 71–78.
[3] *Note on two theorems about function algebras*, Mathematika, London **4** (1957) 138–139.
[4] *Bounded functions and Fourier transforms*, Proc. Amer. Math. Soc. **9** (1958) 440–446.
[5] *Bipositive and isometric isomorphisms of some convolution algebras*, Canad. J. Math. **17** (1965) 839–846.
[6] *Operators commuting with translations*, Pacif. J. Math. **16** (1966) 259–265.
EDWARDS, R. E. and E. HEWITT
[1] *Pointwise limits for sequences of convolution operators*, Acta Math. **113** (1965) 181–218.
EFFROS, E. G.
[1] *A decomposition theory for representations of C*-algebras*, Trans. Amer. Math. Soc. **107** (1963) 83–106.
[2] *Order ideals in a C*-algebra and its dual*, Duke Math. J. **30** (1963) 391–411.
[3] *The Borel space of von Neumann algebras on a separable Hilbert space*, Pacif. J. Math. **15** (1965) 1153–1164.
EHRENPREIS, L. and F. MAUTNER
[1] *Some properties of the Fourier transform on semi-simple Lie groups*. I, Ann. Math. **61** (1955) 406–439; II, Trans. Amer. Math. Soc. **84** (1957) 1–55.
[2] *Uniformly bounded representations of groups*, Proc. Nat. Acad. Sci. USA **41** (1955) 231–233.
EIDELHEIT, M.
[1] *On isomorphisms of rings of linear operators*, Studia Math. **9** (1940) 97–105 (Ukrainian summary).
ÈIDLIN, V. L.
[1] *On the topological characteristics of the space of maximal ideals of a Banach algebra*, Vestn. Leningr. un-ta **22** (1967) 173–174 (Russian).
ELLIOTT, R. J.
[1] *Some results in spectral synthesis*, Proc. Cambridge Phil. Soc. **61** (1965) 395–424.
[2] *Two notes on spectral synthesis of discrete Abelian groups*, Proc. Cambridge Phil. Soc. **61** (1965) 617–420.
[3] *Analytic functions on locally convex algebras*, Proc. London Math. Soc. **16** (1966) 321–341.
ERNEST, J. A.
[1] *A decomposition theory for unitary representations of locally compact groups*, Trans. Amer. Math. Soc. **104** (1962) 252–277.
[2] *Notes on the duality theorem of non-commutative noncompact topological groups*, Tôhoku Math. J. **15** (1963) 182–186.
[3] *A new group algebra for locally compact groups. I, II*, Amer. J. Math. **86** (1964) 467–492; Canad. J. Math. **17** (1965) 604–615.
[4] *The representation lattice of a locally compact group*, Illinois J. Math. **10** (1966) 127–133.

[5] *The big group algebra*, Proceedings of the Internat. Congress of Mathematicians, Moscow, 1966, Functional Analysis, 11.

FAGE, M. K.

[1] *Spectral manifolds of a bounded linear operator in Hilbert space*, DAN SSSR **58** (1947) 1609–1612 (Russian).

FANTAPPIÈ, L.

[1] *Calcolo degli autovalori e delle autofunzioni degli operatori osservabili su un gruppo compatto*, Archiv Math. **5** (1954) 292–300.

[2] *Calcolo degli autovalori e autofunzioni degli operatori "fisici" su un gruppo topologico compatto*, Proc. Internat. Congress Math., Amsterdam **2** (1954) 98–100.

FARRELL, R. N.

[1] *Dense algebras of functions in L_p*, Proc. Amer. Math. Soc. **13** (1962) 324–328.

FELDMAN, C.

[1] *The Wedderburn principal theorem in Banach algebras*, Proc. Amer. Math. Soc. **2** (1951) 771–777.

FELDMAN, J.

[1] *Embedding of AW^*-algebras*, Duke Math. J. **23** (1956) 303–307.

[2] *Some connections between topological and algebraic properties in rings of operators*, Duke Math. J. **23** (1956) 365–370.

[3] *Isomorphisms of finite type II rings of operators*, Ann. Math. **63** (1956) 565–571.

[4] *Nonseparability of certain finite factors*, Proc. Amer. Math. Soc. **7** (1956) 23–26.

FELDMAN, J. and J. M. G. FELL

[1] *Separable representations of rings of operators*, Ann. Math. **65** (1957) 241–249.

FELDMAN, J. and R. KADISON

[1] *The closure of the regular operators in a ring of operators*, Proc. Amer. Math. Soc. **5** (1954) 909–916.

FELL, J. M. G.

[1] *Representations of weakly closed algebras*, Math. Ann. **133** (1957) 118–126.

[2] *The structure of algebras of operator fields*, Acta Math. **106** (1961) 233–280.

[3] *The dual spaces of Banach algebras*, Trans. Amer. Math. Soc. **114** (1965) 227–250.

FELL, J. M. G. and M. GARDNER

[1] *Non-unitary dual spaces of semisimple Lie groups*, Proceedings of the Internat. Congr. Math., Moscow, 1966, Functional Analysis, 11.

FELL, J. M. G. and J. L. KELLEY

[1] *An algebra of unbounded operators*, Proc. Nat. Acad. Sci. USA **38** (1952) 592–598.

FELL, J. M. G. and E. THOMA

[1] *Einige Bemerkungen über vollsymmetrische Banachsche Algebren*, Archiv Math. **12**, 1 (1961) 69–70.

FIGÁ-TALAMANCA, A.

[1] *Multipliers of p-integrable functions*, Bull. Amer. Math. Soc. **70** (1967) 666–699.

[2] *Translation invariant operators in L^p*, Duke Math. J. **32** (1965) 495–501.

FIGÁ-TALAMANCA, A. and D. RIDER

[1] *A theorem of Littlewood and lacunary series for compact groups*, Pacif. J. Math. **16** (1966) 505–514.

FILIPPI, M.

[1] *Variété de non-synthèse spectrale sur un groupe abélien localement compact*, Israel J. Math. **3** (1965) 43–60.

FOIAŞ, C.

[1] *Elementi completamente continui e quasi-completamente continui di un'algebra di Banach*, Atti Accad. Naz. Lincie, Rend. Cl. Sci. Fis. Mat. Nat. **20** (1956) 155–160.

FØLNER, E.

[1] *Besicovitch almost periodic functions in arbitrary groups*, Math. Scand. **5** (1957) 47–53.

FORELLI, P.

[1] *Homomorphisms of ideals in group algebras*, Illinois J. Math. **9** (1965) 410–417.

FREUNDLICH, M.
[1] *Completely continuous elements of a normed ring*, Duke Math. J. **16** (1949) 273–283.
FRISCH, J.
[1] *Fonctions analytiques sur un ensemble semianalytique*, C. R. Acad. Sci. Paris, **260**, 11 (1965) 2974–2976.
FUGLEDE, B. and R. V. KADISON
[1] *On a conjecture of Murray and von Neumann*, Proc. Nat. Acad. Sci. USA **37** (1951) 420–425.
[2] *On determinants and a property of a trace in finite factors*, Proc. Nat. Acad. Sci. USA **37** (1951) 425–431.
[3] *Determinant theory in finite factors*, Ann. Math. **55** (1952) 510–530.
[4] *Function Algebras*, Proc. Int. Symposium on Function Algebras, Tulane Univ., 1965.
FUJIWARA, K.
[1] *Sur les anneaux des fonctions continues à support compact*, Math. J. Okayama Univ. **3** (1954) 175–184.
FUKAMIYA, M.
[1] *On B*-algebras*, Proc. Japan Acad. **27** (1951) 321–327.
[2] *On a theorem of Gelfand and Neumark and the B*-algebra*, Kumamoto J. Sci., ser. A, **1** (1952) 17–22.
FUKAMIYA, M., Y. MISONOU and Z. TAKEDA
[1] *On order and commutativity of B*-algebras*, Tôhoku Math. J. **6** (1954) 89–93.
FURSTENBERG, H.
[1] *Translation invariant cones of functions on semi-simple Lie groups*, Bull. Amer. Math. Soc. **71** (1965) 271–326.
GAMELIN, T.
[1] *Restrictions of subspaces of $C(X)$*, Trans. Amer. Math. Soc. **112** (1964) 278–286.
GAMELIN, T. and H. ROSSI
[1] *Jensen measures and algebras of analytic functions*, Function Algebras, Proc. Int. Symposium on Function Algebras, Tulane Univ., 1965, Scott-Foresman and Co., Palo Alto (1966).
GARDNER, L. T.
[1] *On duality for locally compact groups*, Preliminary report, abstract 622–18, Notices Amer. Math. Soc. **12** (1965) 330.
GARNET, J.
[1] *A topological characterization of Gleason parts*, Pacif. J. Math. **30** (1967) 59–63.
GELFAND, I. M.
[1] *On normed rings*, DAN SSSR **23** (1939) 430–432 (Russian).
[2] *On absolutely convergent trigonometric series and integrals*, DAN SSSR **25** (1939) 571–574 (Russian).
[3] *On the ring of almost periodic functions*, DAN SSSR **25** (1939) 575–577 (Russian).
[4] *Normierte Ringe*, Matem. Sbornik **9** (1941) 3–24 (Russian summary).
[5] *Ideale und primäre Ideale in normierten Ringen*, Matem. Sbornik **(51) 9** (1941) 41–48 (Russian summary).
[6] *Zur Theorie der Charaktere der Abelschen topologischen Gruppen*, Matem. Sbornik **(51) 9** (1941) 49–50 (Russian summary).
[7] *Über absolut konvergente trigonometrische Reihen und Integrale*, Matem. Sbornik **(51) 9** (1941) 51–66 (Russian summary).
[8] *Spherical functions on symmetric Riemann spaces*, DAN SSSR **70** (1950) 5–8 (Russian).
[9] *Center of an infinitesimal group ring*, Matem. Sbornik **68** (1950) 103–112 (Russian).
[10] *Certain problems of functional analysis*, UMN **11** (1956) 3–12 (Russian).
[11] *On subrings of rings of continuous functions*, UMN **12** (1957) 249–251 (Russian).
GELFAND, I. M. and M. I. GRAEV
[1] *Analogue to Plancherel's formula for classical groups*, Trudy Mosk. matem. o-va **4** (1955) 375–404 (Russian).
[2] *Traces of unitary representations of the real unimodular group*, DAN SSSR **100** (1955) 1037–1040 (Russian).

[3] *Geometry of homogeneous spaces, representations of groups in homogeneous spaces and problems of integral geometry related to them*, Trudy Mosk. matem. o-va **8** (1959) 321–390; **9** (1960) 562 (Russian).

GELFAND, I. M., M. I. GRAEV, M. A. NAIMARK and D. A. BEREZIN

[1] *Representations of Lie groups*, Zap. 3-go matem. vsesoyuznogo syezda **37** (1956) (Russian).

GELFAND, I. M. and A. N. KOLMOGOROV

[1] *On rings of continuous functions on topological spaces*, DAN SSSR **22** (1939) 11–15 (Russian).

GELFAND, I. M. and M. A. NAIMARK

[1] *On the embedding of normed rings into the ring of operators in Hilbert space*, Matem. Sbornik **12** (**54**) (1943) 197–213 (Russian).

[2] *Unitary representations of the Lorentz group*, J. Phys. **10** (1946) 93–94.

[3] *Unitary representations of the group of linear transformations of the real line*, DAN SSSR **55** (1947) 571–574 (Russian).

[4] *Unitary representations of the Lorentz group*, Izvestiya AN SSSR, ser. matem. **11** (1947) 411–504 (Russian).

[5] *Analogue of the Plancherel formula for the complex unimodular group*, DAN SSSR **63** (1948) 609–612 (Russian).

[6] *Rings with involution and their representations*, Izvestiya AN SSSR, ser. matem., **12** (1948) 445–480 (Russian).

[7] *Unitary representations of the classical groups*, Trudy Matem. in-ta im. V. A. Steklova **36** (1950) 1–228 (Russian).

[8] *Unitary representations of the unimodular group, containing the identity representation of a unitary subgroup*, Trudy Mosk. matem. o-va **1** (1952) 423–475 (Russian).

GELFAND, I. M. and D. A. RAIKOV

[1] *On the theory of characters of commutative topological groups*, DAN SSSR **28** (1940) 195–198 (Russian).

[2] *Irreducible unitary representations of locally compact groups*, Matem. Sbornik **13** (**55**) (1943) 301–316 (Russian).

GELFAND, I. M., D. A. RAIKOV and G. E. SHILOV

[1] *Commutative normed rings*, Fizmatgiz, Moscow, 1960 (Russian); UMN **1**, 2 (**12**) (1946) 48–146 (Russian).

GELFAND, I. M. and G. E. SHILOV

[1] *On various methods of introducing a topology in the set of maximal ideals of a normed ring*, Matem. Sbornik (**51**) **9** (1941) 25–38 (Russian); *Über verschiedene Methoden der Einführung der Topologie in die Menge der maximalen Ideale eines normierten Ringes*, Matem. Sbornik **9** (**51**) (1941) 25–39 (Russian summary).

GELFAND, I. M., N. YA. VILENKIN and M. I. GRAEV

[1] *Integral geometry and problems of representation theory connected with it*, Moscow (1962) (Russian).

GHIKA, A.

[1] *Algebre de transformări liniare și continue ale unui spațiu hilbertian în altul*, Comun. Acad. Rep. Pop. Rom. **7** (1957) 831–834.

GILLMAN, L. and M. HENRIKSEN

[1] *Concerning rings of continuous functions*, Trans. Amer. Math. Soc. **77** (1954) 340–362.

GILLMAN, L., M. HENRIKSEN and M. JERISON

[1] *On a theorem of Gelfand and Kolmogoroff concerning maximal ideals in rings of continuous functions*, Proc. Amer. Math. Soc. **5** (1954) 447–455.

GINZBURG, YU. and I. IOKHVIDOV

[1] *Investigations on the geometry of infinite-dimensional spaces with bilinear metric*, UMN **17**, 4 (1962) 3–56 (Russian).

GLAESER, G.

[1] *Sur le théorème du prolongement de Whitney*, C. R. Acad. Sci. Paris **245** (1957) 617–619.

[2] *Étude de quelques algèbres tayloriennes*, J. d'Analyse Math., Jerusalem, VI (1958) 1–125.

[3] *Multiplicateurs rugueux des fonctions différentielles et synthèse spectrale*, Ann. École Norm. Sup. **79** (1962) 243–261.

GLEASON, A. M.

[1] *A note on locally compact groups*, Bull. Amer. Math. Soc. **55** (1949) 744–745.

[2] *Function algebras*, "Matematika" **5**, 2 (1961) 161–166 (Russian); Seminars on Analytic Functions, II, Institute for Advanced Study, Princeton, 1957.

[3] *Finitely generated ideals in Banach algebras*, J. Math. and Mech. **13** (1964) 125–332.

GLEASON, A. M. and H. WHITNEY

[1] *The extension of linear functionals defined on H^∞*, Pacif. J. Math. **12** (1962) 163–183.

GLICKFELD, B. W.

[1] *Contributions to the theory of holomorphic functions in commutative Banach algebras with identity*, Columbia Univ., 1964.

GLICKSBERG, I.

[1] *Measures orthogonal to algebras and sets of antisymmetry*, Trans. Amer. Math. Soc. **105** (1962) 415–435.

[2] *Function algebras with closed restrictions*, Proc. Amer. Math. Soc. **14** (1963) 158–161.

[3] *On convex hulls of translates*, Pacif. J. Math. **13** (1963) 97–113.

[4] *A remark on analyticity of function algebras*, Pacif. J. Math. **13** (1963) 1181–1185.

[5] *Some uncomplemented function algebras*, Trans. Amer. Math. Soc. **111** (1964) 121–137.

[6] *Maximal algebras and a theorem of Radó*, Pacif. J. Math. **14** (1964) 919–941.

GLICKSBERG, I. and K. DE LEEUW

[1] *Quasi-invariance and analyticity of measures on compact groups*, Acta Math. **109** (1963) 179–205.

GLICKSBERG, I. and J. WERMER

[1] *Measures orthogonal to a Dirichlet algebra*, Duke Math. J. **30** (1963) 661–666.

GLIMM, J.

[1] *A Stone-Weierstrass theorem for C*-algebras*, Ann. Math. **72** (1960) 216–244.

[2] *On a certain class of operator algebras*, Trans. Amer. Math. Soc. **95** (1960) 318–340.

[3] *Type I C*-algebras*, Ann. Math. **73** (1961) 572–612.

GODEMENT, R.

[1] *Extension à un groupe abélien quelconque des théorèmes taubériens de N. Wiener et d'un théorème de A. Beurling*, C. R. Acad. Sci. Paris **223** (1946) 16–18.

[2] *Théorèmes taubériens et théorie spectrale*, Ann. École Norm. Sup. **64** (1948) 119–138.

[3] *Les fonctions de type positif et la théorie des groupes*, Trans. Amer. Math. Soc. **63** (1948) 1–84.

[4] *Théorie générale des sommes continues d'espaces de Banach*, C. R. Acad. Sci. Paris **228** (1949) 1321–1323.

[5] *L'analyse harmonique dans les groupes non abéliens*, Analyse Harmonique, Colloques Intern. **15**, Paris, 1949.

[6] *Some unsolved problems in the theory of group representations*, Proc. Internat. Congress Math., Cambridge **2** (1950) 106–111.

[7] *Sur la théorie des représentations unitaires*, Ann. Math. **53** (1951) 68–124.

[8] *Mémoire sur la théorie des caractères dans les groupes localement compact unimodulaires*, J. Math. Pures et Appl. **30** (1951) 1–110.

[9] *A theory of spherical functions. I*, Trans. Amer. Math. Soc. **73** (1952) 496–556.

[10] Math. Reviews **13** (1952) 11.

[11] *Théorie des caractères, I. Algèbres unitaires*, Ann. Math. **59** (1954) 47–62.

[12] *Théorie des caractères, II. Définition et propriétés générales des caractères*, Ann. Math. **59** (1954) 63–85.

GOKHBERG, I. Ts.

[1] *Criteria for one-sided invertibility of elements of normed rings and their applications*, DAN SSSR **145**, 5 (1962) 971–974 (Russian).

[2] *The problem of factorization in normed rings, functions of cylindrical and symmetric operators and singular integral equations*, UMN **19**, 1 (1964) 71–124 (Russian).

[3] *On a generalization of theorems of M. G. Krein of the Wiener-Levi type*, Mathematical investigations **1**, 1, AN Mold. SSR, Kishinev, 1966, 110–130 (Russian).

GOKHBERG, I. TS. and M. G. KREIN

[1] *Introduction to the theory of linear non-self-adjoint operators in Hilbert space*, "Nauka", Moscow, 1965 (Russian).

GOLDHABER, J. K. and E. S. WOLK

[1] *Maximal ideals in rings of bounded continuous functions*, Duke Math. J. **21** (1954) 565–569.

GOLDMAN, M.

[1] *Structure of AW*-algebras*, Duke Math. J. **23** (1956) 23–34.

GOLEMA, K.

[1] *Free products of compact general algebras*, Colloq. Math. **13**, 2 (1965) 165–166.

GOLODETS, V. YA.

[1] *On the problem of irreducible representations of commutation and anticommutation relations*, UMN **20** (1965) 175–182 (Russian).

[2] *On factor-representations of type II for commutative and anticommutative relations*, UMN **20** (1965) 68–72 (Russian).

[3] *On factor-representations of anticommutation relations*, DAN SSSR **167** (1966) 19–22 (Russian).

GONCHAR, A. A.

[1] *On the minimal boundary of the algebra A(E)*, Izv. AN SSSR, ser. matem. **27**, 4 (1963) 949–955 (Russian).

GORDON, H.

[1] *The maximal ideal space of a ring of measurable functions*, Amer. Math. J. **88** (1966) 827–843.

GORIN, E. A.

[1] *Commutative Banach algebras, generated by the group of unitary elements*, Funktsional'nyi analiz i ego prilozheniya, I, 3 (1967) 86–87 (Russian).

[2] *Spaces of maximal ideals of some normed rings with uniform convergence*, Vestn. Mosk. u-ta, ser. mekh., **3** (1966) 14–19 (Russian).

[3] *Maximal subalgebras of Banach algebras with involution*, Matem. zametki **1** (2) (1967) 161–167 (Russian).

GORIN, E. and V. LIN

[1] *On a condition on the radical of a Banach algebra ensuring strong decomposability*, Matematich. zametki **2**, 6 (1967) 589–592 (Russian).

GRAEV, M. I.

[1] *On a general method of calculating the traces of infinite-dimensional unitary representations of real simple Lie groups*, DAN SSSR **103** (1955) 357–360 (Russian).

[2] *Unitary representations of real simple Lie groups*, UMN **12** (1957) 179–182 (Russian).

[3] *Irreducible unitary representations of third order matrices which leave invariant an indefinite Hermitian form*, DAN SSSR **113** (1957) 966–969 (Russian).

GRAMSCH, B.

[1] *Integration und holomorphe Funktionen in lokalbeschränkte Räumen*, Math. Ann. **162** (1965) 190–210.

[2] *Funktionalkalkül mehrer komplexer Veränderlichen in lokalbeschränkten Algebren*, Habilitationsschrift, Mainz, 1966 (see also Proceedings of the Intern. Congress Math., Moscow, 1966, Functional Analysis, 14).

[3] *Eine Idealstruktur Banachscher Operatorenalgebren*, J. reine und angew. Math. **225** (1967) 97–115.

GREENLEAF, F. P.

[1] *Norm decreasing homomorphisms on group algebras*, Pacif. J. Math. **15** (1965) 1187–1219.

GRIFFIN, E.

[1] *Some contributions to the theory of rings of operators, I, II*, Trans. Amer. Math. Soc. **75** (1953) 471–504; **79** (1955) 389–400.

GROTHENDIECK, A.

[1] *Sur certains espaces de fonctions holomorphes*, J. reine und angew. Math. **192** (1953) 35–64.

[2] *Un résultat sur le dual d'une C*-algèbre*, J. Math. Pures et Appl. **36** (1957) 97–108.

GROVE, L. C.

[1] *A generalized group algebra for compact groups*, Studia Math. **26** (1965) 73–90.

GRUSHIN, V. V.

[1] *On the structure of the closed ideals in the ring of doubly periodic vector-smooth functions*, Vestn. Mosk. u-ta, ser. matem., mekh., astr., **1** (1961) (Russian) 17–23.

GUENIN, M. and B. MISRA

[1] *On the von Neumann algebras generated by field operators*, Nuovo Cimento **30**, 5 (1963) 1272–1290.

GUICHARDET, A.

[1] *Une caractérisation des algèbres de von Neumann de type I*, C. R. Acad. Sci. Paris **248** (1959) 3398–3400.

[2] *Sur un problème posé par G. W. Mackey*, C. R. Acad. Sci. Paris **250**, 6 (1960) 962–963.

[3] *Caractères des algèbres de Banach involutives*, Ann. Inst. Fourier **13**, 1 (1963) 1–81.

[4] *Algèbres de Banach commutatives* (*Séminaire*), Poitiers, 1963–64.

[5] *Sur la decomposition des représentations des C*-algèbres*, C. R. Acad. Sci. Paris **258**, 3 (1964) 768–770.

[6] *Utilisation des sous-groupes distinguées dans l'étude des représentations unitaires des groupes localement compact*, Compositio Math. **17** (1965) 1–35.

[7] *On the tensor product of C*-algebras*, DAN SSSR **160** (1965) 986–989 (Russian).

[8] *Sur la catégorie des algèbres de von Neumann*, Bull. sci. math. **90**, 1–2 (1966) 41–64.

[9] *Produits tensoriels infinis et représentation des relations d'anticommutation*, Ann. École Norm. Sup. **83** (1966) 1–52.

[10] *Analyse harmonique commutative*, Dunod, Paris, 1968.

GULICK, S. L.

[1] *The bidual of a locally multiplicatively-convex algebra*, Pacif. J. Math. **17** (1966) 71–96.

GULICK, S., T. LIN and A. VAN ROOIJ

[1] *Group algebra modules, I, II*, Canad. J. Math. **19**, 1 (1967) 133–150; 151–173.

GUNNING, R. and H. ROSSI

[1] *Analytic functions of several complex variables*, Prentice-Hall, Inc., Englewood Cliffs, N. J., 1965.

GURARY, V. P.

[1] *On primary ideals of the ring* $L_{\varphi}(0,_{\infty})$, Proceedings of the Internat. Congress Math., Moscow, 1966, Functional Analysis, 45 (Russian).

GUREVICH, A.

[1] *Unitary representation in Hilbert space of compact topological groups*, Matem. Sbornik **13 (55)** (1943) 79–86 (Russian).

HAAR, A.

[1] *Der Massenbegriff in der Theorie der kontinuierlichen Gruppen*, Ann. Math. **34** (1933) 147–169.

HAHN, H.

[1] *Über lineare Gleichungssysteme in linearen Räumen*, J. f. d. reine u. angew. Math. **157** (1927) 214–229.

HALMOS, P. R.

[1] *Theory of measure*, IIL, Moscow, 1953 (Russian); Van Nostrand, N. Y., 1950.

HALPERIN, I.

[1] *Von Neumann's arithmetics of continuous rings*, Acta Sci., Szeged **23** (1962) 1–17.

[2] *Transcendental elements in continuous rings*, Canad. J. Math. **14** (1962) 39–44.

HARISH-CHANDRA

[1] *Representations of semisimple Lie groups on a Banach space. I–VI*, Proc. Nat. Acad. Sci. USA **37** (1951) 170–173, 362–365, 366–369, 691–694; **40** (1954) 1076–1077, 1078–1080.

[2] *Plancherel formula for complex semisimple Lie groups*, Proc. Nat. Acad. Sci. USA **37** (1951) 813–818.

[3] *Plancherel formula for the* 2×2 *real unimodular group*, Proc. Nat. Acad. Sci. USA **38** (1952) 337–342.

[4] *Representations of semisimple Lie groups. I, II, III*, Trans. Amer. Math. Soc. **75** (1953) 185–243; **76** (1954) 26–65, 234–253.

[5] *The Plancherel formula for complex semisimple Lie groups*, Trans. Amer. Math. Soc. **76** (1954) 485–528.

[6] *On the Plancherel formula for the right-invariant functions on a semisimple Lie group*, Proc. Nat. Acad. Sci. USA **40** (1954) 200–204.

[7] *On the characters of a semisimple Lie group*, Bull. Amer. Math. Soc. **61** (1955) 389–396.

[8] *Integrable and square-integrable representations of a semisimple Lie group*, Proc. Nat. Acad. Sci. USA **41** (1955) 314–317.

[9] *The characters of semisimple Lie groups*, Trans. Amer. Math. Soc. **83** (1956) 98–163.

[10] *Invariant differential operators on a semisimple Lie algebra*, Proc. Nat. Acad. Sci. USA **42** (1956) 252–253.

[11] *A formula for semisimple Lie groups*, Proc. Nat. Acad. Sci. USA **42** (1956) 538–540.

[12] *Representations of semisimple Lie groups, VI: Integrable and square-integrable representations*, Amer. J. Math. **78** (1956) 564–628.

[13] *Differential operators on a semisimple Lie algebra*, Amer. J. Math. **79** (1957) 87–120.

[14] *Fourier transforms on a semisimple Lie algebra, I–II*, Amer. J. Math. **79** (1957) 193–257, 653–686.

[15] *Spherical functions on a semisimple Lie group*, Proc. Nat. Acad. Sci. USA **43** (1957) 408–409.

HARTMAN, S.

[1] *Quelques remarques sur les expansions de Fourier*, Studia Math. **17** (1957) 200–208.

HAUSDORFF, F.

[1] *Theory of sets*, GTTI, Moscow-Leningrad, 1937 (Russian); Chelsea, N. Y., 1957; *Mengenlehre*, 2nd ed., Berlin-Leipzig, de Gruyter, 1927.

HAUSNER, A.

[1] *Ideals in a certain Banach algebra*, Proc. Amer. Math. Soc. **8** (1957) 246–249.

HEBLE, M. P.

[1] *The semigroup algebra and representations of a compact simple semigroup*, J. Math. Analysis and Appl. **12** (1965) 263–300.

HEIDER, L. J.

[1] *Directed limits on rings of continuous functions*, Duke Math. J. **23** (1956) 293–296.

HELGASON, S.

[1] *The derived algebra of a Banach algebra*, Proc. Nat. Acad. Sci. USA **40** (1954) 994–995.

[2] *Multipliers of Banach algebras*, Ann. Math. **64** (1956) 240–254.

HELSON, H.

[1] *Spectral synthesis of bounded functions*, Arkiv Mat. **1** (1952) 497–502.

[2] *On the ideal structure of group algebras*, Arkiv Mat. **2** (1952) 83–86.

[3] *Isomorphisms of abelian group algebras*, Arkiv Mat. **2** (1953) 475–487.

[4] *Lectures on invariant subspaces*, Academic Press, N. Y.-London, 1964.

[5] *Compact groups with ordered duals*, Proc. London Math. Soc. A, **14** (1965) 144–156.

HELSON, H. and J. P. KAHANE

[1] *Sur les fonctions opérant dans les algèbres de transformées de Fourier de suites ou de fonctions sommables*, C. R. Acad. Sci. Paris, **247** (1958) 626.

HELSON, H. and F. QUIGLEY

[1] *Maximal algebras of continuous functions*, Proc. Amer. Math. Soc. **8** (1957) 111–114.

[2] *Existence of maximal ideals in algebras of continuous functions*, Proc. Amer. Math. Soc. **8** (1957) 115–119.

HENRIKSEN, M.

[1] *On the ideal structure of the ring of entire functions*, Pacif. J. Math. **2** (1952) 179–184.

HERGLOTZ, G.

[1] *Über Potenzreihen mit positivem reellem Teil im Einheitskreis*, Leipz. Berichte **63** (1911) 501–511.

HERSTEIN, I. N.

[1] *Group-rings as *-algebras*, Publ. Math. Debrecen **1** (1950) 201–204.

[2] *Une note sur un article de M. Turumaru*, Portugaliae Math. **12** (1953) 113–114.

[3] *A theorem on rings*, Canad. J. Math. **5** (1953) 238–241.

HERZ, C. S.
[1] *Spectral synthesis of bounded functions*, Trans. Amer. Math. Soc. **94** (1960) 181–232.
[2] *Remarques sur la Note précédente de N. Varopoulos*, C. R. Acad. Sci. Paris **260** (1965) 6001–6004; **262** (1966) 384–387.

HEWITT, E.
[1] *Certain generalizations of the Weierstrass approximation theorem*, Duke Math. J. **14** (1947) 419–427.
[2] *Rings of real-valued continuous functions. I*, Trans. Amer. Math. Soc. **64** (1948) 45–99.
[3] *A note on normed algebras*, Ann. Acad. Brasil Ci. **22** (1950) 171–174.
[4] *Fourier transforms of the class* $\mathfrak{L}_p$, Ark. Mat. **2** (1954) 571–574.
[5] *A survey of abstract harmonic analysis*, in "Some aspects of analysis and probability", *Surveys in Applied Mathematics*, Vol. 4, John Wiley & Sons, N. Y. (1958) 105–168 (Russian translation in "Matematika" **4**, 4 (1960) 75–133).

HEWITT, E. and K. ROSS
[1] *Abstract harmonic analysis*, Vol. I, Springer, Berlin, 1963.
[2] *Abstract harmonic analysis*, Vol. II, Springer, Berlin, 1970.

HEWITT, E. and E. P. WIGNER
[1] *On a theorem of Magnus*, Proc. Amer. Math. Soc. **8** (1957) 740–744.

HEWITT, E. and H. ZUCKERMAN
[1] *The* l_1*-algebra of a commutative semigroup*, Trans. Amer. Math. Soc. **83** (1956) 70–97.

HILLE, E.
[1] *On the theory of characters of groups and semi-groups in normed vector rings*, Proc. Nat. Acad. Sci. USA **30** (1944) 58–60.
[2] *Functional analysis and semi-groups*, IIL, Moscow, 1951 (Russian); New York, AMS Colloq. Publ. Vol. XXI, 1948.
[3] *On roots and logarithms of elements of a complex Banach algebra*, Math. Ann. **136** (1958) 46–57.

HILLE, E. and R. S. PHILLIPS
[1] *Functional analysis and semi-groups*, IIL, Moscow, 1962 (Russian); AMS Colloq. Publ., N. Y., Vol. XXXI, 1957.

HINRICH, L. H.
[1] *Open ideals in* $C(X)$, Pacif. J. Math. **14** (1964) 1255–1263.

HIRSCHFELD, R.
[1] *Sur l'analyse harmonique dans les groupes localement compacts*, C. R. Acad. Sci. Paris **246** (1958) 1138–1140.
[2] *Sur les algèbres de fonctions dans un groupe compact*, C. R. Acad. Sci. Paris **261**, 16 (1965) 3029–3031.

HOCHSCHILD, G.
[1] *On the cohomology groups of an associative algebra*, Ann. Math. **46** (1945) 58–68.

HOFFMAN, K.
[1] *Boundary behavior of generalized analytic functions*, Trans. Amer. Math. Soc. **87** (1958) 447–466.
[2] *A note on the paper of I. J. Schark*, J. Math. and Mech. **10** (1961) 747–748.
[3] *Analytic functions and logmodular Banach algebras*, Acta Math. **108** (1962) 271–317.
[4] *Banach spaces of analytic functions*, IIL, Moscow, 1963 (Russian); Prentice-Hall, Inc., Englewood Cliffs, 1962.
[5] *Lectures on sup norm algebras*, Summer school on topological algebra theory, Bruges, 1966, 1–74.

HOFFMAN, K. and H. ROSSI
[1] *Function theory and multiplicative linear functionals*, Trans. Amer. Math. Soc. **116** (1965) 536–543.

HOFFMAN, K. and I. M. SINGER
[1] *Maximal subalgebras of* $C(\Gamma)$, Amer. J. Math. **79** (1957) 295–305.
[2] *On some problems of Gelfand*, UMN **14**, 3 (1959) 99–114 (Russian).
[3] *Maximal algebras of continuous functions*, Acta Math. **103** (1960) 217–241.

HOFFMAN, K. and J. WERMER
[1] *A characterization of $C(X)$*, Pacif. J. Math. **12** (1962) 941–944.
HOLLADAY, J. C.
[1] *A note on the Stone-Weierstrass theorem for quaternions*, Proc. Amer. Math. Soc. **8** (1957) 656–657.
HONGO, E.
[1] *On quasi-unitary algebras with semi-finite left rings*, Bull. Kyusyu Inst. Technol., Math. Nat. Sci. **3** (1957) 1–10.
HONGO, E. and M. ORIHARA
[1] *A remark on a quasi-unitary algebra*, Yokohama Math. J. **2** (1957) 69–72.
VAN HOVE, L.
[1] *L'ensemble des fonctions analytiques sur un compact en tant qu'algèbre topologique*, Bull. Soc. Math. Belgique (1952) 8–17.
HSIA, TAO-HSING
[1] *On locally bounded topological algebras*, Acta Math. Sinica **14** (1964) 238–251; Chinese Math. **5** (1964) 261–276.
[2] *Positive functionals on algebras*, DAN SSSR **121** (1958) 233–235 (Russian).
[3] *On semi-normed rings with involution*, DAN SSSR **124**, 6 (1959) 1223–1225 (Russian).
[4] *On semi-normed rings with involution*, Izvestiya AN SSSR **23**, 4 (1959) 509–528 (Russian).
HULANICKI, A.
[1] *Groups whose regular representation weakly contains all unitary representations*, Studia Math. **24**, 1 (1964) 37–59.
[2] *On the spectral radius of Hermitian elements in group algebras*, Pacif. J. Math. **18** (1966) 277–287.
IGARI, S.
[1] *Sur les fonctions qui opèrent sur l'anneau de Dirichlet $D(G)$*, Tôhoku Math. J. **17** (1965) 200–205.
INGELSTAM, L.
[1] *Real Banach algebras*, Arkiv Math. **5** (1965) 239–270.
IOKHVIDOV, I. S.
[1] *On the spectra of Hermitian and unitary operators in a space with indefinite metric*, DAN SSSR **71** (1950) 225–228 (Russian).
IOKHVIDOV, I. S. and M. G. KREIN
[1] *Spectral theory of operators in a space with indefinite metric, I, II*, Trudy Mosk. matem. o-va, **5** (1956) 367–432; **8** (1959) 413–496; **15** (1966) 451–454 (Russian).
IONESCU, T. C.
[1] *Fonctions de type positif*, C. R. Acad. Sci. Paris, **243** (1956) 1389–1392.
ISÉKI, K.
[1] *On B*-algebras*, Nederl. Akad. Wetensch. Proc. ser. A **15** (1953) 12–14.
ISHII, T.
[1] *On homomorphisms on the ring of continuous functions onto the real numbers*, Proc. Japan Acad. **33** (1957) 419–423.
ISMAGILOV, R. C.
[1] *On rings of algebras in a space with indefinite metric*, DAN SSSR **171** (1966) 269–271 (Russian).
[2] *Elementary spherical functions on the group $SL(2,P)$ over the field P, which is not locally compact, with respect to the subgroup of matrices with entire elements*, Izvestiya AN SSSR ser. matem. **31** (1967) 361–390 (Russian).
ITÔ, S.
[1] *Positive definite functions on homogeneous spaces*, Proc. Japan Acad. **26** (1950) 17–28.
[2] *Unitary representations of some linear groups. I, II*, Nagoya Math. J. **4** (1952) 1–13; **5** (1953) 79–96.
IWASAWA, K.
[1] *On group rings of topological groups*, Proc. Imp. Acad. Tokyo **20** (1944) 67–70.

JACOBSON, N.
[1] *The theory of rings*, Math. Surveys II, AMS, No. 4, 1943, New York.
[2] *The radical and semi-simplicity for arbitrary rings*, Amer. J. Math. **67** (1945) 300–320.
[3] *A topology for the set of primitive ideals in an arbitrary ring*, Proc. Nat. Acad. Sci. USA **31** (1945) 333–338.
[4] *On the theory of primitive rings*, Ann. Math. **48** (1947) 8–21.

JACOBSON, N. and C. RICKART
[1] *Jordan homomorphisms of rings*, Trans. Amer. Math. Soc. **69** (1950) 479–502.

JEWETT, R. I.
[1] *A variation of the Stone-Weierstrass theorem*, Proc. Amer. Math. Soc. **14** (1963) 690–693.

JOHNSON, B. E.
[1] *A commutative semisimple annihilator Banach algebra which is not dual*, Bull. Amer. Math. Soc. **73** (1967) 407–409.
[2] *The uniqueness of the (complete) norm topology*, Bull. Amer. Math. Soc. **73**, 4 (1967) 537–539.

JOHNSON, B. E. and A. M. SINCLAIR
[1] *Continuity of derivations and a problem of Kaplansky*, Amer. J. Math. **90** (1968) 1067–1073.

JORDAN, P., J. VON NEUMANN and E. WIGNER
[1] *On an algebraic extension of the quantum mechanical formalism*, Ann. Math. **35** (1934) 29–64.

KADISON, R.
[1] *A representation theory for commutative topological algebra*, Memoirs Amer. Math. Soc. **7** (1951), 39 pages.
[2] *Isometries of operator algebras*, Ann. Math. **54** (1951) 325–338.
[3] *A generalized Schwarz inequality and algebraic invariants for operator algebras*, Ann. Math. **56** (1952) 494–503.
[4] *Infinite unitary groups*, Trans. Amer. Math. Soc. **72** (1952) 386–399.
[5] *Infinite general linear groups*, Trans. Amer. Math. Soc. **76** (1954) 66–91.
[6] *On the general linear group of infinite factors*, Duke Math. J. **22** (1955) 119–122.
[7] *Multiplicity theory for operator algebras*, Proc. Nat. Acad. Sci. USA **41** (1955) 169–173.
[8] *On the additivity of the trace in finite factors*, Proc. Nat. Acad. Sci. USA **41** (1955) 385–387.
[9] *On the orthogonalization of operator representations*, Amer. J. Math. **77** (1955) 600–620.
[10] *Report on operator algebras*, Publ. Nat. Acad. Sci., Nat. Res. Council No. **387** (1955) 4–10.
[11] *Irreducible operator algebras*, Proc. Nat. Acad. Sci. USA **43** (1957) 273–276.
[12] *Theory of operators, II: Operator algebras*, Bull. Amer. Math. Soc. **64** (1958) 61–85.
[13] *The trace in finite operator algebras*, Proc. Amer. Math. Soc. **12** (1961) 973–977.

KAHANE, J. P.
[1] *Algebras de convolucion de sucesiones, funciones y medidas sumables*, Cursos y Semin. Mat. Univ. Buenos Aires **6** (1961) 91 pages.
[2] *Algèbres tensorielles et analyse harmonique*, Semin. Bourbaki, Secrèt. math. 1964–1965, 17, 30 (1966) 291-01–291-10.

KAHANE, J. P. and R. SALEM
[1] *Ensembles parfaits et séries trigonométriques*, Hermann, Paris, 1963.

KAKUTANI, S.
[1] *Rings of analytic functions: Lectures on functions of a complex variable*, Univ. Michigan Press, Ann Arbor, Michigan, 1955.

KALLIN, E.
[1] *A nonlocal function algebra*, Proc. Nat. Acad. Sci. USA **49** (1963) 821–824.

KAMOWITZ, G.
[1] *Cohomology groups of commutative Banach algebras*, Trans. Amer. Math. Soc. **102** (1962) 352–372.

KANTOROVITZ, S.
[1] *The annihilator of a closed ideal in a function algebra*, Bull. Res. Council Israel Sect. **9** F (1960) 132–134.

KAPLANSKY, I.

[1] *Topological rings*, Amer. J. Math. **69** (1947) 153–183.

[2] *Topological rings*, Bull. Amer. Math. Soc. **54** (1948) 809–826.

[3] *Dual rings*, Ann. Math. **49** (1948) 689–701.

[4] *Rings with polynomial identity*, Bull. Amer. Math. Soc. **54** (1948) 575–580.

[5] *Regular Banach algebras*, J. Indian Math. Soc. (N. S.) **12** (1948) 57–62.

[6] *Locally compact rings*, Amer. J. Math. **70** (1948) 447–459.

[7] *Groups with representations of bounded degree*, Canad. J. Math. **1** (1949) 105–112.

[8] *Primary ideals in group algebras*, Proc. Nat. Acad. Sci. USA **35** (1949) 133–136.

[9] *Normed algebras*, Duke Math. J. **16** (1949) 309–418.

[10] *Topological representation of algebras, II*, Trans. Amer. Math. Soc. **68** (1950) 62–75.

[11] *Quelques résultats sur les anneaux d'opérateurs*, C. R. Sci. Paris **231** (1950) 485–486.

[12] *Topological algebras*, Proc. Internat. Congress Math., Cambridge **2** (1950) 112–113.

[13] *The structure of certain operator algebras*, Trans. Amer. Math. Soc. **70** (1951) 219–255.

[14] *Locally compact rings, II*, Amer. J. Math. **73** (1951) 20–24.

[15] *Group algebras in the large*, Tôhoku Math. J. **3** (1951) 249–256.

[16] *Projections in Banach algebras*, Ann. Math. **53** (1951) 235–249.

[17] *A theorem on rings of operators*, Pacif. J. Math. **1** (1951) 227–232.

[18] *Symmetry of Banach algebras*, Proc. Amer. Math. Soc. **3** (1952) 396–399.

[19] *Algebras of type I*, Ann. Math. **56** (1952) 460–472,

[20] *Modules over operator algebras*, Amer. J. Math. **75** (1953) 839–858.

[21] *Ring isomorphisms of Banach algebras*, Canad. J. Math. **6** (1954) 374–381.

[22] *Functional analysis*, in "Some aspects of analysis and probability", Surveys in Appl. Math., Vol. 4, J. Wiley & Sons, N. Y. (1958) 3–36.

KASTLER, D.

[1] *The C*-algebras of a free boson field*, Commun. Math. Phys. **1** (1965) 14–48.

KATS, G. I.

[1] *Generalization of the group duality principle*, DAN SSSR **138** (1961) 275–278 (Russian).

[2] *Generalized functions on a locally compact group and decompositions of unitary representations*, UMN **16** (1961) 190–191 (Russian).

[3] *Representations of commutative ring groups*, DAN SSSR **145** (1962) 982–992 (Russian).

[4] *Ring groups and the principle of duality, I, II*, Trudy Mosk. mat. o-va, Volumes 12, 13 (1965).

KATZNELSON, Y.

[1] *Sur les fonctions opérant sur l'algèbre des séries de Fourier absolument convergentes*, C. R. Acad. Sci. Paris **247** (1958) 404–406.

[2] *Algèbres characterisées par les fonctions qui opèrent sur elles*, C. R. Acad. Sci. Paris, **247** (1958) 903–905.

[3] *Sur le calcul symbolique dans quelques algèbres de Banach*, Ann. École Norm. Sup., **76** (1959) 83–123.

[4] *Calcul symbolique dans les algèbres homogènes*, C. R. Acad. Sci. Paris, **254**, 15 (1962) 2700–2702.

KAWADA, Y.

[1] *Über den Dualitätssatz der Charaktere nichtkommutativer Gruppen*, Proc. Physico-Math. Soc. Japan **24** (1942) 97–109.

[2] *Über die Erweiterung der maximalen Ideale in normierten Ringen*, Proc. Imp. Acad. Tokyo **19** (1943) 267–268.

[3] *Über den Operatorenring Banachscher Räume*, Proc. Imp. Acad. Tokyo **19** (1943) 616–621.

[4] *On the group ring of a topological group*, Math. Japonicae **1** (1948) 1–5.

KELDYSH, M. V.

[1] *Sur les suites de polynômes bornés dans leur ensemble*, Matem. Sbornik **42** (1935) 719–724.

[2] *Sur l'approximation des fonctions analytiques dans les domains fermés*, Matem. Sbornik **8** (1940) 137–148.

[3] *On the representation of functions of a complex variable by series of polynomials in closed domains*, Matem. Sbornik **16** (1945) 249–258 (Russian).

KELLEY, J. L.
[1] *Convergence in topology*, Duke Math. J. **17** (1950) 277–283.
[2] *Commutative operator algebras*, Proc. Nat. Acad. Sci. USA **38** (1952) 598–605.
[3] *General topology*, Van Nostrand, Princeton, N. J., 1955; Nauka, Moscow, 1968.
[4] *Duality for compact groups*, Proc. Nat. Acad. Sci. USA **49** (1963) 457–458.
KELLEY, J. L. and A. P. MORSE
[1] *Invariant measures*, Math. Ann. **148** (1962) 98–124.
KELLEY, J. L. and R. L. VAUGHT
[1] *The positive cone in Banach algebras*, Trans. Amer. Math. Soc. **74** (1953) 44–45.
KELLOGG, C. N.
[1] *Centralizers and H*-algebras*, Pacif. J. Math. **17** (1966) 121–129.
KEOWN, E. R.
[1] *Reflexive Banach algebras*, Proc. Amer. Math. Soc. **6** (1955) 252–259.
KHALILOV, Z. I.
[1] *Linear singular equations in normed rings*, DAN SSSR **58** (1947) 1613–1616 (Russian).
KHELEMSKY, A. YA.
[1] *On commutative normed rings with finite-dimensional radical*, Vestn. Mosk. un-ta, ser. matem.-mekh., **6** (1964) 7–16 (Russian).
[2] *Description of annihilator commutative Banach algebras*, DAN SSSR **157**, 1 (1964) 60–62 (Russian).
[3] *Annihilator extensions of commutative Banach algebras*, Izvestiya AN SSSR, ser. matem., **29**, 4 (1965) 945–956 (Russian).
[4] *On an analytic condition on the radical of a commutative Banach algebra and questions of decomposability related to it*, DAN SSSR **167**, 3 (1966) 525–527 (Russian).
KHURGIN, YA. I.
[1] *On subrings of the ring of continuous functions in a circle*, Uch. zap. MGU, ser. matem., **3** (1950) (Russian).
KHURGIN, YA. I. and N. I. SHCHETININ
[1] *On closed subrings of the ring of functions with n continuous derivatives*, DAN SSSR **29** (1940) 288–291 (Russian).
KLEINECKE, D. G.
[1] *On operator commutators*, Proc. Amer. Math. Soc. **8** (1957) 535–536.
KODAIRA, K. and S. KAKUTANI
[1] *Normed ring of a locally compact Abelian group*, Proc. Imp. Acad. Tokyo **19** (1943) 360–365.
KOEHLER, F.
[1] *Note on a theorem of Gelfand and Šilov*, Proc. Amer. Math. Soc. **2** (1951) 541–543.
KOLMOGOROV, A. N.
[1] *Zur Normierbarkeit eines allgemeinen topologischen linearen Raumes*, Studia Math. **5** (1934) 29–33.
KOLMOGOROV, A. N. and S. V. FOMIN
[1] *Elements of the theory of functions and functional analysis*, MGU, Moscow, 1954 (Russian); Graylock Press, Baltimore, 1957; *Introduction to real analysis*, Prentice-Hall, Inc., 1970.
KONDÔ, M.
[1] *Les anneaux des opérateurs sur un espace de S. Banach et quelques problèmes qui s'y rattachent*. I, J. Math. Tokyo **1** (1951) 35–54.
KÖNIG, H.
[1] *Zur abstrakten Theorie der analytischen Funktionen*, I, II, Math. Zeit. **88** (1965) 136–165; Math. Ann. **163** (1966) 9–17.
[2] *Lectures on abstract H^p theory*, Summer school on topological algebra theory, Bruges, 1966, 75–127.
KOOSIS, P.
[1] *An irreducible unitary representation of a compact group is finite dimensional*, Proc. Amer. Math. Soc. **8** (1957) 712–715.

KORENBLYUM, B. I.

[1] *On some special commutative normed rings*, DAN SSSR **64** (1949) 281–284 (Russian).

[2] *On a normed ring of functions with convolution*, DAN SSSR **115** (1957) 226–229 (Russian).

[3] *Generalization of Wiener's Tauberian theorem and harmonic analysis of rapidly growing functions*, Trudy Mosk. matem. o-va **7** (1958) 121–148 (Russian).

KOSHI, SH.

[1] *On Weierstrass-Stone's theorem*, J. Math. Soc. Japan **5** (1953) 351–352.

KOSTYUCHENKO, A. G. and B. S. MITYAGIN

[1] *Positive definite functionals on nuclear spaces*, Trudy Mosk. matem. o-va **9** (1960) 283–316 (Russian).

[2] *Positive definite functionals on nuclear spaces*, UMN **15**, 1 (1960) 244–246 (Russian).

KÖTHE, G.

[1] *Abstrakte Theorie nichtkommutativen Ringe mit einer Anwendung auf die Darstellungstheorie kontinuierlicher Gruppen*, Math. Ann. **103** (1930) 545–572.

[2] *Dualität in der Funktionentheorie*, J. reine angew. Math. **191** (1953) 29–49.

[3] *Topologische lineare Räume*, Springer Verlag, 1960; *Topological linear spaces*,

KOWALSKY, H.

[1] *Beiträge zur topologischen Algebra*, Math. Nachr. **11** (1954) 143–185.

KRABBE, G. L.

[1] *Abelian rings and spectra of operators on l_p*, Proc. Amer. Math. Soc. **7** (1956) 783–790.

[2] *Spectral isomorphisms of some rings of infinite matrices on a Banach space*, Amer. J. Math. **78** (1956) 42–50.

[3] *Spectra of convolution operators on L^p and rings of factor-sequences*, Quart. J. Math. **8** (1957) 1–12.

KREIN, M. G.

[1] *Sur les fonctionnelles positives additives dans les espaces linéaires normées*, Zap. Nauk. dosl. inst. matem. i mekh. KhDU i Kharkivsk. matem. tov. **14** (1937) 227–237 (Ukrainian summary).

[2] *Propriétés fondamentales des ensembles coniques normaux dans l'espace de Banach*, DAN SSSR **28** (1940) 13–17 (Russian summary).

[3] *On a ring of functions, defined on a topological group*, DAN SSSR **29** (1940) 275–280 (Russian).

[4] *On a special ring of functions*, DAN SSSR **29** (1940) 355–359 (Russian).

[5] *On the theory of almost periodic functions on a topological group*, DAN SSSR **30** (1941) 5–8 (Russian).

[6] *On positive functionals on almost periodic functions*, DAN SSSR **30** (1941) 9–12 (Russian).

[7] *Sur une généralisation du théorème de Plancherel au cas des intégrales de Fourier sur les groupes topologiques commutatifs*, DAN SSSR **30** (1941) 482–486 (Russian summary).

[8] *A principle of duality for bicompact groups and quadratic block algebras*, DAN SSSR **69** (1949) 725–728 (Russian).

[9] *Hermitian-positive kernels on homogeneous spaces, I, II*, Ukr. matem. zh. **1** (1949) 64–98; **2** (1950) 10–59 (Russian).

[10] *Integral equations on semigroups with kernel depending on the difference of arguments*, UMN **13**, 5 (1958) 3–120 (Russian).

[11] *On some new Banach algebras and theorems of Wiener-Levi type for Fourier series and integrals*, Matematicheskie issledovaniya 1 : 1, AN Mold. SSSR, Kishinev (1966) 82–109 (Russian).

KREIN, M. G. and D. MILMAN

[1] *On extreme points of regularly convex sets*, Studia Math. **9** (1940) 133–138 (Ukrainian summary).

KREIN, M. G. and M. RUTMAN

[1] *Linear operators which leave invariant a cone in Banach space*, UMN **3** **(23)** (1948) 3-95 (Russian); AMS Translation No. 26 (1950).

KREIN, M. G. and V. SHMULYAN

[1] *On regularly convex sets in the space conjugate to a Banach space*, Ann. Math. **41** (1940) 556–583.

KURATOWSKI, K.
[1] *Topology*, Vol. I, "Mir", Moscow, 1965 (Russian); *Topologie*, 3rd ed., Warsaw, 1952; English ed., Academic Press, N. Y., 1966.
[2] *Wstęp do teorii mnogości i topologii*, Państwowe Wydawnictwo Naukowe, Warsaw, 1955; *Introduction to set theory and topology*, PWN, Warsaw, 1961.

LAMPERTI, J.
[1] *On the isometries of certain function spaces*, Pacif. J. Math. **8** (1958) 459–466.

LANGER, H.
[1] *On J-Hermitian operators*, DAN SSSR **134** (1960) 263–266 (Russian).
[2] *Zur Spektraltheorie J-selbstadjungierter Operatoren*, Math. Ann. **146** (1962) 60–85.

LARSON, R. G.
[1] *Commutative Hopf algebras*, Canad. J. Math. **19** (1967) 350–360.

LAVRENTEV, M. A.
[1] *Sur les fonctions d'une variable complex représentables par des séries de polynômes*, Hermann, Paris, 1936.

LAX, P.
[1] *Translation invariant subspaces*, Acta Math. **101** (1959) 163–178.

DE LEEUW, K.
[1] *A type of convexity on the space of n complex variables*, Trans. Amer. Math. Soc. **83** (1956) 193–204.
[2] *Functions of circular subsets of the space of n complex variables*, Duke Math. J. **24** (1957) 415–432.
[3] *The isometries of H^1*, mimeographed notes, Stanford, 1960.

DE LEEUW, K. and C. HERZ
[1] *An invariance property of spectral synthesis*, Illinois J. Math. **9** (1965) 220–229.

DE LEEUW, K. and Y. KATZNELSON
[1] *Functions that operate on non self-adjoint algebras*, Iton leanalisa matematut. J. analyse math., 1963, 11.

DE LEEUW, K. and H. MIRKILL
[1] *Intrinsic algebras on the torus*, Trans. Amer. Math. Soc. **81** (1956) 320–330.

DE LEEUW, K. and W. RUDIN
[1] *Extreme points and extremum problems in H^1*, Pacif. J. Math. **8** (1958) 467–485.

DE LEEUW, K., W. RUDIN and J. WERMER
[1] *The isometries of some function spaces*, Proc. Amer. Math. Soc. **11** (1960) 694–698.

LEIBENZON, Z. L.
[1] *On the ring of continuous functions on a circle*, UMN **7**, 4 (**50**) (1952) 163–164 (Russian).
[2] *On the ring of functions with absolutely convergent Fourier series*, UMN **9**, 3 (1954) 157–162 (Russian).
[3] *On homomorphisms of the rings $A(\alpha_n)$*, UMN **20**, 4 (1965) 201–203 (Russian).

LEPTIN, H.
[1] *Reduktion linearer Funktionale auf Operatorenringen*, Abh. Math. Sem. Univ. Hamburg **22** (1958) 98–113.
[2] *Zur Reduktionstheorie Hilbertschen Räume*, Math. Zeit. **69** (1958) 40–58.

LERAY, J.
[1] *Fonction de variable complexe. Sa représentation comme somme de puissances négatives de fonctionnelles linéaires*, R. C. Acad. Lincei (1956) 589–590.

LERESCHE, G.
[1] *Algèbres d'opérateurs non bornés sur un espace de Hilbert*, Comment. math. helv. **40**, 4 (1966) 281–324.

LEVITAN, B. M.
[1] *Normed rings generated by a generalized translation operation*, DAN SSSR **47** (1945) 3–6 (Russian).
[2] *A theorem on the representation of positive definite functions for a generalized translation operation*, DAN SSSR **47** (1945) 163–165 (Russian).

[3] *Plancherel's theorem for a generalized translation operation*, DAN SSSR **47** (1945) 323–326 (Russian).
[4] *Duality theorem for a generalized translation operation*, DAN SSSR **47** (1945) 401–403 (Russian).
[5] *A generalization of the operation of translation and infinite hypercomplex systems, I, II*, Matem. Sbornik **16 (58)** (1945) 259–280; **17** (1945) 9–44 and 163–192 (Russian summaries).
[6] *Rings of operators and the generalized translation operation*, DAN SSSR **52** (1946) 99–102 (Russian).
[7] *Concerning the theory of unitary representations of locally compact groups*, Matem. Sbornik **19 (61)** (1946) 407–428 (Russian).
[8] *Correction to the paper "A generalization of the operation of translation and infinite hypercomplex systems"*, Matem. Sbornik **24 (66)** (1949) 501–502 (Russian).
[9] *Application of generalized translation operators to linear differential equations of the second order*, UMN **4**, 1 (1949) 3–112 (Russian).
[10] *Generalized translation operators and some of their applications*, Moscow, 1962 (Russian).

LEVITAN, B. M. and A. YA. POVZNER
[1] *Sturm-Liouville differential equations on the half-axis and Plancherel's theorem*, DAN SSSR **52** (1946) 483–486 (Russian).

LÉVY, P.
[1] *Sur la convergence absolue des séries de Fourier*, Compositio Math. **1** (1934) 1–14.

LITTLEWOOD, D. E.
[1] *The theory of group characters and matrix representations of groups*, Oxford Univ. Press, N. Y., 1940.

LJASTAD, O.
[1] *Uniform approximation in various function spaces*, Math. Scand. **14**, 1 (1964) 5–18.

LOOMIS, L. H.
[1] *Haar measure in uniform structures*, Duke Math. J. **16** (1949) 193–208.
[2] *An introduction to abstract harmonic analysis*, IIL, Moscow, 1956 (Russian); Van Nostrand, N. Y., 1953.
[3] *The lattice theoretic background of the dimension theory of operator algebras*, Memoirs Amer. Math. Soc. **18** (1955) 36 pages.

LOOMIS, L. H. and D. WIDDER
[1] *The Poisson integral representation of functions which are positive and harmonic in a half-plane*, Duke Math. J. **9** (1942) 643–645.

LORCH, E. R.
[1] *The spectrum of linear transformations*, Trans. Amer. Math. Soc. **52** (1942) 238–248.
[2] *The theory of analytic functions in normed abelian vector rings*, Trans. Amer. Math. Soc. **57** (1943) 414–425.
[3] *The structure of normed abelian rings*, Bull. Amer. Math. Soc. **50** (1944) 447–463.
[4] *L'integrazione e gli ideali massimi*, Rend. Sem. Mat. Univ., Torino, **13** (1953/54) 33–38.
[5] *Normed rings—the first decade*, Proc. Symp. Spectral Theory and Differential Problems, Oklahoma (1955) 249–258.

LOUPIAS, G. and S. MIRACLE-SOLE
[1] *C*-algèbres des systèmes canoniques. I*, Comm. Math. Phys. **2** (1966) 31–48.

LOWDENSLAGER, D. B.
[1] *Potential theory in bounded symmetric homogeneous complex domains*, Ann. Math. **67** (1958) 467–484.

LUMER, G.
[1] *The range of the exponential function*, Publ. Inst. Math. y Estadist. Fac. Ing. y Agrimens **3** (1957) 53–55.
[2] *Commutadores en álgebras de Banach*, Publ. Inst. Math. y Estadist. Fac. Ing. y Agrimens **3** (1957) 57–63.
[3] *Semi-inner product spaces*, Trans. Amer. Math. Soc. **100** (1961) 29–43.
[4] *Analytic functions and the Dirichlet problem*, Bull. Amer. Math. Soc. **70** (1964) 98–104.
[5] *Herglotz transformation and H^p theory*, Bull. Amer. Math. Soc. **71** (1965) 725–730.

[6] *Intégralité uniforme dans es algèbres de fonctions classes H^p et classe the Hardy universelle*, C. R. Acad. Sci. Paris, **262** (1966) 1046–1049.

LUZIN, N. N. and I. I. PRIVALOV

[1] *Sur l'unicité et la multiplicité des fonctions analytiques*, Ann. Sci. École Norm. Sup. **42** (1925).

LYANTSE, V. È.

[1] *On a generalization of the concept of spectral measure*, Matem. Sbornik **61** (1961) 80–120 (Russian).

[2] *On a generalization of the concept of a spectral operator*, DAN SSSR **142**, 2 (1962) 278–281 (Russian).

LYAPUNOV, A. A.

[1] *Problème général de la stabilité du mouvement*, Annals Math. Studies, Princeton, N. J., 1949.

LYUBARSKY, G. YA.

[1] *Harmonic analysis on a topological manifold with transitive group*, Dissertation, Kazan, 1945 (Russian).

LYUSTERNIK, L. A., and V. I. SOBOLEV

[1] *Elements of functional analysis*, "Nauka", Moscow, 1965 (Russian); F. Ungar, New York, 1960; *Elemente der Funktionalanalysis*, Akademie-Verlag, Berlin, 1955.

MAAK, W.

[1] *Darstellungstheorie unendlicher Gruppen und fastperiodische Funktionen*, Enzyklopädie Math. Wiss., Bd. I, 1. Teil, Heft 7, Artikel 16, B. G. Teubner, Leipzig, 1953.

[2] *Fastperiodische Funktionen auf der Modelgruppe*, Math. Scand. **3** (1955) 44–48.

MACDOWELL, R.

[1] *Banach spaces and algebras of continuous functions*, Proc. Amer. Math. Soc. **6** (1955) 67–78.

MACKEY, G. W.

[1] *The Laplace transform for locally compact Abelian groups*, Proc. Nat. Acad. Sci. USA **34** (1948) 156–162.

[2] *Imprimitivity for representations of locally compact groups*, I, Proc. Nat. Acad. Sci. USA **35** (1949) 537–545.

[3] *Imprimitivité pour les représentations des groupes localement compacts.* II: *Nombres d'entrelacement pour les représentations imprimitives*; III: *Produits de Kronecker et nombres d'entrelacement fortes*, C. R. Acad. Sci. Paris **230** (1950) 808–809, 908–909.

[4] *Functions on locally compact groups*, UMN **8**, 4 (**56**) (1953) 95–129 (Russian); Bull. Amer. Math. Soc., **56** (1950) 385–412.

[5] *On induced representations of groups*, Amer. Journ. Math. **73** (1951) 576–592.

[6] *Induced representations of locally compact groups, I, II: The Frobenius reciprocity theorem*, Ann. Math. **55** (1952) 101–139; **58** (1953) 193–231.

[7] *Borel structure in groups and their duals*, Trans. Amer. Math. Soc. **85** (1957) 134–165.

[8] *Unitary representations of group extensions, I*, Acta Math. **99** (1958) 265–311.

[9] *Infinite-dimensional representations of groups*, Matematika **6**, 6 (1962) 57–103 (Russian).

[10] *Infinite-dimensional group representations*, Colloquium lecture given at Stillwater, Oklahoma, August 29–September 1, 1961 (66th summer meeting of the AMS).

[11] *Imprimitivity for representations of locally compact groups*, I, Proc. Nat. Acad. Sci. USA **35** (1949) 537–545.

[12] *On induced representations of groups*, Amer. J. Math. **73** (1951) 576–592.

[13] *Induced representations of locally compact groups*, I, II: *The Frobenius reciprocity theorem*, Ann. Math. **55** (1952) 101–139; **58** (1953) 193–231.

[14] *Borel structure in groups and their duals*, Trans. Amer. Math. Soc. **85** (1957) 134–165.

MAEDA, F.

[1] *Relative dimensionality in operator rings*, J. Sci. Hiroshima Univ. **11** (1941) 1–6.

MAEDA, SH.

[1] *Lengths of projections in rings of operators*, J. Sci. Hiroshima Univ., A, **20** (1956) 5–11.

MALGRANGE, B.
[1] *Ideals of differentiable functions*, Tata Inst., Bombay, 1966.
MALLIAVIN, P.
[1] *Théorèmes d'adhérence pour certaines séries de Dirichlet. Procédés d'extrapolation en analyse fonctionnelle*, C. R. Acad. Sci. Paris **239** (1954) 20–22.
[2] *Sur l'impossibilité de la synthèse spectrale sur la droit*, C. R. Acad. Sci. Paris **248** (1959) 2155–2157.
MARKOV, A. A.
[1] *On mean values and exterior densities*, Matem. Sbornik **4**, 1 (**46**) (1938) 165–191 (Russian summary).
MARKUSHEVICH, A. I.
[1] *Theory of analytic functions*, Gostekhizdat, Moscow-Leningrad, 1950 (Russian).
MARKUSHEVICH, L. A.
[1] *Structure of rings of continuous functions on the circle with two generators*, DAN SSSR **132**, 6 (1960) 1265–1269 (Russian).
[2] *Approximation by polynomials of continuous functions on Jordan arcs in the space of n complex variables*, Sib. matem. zh. **2**, 2 (1961) 260–273 (Russian).
MASANI, P. and N. WIENER
[1] *The prediction theory of multivariate stochastic processes, I, II*, Acta Math. **98** (1957) 111–150, **98** (1958) 93–137.
MATAGNE, R.
[1] *Les espaces de Silva*, Bull. Soc. Roy. Sci. Liège (1964) 754–768.
MATÉ, L.
[1] *Embedding multiplier operators of a Banach algebra B into its second conjugate space B***, Bull. Acad. Polon. Sci. Ser. Math. Astr. Phys. **13**, 11–12 (1965) 809–812.
MATSUSHITA, S.
[1] *Positive linear functionals in self-adjoint B-algebras*, Proc. Japan Acad. **29** (1953) 427–430.
[2] *Analyse harmonique dans les groupes localement compacts*, I-II, C. R. Acad. Sci. Paris **237** (1953) 955–957, 1056–1057.
[3] *Über einen Satz von K. Iwasawa*, J. Inst. Polytechn. Osaka City Univ., A, **4** (1953) 59–61.
[4] *Plancherel's theorem on general locally compact groups*, J. Inst. Polytechn. Osaka City Univ., A, **4** (1953) 63–70.
[5] *Sur le théorème de Plancherel*, Proc. Japan Acad. **30** (1954) 557–561.
[6] *Positive functionals and representation theory on Banach algebras, I*, J. Inst. Polytech. Osaka City Univ., A, **6** (1955) 1–18.
MATTHES, K.
[1] *Über eine Verallgemeinerung eines Satzes von Gelfand und Kolmogoroff*, Math. Nachr. **15** (1956) 117–121.
MAURIN, K.
[1] *Allgemeine Eigenfunktionsentwicklungen Spektraldarstellung abstrakter Kerne, Eine Verallgemeinerung der Distributionen auf Lie'schen Gruppen*, Bull. Acad. Polon. Sci. Ser. Math. Astr. Phys. **7**, 8 (1959) 471–479.
[2] *Allgemeine Eigenfunktionsentwicklungen, unitäre Darstellungen lokalkompakter Gruppen und automorphe Funktionen*, Math. Ann. **165** (1966) 204–222.
[3] *Methods of Hilbert space* "Mir", Moscow, 1965 (Russian).
MAURIN, K. and L. MAURIN
[1] *Universelle umhüllende Algebra einer l.k. Gruppe und ihre s.a. Darstellungen*, Anw. Studia Math. **24** (1964) 227–253.
[2] *Automorphic forms, spectral theory and group representations*, Bull. Acad. Polon. Sci. Sér. Math. Astro. Phys. **13** (1965) 199–203.
MAUTNER, F. I.
[1] *The completeness of the irreducible unitary representations of a locally compact group*, Proc. Nat. Acad. Sci. USA **34** (1948) 52–54.
[2] *Unitary representations of locally compact groups*, I, II, Ann. Math. **51** (1950) 1–25; **52** (1950) 528–556.

[3] *On the decomposition of unitary representations of Lie groups*, Proc. Amer. Math. Soc. **2** (1951) 490–496.

[4] *A generalization of the Frobenius reciprocity theorem*, Proc. Nat. Acad. Sci. USA **37** (1951) 431–435.

[5] *Fourier analysis and symmetric spaces*, Proc. Nat. Acad. Sci. USA **37** (1951) 529–533.

[6] *Induced representations*, Amer. J. Math. **74** (1952) 737–758.

[7] *Geodesic flows and unitary representations*, Proc. Nat. Acad. Sci. USA **40** (1954) 33–36.

[8] *Note on the Fourier inversion formula for groups*, Trans. Amer. Math. Soc. **78** (1955) 371–384.

MAZUR, S.

[1] *Sur les anneaux linéaires*, C. R. Acad. Sci. Paris, **207** (1938) 1025–1027.

McDUFF, D.

[1] *On the structure of II_1-factors*, Uspekhi mat. nauk, **6** (1970) 29–51 (Russian).

McKISSICK, R.

[1] *A non trivial normal sup norm algebra*, Bull. Amer. Math. Soc. **69** (1963) 391–395.

MERGELYAN, S. N.

[1] *On the representation of functions by series of polynomials on closed sets*, DAN SSSR **78** (1951) 405–408 (Russian).

[2] *Uniform approximation of functions of a complex variable*, UMN **7**, 2 (1952) 31–128 (Russian); AMS Translation **101** (1954).

[3] *Appendix to the book by Walsh "Interpolation and approximation by rational functions in the complex domain"*, Moscow, 1961 (Russian).

MEYER, M. Y.

[1] *Prolongement des multiplicateurs d'idéaux fermés de $L^1(R^n)$*, C. R. Acad. Sci. Paris **262** (1966) 744–745.

MICHAEL, E. A.

[1] *Locally multiplicatively-convex topological algebras*, Memoirs Amer. Math. Soc. **11** (1952) 79 pages.

MIKUSIŃSKI, J. G.

[1] *L'anneau algébrique et ses applications dans l'analyse fonctionnelle, I, II*, Ann. Univ. Mariae Curie-Skłodowska, Sec. A **2** (1947) 1–48; **3** (1949) 1–84 (Polish summary).

[2] *Sur les fondements du calcul opératoire*, Studia Math. **11** (1950) 41–70.

[3] *Une nouvelle justification du calcul de Heaviside*, Atti Lincei, Fis. Mat. Nat. (8), Ser. **1**, 2 (1950) 113–121.

[4] *Operator calculus*, IIL, Moscow, 1956 (Russian); *Rachunek operatorów*, Warszawa, 1953.

MILLER, J. B.

[1] *Generalized-function calculus for the Laplace transformation*, Arch. Rat. Mech. and Analys. **12** (1963) 409–419.

[2] *Averaging and Reynolds Operators in Banach Algebras, I. Representation by Derivations and Antiderivations*, J. Math. Analysis and Appl. **17** (1966) 527–548.

MILMAN, D. P.

[1] *On the normability of topological rings*, DAN SSSR **47** (1945) 162–164 (Russian).

[2] *Characterization of extremal points of regularly convex sets*, DAN SSSR **57** (1947) 119–122 (Russian).

[3] *Accessible points of a functional compactum*, DAN SSSR **59** (1948) 1045–1048 (Russian).

[4] *On the theory of rings with involution*, DAN SSSR **76** (1951) 349–352 (Russian).

[5] *On integral representations of functions of many variables*, DAN SSSR **87** (1952) 9–10 (Russian).

MISONOU, Y.

[1] *On weakly central operator algebra*, Tôhoku Math. J. **4**, 2 (1952) 194–202.

[2] *Operator algebras of type I*, Kōdai Math. Sem. Reports, No. 3 (1953) 87–90.

[3] *Unitary equivalence of factors of type III*, Proc. Japan Acad. **29** (1953) 482–485.

[4] *On the direct product of W^*-algebras*, Tôhoku Math. J. **6** (1954) 189–204.

[5] *Generalized approximately finite W^*-algebras*, Tôhoku Math. J. **7** (1955) 192–205.

[6] *On divisors of factors*, Tôhoku Math. J. **8** (1956) 63–69.

MISONOU, Y. and M. NAKAMURA

[1] *Centering of an operator algebra*, Tôhoku Math. J. **3**, 2 (1951) 243–248.

MITYAGIN, B. S.

[1] *Maximal ideals of some normed rings*, Sib. matem. zh. **2** (1960) (Russian) 82–88.

MITYAGIN, B. S., S. ROLEWICZ and W. ŻELAZKO

[1] *Entire functions in B_0-algebras*, Studia Math. **21** (1962) 291–306.

MIYANAGA, Y.

[1] *A note on Banach algebras*, Proc. Japan Acad. **32** (1956) 176.

MLAK, W.

[1] *Representations of some algebras of generalized analytic functions*, Bull. Acad. Polon. Sci. math. astr. phys. **13** (1965) 211–214.

MOCHIZUKI, N.

[1] *A note on the Choquet boundary of a restricted function algebra*, Tôhoku Math. J. **18** (1966) 316–317.

MOLLIOS, A.

[1] *Heredity of tensor products of topological algebra*, Math. Ann. **162** (1966) 246–257.

MOORE, E. H. and H. L. SMITH

[1] *A general theory of limits*, Amer. J. Math. **44** (1922) 102–121.

MOSTOW, G. D.

[1] *On the L^2-space of a Lie group*, Amer. J. Math. **74** (1952) 920–928.

MULLINS, R. E.

[1] *The essential set of function algebras*, Proc. Amer. Math. Soc. **18** (1967) 271–273.

MÜNTZ, C.

[1] *Über den Approximationssatz von Weierstrass*, H. A. Festschrift, Berlin, 1914.

MURRAY, F. J.

[1] *Linear transformations between Hilbert spaces and the application of this theory to linear partial differential equations*, Trans. Amer. Math, Soc. **37** (1935) 301–338.

MURRAY, F. J. and J. VON NEUMANN

[1] *On rings of operators. I*, Ann. Math. **37** (1936) 116–229; *II*, Trans. Amer. Math. Soc. **41** (1937) 208–248; *IV*, Ann. Math. **44** (1943) 716–808.

MYERS, S. B.

[1] *Algebras of differentiable functions*, Proc. Amer. Math. Soc. **5** (1954) 917–922.

NACHBIN, L.

[1] *A generalization of Whitney's theorem on ideals of differentiable functions*, Proc. Nat. Acad. Sci. USA **43** (1957) 935–937.

NAGASAWA, M.

[1] *Isomorphisms between commutative Banach algebras with an application to rings of analytic functions*, Kōdai Math. Sem. Rep. **11** (1959).

NAGUMO, M.

[1] *Einige analytische Untersuchungen in linearen metrischen Ringen*, Japan J. Math. **13** (1936) 61–80.

NAIMARK, M. A.

[1] *Positive definite operator functions on a commutative group*, Izvestiya Akad. Nauk SSSR, ser. matem. **7** (1943) 237–244 (Russian).

[2] *Rings with involution*, UMN **3** (1948) 52–145 (Russian); AMS Translation **25** (1950) 131 pp.

[3] *Rings of operators in Hilbert space*, UMN **4** (1949) 83–147 (Russian).

[4] *On a problem in the theory of rings with involution*, UMN **6** (1951) 162–164 (Russian).

[5] *Description of all irreducible unitary representations of classical groups*, DAN SSSR **84** (1952) 883–886 (Russian).

[6] *On irreducible linear representations of the proper Lorentz group*, DAN SSSR **97** (1954) 969–972 (Russian).

[7] *On the continuous analogue of the Schur lemma*, DAN SSSR **98** (1954) 185–188 (Russian).

[8] *Linear representations of the Lorentz group*, UMN **9**, 4 (**62**) (1954) 19–63 (Russian); AMS Translation (1957) 379–458.

[9] *On the description of all unitary representations of the complex classical groups, I*, Matem. Sbornik **35** (**77**) (1954) 317–356 (Russian).

[10] *Normed rings*, Gostekhizdat, Moscow, 1956 (Russian); P. Noordhoff N.V., Groningen, 1959, 1964.

[11] *On irreducible linear representations of a complete Lorentz group*, DAN SSSR **112** (1957) 583–586 (Russian).

[12] *Linear representations of the Lorentz group*, Fizmatgiz, Moscow, 1958 (Russian). Macmillian, N. Y. (1964) (English); Dunod, Paris (1962) (French).

[13] *On the decomposition of the tensor product of the representations of the principal series of the proper Lorentz group into irreducible representations*, DAN SSSR **119** (1958) 872–876 (Russian).

[14] *On the decomposition of irreducible representations of the principal series of the nth order complex unimodular group into representations of the second order complex unimodular group*, DAN SSSR **121** (1958) 590–593 (Russian).

[15] *Decomposition of the tensor product of irreducible representations of the proper Lorentz group into irreducible representations*, Trudy Mosk. matem. o-va **8** (1959) 121–153; **9** (1960) 237–282; **10** (1961) 181–216 (Russian).

[16] *On factor-representations of a locally compact group*, DAN SSSR **134**, 2 (1960) 275–277 (Russian).

[17] *On the decomposition into factor-representations of a unitary representation of a locally compact group*, Sib. matem. zh. **2**, 1 (1961) 89–99 (Russian).

[18] *Lineare Darstellungen der Lorentzgruppe*, VEB Deutscher Verlag der Wissenschaften, Berlin, 1963 (German edition of [12]).

[19] *On the structure of factor-representations of a locally compact group*, DAN SSSR **148** (1963) 775–778 (Russian).

[20] *On commuting unitary operators in spaces with indefinite metric*, Acta Sci. Szeged **24**, 3–4 (1963) 177–189.

[21] *Infinite-dimensional representations of groups and related problems*, Itogi nauki, AN SSSR, Inst. nauchn. informatsii, Moscow, 1964 (Russian).

[22] *Kommutative symmetrische operatoren Algebra in Pontryaginschen Räumen Π_κ*, Math. Ann. **162** (1965) 147–171.

[23] *On unitary group representations in spaces with indefinite metric*, Acta Sci. Szeged **26**, 3–4 (1965) 201–209.

[24] *On the structure of unitary representations of locally compact groups and symmetric representations of algebras in the Pontryagin spaces Π_κ*, Izv. AN SSSR, ser. matem., **30**, 5 (1966) 1111–1132 (Russian).

[25] *Degenerate algebras of operators in the Pontryagin space Π_κ*, Izv. AN SSSR, ser. matem., **30** (1966) 1229–1256 (Russian).

NAIMARK, M. A. and S. V. FOMIN

[1] *Continuous direct sums of Hilbert spaces and some of their applications*, UMN **10** (1955) 111–142 (Russian).

NAKAMURA, M.

[1] *The two-sided representations of an operator algebra*, Proc. Japan Acad. **27** (1951) 172–176.

[2] *On the direct product of finite factors*, Tôhoku Math. J. **6** (1954) 205–207.

NAKAMURA, M. and Z. TAKEDA

[1] *The Radon-Nikodým theorem of traces for a certain operator algebra*, Tôhoku Math. J. **4** (1952) 275–283.

[2] *Normal states of commutative operator algebras*, Tôhoku Math. J. **5** (1953) 109–121.

[3] *On some elementary properties of the crossed products of von Neumann algebras*, Proc. Japan Acad. **34** (1958) 489–494.

[4] *On certain examples of the crossed product of finite factors I-II*, Proc. Japan Acad. **34** (1958) 495–499, 500–502.

[5] *A Galois theory for finite factors*, Proc. Japan Acad. **36** (1960) 258–260.

[6] *On the fundamental theorem of the Galois theory for finite factors*, Proc. Japan Acad. **36** (1960) 313–318.

NAKAMURA, M., M. TAKESAKI and H. UMEGAKI
[1] *A remark on the expectations of operator algebras*, Kōdai Math. Sem. Rep. **12** (1960) 82–90.
NAKAMURA, M. and T. TURUMARU
[1] *On extensions of pure states of an Abelian operator algebra*, Tôhoku Math. J. **6** (1954) 253–257.
NAKAMURA, M. and H. UMEGAKI
[1] *A remark on theorems of Stone and Bochner*, Proc. Japan Acad. **27** (1951) 506–507.
[2] *On a proposition of von Neumann*, Kōdai Math. Sem. Rep. **8** (1956) 142–144.
NAKANO, H.
[1] *Hilbert algebras*, Tôhoku Math. J. **2** (1950) 4–23.
NAKAYAMA, T.
[1] *Remark on the duality for noncommutative compact groups*, Proc. Amer. Math. Soc. **2** (1951) 849–854.
NATANSON, I. P.
[1] *Theory of functions of a real variable*, Gostekhizdat, Moscow-Leningrad, 1950, 1957. (Chapters I-IX), F. Ungar, New York, 1955; (Chapters X-XVII), F. Ungar, New York, 1960.
NELSON, E.
[1] *Analytic vectors*, Ann. Math. **70** (1959) 572–615.
VON NEUMANN, J.
[1] *Zur Algebra der Funktionaloperationen und Theorie der normalen Operatoren*, Math. Ann. **102** (1929) 370–427.
[2] *Über adjungierte Funktionaloperatoren*, Ann. Math. **33** (1932) 294–310.
[3] *Almost periodic functions in a group. I*, Trans. Amer. Math. Soc. **36** (1934) 445–492.
[4] *On a certain topology for rings of operators*, Ann. Math. **37** (1936) 111–115.
[5] *On regular rings*, Proc. Nat. Acad. Sci. USA **22** (1936) 707–713.
[6] *Continuous rings and their arithmetics*, Proc. Nat. Acad. Sci. USA **23** (1937) 341–349.
[7] *On infinite direct products*, Compositio Math. **6** (1938) 1–77.
[8] *On rings of operators. III*, Ann. Math. **41** (1940) 94–161.
[9] *On some algebraic properties of operator rings*, Ann. Math. **44** (1943) 709–715.
[10] *On rings of operators, Reduction theory*, Ann. Math. **50** (1949) 401–485.
[11] *Functional operators. I, II*, Princeton, Princeton Univ. Press, 1950.
[12] *The non-isomorphism of certain continuous rings*, Ann. Math. **67** (1958) 485–496.
[13] *Continuous geometry*, Princeton, 1960.
VON NEUMANN, J. and R. SCHATTEN
[1] *The cross-space of linear transformations, I-III*, Ann. Math. **47**, 2 (1946) 73–84; 608–630; **49**, 2 (1949) 557–582.
NEWMAN, D. J.
[1] *Some remarks on the maximal ideal structure of* H^∞, Ann. Math. **70** (1959) 438–445.
[2] *Interpolation in* H^∞, Trans. Amer. Math. Soc. **92** (1959) 501–507.
[3] *The nonexistence of projections from* L^1 *to* H^1, Proc. Amer. Math. Soc. **12** (1961) 98–99.
NORGUET, F.
[1] *Intégration de formes différentielles non fermées*, Rendiconti di Math. **20** (1961) 335–372.
NORTHCOTT, D.
[1] *Ideal theory*, Cambridge Univ. Press, Cambridge, 1953.
NOVODVORSKY, M. E.
[1] *On some homological invariants of the space of maximal ideals*, Matem. zametki I, 4 (1967) 487–494 (Russian).
OGASAWARA, T.
[1] *Finite-dimensionality of certain Banach algebras*, J. Sci. Hiroshima Univ., A, **17** (1954) 359–364.
[2] *A theorem on operator algebras*, J. Sci. Hiroshima Univ., A, **18** (1955) 307–309.
[3] *A structure theorem for complete quasi-unitary algebras*, J. Sci. Hiroshima Univ., A, **19** (1955) 79–85.

[4] *Topologies on rings of operators*, J. Sci. Hiroshima Univ., A, **19** (1955) 255–272.

OGASAWARA, T. and K. YOSHINAGA

[1] *Weakly completely continuous Banach *-algebras*, J. Sci. Hiroshima Univ., A, **18** (1954) 15–36.

[2] *A non-commutative theory of integration for operators*, J. Sci. Hiroshima Univ., A, **18** (1955) 311–347.

OKA, K.

[1] *Sur les fonctions de plusieurs variables, II. Domaines d'holomorphie*, J. Sci. Hiroshima Univ. A7 (1937) 115.

[2] *Sur les fonctions analytiques de plusieurs variables, VII. Sur quelques notions arithmétiques*, Bull. Soc. Math. France **78** (1950) 1 27.

OLUBUMMO, A.

[1] *Left completely continuous B*-algebras*, J. London Math. Soc. **32** (1957) 270–276.

O'NEILL, B.

[1] *Parts and one-dimensional analytic spaces*, Thesis, Brown Univ., 1965.

ONO, T.

[1] *A generalization of the Hahn-Banach theorem*, Nagoya Math. J. **6** (1953) 171–176.

[2] *Local theory of rings of operators, I, II*, J. Math. Soc. Japan **10** (1958) 184–216; 438–458.

[3] *Note on a B*-algebra*, J. Math. Soc. Japan **11** (1959) 146–158.

ORIHARA, M.

[1] *Correction to my paper "Rings of operators and their traces"*, Memoirs Fac. Sci. Kyusyu Univ., A, **8** (1953) 89–91.

ORIHARA, M. and T. TSUDA

[1] *The two-sided regular representations of a locally compact group*, Memoirs Fac. Sci. Kyusyu Univ., A, **6** (1951) 21–29.

OTOBE, Y.

[1] *Note on locally compact simple rings*, Proc. Imp. Acad. Tokyo **20** (1944) 283.

[2] *On locally compact fields*, Japan J. Math. **19** (1945) 189–202.

PADMANABHAN, A. R.

[1] *Some dominated convergence theorems in a von Neumann algebra*, Proc. Japan Acad. **42** (1966) 347–350.

PALEY, R. E. A. C. and N. WIENER

[1] *Notes on the theory and application of Fourier transforms, I-II*, Trans. Amer. Math. Soc. **35** (1933).

[2] *Fourier transforms in the complex domain*, "Nauka", Moscow, 1964 (Russian); AMS Colloq. Publ., New York, Vol. XIX, 1934.

PALLU DE LA BARRIÈRE, R.

[1] *Algèbres auto-adjointes faiblement fermées et algèbres hilbertiennes de classe finie*, C. R. Acad. Sci. Paris, **232** (1951) 1994–1995.

[2] *Algèbres unitaires et espaces d'Ambrose*, C. R. Acad. Sci. Paris, **233** (1951) 997–999.

[3] *Isomorphisme des *-algèbres faiblement fermées d'opérateurs*, C. R. Acad. Sci. Paris, **234** (1952) 795–797.

[4] *Algèbres unitaires et espaces d'Ambrose*, Ann. École Norm. Sup., **70**, 3 (1953) 381–401.

[5] *Sur les algèbres d'opérateurs dans les espaces hilbertiens*, Bull. Soc. Math. France **82** (1954) 1–52.

PALYUTKIN, V. G.

[1] *Invariant measure on a compact ring group*, Ukr. matem. zh. **18**, 4 (1966) 49–59 (Russian).

PEDERSEN, E. A.

[1] *A decomposition theory for rings of operators*, Diss., Illinois Inst. Technol., 1966.

PEŁCZYŃSKI, A.

[1] *Uncompleted function algebras with separable annihilators*. Duke Math. J. **33** (1966) 605–612.

[2] *A generalization of Stone's theorem on approximation*, Bull. Acad. Polon. Sci., Cl. 3, **5** (1957) 105–107.

PETER, F. and H. WEYL

[1] *The completeness of the primitive representations of a closed continuous group*, UMN **2** (1936) 144–160 (Russian); *Die Vollständigkeit der primitiven Darstellungen einer geschlossenen kontinuierlichen Gruppe*, Math. Annalen, **97** (1927) 737–755.

PHELPS, R. R.

[1] *Lectures on Choquet's theorem*, N.Y., 1966.

PIETSCH, A.

[1] *Nuclear locally convex spaces*, "Mir", Moscow, 1967.

PITCHER, T.

[1] *Sets of "positive" functions in H-systems*, Trans. Amer. Math. Soc. **77** (1954) 481–489.

PITT, H.

[1] *General Tauberian theorems*, Proc. London Math. Soc. **47**, 2 (1938) 243–288.

[2] *A theorem on absolutely convergent trigonometric series*, J. Math. and Phys. **16** (1938) 191–195.

PLESNER, A. I.

[1] *Spectral theory of linear operators*, UMN **9** (1941) 3–125 (Russian).

[2] *Spectral theory of linear operators*, "Nauka", Moscow, 1965 (Russian).

PLESNER, A. I. and V. L. ROKHLIN

[1] *Spectral theory of linear operators, II*, UMN **1**, 1 (**11**) (1946) 71–191 (Russian).

POLISHCHUK, E. M.

[1] *Volterra composition and normed rings*, Nauchn. zap. Mag. ped. inst. (1954) 263–269 (Russian).

PONTRYAGIN, L. S.

[1] *Über stetige algebraische Körper*, Ann. Math. **33** (1932) 163–174.

[2] *The theory of topological commutative groups*, Ann. Math. **35** (1934) 361–388.

[3] *Linear representations of compact topological groups*, Matem. Sbornik **1** (**43**) (1936) 267–271; UMN (1936) 177–195 (Russian).

[4] *Linear representations of compact topological groups*, UMN (1936) 177–195 (Russian translation of article [3]).

[5] *Hermitian operators in a space with indefinite metric*, Izv. AN SSSR, ser. matem., **8** (1944) 243–280 (Russian).

[6] *Continuous groups*, 2nd edition, Moscow, Gostekhizdat, 1954 (Russian); *Topological groups*, Princeton Univ. Press, Princeton, 1939; *Topologische Gruppen*, Teubner.

POOL, J. C. T.

[1] *Mathematical aspects of the Weyl correspondence*, Mathem. Physics of the Weyl correspondence **7**, 1 (1966) 66–76.

PORTA, H. and J. T. SCHWARTZ

[1] *Representations of the algebra of all operators in Hilbert space and related analytic function algebras*, June 1966, Reprint.

POTAPOV, V. P.

[1] *On holomorphic matrix-functions bounded in the unit disc*, DAN SSSR **72** (1950) 849–852 (Russian).

POVZNER, A. YA.

[1] *On positive functions on an abelian group*, DAN SSSR **28** (1940) 294–295 (Russian).

[2] *Concerning equations of Sturm-Liouville type and positive functions*, DAN SSSR **43** (1944) 387–391 (Russian).

[3] *On equations of Sturm-Liouville type on a half-axis*, DAN SSSR **53** (1946) 295–298 (Russian).

[4] *On some general inversion formulas of Plancherel type*, DAN SSSR **57** (1947) 123–125 (Russian).

[5] *On the spectrum of bounded functions*, DAN SSSR **57** (1947) 755–758 (Russian).

[6] *On the spectrum of bounded functions and the Laplace transform*, DAN SSSR **57** (1947) 871–874 (Russian).

[7] *On differential equations of Sturm-Liouville type on a half-axis*, Matem. Sbornik **23** (**65**) (1948) 3–52 (Russian).

POWERS, R. T.
[1] *Representations of uniformly hyperfinite algebras and their associated von Neumann rings*, Ann. Math. **86** (1967) 138–171.
PRÉKOPA, A.
[1] *Extension of multiplicative set functions with values in a Banach algebra*, Acta Math. Acad. Sci. Hung. **7** (1956) 201–213.
PRICE, J. J.
[1] *Some duality theorems*, Illinois J. Math. **1** (1957) 433–445.
PUKÁNSZKY, L.
[1] *On a theorem of Mautner*, Acta Sci. Math. **15** (1954) 145–148.
[2] *The theorem of Radon-Nikodým in operator rings*, Acta Sci. Math. **15** (1954) 149–156.
[3] *On the theory of quasi-unitary algebras*, Acta Sci. Math. **16** (1955) 103–121.
[4] *Some examples of factors*, Publicationes Mathematicae, Debrecen, **4** (1955–1956) 135–156.
[5] *On the Kronecker products of irreducible representations of the* 2×2 *real unimodular group, I*, Trans. Amer. Math. Soc. **100** (1961) 116–152.
PURSELL, L. E.
[1] *A note on isomorphism of* $C(X,R)$ *and* $C^*(X,R)$, Bull. Calcutta Math. Soc. **49** (1957) 47–48.
[2] *The ring* $C(X,R)$ *considered as the subring of the ring of all real-valued functions*, Proc. Amer. Math. Soc. **8** (1957) 820–821.
QUIGLEY, F.
[1] *Approximation by algebras of functions*, Math. Ann. **135** (1958) 81–92.
RABINOVICH, V. S.
[1] *On the group of unitary operators in Hilbert space*, Uch. zap., Tashkent, **37** (1954) 125–130 (Russian).
RAIKOV, D. A.
[1] *On positive definite functions*, DAN SSSR **26** (1940) 857–862 (Russian).
[2] *Positive definite functions on discrete commutative groups*, DAN SSSR **27** (1940) 325–329 (Russian).
[3] *Positive definite functions on commutative groups with invariant measure*, DAN SSSR **28** (1940) 296–300 (Russian).
[4] *Generalized duality theorem for commutative groups with invariant measure*, DAN SSSR **30** (1941) 589–591 (Russian).
[5] *A new proof of the uniqueness of Haar's measure*, DAN SSSR **34** (1942) 231–233 (Russian).
[6] *Harmonic analysis on commutative groups with Haar measure and the theory of characters*, Trudy Matem. in-ta im. V. A. Steklova AN SSSR **14** (1945) 1–86 (Russian).
[7] *On the theory of normed rings with involution*, DAN SSSR **54** (1946) 387–390 (Russian).
[8] *On various types of convergence of positive definite functions*, DAN SSSR **58** (1947) 1279–1282 (Russian).
RAINWATER, J.
[1] *A remark on regular Banach algebras*, Proc. Amer. Math. Soc. **18** (1967) 255–256.
RAJAGOPALAN, M.
[1] L^p*-conjecture for locally compact groups, I, II*, Trans. Amer. Math. Soc. **125** (1966) 216–222; Math. Ann. **169** (1967) 331–339.
RAJESWARA, A. and K. V. RAO
[1] *On a generalized corona problem*, J. Analyse Math. **18** (1967) 277–278.
RAMASVAMI, V.
[1] *Normed algebras, isomorphism and the associative postulate*, J. Indian Math. Soc. **14** (1950) 47–64.
[2] *On a theorem of Gelfand and Hille*, J. Indian Math. Soc. **14** (1950) 129–138.
REID, G. A.
[1] *A generalization of* W^**-algebras*, Pacif. J. Math. **15** (1965) 1019–1026.
REITER, H.
[1] *Investigations in harmonic analysis*, Trans. Amer. Math. Soc. **73** (1952) 401–427.
[2] *On a certain class of ideals in the* L^1*-algebra of a locally compact abelian group*, Trans. Amer. Math. Soc. **75** (1953) 505–509.

[3] *Über L^1-Räume auf Gruppen, I, II*, Monatshefte f. Math. **58** (1954) 73–76, 172–180.
[4] *Contributions to harmonic analysis*, Acta Math. **96** (1956) 253–263.
[5] *Beiträge zur harmonischen Analyse, II*, Math. Ann. **133** (1957) 298–302.
[6] *Contributions to harmonic analysis, III*, J. London Math. Soc. **32** (1957) 477–483.

RENNISON, J. F.
[1] *A note on the extension of associative products*, Proc. Amer. Math. Soc. **17** (1966) 1375–1377.

RICKART, C. E.
[1] *Banach algebras with an adjoint operation*, Ann. Math. **47** (1946) 528–550.
[2] *The singular elements of a Banach algebra*, Duke Math. J. **14** (1947) 1066–1077.
[3] *The uniqueness of norm problem in Banach algebras*, Ann. Math. **51** (1950) 615–628.
[4] *Representations of certain Banach algebras of Hilbert space*, Duke Math. J. **18** (1951) 27–39.
[5] *On spectral permanence for certain Banach algebras*, Proc. Amer. Math. Soc. **4** (1953) 191–196.
[6] *General theory of Banach algebras*, Van Nostrand, Princeton, N. J., 1960.

RIEFFEL, M. A.
[1] *Burnside's theorem for representations of Hopf algebras*, J. Algebra **6** (1967) 123–130.

RIESZ, F.
[1] *Über Sätze von Stone und Bochner*, Acta Univ. Szeged **6** (1933) 184–198.

RIESZ, F. and B. SZÖKEFALVI-NAGY
[1] *Lectures on functional analysis*, IIL, Moscow, 1954 (Russian); *Functional analysis*, F. Ungar Publ. Co., N.Y., 1955; *Funktionalanalysis*, VEB, Berlin, 1956; *Leçons d'analyse fonctionnelle*, Gauthier-Villars, Paris and Akad. Kiadó, Budapest, 1955.

RINGROSE, J. R.
[1] *On subalgebras of a C*-algebra*, Pacif. J. Math. **15** (1965) 1377–1382.

ROKHLIN, V. A.
[1] *Unitary rings*, DAN SSSR **59** (1948) 643–649 (Russian).

ROLEWICZ, S.
[1] *Entire functions in B_0-algebras containing a dense division algebra*, Studia Math. **23** (1963) 181–187.
[2] *Example of a semi-simple m-convex B_0-algebra which is not a projective limit of semi-simple B-algebras*, Bull. Acad. Polon. Sci., Ser. math. astr. phys., **11**, 7 (1963) 459–462.

ROLEWICZ, S. and W. ŻELAZKO
[1] *Some problems concerning B_0-algebras*, "Tensor" **13** (1963) 269–276.

ROMM, B. D.
[1] *Decomposition into irreducible representations of the tensor product of two irreducible unitary representations of the real unimodular group of the second order*, DAN SSSR **153**, 2 (1963) 276–277 (Russian).

ROSENBERG, A.
[1] *The number of irreducible representations of simple rings with no minimal ideals*, Amer. J. Math. **75** (1953) 523–530.

ROSENBLUM, M.
[1] *On the operator equation $BX—XA = Q$*, Duke Math. J. **23** (1956) 263–269.

ROSENTHAL, H. P.
[1] *Sur les ensembles de Ditkin forts*, C. R. Acad. Sci. Paris **262** (1966) 873–876.

ROSSI, H.
[1] *The local maximum modulus principle*, Ann. Math. **72**, 2 (1960) 1–11.
[2] *Algebras of holomorphic functions on one-dimensional varieties*, Trans. Amer. Math. Soc. **100** (1961) 439–458.

ROSSI, H. and G. STOLZENBERG
[1] *Analytic function algebras*, Function Algebras, Scott-Foresman, Chicago, 1965.

ROYDEN, H.
[1] *Some remarks on function algebras*, Stanford Univ. Techn. Report 5 (1957).
[2] *On a theorem of Wermer's*, Stanford Univ. Techn. Report 9 (1959).

[3] *Algebras of bounded analytic functions on Riemann surfaces*, Acta Math. **114**, 1–2 (1965) 113–142.

Rudin, W.

[1] *Analyticity and the maximum modulus principle*, Duke Math. J. **20** (1953) 449–458.

[2] *Some theorems on bounded analytic functions*, Trans. Amer. Math. Soc. **78** (1955) 333–342.

[3] *The automorphisms and the endomorphisms of the group algebra of the unit circle*, Acta Math. **95** (1956) 39–55.

[4] *Subalgebras of the space of continuous functions*, Collection "Some problems of approximation theory", IIL, Moscow, 1963 (Russian); Proc. Amer. Math. Soc. **7** (1956) 825–830.

[5] *Factorization in the group algebra of the real line*, Proc. Nat. Acad. Sci. USA **43** (1957) 339–340.

[6] *Continuous functions on compact spaces without perfect subsets*, Proc. Amer. Math. Soc. **8** (1957) 39–42.

[7] *The closed ideals in an algebra of analytic functions*, Canad. J. Math. **9** (1957) 426–434.

[8] *Les idéaux fermées dans un anneau de fonctions analytiques*, C. R. Acad. Sci. Paris **244** (1957) 997–998.

[9] *On the structure of maximum modulus algebras*, Proc. Amer. Math. Soc. **9** (1958) 708–712.

[10] *Independent perfect sets in groups*, Michigan Math. J. **5** (1958) 159–161.

[11] *On isomorphisms of group algebras*, Bull. Amer. Math. Soc. **64** (1958) 167–169.

[12] *Fourier analysis on groups*, Interscience, N.Y., 1962.

[13] *Ideals with small automorphisms*, Bull. Amer. Math. Soc. **72** (1965) 339–341.

Ruelle, D.

[1] *States of physical systems*, Comm. Math. Phys. **3** (1966) 133–150.

Rumshisky, L. Z.

[1] *On some classes of positive functions*, DAN SSSR **33** (1941) 105–108 (Russian).

Russo, B. and H. A. Dye

[1] *A note on unitary operators in C^*-algebras*, Duke Math. J. **33** (1966) 413–416.

Sakai, S.

[1] *A remark on Mautner's decomposition*, Kōdai Math. Sem. Reports (1952) 107–108.

[2] *On the representations of semi-simple Lie groups*, Proc. Japan Acad. **30** (1954) 14–18.

[3] *On infinite-dimensional representations of semi-simple Lie algebras and some functionals on the universal enveloping algebras, I*, Proc. Japan Acad. **30** (1954) 305–312.

[4] *On the group isomorphism of unitary groups in AW^*-algebras*, Tôhoku Math. J. **7** (1955) 87–95.

[5] *A characterization of W^*-algebras*, Pacif. J. Math. **6** (1956) 763–773.

[6] *The absolute value of W^*-algebras of finite type*, Tôhoku Math. J. **8** (1956) 70–85.

[7] *On the σ-weak topology of W^*-algebras*, Proc. Jap. Acad. **32** (1956) 329–332.

[8] *On topological properties of W^*-algebras*, Proc. Japan Acad. **33** (1957) 439–444.

[9] *On a characterization of type I C^*-algebras*, Bull. Amer. Math. Soc. **72** (1966) 508–512.

[10] *On a problem of Calkin*, Amer. J. Math. **88** (1966) 935–941.

[11] *The theory of W^*-algebras*, Lecture Notes, Yale Univ., 1962.

[12] *An uncountable number of II_1 and II_∞ factors*, Journ. Functional Analysis **5** (1970) 236–246.

[13] *An uncountable family of non-hyperfinite type III factors*, Functional Analysis, Proceedings of a symposium, Academic Press, N.Y., 1970.

[14] *C^*-algebras and W^*-algebras*, Springer-Verlag, Berlin-Heidelberg-New York, 1971.

Saks, S.

[1] *Theory of the integral*, IIL, Moscow, 1949 (Russian); Monografie Matematyczne, Warsaw, 1939.

Sâmboan, G.

[1] *Asupra integralei produs*, Bull. Ştiinţ. Acad. Rep. Pop. Rom., Sect. Mat. Fiz. **9** (1957) 241–246.

Sarason, D.

[1] *Invariant subspaces and unstarred operator algebras*, Pacif. J. Math. **17** (1966) 511–517.

Sasaki, V.

[1] *Lattice of projections in AW^*-algebras*, J. Sci. Hiroshima Univ. A, **19** (1955) 1–30.

SAWOROTNOW, P. P.

[1] *On a generalization of the notion of H*-algebras*, Proc. Amer. Math. Soc. **8** (1957) 49–55.

[2] *On the imbedding of a right complemented algebra into Ambrose's H*-algebra*, Proc. Amer. Math. Soc. **8** (1957) 56–62.

[3] *Discrete complemented algebras*, Proc. Amer. Math. Soc. **17** (1966) 499–501.

[4] *On two-sided H*-algebras*, Pacif. J. Math. **16** (1966) 365–370.

SCHARK, I. J.

[1] *The maximal ideals in an algebra of bounded analytic functions*, J. Math. and Mech. **10** (1961) 735–746.

SCHATTEN, R.

[1] *A theory of cross-spaces*, Ann. Math. Studies 26, Princeton Univ. Press, Princeton, 1950.

SCHATZ, J. A.

[1] *Representation of Banach algebras with involution*, Canad. J. Math. **9** (1957) 435–442.

SCHWARTZ, J. T.

[1] *Two finite, non-hyperfinite, non-isomorphic factors*, Comm. Pure and Appl. Math. **16** (1963) 19–26.

[2] *Non-isomorphism of a pair of factors of type III*, Comm. Pure and Appl. Math. **16** (1963) 111–120.

SCHWARTZ, L.

[1] *Sur une propriété de synthèse spectrale dans les groups non compacts*, C. R. Acad. Sci. Paris, **227** (1948) 424–426.

[2] *Théorie des distributions*, *I*, 1950; *II*, 1951, Hermann, Paris. New edition, 1966 (volumes I and II combined).

SEGAL, I. E.

[1] *The group ring of a locally compact group. I*, Proc. Nat. Acad. Sci. USA **27** (1941) 348–352.

[2] *Representation of certain commutative Banach algebras*, Bull. Amer. Math. Soc. **52** (1946) 421–422.

[3] *Irreducible representations of operator algebras*, Bull. Amer. Math. Soc. **53** (1947) 73–88.

[4] *The group algebra of a locally compact group*, Trans. Amer. Math. Soc. **61** (1947) 69–105.

[5] *Postulates for general quantum mechanics*, Ann. Math. **48** (1947) 930–948.

[6] *Two-sided ideals in operator algebras*, Ann. Math. **50** (1949) 856–865.

[7] *A kind of abstract integration pertinent to locally compact groups. I*, Bull. Amer. Math. Soc. **55** (1949) 46.

[8] *The two-sided regular representation of a unimodular locally compact group*, Ann. Math. **51** (1950) 293–298.

[9] *An extension of Plancherel's formula to separable unimodular locally compact groups*, Ann. Math. **52** (1950) 272–292.

[10] *A class of operator algebras which are determined by groups*, Duke Math. J. **18** (1951) 221–265.

[11] *Decomposition of operator algebras. I, II*, Memoirs Amer. Math. Soc. **9** (1951), 67 pp. and 66 pp. respectively.

[12] *Hypermaximality of certain operators on Lie groups*, Proc. Amer. Math. Soc. **3** (1952) 13–15.

[13] *A non-commutative extension of abstract integration*, Ann. Math. **57** (1953) 401–457.

[14] *Tensor algebras over Hilbert spaces, I-II*, Trans. Amer. Math. Soc. **81** (1956) 106–134; Ann. Math. **63** (1956) 160–175.

[15] *The structure of a class of representations of the unitary group on a Hilbert space*, Proc. Amer. Math. Soc. **8** (1957) 197–203.

SEGAL, I. E. and J. VON NEUMANN

[1] *A theorem on unitary representations of semisimple Lie groups*, Ann. Math. **52** (1950) 509–517.

SEMADENI, Z.

[1] *Spaces of continuous functions on compact sets*, Adv. in Math. **1,** 3 (1965) 320–382.

[2] *Banachspaces of continuous functions*, Vol. I, PWN, Warsaw, 1971.

SERRE, J. P.

[1] *Un théorème de dualité*, Comm. Math. Helv. **29** (1955) 9–26.

SHERMAN, S.

[1] *The second adjoint of a C*-algebra*, Proc. Int. Congress Math., Cambridge, **1** (1950) 470.

[2] *Order in operator algebras*, Amer. J. Math. **73** (1951) 227–232.

SHERMAN, T.

[1] *A weight theory for unitary representations*, Canad. J. Math. **18** (1960) 159–168.

SHIGA, K.

[1] *Representations of a compact group on a Banach space*, J. Math. Soc. Japan **7** (1955) 224–248.

[2] *Bounded representations on a topological vector space and weak almost periodicity*, Japan J. Math. **25** (1955) 21–35.

SHILOV, G. E.

[1] *Ideals and subrings of the ring of continuous functions*, DAN SSSR **22** (1939) 7–10 (Russian).

[2] *Sur la théorie des idéaux dans les anneaux normés de fonctions*, DAN SSSR **27** (1940) 900–903 (Russian summary).

[3] *On the extension of maximal ideals*, DAN SSSR **29** (1940) 83–85 (Russian summary).

[4] *On certain normed rings*, Sb. Nauchn. Stud. Rabot. MGU **18** (1940) 5–25 (Russian).

[5] *On regular normed rings*, Trudy Matem. in-ta im. V. A. Steklova **21** (1947) 1–118 (Russian with English summary).

[6] *On normed rings with one generator*, Matem. Sbornik **21** **(63)** (1947) 25–47 (Russian with English summary).

[7] *Example of a continuous system of primary ideals in a ring of functions*, Uch. zap. MGU (1947) (Russian).

[8] *On a property of rings of functions*, DAN SSSR **58** (1947) 985–988 (Russian).

[9] *Rings of type $\mathscr{C}$*, DAN SSSR **66** (1949) 813–816 (Russian).

[10] *Rings of type $\mathscr{C}$ on the line and on the circle*, DAN SSSR **66** (1949) 1063–1066 (Russian).

[11] *On a theorem of I. M. Gelfand and its generalizations*, DAN SSSR **72** (1950) 641–644 (Russian).

[12] *Description of a class of normed rings of functions*, Matem. Sbornik **26** **(68)** (1950) 291–310 (Russian).

[13] *On continuous sums of finite-dimensional rings*, Matem. Sbornik **27** **(69)** (1950) 471–484 (Russian).

[14] *Homogeneous rings of functions*, UMN **6** (1951) 91–137 (Russian).

[15] *On rings of functions with uniform convergence*, Ukr. Matem. Zh. **3** (1951) 404–411 (Russian).

[16] *On homogeneous rings of functions on the torus*, DAN SSSR **82** (1952) 681–684 (Russian).

[17] *On rings of functions on the n-dimensional torus*, Matem. zb. Kiivsk. derzh. u-ta **6** (1952) 17–23 (Russian).

[18] *On the decomposition of a commutative normed ring into the direct sum of ideals*, Matem. Sbornik **32** **(74)** (1953) 353–364 (Russian).

[19] *Criteria for compactness in a homogeneous space of functions*, DAN SSSR **92** (1953) 11–12 (Russian).

[20] *On certain problems of the general theory of commutative normed rings*, UMN **12**, 1 (1957) 246–249 (Russian).

[21] *On analytic functions in a normed ring*, UMN **15**, 3 **(93)** (1960) 181–183 (Russian); AMS Translation **28** (1963) 207–209.

[22] *On locally analytic functions* [To the author's article "On analytic functions in a normed ring" in the journal "Uspekhi Matem. Nauk", Vol. 15, No. 3, 1960. Letter to the Editor.] UMN **21**, 6 (1966), 177–182 (Russian).

SHIROTA, T.

[1] *On ideals in rings of continuous functions*, Proc. Japan Acad. **30** (1954) 85–89.

SHMULYAN, V. L.

[1] *On multiplicative linear functionals in certain special normed rings*, DAN SSSR **26** (1940) 13–16 (Russian).

SHNOL, È. È.
[1] *Structure of ideals in rings R_α*, Matem. Sbornik **27** (**69**) (1950) 143–146 (Russian).
[2] *Closed ideals in the ring of continuously differentiable functions*, Matem. Sbornik **27** (**69**) (1950) 281–284 (Russian).

SHREIDER, YU. A.
[1] *Structure of maximal ideals in rings of measures with convolution*, Matem. Sbornik, **27** (**69**) (1950) 297–318 (Russian); AMS Translation No. 81 (1953), 28 pp.

SINGER, I. M.
[1] *Uniformly continuous representations of Lie groups*, Ann. Math. **56** (1952) 242–247.
[2] *Automorphisms of finite factors*, Amer. J. Math. **77** (1955) 117–133.
[3] *Report on group representations*, Publ. Nat. Acad. Sci. Nat. Res. Council, No. **387** (1955) 11–26.

SINGER, I. M. and J. WERMER
[1] *Derivations on commutative normed algebras*, Math. Ann. **129** (1955) 260–264.

SMILEY, M. F.
[1] *Right H*-algebras*, Proc. Amer. Math. Soc. **4** (1953) 1–4.
[2] *Right annihilator algebras*, Proc. Amer. Math. Soc. **6** (1955) 698–701.

SMIRNOV, V. I.
[1] *Course in higher mathematics*, Vol. V, Gostekhizdat, Moscow-Leningrad, 1959 (Russian).

SMIRNOV, YU. S.
[1] *A necessary and sufficient condition for metrizability of a topological space*, DAN SSSR **77** (1951) 197–210 (Russian); AMS Translation No. 26.

"Some problems of approximation theory", IIL, Moscow, 1963.

SONIS, M. G.
[1] *On the Wiener relation in commutative rings, I*, resp. Matem. Konf. Molodykh Issledovatelei, Proceedings, vyp. II, Kiev, 1965, 616–621 (Russian).
[2] *On positive functionals in completely symmetric rings*, Vestn. Mosk. un-ta **4** (1966) 58–65 (Russian).

SRINIVASAN, T. P. and JU-KWEI WANG
[1] *Weak-star Dirichlet algebras*, Functional Algebra, Proc. Int. Symposium, Tulane Univ., 1965, 216–249.

STEGEMAN, J.
[1] *Extension of a theorem of H. Helson*, Proceedings of the Int. Congr. Math., Moscow, 1966, Functional Analysis, 28.

STINESPRING, W. F.
[1] *Positive functions on C*-algebras*, Proc. Amer. Math. Soc. **6** (1955) 211–216.
[2] *Integration theorems for gages and duality for unimodular groups*, Trans. Amer. Math. Soc. **90** (1959) 15–56.

STOLZENBERG, G.
[1] *A maximal ideal space with no analytic structure*, J. Math and Mech. **12** (1963) 103–111.
[2] *Polynomially and rationally convex sets*, Acta Math. **109** (1963) 259–289.
[3] *Uniform approximation on smooth curves*, Acta Math. **115** (1966) 185–198.

STONE, M. H.
[1] *Linear transformations in Hilbert space and their applications to analysis*, Amer. Math. Soc., Coll. Publ. XV, N.Y., 1932.
[2] *On one-parameter unitary groups in Hilbert space*, Ann. Math. **33** (1932) 643–648.
[3] *Applications of the theory of Boolean rings to general topology*, Trans. Amer. Math. Soc. **41** (1937) 375–481.
[4] *A general theory of spectra, I*, Proc. Nat. Acad. Sci. USA **26** (1940) 280–283.
[5] *The generalized Weierstrass approximation theorem*, Math. Mag. **21** (1948) 167–187, 237–254.
[6] *Notes on integration*, Proc. Nat. Acad. Sci. USA **34** (1948) 447–455.
[7] *On a theorem of Pólya*, J. Indian Math. Soc. **12** (1948) 1–7.
[8] *On the theorem of Gelfand-Mazur*, Ann. Soc. Polon. Math. **25** (1952) 238–240.

STØRMER, E.

[1] *On the Jordan structure of C*-algebras*, Trans. Amer. Math. Soc. **120**, 3 (1965) 438–447.

[2] *Jordan algebras of type I*, Acta Math. **115**, 3–4 (1966) 165–184.

STRICHARTZ, R. S.

[1] *Isomorphisms of group algebras*, Proc. Amer. Math. Soc. **17** (1966) 858–862.

SUKHOMLINOV, G. A.

[1] *On the extension of linear functionals in complex and quaternion linear space*, Matem. Sbornik **3** (**45**) (1938) 353–358 (Russian with German summary).

Summer school on topological algebra theory, Bruges, September 6–16, 1966.

SUNOUCHI, H.

[1] *A characterization of the maximal ideal in a factor of the case* (II_∞), Kōdai Math. Sem. Reports **6** (1957) 7.

[2] *A characterization of the maximal ideal in a factor, II*, Kōdai Math. Sem. Reports **7** (1955) 65–66.

[3] *Infinite Lie rings*, Tôhoku Math. J. **8** (1956) 291–307.

SUZUKI, N.

[1] *On the invariants of W*-algebras*, Tôhoku Math. J. **7** (1955) 177–185.

[2] *On the automorphisms of W*-algebras leaving the center elementwise invariant*, Tôhoku Math. J. **7** (1955) 186–191.

SUZUKI, N. and T. SAITÔ

[1] *On the operators which generate continuous von Neumann algebras*, Tôhoku Math. J. **15** (1963) 277–280.

SZÖKEFALVI-NAGY, B.

[1] *Transformations de l'espace de Hilbert, fonctions de type positif sur un groupe*, Acta Sci. Math. **15** (1954) 104–114.

[2] *Note on sums of almost orthogonal operators*, Acta Sci. Math. **18** (1957) 189–191.

TAKEDA, Z.

[1] *On a theorem of R. Pallu de la Barrière*, Proc. Japan Acad. **28** (1952) 558–563.

[2] *Note on Fourier-Stieltjes integral, II*, Kōdai Math. Sem. Rep. **2** (1953) 33–36.

[3] *Conjugate spaces of operator algebras*, Proc. Japan Acad. **30** (1954) 90–95.

[4] *On the representations of operator algebras*, Proc. Japan Acad. **30** (1954) 299–304.

[5] *On the representations of operator algebras, II*, Tôhoku Math. J. **6** (1954) 212–219.

TAKENOUCHI, O.

[1] *On the maximal Hilbert algebras*, Tôhoku Math. J. **3** (1951) 123–131.

[2] *Sur une classe de fonctions continues de type positif sur un groupe localement compact*, Math. J. Okayama Univ. (1955) 143–173.

[3] *Families of unitary operators defined on groups*, Math. J. Okayama Univ. **6** (1957) 171–179.

TAKESAKI, M.

[1] *A duality in representation theory of C*-algebras*, Ann. Math. **85** (1967) 370–382.

TANNAKA, S.

[1] *Über den Dualität der nichtkommutativen topologischen Gruppen*, Tôhoku Math. J. **45** (1938) 1–12.

TATSUUMA, N.

[1] *A duality theorem for locally compact groups, I, II, III*, Proc. Japan Acad. **41** (1965) 878–882; **42** (1966) 46–49, 87–90.

TAYLOR, A.

[1] *Introduction to functional analysis*, Wiley, New York, 1958.

TAYLOR, J. L.

[1] *The Tomita decomposition of rings of operators*, Trans. Amer. Math. Soc. **113** (1967) 30–39.

TELEMAN, S.

[1] *Sur les algèbres de J. von Neumann*, Bull. Sci. Math. **82** (1958) 117–126.

[2] *Contribution à l'analyse harmonique sur un groupe topologique quelconque*, C. R. Acad. Sci. Paris **257**, 16 (1963) 2227–2230.

THOMA, E.

[1] *Zur Reduktionstheorie in separablen Hilbert-Räumen*, Math. Zeit. **67** (1957) 1–9.

[2] *Zur Reduktionstheorie in allgemeinen Hilbert-Räumen*, Math. Zeit. **68** (1957) 153–188.

[3] *Die unitären Darstellungen der Bewegungsgruppe des R^2*, Abst. Short Comm. Int. Congr. Math., Univ. Edinburgh, (1958) 24–25 (also see Math. Ann. **134** (1958) 428–452).

TIKHONOV, A. N.

[1] *Über die topologische Erweiterung von Räumen*, Math. Ann. **102** (1929) 544–561.

TOMITA, M.

[1] *On rings of operators in non-separable Hilbert spaces*, Memoirs Fac. Sci., Kyusyu Univ. A, **7** (1953) 129–168.

[2] *On the regularly convex hull of a set in a conjugate Banach space*, Math. J. Okayama Univ. **3** (1954) 143–145.

[3] *Representations of operator algebras*, Math. J. Okayama Univ. **3** (1954) 147–173.

[4] *Banach algebras generated by a bounded linear operator*, Math. J. Okayama Univ. **4** (1955) 97–102.

TOMIYAMA, J.

[1] *On the projection of norm one in W^*-algebras*, Proc. Japan Acad. **33** (1957) 608–612.

[2] *Generalized dimension function for W^*-algebras of infinite type*, Tôhoku Math. J. **10** (1958) 121–129.

TSUJI, K.

[1] *N^*-algebras and finite class groups*, Bull. Kyusyu Inst. Technol., Math. Nat. Sci. **1** (1955) 1–9.

[2] *Representation theorems of operator algebra and their applications*, Proc. Japan Acad. **31** (1955) 272–277.

[3] *Harmonic analysis on locally compact groups*, Bull. Kyusyu Inst. Technol., Math. Nat. Sci. **2** (1956) 16–32.

[4] *W^*-algebras and abstract (L)-spaces*, Bull. Kyusyu Inst. Technol., Math. Nat. Sci. **3** (1957) 11–13.

[5] *ω-almost periodic functions on arbitrary groups*, Bull. Kyusyu Inst. Technol., Math. Nat. Sci. **4** (1958) 7–14.

[6] *Tensor product of annihilating spaces*, Bull. Kyusyu Inst. Technol., Math. Nat. Sci. **10** (1963) 15–18.

[7] *Annihilator of W^*-algebras annihilating spaces*, Bull. Kyusyu Inst. Technol., Math. Nat. Sci. **10** (1963) 25–39.

[8] *The J. von Neumann commutation theorem*, Bull. Kyusyu Inst. Technol., Math. Nat. Sci. **13** (1966) 1–3.

TURPIN, P.

[1] *Sur une classe d'algèbres topologiques généralisant les algèbres localement bornées*, Thèse de troisième cycle, Grenoble, 1966.

[2] *Sur une classe d'algèbres topologiques*, C. R. Acad. Sci. Paris, **263**, 14 (1960) 436–439.

TURUMARU, T.

[1] *On the commutativity of the C^*-algebra*, Kōdai Math. Sem. Reports (1951) 51.

[2] *On the direct product of operator algebras, I, II, III, IV*, Tôhoku Math. J. **4** (1953) 242–251; **5** (1953) 1–7; **6** (1954) 208–211; **8** (1956) 281–285.

UMEGAKI, H.

[1] *On some representation theorems in an operator algebra. I-III*, Proc. Japan Acad. **27** (1951) 328–333, 501–505; **28** (1952) 29–31.

[2] *Operator algebra of finite class. I, II*, Kōdai Math. Sem. Reports (1952) 123–129; (1953) 61–63.

[3] *Decomposition theorems of operator algebras and their applications*, Japan J. Math. **22** (1952/53) 27–50.

[4] *Ergodic decomposition of stationary linear functionals*, Proc. J. Acad. **30** (1954) 358–362.

[5] *Note on irreducible decomposition of a positive linear functional*, Kōdai Math. Sem. Rep. **4** (1957) 25–32.

[6] *Positive definite functions and direct product Hilbert space*, Tôhoku Math. J. **7** (1955) 206–211.

[7] *Conditional expectation in an operator algebra, II*, Tôhoku Math. J. **8** (1956) 86–100.

[8] *Weak compactness in an operator space*, Kōdai Math. Sem. Rep. **8** (1956) 145–151.

URBANIK, K.

[1] *On quotient-fields generated by pseudonormed rings*, Studia Math. **15** (1955) 31–33.

VAROPOULOS, N. TH.

[1] *Continuité des formes linéaires positives sur une algèbre de Banach avec involution*, C. R. Acad. Sci. Paris, **258** (1964) 1121.

[2] *Sur les formes positives d'une algèbre de Banach*, C. R. Acad. Sci. Paris, **258** (1964) 2465–2467.

[3] *Sur les ensembles parfaits et les séries trigonométriques*, C. R. Acad. Sci. Paris, **260** (1965), 3831–3834, 4668–4670, 5165–5168, 5997–6000; **262** (1966) 384–387.

[4] *Caractérisations des ensembles de Sidon d'une algèbre tensorielle*, C. R. Acad. Sci. Paris, **262** (1966) 447–449.

[5] *Algèbres tensorielles et applications à l'analyse harmonique*, Summer school on topological algebra theory, Bruges, 1966, 252–283.

VASILEV, N. B.

[1] *C*-algebras with finite-dimensional irreducible representations*, UMN **21**, 1 (1966) 135–154.

VERNIKOV, I. KH., S. G. KREIN and A. V. TOVBIN

[1] *On semi-ordered rings*, DAN SSSR **30** (1941) 778–780 (Russian).

VIDAV, I.

[1] *Über eine Vermutung von Kaplansky*, Math. Z. **62** (1955) 330.

[2] *Quelques propriétés de la norme dans les algèbres de Banach*, Publ. Inst. Math. Acad. Serbe Sci. **10** (1956) 53–58.

[3] *Über die Darstellung der positiven Funktionale*, Math. Z. **68** (1958) 362–366.

[4] *Sur une système d'axiomes caractérisant les C*-algèbres*, Glasnik math. fiz. i astron. **16**, 3–4 (1961) 189–193.

VILENKIN, N. YA.

[1] *On the theory of associated spherical functions*, DAN SSSR **111** (1956) 742–744 (Russian).

[2] *Matrix elements of irreducible unitary representations of the group of real orthogonal matrices and motion groups in $(n-1)$-dimensional Euclidean space*, DAN SSSR **113** (1957) 16–19 (Russian).

[3] *Special functions and the theory of group representations*, "Nauka", Moscow, 1965.

VILENKIN, N. YA., È. L. AKIM and A. A. LEVIN

[1] *Matrix elements of irreducible unitary representations of groups of Euclidean motions in three-dimensional space and their properties*, DAN SSSR **112** (1957) 987–989 (Russian).

VITUSHKIN, A. G.

[1] *Some theorems on the possibility of uniformly approximating continuous functions by analytic functions*, DAN SSSR **123** (1958) 959–962 (Russian).

[2] *Conditions on a set which are necessary and sufficient for the uniform approximation by analytic (or rational) functions of every function which is continuous on this set*, DAN SSSR **128** (1959) 17–20 (Russian).

VOWDEN, B. J.

[1] *On the Gelfand-Neumark theorem*, J. London Math. Soc. **42** (1967) 725–731.

VULIKH, B. Z.

[1] *Partially ordered rings*, zap. 3-go Vsesoyuznogo Mat. syezda (1956) 20–21 (Russian).

[2] *Introduction to functional analysis*, Gosudar. Izd. Fiziko-Matem. Lit., Moscow, 1958 (Russian).

[3] *Introduction to the theory of partially ordered spaces*, Wolters-Noordhoff N.V., Groningen, 1967.

WADELL, M. C.

[1] *Properties of regular rings*, Duke Math. J. **19** (1952) 623–627.

WAELBROECK, L.

[1] *Le calcule symbolique dans les algèbres commutatives*, J. Math. Pures et Appl. **33** (1954) 147–186.

[2] *Les algèbres à inverse continu*, C. R. Acad. Sci. Paris **238** (1954) 640–641.

[3] *Structure des algèbres à inverse continu*, C. R. Acad. Sci. Paris, **238** (1954) 762–764.

[4] *Étude spectrale des algèbres complètes*, Brussels, 1960.
[5] *Théorie des algèbres de Banach et des algèbres localement convexes*, Univ. de Montréal, 1962.
[6] *On the analytic spectrum of Arens*, Pacif. J. Math. **13** (1963) 317–319.
[7] *Lectures in spectral theory*, Yale Univ. Dept. Math., 1963.
[8] *About a spectral theorem*, Proc. Symp. of Functional Algebra Theory at Tulane, 1965. Scott Foresman, 1966.
[9] *Continuous inverse of locally pseudo-convex algebras*. Summer school on topological algebra theory, Bruges (1966) 128–185.

WALSH, B.
[1] *Banach algebras of scalar-type elements*, Proc. Amer. Math. Soc. **16** (1955) 1167–1170.

WALSH, J. L.
[1] *Interpolation and approximation by rational functions in the complex domain*, IIL, Moscow, 1961 (Russian).

WARNER, S.
[1] *Weak locally multiplicatively-convex algebras*, Pacif. J. Math. **5**, Suppl. 2 (1955) 1025–1032.
[2] *Inductive limits of normed algebras*, Trans. Amer. Math. Soc. **82** (1956) 190–216.
[3] *Polynomial completeness in locally multiplicatively-convex algebras*, Duke Math. J. **23** (1956) 1–11.
[4] *Weakly topologized algebras*, Proc. Amer. Math. Soc. **8** (1957) 314–316.

WEIERSTRASS, K.
[1] *Über die analytische Darstellbarkeit sogenannter willkürlicher Funktionen einer reellen Veränderlichen*, Sitzber. Akad. Berlin (1885) 633–640, 789–806.

WEIL, A.
[1] *L'intégration dans les groupes topologiques et ses applications*, Actualités Sci. et Industr., 869, Paris, 1938; *Integration on topological groups and its applications*, IIL, Moscow, 1950 (Russian).

WENDEL, J. G.
[1] *On isometric isomorphism of group algebras*, Pacif. J. Math. **1** (1951) 305–311.
[2] *Left centralizers and isomorphisms of group algebras*, Pacif. J. Math. **2** (1952) 251–261.

WERMER, J.
[1] *On algebras of continuous functions*, Proc. Amer. Math. Soc. **4** (1953) 866–869.
[2] *On a class of normed rings*, Arkiv för Math. **2** (1954) 537–551.
[3] *Algebras with two generators*, Amer. J. Math. **76** (1954) 853–859.
[4] *Subalgebras of the algebra of all complex-valued continuous functions on the circle*, Amer. J. Math. **78** (1956) 225–242.
[5] *Maximal subalgebras of group-algebras*, Proc. Amer. Math. Soc. **6** (1955) 692–694.
[6] *Approximation by polynomials on an arc in C^3*, Collection "Some problems of approximation theory", IIL, Moscow, 1963 (Russian); Ann. Math. **62** (1955) 269–270.
[7] *Function rings and Riemann surfaces*, Ann. Math. **67** (1958) 45–71.
[8] *Rings of analytic functions*, Ann. Math. **67** (1958) 497–516.
[9] *The hull of a curve in C^n*, Ann. Math. **68** (1958) 550–561.
[10] *An example concerning polynomial convexity*, Ann. Math. **139** (1959) 147–150.
[11] *The maximum principle for bounded functions*, Ann. Math. **69** (1959) 598–604.
[12] *Dirichlet algebras*, Duke Math. J. **27** (1960) 373–382.
[13] *Banach algebras and analytic functions*, Collection "Some problems in approximation theory", IIL, Moscow, 1963 (Russian); Advances Math. **1** (1961) 51–102.
[14] *Uniform approximation and maximal ideal spaces*, Bull. Amer. Math. Soc. **68** (1962) 298–305.
[15] *Maximal ideal spaces*, Proc. Int. Congr. Math., 1962.
[16] *The space of real parts of a function algebra*, Pacif. J. Math. **13** (1963) 1423–1426.
[17] *Approximation on a disc*, Math. Ann. **155** (1964) 331–333.
[18] *Seminar über Funktionen-Algebren*, Lecture Notes in Math., Springer Verlag, Berlin, 1964.
[19] *Polynomially convex discs*, Math. Ann. **158** (1965) 6–10.

WERNER, R. G.
[1] *Closed ideals in the group algebra $L^1(G) \cap L^2(G)$*, Trans. Amer. Math. Soc. **121** (1966) 409–423.
WHITNEY, H.
[1] *On ideals of differentiable functions*, "Mathematika" **10**, 4 (1966) (Russian); Amer. J. Math. **70** (1948) 635–658.
WIDOM, H.
[1] *Approximately finite algebras*, Trans. Amer. Math. Soc. **83** (1956) 170–178.
[2] *Nonisomorphic approximately finite factors*, Proc. Amer. Math. Soc. **8** (1957) 537–540.
WIENER, N.
[1] *Tauberian theorems*, Ann. Math. **33** (1932) 1–100.
[2] *The Fourier integral and certain of its applications*, Fizmatgiz, Moscow, 1963 (Russian); Cambridge Univ. Press, Cambridge, 1933.
WIENER, N. and H. R. PITT
[1] *On absolutely convergent Fourier-Stieltjes transforms*, Duke Math. J. **4** (1938) 420–436.
WILFSOHN, A.
[1] *The reduced dual of a direct product of groups*, Proc. Cambridge Philos. Soc. **62**, 1 (1962) 5–6.
WILKEN, D. R.
[1] *Local behavior in function algebras*, Doct. diss., Tulane Univ. 1965, 56 pp.; Dissert. Abstr. 26, 4 (1965) 2243–2244.
WILLCOX, A. B.
[1] *Note on certain group algebras*, Proc. Amer. Math. Soc. **7** (1956) 874–879.
WOLFSON, K. G.
[1] *The algebra of bounded operators on Hilbert space*, Duke Math. J. **20** (1953) 533–538.
[2] *The algebra of bounded functions*, Proc. Amer. Math. Soc. **5** (1954) 10–14.
[3] *A class of primitive rings*, Duke Math. J. **22** (1955) 157–163.
[4] *A note on the algebra of bounded functions*, Proc. Amer. Math. Soc. **7** (1956) 852–855.
WRIGHT, F. B.
[1] *A reduction for algebras of finite type*, Ann. Math. **60** (1954) 560–570.
YANKOV, V.
[1] *On the uniformization of A- and B-sets*, DAN SSSR **30** (1941) 591–592 (Russian).
YEN, TI
[1] *Trace of finite AW*-algebras*, Duke Math. J. **22** (1955) 207–222.
[2] *Isomorphism of unitary groups in AW*-algebras*, Tôhoku Math. J. **8** (1956) 275–280.
[3] *Quotient algebra of a finite AW*-algebra*, Pacif. J. Math. **6** (1956) 389–395.
[4] *Isomorphism of AW*-algebras*, Proc. Amer. Math. Soc. **8** (1957) 345–349.
YOOD, B.
[1] *On ideals in operator rings over Banach spaces*, Bull. Amer. Math. Soc. **53** (1947) 281.
[2] *Banach algebras of bounded functions*, Duke Math. J. **16** (1949) 151–163.
[3] *Transformations between Banach spaces in the uniform topology*, Ann. Math. **50** (1949) 486–503.
[4] *Banach algebras of continuous functions*, Amer. J. Math. **73** (1951) 30–42.
[5] *Topological properties of homomorphisms between Banach algebras*, Amer. J. Math. **76** (1954) 155–167.
[6] *Difference algebras of linear transformations on a Banach space*, Pacif. J. Math. **4** (1957) 615–636.
[7] *Periodic mappings on Banach algebras*, Amer. J. Math. **77** (1955) 17–28.
[8] *Faithful representations of normed algebras*, Pacif. J. Math. **10** (1960) 345–363.
YOSHIZAWA, H.
[1] *Unitary representations of locally compact groups*, Osaka Math. J. **1** (1949) 81–89.
[2] *A proof of the Plancherel theorem*, Proc. Japan Acad. **30** (1954) 276–281.
YOSIDA, K.
[1] *On the group imbedded in the metrical complete ring*, Japan J. Math. **13** (1936) 7–26, 459–472.

[2] *On the exponential formula in the metrical complete ring*, Proc. Imp. Acad. Tokyo **13** (1937) 301–304.
[3] *On the duality theorem on non-commutative compact groups*, Proc. Imp. Acad. Tokyo **19** (1943) 181–183.
[4] *Normed rings and spectral theorems. I-VI*, Proc. Imp. Acad. Tokyo **19** (1943) 356–359, 466–470; **20** (1943) 71–73, 183–185, 269–273, 580–583.

YOSIDA, K. and T. NAKAYAMA
[1] *On the semi-ordered ring and its application to the spectral theorem*. I-II, Proc. Imp. Acad. Tokyo **18** (1942) 555–560; **19** (1943) 144–147.

ZAANEN, A. A.
[1] *Linear analysis. Measure and integral, Banach and Hilbert space, linear integral equations*, Noordhoff, Groningen, 1953.

ZAMFORESCU, I.
[1] *Une généralisation du théorème de Weierstrass-Stone*, C. R. Acad. Sci. Paris, **246** (1958) 524–525.

ŻELAZKO, W.
[1] *On the locally bounded and m-convex topological algebras*, Studia Math. **19** (1960) 333–356.
[2] *A theorem on the B_0-division algebra*, Bull. Acad. Pol. Sci. **8** (1960) 373–375.
[3] *On the radicals of p-normed algebras*, Studia Math. **21** (1962) 203–206.
[4] *On the analytic functions in p-normed algebras*, Studia Math. **21** (1962) 345–350.
[5] *On decomposition of commutative p-normed algebra into a direct sum of ideals*, Colloq. Math. **10**, 1 (1963) 57–60.
[6] *Metric generalizations of Banach algebras*, Rozprawy Mat. **47** (1965) 70.
[7] On *generalized topological divisiors of zero*, Proceedings of the Int. Congr. Math., Moscow, 1966, Functional Analysis, 33.

ZELINSKY, D.
[1] *Raising idempotents*, Duke Math. J. **21** (1954) 315–322.

ZELLER-MEIER, G.
[1] *Produit croisés d'une C*-algèbre par un groupe d'automorphismes*, C. R. Acad. Sci. Paris **263** (1966) 20–23.

ZERNER, M.
[1] *Fonctionnelles propres des opérateurs sur les espaces nucléaires*, Summer school on topological algebra theory, Bruges, 1966, 186–232.

ZHELOBENKO, D. D.
[1] *Description of a class of representations of the Lorentz group*, DAN SSSR **121**, 4 (1958) 586–590 (Russian).
[2] *Structure of the group ring of the Lorentz group*, DAN SSSR **126**, 3 (1959) 482–485 (Russian).
[3] *Linear representations of the Lorentz group*, DAN SSSR **126**, 5 (1959) 935 (Russian).

ZORN, M.
[1] *A remark on method in transfinite algebra*, Bull. Amer. Math. Soc. **41** (1935) 667–670.

INDEX